Biogeochemistry of WETLANDS

Science and Applications

Biogeochemistry of WETLANDS

Science and Applications

K. Ramesh Reddy and **Ronald D. DeLaune**

CRC Press
Taylor & Francis Group
Boca Raton London New York

CRC Press is an imprint of the
Taylor & Francis Group, an **informa** business

Front cover image (clockwise from top right): (1) Rice paddy under cultivation, Wargal (near Hyderabad), Andhra Pradesh, India (Photo by K. Ramesh Reddy); (2) pitcher plants (*Sarracenia flava*) in wet prairie of Tate's Hell Swamp, Florida (Photo by Todd. Z. Osborne); (3) mangrove swamp, Lost Man's River, Everglades National Park, Florida (photo by Todd Z. Osborne); and (4) shore of Blue Cypress Lake, Blue Cypress Marsh, Florida (Photo by Todd Z. Osborne).

Back cover image (left to right): (1) Garnet Strand, Big Cypress National Preserve, Florida (Photo by Todd Z. Osborne); (2) salt marsh on Cabretta, Sapelo Island, Georgia (photo by Todd Z. Osborne); and (3) sea grasses and mangroves, Florida Bay, Everglades National Park, Florida (photo by Todd Z. Osborne).

CRC Press
Taylor & Francis Group
6000 Broken Sound Parkway NW, Suite 300
Boca Raton, FL 33487-2742

© 2008 by Taylor & Francis Group, LLC
CRC Press is an imprint of Taylor & Francis Group, an Informa business

No claim to original U.S. Government works
Printed in the United States of America on acid-free paper
10 9 8 7 6 5 4 3 2 1

International Standard Book Number-13: 978-1-56670-678-0 (Hardcover)

This book contains information obtained from authentic and highly regarded sources. Reasonable efforts have been made to publish reliable data and information, but the author and publisher cannot assume responsibility for the validity of all materials or the consequences of their use. The authors and publishers have attempted to trace the copyright holders of all material reproduced in this publication and apologize to copyright holders if permission to publish in this form has not been obtained. If any copyright material has not been acknowledged please write and let us know so we may rectify in any future reprint.

Except as permitted under U.S. Copyright Law, no part of this book may be reprinted, reproduced, transmitted, or utilized in any form by any electronic, mechanical, or other means, now known or hereafter invented, including photocopying, microfilming, and recording, or in any information storage or retrieval system, without written permission from the publishers.

For permission to photocopy or use material electronically from this work, please access www.copyright.com (http://www.copyright.com/) or contact the Copyright Clearance Center, Inc. (CCC), 222 Rosewood Drive, Danvers, MA 01923, 978-750-8400. CCC is a not-for-profit organization that provides licenses and registration for a variety of users. For organizations that have been granted a photocopy license by the CCC, a separate system of payment has been arranged.

Trademark Notice: Product or corporate names may be trademarks or registered trademarks, and are used only for identification and explanation without intent to infringe.

Library of Congress Cataloging-in-Publication Data
Reddy, Ramesh. 　Biogeochemistry of wetlands: science and applications / Ramesh Reddy and Ronald D. DeLaune. 　　p. cm. 　Includes bibliographical references and index. 　ISBN 978-1-56670-678-0 (alk. paper) 　1. Biogeochemistry. 2. Biogeochemical cycles. 3. Wetlands. I. DeLaune, R. D. II. Title. QH344.R43 2008 577.68--dc22　　　2007048390

Visit the Taylor & Francis Web site at
http://www.taylorandfrancis.com

and the CRC Press Web site at
http://www.crcpress.com

*This book is dedicated to the memories of
William H. Patrick, Jr. and Robert G. Wetzel*

Table of Contents

Preface xxi
Acknowledgments xxiii

Chapter 1 Introduction 1

Chapter 2 Basic Concepts and Terminology 7
2.1 Introduction 7
2.2 Chemistry 7
    2.2.1 Aqueous Chemistry 7
        2.2.1.1 Concentration Units 7
    2.2.2 Acids and Bases 8
    2.2.3 Equilibrium Constant 9
    2.2.4 Thermodynamics 9
        2.2.4.1 Influence of pH 12
    2.2.5 Oxidation–Reduction Reactions 13
        2.2.5.1 Oxidation–Reduction 13
        2.2.5.2 Oxidation State or Number 13
    2.2.6 Balancing Oxidation–Reduction Reactions 13
2.3 Microbiology and Biochemistry 15
    2.3.1 Microbial Cell 15
    2.3.2 Microbial Classification 16
    2.3.3 Chemistry of Biological Molecules 16
    2.3.4 Metabolic Reactions 16
    2.3.5 Enzymes 17
    2.3.6 Biochemical Kinetics 17
2.4 Isotopes 18
    2.4.1 Radioactive Isotopes and Decay 18
    2.4.2 Half-Life 18
    2.4.3 Stable Isotopes 19
2.5 Terminology in Soil Science 19
    2.5.1 Master Soil Horizon 19
    2.5.2 Properties Used in Soil Description 20
    2.5.3 Soil Taxonomy 21
    2.5.4 Physical Properties 23
    2.5.5 Chemical Properties 24
2.6 Units 24
Study Questions 25
Further Readings 26

Chapter 3 Biogeochemical Characteristics 27
3.1 Introduction 27
3.2 Types of Wetlands 31
    3.2.1 Coastal Wetlands 31
    3.2.2 Inland Wetlands 32
3.3 Wetland Hydrology 34

3.4	Wetland Soils	35
	3.4.1 Physical Characteristics	36
	3.4.2 Chemical Characteristics	38
	3.4.3 Biological Characteristics	40
3.5	Wetland Vegetation	41
3.6	Biogeochemical Features of Wetlands	41
	3.6.1 Presence of Molecular Oxygen in Restricted Zones	42
	3.6.2 Sequential Reduction of Other Inorganic Electron Acceptors	42
	3.6.3 Oxidized Soil–Floodwater Interface	43
	3.6.4 Exchanges at the Soil–Water Interface	45
	3.6.5 Presence of Hydrophytic Vegetation	46
3.7	Types of Wetland/Hydric Soils	46
	3.7.1 Waterlogged Mineral Soils	46
	3.7.2 Organic Soils (Histosols)	51
	3.7.3 Marsh Soils	52
	3.7.4 Paddy Soils	53
	3.7.5 Subaqueous Soils	53
3.8	Field Indicators of Hydric Soils	54
	3.8.1 All Soils	55
	3.8.2 Sandy Soils	58
	3.8.3 Loamy and Clayey Soils	60
3.9	Summary	63
Study Questions		64
Further Readings		65

Chapter 4   Electrochemical Properties 67

4.1	Introduction	67
4.2	Theoretical Relationships	69
	4.2.1 $E°$ vs. log K	76
	4.2.2 pe vs. Eh	77
4.3	Measurement of Eh	79
4.4	Eh–pH Relationships	81
4.5	Buffering of Redox Potential (Poise)	84
4.6	Measurement of Redox Potentials	85
	4.6.1 Construction of Platinum Electrodes	85
	4.6.2 Standardization of Electrodes	86
	4.6.3 Redox Potentials in Soils	88
4.7	pH	92
	4.7.1 Soil pH	93
	4.7.2 Floodwater pH	96
	4.7.3 pH Effects	97
4.8	Redox Couples in Wetlands	98
	4.8.1 Intensity	99
	4.8.2 Capacity	100
4.9	Redox Gradients in Soils	101
4.10	Microbial Fuel Cells	106
4.11	Specific Conductance	106
4.12	Summary	107
Study Questions		108
Further Readings		109

Table of Contents

Chapter 5 Carbon 111

5.1 Introduction 111
5.2 Major Components of Carbon Cycle in Wetlands 114
    5.2.1 Plant Biomass Carbon (Net Primary Productivity) 114
    5.2.2 Particulate Organic Matter (Detrital and Soil) 114
    5.2.3 Microbial Biomass Carbon 115
    5.2.4 Dissolved Organic Matter 116
    5.2.5 Gaseous Forms of Carbon 117
5.3 Organic Matter Accumulation 118
5.4 Characteristics of Detritus and Soil Organic Matter 119
    5.4.1 Nonhumic Substances 121
        5.4.1.1 Carbohydrates 122
    5.4.2 Phenolic Substances 122
    5.4.3 Humic Substances 125
5.5 Decomposition 127
    5.5.1 Leaching and Fragmentation 129
    5.5.2 Extracellular Enzyme Hydrolysis 129
    5.5.3 Catabolic Activity 130
        5.5.3.1 Aerobic Catabolism 136
        5.5.3.2 Anaerobic Catabolism 138
        5.5.3.3 Aerobic vs. Anaerobic Catabolism 141
5.6 Organic Matter Turnover 150
    5.6.1 Abiotic Decomposition 151
    5.6.2 Biotic Decomposition 151
5.7 Regulators of Organic Matter Decomposition 153
    5.7.1 Quality and Quantity of Organic Matter 157
    5.7.2 Microbial Communities and Biomass 158
    5.7.3 Water Table or Soil Aeration Status 160
    5.7.4 Availability of Electron Acceptors
           with Higher Reduction Potentials 162
    5.7.5 Nutrient Availability 164
    5.7.6 Temperature 167
5.8 Environmental and Ecological Significance 170
5.9 Functions of Organic Matter in Soils 173
5.10 Summary 178
Study Questions 180
Further Readings 183
   184

Chapter 6 Oxygen 185

6.1 Introduction 185
6.2 Oxygen–H_2O Redox Couple 186
6.3 Soil Gases 187
    6.3.1 Redox Potential 192
    6.3.2 Oxygen Diffusion Rate 195
    6.3.3 Soil Oxygen Content 198
6.4 Sources of Oxygen 199
6.5 Aerobic–Anaerobic Interfaces 200
6.6 Oxygen Consumption 204
    6.6.1 Oxygen as Reactant 204
    6.6.2 Oxygen as an Electron Acceptor 205

6.7 Summary	211
Study Questions	212
Further Readings	212

Chapter 7 Adaptation of Plants to Soil Anaerobiosis 215

7.1	Introduction		215
7.2	Distribution of Wetland Plants		217
7.3	Mechanisms of Flood Tolerance		218
	7.3.1	Metabolic Adaptations	219
	7.3.2	Morphological/Anatomical Adaptations	221
		7.3.2.1 Roots	221
		7.3.2.2 Pneumatophores	221
		7.3.2.3 Lenticels	222
		7.3.2.4 Intercellular Airspaces	222
	7.3.3	Aerenchyma Formation	223
	7.3.4	Intercellular Oxygen Concentration	226
7.4	Mechanisms of Oxygen Movement in Wetland Plants		227
	7.4.1	Diffusion	228
	7.4.2	Mass Flow	229
7.5	Oxygen Release by Plants		237
7.6	Measurement of Radial Oxygen Loss		239
7.7	Soil Phytotoxic Accumulation Effects on Plant Growth		241
7.8	Oxidizing Power of Plant Roots		245
	7.8.1	Root Iron Plaque Formation	246
7.9	Effect of Intensity and Capacity of Soil Reduction on Wetland Plant Functions		247
	7.9.1	Effect of Soil Reduction Intensity	249
	7.9.2	Relationship of Reduction Intensity with Root Porosity and Radial Oxygen Loss	250
	7.9.3	Effect of Soil Reduction Intensity on Nutrient Uptake	252
	7.9.4	Soil Reduction Capacity Effects on Carbon Assimilation and Radial Oxygen Loss	253
7.10	Summary		255
Study Questions			256
Further Readings			256

Chapter 8 Nitrogen 257

8.1	Introduction		257
8.2	Forms of Nitrogen		257
	8.2.1	Inorganic Nitrogen	257
	8.2.2	Organic Nitrogen	258
8.3	Major Storage Compartments		258
	8.3.1	Plant Biomass Nitrogen	260
	8.3.2	Particulate Organic Nitrogen	260
	8.3.3	Microbial Biomass Nitrogen	261
	8.3.4	Dissolved Organic Nitrogen	261
	8.3.5	Inorganic Forms of Nitrogen	261
	8.3.6	Gaseous End Products	261
8.4	Redox Transformations of Nitrogen		262
8.5	Mineralization of Organic Nitrogen		264

	8.5.1 C:N Ratio Concept 265
	8.5.2 Chemical Composition of Organic Nitrogen 267
	8.5.3 Microbial Degradation of Organic Nitrogen 274
	8.5.4 Regulators of Organic Nitrogen Mineralization 278

8.6 Ammonia Adsorption–Desorption 280
8.7 Ammonia Fixation 283
8.8 Ammonia Volatilization 284
    8.8.1 Physicochemical Reaction 284
    8.8.2 Regulators of Ammonia Volatilization 286
8.9 Nitrification 289
    8.9.1 Chemoautotrophic Bacteria 289
    8.9.2 Methane-Oxidizing Bacteria 291
    8.9.3 Heterotrophic Bacteria and Fungi 292
    8.9.4 Regulators of Ammonium Oxidation 292
8.10 Anaerobic Ammonium Oxidation 294
8.11 Nitrate Reduction 296
    8.11.1 Denitrification 297
    8.11.2 Nitrifier Denitrification 298
    8.11.3 Aerobic Denitrification 301
    8.11.4 Chemodenitrification 301
    8.11.5 Dissimilatory Nitrate Reduction to Ammonia (DNRA) 302
    8.11.6 Regulators of Nitrate Reduction 303
    8.11.7 Nitrate Reduction Rates in Wetlands and Aquatic Systems 307
8.12 Nitrogen Fixation 309
    8.12.1 Regulators of Dinitrogen Fixation 310
    8.12.2 Nitrogen Fixation Rates 313
8.13 Nitrogen Assimilation by Vegetation 314
8.14 Nitrogen Processing by Wetlands 317
    8.14.1 Ammonium Flux 318
    8.14.2 Nitrate Flux 320
8.15 Summary 322
Study Questions 322
Further Readings 323

Chapter 9 Phosphorus 325

9.1 Introduction 325
9.2 Phosphorus Accumulation in Wetlands 328
    9.2.1 Why Does Phosphorus Added to Wetlands
         Accumulate in Soils? 328
9.3 Phosphorus Forms in Water Column and Soil 330
    9.3.1 Water Column 332
    9.3.2 Soil 334
9.4 Inorganic Phosphorus 335
9.5 Phosphorus Sorption by Soils 340
    9.5.1 Adsorption—Desorption 343
    9.5.2 Phosphorus Sorption Isotherms 346
         9.5.2.1 Linear Equation 347
         9.5.2.2 Freundlich Equation 348
         9.5.2.3 Langmuir Equation 349

		9.5.2.4	Single-Point Isotherms	349

| | 9.5.2.5 Quantity (*Q*)/Intensity (*I*) Relationships | 350 |

9.5.3 Precipitation and Dissolution ... 350
9.5.4 Regulators of Phosphorus Retention and Release ... 353
9.6 Organic Phosphorus ... 357
    9.6.1 Forms of Organic Phosphorus ... 357
    9.6.2 Chemical Characterization of Organic Phosphorus ... 367
9.7 Phosphorus Uptake and Storage in Biotic Communities ... 370
    9.7.1 Microorganisms ... 370
    9.7.2 Periphyton ... 371
    9.7.3 Vegetation ... 372
9.8 Mineralization of Organic Phosphorus ... 376
    9.8.1 Abiotic Degradation and Stabilization of Organic Phosphorus ... 377
        9.8.1.1 Leaching of Soluble Organic Phosphorus ... 377
        9.8.1.2 Noncatalyzed Hydrolysis of Phosphate Esters ... 377
        9.8.1.3 Photolysis ... 378
        9.8.1.4 Stabilization of Organic Phosphorus ... 378
    9.8.2 Enzymatic Hydrolysis of Organic Phosphorus ... 378
        9.8.2.1 Phosphatases or Monoesterases ... 379
        9.8.2.2 Phosphodiesterases ... 381
    9.8.3 Microbial Activities and Phosphorus Release ... 381
        9.8.3.1 Litterbag Method ... 384
        9.8.3.2 Basal Mineralization of Organic Phosphorus ... 384
        9.8.3.3 Potentially Mineralizable Phosphorus ... 384
        9.8.3.4 Mineralization of Added Organic Phosphorus ... 385
        9.8.3.5 Substrate-Induced Organic Phosphorus Mineralization ... 385
    9.8.4 Regulators of Organic Phosphorus Mineralization ... 387
9.9 Biotic and Abiotic Interactions on Phosphorus Mobilization ... 388
    9.9.1 Phosphorus–Iron–Sulfur Interactions ... 388
    9.9.2 Periphyton–Phosphate Interactions ... 391
    9.9.3 Biotic and Abiotic Interactions of Fe and Ca with Phosphorus ... 394
    9.9.4 Gaseous Loss of Phosphorus ... 395
9.10 Phosphorus Exchange between Soil and Overlying Water Column ... 395
9.11 Phosphorus Memory by Soils and Sediments ... 397
9.12 Summary ... 401
Study Questions ... 403
Further Readings ... 403

Chapter 10 Iron and Manganese ... 405

10.1 Introduction ... 405
10.2 Storage and Distribution ... 405
10.3 Eh–pH Relationships ... 407
    10.3.1 Iron ... 409
    10.3.2 Manganese ... 411
10.4 Reduction of Iron and Manganese ... 411
    10.4.1 Microbial Communities ... 413
    10.4.2 Biotic and Abiotic Reduction ... 415
        10.4.2.1 Biotic Reduction ... 415
        10.4.2.2 Abiotic Reduction ... 417

Table of Contents

- 10.4.3 Forms of Iron and Manganese 421
  - 10.4.3.1 Iron 422
  - 10.4.3.2 Manganese 423
  - 10.4.3.3 Complexation of Iron and Manganese with Dissolved Organic Matter 425
  - 10.4.3.4 Mobile and Immobile Pools of Iron and Manganese 425
- 10.5 Oxidation of Iron and Manganese 427
  - 10.5.1 Microbial Communities 428
  - 10.5.2 Biotic and Abiotic Oxidation 429
    - 10.5.2.1 Iron 430
    - 10.5.2.2 Manganese 432
- 10.6 Mobility of Iron and Manganese 432
- 10.7 Ecological Significance 435
  - 10.7.1 Nutrient Regeneration/Immobilization 435
    - 10.7.1.1 Organic Matter Decomposition and Nutrient Release 435
    - 10.7.1.2 Phosphorous Release or Retention 438
    - 10.7.1.3 Coprecipitation of Trace Elements with Iron and Manganese Oxides 439
  - 10.7.2 Ferromanganese Nodules 439
  - 10.7.3 Root Plaque Formation 440
  - 10.7.4 Ferrolysis 441
  - 10.7.5 Methane Emissions 441
- 10.8 Summary 443
- Study Questions 444
- Further Readings 445

Chapter 11 Sulfur 447

- 11.1 Introduction 447
- 11.2 Major Storage Compartments 447
- 11.3 Forms of Sulfur 448
- 11.4 Oxidation–Reduction of Sulfur 451
- 11.5 Assimilatory Sulfate and Elemental Sulfur Reduction 454
- 11.6 Mineralization of Organic Sulfur 454
- 11.7 Electron Acceptor—Reduction of Inorganic Sulfur 457
  - 11.7.1 Dissimilatory Sulfate Reduction 457
  - 11.7.2 Role of Sulfur in Energy Flow 461
  - 11.7.3 Measurement of Sulfate Reduction in Wetland Soils 462
  - 11.7.4 Regulators of Sulfate Reductions 464
- 11.8 Electron Donor—Oxidation of Sulfur Compounds 466
- 11.9 Biogenic Emission of Reduced Sulfur Gases 470
- 11.10 Sulfur–Metal Interactions 471
- 11.11 Sulfide Toxicity 473
- 11.12 Exchange between Soil and Water Column 473
- 11.13 Sulfur Sinks 474
- 11.14 Summary 475
- Study Questions 475
- Further Readings 476

Chapter 12 Metals/Metalloids 477

- 12.1 Introduction 477
- 12.2 Factors Governing Metal Availability and Transformation 477
  - 12.2.1 Soil/Sediment Redox–pH Conditions 480

12.3	Mercury—Methyl Mercury		482
12.4	Arsenic		485
	12.4.1	Sources of Arsenic	485
	12.4.2	Dissolution of Primary Minerals	485
	12.4.3	Biotransformation	486
		12.4.3.1 Thermodynamics	486
	12.4.4	Oxidation–Reduction	486
	12.4.5	Reductive Dissolution of Metal Oxides	487
	12.4.6	Kinetics of Arsenic Oxidation–Reduction in Soils	487
	12.4.7	Importance of As Speciation	487
		12.4.7.1 Competition with Other Anions	488
		12.4.7.2 Coprecipitation with Metal Oxides and Sulfide	489
	12.4.8	Chemical Oxidation and Reduction of Arsenic	489
		12.4.8.1 Oxidation by Metal Oxides	489
		12.4.8.2 Reduction by Sulfides	489
12.5	Copper		489
12.6	Zinc		493
	12.6.1	Distribution in Soils and Sediments	493
12.7	Selenium		494
12.8	Chromium		496
12.9	Cadmium		499
12.10	Lead		501
12.11	Nickel		503
12.12	Summary		505
Study Questions			505
Further Readings			506

Chapter 13 Toxic Organic Compounds 507

13.1	Introduction		507
	13.1.1	Pharmaceuticals	511
13.2	Biotic Pathways		513
	13.2.1	Acclimation	514
	13.2.2	Biodegradation	514
	13.2.3	Cometabolism	514
	13.2.4	Microbial Accumulation	514
	13.2.5	Polymerization and Conjugation	514
13.3	Metabolism of Organic Compounds		515
	13.3.1	Hydrolysis	515
	13.3.2	Oxidation	516
		13.3.2.1 Hydroxylation	516
		13.3.2.2 Dealkylation	517
		13.3.2.3 β-Oxidation	517
		13.3.2.4 Decarboxylation	517
		13.3.2.5 Cleavage of Ether Linkage	518
		13.3.2.6 Epoxidation	518
		13.3.2.7 Oxidative Coupling	518
		13.3.2.8 Aromatic Ring Cleavage	518
		13.3.2.9 Heterocyclic Ring Cleavage	518
		13.3.2.10 Sulfoxidation	518

| | 13.3.3 | Reduction | 519 |

        13.3.3   Reduction 519
                13.3.3.1   Reductive Dehalogenation 519
        13.3.4   Synthesis 520
13.4   Plant and Microbial Uptake 521
13.5   Abiotic Pathways 521
        13.5.1   Redox–Potential–pH 521
        13.5.2   Hydrolysis 522
        13.5.3   Sorption to Suspended Solids and the Substrate Bed 522
                13.5.3.1   Effect of Colloidal Organic Matter in Surface Water
                              on Sorption in Wetlands 524
        13.5.4   Exchange between Soil and Water Column 525
        13.5.5   Settling and Burial of Particulate Contaminants 525
        13.5.6   Photolysis 525
        13.5.7   Volatilization 526
        13.5.8   Runoff and Leaching 527
13.6   Regulators 528
        13.6.1   Effect of Electron Acceptors on Toxic Organic Degradation 528
        13.6.2   Denitrifying Bacteria 528
        13.6.3   Effect of Sediment Redox–pH Conditions on Degradation 529
        13.6.4   Burial 533
13.7   Summary 533
Study Questions 534
Further Readings 535

Chapter 14   Soil and Floodwater Exchange Processes 537

14.1   Introduction 537
14.2   Advective Flux 539
        14.2.1   Advective Flux Processes 539
        14.2.2   Measurement of Advective Flux 540
                14.2.2.1   Seepage Meters 541
                14.2.2.2   Piezometer 542
                14.2.2.3   Salinity/Conductivity 542
                14.2.2.4   Radium/Radon Isotopes 542
                14.2.2.5   Dyes 543
14.3   Diffusive Flux 543
        14.3.1   Diffusive Flux Processes 543
                14.3.1.1   Ammonium Flux 545
                14.3.1.2   Phosphate Flux 546
                14.3.1.3   Sulfate Flux 546
14.4   Bioturbation 547
        14.4.1   Macrobenthos Communities 548
        14.4.2   Benthic Invertebrates and Sediment–Water Interactions 549
14.5   Wind Mixing and Resuspension 550
14.6   Exchange of Dissolved Solutes between Soil/Sediment and the Water Column 551
        14.6.1   Gradient-Based Measurements 552
        14.6.2   Overlying Water Incubations 552
                14.6.2.1   Benthic Chambers 552
                14.6.2.2   Intact Cores 554

14.7	Sediment Transport Processes	556
	14.7.1 Sediment/Organic Matter Accretion in Wetlands	557
	14.7.2 Measurement of Sedimentation or Accretion Rates	560
	14.7.2.1 Filter Pad Traps	562
	14.7.2.2 Artificial Marker Horizons	562
	14.7.2.3 Sedimentation–Erosion Table	563
	14.7.2.4 Beryllium-7 Dating	564
	14.7.2.5 Lead-210 Dating	565
	14.7.2.6 Cesium-137 Dating	566
	14.7.2.7 Carbon-14 Dating	567
	14.7.2.8 Application of Sediment Dating	568
14.8	Vegetative Flux/Detrital Export	568
14.9	Air–Water Exchange	569
14.10	Biogeochemical Regulation of Exchange Processes	570
14.11	Summary	572
Study Questions		573
Further Readings		573

Chapter 15 Biogeochemical Indicators 575

15.1	Introduction	575
15.2	Concept of Indicators	577
15.3	Guidelines for Indicator Development	578
	15.3.1 Conceptual Relevance	578
	15.3.2 Feasibility of Implementation	578
	15.3.3 Response Variability	579
	15.3.4 Interpretation and Utility	579
15.4	Levels of Indicators	579
15.5	Wetland Ecosystem Reference Conditions	581
15.6	Sampling Protocol and Design	582
	15.6.1 Water Quality Indicators	588
	15.6.2 Soil Quality Indicators	588
	15.6.3 Minimum Monitoring Requirements	590
15.7	Data Analysis	590
	15.7.1 Impact/Recovery Indices	594
15.8	Summary	597
Study Questions		597
Further Readings		598

Chapter 16 Wetlands and Global Climate Change 599

16.1	Introduction	599
16.2	Potential Impact of Global Change to Wetlands	601
16.3	Methane	602
	16.3.1 Wetlands as a Source of Methane	602
	16.3.2 Methane Production in Wetlands	603
	16.3.3 Methane Emission	604
	16.3.4 Regulators of Methane Emission	607
	16.3.5 Methane Sinks	608

16.4	Nitrous Oxide		609
	16.4.1	Wetlands as a Source of Nitrous Oxide	609
	16.4.2	Nitrous Oxide Production in Wetlands	609
	16.4.3	N_2O Emission from Wetlands	611
	16.4.4	Production and Emissions from Natural Wetlands	611
	16.4.5	Regulators of N_2O Production and Emissions	613
	16.4.6	Nitrous Oxide Consumption	614
16.5	Carbon Sequestration		615
16.6	Impact of Sea-Level Rise on Coastal Wetlands		616
	16.6.1	Marsh Accretion	619
16.7	Summary		620
Study Questions			620
Further Readings			621

Chapter 17 Freshwater Wetlands: The Everglades 623

17.1	Introduction			623
17.2	Everglades Wetlands			625
	17.2.1	Historical Perspective		626
	17.2.2	Hydrologic Units		627
		17.2.2.1	Everglades Agricultural Area and C-139 Basin	627
		17.2.2.2	Stormwater Treatment Areas	629
		17.2.2.3	Water Conservation Areas	629
		17.2.2.4	Holeyland and Rotenberger Wildlife Management Areas	630
		17.2.2.5	Everglades National Park	630
17.3	Nutrient Loads and Ecological Alternations			630
	17.3.1	Surface Water Quality and Loads		631
	17.3.2	Soil Nutrient Distribution and Storage		633
	17.3.3	Vegetation		639
	17.3.4	Periphyton		641
	17.3.5	Microbial Communities and Biomass		643
		17.3.5.1	Microbial Communities	643
		17.3.5.2	Microbial Biomass	645
17.4	Biogeochemical Cycles			647
	17.4.1	Enzymes		647
	17.4.2	Carbon Cycling		649
		17.4.2.1	Decomposition of Organic Matter	649
		17.4.2.2	Microbial Respiration	649
		17.4.2.3	Methane Emissions	651
	17.4.3	Nitrogen Cycling		653
		17.4.3.1	Organic Nitrogen Mineralization	653
		17.4.3.2	Nitrification–Denitrification	655
		17.4.3.3	Biological Nitrogen Fixation	656
	17.4.4	Phosphorus Cycling		657
		17.4.4.1	Biotic Processes	658
		17.4.4.2	Abiotic Processes	659
	17.4.5	Sulfur Cycling		660
		17.4.5.1	The Methylmercury–Sulfate Link	663

17.5	Restoration and Recovery	663
17.6	Summary	666

Study Questions ... 667
Further Readings ... 667

Chapter 18 Coastal Wetlands: Mississippi River Deltaic Plain Coastal Marshes, Louisiana ... 669

18.1	Introduction	669
18.2	Biogeography and Geology of Louisiana Coastal Wetlands	669
18.3	Coastal Wetland Loss	670
18.4	Case Studies	672
	18.4.1 Processes Governing Coastal Marsh Stability	672
	18.4.2 Comparison of Vertical Accretion of Louisiana Marsh to Other Gulf Coast Marsh	673
	18.4.3 Influence of Sediment Addition to a Deteriorating Louisiana Salt Marsh	676
	18.4.4 Impact of Mississippi River Diversion on Enhancing Marsh Accretion	676
18.5	Impact of Flooding and Saltwater Intrusion on Louisiana Coastal Vegetation	680
18.6	Carbon Cycling	684
	18.6.1 Primary Production	684
	18.6.2 Methane and Carbon Dioxide Emission along a Salinity Gradient in Louisiana Coastal Marshes	685
	18.6.3 Carbon Sinks	686
	18.6.4 Decomposition of Surface Peat	686
	18.6.5 Carbon Losses Resulting from Wetland Deterioration	686
18.7	Nitrogen Cycling	687
	18.7.1 Nitrogen Inputs	688
	18.7.2 Nitrogen Regeneration and Uptake	688
	18.7.3 Nitrogen Losses	689
	18.7.4 Nitrogen Budget	690
	18.7.5 Processing Capacity of Added Nitrogen Entering Louisiana Wetland	691
	18.7.6 Capacity of Freshwater Marsh to Process Nitrate in Diverted Mississippi River Water	692
18.8	Sulfur Cycling	693
	18.8.1 Forms of Sulfur in Louisiana Marsh Soil	693
	18.8.2 Sulfate Reduction Rates in Louisiana Marsh Soils	694
	18.8.3 Flux of Reduced Sulfur Gases	694
	18.8.3.1 Salt Marsh	696
	18.8.3.2 Brackish Marsh	696
	18.8.3.3 Freshwater Marsh	697
18.9	Case Studies of Factors Governing the Fate of Toxic Organic Compounds and Pollutants in the Louisiana Coastal Wetland	697
	18.9.1 Toxic Organic Compounds	697
	18.9.2 Mercury	700
18.10	Summary	701

Study Questions ... 701
Further Readings ... 702

Chapter 19 Advances in Biogeochemistry 703

19.1 Introduction 703
19.2 Biogeochemical Processes 705
19.3 Algal and Microbial Interactions 708
19.4 Vegetation and Microbial Interactions 709
19.5 Modern Tools to Study Biogeochemical Cycles 710
    19.5.1 Microbial Communities and Diversity 710
    19.5.2 Nuclear Magnetic Resonance Spectroscopy 711
    19.5.3 Diffuse Reflectance Spectroscopy 711
    19.5.4 Stable Isotopes 711
19.6 Synthesis: Mechanistic and Statistical Models 712
    19.6.1 Mechanistic Models 712
    19.6.2 Stochastic (Statistical) Models 713
    19.6.3 Geospatial Models 715
19.7 Future Directions and Perspectives 716
Further Readings 717

References 719

Index 757

Preface

Wetland science is now emerging as an interdisciplinary subject. Hydrologists, biogeochemists, pedologists, ecologists, microbiologists, and scientists from various disciplines are working individually or together to improve our understanding of the functions and ecosystem services of the wetlands. The idea for this book was conceptualized some 25 years back, when the first author of the book started teaching a new course entitled "Biogeochemistry of Wetlands" at the University of Florida. Since then, there has been steady need expressed by more than 500 students who took this course at the University and by several colleagues working with other universities, governmental agencies, and industries. At present, both of us are in the fourth quarter of our professional career, and, about three years ago, we felt that the time has come to synthesize the work on wetland biogeochemistry into a book by including examples of the research conducted at our respective institutions. Our approach is to view biogeochemistry as the key "operating system" that regulates the physical, chemical, and biological processes of elemental cycles within a wetland, thereby affecting large-scale ecosystem services.

This book focuses on "organic matter" as a hub of biogeochemistry, and on oxidation–reduction reactions as primary drivers of biogeochemical processes. Wetlands are unique in that a range of soil–sediment conditions, from strongly reducing (anaerobic) to oxidizing (aerobic), can be found at a range of spatial and temporal scales. These environments include forested wetlands, tidal freshwater and salt marsh wetlands, inland freshwater marshes and northern peat lands, swamp forests, and riparian wetlands and estuaries. In certain topical areas, we relied heavily on the biogeochemistry of aquatic systems because a limited amount of information was available on wetlands. This book was written as both a reference and a text for a graduate-level course. Individuals with an interest in environmental science, biology, chemistry, ecology, and environmental engineering would also find this book useful. The impact of soil redox processes on elemental cycling, biotransformation, and heavy metal chemistry is emphasized. In this context we present 19 chapters involving science and application of biogeochemical principles in wetlands and aquatic systems. The book includes chapters dealing with terminology (Chapters 2 and 3) and electrochemical properties (Chapter 4) describing basic biogeochemical processes that drive transformation processes in wetlands. In addition, there are individual chapters dealing with carbon, oxygen, nitrogen, phosphorus, and sulfur cycles (Chapters 5, 6, 8–11). Attention is given to microbially mediated process in soil and water and atmospheric exchange as related to elemental biogeochemical cycling. One chapter describes plant adaptation (metabolic and morphological) to soil–sediment redox conditions including oxygen transport mechanisms (Chapter 7). This book also includes chapters on iron, manganese, and other heavy metals with a strong emphasis on the role that soil redox–pH plays on metal speciation, availability, and transformations (Chapters 11–12). In addition, the fate and transport of toxic organic compounds in wetland environments are also included (Chapter 13). Chapter 14 describes the exchange processes regulating the movement of elements between the soil and water column. Chapter 15 describes various biogeochemical indicators that can be used to determine contaminant impacts on wetlands and use of these indicators to determine the successes/failures of various restoration programs. Chapter 16 covers the role of wetlands in climate change. In addition, wetland case studies for research conducted in freshwater wetlands (Florida Everglades) and coastal wetlands (Louisiana Mississippi River deltaic plain) are also presented. Chapter 19 provides a glimpse of some of the recent advances in biogeochemistry pertinent to wetland environments.

Acknowledgments

We extend our sincere thanks to many of our colleagues who reviewed the content, provided information to support the chapters individuals, and helped to improve the quality of the book. These include M. Clark, R. Corstanje, E. D'Angelo, W. F. DeBusk, E. Dunne, M. Fisher, W. Hurt, R. Gambrell, S. Grunwald, P. Inglett, K. S. Inglett, J. Jawitz, A. Jugsujinda, A. Keppler, L. Mantini, S. Newman, A. Ogram, T. Osborne, R. Pezeshki, B. Turner, J. White, Y. Wang, K. Yu, and many graduate students and post-doctoral fellows of the Wetland Biogeochemistry Laboratory, University of Florida. Many illustrations presented in this book were originally developed as a part of a graduate course taught at the University of Florida by the first author. For consistency, these figures were redrawn. Special thanks to both Patrick Inglett and Kanika Sharma Inglett for working long hours with both authors during various stages in completing the book. Patrick prepared all illustrations and figures presented in the book. Kanika worked with authors on a regular basis by reading all the chapters and keeping authors on track. The authors thank Jeremy Bright for his assistance in proof reading all chapters. The authors also acknowledge their mentor William H. Patrick, Jr. (deceased), Boyd Professor, Louisiana State University, for providing inspiration and support during early stages of their careers. Special contributions of the following are duly acknowledged: Wade Hurt provided write-up on hydric soils (Chapter 3); Mark Clark and Sabine Grunwald made significant contributions to Chapter 15, especially providing the text and interpretation for Sections 15.6 (Clark) and 15.7 (Grunwald). R. Corstanje, S. Grunwald, and P. W. Inglett significantly contributed to Chapter 19. Corstanje and Grunwald provided a discussion on mechanistic and statistical models (Section 19.6) and Inglett provided input on discussions related to biogeochemical cycles (Sections 19.3 and 19.4). The authors thank Ramona Smith her assistance in developing index words for the book.

Ron DeLaune (one of the authors) thanks his wife, Carole, children, and grandchildren for their support and understanding, which enabled him to devote long hours to complete the book. Ramesh Reddy would also like to thank his wife, Sulochana, and children (Rony and Sameer) for their support over the years in allowing him to use evenings and weekends, and lots of space in the house to complete the book.

K. Ramesh Reddy
R. DeLaune

1 Introduction

Globally, wetlands can be found in all climates, from tropical to tundra, with the exception of Antarctica. Approximately 6% of Earth's land surface, which equals about 2 billion acres (approximately 800 million ha), is covered by wetlands. The United States alone contains about 14% of the world's wetlands, or about 274 million acres (111 million ha). The Convention on Wetlands, signed in Ramsar, Iran, in 1971, is an intergovernmental treaty that provides the framework for national action and international cooperation for the conservation and wise use of wetlands and their resources. There are presently 158 contracting parties to the convention, with 1723 wetland sites, totaling 160 million ha, designated for inclusion in the Ramsar List of Wetlands of International Importance (http://www.ramsar.org/).

Wetlands are complex ecosystems. Functions in these systems are driven by many physical, chemical, and biological processes. No one discipline or specialization can describe these complex processes. In the past, scientists often described wetland ecosystems on the basis of their disciplinary bias, with an emphasis on one of the specialties such as hydrology, chemistry, wildlife, microbiology, or vegetation. This was like the fabled group of blind men trying to describe an elephant by each touching a different part, such as the side or the trunk or the tail. When the blind men compared notes, they found that they were in complete disagreement. The one who had touched the elephant's side claimed an elephant was like a wall; the man who had felt the tail said an elephant was like a rope. It took the whole group of blind men to accurately describe the elephant. Similarly, it has become very clear that no single discipline can adequately describe a complex ecosystem such as wetlands. Describing a wetland ecosystem requires an interdisciplinary approach linking various specializations—biology, ecology, environmental science, hydrology, and so on. Understanding must also draw on disciplines outside these fields. For example, much of the chemical and microbiological processes measured in wetland soils is based on studies of saturated soils, which began around the turn of the twentieth century with research into nutrient behavior in paddy soils and processes measured in lake and marine sediments.

Wetlands are a critical feature of the global landscape because of their unique role in regulating global biogeochemical cycles. The value and function of wetlands are well recognized, as evidenced by national and international policies to preserve wetland ecosystems. Wetlands are some of the most biologically productive ecosystems on earth; their productivity can exceed that of terrestrial and aquatic systems. Wetlands not only serve to promote and sustain biota in many forms, but also serve as living filters that process pollutants from terrestrial runoff and atmospheric deposition. Biodegradation of organic compounds, elemental cycling, atmospheric exchange, processing capacity, and plant response are controlled by the unique conditions found in the wetland environment (Figure 1.1).

Wetlands are located in areas with low elevation and a high water table. Wetlands can include marshes, swamps, bogs, and similar areas. These areas are poorly drained and retain water during rainy periods. Wetlands provide a unique habitat for plants and animals, provide flood control, offer a habitat for commercial fisheries, recharge groundwater, and can improve water quality. Wetlands typically occur in landscapes between upland and aquatic ecosystems. Because uplands are often the source of water to wetlands, components within runoff water are also supplied by uplands. In the absence of wetlands, contaminants added to or generated within upland areas are directly transported into receiving aquatic ecosystems. Several physical, chemical, and biological processes functioning in the soil of uplands and wetlands are involved in regulating the fate (availability)

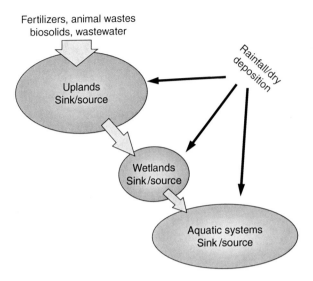

FIGURE 1.1 Linkages between uplands, wetlands, and aquatic systems.

of contaminants. For example, uplands and wetlands can serve as both "sink" and "source" for contaminants:

- *Sink:* Contaminants are transformed to biologically unavailable forms within the system. For example, wetlands can convert nitrate to N_2 gas through a biological reaction called denitrification (this process is discussed in detail in Chapter 8).
- *Source:* Contaminants are transported from one ecosystem to another. For example, uplands can serve as a "source" for suspended solids, nutrients, and other contaminants to wetlands. Similarly, eutrophic wetlands can be a "source" of contaminants or nutrients to adjacent aquatic systems such as streams, rivers, lakes, and estuaries.
- *Transformer:* Contaminants added to a wetland can also be transformed and released as different or complexed compounds, or as new compounds to the aquatic ecosystem downstream. Because wetlands receive runoff from upland ecosystems, the changes in wetlands can be used as an indicator of upland ecosystem's "health."

Because wetland soils can serve as sinks, sources, and transformers of nutrients and other chemical contaminants, they have a significant impact on water quality and ecosystem productivity. The primary driver of wetland processes is ecosystem biogeochemistry. Biogeochemistry is defined as the study of the exchange or flux of materials between living and nonliving components of the biosphere. Biogeochemistry is an interdisciplinary science, involving the interaction of complex processes regulated by physical, chemical, and biological processes in various components of the ecosystem, including the exchange of materials between biotic and abiotic components of the ecosystem. As defined, biogeochemistry encompasses interactions from the smallest scale to the global scale, the biosphere. Wetlands, as the term is used in this book, principally relate to small-scale exchanges at the particle and microbial scale to the field-scale. However, the impact of these small-scale processes on a landscape and on global reservoirs can be significant and will also be addressed.

Living pools in wetland soil reservoirs can act as exchange or cycling pools, with rapid turnover and cycling between the organisms in the pool and their immediate environment. Reservoir pools, which are larger with slower turnover, provide long-term storage. The amount of a given constituent in these pools depends on its residence time, which is simply the amount of that material in the reservoir divided by the rate at which it is removed or added to the reservoir or the rate at which it is transformed. Biogeochemical cycles are influenced by various processes that result in exchange

Introduction

of materials between two storage pools. The exchange or cycling pool can encompass up to 20% of the total amount of a given compound of a system and turn over rapidly immobilizing and remobilizing compounds in a short time. The reservoir pool typically contains the majority of a given compound in a system, is less reactive, and provides long-term storage. When wetlands are used for wastewater treatment, designs that increase the percentage of contaminants in the reservoir pool are more desirable because this provides long-term removal of the contaminant.

Wetland biogeochemistry involves processes by which an element or a compound is transformed within wetlands, including means by which various forms are interchanged between the solid, liquid, and gaseous phases. Thus, the broader ecosystem biogeochemistry definition is also applied to wetland biogeochemistry, which focuses on surface or near-surface processes in wetlands that govern biogeochemical cycles, plant production, microbial transformations, nutrient availability, pollutant removal, heavy metal chemistry, atmospheric exchange, and sediment transport. Wetland biogeochemistry has its parentage in biology, soil science, chemistry, and geology. Wetland biogeochemistry is taking its place along basic research efforts in bioscience, geosciences, and atmospheric sciences in providing the knowledge needed to develop solutions to the challenges faced in wetland ecosystems. Biogeochemical processes in wetlands involve a host of complex microbial communities, periphyton, and vegetation, and interaction and mutual dependency among these communities. Biogeochemical processes quantify exchange and transport of elements or compounds within wetlands, including exchange or transport to other systems (e.g., the atmosphere). Wetland biogeochemical processes provide overviews of the environmental factors and alterations in wetlands that control or impact functioning at the local, regional and global level. Nitrogen and carbon cycles are good examples of biogeochemical processes that impact processes at the local level (e.g., plant growth and soil accretion, regional level (water quality), and global level (greenhouse gas emissions and carbon storage). Wetland biogeochemical cycles can have global significance as follows (Figure 1.2):

- Eutrophication of oligotrophic wetlands resulting from increased nutrient loads can enhance the primary productivity.

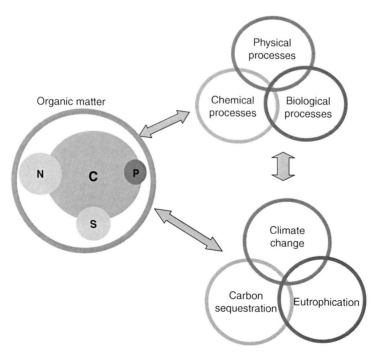

FIGURE 1.2 Linkages between physical, chemical, and biological processes and global scale processes in the biosphere.

- High primary productivity can result in increased rates of organic matter accumulation, providing sink for carbon (increased carbon sequestration).
- High rates of carbon accumulation can enhance microbial activities in soil and the water column.
- Increased rates of microbial activities can increase the production of greenhouse gases, and increased levels of greenhouse gases can have a negative impact on climate.

Thus, biogeochemical cycles within wetlands can have both positive and negative feedback.

Knowledge of wetland biogeochemical processes is useful for predicting the environmental fate and transport of elements and compounds that occur naturally in wetlands and those that enter the system through anthropogenic sources. The type of biogeochemical transformation occurring in wetlands, in contrast to upland systems, is strongly governed by hydrology. Transformation that

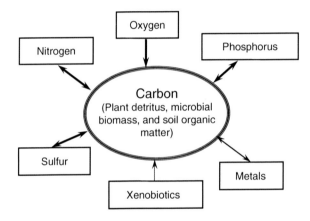

FIGURE 1.3 Linkages between organic carbon and other chemical constituents discussed in the book.

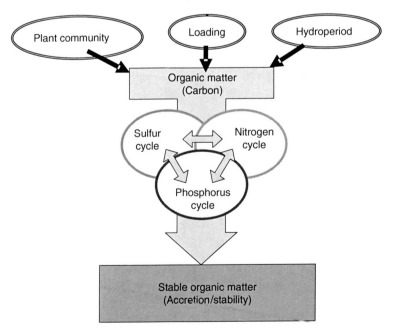

FIGURE 1.4 Relationship between coupled biogeochemical cycles and organic matter accretion in wetlands.

Introduction

occurs in wetlands involves both anaerobic and aerobic processes. The chapters presented in this book focus on the role excess water and soil oxidation–reduction processes play in elemental cycling, heavy metal transformation, wetland plant response, and toxic organic transformation (Figure 1.3).

Biogeochemical processes in the soil and water column are key drivers of several ecosystem functions associated with wetland values (e.g., water quality improvement through denitrification and long-term nutrient storage in the organic matter) (Figure 1.4). The hub for biogeochemistry is organic matter, and its cycling in the soil and water column is far more interesting and important than any other constituent of wetlands. Nutrients such as nitrogen, phosphorus, and sulfur are primary components of soil organic matter, and the cycling of these nutrients is always coupled with carbon cycling. The rate and extent of many of these reactions in the soil and the water column, involving carbon, nitrogen, phosphorus, and sulfur, are mediated by microbial communities and associated physicochemical reactions.

Biogeochemical processes occurring in wetland ecosystems are also important in global biogeochemical cycles, including global warming, carbon sequestration, and water quality. Wetlands are atmospheric sources of carbon dioxide, methane, and nitrous oxide. Wetlands, due to flooded or reducing conditions, can also support biogeochemical processes that limit organic matter turnover, thus serving as important global carbon sinks.

2 Basic Concepts and Terminology

2.1 INTRODUCTION

Understanding the biogeochemical cycles in wetland ecosystems requires some fundamental knowledge of the basic terminology and concepts used in chemistry and biology. In addition, the reader needs to be familiar with basic soil science terms. The reader should refer to any basic chemistry or biology book to review some of the concepts used throughout this book. This chapter briefly reviews some basic concepts and terminology used in the book. The information was obtained from textbooks by Madigan and Martinko (2006), Segal (1976), and Brady and Weil (2003). A recent book by Coyne and Thompson (2006) on basic concepts for soil scientists is also available.

2.2 CHEMISTRY

Atoms. Extremely small, invisible particles comprising an element. Atoms consist of particles called protons (positively charged), electrons (negatively charged), and neutrons (uncharged).

Molecule. A group of two or more atoms held together by chemical bonds. Molecules are basic building blocks of elements. Types of chemical bonds are ionic bonds, covalent bonds, and hydrogen bonds.

Mole. Mass of a mole of a substance, in grams, numerically equals to the formula mass of the substance.

Ionic bonds. Chemical bonds in which electrons are not shared equally between atoms, resulting in charged molecules.

Covalent bonds. Chemical bonds in which electrons are shared equally between two atoms.

Hydrogen bonds. Chemical bonds between hydrogen atoms and electronegative elements such as nitrogen and oxygen.

2.2.1 Aqueous Chemistry

Many biochemical reactions occur in an aqueous medium, thus it is useful to briefly review some of the terminology related to solution chemistry.

2.2.1.1 Concentration Units

Concentrations of dissolved solutes are expressed based on volume or by weight. The most common method is to express on volume basis.

Molarity (M). Number of moles of solute per liter of solution (1,000 mL)
Moles of solute. Weight (g) of solute/molecular weight (MW) of the solute
Molality (m). Number of moles of solute per 1,000 g of solution

$$1 \text{ M solution} = \text{one Avogadro's number of molecules}$$

where Avogadro's number is 6.023×10^{23} or number of molecules per g mole. For dilute solutions of monovalent and divalent ions present in natural waters M = m.

- 1 mmol = 10^{-3} mol or 1 mM = 10^{-3} M = 1 mmol L^{-1}
- 1 μmol = 10^{-6} mol or 1 μM = 10^{-6} M = 1 μmol L^{-1}
- 1 nmol = 10^{-9} mol or 1 nM = 10^{-9} M = 1 nmol L^{-1}

Normality (N). Number of equivalents of solutes per liter of solution

Equivalents of solute. Weight (g) of solute/equivalent weight (EW) of the solute, where EW = MW/n; n = number of H^+ (for acids) and OH^- (for bases) per molecule of oxidation–reduction reactions, number of electrons gained per molecule.

$$Molarity(M) = \frac{Normality}{n}$$

Mole fraction is defined as the ratio of the number of moles of a compound to the total number of moles of all compounds.

$$\text{Mole fraction of } C = \frac{C_1}{C_2 + C_3 + C_4 + \cdots + C_n}$$

where C is the number of moles of a compound and 1, 2, 3, …, n represents different compounds.

Activity. Effective concentration of a solute in a reaction and is expressed as

$$A = \gamma[M]$$

Here γ is the activity coefficient, which is defined as the active fraction of the actual concentration. The value of γ is less than 1. For dilute solutions $A = [M]$ with $\gamma = 1$. For simplicity purposes all examples presented in this book will assume $\gamma = 1$, i.e., molar concentration is equal to their activities.

2.2.2 Acids and Bases

Acid is a substance that donates protons (H^+), that is, its capacity to neutralize OH^-. Consider an acid HA, when dissolved in water ionizes to yield $H^+ + A^-$ where A^- is its conjugate base:

$$HA + H_2O \rightleftharpoons H_3O^+ + A^-$$

$$K_a = \frac{[H_3O^+][A^-]}{[HA][H_2O]}$$

where K_a is the dissociation constant, which is directly proportional to $[H^+]$, suggesting that greatest K_a $[A^-]$ will be its acidity. Since $[H_2O]$ is constant and $[H_3O^+]$ is the same as $[H^+]$, the above equation is usually written as

$$K_a = \frac{[H^+][A^-]}{[HA]}$$

A strong acid ionizes almost 100% in aqueous solution. For example HCl ionization in water:

$$HCl + H_2O \rightarrow H_3O^+ + Cl^-$$

Basic Concepts and Terminology

Base is a substance that accepts protons, that is, indicate the alkalinity of water. A strong base ionizes in solution and yields OH⁻ ions. Sodium and potassium hydroxides are examples of strong inorganic bases:

$$KOH \rightarrow K^+ + OH^-$$

Atmospheric CO_2 or the CO_2 produced during plant or microbial respiration in equilibrium with water can result in the formation of weak acids:

$$CO_{2aq} + H_2O \rightleftharpoons HCO_3^- + H^+ \rightleftharpoons CO_3^{2-} + H^+$$

$$K_a = \frac{[HCO_3^-][H^+]}{[CO_{2aq}]}$$

$$K_{a2} = \frac{[CO_3^{2-}][H^+]}{[HCO_3^-]}$$

2.2.3 Equilibrium Constant

Many biogeochemical reactions that occur in nature are reversible, and under most conditions reactions never reach 100% completion, thus reaching "equilibrium." The stage at which reaction approaches steady state is usually expressed as "equilibrium constant (K_{eq})." For a system at equilibrium, the rates of forward and backward reactions are equal.

For generalized reaction

$$aA + bB = cC + dD$$

where a, b, c, and d represent the moles of the reactants (A, B) and the products (C, D), respectively:

$$\frac{cC + dD}{aA + bB} = K_{eq}$$

2.2.4 Thermodynamics

All biogeochemical processes that occur in the biosphere are subject to the basic laws of thermodynamics. Thermo (heat) refers to "energy in transit" and dynamics relates to "movement." Thus, thermodynamics studies the movement of energy and how energy supports movement. The laws of thermodynamics postulate that energy can be exchanged between physical systems as heat or work. The laws also postulate the existence of a quantity named entropy, which can be defined for any system. Thermodynamics can describe how systems respond to changes in their surroundings. This can be applied to a wide variety of topics in science and engineering, such as phase transitions, chemical reactions, and transport phenomena.

The *first law of thermodynamics* deals with the conservation of energy. It states that energy can be neither created nor destroyed. Energy can only be changed from one form to another. The concept of *enthalpy* (heat flow) evolved from the first law, and is defined as a measure of change in heat content of reactants and products. Change in enthalpy refers to heat being released or absorbed during a reaction, but it does not predict whether a reaction is favorable.

The *second law of thermodynamics* deals with the concept of *entropy*. It states that for any spontaneous process, the reaction always proceeds in the direction of increasing order. Generally, spontaneous reactions release energy as they progress toward equilibrium. This energy can be harnessed and is useful to do work. Entropy is a measure of randomness or disorder of reactants and products.

The following equation describes the relationship between enthalpy and entropy. This relation is known as the Gibbs-Helmholtz equation and is one of the most important in the field of thermodynamics:

$$\Delta G = \Delta H - T\Delta S$$

where Δ is the denotion of the "change in"; G is the free energy released that is available to do the work, units are cal mol^{-1} or J mol^{-1} (1 cal = 4.18 J); H (enthalpy) is the total amount of the energy released during a reaction, units are cal mol^{-1} or J mol^{-1}; S (entropy) is the measure of degree of randomness of the system, unit is cal mol^{-1} K^{-1} (1 cal = 4.18 J); T is the absolute temperature (1 K = 1°C + 273.15).

Standard state of a substance is referred in terms of the reference conditions. Standard state conditions are 1 atmospheric pressure and specific temperature (K = 298.15 or 25°C), pH 7 and all reactants and products initially at 1 M concentration pure substances at standard state have unit activity.

Consider the generalized chemical reactions:

$$\underset{\text{(reactants)}}{a\text{A} + b\text{B}} \rightleftharpoons \underset{\text{(products)}}{c\text{C} + d\text{D}}$$

The energy released or utilized by this reaction can be expressed as follows:

$$\Delta G = \sum \text{free-energy products} - \sum \text{free-energy reactants}$$

At constant pressure and temperature, this energy difference is expressed as the "Gibbs free-energy change" or "ΔG."

For a system in true equilibrium $\Delta G = 0$, thus any reaction not at equilibrium will proceed spontaneously and release energy. For reactants A and B to yield products C and D, the free-energy content of the reactants should be higher than that of the products, that is, ΔG is negative. If the above reaction occurs under standard state conditions, the free-energy change is expressed as $\Delta G°'$. This reaction is *exergonic* when ΔG is negative. This indicates that the reaction will proceed spontaneously. When ΔG is positive, the product formation requires energy and this reaction is called *endergonic*.

When ΔG is negative, the reaction proceeds until the products accumulate with the release of energy. The reaction then proceeds in the reverse direction until the rates of forward and backward reactions are approximately the same and the reaction is balanced.

The concentration of the reactants and the products may not be saved and is related to the free energy of the reaction, that is, high negative value of ΔG suggests that the concentrations of the products will be greater than that of the reactants. On the other hand, the concentration may be similar if ΔG is minimal; however, in the natural system most of the reactions that occur never approach equilibrium, but the reactants and the products approach steady-state levels that may be different from equilibrium levels.

To relate ΔG of the reactions, we need to know the free energy of formation for the substances involved in the reaction. *Free-energy formation* for a substance is defined as the free energy released or used to form one mole of the substance in its standard state and is denoted as $G_f°$, where "f" stands for formation, and additional subscripts such as T and t can be used to indicate whether temperature is in Kelvin or °C, respectively. By convention $G_f°$ is for stable configuration of elements in their standard states. For example, $G_f°$ for C, H_2O, N_2, and O_2 is set at zero. Examples of free energies of formation for selected compounds involved in biogeochemical cycles of elements in wetlands ($G°$, kJ mol^{-1}) are shown in Table 2.1 (Lindsay, 1979; Madigan and Martinko, 2006).

TABLE 2.1
Free Energies of Formation for Selected Compounds Involved in Biogeochemical Cycles of Elements in Wetlands ($G°$, kJ mol^{-1})

Compounds	($G°$, kJ mol^{-1})	Compounds	($G°$, kJ mol^{-1})
CO	−137.3	HPO_4^{2-}	−1,095.3
CO_2	−394.9	$H_2PO_4^-$	−1,136.3
CH_4	−50.8	H_3PO_4	−1,148.6
H_2CO_3	−624.1	PH_3	−9.0
HCO_3^-	−587.7	e^- (electron)	0
CO_3^-	−528.7	S^0	0
Formate	−351.5	SO_4^{2-}	−745.8
Acetate	−369.9	H_2S	−27.2
Alanine	−372.1	HS^-	+12.1
Butyrate	−352.2	S^{2-}	+85.9
Ethanol	−181.8	Fe^{2+}	−85.1
Glucose	−917.6	Fe^{3+}	−10.5
Lactate	−519.6	$FeCl_3$	−414.6
Propionate	−361.1	$Fe(OH)_3$	−652.9
Pyruvate	−473.5	FeOOH (goethite)	−490.8
Urea	−204.1	Fe_2O_3 (hematite)	−743.4
C_2H_2	209	$FePO_4$	−1,183.7
C_2H_4	68.1	$FePO_4$ $2H_2O$ (stregite)	−1,666.1
H_2	0	$(Fe)_3(PO_4)$ $8H_2O$ (vivianite)	−4,423.9
H^+	0 at pH 0; −5.7 per pH unit	FeS_2	−162.1
		Mn^{2+}	−227.9
O_2	0	Mn^{3+}	−82.1
OH^-	−157.5 at pH 14; −198.8 at pH 7	MnO_4^{2-}	−506.6
		MnO_2	−456.7
N_2	0	$AlPO_4$ $2H_2O$ (variscite)	−2,114.9
NO	+86.7	$CaCO_3$ (calcite)	−1,129.4
NO_2^-	−37.3	Cu^+	50.3
NO_3^-	−111.5	Cu^{2+}	64.9
NH_4^+	−79.6	CuS	−49.0
N_2O	+104.3	ZnS	−198.6
H_2O	−237.6	HgS	−49.0
H_2O_2	−134.3	Hg	37.2
PO_4^{3-}	−1,026.6	Hg^{2+}	−164.5

Source: Lindsay (1979) and Madigan and Martinko (2006).

For generalized chemical reaction

$$\Delta G_r^° = \sum[\Delta G_f^° C + \Delta G_f^° D] - \sum[\Delta G_f^° A + \Delta G_f^° B]$$

$\Delta G_r^°$ of the reaction is related to the equilibrium constant as follows:

$$\Delta G_r^° = -RT \ln K$$

where $K = aC^c \, aD^d / aA^a \, aB^b$, R is the universal gas constant (0.001987 kcal mol^{-1} deg^{-1} or 8.314 J mol^{-1} deg^{-1} [1 cal = 4.184 J]), and T is the absolute temperature in Kelvin.

$$\Delta G_r^° (\text{kcal}) = -0.001987 \times 298.15 \times 2.303 \log K$$
$$= -1.364 \log K$$

$$\Delta G_r^\circ(J) = -8.314 \times 298.15 \times 2.303 \log K$$
$$= -5{,}709 \log K$$

As mentioned earlier, many biogeochemical reactions seldom approach equilibrium conditions. As these reactions approach near equilibrium conditions, the ΔG_r° can be expressed as

$$\Delta G_r = \Delta G_r^\circ + RT \ln Q$$

where Q is the reaction quotient $aC^c aD^d / aA^a aB^b$.

aC, aD, aA, aB refer to the activities of substances A, B, C, D involved in the reaction. If the activities of the substances is unity, then $Q = 1$.

$$\Delta G_r = \Delta G_r^\circ + RT \ln 1$$
$$\Delta G_r = \Delta G_r^\circ$$

For a system in equilibrium $\Delta G_r = 0$ and

$$\Delta G_r = -RT \ln K$$

or

$$\log K = \frac{-\Delta G_r^\circ}{RT \times 2.303}$$
$$= \frac{-\Delta G_r^\circ}{1.364}$$

2.2.4.1 Influence of pH

In many biogeochemical reactions, H^+ appears either as a reactant or a product. Easy experimental tools are available to measure the activity of H^+, thus scientists routinely use aH^+ rather than molar concentration. The logarithmic expression of aH^+ commonly used is

$$pH = -\log aH^+$$

For a standard state, H^+ concentration is set as 1 M, which is equal to pH = 0. However, biological reactions do not occur at pH = 0, thus for standard state reactions, scientists adopted pH of 7 (10^{-7} M). For a generalized reaction involving H^+:

$$A \rightarrow B + H^+$$

$$\Delta G_r = \Delta G^\circ + RT \ln \frac{aB\, aH^+}{aA}$$

If

$$aA = aB = 1$$

$$\Delta G_r = \Delta G^\circ + 1.364 \log aH^+$$
$$= \Delta G^\circ + 1.364\, pH$$

Basic Concepts and Terminology

In biogeochemistry, pH is a master variable that can be used in reactions involving oxides, hydroxides, sulfides, carbonates, and many enzymatically mediated biological reactions.

2.2.5 Oxidation–Reduction Reactions

2.2.5.1 Oxidation–Reduction

Many biogeochemical reactions in the natural systems involve oxidation–reduction. The biogeochemical cycles of C, N, O, S, P, and trace metals are regulated by oxidation–reduction reaction. Thus, we have dedicated Chapter 4 primarily to discuss oxidation–reduction reactions in wetlands. In this chapter we will primarily introduce the terminology. For details, the reader should refer to Chapter 4.

Oxidation refers to the tendency of a reaction to lose electrons.
Reduction refers to the tendency of a reaction to gain electrons.

2.2.5.2 Oxidation State or Number

The following are few simple rules for determining the oxidation state or number of elements in chemical reactions (Madigan and Martinko, 2006).

1. The oxidation state of an element in an elementary substance (H_2, O_2) is 0.
2. The oxidation state of an ion is equal to its charge (e.g., $K^+ = +1$; $Ca^{2+} = +2$; $Fe^{3+} = +3$; $Cl^- = -1$; $O^{2-} = -2$).
3. The oxidation state of H^+ is always +1 and that of O is always −2.
4. The sum of oxidation numbers in a neutral molecule is 0. For example, H_2O is neutral because the net charge equals 0: $O(-2 \times 1 = -2) + H^+(+1 \times 2 = +2)$.
5. In an ion, the sum of oxidation numbers of all atoms is equal to the charge on that ion. For example, in a NO_3^- ion, $O(-2 \times 3 = -6) + N(+5) = -1$.
6. The oxidation state of C in simple compounds can be calculated by adding the oxidation numbers of the O and H atoms present in the compound. According to rule number 4, the sum of all oxidation numbers must equal to 0. For example, the oxidation state of C in CO_2 can be determined as follows: $O(-2 \times 2 = -4) + C(+4) = 0$.
7. In organic compounds, it may not be possible to assign a specific oxidation number for C atoms, but the oxidation state of a whole compound can be calculated. For example, the oxidation state of C in the glucose molecule $C_6H_{12}O_6$ is 0, because 6O at −2 = −12 and $12H^+$ at +1 = +12.
8. In oxidation–reduction reactions, there is a balance between reduced products and oxidized reactants. To calculate this balance, the number of molecules of each reactant or product must be multiplied by their oxidation number or state.

2.2.6 Balancing Oxidation–Reduction Reactions

Most reactions that occur in living cells are some form of oxidation–reduction reactions. *Oxidation–reduction reactions* must occur together, since no substances can lose electrons without another substance gaining electrons. In biological systems, oxidation–reduction reactions involve not only transfer of electrons but also transfer of hydrogen that has both one proton (H^+) and one electron.

Oxidation of H_2 gas will result in the release of H^+ and one electron:

$$2H_2 \rightarrow 4H^+ + 4e^-$$

This oxidation reaction must be coupled with reduction of another substance that can accept electrons:

$$O_2 + 4e^- + 4H^+ \rightarrow 2H_2O$$

Thus, coupled reaction can be written as

$$O_2 + 2H_2 \rightarrow 2H_2O$$

Electron donor is the substance undergoing oxidation or the substance that loses electrons during oxidation. The substance is also referred to as "reductant" or "reducing agents." Common electron donors are organic carbon compounds and reduced inorganic compounds.

Electron acceptor is the substance undergoing reduction or the substance that gains electrons during reduction. The substance is also referred to as "oxidant" or the "oxidizing agents." The most common electron acceptor is oxygen, which is used in aerobic respiration of living organisms: plants, animals, and microorganisms. Under oxygen-free environments, selected microorganisms can use alternate electron acceptors such as NO_3^-, Fe^{3+}, Mn^{4+}, and SO_4^{2-}.

Reduction potential ($E°$) is the tendency of a substance undergoing oxidation to give up electrons and by the substance undergoing reduction to gain electrons. These potentials are measured in reference to standard hydrogen electrodes. Substances with positive reduction potentials are usually good oxidizing agents while the substances with highly negative reduction potentials are good reducing agents. The reduction potentials of the various organic and inorganic substances are published in chemistry and biochemistry textbooks.

Many chemical reactions can be easily balanced. However, balancing oxidation–reduction reactions can be a bit complex. The following general approach can be used in balancing redox reactions.

1. Divide the equation into two half-equations, as oxidation and reduction reactions. Identify the two half-reactions (oxidation and reduction) and balance them separately. Combine the balanced reactions to yield a single balanced equation for the overall reaction.
2. Balance the half-equations, first with respect to the number of atoms. Balance the elements on both sides of the equations, except hydrogen and oxygen. Because all reactions take place in aqueous solutions, H_2O molecules can be added as needed to balance oxygen atoms in the equation.
3. Balance the half-equations with respect to charge. In most biogeochemical reactions, ionic balance can be achieved by adding either H^+ or OH^- to the left or right side of equations. Mostly, H^+ is added to balance hydrogens in the equation.
4. Balance oxidation and reduction by balancing electrons. Balance electrons in both the half-equations, and then combine them so that electrons are canceled, leaving none on either side of equation.

EXAMPLE: OXIDATION OF SULFIDE TO SULFATE WITH OXYGEN

Oxidation Reaction

$$H_2S \Rightarrow SO_4^{2-}$$

Sulfur is balanced on both sides of the equation. Now add the H_2O molecule on the left side of the equation to balance the oxygen:

$$H_2S + 4H_2O \Rightarrow SO_4^{2-}$$

Now, add H^+ on the right side of the equation to balance the hydrogen:

$$H_2S + 4H_2O \Rightarrow SO_4^{2-} + 10H^+$$

Now, determine the number of electrons given off in the preceeding oxidation reaction.

Left side of equation:

$$H_2(+2) \; S(-2) = \text{net charge is } 0$$

$$4H_2(+8) \; O(-8) = \text{net charge is } 0$$

Right side of the equation:

$$S(+6) \; O_4^{2-}(-8) = \text{net charge is } -2$$

$$10H^+(+10) = \text{net charge is } +10$$

To balance the right side of the equation, −8 negative charges are needed to maintain net charge of 0. This can be obtained by adding 8 electrons to the right side of the equation:

$$H_2S + 4H_2O \Rightarrow SO_4^{2-} + 10H^+ + 8e^-$$

Now, the half-reaction is balanced. In the oxidation reaction 10 protons and 8 electrons are released.

Reduction Reaction

Repeat all the previous steps discussed to balance the reduction reaction.

$$O_2 \Rightarrow H_2O$$

$$O_2 \Rightarrow 2H_2O$$

$$O_2 + 4H^+ \Rightarrow 2H_2O$$

$$O_2 + 4H^+ + 4e^- \Rightarrow 2H_2O$$

Oxidation–Reduction Reaction

$$H_2S + 4H_2O \Rightarrow SO_4^{2-} + 10H^+ + 8e^- \text{ (oxidation)}$$

$$2(O_2 + 4H^+ + 4e^- \Rightarrow 2H_2O) \text{ (reduction)}$$

$$H_2S + 2O_2 \Rightarrow SO_4^{2-} + 2H^+ \text{ (oxidation–reduction)}$$

2.3 MICROBIOLOGY AND BIOCHEMISTRY

2.3.1 MICROBIAL CELL

A typical feature that separates a prokaryotic cell from a eukaryotic cell is the lack of membrane-bound nuclear material. This region, referred to as nucleoid, has the nucleic acid DNA present in free form. Prokaryotic cells also lack well-defined organelles. Prokaryotes can be divided into eubacteria (bacteria) or archaebacteria (archaea). Archaea are distinct from bacteria in the unique structural components of the cell membrane. Unlike bacterial cells, which have ester linkages between fatty acids and glycerol molecule, archaeal cells have ether linkages between glycerol and hydrophobic chains. Typically, a prokaryotic (bacterial) cell has a rigid outer cell wall that envelops a cell membrane. The major constituent of cell walls of eubacteria is peptidoglycan. The archael cell wall consists of pseudopeptidoglycan, glycoprotein, and polysaccharide. Based on their reaction to Gram's stain, bacteria can be divided into two classes: gram-positive and gram-negative. The difference in reaction to the stain is due to the differences in cell wall structure. Prokaryotic cell size ranges between 0.5 and 2.0 µm.

2.3.2 Microbial Classification

Gram-positive. Prokaryotic cells with thick peptidoglycan layers lacking the outer membrane.

Gram-negative. Prokaryotic cells with thin layers of peptidoglycan which is enveloped by an outer membrane of lipopolysaccharide (LPS) and other complex macromolecules.

Autotroph. Organism capable of biosynthesizing cell material from carbon dioxide as a sole carbon source.

Chemolithotroph. Organism that uses inorganic compounds as energy sources (electron donor).

Chemoorganotroph. Organism that uses organic compounds as energy sources (electron donor).

Heterotroph. Organism requiring organic compounds as energy sources.

Phototroph. Organism capable of using light as an energy source.

2.3.3 Chemistry of Biological Molecules

Cells are comprised of several types of polymeric macromolecules formed by joining several monomers. Some of the essential macromolecules are considered here.

Carbohydrates are organic compounds containing C, H, and O in the ratio of 1:2:1. Polysaccharides are defined as high-molecular-weight carbohydrates (sugars) that are formed when multiple monosaccharides are linked with glycosidic bonds. Polysaccharides can also combine with lipids or proteins to form glycol lipids or glycoproteins. Examples of polysaccharides include cellulose as structural component of cell walls of plant and algae, and storage compounds such as starch and glycogen.

Lipids are composed of fatty acids that contain both highly hydrophobic and hydrophilic regions. Simple lipids, also referred to as glycerides, are made of three fatty acids bonded to C_3 alcohol glycerol. Complex lipids are simple lipids with nitrogen, phosphorus, sulfur, or other small hydrophilic compounds. One example is phospholipids, which have a major structural role in cellular membranes.

Nucleotides are monomers that link together to form polymers called *nucleic acids*: deoxyribonucleic acid (DNA) and ribonucleic acid (RNA). Each nucleotide comprises a C_5 sugar, a phosphate group, and a nitrogen base. Nucleotides are major constituents of the hereditary material and can act as energy currency (ATP) in cells. Some derivatives of nucleotides can function in oxidation–reduction reactions in the cell.

Proteins are polymers of amino acids that are linked by peptide bonds. All amino acids contain an amino group ($-NH_2$) and a carboxyl group ($-COOH$). Dipeptides are formed when two amino acids are joined together by peptide bonds. Several covalently linked amino acids are referred to as polypeptides. Proteins can have one or more polypeptides.

In cells, proteins are present as structural proteins or catalytic proteins (enzymes). Structural proteins are a part of cell membranes, walls, and cytoplasmic components. The primary structure of a protein is determined by its amino acid sequence, and the folding decides its functional ability within the cell. Denaturation of proteins can be caused by extreme conditions, such as pH, heat, or chemical reagents, which affect the folding properties.

2.3.4 Metabolic Reactions

Anabolism. The sum of all biosynthetic reactions in cells that are used to build the chemical substances that the cells are composed of. It results in conservation of energy.

Catabolism. Biochemical reactions leading to a breakdown of compounds and the production of usable energy (usually ATP) by the cell.

Biochemical pathway. Series of reactions involved in chemical oxidation of biological compounds.

Aerobic respiration. Process in which compounds are oxidized with oxygen serving as the terminal electron acceptor, usually accompanied by production of adenosine triphosphate (ATP) by oxidative phosphorylation.

Anaerobic respiration. Process in which electron acceptors other than molecular oxygen are used.

Substrate-level phosphorylation. Production of adenosine triphosphate (ATP) by the direct transfer of a high-energy phosphate molecule to adenosine diphosphate (ADP) during catabolism of a phosphorylated organic compound. It occurs under both aerobic and anaerobic conditions.

Oxidative phosphorylation. Process that uses the energy derived from the flow of electrons through the electron transport system (ETS) to drive the synthesis of ATP from ADP and Pi.

Fermentation. Anaerobic catabolism of an organic compound in which the compound serves as both electron donor and electron acceptor.

2.3.5 Enzymes

Enzymes are proteins that contain catalytic sites for biochemical reactions and act as biological catalysts. Thus, enzymes direct all metabolic events. Enzymes increase the velocity of chemical reactions and are not consumed during the reaction they catalyze. They are often referred by common names created by adding the suffix "-ase" to the name of the substrate or to the reaction they catalyze. The International Union of Biochemistry and Molecular Biology (IUBMB) developed a system of nomenclature and grouped enzymes into six main classes:

1. Oxidoreductases: Catalyze oxidation–reduction reactions
2. Transferases: Catalyze the transfer of carbon-, nitrogen-, or phosphorus-containing groups
3. Hydrolases: Catalyze the cleavage of bonds through the addition of water
4. Lyases: Catalyze cleavage of C–C, C–S, and certain C–N bonds
5. Isomerases: Catalyze the racemization of optical geometric isomers
6. Ligases: Catalyze the formation of bonds between carbon and O, S, N coupled to hydrolysis of high-energy phosphates

Enzyme (E) molecules contain special pockets called active sites. The active site binds the substrate (S) forming an enzyme–substrate (ES) complex. The ES is converted to an enzyme–product (EP), which subsequently dissociates to products (P) and enzymes (E). Enzymes can be highly substrate specific and catalyze only one group of chemical reactions. Depending on the needs of living cells, some enzyme production can be regulated; that is, enzymes can be induced or repressed. Enzyme activity can be influenced by a number of factors, including substrate concentration, temperature, pH, and the concentration of end products of a reaction.

2.3.6 Biochemical Kinetics

The concentration of the enzyme–substrate complex influences the velocity of enzymatic reactions. The relationship between the velocity of a reaction and the concentration of substrates is described by the Michaelis–Menton equation:

$$v = \frac{V_m[S]}{K_m + [S]}$$

where v is the initial velocity, V_m the maximum velocity of the reaction, $[S]$ the substrate concentration, and K_m the Michaelis constant.

K_m is the substrate concentration at which $v = 1/2V_m$. A higher value of K_m indicates a low affinity of the enzyme for substrate, whereas a lower K_m value indicates a higher affinity for the substrate.

Order of reaction. At very high substrate concentrations [S] $\gg K_m$, the velocity of reaction is essentially independent of the substrate concentration. The velocity is constant and equal to V_{max}. The rate of reaction is a *zero-order* reaction. At very low substrate concentrations [S] $\ll K_m$, the velocity of reaction is directly proportional to the substrate concentration. The rate of reaction is *first order* with respect to the substrate.

Activation energy. The minimum energy required to carry out the reaction is called the energy of activation, E_a. If a reaction requires higher activation energy, the rate of reaction is lowered. The presence of a *catalyst* lowers the activation energy and increases the rate of reaction. In biological systems, enzymes act as catalysts.

Arrhenius equation. Explains the relationship between the rate constant of a reaction, k, and the activation energy, E_a:

$$k = Ae^{-E_a/RT}$$

$$E_a = \frac{2.3RT_2T_1}{T_2 - T_1} \log \frac{k_2}{k_1}$$

where k_2 and k_1 are the specific reaction rate constants at two different temperatures, T_2 and T_1, respectively.

The effect of temperature on the rate of reaction is frequently expressed in terms of a temperature coefficient, Q_{10}, which is the factor by which the rate of reaction increases when the temperature is raised by 10°C.

2.4 ISOTOPES

Isotopes are atoms that contain the same number of protons (have the same atomic number) but a different number of neutrons (i.e., have different atomic weights).

2.4.1 RADIOACTIVE ISOTOPES AND DECAY

A radioactive isotope radionuclide is an atom with an unstable nucleus, which is a nucleus characterized by excess energy available to be imparted either to a newly created radiation particle within the nucleus or to an atomic electron. The radionuclide, in this process, undergoes radioactive decay and emits gamma ray(s), subatomic particles, or both. Radionuclides occur naturally, but can also be artificially produced.

2.4.2 HALF-LIFE

The radioactive half-life for a given radioisotope is the time it takes for half of the radioactive nuclei in any sample to undergo radioactive decay. After two half-lives, one-fourth of the original sample will be left, after three half-lives one-eighth of the original sample will be left, and so forth.

The rate of radioactive decay is typically expressed in terms of either the radioactive half-life or the radioactive decay constant. They are related as follows:

$$T_{1/2} = \frac{\ln 2}{\lambda} \approx \frac{0.693}{\lambda} \approx 0.693\tau$$

- $T_{1/2}$: Radioactive half-life
- λ: Radioactive decay constant
- τ: Mean lifetime

Basic Concepts and Terminology

The decay constant is sometimes also called the disintegration constant. The half-life and the decay constant give the same information, so either may be used to characterize decay. Another useful concept in radioactive decay is the average lifetime. Average lifetime is the reciprocal of decay constant.

2.4.3 Stable Isotopes

Stable isotopes are chemical isotopes that are not radioactive. Stable isotopes of the same element have the same chemical characteristics and behave chemically almost identically. Mass differences, due to a difference in the number of neutrons, result in partial separation of the light from the heavy isotopes during chemical reactions (isotope fractionation).

Commonly analyzed stable isotopes include oxygen, carbon, nitrogen, hydrogen, and sulfur. These isotope systems have been used in research for many years because they are relatively simple to measure. Recent advances in mass spectrometry now enable the measurement of heavier stable isotopes, such as iron, copper, zinc, and molybdenum.

2.5 TERMINOLOGY IN SOIL SCIENCE

For soil science terms, the reader is referred to the *Glossary of Soil Science Terms* published by the Soil Science Society of America, Madison, Wisconsin, http://www.soils.org. In this section, few of these terms will be defined. For details on many soil science concepts, the reader is directed to Brady and Weil (2003).

Pedon. The smallest three-dimensional unit of soil that has all the primary characteristics of that soil type and can be used to characterize that soil individual. Pedons may occupy approximately 1–10 m^2 of the landscape.

Polypedon. A group of pedons closely associated in the field and having similar soil properties. They are used to serve as the basic classification unit or soil individual.

Epipedon. The diagnostic horizons that occur at the soil surface. This represents the surface horizon of soil profile.

2.5.1 Master Soil Horizon

Soil horizons refers to the layers approximately parallel to the land surface and differing from adjacent layers in their physical, chemical, and biological properties, produced during soil-forming processes. Soil taxonomists use the capital letters O, A, E, B, and C to represent master soil horizons (Figure 2.1).

O horizon. The surface organic horizon in mineral soils. This layer is dominated by fresh or partly decomposed organic material, including plant detrital matter or litter, moss, and lichens. Some are saturated with water for long periods, some were once saturated but drained, and some were never saturated with water. The mineral fraction of this horizon is usually small.

A horizon. Mineral horizons that formed at the surface or below an O horizon. This layer is distinguished by an accumulation of humified organic matter mixed with mineral fraction and is not dominated by properties characteristic of E or B horizons.

E horizon. The mineral horizon whose main feature is loss of silicate clay, iron, or aluminum, leaving sand and silt particles. This horizon is also referred as the "eluviation zone."

B horizon. Soil layers where illuviation has taken place from above. This includes (1) illuvial concentrations of silicate clay, iron, aluminum, humus, carbonates, gypsum, or silica, alone or in combination; (2) evidence of accumulation or removal of carbonates; and (3) residual concentration of iron and aluminum oxides.

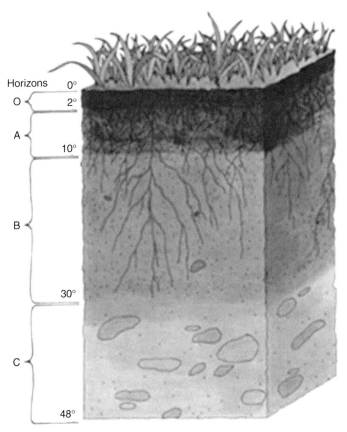

FIGURE 2.1 Soil profile with distinct horizons. (From http://soils.usda.gov/education/resources/k_12/lessons/profile/profile.jpg.)

C horizon. Unconsolidated material underlying the B horizon. This material is not affected by soil-forming processes and lacks the properties of O, A, E, and B horizons.

R horizon. Hard bed rock.

2.5.2 Properties Used in Soil Description

Color. A color designation system (Munsell color system) that specifies the relative degrees of three simple variables: hue, value, and chroma, as follows:
- Hue: Specific color (red, orange, yellow, green, blue, indigo, violet)
- Value: The lightness or darkness of the color, 0 being the darkest and 10 being the lightest
- Chroma: The light intensity, 0 being the least intense and 10 being the most intense

For example, 10YR 6/4 is a color (of soil) with a hue = 10YR, value = 6, and chroma = 4.

Soil texture. The relative proportions of various soil separates in a soil as described by the classes of soil texture of mineral soils (Figure 2.2). These include sand, loamy sand, sandy loam, loam, silt loam, silt, sandy clay loam, and clay. The relative proportion of these particles can be determined by particle size analysis. Soil texture is an expression of both qualitative and quantitative properties including the predominant size or size range of particles. The qualitative aspect is the "feel" of soil material (coarse, gritty, fine, and smooth),

Basic Concepts and Terminology

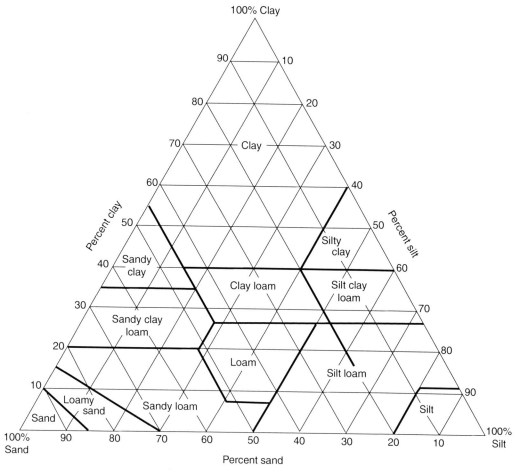

FIGURE 2.2 Soil textural triangle used to determine relative proportion of sand, silt, and clay. (From Soil Survey Staff, 1975.)

and the quantitative aspect refers to relative proportions of various sizes of particles of a given soil.

Structure. Soil structure is defined as the mutual arrangement, orientation, and organization of particles in the soil. In other words, structure refers to the combination or arrangement of primary soil particles into secondary units or peds. The secondary units are characterized on the basis of size, shape, and grade (degree of distinctness).

2.5.3 Soil Taxonomy

Soil classification is the arrangement of soils into groups of soils on the basis of soil characteristics. There are six categories of classification in soil taxonomy.

1. Order (12)
2. Suborder (55)
3. Great group (238)
4. Subgroup (1,243)
5. Family (7,504)
6. Series (18,807)

There are 12 orders of soils, which are sorted by degree of horizon development and the kinds of horizon present. The description of soil orders presented here was obtained from http://soils.usda.gov/technical/classification/orders/, http://soils.ag.uidaho.edu/soilorders/orders.htm, and https://www.soils.org/sssagloss/

Andisols. Soils of volcanic origin. The colloidal fraction is dominated by allophone or Al-humus compounds. Andisols are dominated by short-range-order minerals. They include weakly weathered soils with much volcanic material as well as more strongly weathered soils. Hence, the content of volcanic material is one of the characteristics used in defining andic soil properties. Materials with andic soil properties comprise 60% or more of the thickness between the mineral soil surface and the top of an organic layer with andic soil properties. Andisols are divided into eight suborders: *Aquands, Gelands, Cryands, Torrands, Xerands, Vitrands, Ustands,* and *Udands.*

Alfisols. Alfisols are characterized by translocation of silicate clays to form an argillic horizon (a mineral soil horizon that is characterized by the illuvial accumulation of phyllosilicate clays). These are soils with gray to brown surface horizons and B horizon of illuvial clay accumulation and a base saturation of 35% or greater. These soils are well developed and contain a subsurface horizon in which clays have accumulated. Alfisols are mostly found in temperate, humid, and subhumid regions of the world. They typically have an ochric epipedon, but may have an umbric epipedon. Alfisols are divided into five suborders: *Aqualfs, Cryalfs, Udalfs, Ustalfs,* and *Xeralfs.*

Aridisols. Mineral soils that have an aridic moisture regime, an ochric epipedon, and other pedogenic horizons but no oxic horizon. Aridisols are too dry for mesophytic plants to grow. They have either (1) an aridic moisture regime and an ochric or anthropic epipedon and one or more of the following with an upper boundary within 100 cm of the soil surface: a calcic, cambic, gypsic, natric, petrocalcic, petrogypsic, salic, duripan, or argillic horizon; or (2) a salic horizon and saturation with water within 100 cm of the soil surface for 1 month or more in normal years. Aridisols are divided into seven suborders: *Cryids, Salids, Durids, Gypsids, Argids, Calcids,* and *Cambids.*

Entisols. Mineral soils that have no distinct subsurface diagnostic horizons within 1 m of the soil surface. Entisols have little or no evidence of the development of pedogenic horizons. Most Entisols have no diagnostic horizons other than an ochric epipedon. Entisols are divided into five suborders: *Aquents, Arents, Psamments, Fluvents,* and *Orthents.*

Gelisols. Gelisols have permafrost within 100 cm of the soil surface or have gelic materials within 100 cm of the soil surface and permafrost within 200 cm. Gelic materials are mineral or organic soil materials that have evidence of cryoturbation (frost churning) or ice segregation in the active layer (seasonal thaw layer) or the upper part of the permafrost. Gelisols are divided into three suborders: *Histels, Turbels,* and *Orthels.*

Histosols. Organic soils that have organic soil materials in more than half of the upper 80 cm, or that are of any thickness if overlying rock or fragmental material that has interstices filled with organic soil materials. Histosols are dominantly organic. They are mostly soils that are commonly called bogs, moors, or peats and mucks. They contain at least 20–30% organic matter by weight and are more than 40 cm thick. Bulk densities are quite low, often less than 0.3 g cm^{-3}. Histosols are divided into four suborders: *Folists, Fibrists, Saprists,* and *Hemists.*

Inceptisols. Mineral soils that have one or more pedogenic horizons in which mineral materials other than carbonates or amorphous silica have been altered or removed, but not accumulated to a significant degree. Under certain conditions, Inceptisols may have an ochric, umbric, histic, plaggen, or mollic epipedon. Water is available to plants more than half of the year or more than 90 consecutive days during a warm season. They are more developed than Entisols, but still lack the features that are characteristic of other soil orders.

Inceptisols have altered horizons that have lost bases or iron and aluminum but retain some weatherable minerals. They do not have an illuvial horizon enriched with either silicate clay or with an amorphous mixture of aluminum and organic carbon. Inceptisols are divided into seven suborders: *Aquepts, Anthrepts, Gelepts, Cryepts, Ustepts, Xerepts,* and *Udepts.*

Mollisols. Mineral soils that have a mollic epipedon overlying mineral material with a base saturation of 50% or more when measured at pH 7. Mollisols may have argillic, natric, albric, cambic, gypsic, or petrocalcic horizon, a hystic epipedon, or a duripan, but not an oxic or spodic horizon. They are characterized by a thick, dark surface horizon. This fertile surface horizon, known as a mollic epipedon, results from the long-term addition of organic materials derived from plant roots. Mollisols are divided into eight suborders: *Albolls, Aquolls, Rendolls, Gelolls, Cryolls, Xerolls, Ustolls,* and *Udolls.*

Oxisols. Mineral soils that have an oxic horizon within 2 m of the surface or plinthite as a continuous phase within 30 cm of the surface, and that do not have a spodic or argillic horizon above the oxic horizon. Oxisols are very highly weathered soils that are found primarily in the intertropical regions of the world. These soils contain few weatherable minerals and are often rich in Fe and Al oxide minerals. Oxisols are divided into five suborders: *Aquox, Torrox, Ustox, Perox,* and *Udox.*

Spodosols. Mineral soils that have spodic horizon or a placic horizon that overlies a fragipan. These are acid soils and are characterized by a subsurface accumulation of humus that is complexed with Al and Fe. These photogenic soils typically form in coarse-textured parent material and have a light-colored E horizon overlying a reddish-brown spodic horizon. In undisturbed soils there is normally an overlying eluvial horizon. Spodosols are divided into five suborders: *Aquods, Gelods, Cryods, Humods,* and *Orthods.*

Ultisols. Mineral soils that have an argillic horizon with a base saturation of <35% when measured at pH 8.2. Ultisols are strongly leached, acid forest soils with relatively low native fertility. They are found primarily in humid temperate and tropical areas of the world, typically on older, stable landscapes. These soils are characterized by an accumulation of clays in the subsurface horizon and exhibit strong yellowish or reddish colors resulting from the presence of iron oxides. Ultisols are divided into five suborders: *Aquults, Humults, Udults, Ustults,* and *Xerults.*

Vertisols. Mineral soils that have 30% or more clay, with deep, wide cracks when dry, and either gilgai microrelief, intersecting slickensides, or wedge-shaped structural aggregates tilted at an angle from the horizon. Vertisol soils have a high content of expanding clay and at some times of the year develop deep wide cracks. Vertisols shrink when drying and swell when they become wetter. Vertisols are divided into six suborders: *Aquerts, Cryerts, Xererts, Torrerts, Usterts,* and *Uderts.*

2.5.4 Physical Properties

Soil bulk density. The mass of dry soil per unit bulk volume, including any air-filled spaces. The bulk volume is determined before drying to a constant weight.

$$\text{Bulk density} = \frac{\text{Dry weight of soil (g)}}{\text{Volume of soil solids and pores (cm}^3\text{)}}$$

Bulk density values range from <0.5 g cm^{-3} in organic soils to 1.8 g cm^{-3} in mineral soils. Bulk density values are needed to calculate the total storage of a given nutrient per unit area in a given depth of soil.

Soil particle density. The density of the soil particle, which is the dry mass of the particles expressed on volume of soil solids (not the bulk volume of the particles).

Soil porosity. Refers to the volume fractions of pores in soil. Total porosity does not reflect the pore size distribution.

Soil water. The water loss from soil on drying to a constant mass at 105°C: expressed as the mass of water per unit mass of dry soil or as the volume of water per unit bulk volume of soil.

Soil air. The fraction of the bulk volume of soil that is filled with air at any given time or under any given condition.

2.5.5 CHEMICAL PROPERTIES

Cations. Atom or group of atoms carrying a positive charge. The charge is a result of the presence of more protons than electrons.

Anions. Anions are atoms or group of atoms having more negatively charged electrons than positively charged protons.

Cation exchange capacity (CEC). The sum of *exchangeable bases* plus the total soil acidity at a specific pH value, usually 7.0 or 8.0. When acidity is expressed as salt extractable acidity, the CEC is called the effective cation exchange capacity (ECEC) because this is considered to be the CEC of the exchanger at the native pH value. It is usually expressed in centimoles of charge per kilogram of exchanger.

Anion exchange capacity. The sum of exchangeable anions that a soil can adsorb. Usually expressed as centimoles, or millimoles, of charge per kilogram of soil (or of other adsorbing material such as clay).

Soil pH. The pH of a solution in equilibrium with soil. It is determined by means of a glass, pH electrode or other suitable electrode or indicator at a specific soil–solution ratio in a specified solution, usually distilled water, 0.01 M $CaCl_2$, or 0.01 M KCl.

pH is defined in terms of the activity of the hydrogen ion:

$$pH = -\log\{H^+\}$$

Redox potential (Eh). The potential that is generated between an oxidation or reduction half-reaction and the standard hydrogen electrode (SHE) (0.0 V at pH = 0). In soils, it is the potential created by oxidation–reduction reactions that take place on the surface of a platinum electrode measured against a reference electrode minus the Eh of the reference electrode. This is a measure of oxidation–reduction potential of redox active components in the soil (see Chapter 4).

2.6 UNITS

The following are some of the common units used in various chapters of this book:

Non-SI Unit	Multiply By	To SI Unit
Length		
Inch (in.)	2.54	Centimeter, cm (10^{-2} m)
Foot (ft)	0.304	Meter, m
Mile	1.609	Kilometer, km (10^3 m)
Micron	1.0	Micrometer, μm (10^{-6} m)
Angstrom unit (Å)	0.1	Nanometer, nm (10^{-9} m)
Area		
Acre	0.405	Hectare, ha (10^4 m^2)
Square foot	9.29×10^{-2}	Square meter, m^2
Square inch	645	Square millimeter, mm^2
Square mile	2.59	Square kilometer, km^2

Non-SI Unit	Multiply By	To SI Unit
Volume		
Acre-inch	102.8	Cubic meter, m^3
Cubic foot	2.83×10^{-2}	Cubic meter, m^3
Cubic inch	1.64×10^{-5}	Liter, L
Quart	0.946	Liter, L
Gallon	3.78	Liter, L
Ounce	2.96×10^{-2}	Liter, L
Pint	0.473	Liter, L
Gallon per minute	0.06308	Liter per second, $L\,s^{-1}$
Cubic foot per second (cfs)	28.32	Liter per second, $L\,s^{-1}$
Mass		
Ounce	28.4	Gram, g
Pound (lb)	0.454	Kilogram, kg (10^3 g)
Ton (2000 pounds)	0.907	Megagram, Mg (10^6 g)
Tonne (metric)	1000	Kilogram, kg
Pressure		
Atmosphere	0.101	Megapascal, MPa (10^6 Pa)
Bar	0.1	Megapascal, MPa
Pound per square foot	47.9	Pascal, Pa
Pound per square inch	6.9×10^3	Pascal, Pa
Temperature		
Degrees Fahrenheit (°F −32)	0.556	Degrees, °C
Degrees Celsius (°C +273)	1	Kelvin, K
Energy		
Calories	4.184	Joule, J
British thermal unit (btu)	1.05×10^3	Joule, J
Erg	10^{-7}	Joule, J
Dyne	10^{-5}	Newton, N
Concentrations		
Percent (%)	10	Gram per kilogram, $g\,kg^{-1}$
Parts per million	1	Milligram per kilogram, $mg\,kg^{-1}$
Milliequivalents per 100 g	1	Centimole per kilogram, $cmol\,kg^{-1}$
Radioactivity		
Curie (Ci)	3.7×10^{10}	Becquerel, Bq
Rad	100	Gray, Gy

STUDY QUESTIONS

1. What is the difference between molality and normality?
2. What are the laws of thermodynamics?
3. Define equilibrium constant.
4. Explain the difference between acids and bases.
5. Describe oxidation–reduction reactions.
6. Describe the basic difference between gram-positive and gram-negative bacteria.
7. List the differences between chemolithotrophic and chemorganotrophic bacteria.
8. Define aerobic and anaerobic respiration.
9. List the differences in the composition of polysaccharide, lipids, and proteins.
10. Define anabolism and catabolism.
11. What is an electron donor? How is it different from electron acceptors?
12. Explain and provide the reaction for free-energy change (ΔG).
13. What is enthalpy change? How does it relate to free-energy change?

14. What is an enzyme? List the enzyme classes found in soils.
15. Describe the Michaelis–Menton equation as it relates to enzyme kinetics.
16. What is an activation energy?
17. Define "isotope."
18. Define "half-life."
19. What is the difference between a radioactive isotope and a stable isotope?
20. What are soil horizons? List the factors or physical, chemical, and biological properties used to distinguish individual horizons.
21. Describe soil texture.
22. What are soil orders? List and describe some of the orders.
23. Describe the difference between soil bulk density and soil particle density.
24. What is soil porosity?
25. What is the cation exchange capacity of a soil?

FURTHER READINGS

Brady, N. C. and R. R. Weil. 2003. *The Nature and Properties of Soils*. Prentice Hall, New Jersey. 960 pp.

Madigan, M. T. and J. M. Martinko. 2006. *Brock Biology of Microorganisms*. 11th Edition. Pearson Prentice Hall, Upper Saddle River, NJ.

Segal, I. H. 1976. *Biochemical Calculations*. 2nd Edition. Wiley, New York.

3 Biogeochemical Characteristics

3.1 INTRODUCTION

Wetlands are a key component of the landscape. Although wetlands are one of the most productive ecosystems on earth, their functions and values have only been recognized by society in the past three decades. This is clearly evident as interest in understanding wetland science and protection and conservation of these ecosystems has increased. The Clean Water Act of the United States requires that wetlands be protected from degradation because of their multiple functions including water quality improvement and wildlife habitat. The values and functions of wetlands are now well recognized, as evidenced by public awareness and implementation of national policy to protect and preserve these fragile ecosystems. Economic analysis of various ecosystems suggests that freshwater wetlands and estuaries are more valuable than other ecosystems of the biosphere (Constanza et al., 1997). The estimated economic value is based on the services (elemental cycling, water storage and supply, flood regulation, and water treatment) provided by these ecosystems (Table 3.1).

Although wetlands occupy only a small portion of the total landscape, their overall role in the whole ecosystem is much greater than their area. Wetlands exist at the interface between terrestrial and aquatic environments. They serve as sources, sinks, and transformers of materials (Figure 3.1a,b). Wetlands are divided into two broad categories: freshwater wetlands and coastal wetlands. Freshwater wetlands are situated within interior landscapes, such as floodplains along rivers and streams, wet prairies and hardwood swamps, prairie potholes, picosin wetlands, and marshy areas around lakes and ponds. Coastal wetlands are linked to estuaries that support coastal marshes and mangroves. The biological productivity of wetlands can often exceed that of terrestrial and aquatic systems. Although there is broad range of wetland types on the landscape, they have common characteristics, some structural (water, soils, biota) and others functional (nutrient cycling, water balance, organic matter production and accretion) (Lewis, 1995).

"Wetlands," as the name implies, are the lands located in wet areas. Three major components constitute wetlands:

- Hydrology (presence of water at or near the surface for a period of time)
- Hydrophytic vegetation (wetland plants adapted to saturated soil conditions)
- Hydric soils (saturated soil conditions exhibiting temporary or permanent anaerobiosis)

Wetlands can be very diverse with very high internal spatial heterogeneity with respect to vegetation, soils, and hydrology. Thus, the characteristics and functions of any given wetland can be determined by the position on the landscape, climate, hydrology, vegetation, and soils.

Defining a wetland is complex and difficult, because many wetlands are transient and have distinct moisture gradients ranging from dry soil to flooded. Although it is relatively easy to describe some wetland properties, difficulty can be encountered when attempts are made to precisely define wetlands, especially in establishing boundaries with adjacent uplands. Because of this, several definitions (legal and nonlegal) have been reported, all with three components: water, soils, and vegetation.

> Definition: "Land or areas (such as tidal flats or swamps) containing much soil moisture." Source: Webster's Collegiate Dictionary.

TABLE 3.1
Estimated Economic Value of Selected Ecosystems of the Biosphere

Ecosystem	US $ ha^{-1} year^{-1}
Estuaries	22,832
Swamps and floodplains	19,580
Coastal sea grass/algae beds	19,004
Tidal marsh/mangroves	9,990
Lakes/rivers	8,498
Coral reefs	6,075
Tropical forests	2,007
Coastal continental shelf	1,610
Temperate/boreal forests	302
Open oceans	252

Source: Constanza et al. (1997) and Batzer and Sharitz (2006).

Definition: "Those areas that are inundated or saturated by surface or ground water at a frequency and duration sufficient to support, and that under normal circumstances do support, prevalence of vegetation typically adapted for life in saturated soil conditions. Wetlands generally include swamps, marshes, bogs and similar areas." Source: EPA (40 CFR 230.3(t)) [45 FR 85344, December 24, 1980, as amended at 58 FR 45037, August 25, 1993]. This is a regulatory definition of wetlands used by the U.S. Army Corps of Engineers and the U.S. Environmental Protection Agency in administering dredge and fill permitting under Section 404 of the Federal Clean Water Act. This definition places major emphasis on hydrophytic vegetation, as compared to other two criteria. In practice, this allows for a rapid identification of a wetland, however, precise boundaries cannot be identified using this simple approach.

Definition: "Wetlands are lands transitional between terrestrial and aquatic systems where the water table is usually at or near the surface or the land is covered by shallow water. For purposes of this classification, wetlands must have one or more of the following three attributes: (i) at least periodically, the land supports predominately hydrophytes, (ii) the substrate is predominately undrained hydric soil, and (iii) the substrate is nonsoil and is saturated with water or covered by shallow water at some time during the growing season of each year." Source: Cowardin et al. (1979). U.S. Department of Interior—Fish and Wildlife Service. This definition places emphasis on all three major attributes of wetlands, and is difficult to apply as it requires a comprehensive study of the site. The boundaries identified by this criteria are much more reliable than the EPA definition.

Parallel to federal efforts, many state agencies developed their own definition of wetlands to protect these areas from development. For example, the State of Florida defined wetlands in subsection 373.019(17), F.S. "as those areas that are inundated or saturated by surface water or ground water at a frequency and a duration sufficient to support, and under normal circumstances do support, a prevalence of vegetation typically adapted for life in saturated soils." Soils present in wetlands generally are classified as hydric or alluvial, or possess characteristics that are associated with anaerobic soil conditions. The prevalent vegetation in wetlands generally consists of facultative or obligate hydrophytic macrophytes that are typically adapted to areas having the soil conditions described above. These species, due to morphological, physiological, or reproductive adaptations, have the ability to grow, reproduce, or persist in aquatic environments or anaerobic soil conditions. Florida wetlands generally include swamps, marshes, bayheads, bogs, cypress domes and strands, sloughs, wet prairies, riverine swamps and marshes, hydric seepage slopes, tidal marshes, mangrove swamps, and other similar areas. Florida wetlands generally do not include longleaf or slash pine flatwoods with an understory dominated by saw palmetto.

Biogeochemical Characteristics

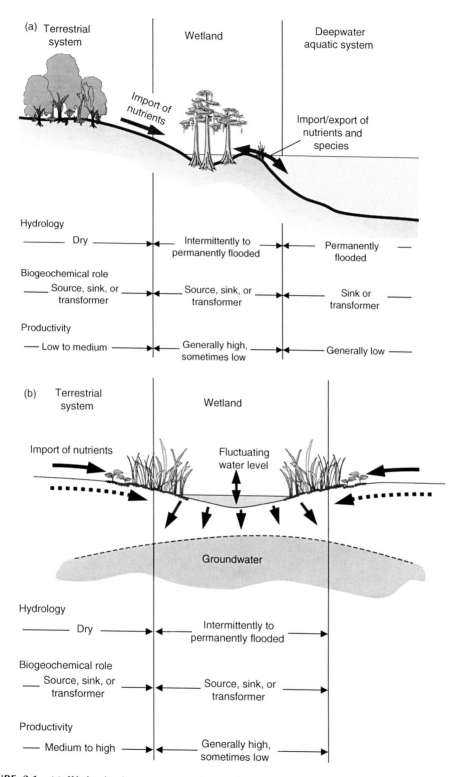

FIGURE 3.1 (a) Wetlands shown as a continuum between terrestrial and deepwater aquatic systems. (b) Wetlands can exist as isolated from other water bodies. (From Mitsch and Gosselink, 2000. With permission.)

Definition: "Hydric soil is a soil that in its undrained conditions is saturated, flooded, or ponded long enough during the growing season to develop anaerobic conditions that favor the growth and regeneration of hydrophytic vegetation." Source: United States Department of Agriculture (USDA), Natural Resource Conservation Service—Food Security Act of 1985 (USDA, NRCS, 2006). This definition was revised by deleting phrase "in its undrained condition" and any direct reference to hydrophytic vegetation (USDA, NRCS, 2006). This was necessary because the phrase "in its undrained condition" implies that drained soils are not hydric soils. "A hydric soil is a soil that is saturated, flooded, or ponded long enough during the growing season to develop anaerobic conditions in the upper part." This definition implies that soil without hydrophytic vegetation can be hydric soil. Saturated soil conditions associated with altered microbial activity and depletion of oxygen can result in changes in various biogeochemical properties including promoting various oxidation–reduction reactions, and movement and accumulation of various reduced compounds such as iron. These processes can result in distinct changes in morphological patterns that may persist in wetland soils during dry conditions, making them useful for identification of hydric soils. For details on field indicators, the reader is referred to a document that provides guidelines for identification and delineating hydric soils (USDA NRCS, 2006).

The National Technical Committee for Hydric Soils (NTCHS) has developed the following criteria for hydric soil identification (USDA NRCS, 2006):

1. All Histosols except Folists
2. Soils in Aquic suborders, Aquic subgroups, Albolls suborder, Salorthids great group, or Pell great groups of Vertisols that are
   a. Somewhat poorly drained and have water table less than 0.5 ft (15 cm) from the surface for a significant period (usually more than 2 weeks) during the growing season
   b. Poorly or very poorly drained and have any of the following:
      i. A frequently occurring water table at less than 0.5 ft (15 cm) from the surface for a significant period (usually more than 2 weeks) during growing season and if textures are coarse sand, sand, or fine sand in all layers within 2 in. or for other soils
      ii. Water table at less than 1.0 ft (30 cm) from the surface for a significant period (usually more than 2 weeks) during the growing season if permeability is equal to or greater than 6 in. h^{-1} (50 cm), in all layers within 20 in. (50 cm)
      iii. A frequently occurring water table at less than 1.5 ft (45 cm) from the surface for a significant period (usually more than 2 weeks) during the growing season if permeability is less than 6.0 in. h^{-1} (15 cm h^{-1}) in any layer within 20 in. (50 cm)
3. Soils that are ponded for long or very long duration during the growing season
4. Soils that are frequently flooded for long or very long duration during the growing season

The Committee on Wetlands Characterization, Water Science and Technology Board, National Academy of Sciences, developed a reference definition for wetland that stands outside the interests of any private or public agency (Lewis, 1995). "A wetland is an ecosystem that depends on constant or recurrent, shallow inundation or saturation at or near the surface of the substrate. The minimum essential characteristics of a wetland are recurrent, sustained inundation or saturation at or near the surface and the presence of physical, chemical, and biological features reflective of recurrent, sustained inundation or saturation. Common diagnostic features of wetlands are hydric soils and hydrophytic vegetation. These features will be present except where specific physicochemical, biotic, or anthropogenic factors have removed them or prevented their development."

Wetlands, as defined by various groups, are limited to areas in which there are emergent plants. At present this definition is the most commonly accepted, although there is considerable disagreement on the boundaries of developing wetlands from upland areas.

In the context of soils and biogeochemistry, we provide a much broader definition of wetlands: "Wetlands consist of a biologically active soil or sediment in which the content of water in or the overlying floodwater is great enough to inhibit oxygen diffusion into the soil/sediment and stimulate anaerobic (oxygen-free) biogeochemical processes and support hydrophytic vegetation." This definition not only makes reference to plants, but also includes biologically active soil or sediment that supports a consortium of microbial communities that regulate a range of oxidation–reduction reactions and direct feedback to vegetation communities.

A narrower scientific definition (and the most commonly accepted one) would limit wetlands to areas in which there are plants. This type of wetland is usually much higher in soil organic matter than those without emergent plants. This is due to the high production of plant biomass and the slow rate of organic matter decomposition, because of the limited supply of oxygen in the wet soil.

3.2 TYPES OF WETLANDS

In North America, common terms used for wetlands are swamps and marshes. Swamps have primarily woody vegetation, whereas marshes are dominated by herbaceous vegetation. However, some of this terminology is interchanged in other parts of the world. In North America, wetlands constitute approximately 110 million ha in the United Sates and 127 million ha in Canada (Mitsch and Gosselink, 2007). Not all wetlands are the same. Depending on landscape position, wetlands differ based on soil types, climate, hydrologic regime, vegetation, physicochemical properties of water or water quality, and anthropogenic disturbance. Broadly, wetlands can be grouped into the following categories (values in parentheses are area in the United States): (Mitsch and Gosselink, 2007):

Coastal wetlands:
- Tidal salt marshes (1.9 million ha)
- Tidal freshwater marshes (0.8 million ha)
- Mangrove wetlands (0.5 million ha)

Inland wetlands:
- Freshwater marshes (27 million ha)
- Peatlands (55 million ha)
- Freshwater swamps and riparian wetlands (25 million ha)

3.2.1 COASTAL WETLANDS

Tidal salt marshes occur along coastlines in middle and high latitudes worldwide. These wetlands are influenced by tides and by freshwater (surface water and groundwater) inputs from adjacent watersheds. In the United States, these marshes are prevalent along the coast of the Gulf Mexico and along the eastern coast from Maine to Florida. These wetlands are also characterized by salt-tolerant plants such as *Spartina* and others. Soils can vary from mineral to organic, with distinct features including sulfide accumulation. As a result of nutrient loading from adjacent watersheds and from tidal water, these ecosystems exhibit high rates of primary productivity.

Tidal freshwater marshes are typically found upstream of estuaries. Water levels in these ecosystems are influenced by tides. These marshes are characterized by emergent macrophytes that are not tolerant to salinity. Vegetation diversity is typically high, with common species including cattails, pickerel weed, wild rice, arrowhead, and others.

Mangrove wetlands are also found along the coastline in subtropical and tropical regions of the world. These ecosystems are characterized by salt-tolerant trees, shrubs, and other plants adapted to brakish and saline tidal waters. These wetlands are also found at the interface where saltwater meets freshwater. In the United States, the Gulf coasts of Texas and Florida contain mangrove swamps. Florida's southwest coast supports one of the largest mangrove swamps in the world. Three types of mangroves have been identified in the United States: *Black mangroves* (*Avicennia* sp.) are characterized by their ability to grow inland. These trees possess root modifications called "pneumatophores," which are used as conduits for gas exchange, including the transport of oxygen to the root zone. *Red mangroves (Rhizophera mangle)* are characterized by their distinctive arching roots. *White mangroves* are characterized by their ability to grow farther inland with no visible modified root structure.

3.2.2 INLAND WETLANDS

Freshwater marshes are widely distributed all over the world, although a significant portion of these are drained for alternative land uses. These ecosystems are characterized by either mineral or organic soils and the presence of emergent macrophytes. They frequently occur along streams in poorly drained depressions, in shallow water along the boundaries of lakes and ponds. Hydrology in these systems can vary with water levels from a few centimeters to one meter and in some marshes complete dry out can occur as a result of extensive drought. Some examples of freshwater marshes are wet meadows, prairie potholes, playa lakes, and vernal pools.

Wet meadows occur in low-lying areas of the landscape, such as farm lands, transitional areas, poorly drained soils in mountain regions, and shallow lake basins. Even though there are periods of high water levels, these marshes can also dry out. Vegetation is dominated by water-tolerant grasses, sedges, and rushes, and often resembles grasslands.

Prairie potholes are characterized as depressional wetlands often found in the upper Midwest including North Dakota, South Dakota, Minnesota, Iowa, and Wisconsin. Hydrology is controlled by snowmelt and rain during spring, which fills low-lying areas forming concentric circles.

Playas are round depressional areas on the landscape fed by fresh water from rainfall. These wetlands occur in southern high plains, including areas in West Texas, Oklahoma, New Mexico, Colorado, and Kansas. Some of these wetlands accumulate salts as water from underlying aquifers supplying salt percolates upward through the soil profile. These wetlands are formed seasonally and may stay flooded during the wet season and dry out during the remainder of the year.

Vernal pools are small depressional wetlands that range in size from small pockets saturated with water to shallow lakes. These wetlands are found in gently sloping grassland landscapes with Mediterranean climate, such as in the West Coast. Many of these wetlands in California have been drained for development. During any given single season, these wetlands may fill and dry several times. The subsurface is either bedrock or a clay layer that helps to hold the water.

Freshwater swamps are dominated by woody plants ranging from the forested red maple (*Acer rubrum*) swamps of the northeastern United States to bottomland hardwood forests in the southeastern United States consisting of tupelo gum (*Nyssa* sp.), oak (*Quercus* sp.), and bald cypress (*Taxodium distichum*). These wetlands also occur in isolated cypress domes fed primarily by rain water. Soils in these wetlands can be mineral and organic, and are characterized by high organic matter content. Soils are always saturated or flooded, creating highly anaerobic conditions.

Riparian wetlands are land areas adjacent to perennial, intermittent, and ephemeral streams, lakes, or rivers. In addition to precipitation, riparian areas receive water from three sources: (1) groundwater discharge, (2) overland and shallow subsurface flow from adjacent uplands, and (3) flow from adjacent surface water body (NRC, 2002). These areas have high water tables and periodic flooding. These areas support a wide range of wetland vegetation including emergent macrophytes, grasses, and trees. These floodplain wetlands have alluvial mineral soils.

Bogs are characterized by spongy peat deposits, acidic waters, sphagnum moss, and low nutrient status. The primary source of water to bogs is precipitation. In the United States, bogs are found in the glaciated northeast and Great Lakes regions and in the southeast (pocosins). Bogs are formed as a result of (1) sphagnum growth over ponds and lakes that slowly fills them and (2) sphagnum growth on wet areas of uplands that creates poorly drained conditions by holding water. As a result of these conditions, several feet of acidic peat deposits can build up. In the southeast, bogs known as pocosins (swamps on the hill) are found on the Atlantic coastal plain with dominant areas in North Carolina. Soils range from mineral to organic, with acidic conditions.

Fens are peat-forming wetlands that receive nutrients from surrounding watershed through drainage and surface runoff. Fens are less acidic and often more nutrient enriched and eutrophic, and support more diverse vegetation than bogs. These wetlands are permanently saturated with water. Fens support sedges, grasses, shrubs, and trees.

Constructed wetlands are built for specific use such as for water treatment, water storage, and wildlife habitat. Depending on the use, constructed wetlands can vary in size, location, soil types, and vegetation. One group of constructed wetlands are those established on a former upland environment for wildlife habitat and water quality improvement. Once created, some of these wetlands are managed to optimize their utilization. For example, some constructed wetlands are managed to optimize water quality improvement. These wetlands are also known as treatment wetlands. Depending on location, the soils in these wetlands can vary from mineral to organic. Although many wetlands are created as a part of mitigating wetland habitat loss, it is not clear how well these systems function as compared to their natural counterparts. Life expectancy of wetlands created for water quality improvement may vary depending on hydraulic and nutrient loading. Often these wetlands require management to maintain long-term sustainability. The effect of original soil type on overall function of a wetland may last only a few years until a significant amount of organic matter is accumulated. Wetlands accumulate organic matter at a much faster rate than other ecosystems, altering the ecosystem hydrology and biogeochemistry. The reader is referred to two books, Mitsch and Gosselink (2007) and Kadlec and Knight (1996), for details on the overall function and management of wetlands.

For the past three decades, constructed wetlands have been heavily used to treat a wide variety of wastewater, including domestic (ranging from individual homes to small towns), agricultural, mine drainage, landfill leachate, urban stormwater, and agricultural drainage or surface runoff water. Constructed wetlands can be grouped into two broad categories:

- Freewater surface (FWS) wetlands
- Subsurface flow (SF) wetlands

The basic biogeochemical processes involved in removing contaminants from wastewaters are the same as those encountered in natural systems. These may include filtration, sedimentation, and microbial degradation. For example, total suspended solids are removed by filtration and sedimentation, biological oxygen demand (BOD) by microbial degradation, nitrogen by nitrification–denitrification, and phosphorus by adsorption and precipitation reactions. These processes are discussed in detail in various chapters of this book.

In FWS wetlands, oxygen exchange is primarily through air–water interactions. Physical, chemical, and biological processes in the water column are important in regulating the contaminant removal. In addition to removal of contaminants through uptake, vegetation can influence physical filtration of suspended solids. Major contaminant removal processes in these systems are accumulation of organic matter, sediments, and associated contaminants. The SF wetlands are designed for water to stay subsurface. Because the water flows through the media (such as soils or artificial substrates), the interaction between water and substrate is much greater than that in FWS wetlands. Surface air exchange in these systems is of minor importance. The root zone of vegetation supports a wide range of microbial communities, which may be involved in removal of contaminants. The design criteria, operation, and performance of these systems are described in various manuals published by governmental agencies such as the U.S. Environmental Protection Agency. For details, the reader is referred to Kadlec and Knight (1996).

Paddy fields can be classified as managed wetlands. Paddy (rice) fields are typically surrounded by earthen levees to retain irrigation and rain water and ensure soil saturation and flooding. Paddy fields are distributed across all continents, with the greatest area in Asia. During the growing season, these systems exhibit characteristics similar to wetlands, with respect to morphology, chemistry, and biology. Paddy fields are the major agroecosystem in the world. Rice is the staple food for nearly half of the world's population, and approximately 95% of the global rice production occurs in fields with soils flooded during at least a part of the growing season. These systems are highly managed with respect to land preparation and water and nutrient management. Thus, paddy fields can be viewed as constructed wetlands used for food production. Rice paddies contribute substantially to the increasing atmospheric methane concentration.

3.3 WETLAND HYDROLOGY

The hydrology of wetlands is defined by hydroperiod (depth, duration, and frequency of inundation or soil saturation), hydrodynamics (direction and velocity of water movement), and source of water (groundwater or surface water). Hydrology controls the biogeochemical characteristics of wetlands, including physical, chemical, and biological properties of soil, productivity of biotic communities, and water quality. Water level and flow in wetlands vary considerably depending on their location on the landscape and are specific for a given wetland. For example, in coastal wetlands water level can fluctuate daily, whereas in freshwater wetlands seasonal fluctuations can be observed in water levels.

The hydroperiod of wetlands is dependent on flow, depth, frequency, duration (the amount of time wetlands are in standing water), seasonality, and frequency of flooding (average number of times wetlands are flooded or saturated). Thus, the hydroperiod of a wetland includes all aspects of water budget (rainfall, evapotranspiration, and subsurface and surface flow) irrespective of the source of the water. The following qualitative definitions have been presented for wetland hydroperiods in tidal and nontidal wetlands (Cowardin et al., 1979; summarized by Mitsch and Gosselink, 2000).

Major sources of water to wetlands can be grouped into

- Precipitation
- Surface water inflow
- Groundwater (inflow through soils)

Relative importance of the three sources of water to wetlands is illustrated in Figure 3.2 (Brinson, 1993). For example, all wetlands are influenced by precipitation. Areas with high precipitation are

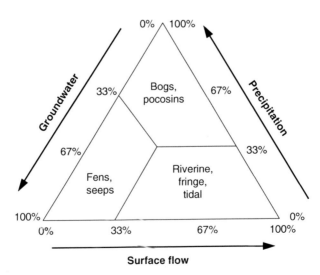

FIGURE 3.2 Relationship between water source and wetland vegetation. (Modified from Brinson, 1993.)

favorable for many wetlands, especially in areas where rainfall/snowfall is in excess of surface runoff and evapotranspiration. However, some wetlands are not influenced significantly by precipitation. These include coastal wetlands, which receive water from tides, and riparian or riverine wetlands, which receive water from rivers and streams. The input of water from precipitation depends on climate and region. Many areas have distinct wet and dry seasons. If the source of water is primarily precipitation, the wetland would likely be an ombrotrophic bog.

Surface inflow and outflow are seasonally dependent on the position of wetlands on the landscape and precipitation. If the source of water is from surface inflow, the wetland would likely to be a riparian or riverine wetland. If the groundwater is the major input, the wetland would likely be a depressional wetland.

Tidal wetlands:
- Subtidal—permanently flooded with tidal water
- Irregularly exposed—surface exposed by tides less often than daily
- Regularly flooded—alternately flooded and exposed at least once daily
- Irregularly exposed—flooded less often than daily

Nontidal wetlands:
- Permanently flooded—flooded throughout the year in all years
- Intermittently exposed—flooded throughout the year except in years of extreme drought
- Semipermanently flooded—flooded during growing season in most years
- Seasonally flooded—flooded for extended periods during growing season, but usually no surface water by the end of growing season
- Saturated—soil is saturated for extended periods during growing season, but standing water is rarely present
- Temporarily flooded—flooded for brief periods during the growing season, but water table during most of the period below the soil surface
- Intermittently flooded—surface is usually exposed with surface water present for variable periods without detectable seasonal pattern

Hydroperiod has major influences on the relative rates and extent of various biogeochemical cycles in wetlands. These effects are discussed in various chapters in this book.

3.4 WETLAND SOILS

The *Glossary of Soil Science Terms* defines soil as "(1) The unconsolidated mineral or organic material on the immediate surface of the earth that serves as a natural medium for the growth of plants. (2) The unconsolidated mineral or organic material on the immediate surface of the earth that has been subjected to and shows the effects of genetic and environmental factors of climate (including moisture and temperature effects), and macro- and microorganisms, conditioned by relief, acting on parent material over a period of time. The soil produced from the combination of these factors differs from the parent material in many physical, chemical, and morphological properties and characteristics."

Although this definition primarily focuses on uplands, in a broader sense, it does include soils that undergo periodic or continuous flooding. Depending on scientific disciplines and ecosystems, soils saturated with water are often called flooded soils, wetland soils, waterlogged soils, and marsh soils. Soil scientists have used terms such as flooded soils, waterlogged soils, and paddy soils. Ecologists refer to these systems as wetland soils. Now, wetland soils have been defined as hydric soils.

Wetland soils are widely distributed throughout the world. Large areas of wetlands suitable for food production have been developed. Many of these soils have already lost wetland soil

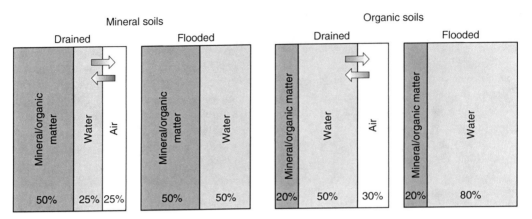

FIGURE 3.3 Composition of soil volume in soils under drained and flooded conditions.

characteristics and are classified as upland soils. The drainage of some of these former wetlands is artificially controlled by water-table manipulation for growing crops. The physical, chemical, and biological characteristics of wetland soils are important in determining the properties and functioning of wetlands.

3.4.1 Physical Characteristics

Soil volume primarily comprises solid matter, water, and air (Figure 3.3). When soils are flooded, most of their nonsolid or pore volume is occupied by water. The solid matter of soil is composed of inorganic material and organic matter. The density of solid mineral matter is approximately 2.6 g cm^{-3}. The density of organic matter is approximately 1 g cm^{-3}. The hydroperiod and water-table fluctuations influence the air-filled pore space of soils, which are extremely important for oxygen diffusion into the soil from the atmosphere (soil aeration).

Upland mineral soils generally consist of about 50% volume solid, 25% water, and 25% air. Oxygen governs most of the biogeochemical reactions in upland soils. For organic soils, approximately one-fifth of the volume is occupied by mineral plus organic matter, whereas the remaining volume by water and air. Generally, reduced compounds are not found in upland soils. Gaseous exchange is not restricted because of continuing of air spaces in upland soils. Gaseous composition of soil pores is about 10–21% oxygen, 0.03–1% carbon dioxide, and trace amounts of nitrous oxide and ammonia.

In wetland soils, oxygen is curtailed, because soil pores are filled with water. In wet mineral soils about 50% of the soil volume is solids, whereas the remaining 50% is occupied by water. In wetland organic soils, a large proportion (up to 90%) of soil volume is occupied by water, with soil organic matter and mineral matter occupying <20%. In the absence of oxygen, facultative anaerobes and obligate anaerobes predominate. Poorly drained soils and soils receiving high carbonaceous waste can also have characteristics similar to wetland soils because of the rapid depletion of oxygen. The presence of wetland vegetation suggests the degree of soil wetness and intensity of anaerobic conditions.

Relative proportion of air and water per unit volume of soil depends on soil type and hydrologic conditions (such as rainfall, irrigation, or water table). For example, the better-drained, coarser-textured soils (Bruin and Dundee) in the Mississippi River floodplain (Figure. 3.4) had high oxygen content in the soil pores at all depths in the profile, whereas poorly drained soils (fine-textured soils such as Mhoon and Tunica) were usually low in oxygen content at lower depths. These conditions can result in a temporary waterlogging condition in the soil, and depending on the retention time of

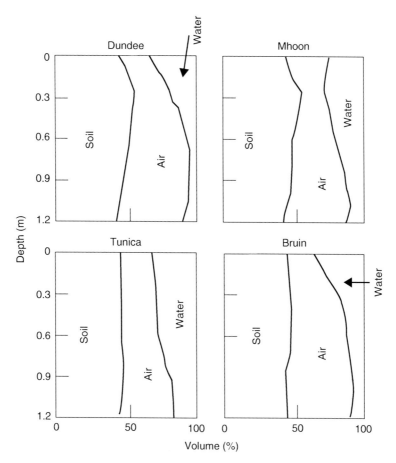

FIGURE 3.4 Soil type and the hydrologic conditions influence the relative proportion of air and water per unit volume of soil (Patrick, 1973).

water these soils can assume the characteristics of wetland soils. However, many of these soils are artificially drained to create optimal field moisture conditions for plant growth.

In upland soils, low oxygen content of heavy-textured soils is shown to decrease root density of plants. Thus, in these systems, internal drainage is necessary to improve soil aeration and plant growth. Similarly, organic soils have high water-holding capacity, and during heavy rainfall events these soils become completely saturated with water. To obtain optional moisture content for growing crops, organic soils are artificially drained with an extensive system of drainage canals, such as those observed in the Everglades Agriculture Area (EAA) of South Florida.

Depending on hydrologic conditions, wetland soils can be characterized as (i) flooded with defined water depth above soil surface, (ii) saturated soil conditions with no excess floodwater, and (iii) water table below the soil surface at a certain depth depending on soil characteristics (Figure 3.5). Under the first two conditions, wetland soils can be classified as hydric soils, whereas the third group can mimic the characteristics of both wetland and upland soils, depending on soil type and hydrologic conditions. Soil taxonomy classifies soils with these characteristics into a suborder *Aquic*, which implies that soil pores are filled with water (from soil surface to a depth of 2 m), and many of the oxidized compounds are enzymatically reduced, with end products of these reductive processes accumulating in the soil. Soil taxonomists classify soils under *Aquic* moisture on the basis of soil color and not accumulation of reduced products. Generally, gray colors or low chroma (2) are used as indicators of soil anaerobiosis (Vepraskas, 1992).

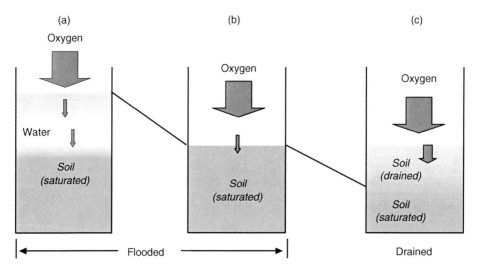

FIGURE 3.5 Wetland soils under different hydrologic conditions.

Besides restricting the supply of oxygen, excess water causes other important effects in wetland soils. Some of these effects are:

1. Softening of the soil material as a result of the weakening effect of water on the bonds holding soil particles together as stable aggregates. This physical effect can have several consequences:
   - Root penetration by wetland plants is made easier, and the soil is much easier to manipulate when wet, an advantage that rice farmers have used for centuries.
   - Trafficability of land is much poorer if it is flooded, and permeability of water is usually decreased.
   - Soil structure may disintegrate under flooded conditions as a result of the reduction in cohesive forces, deflocculation of clay particles, pressure of entrapped air, and destruction of cementing agents.
2. Flooding alters soil temperature by changing the soil color, thus altering the absorption of heat, and affecting the heat of conductivity. Soil is usually cooler at the surface as a result of evaporation.
3. Soil bulk density (weight of dry soil per unit volume) usually decreases as a result of flooding. This is due to the high water-absorption capacity of organic matter and the destruction of soil aggregates.

3.4.2 Chemical Characteristics

Upland soils can be transformed into wetland soils as a result of excessive rainfall, poor drainage, and high oxygen demand in the soil. Under these conditions, oxidized forms are reduced as a result of the respiratory requirements of anaerobic bacteria. Similarly, when wetland soils are drained, they function as upland soils, and under these conditions many of the reduced compounds are oxidized by either chemical or biochemical reactions.

In upland/drained soils, oxidized forms of chemical species dominate the system, whereas in wetland soils reduced forms dominate the system (Figure 3.6). During flooding or in the absence of molecular oxygen, the oxidized forms are converted into reduced forms, through several microbially mediated catabolic processes. Under drained conditions, many of the reduced forms are converted into oxidized forms through chemical and biological processes. Presence of reduced forms indicates soil wetness or anaerobic soil conditions, which are used as indicators of hydric soil identification.

Biogeochemical Characteristics

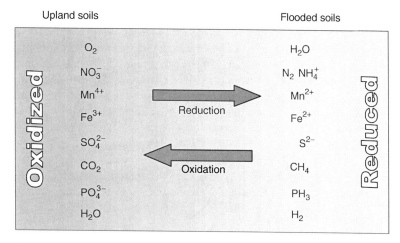

FIGURE 3.6 Redox forms of chemical compounds dominating in drained/flooded soils.

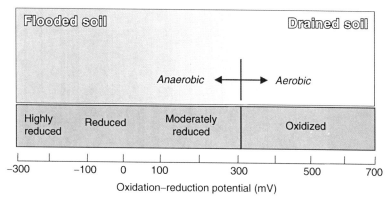

FIGURE 3.7 Relationship among soil hydrologic conditions, E_h and microbial metabolic activities.

Alterations in chemical reactions as a result of flooding or soil anaerobiosis affect soil pH, redox potential, electrical conductivity, CEC, and sorption and desorption of ions. In general, saturated soil conditions result in an increase in pH, electrical conductivity, and ionic strength, but a decrease in soil redox potential (Eh). The pH of most soils tends to approach the neutral point under flooded conditions, with acid soils increasing and alkaline soil decreasing in pH. The increase in the pH of acid soils depends on the activities of oxidants (such as nitrate, iron and manganese oxides, and sulfate) and proton consumption during reduction of these oxidants under flooded conditions. In alkaline soils, pH is controlled by the accumulation of dissolved carbon dioxide and organic acids (see Chapter 4 for detailed discussion).

Redox potential reflects the intensity of reduction or a measure of electron (e^-) activity analogous to pH (which measures H^+ activity). Depending on soil characteristics, upon flooding, Eh generally decreases with time and approaches a steady value. Redox potential is the most common parameter used to measure the degree of soil wetness or intensity of soil anaerobic conditions. A detailed discussion on redox potential is presented in Chapter 4. The relationship among soil hydrologic conditions, Eh, and microbial metabolic activities is depicted in Figure 3.7. The range of Eh values observed in wetland soils is from +700 to −300 mV. Negative values represent high electron activity and intense anaerobic conditions typical of permanently waterlogged soils. Positive values represent low electron activity and aerobic conditions or moderately anaerobic conditions typical of wetlands in a transition zone.

A comparison of selected soil physicochemical properties of freshwater wetland soils is shown in Table 3.2. Numerous examples on soil properties of various wetlands are presented in several chapters of this book. High degree variability exists among wetland types with respect to soil physicochemical properties. Table 3.2 shows a broad range among wetland types. Depending on the geographic location and impacts, a broad range of chemical characteristics can be noted even within same type of wetland. The pH of wetland soils ranged from 3.9 to 6.0, but even a wider range can be observed, if one includes freshwater wetlands located in limestone areas (such as Florida Everglades), which may have pH values of up to 8.0. Similarly, coastal wetlands may have pH in the range of 7–8. The organic matter content varies among wetland types with mineral wetland soils with lower organic matter and higher bulk densities, as compared to organic wetland soils with lower bulk densities. As shown in Table 3.2, bulk densities ranged from 0.07 to 0.55 g cm^{-3}. Significance of these physicochemical properties in biogeochemical cycling of elements is discussed in several chapters of this book.

3.4.3 BIOLOGICAL CHARACTERISTICS

Saturated soil conditions support microbial populations adapted to anaerobic environments. Aerobic microbial populations are restricted to zones where oxygen is available. Most of the aerobic organisms become quiescent, or die, and new inhabitants, largely facultative (these organisms can function under both aerobic and anaerobic environments) and obligate anaerobic bacteria, take over (Figure 3.8).

TABLE 3.2
Comparison of Selected Soil Physical and Chemical Properties (0–20 cm) among Wetland Types

Site	BD (g cm^{-3})	pH	OM (%)	N (mg g^{-1})	P (mg kg^{-1})	Ex-Ca (mg kg^{-1})	Ex-Mg (mg kg^{-1})	Ox-Fe (mg kg^{-1})	Ox-Al (mg kg^{-1})
Fen (MI)	0.22	6.0	55	22.0	900	8,120	906	4,924	2,295
Pocosin (NC)	0.07	3.9	77	14.0	300	1,033		2,370	814
Bog (MD)	0.11	4.5	68	16.9	1,000	710	477	5,710	6,400
Forested swamp (MD)	0.25	4.5	59	14.1	1,400	706	517	5,410	7,600
Marsh (WI)		6.5	41	17.7		12,730	2,300		
Swamp (NC)	0.55	4.1	17	5.7	800	4,630	499	1,301	2,280

Note: BD = bulk density; OM = organic matter; N = nitrogen; P = phosphorus; Ex-Ca = exchangeable calcium; Ex-Mg = exchangeable magnesium; Ox-Fe = oxalate extractable iron; Ox-Al = oxalate extractable aluminum (summarized by Faulkner and Richardson, 1989).

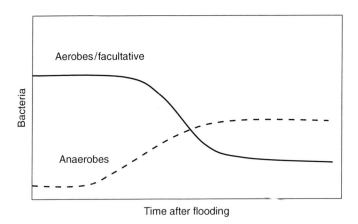

FIGURE 3.8 Schematic of changes in microbial communities after a flooding event.

Fungi, which are active in upland environments, cease to exist in wetland soils. This is primarily due to the absence of oxygen and alteration in soil pH (acid to neutral) under anaerobic conditions. Overall, microbial biomass decreases under saturated soil conditions. The metabolic activities of anaerobic bacteria depend on alternate electron acceptors, such as oxidized forms of nitrogen, iron, manganese, and sulfur. Under wetland soil conditions, rates of many microbially mediated reactions decline, and some reactions may be eliminated and replaced by new ones. New microbial reactions are involved in the reduction of oxidized compounds during respiratory processes, resulting in the production of reduced compounds.

Saturated soil conditions support the growth of microphytic communities including a variety of planktonic and epiphytic in the water column and benthic algae at the soil–floodwater interface. The species composition varies with physicochemical conditions within the wetland. The chlorophyll-bearing organisms are indigenous to most soils and are favored by an environment with adequate moisture and sunlight. These organisms obtain small amounts of various nutrients from soil and water in support of growth. As photosynthetic process liberates oxygen into the water column, these organisms have special significance in wetlands in regulating biogeochemical cycles of nutrients. Diel fluctuations in dissolved oxygen produced as a consequence of photosynthesis often increase the oxygen levels in the floodwater beyond saturation levels during the day and to low levels during the night (see Chapter 6 for discussion).

3.5 WETLAND VEGETATION

Saturated soil conditions support hydrophytic vegetation, one of the criteria used to identify wetlands. These plants have unique characteristics that allow them to adapt to the oxygen-deficient conditions of saturated soils, including physiological adaptations (such as capability to respire anaerobically), anatomic adaptations (such as development of intercellular air spaces), and morphological adaptations (such as water roots and adventitious roots) (see Chapter 6 for details). Approximately 7,000 vascular plant species that grow in wetland environments have been identified. Of those identified, approximately 27% are obligate wetland plants (i.e., plants that only occur under oxygen-deficient soil conditions) (Reed, 1988). The remaining 73% of wetland plant species that occur both in wetlands and uplands are grouped under "facultative wetland plants." The majority of plant species that grow in wetlands also grow in transitional areas adjacent to wetlands.

The Fish and Wildlife Service divides wetland plants into the following groups:

- Obligate wetland plants that occur only in wetland environments
- Facultative wetland plants that occur in wetlands, but are occasionally found in nonwetlands
- Facultative plants that occur in wetlands or nonwetlands
- Facultative upland plants that occur in upland environments, but are occasionally found in wetlands
- Obligate upland plants that occur only in upland environments

The grouping is based on the probability of occurrence of these plants in wetlands.

3.6 BIOGEOCHEMICAL FEATURES OF WETLANDS

Wetland soils have unique characteristics compared to upland soils (Ponnamperuma, 1972) and can be identified by the presence of

- Molecular oxygen in restricted zone
- Reduction of inorganic electron acceptors and accumulation of reduced compounds
- Oxidized soil–floodwater interface
- Exchange of dissolved species between soil and floodwater

- Hydrophytic vegetation
- Accumulation of organic matter (source of electron donor)

3.6.1 Presence of Molecular Oxygen in Restricted Zones

In wetlands, the gaseous exchange between air and soil is severely curtailed because soil pores are filled with water. Oxygen diffusion through water is approximately 10,000 times slower than diffusion in air. Similarly, poorly drained soils or heavy-textured soils or organic soils can become oxygen deficient during short-term flooding as a result of heavy rainfall.

Oxygen is introduced into wetland soils by

- Diffusion and mass flow through floodwater
- Diffusion and mass flow through plants
- Alternate wet/dry cycles or water-table fluctuations

Dissolved oxygen present in the soil pore water is rapidly consumed during aerobic respiration and approaches zero within a few hours depending on soil oxygen demand. Similarly, as the soil water content increases due to rainfall or irrigation, oxygen concentration can rapidly decrease in the soil profile. Although not related to wet soil conditions, application of oxygen-demanding wastes (animal wastes, composts, sewage sludge) or fertilizer (ammoniacal fertilizer) can also result in oxygen depletion in the soil.

Aerobic organisms use oxygen as their primary electron acceptor during their catabolic activities. In the absence of oxygen, aerobic respiration is curtailed and the microbial catabolic activity switches to anaerobic respiration. During this process, aerobic bacteria obtain energy through oxidation of several organic and inorganic compounds, through several intermediate steps and the release of electrons. The electrons released from reduced organic compounds pass through an electron transport chain containing several electron carriers. This ETS is capable of moving electrons from organic substrates to oxygen.

Oxygen is needed in soils for two important purposes:

- Microbial respiration
- Plant root respiration

When oxygen is limited, as is the case in wetland soils, unique conditions are set in motion that differentiate wetlands from uplands in such a way as to increase organic matter in the soil, which may even result in the formation of thick layers of peat, and a change in the distribution of microorganisms (with anaerobic bacteria being more active) and chemical properties of wetland soil.

In wetlands, oxygen cannot always move into the soil rapidly enough to take care of the BOD of organisms because of the blockage of soil pores by water. The fate of oxygen in wetlands is discussed in Chapter 6.

3.6.2 Sequential Reduction of Other Inorganic Electron Acceptors

Oxygen, rather than any other electron acceptor, is the most preferred electron acceptor for microorganisms. As a result, oxygen is always used first by microorganisms. When the oxygen supply is curtailed, specialized microorganisms in the soil have the capacity to switch to other oxidants that replace oxygen in supporting biological oxidation of organic substrates. The other oxidants that can substitute for oxygen are shown in Table 3.3.

Once the oxygen is depleted in the soil, selected aerobic bacteria assumes a facultative role and functions like anaerobic bacteria. During this process, these bacteria use other electron acceptors

TABLE 3.3
Inorganic Redox System in Flooded Soils and Sediments

Sequence of Reduction	System	Oxidized Form	Properties	Reduced Form	Properties
1	Oxygen	O_2	Soluble	H_2O	—
2	Nitrogen	NO_3	Very soluble	NH_4	Soluble
		NO_2	Not adsorbed on soil	NH_4	Adsorbed on exchange complex
		N_2O	Soluble	N_2O	Soluble
				N_2	Slightly soluble
2	Manganese	MnO_2	Insoluble pH dependent	Mn^{2+}	Slightly soluble pH dependent
3	Iron	Fe_2O_3	Insoluble	Fe^{2+}	Slightly soluble pH dependent
		$Fe(OH)_3$	pH dependent		
4	Sulfur	SO_4^{2-}	Relatively soluble pH dependent	S^{2-}	Slightly soluble
				HS^-	Precipitates with metallic cations
				H_2S	Precipitates with metallic cations
5	Carbon dioxide	$CO_2(g)$	Soluble	CH_4	Slightly soluble
		CO_3^{2-}	pH dependent		
		HCO_3	pH dependent		
5	Hydrogen	H_2O		H_2	
6	Phosphate	PO_4^{3-}	Soluble	PH_3	Slightly soluble
		HPO_4^{2-}	pH dependent		
		H_2PO	pH dependent		

for respiration, while using organic or inorganic substrates as their energy source. These electron acceptors are derived from both organic and inorganic sources. The sequential reduction of these compounds is dependent on the electron affinity of the electron acceptor and energy yield, and related enzyme systems in the bacteria (see Chapter 5 for details). In a biological system, oxidation usually proceeds by the removal of electrons and not by the addition of oxygen.

Anaerobic bacteria have ETSs similar to those of aerobic bacteria. The order of reduction starts with oxygen followed by other electron acceptors. Oxygen is the most preferred electron acceptor, followed by oxides of nitrogen such as nitrate, oxides of iron and manganese, sulfate and elemental sulfur, and carbon dioxide. The rate at which these electron acceptors are consumed in soil systems depends on their concentration, readily biodegradable organic compounds, and the microbial population involved in the reductive processes. Relative reduction rates of electron acceptors as a function of time after flooding a soil are shown in Figure 3.9. A detailed discussion of oxidation of organic substrates as influenced by electron acceptors is presented in Chapter 6.

3.6.3 Oxidized Soil–Floodwater Interface

Although oxygen supply to the soil is restricted as a result of flooding, the demand for oxygen remains high. These conditions result in the development of two distinctly different soil layers: (i) an oxidized or aerobic surface soil layer where oxygen is present and (ii) an underlying reduced anaerobic soil layer where no free oxygen is present (Figure 3.10).

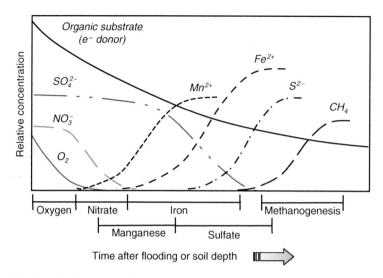

FIGURE 3.9 Relative reduction rates of electron acceptors as a function of time after flooding a soil.

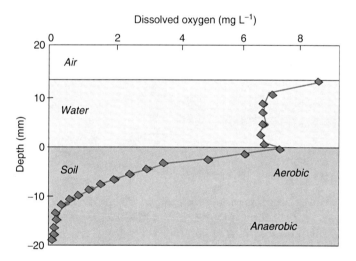

FIGURE 3.10 Schematic showing aerobic and anaerobic regions in a flooded system.

The thickness of this oxidized soil layer and associated processes are discussed in Chapter 5. Ecologically, the thickness of this oxidized layer has a significant effect on regulating nitrogen reactions (as discussed in Chapter 8) and functions as a sink for reduced compounds diffusing from the underlying anaerobic soil layer.

The depth of the oxidized soil–floodwater interface where a given electron acceptor is stable depends on the inflow of electron acceptors in the system, the availability of organic substrates, and hydrologic conditions. The sequential reduction of electron acceptors as a function of depth follows the order of oxygen reduction at the soil–floodwater interface, followed by other electron acceptors. The soil depth at which these reductions occur are depicted in Figure 3.9. It should be noted that the electron acceptors (listed in Table 3.3) are utilized in the soil simultaneously at different soil depths. Thus, wetlands can be characterized by measuring the concentration of reduced species such as manganous manganese (Mn^{2+}), ferrous iron (Fe^{2+}), sulfide, and methane or by the concentration of oxidized species such as oxygen, nitrate, and sulfate.

3.6.4 Exchanges at the Soil–Water Interface

The accumulation of reduced compounds in the anaerobic or reduced soil layer results in the establishment of concentration gradients across the aerobic–anaerobic interface. The concentration of reduced compounds is usually higher in the anaerobic layer, which results in upward diffusion into the aerobic soil or floodwater where they are oxidized. Similarly, some of the dissolved oxidized compounds diffuse downward, that is, from the floodwater or aerobic soil layer into the underlying anaerobic soil layer, where they will be reduced. For example, the steep gradients in ammonium concentrations in the soil profile are due to diffusion into aerobic soil layer or floodwater, and subsequent oxidation to nitrate (Figure 3.11). The nitrate formed diffuses downward into the anaerobic soil layer and is consumed as electron acceptor by microorganisms (Figure 3.12). Similar oxidation

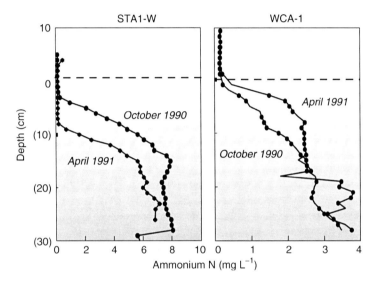

FIGURE 3.11 Concentration of ammonium N with depth in a flooded system. STA1-W = stormwater treatment Area-1-west; WCA-1 = water conservation Area-1. Both welands are located in northern Everglades of Florida. (Reddy, K. R., Unpublished Results, University of Florida.)

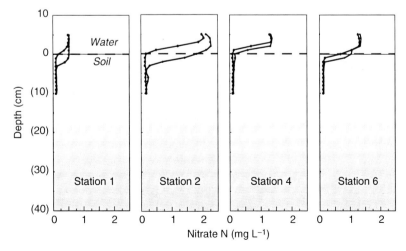

FIGURE 3.12 Concentration of nitrate N with depth in a wetland (Central Florida) receiving sewage effluent. (Reddy, K. R., Unpublished Results, University of Florida.)

and reduction reactions occur in soil profiles for carbon, sulfur, iron, and manganese (see discussions in respective chapters). Another example of the exchange of dissolved methane and sulfate between the soil and overlying water column of a constructed wetland used for treating eutrophic lake water is shown in Figure 3.13.

These are typical characteristics of wetlands with overlying floodwater and lake and marine sediments. The exchange rates between soil and water column determine whether wetland soils or sediments are functioning as a sink or source for nutrients. The rate of exchange of dissolved species depends on:

- Concentration of dissolved species in soil pore water
- Soil type and other related physicochemical properties (pH, CEC, organic matter content, and bulk density)
- Concentration of dissolved species in the floodwater
- Kinetics of related biogeochemical processes in soil and floodwater

3.6.5 Presence of Hydrophytic Vegetation

One of the unique characteristics of wetlands is the presence of hydrophytic vegetation. These characteristics are used in delineating wetlands and upland ecosystems. The oxygen requirements of wetland plant roots are largely met by oxygen transport through the aerenchyma tissue of these plants (Figure 3.14). Root oxygenation is an important characteristic that helps wetland plants adapt to intense soil anaerobiosis. Oxygen transport through plants occurs through diffusion and mass flow.

Under most conditions, oxygen release into rhizosphere should be adequate to oxidize excessive levels of reduced compounds in order for wetland plants to survive soil anaerobiosis. Release of oxygen into the rhizosphere is demonstrated by the observation of oxidation of Fe^{2+} to Fe^{3+} and precipitation on the root surface, oxidation of carbonaceous compounds, and nitrification of ammonium nitrogen. A detailed discussion on the fate of oxygen in the rhizosphere is presented in Chapter 6.

3.7 TYPES OF WETLAND/HYDRIC SOILS

By definition, all wetlands have hydric soils. However, land areas with hydric soil features may not support all functions of wetlands. With a broader definition, wetland soils can be divided into the following major groups:

- Waterlogged mineral soils
- Organic soils
- Marsh soils
- Paddy soils
- Subaqueous soils

3.7.1 Waterlogged Mineral Soils

Waterlogged mineral soils are those in which part or the whole soil profile is saturated for a sufficient period of time to create distinctive gley horizons in the profile. These soils can be sandy, loamy, or clay soils. The other most commonly referred term is flooded soils. We confine our discussion in this group primarily to mineral soils. The distinctive characteristics of these soils are

- Increase in organic matter accumulation in surface horizon
- Mottled zone (gley horizon) where iron and manganese accumulate
  - Hydrated geothite (α-FeOOH)
  - Lepidocrocite (γ-FeOOH)
  - Pure oxides

Biogeochemical Characteristics

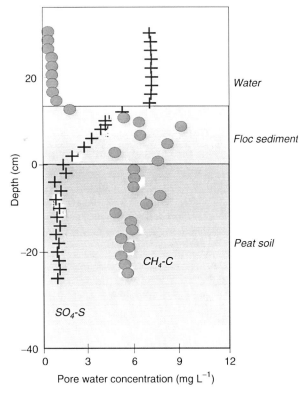

FIGURE 3.13 Concentration of methane and sulfate with depth in a flooded system. (Redrawn from D'Angelo and Reddy, 1994b.)

FIGURE 3.14 Micrograph of a cross-sectional view of aerenchyma tissue of plants. (Courtesy: Brix, H., University of Aarhus, Denmark.)

- Permanently reduced zone (gray color or bluish-green color with B_g or C_g horizon)
- Formation of secondary minerals:
  - Hydrated magnetite
  - Pyrite (FeS_2)
  - Siderite ($FeCO_3$)
  - Vivianite ($Fe_3(PO_4)_2 \cdot 8H_2O$).
  - Ferrous silicates
  - Jarosite ($KFe_3(SO_4)_2(OH)_6$)

The change in color from a well-oxidized upland soil (red, orange, or yellow) to a wetland soil proceeds along a continuum until the soil is "gleyed" or assumes gray color (Figures 3.15a and 3.15b). The soil profile in Figure 3.15a has been in an oxidized state for most of its history in the upper part of its profile and in a moderately reduced state in the lower part of its profile. The soil profile in Figure 3.15b has been in a reduced state throughout the profile for most of its history. Colors are mostly chroma 2 or less. Soils with a dominate chroma of 3 may be saturated but not reduced for much of the time.

When the water table is at the surface, anaerobic or reducing conditions will create an ideal environment for reduction of iron and manganese oxides to reduced iron and manganese, respectively, resulting in accumulation of reduced cations in soil pore water (Figure 3.16). Under these conditions, the cation exchange sites are usually dominated by reduced iron and manganese ions displacing base cation such as calcium, magnesium, and ammonium into soil pore water.

As the water table is lowered as a result of drainage, the reduced cations and other base cations present in soil pore water are transported downward. The center of the soil aggregate in the drained portion of the soil may remain anaerobic, whereas the surface of the soil aggregate becomes aerobic, as a result of oxygen diffusion. The dissolved Fe^{2+} and Mn^{2+} are transported through the soil matrix until they encounter another aerobic zone, where they are oxidized and precipitated. This process is generally heterogeneous, and leads to soils with a matrix of one color and mottles of another (Schwertmann, 1993).

In the transition zone, where soil is alternating from unsaturated and well-aerated to saturated, soil mottles will form. During this period, Fe^{2+} and Mn^{2+} present in the outer layer of soil are oxidized to iron and manganese oxides, respectively. This oxidation reaction can also result in the formation of ferric and manganic oxyhydroxides, respectively. The center of the aggregate will remain anaerobic with high concentrations of Fe^{2+} and Mn^{2+}, assuming a gray color as a result of reduction, whereas the outer layers assume a brown color as a result of oxidation. These conditions result in a mottled appearance with a predominantly red to yellow soil matrix with gleyed mottles. Subsequently, if the soil is allowed to drain for a sufficiently long period, the Fe^{2+} and Mn^{2+} present in the center of the soil aggregate are completely oxidized. The depth at which these mottle formations appear can be used as a defining characteristic of water-table depth. The closer these mottle formations are to the surface, the closer the water table will be to the surface, thus developing wetland soil characteristics. When drained soil is flooded, the pore spaces between soil aggregates are saturated with water and become anaerobic, whereas the center of the aggregate may remain aerobic. The outer layers may have Fe^{2+} and Mn^{2+}, whereas in inner layers iron and manganese are present in oxidized forms. These conditions are very short-term and transitionary.

The mechanism of mottle formation is governed by oxidation–reduction reactions of iron and manganese as a result of the fluctuating water table. Soil mottles vary in size, but generally small patches of soil are colored as a result of oxidation of ferrous iron. Mottles form in areas of the soil profile that undergo intermittent wetting and drying, and consequently intermittent periods of oxidation and reduction. The color of mottles is regulated by the amount of iron oxides present in the soil. Mottles with a chroma of <2 infer a seasonally high water table, which is indicative of a more "developed" Aquic moisture regime.

Biogeochemical Characteristics

FIGURE 3.15 (See color insert following page 392.) Soil profiles showing (a) oxidized and (b) reduced forms of soils.

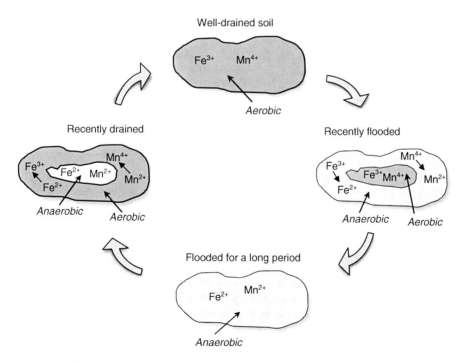

FIGURE 3.16 Oxidation–reduction of iron and manganese under flooded and drained soil conditions.

In soils with low permeability and poor drainage, flooding can result in a ponding condition. Gleying occurs rapidly in the surface horizon as a result of reduction processes, whereas the lower horizons still remain aerobic. As the water moves down through the profile through channels, it carries reduced Fe and Mn, resulting in coating of soil aggregates, and thus mottle formation. These conditions result in B_g horizon.

In A horizons, some Fe^{2+} and Mn^{2+} can also be precipitated around plant roots. In many wetland plants, reddish-brown layers are seen around the roots, as a result of Fe^{2+} oxidation and subsequent precipitation. This can also result in mottle formation along the channels of plant roots with surrounding gleyed soil. Root zone oxidation is most visible in completely reduced and gleyed soils, where they appear on red to yellow mottles.

Pedologists have used the morphological features of oxidation–reduction characteristics of iron and manganese to characterize hydric soils (see Section 3.8) and refer to these as redoximorphic features. These features are referred to as redox concentrations (Fe and Mn concentrations), redox depletion (movement of reduced Fe and Mn), and redox matrix. Redox concentrations are the zones (usually reddish in color) where Fe/Mn are concentrated. They are of three types (Figure 3.17): (1) nodules and concretions, which are extremely firm (hard when dry) irregular-shaped bodies and most often are not reliable evidence of current wetness but reflect prior wetness; (2) masses, which are soft bodies within the soil matrix; and (3) pore linings, which are coatings or impregnations on a natural soil surface (root pore, air pore, ped face, crack face). Redox depletions are zones (usually grayish in color) where Fe/Mn oxides have been removed. Redox depletions usually occur when reduced Fe and Mn move along with water either by diffusion or mass flow to zones of lower concentrations. Some of these depletions may occur along the root channels. Reduced iron oxidizes when exposed to air (see Richardson and Vepraskas, 2001). We prefer not to use "redoximorphic features" or "redox concentrations" to describe the role of iron in altering morphological features of soils. We believe the use of "redox" in this way is not proper, as redox refers to broad group of biogeochemical reactions as described in this book.

FIGURE 3.17 (See color insert following page 392.) Redox concentrations shown as (a) nodules and concretions, (b) pore linings, and (c) soft bodies within the soil matrix. (USDA–NRCS, 2006.)

FIGURE 3.18 Soil subsidence due to oxidation of organic matter. (Location: Everglades Research and Education Center, University of Florida, Belle Glade, FL.)

3.7.2 Organic Soils (Histosols)

Histosols (also known as peatlands or organic soils) are characterized with high organic matter (>12% total carbon) in the upper 1 m of the profile. Soil formation is due to accumulation of partially decomposed organic matter, where peat decomposition processes cannot keep up with primary productivity. Extensive areas of peatlands are found in northern parts of the United States (e.g., in Alaska and Minnesota) and in Canada. In southern regions of the United States, large areas of peatlands are found in the pocosins and Carolina bays of the southeast and the Everglades in south Florida. Surface horizons are usually well decomposed, underlain by fibrous undecomposed peat. This differentiation is based on drained organic soils. Organic soils have high water-holding capacity and poor drainage. Upon flooding, these soils are rapidly reduced assuming intense anaerobic soil conditions, with E_h dropping to <-200 mV within a few days of flooding. Under natural conditions, as mentioned earlier, peat forms as a result of plant litter accumulation from hydrophytic vegetation. Several million hectares of peatlands in North America, Europe, the USSR, and Southeast Asia have been drained for farming. For example, the peat deposits in the Everglades of south Florida were formed about 5,000 years ago from sawgrass. When these flooded organic soils in the Everglades were drained for agricultural purposes, the process of organic matter accumulation was completely reversed. Under drained conditions, Everglades peat soils are currently oxidizing at a rate of about 2 cm year^{-1}. As shown in Figure 3.18, soil below the houses constructed at ground level with pilings resting on the bedrock is lost as a result of subsidence. At current rate of subsidence, by the year 2050, a majority of the agricultural land in the EAA will have peat depths of few cm, making them less attractive for productive agriculture.

Soil subsidence is governed by

- Rate of microbial oxidation
- Climatic conditions (temperature and rainfall)
- Water management (frequency of draining and flooding, and water-table fluctuations)
- Cropping system
- Geological subsidence
- Compaction
- Nature of organic matter
- Fire and wind erosion

Soil subsidence through biological oxidation can be reduced by creating anaerobic conditions, such as flooding organic soils for short periods (1–2 months) in Florida, primarily to control weeds and nematodes. This practice along with high water-table management can reduce subsidence due to biological oxidation.

Organic soils are classified on the basis of the presence of identifiable plant material:

- *Saprists*. About two-thirds of the material is well decomposed and <1/3 of the plant material is identifiable.
- *Fibrists*. About one-third of the material is well decomposed and >2/3 of the plant material is identifiable.
- *Hemists*. About half of the material is well decomposed and the other half contains identifiable plant material.
- Muck = well-decomposed peat material
- Peat = fibrous, with identifiable plant material

3.7.3 Marsh Soils

There are two types of marsh soils: freshwater marsh soils and saltwater marsh soils. These soils are generally permanently waterlogged. Freshwater marshes occur in the areas adjacent to lakes and streams, whereas saltwater marshes occur in estuaries. Typical characteristics of these soils are

- Accumulation of plant residues
- Permanently reduced B_g or C_g horizon

The subscript "g" is used to indicate a gleyed horizon resulting from the reduction of Fe^{2+}. Use of "g" with "B" indicates pedogenic changes in the horizon, and use of "g" with "C" suggests no pedogenic changes in the horizon.

Freshwater marshes can be divided into

- Upland marshes (ombrotrophic):
  - Receive rain water low in bases and hydrodynamically isolated
  - Have pH range 3.5–4.5
- Lowland marshes (rheotrophic):
  - Receive inflow water and bases from surrounding areas
  - Have pH range 5–6

These soils closely resemble Histosols except that marsh soils have a wider range of organic carbon content when compared to that of 30–45% for Histosols.

Saltwater marsh soils under flooded conditions have neutral pH and support salt-tolerant plants. Many of these soils contain large amounts of FeS_2. Pyrite formation occurs as a result of high sulfate concentrations of seawater associated with high concentrations of Fe^{2+} in the sediments and rapid accumulation of organic matter (see Chapter 11 for details).

Flooded conditions:

$$Fe(OH)_2 + H_2S = FeS + 2H_2O$$

$$FeS + S = FeS_2$$

After these soils are drained, pyrite is oxidized to ferric hydroxide resulting in severe acidity to pH less than 2. The bacteria involved in oxidizing FeS_2 are *Thiobacillus ferroxidans* and *T. thioxidans*.

Drained conditions:

$$FeS_2 + O_2 = Fe(OH)_3 + H_2SO_4$$

3.7.4 Paddy Soils

Paddy soils are managed for the purpose of growing rice. Rice is grown in flooded soils, poorly drained soils, and well-drained soils. In this section, we primarily discuss the flooded paddy soils (lowland) used for rice production. Management practices imposed on paddy soils include

- Plowing and land leveling
- Flooding: 5–15 cm water
- Alternate flooding/draining during growing season
- Draining after a rice harvest
- Leaving fallow for 4–6 months during a nongrowing season
- Developing hard pan or plow pan

Continuous puddling of soil in the surface 15–20 cm soil layers creates a hard pan below the plow layer, which is highly compacted and has very low permeability. The hard pan development does not occur in recently developed paddy soils, but in soils that have been under cultivation for several years.

In paddy soils, oxygen is introduced through the floodwater and consumed at the soil–water interface, and to some extent oxygen is also introduced through the plants into the root zone. Manganous manganese and ferrous iron formed in the anaerobic zone of surface layer diffuses in two directions: (1) to the surface layer, where it is oxidized and (2) to the subsurface layer, where it is oxidized.

Oxidation of Mn^{2+} to Mn^{4+} and Fe^{2+} to Fe^{3+} occurs

- At the soil–water interface
- Around rice roots
- In the B horizon just below the hard pan
- In the lower G horizon near the groundwater surface

3.7.5 Subaqueous Soils

These soils are formed from river, lake, and ocean sediments. Ponnamperuma (1972) uses "soil" terms to describe the uppermost layers of unconsolidated aqueous sediments for the following reasons:

- Sediments are formed from soil components.
- All soil-forming processes proceed in the upper layer of sediments.
- Sediments are high in organic matter.
- Bacterial populations are similar to those of soil.
- Surface layers show distinct horizons.
- Sediments, like soils, differ in texture, oxidation–reduction level, clay, mineralogy, and organic matter content.

This concept was adopted by pedologists and identifies these soils as subaqueous soils, which are permanently flooded soils that occur immediately below a water depth of <2.5 m. These soils have been mapped in an estuarine environment (Demas et al., 1996; Demas and Rabenhorst, 1999). The Natural Resource Conservation Service has amended the definition of soil to include sediments under as much as 2.5 m water (Soil Survey Staff, 1999). The justification used for including sediments of shallow water environments is that these sediments undergo soil-forming processes and are capable of supporting rooted plants, and meet the definition of soil according to the criteria defined in *Soil Taxonomy*.

3.8 FIELD INDICATORS OF HYDRIC SOILS

Hydric soils are defined as soils that formed under conditions of saturation, flooding, or ponding during the growing season long enough to develop anaerobic conditions in the upper part of the soil (Federal Register, July 13, 1994 and 1995). Most of the hydric soils exhibit morphological characteristics that result from repeated periods of saturation or flooding, for more than few days. As discussed previously, flooding or saturation creates oxygen-deficient conditions promoting anaerobic biogeochemical processes. These processes may include accumulation of organic matter, reduction, translocation, and accumulation of reduced iron and other elements, which may result in morphological characteristics that may persist in the soil during wet and dry seasons, making them useful for identifying hydric soils (USDA NRCS, 2006). The field indicators identified are regionally specific and are used as a guide to help identify and delineate hydric soils under field conditions. Soils function as long-term integrators and any morphological indicator developed may be more reliable in areas where consistent information on hydrology and vegetation is not available. For details on hydric soil field indicators, the reader is referred to USDA NRCS document. A brief summary on field indicators for hydric soils is provided in this chapter.

The proposed indicators are to be used for a specific Land Resource Region (LRR) or the Major Land Resource Areas (MLRA) (Figure 3.19). To document a hydric soil, dig a hole and describe the soil profile to a depth of approximately 50 cm (20 in.). Using the completed soil description specify which, if any, of the indicators have been met. Deeper examination of soil may be required where field indicators are not easily seen within 50 cm (20 in.) of the surface. It is always recommended that soils be excavated and a profile described as deep as necessary to make reliable interpretations. For example, examination to less than 50 cm (20 in.) may suffice in soils with surface horizons of organic material or mucky mineral material because these shallow organic accumulations only occur in hydric soils. Conversely, the depth of excavation should be greater than 50 cm (20 in.) in Mollisols because the upper horizons of these soils, due to the masking effect of organic material, often contain no visible redoximorphic features. In many locations, it is necessary to make exploratory observations to a meter or more. These observations should

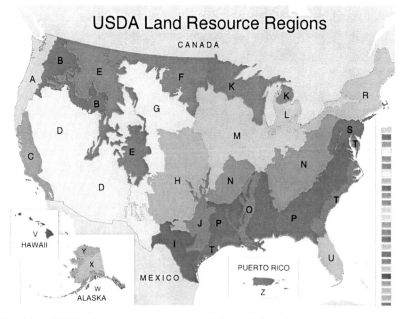

FIGURE 3.19 Map of USDA land resource regions. (USDA–NRCS, 2006.)

Biogeochemical Characteristics

be made with the intent of documenting and understanding the variability in soil properties and hydrologic relationships on the site.

Many of the hydric soil indicators were developed for delineation purposes. During the development of these hydric soil indicators, observations were concentrated near the edge of the wetlands and in the interior of wetlands; therefore, there are wetlands that lack any of the approved hydric soil indicators in their wettest portions. Delineators and other users of the hydric soil indicators should concentrate their observation efforts at the wetland edge when these conditions are suspect.

To determine whether an indicator is present or not, it is critical to know exactly where to look. Depths used in the indicators are measured from the muck or mineral soil surface in most of the United States. We should look for an indicator at the soil surface nationwide when applying indicators A1 and A2 and in LRRs F, G, H, and M if the material beneath any mucky peat or peat is sandy.

In LRRs R, W, X, and Y, our observations should begin at the top of the mineral surface (underneath any or all fibric, hemic, or sapric material) except for the application of indicators A1 and A2. In the remaining LRRs and in LRRs F, G, H, and M, if the material beneath any mucky peat or peat is not sandy, our observations should begin at the top of the muck or mineral surface (underneath any fibric or hemic material) except for the application of indicators A1 and A2.

The following list of indicators was obtained from the USDA NRCS (2006) report on field indicators of hydric soils. Field indicators of hydric soils are grouped as follows:

- All soils
- Sandy soils
- Loamy and clayey soils

3.8.1 ALL SOILS

All mineral layers above any of the A indicators except for indicator A16 have dominant chroma 2 or less, or the layer(s) with dominant chroma of more than 2 is less than 15 cm (6 in.) thick.

A1. Histosol (except Folist). For use in all LRRs or *Histel* (except Folistel); for use in LRRs with permafrost (Figure 3.20).

A2. Histic Epipedon. For use in all LRRs. A Histic Epipedon underlain by mineral soil material with chroma 2 or less.

A3. Black Histic. For use in all LRRs. A layer of peat, mucky peat, or muck 20 cm (8 in.) or more thick starting within the upper 15 cm (6 in.) of the soil surface having hue 10YR or yellower, value 3 or less, and chroma 1 or less underlain by mineral soil material with chroma 2 or less.

A4. Hydrogen Sulfide. For use in all LRRs. A hydrogen sulfide odor within 30 cm (12 in.) of the soil surface.

A5. Stratified Layers. For use in LRRs C, F, K, L, M, N, O, P, R, S, T, and U; for testing in LRRs V and Z. Several stratified layers starting within the upper 15 cm (6 in.) of the soil surface. One or more of the layers has value 3 or less with chroma 1 or less, or it is muck, mucky peat, peat, or mucky-modified mineral texture. The remaining layers have chroma 2 or less (Figure 3.21).

A6. Organic Bodies. For use in LRRs P, T, U, and Z. Presence of 2% or more organic bodies of muck or a mucky-modified mineral texture, approximately 1–3 cm (0.5–1 in.) in diameter, starting within 15 cm (6 in.) of the soil surface. In some soils, the organic bodies are smaller than 1 cm (Figure 3.22).

A7. 5 cm Mucky Mineral. For use in LRRs P, T, U, and Z. A layer of mucky-modified mineral soil material 5 cm (2 in.) or more thick starting within 15 cm (6 in.) of the soil surface.

A8. Muck Presence. For use in LRRs U, V, and Z. A layer of muck with value 3 or less and chroma 1 or less within 15 cm (6 in.) of the soil surface.

FIGURE 3.20 (See color insert following page 392.) This soil meets the requirements of indicator A1. Muck (sapric soil) material is about 0.5 m thick. The shovel is approximately 1 m. (USDA–NRCS, 2006.)

FIGURE 3.21 (See color insert following page 392.) Indicator A5 (Stratified Layers) in sandy soil material. (USDA–NRCS, 2006.)

- *A9. 1 cm Muck.* For use in LRRs D, F, G, H, P, and T; for testing in LRRs C, I, J, and O. A layer of muck 1 cm (0.5 in.) or more thick with value 3 or less and chroma 1 or less starting within 15 cm (6 in.) of the soil surface.
- *A10. 2 cm Muck.* For use in LRRs M and N; for testing in LRRs A, B, E, K, L, S, W, X, and Y. A layer of muck 2 cm (0.75 in.) or more thick with value 3 or less and chroma 1 or less starting within 15 cm (6 in.) of the soil surface.

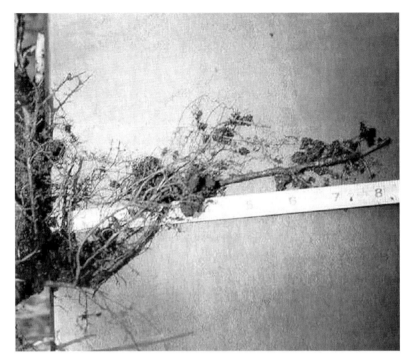

FIGURE 3.22 (See color insert following page 392.) Indicator A6 (Organic Bodies). Scale is in inches. (USDA–NRCS, 2006.)

A11. Depleted Below Dark Surface. For use in all LRRs except W, X, and Y; for testing in LRRs W, X, and Y. A layer with a depleted or gleyed matrix that has 60% or more chroma 2 or less starting within 30 cm (12 in.) of the soil surface that has a minimum thickness of either of the following:
  a. 15 cm (6 in.)
  b. 5 cm (2 in.) if the 5 cm (2 in.) consists of fragmental soil material

Loamy/clayey layer(s) above the depleted or gleyed matrix must have value 3 or less and chroma 2 or less. Any sandy material above the depleted or gleyed matrix must have value 3 or less and chroma 1 or less, and at least 70% of the visible soil particles must be covered, coated, or similarly masked with organic material.

A12. Thick Dark Surface. For use in all LRRs. A layer at least 15 cm (6 in.) thick with a depleted or gleyed matrix that has 60% or more chroma 2 or less starting below 30 cm (12 in.) of the surface. The layer(s) above the depleted or gleyed matrix must have value 2.5 or less and chroma 1 or less to a depth of at least 30 cm (12 in.) and value 3 or less and chroma 1 or less in any remaining layer above the depleted or gleyed matrix. Any sandy material above the depleted or gleyed matrix must have at least 70% of the visible soil particles covered, coated, or similarly masked with organic material.

A13. Alaska Gleyed. For use in LRRs W, X, and Y. A mineral layer with a dominant hue of N, 10Y, 5GY, 10GY, 5G, 10G, 5BG, 10BG, 5B, 10B, or 5PB, with value 4 or more in more than 50% of the matrix. The layer starts within 30 cm (12 in.) of the mineral surface, and is underlain within 1.5 m (60 in.) by soil material with hue 5Y or redder in the same type of parent material.

A14. Alaska Redox. For use in LRRs W, X, and Y. A mineral layer that has dominant hue 5Y with chroma of 3 or less, or a gleyed matrix, with 10% or more distinct or prominent redox

FIGURE 3.23 (See color insert following page 392.) Indicator A14 (Alaska Redox) in a gleyed matrix with reddish redox concentrations around pores and root channels. (USDA–NRCS, 2006.)

concentrations as pore linings with value and chroma 4 or more. The layer occurs within 30 cm (12 in.) of the soil surface (Figure 3.23).

A15. Alaska Gleyed Pores. For use in LRRs W, X, and Y. A mineral layer that has 10% or more hue N, 10Y, 5GY, 10GY, 5G, 10G, 5BG, 10BG, 5B, 10B, or 5PB with value 4 or more along root channels or other pores starting within 30 cm (12 in.) of the soil surface. The matrix has dominant hue of 5Y or redder.

A16. Coast Prairie Redox. For use in MLRA 150A of LRR T. A layer starting within 15 cm (6 in.) of the soil surface that is at least 10 cm (4 in.) thick and has a matrix chroma 3 or less with 2% or more distinct or prominent redox concentrations as soft masses or pore linings.

3.8.2 Sandy Soils

Sandy soils include all sandy material except loamy, very fine sand. All mineral layers above any of the S indicators except for indicators S6 and S9 have dominant chroma 2 or less, or the layer(s) with dominant chroma of more than 2 is less than 15 cm (6 in.) thick.

S1. Sandy Mucky Mineral. For use in all LRRs except W, X, and Y and those LRRs that use indicator A7 (P, T, U, and Z). A layer of mucky-modified sandy soil material 5 cm (2 in.) or more thick starting within 15 cm (6 in.) of the soil surface.

S2. 2.5 cm Mucky Peat or Peat. For use in LRRs G and H. A layer of mucky peat or peat 2.5 cm (1 in.) or more thick with value 4 or less and chroma 3 or less starting within 15 cm (6 in.) of the soil surface underlain by sandy soil material.

S3. 5 cm Mucky Peat or Peat. For use in LRRs F and M; for testing in LRR R. A layer of mucky peat or peat 5 cm (2 in.) or more thick with value 3 or less and chroma 2 or less starting within 15 cm (6 in.) of the soil surface underlain by sandy soil material.

S4. Sandy Gleyed Matrix. For use in all LRRs except W, X, and Y. A gleyed matrix that occupies 60% or more of a layer starting within 15 cm (6 in.) of the soil surface.

S5. Sandy Redox. For use in all LRRs except V, W, X, and Y. A layer starting within 15 cm (6 in.) of the soil surface that is at least 10 cm (4 in.) thick, and has a matrix with 60% or

FIGURE 3.24 (See color insert following page 392.) Indicator S5 (Sandy Redox) with redox concentrations occurring almost at the surface. The soil slice is about 40 cm long. (USDA–NRCS, 2006.)

more chroma 2 or less with 2% or more distinct or prominent redox concentrations as soft masses or pore linings (Figure 3.24).

S6. Stripped Matrix. For use in all LRRs except V, W, X, and Y. A layer starting within 15 cm (6 in.) of the soil surface in which iron/manganese oxides or organic matter have been stripped from the matrix, exposing the primary base color of soil materials. The stripped areas and translocated oxides or organic matter form a faint diffuse splotchy pattern of two or more colors. The stripped zones are 10% or more of the volume; they are rounded and approximately 1–3 cm (0.5–1 in.) in diameter (Figure 3.25).

S7. Dark Surface. For use in LRRs N, P, R, S, T, U, V, and Z. A layer 10 cm (4 in.) or more thick starting within the upper 15 cm (6 in.) of the soil surface with a matrix value 3 or less and chroma 1 or less. At least 70% of the visible soil particles must be covered, coated, or similarly masked with organic material. The matrix color of the layer immediately below the dark layer must have chroma 2 or less.

S8. Polyvalue Below Surface. For use in LRRs R, S, T, and U; for testing in LRRs K and L. A layer with value 3 or less and chroma 1 or less starting within 15 cm (6 in.) of the soil surface underlain by a layer(s) where translocated organic matter unevenly covers the soil material forming a diffuse splotchy pattern. At least 70% of the visible soil particles in the upper layer must be covered, coated, or masked with organic material. Immediately below this layer, the organic coating occupies 5% or more of the soil volume and has value 3 or less and chroma 1 or less. The remainder of the soil volume has value 4 or more and chroma 1 or less to a depth of 30 cm (12 in.) or to the spodic horizon, whichever is less.

S9. Thin Dark Surface. For use in LRRs P, R, S, T, and U; for testing in LRRs K and L. A layer 5 cm (2 in.) or thicker within the upper 15 cm (6 in.) of the surface, with value 3 or less and chroma 1 or less. At least 70% of the visible soil particles in this layer must be covered, coated, or masked with organic material. This layer is underlain by a layer(s) with value 4 or less and chroma 1 or less to a depth of 30 cm (12 in.) or to the spodic horizon, whichever is less.

S10. Alaska Gleyed. This indicator is now indicator A13 (Alaska Gleyed).

FIGURE 3.25 (See color insert following page 392.) Indicator S6 (Stripped Matrix) occurs at a depth of about 12 cm. The knife blade is 15 cm long. (USDA–NRCS, 2006.)

3.8.3 LOAMY AND CLAYEY SOILS

All mineral layers above any of the F indicators except for indicators F8, F12, F19, and F20 have dominant chroma 2 or less, or the layer(s) with dominant chroma of more than 2 is less than 15 cm (6 in.) thick.

> *F1. Loamy Mucky Mineral.* For use in all LRRs except N, R, S, V, W, X, and Y, those using A7 (LRRs P, T, U, and Z) and MLRA 1 of LRR A. A layer of mucky-modified loamy or clayey soil material 10 cm (4 in.) or more thick starting within 15 cm (6 in.) of the soil surface.
>
> *F2. Loamy Gleyed Matrix.* For use in all LRRs except W, X, and Y. A gleyed matrix that occupies 60% or more of a layer starting within 30 cm (12 in.) of the soil surface (Figure 3.26).
>
> *F3. Depleted Matrix.* For use in all LRRs except W, X, and Y. A layer with a depleted matrix that has 60% or more chroma 2 or less that has a minimum thickness of either of the following:
> a. 5 cm (2 in.) if 5 cm (2 in.) is entirely within the upper 15 cm (6 in.) of the soil
> b. 15 cm (6 in.) and starts within 25 cm (10 in.) of the soil surface
>
> *F4. Depleted Below Dark Surface.* This indicator is now indicator A11 (Depleted Below Dark Surface).
>
> *F5. Thick Dark Surface.* This indicator is now indicator A12 (Thick Dark Surface).
>
> *F6. Redox Dark Surface.* For use in all LRRs except LRRs W, X, and Y; for testing in LRRs W, X, and Y. A layer at least 10 cm (4 in.) thick entirely within the upper 30 cm (12 in.) of the mineral soil that has either of the following:
> a. Matrix value 3 or less and chroma 1 or less and 2% or more distinct or prominent redox concentrations as soft masses or pore linings

FIGURE 3.26 (See color insert following page 392.) Indicator F2 (Loamy Gleyed Matrix) occurs at the soil surface in this example of the indicator. (USDA–NRCS, 2006.)

  b. Matrix value 3 or less and chroma 2 or less and 5% or more distinct or prominent redox concentrations as soft masses or pore linings
F7. Depleted Dark Surface. For use in all LRRs except W, X, and Y; for testing in LRRs W, X, and Y. Redox depletions, with value 5 or more and chroma 2 or less, in a layer at least 10 cm (4 in.) thick entirely within the upper 30 cm (12 in.) of the mineral soil that has either of the following:
  a. Matrix value 3 or less and chroma 1 or less and 10% or more redox depletions
  b. Matrix value 3 or less and chroma 2 or less and 20% or more redox depletions
F8. Redox Depressions. For use in all LRRs except LRRs W, X, and Y; for testing in LRRs W, X, and Y. In closed depressions subject to ponding, 5% or more distinct or prominent redox concentrations as soft masses or pore linings in a layer 5 cm (2 in.) or thicker entirely within the upper 15 cm (6 in.) of the soil surface (Figure 3.27).
F9. Vernal Pools. For use in LRRs B, C and D. In closed depressions subject to ponding, presence of a depleted matrix with 60% or more chroma 2 or less in a layer 5 cm (2 in.) thick entirely within the upper 15 cm (6 in.) of the soil surface.
F10. Marl. For use in LRR U. A layer of marl with a value of 5 or more starting within 10 cm (4 in.) of the soil surface.
F11. Depleted Ochric. For use in MLRA 151 of LRR T. A layer(s) 10 cm (4 in.) or thicker that has 60% or more of the matrix with value 4 or more and chroma 1 or less. The layer is entirely within the upper 25 cm (10 in.) of the soil surface.
F12. Iron/Manganese Masses. For use in LRRs N, O, P, and T; for testing in LRR M. On floodplains, a layer 10 cm (4 in.) or thicker with 40% or more chroma 2 or less, and 2% or more distinct or prominent redox concentrations as soft iron/manganese masses with diffused boundaries. The layer occurs entirely within 30 cm (12 in.) of the soil surface. Iron/manganese masses have value 3 or less and chroma 3 or less; most commonly they are black. The thickness requirement is waived if the layer is the mineral surface layer.

FIGURE 3.27 (See color insert following page 392.) Indicator F8 (Redox Depression) occurs at the soil surface in this example of the indicator. (USDA–NRCS, 2006.)

- *F13. Umbric Surface.* For use in LRRs P, T, and U and MLRA 122 of LRR N. In depressions and other concave landforms, a layer 25 cm (10 in.) or thicker starting within 15 cm (6 in.) of the soil surface in which the upper 15 cm (6 in.) must have value 3 or less and chroma 1 or less, and the lower 10 cm (4 in.) of the layer must have these colors or any other color that has a chroma 2 or less.
- *F14. Alaska Redox Gleyed.* This indicator is now indicator A14 (Alaska Redox).
- *F15. Alaska Gleyed Pores.* This indicator is now indicator A15 (Alaska Gleyed Pores).
- *F16. High Plains Depressions.* For use in MLRAs 72 and 73 of LRR H; for testing in other MLRAs of LRR H. In closed depressions subject to ponding, a mineral soil that has chroma 1 or less to a depth of at least 35 cm (13.5 in.) and a layer at least 10 cm (4 in.) thick within the upper 35 cm (13.5 in.) of the mineral soil that has either of the following:
  a. 1% or more redox concentrations as nodules or concretions
  b. Redox concentrations as nodules or concretions with distinct or prominent corona
- *F17. Delta Ochric.* For use in MLRA 151 of LRR T. A layer 10 cm (4 in.) or more thick that has 60% or more of the matrix with value 4 or more and chroma 2 or less with no redox concentrations. This layer occurs entirely within the upper 30 cm (12 in.) of the soil surface.
- *F18. Reduced Vertic.* For use in MLRA 150 of LRR T; for testing in all LRRs with Vertisols and Vertic intergrades. In Vertisols and Vertic intergrades, a positive reaction to alpha alpha-dipyridyl that (a) is the dominant (60% or more) condition of a layer at least 4 in. thick within the upper 12 in. (or at least 2 in. thick within the upper 6 in.) of the mineral or muck soil surface, (b) occurs for at least 7 continuous days and 28 cumulative days, and (c) occurs during a normal (within 16–84% of probable precipitation) or drier season and month.
- *F19. Piedmont Floodplain Soils.* For use in MLRAs 149A and 148 of LRR S; for testing on floodplains subject to Piedmont deposition throughout LRRs P, S, and T. On active

floodplains, a mineral layer at least 15 cm (6 in.) thick starting within 25 cm (10 in.) of the soil surface with a matrix (60% or more of the volume) chroma less than 4 and 20% or more distinct or prominent redox concentrations as soft masses or pore linings.

F20. Anomalous Bright Loamy Soils. For use in MLRA 149A of LRR S and MLRA 153C and 153D of LRR T; for testing in MLRA 153B of LRR T. Within 200 m (656 ft) of estuarine marshes or waters and within 1 m (3.28 ft) of mean high water, a mineral layer at least 10 cm (4 in.) thick starting within 20 cm (8 in.) of the soil surface with a matrix (60% or more of the volume) chroma less than 5 and 10% or more distinct or prominent redox concentrations as soft masses, pore linings, or depletions.

3.9 SUMMARY

- Wetland functions and values.
  - Depending on wetland type, hydrologic regime, and nutrient/contaminant inputs, wetlands can serve as a sink, source, and transformer.
  - Hydrologic flux and storage provides flood control, erosion control, groundwater recharge, and serves as a medium to transport nutrients.
  - Biogeochemical cycling controls sediments and nutrients while influencing water quality.
  - Biological productivity affects fisheries as well as timber and shrub crops.
  - Carbon fixing and detrital production/decomposition serve as the basis for C cycle.
  - Wildlife/community habitat provides human recreation, animal harvest, and wildlife refugia.
- Wetlands are usually defined by hydric soils, the presence of vegetation adapted to wet conditions, and hydrology. There are several definitions of wetlands, each stressing different characteristics and wetland values.
- Wetland values are dependent on social perceptions. The valued functions have historically included water storage, flood control, erosion control, sediment control, nutrient removal, protection of general water quality, habitat for crops and fisheries, recreation, and wildlife refugees.
- Wetlands can function as a sink, source, or transformer for nutrients. In addition to functioning as sources or sinks, wetlands act as transformers through processes such as denitrification, methanogenesis, and the microbial breakdown of organic matter.
- When wetlands are drained, the organic matter is oxidized at a much quicker rate. This results in soil subsidence. The soils become acidic because hydrogen ions are released as the organic material is oxidized. For example, much of the organic matter accumulated over a period of 5,000 years in the Everglades was oxidized, on drainage, in less than 100 years.
- Carbon is the primary driver for all biogeochemical processes in wetlands.
- Accumulation of organic matter depends on hydrology, nutrients, vegetation, and fire.
- Reddish layers in the soil are created when ferrous iron is oxidized into the ferric state. These layers are often found around the roots of plants because oxygen is diffusing from the roots. Gray areas (mottles) within the soil indicate the presence of ferrous iron.
- Diffusion of gases through water is 10,000 times slower than in air where oxidized forms act as electron acceptors. Wetland plants pump oxygen to the roots to assist in respiration.
- Reduced forms function as electron donors and oxidized forms function as electron acceptors. Redox potential reflects the intensity of reduction, and it is a measure of electron activity, which is analogous to pH.
- (–) Eh values represent high electron activity and intense anaerobic conditions. (+) Eh values represent low electron activity and aerobic conditions. Eh values >300 mV: aerobic/uplands. Eh values <300 mV: anaerobic/wetlands.

- Wetland soils have the following electron donors and acceptors:
  - Reductants (electron donors): organic matter, ammonium, sulfide, reduced iron and manganese (Fe^{2+}, Mn^{2+}), methane, hydrogen, and fatty acids.
  - Oxidants (electron acceptors): oxygen, sulfate, iron and manganese oxides, nitrogen oxides, carbon dioxide, phosphate, select fatty acids, humic acids.
  - Anaerobic–aerobic interface shows a *gradient* of oxidized to reduced conditions.
  - Potential (mV) ranges from -300 (reduced/anaerobic) to $+700$ (aerobic) and pH range of 4–8.
- Plant interactions:
  - Plants have adapted to the harsh anaerobic conditions of wetland soils. Development of aerenchyma tissues permit oxygen pumping to the roots, to support root respiration and aerobic bacteria in the root zone.
  - Oxidation to form ferric Fe around the roots produces the reddish color.
- Mineral and organic soils in wetlands:
  - Mineral soils have lower organic matter content, higher bulk density, and lower porosity than organic soils.
  - Mineral soils that are waterlogged may have secondary minerals or iron and manganese concretions.
  - Organic soils have high CEC, typically saturated with H^+, as seen in peat bogs.
  - Upland marshes are ombrotrophic (nutrients from atmosphere and ground water) and acidic, whereas lowland marshes are rheotrophic (nutrients from surface and ground water) and have pH 5–6.
  - Saltwater marshes have near-neutral pH and pyrite.
  - Draining salt marshes may lead to very acidic conditions and Al toxicity.
- Anaerobic conditions drive the dominant processes in wetland soils:
  - Anaerobic bacteria sequentially reduce alternate electron acceptors, from nitrate, to ferric and manganic compounds, to sulfate, and finally to carbon dioxide.
  - Anaerobic processes promote slower decomposition, which results in build of organic materials.
  - Generally, soils with higher organic content have a lower pH and lower nutrient availability, and are more highly reduced than mineral soils.
- The soil-floodwater interface in wetland systems is extremely important in cycling of nutrients:
  - High levels of reduced nitrogen compounds such as ammonium in the soils and low levels in the water column cause an upward diffusion of ammonium into the water column where it is oxidized into nitrate, etc.
  - Conversely, high levels of nitrate in water column in aerobic soil layers causes a downward diffusion of nitrate, where it is rapidly reduced by microbes.
  - High amounts of sulfur in coastal wetlands result in the use of sulfate as an alternate electron acceptor.
  - Freshwater wetlands lack substantial amounts of sulfur and tend to depend on methanogenesis for organic matter decomposition.

STUDY QUESTIONS

1. List the three major components that constitute a wetland.
2. Discuss the various definitions of wetlands, including how they differ.
3. List the major criteria developed by the National Technical Committee for hydric soil identification.
4. What are the major types of wetlands? Distinguish or identify the differences.
5. List and describe the various types of (1) inland wetlands and (2) coastal wetlands.

6. What is a constructed wetland? For what purpose are constructed wetlands built?
7. What is a paddy soil? How are these systems similar to wetland soils?
8. Define and discuss hydrology of wetlands. What is hydroperiod dependent upon? What biogeochemical properties does hydrology control?
9. What are the major sources of water to wetlands?
10. What is the difference between tidal and nontidal wetlands?
11. Explain how soil type influences the relative proportion of air and water volume in wetland soils.
12. In addition to restricting oxygen supply, what other important effects does excess water have on wetland soils?
13. What is hydrophytic vegetation? List the unique characteristics that allow these plants to grow in wetland soils.
14. List the major groups that the Fish and Wildlife Service use to identify wetland plant species.
15. What unique characteristics of wetland soils separate them from upland soils?
16. List the ways in which oxygen is introduced into wetland soils.
17. What are the two important purposes for which oxygen is needed in soils? Which electron acceptors can be used by microorganisms in wetland soils?
18. List the inorganic redox system found in flooded soils and sediment. What is the order of sequence of reduction of the inorganic redox system?
19. How are the oxidized surface layers in flooded soils formed? What governs the depth of the surface oxidized layer?
20. What factors govern the exchange of dissolved species between the soil and underlying water?
21. List the major groups into which wetland soils can be divided.
22. What are some distinct characteristics of waterlogged mineral soils?
23. Explain how soil color can be used to describe wetland soil or wetland soil processes.
24. What is a redoxmorphic feature?
25. Describe an organic soil or Histosols. How do these soils form?
26. How are organic soils classified on the basis of identifiable plant materials? List the classifications. How does muck differ from peat?
27. List some important field indicators used in the identification of hydric soils.

FURTHER READINGS

Batzer, D. P. and R. R. Sharitz (eds.). 2006. *Ecology of Freshwater and Estuarine Wetlands.* University of California Press, Berkeley/Los Angeles, CA.

Brinson, M. M. 1993. *A Hydrogeomorphic Classification of Wetlands.* Technical Report; WRP-DE-4, U.S. Army Corps. 79 pp.

Cowardin, L. M., V. Carter, F. C. Golet, and E. T. LaRoe. 1979. *Classification of Wetlands and Deepwater Habitats of the United States.* Fish and Wildlife Service, U.S. Department of Interior, Washington, DC.

Lewis, W. M. 1995. *Wetlands—Characteristics and Boundaries.* National Research Council, National Academy Press, Washington, DC. 306 pp.

Mitsch, W. J. and J. G. Gosselink. 2007. *Wetlands.* Wiley, New York. 920 pp.

Richardson, J. L. and M. J. Vepraskas. 2001. *Wetland Soils.* Lewis Publishers, Boca Raton, FL. 417 pp.

USDA NRCS. 2006. In G. W. Hurt and L. M. Vasilas (eds.) *Field Indicators of Hydric Soils in the United States, Version 6.0.* USDA NRCS in cooperation with the National Technical Committee for Hydric Soils, Ft. Worth, TX.

4 Electrochemical Properties

4.1 INTRODUCTION

The three most important electrochemical properties of the soil that are affected by flooding are oxidation–reduction potential or redox potential (Eh), pH, and ionic strength (electrical conductivity) of soil pore water. Among the three electrochemical properties, Eh undergoes dynamic changes as wetlands are subjected to hydrologic fluctuations. Compared with upland soils, one of the most striking characteristics of a wetland soil is its low Eh, which is a measure of electron activity or potential in the soil. Redox potential of soil is measured using a platinum electrode with a standard calomel reference electrode. In this chapter, major emphasis is placed on changes in Eh and pH in relation to biogeochemical reactions of interest in wetlands.

Oxidation and reduction reactions regulate many of the biogeochemical reactions in wetlands. Electrons are essential to many biogeochemical reactions. Oxidation, the loss of electrons, couples with reduction, the gain of electrons. The Eh of soil is determined by the concentration (activities) of oxidants and reductants. Oxidants include oxygen, nitrate, nitrite, manganese, iron, sulfate, and carbon dioxide, while reductants include various organic substrates and reduced inorganic compounds. From the ecosystem point of view, photosynthesis and respiration are two examples of reduction and oxidation reactions, which regulate energy flow and many biogeochemical reactions. For example, during photosynthesis carbon dioxide is reduced to carbohydrates, and during respiration the reduced organic compounds are oxidized to carbon dioxide. Photosynthesis provides organic matter source to the soil through plant productivity and detrital accumulation.

$$6CO_2 + 24e^- + 24H^+ = C_6H_{12}O_6 + 6H_2O \text{ (Reduction)}$$

$$12H_2O = 6O_2 + 24e^- + 24H^+ \text{ (Oxidation)}$$

$$6CO_2 + 6H_2O = O_2 + C_6H_{12}O_6 \text{ (Photosynthesis)}$$

Respiration by soil microbes involves oxidation of organic matter, a major controller of many biogeochemical reactions in soils.

$$C_6H_{12}O_6 + 6H_2O = 6CO_2 + 24e^- + 24H^+ \text{ (Oxidation)}$$

$$6O_2 + 24e^- + 24H^+ = 12H_2O \text{ (Reduction)}$$

$$C_6H_{12}O_6 + 6O_2 = CO_2 + 6H_2O \text{ (Respiration)}$$

These processes are discussed in detail in Chapter 5.

Oxidation of reduced compounds (such as organic substrates) results in the release of electrons, and these electrons, unless they are removed, create an electron pressure or intensity in the system. If oxygen is present the electrons are easily removed, and no large electron pressure develops. If oxygen is not present, but nitrate is present, a slightly higher electron pressure is developed that enables the facultative bacteria (e.g., denitrifiers) in the soil to use nitrate to remove electrons from the system. If nitrate is completely consumed, more electron pressure (reduction intensity) has to develop to allow manganese reducers to remove electrons during the reduction of manganic manganese. And similarly ferric iron, sulfate, and carbon dioxide, which are electron acceptors, are used sequentially as the electron pressure increases.

Aerobic organisms that oxidize organic substrate using oxygen as an electron acceptor must develop a low reduction intensity so that oxygen accepts electrons produced during the oxidation of organic substrate. Oxygen in this reaction is reduced to water. However, obligate anaerobic organisms such as sulfate reducers must exert a very high electron pressure or reduction intensity to cause the sulfate ion to accept electrons and become reduced to sulfide. These gradients in electron pressures can be characterized by Eh measurements in wetland soils (Figure 4.1).

Eh scale can be used to quantify reduction intensity (or oxidation intensity) in chemical and biological systems, as shown in Figure 4.2. The scale is in electromotive force (EMF) units or volts (or millivolts). We can conveniently divide the redox scale into zones ranging from oxidized at one end (where the aerobes function) to highly reduced at the other end (where the methane producers and sulfate reducers function).

In systems where aerobic organisms function, the Eh range is very narrow, ranging from only about +700 mV down to about +300 mV. Below 300 mV, facultative anaerobes function down to about 0 mV. Below this range, obligate anaerobes function. In wetland (waterlogged or flooded) soils, Eh can be anywhere along the entire scale. Where oxygen is present in a wetland soil, Eh can be as high as in a drained soil, but where oxygen is not present it can be low (−250 to +300 mV). The narrow range and poor reproducibility of redox potentials in well-drained soils limit their value as a tool for characterizing soil aeration. The poor reproducibility is caused primarily by a lack of poising of oxidation–reduction systems dominated by oxygen (discussed in the latter part of this chapter). Redox potential measurements are most useful if limited to wetland soils and sediments. Generally, the oxidation–reduction status of surface waters and upland soils is better characterized

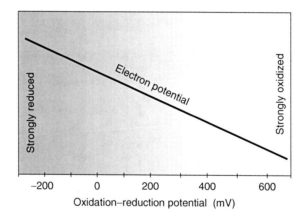

FIGURE 4.1 Schematic showing the relationship between electron pressure and redox potential wetland soils.

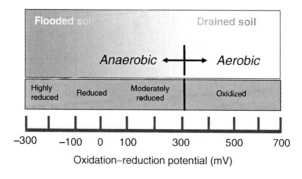

FIGURE 4.2 Schematic showing the range in redox potential in wetland soils.

by dissolved oxygen and soil atmospheric oxygen measurement, respectively. An exception would be beneath the thermocline of a deep-water aquatic system where strongly reduced conditions exist and one may want to know whether the system is moderately reducing (i.e., ferrous iron is stable) or strongly reducing (sulfide is stable). Another exception is wetland delineation studies where a transect from wetlands to uplands is being characterized and Eh values from wetland and transition zones can be compared to Eh values from upland zones. Also, the effect of seasonal changes in hydrology on the oxidation–reduction status of a wetland can be evaluated.

The redox potential of soils is also influenced by frequent additions of organic substrates, or soils high in native organic matter. In these soils Eh values approach −100 mV within a few days after flooding. For example, the Eh values of organic soils decrease rapidly to <−200 mV within a few days after flooding, as compared to mineral soils. From an oxidation–reduction point of view, wetland soils are usually limited in oxidants (electron acceptors) and contain an unlimited supply of reductants, such as organic matter (electron donors). In contrast, drained/upland soils are usually limited in reductants (electron donors) and contain an unlimited supply of oxidants (electron acceptors).

Geochemists, soil scientists, and limnologists have used Eh measurements to characterize the oxidation–reduction status of soils and sediments over the past several decades. As early as 1920, Gillespie (1920) studied electrode potentials in waterlogged soils using platinum electrodes. A number of other soil scientists have used Eh as an operational parameter to characterize paddy soils for the last half century (Ponnamperuma, 1955; Patrick, 1960; Liu, 1993), whereas limnologists used Eh to characterize the oxidation–reduction status of lake sediments (Pearsall and Mortimer, 1939; Mortimer, 1941, 1942). For the past four decades, ecologists and engineers have widely used Eh to characterize wetlands. This wide use of Eh, which was due to the ease of measurement, allowed rapid characterization of the degree of anaerobiosis and reasonable predictions of stability of various compounds regulating nutrient availability and water quality. This intensive variable, along with pH, has been used as a key combination to describe the stability of various minerals (Garrels and Christ, 1965). These stability diagrams are commonly known as Eh–pH phase diagrams, discussed later in this chapter).

4.2 THEORETICAL RELATIONSHIPS

Oxidation and reduction reactions involve the transfer of electrons from one compound to another and play a major role in regulating many reactions in biological systems. Oxidation–reduction reactions are two coupled half reactions involving (i) oxidation and (ii) reduction.

Oxidation is defined as the removal of electrons from a compound (electron donor). This compound is usually referred to as the electron donor or reductant. During this process the compound is oxidized and its oxidation number is increased:

$$\text{Reductant} = \text{Oxidant} + e^- \tag{4.1}$$

Reduction is defined as the addition of electrons to a compound. This compound is usually referred to as the electron acceptor or oxidant. During this process it is reduced and its oxidation number is decreased:

$$\text{Oxidant} + e^- = \text{Reductant} \tag{4.2}$$

The tendency of compounds to accept or donate electrons is expressed as reduction potential or redox potential. The redox potential of a substance depends on the

- Affinity of molecules for electrons
- Concentration of reductants and oxidants (referred to as redox pair)

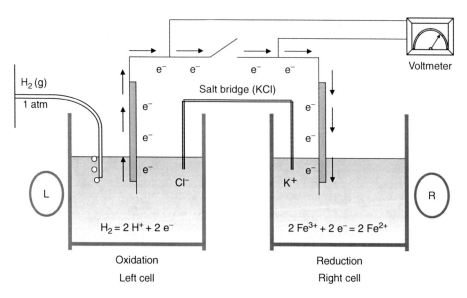

FIGURE 4.3 Electrochemical cell with H_2/H^+ redox couple or standard hydrogen electrode (SHE) in the left cell (L) and Fe^{3+}/Fe^{2+} redox couples in the right cell (R). The left cell is maintained at 1 atm pressure of H_2 and 1 M H^+ activity. In this cell platinum electrode in H_2/H^+ half cell is anode and platinum electrode in Fe^{3+}/Fe^{2+} cell is cathode.

The electron affinity of a substance can be measured using an inert electrode such as platinum. This potential is measured as an electrical potential in reference to a reference or standard electrode, usually a SHE placed in a half cell containing the substance in solution (Figure 4.3).

SHE consists of the following:

- A platinum electrode
- A solution containing H^+ ions at unit activity (1 M)
- Hydrogen gas at 1 atm pressure
- System temperature at 25°C (298.15 K)

The following initial conditions are assigned to SHE:

- The electrical potential is zero
- ΔG_r of $[H^+]$ in solution is zero and the activity of H^+ is 1 M; thus, pH = 0
- ΔG_r of an $[e^-]$ in solution is zero

The SHE constructed with these initial conditions can function as a standard reference, and the potential of any other redox couple can be measured using a second electrode.

The International Union of Pure and Applied Chemistry (IUPAC) recommends that

- The half-cell reactions of substances always be written as reduction reactions.
- The electrical potential of the cell be calculated as the difference between right (reduction) cell voltage and left (oxidation) cell voltage.

Let us consider two half cells, a left cell (oxidation cell) and a right cell (reduction cell), to describe the oxidation–reduction reactions (Figure 4.3).

Assume that the right cell "R" contains a solution with Fe^{3+}/Fe^{2+} redox couple and a platinum electrode. This cell represents the reduction of an oxidant or electron acceptor, and in electrochemistry the electrode in this cell is referred to as cathode. This electrode is connected to another platinum electrode placed in the left cell "L." This cell represents the oxidation of a reductant or electron donor, and is referred to as anode. Hydrogen gas is bubbled into the solution of cell "L" at 1 atm and a hydrogen ion concentration is maintained at 1 M. Both cells are maintained at 25°C and connected with a salt bridge containing saturated KCl. The saturated KCl is mixed with an agar medium and poured into a "U"-shaped salt bridge. The salt bridge completes the electrical circuit and maintains electrical neutrality. For example, the removal of electrons from the cell "L" increases the positive charge, and the Cl^- ions from the salt bridge maintain any deficit in negative charges. Similarly, gain of electrons in the cell "R" increases the negative charge, and the K^+ ions from the salt bridge maintain any deficit in positive charges. In both half cells, platinum electrodes function as a sink or source of electrons to the solution, so that both oxidant and reductant of each half cell can attain equilibrium with it.

In the electrochemical cell shown in Figure 4.3, the force operating to cause electrons to move from oxidation cell to reduction cell is called the EMF. Oxidation of H_2 in the left cell "L" supplies electrons to the platinum electrode, while the platinum electrode in the right cell "R" supplies electrons to the reduction cell. This electron flow is prevented by a switch, and the potential between these two half cells is measured by a voltmeter that responds to EMF or potential. During this process the voltmeter registers the difference in electrical potential between the two cells. Since the potential in cell "L" is zero, the registered electrode potential will represent the potential of half-cell "R." This potential is called the electrode potential, where E represents the difference in electrical potential of two half cells. Since "E" is measured in reference to the hydrogen electrode, the resulting potential is also referred to as Eh. The Eh value is negative when electron activity in cell "L" is greater than that in cell R, and it is positive when electron activity in cell L is less than that in cell R.

The oxidation–reduction half cells presented in Figure 4.3 provide a simplistic way of presenting very complex reactions. We have set SHE at pH = 0 (or 1 M hydrogen ion activity or concentration) but many of the biological systems function at pH = 7 (usually in a pH range of 4–10), and under these conditions the SHE has a redox potential of –0.41 V. The reduced forms of the redox couple with more negative potential are good reducing agents (electron donors or oxidants) (Figure 4.4), while the oxidized forms with more positive potential are good oxidizing agents (electron acceptors or reductants). The redox couple is written with oxidized form always on the left, followed by the reduced form. Examples of some redox couples of interest in wetlands are shown in Table 4.1. The measured potentials assume that the oxidants and reductants are in equilibrium at the electrode surface.

The redox couple with a negative potential has a strong tendency to donate electrons and function as a reductant, while the redox couple with a positive potential has a strong tendency to accept the electrons:

$$2Fe^{3+} + e^- = 2Fe^{2+} \quad (4.3)$$

$$(Oxidant) + electrons = (reductant)$$

In this reaction, Fe^{3+} ions are reduced to Fe^{2+} ions, with the consumption of an electron. This reaction occurs in cell R, and the platinum electrode in this cell functions as a source of electrons:

$$H_2 = 2H^+ + 2e^- \quad (4.4)$$

In the oxidation cell, the platinum electrode functions as a sink for electrons.

The overall oxidation–reduction reaction can be written as

$$2Fe^{3+} + H_2 = 2Fe^{2+} + 2H^+ \quad (4.5)$$

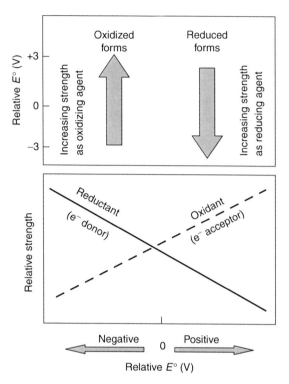

FIGURE 4.4 Relative strength of oxidant (oxidizing agent or electron acceptor) and reductant (reducing agent or electron donor).

The cell potential for standard conditions can be calculated as follows:

$$E°\text{-cell} = [E°\text{-ox}] - [E°\text{-red}] \tag{4.6a}$$

where R is the reduction cell (right) and L the oxidation cell (left). The oxidation cell is SHE where $E°$-ox $= 0$ V. The electrode potential [$E°$-red] for reduction of ferric iron is 0.77 V, when measured in reference to SHE. This suggests that the standard electrode potential for Fe^{3+}/Fe^{2+} redox couple is 0.77 V more positive than SHE.

Equation 4.6a is valid only when standard electrode potentials of the redox reactions in both half cells are written as reduction reactions (as shown in Table 4.1). If the reaction in the oxidation cell "L" is written as an oxidation reaction, the positive $E°$ changes to negative and the negative $E°$ changes to positive. Under these conditions, the following equation applies to obtain the electrode potential for the overall reaction:

$$E°\text{-cell} = [E°\text{-ox}] + [E°\text{-red}] \tag{4.6b}$$

where R is the reduction cell (right) and L the oxidation cell (left). Similarly, H_2 is oxidized in cell L with the release of H^+ and e^-.

Now consider the free energy of a redox reaction. The standard free-energy change of a reaction is the sum of the free energy of formation of products in their standard states *minus* the free energies of formation of the reactants in their standard states. Negative values of ΔG_r represent a spontaneous reaction, and the reaction proceeds as shown below:

$$\Delta G_r° = \sum \Delta G_r°\text{products} - \sum \Delta G_r°\text{reactants} \tag{4.7}$$

TABLE 4.1
Reduction Potentials (Standard at 25°C) of Selected Redox Couples of Significance in Wetlands[a]

Reduction Reaction (Oxidant + e^- + H^+ = Reductant)	$E°$ (V)	$E_7°$ (V) (pH = 7)
(1) $O_2 + 4H^+ + 4e^- = 2H_2O$	+1.23	+0.81
(2) $NO_3^- + 2H^+ + 2e^- = NO_2^- + H_2O$	+0.83	+0.41
(3) $2NO_3^- + 12H^+ + 10e^- = N_2 + 6H_2O$	+1.24	+0.82
(4) $NO_3^- + 10H^+ + 8e^- = NH_4^+ + 3H_2O$	+0.88	+0.46
(5) $NO_2^- + 8H^+ + 6e^- = NH_4^+ + 2H_2O$	+0.89	+0.47
(6) $N_2 + 8H^+ + 6e^- = 2NH_4^+$	+0.28	−0.14
(7) $MnO_2 + 4H^+ + 2e^- = Mn^{2+} + 2H_2O$	+1.29	+0.87
(8) $MnO_2 + HCO_3^- + 3H^+ + 2e^- = MnCO_3 + 2H_2O$	+0.94	+0.52
(9) $Fe(OH)_3 + 3H^+ + e^- = Fe^{2+} + 3H_2O$	+0.80	+0.38
(10) $Fe(OH)_3 + HCO_3^- + 2H^+ + e^- = FeCO_3 + 3H_2O$	+0.78	+0.36
(11) $Fe^{3+} + e^- = Fe^{2+}$	+0.77	+0.35
(12) $NAD^+ + 2H^+ + 2e^- = NADH + H^+$	+0.10	−0.32
(13) Fumarate + $2H^+ + 12e^-$ = Succinate	+0.45	+0.03
(14) Pyruvate + $2H^+ + 12e^-$ = Lactate	+0.23	−0.19
(15) $SO_4^{2-} + 10H^+ + 8e^- = H_2S + 4H_2O$	+0.34	−0.08
(16) $SO_4^{2-} + 9H^+ + 8e^- = HS^- + 4H_2O$	+0.24	−0.18
(17) $SO_4^{2-} + 8H^+ + 8e^- = S^{2-} + 4H_2O$	+0.22	−0.20
(18) $S + 2H^+ + 2e^- = H_2S$	+0.17	−0.25
(19) $CO_2 + 8H^+ + 8e^- = CH_4 + 2H_2O$	+0.17	−0.25
(20) $CO_2 + 4H^+ + 4e^- = CH_2O + H_2O$	−0.07	−0.18
(21) $2H^+ + 2e^- = H_2$	0	−0.42
(22) $H_3PO_4 + 2H^+ + 2e^- = H_3PO_3 + H_2O$	−0.28	−0.70
(23) $H_3PO_3 + 2H^+ + 2e^- = H_3PO_2 + H_2O$	−0.50	−0.92
(24) $H_3PO_2 + H^+ + e^- = P + 2H_2O$	−0.51	−0.93
(25) $P + 3H^+ + 3e^- = PH_3$	+0.06	−0.36
(26) $NAD^+ + 2e^- + 2H^+ = NADH + H^+$	−0.12	−0.54
(27) $FAD + 2H^+ + 2e^- = FADH_2$	+0.20	−0.22

[a] To obtain electrode potentials for oxidation reactions change (+) to (−) or (−) to (+), depending on the reduction potential.

For the oxidation–reduction reaction 4.5:

$$\Delta G_r° = (2\Delta G_f° Fe^{2+} + 2\Delta G_f° H^+) - (2\Delta G_f° Fe^{3+} + \Delta G_f° H_2) \quad (4.8)$$

Since $\Delta G_f° H^+ = 0$ and $\Delta G_f° H_2 = 0$,

$$\Delta G_r° = 2\Delta G_f° Fe^{2+} - 2\Delta G_f° Fe^{3+} \quad (4.9)$$

ΔG_r is expressed in kcal mol^{-1} (1 kcal = 4.184 kJ).

If the value for $\Delta G°$ is negative, then it is likely that the reaction will proceed spontaneously as written, resulting in the release of energy. However, this does not mean that the reaction occurs instantaneously at a measurable rate. This means that if the reaction is complete, energy is released. For example, exposing wood or glucose to oxygen in air will give negative ΔG_r, but under sterile conditions this reaction will not occur at normal temperature. This is because of the high activation energy associated with these reactions. Thus, in natural systems, supply of enzymes and catalysts will activate these reactions.

The equilibrium constant (K_{eq}) expressed for Equation 4.5 is given as follows:

$$K_{eq} = \frac{([Fe^{2+}]^2 [H^+]^2)}{([Fe^{3+}]^2 [P_{H_2}])} \qquad (4.10)$$

Since activity of H^+ unity and $P_{H_2} = 1$ atm, and for many dilute solutions, we assume that the activities of reactants and products are equal to their respective concentration expressed as [M], the equilibrium relationship for Fe^{3+}/Fe^{2+} redox couple can be written as

$$K_{eq} = \frac{[Fe^{2+}]}{[Fe^{3+}]} = \frac{[Reductant]}{[Oxidant]} \qquad (4.11)$$

The standard free-energy change of a reaction is related to the equilibrium constant by

$$\Delta G_r^\circ = -RT \ln K \qquad (4.12)$$

where K is the equilibrium constant, as shown in Equation 4.10, R the universal gas constant (0.001987 kcal mol^{-1} deg^{-1} or 8.314 J mol^{-1} deg^{-1}), and T the absolute temperature (298.15 K or 273.15 + 25°C).

$$\Delta G_r^\circ (\text{kcal}) = (-0.001987 \text{ kcal mol}^{-1} \text{ deg}^{-1})(298.15 \text{ deg})(2.303 \log K) \qquad (4.13)$$

$$\Delta G_r^\circ = -1.364 \log K \qquad (4.14)$$

Reaction (4.13) is valid only when the activities of the products and the reactants for a system are at equilibrium.

In general, for any chemical reaction:

$$\text{Oxidant} + e^- = \text{Reductant}$$

$$\Delta G_r = \Delta G_r^\circ + RT \ln \frac{[Reductant]}{[Oxidant][e^-]} \qquad (4.15)$$

At equilibrium [Reductant] = [Oxidant], then

$$\Delta G_r = \Delta G_r^\circ + RT \ln 1 \qquad (4.16)$$

$$\Delta G_r = \Delta G_r^\circ$$

For a given reaction when $\Delta G_r = 0$

$$\Delta G_r = \Delta G_r^\circ + RT \ln Q \qquad (4.17)$$

where Q is the reaction quotient for nonequilibrium or metastable conditions. It should be noted that Q has the same form as the equilibrium constant. $Q = K$ when the products and reactants are in their standard states and the reaction is at equilibrium.

The amount of energy required to move a charge of F coulombs through a potential difference can be related to the free energy available to do useful work. The relation between the standard free-energy change of a reaction and the electrical potential of the corresponding half cell can be expressed as

$$\Delta G_r^\circ = -nFE^\circ \qquad (4.18)$$

$$\Delta G_r = -nFE \qquad (4.19)$$

where F is the Faraday's constant expressed as charge per mole of electrons (kcal (volt equivalent)$^{-1}$), n the number of electrons involved in a reaction, and E the electrical potential (V). E can be measured directly by separating half cells and a pair of electrodes measuring the electrical potential of each cell. If $E°$ is >0 (positive), the reaction will proceed spontaneously as written; if $E°$ is <0 (negative) the reaction will occur in the opposite direction. When $E° = 0$, the reaction is at equilibrium.

By substituting Equations 4.18 and 4.19 in Equation 4.15, we obtain

$$-nFE = -nFE° + RT \ln \frac{[\text{Reductant}]}{[\text{Oxidant}]} \quad (4.20)$$

$$\frac{-nFE}{-nF} = \frac{-nFE°}{-nF} + \frac{RT}{-nF} \ln \frac{[\text{Reductant}]}{[\text{Oxidant}]} \quad (4.21)$$

which reduces to

$$E = E° - \frac{RT}{nF} \ln \frac{[\text{Reductant}]}{[\text{Oxidant}]} \quad (4.22)$$

where R and T are as described for Equation 4.12, F is the Faraday constant (23.061 kcal (volt equivalent)$^{-1}$), E the cell potential for the reaction (V), $E°$ the standard cell potential (V), and n the number of electrons participating in the reaction (1 kcal = 4.184 kJ).

Since the electrode potential E is measured in reference to SHE, Equation 4.22 can be written as

$$Eh = E° - \frac{RT}{nF} \ln \frac{[\text{Reductant}]}{[\text{Oxidant}]} \quad (4.23)$$

or

$$Eh = E° - \frac{RT}{nF} 2.303 \log \frac{[\text{Reductant}]}{[\text{Oxidant}]}$$

For the reactions at 25°C

$$\frac{RT}{F} = \frac{(0.001987 \text{ kcal deg}^{-1})(298.15 \text{ K})}{(23.06 \text{ kcal (volt equivalent)}^{-1})} \quad (4.24)$$

and

$$\frac{RT}{F} 2.303 = 0.059 \quad (4.25)$$

$$Eh = E° - \frac{0.059}{n} \log \frac{[\text{Reductant}]}{[\text{Oxidant}]} \quad (4.26)$$

For any given reaction, the terms $E°$, n, R, T, and F are constant; thus, Eh can be determined from the activity ratio of the reductant and the oxidant. An increase in oxidant activity will increase Eh because of low electron pressure. Similarly, an increase in reductant activity will decrease Eh because of high electron pressure. Eh values are positive or high in strongly oxidized systems and negative or low in strongly reduced systems. Such a relationship exists between the oxidized and reduced components for any given redox couple in soils (and in other biological and chemical systems). For example, for oxidized iron (Fe^{3+}) and reduced iron (Fe^{2+}), the more oxidized the system, the greater the concentration (or more precisely the chemical activity) of Fe^{3+} in relation to Fe^{2+}, and vice versa. We can express this relationship in another way by stating that the greater the ratio Fe^{3+}/Fe^{2+}, the more oxidized the system, and the greater the ratio Fe^{2+}/Fe^{3+}, the more reduced the system. Using the basic equation: oxidant + e^- = reductant, we can calculate the concentration or activity ratio of oxidized to reduced substances. The greater the ratio, the more oxidized the system, and vice versa. For these reasons, measured Eh values can be viewed as the intensity factor of the system.

This general equation was derived during early 20th century by a German chemist named Walter Nernst. Nernst won the Nobel Prize in 1920 for his contributions to the field of physical chemistry. We use a simplified form of the Nernst equation to show the relationship between the concentrations of oxidized and reduced components and the Eh and pH values of a chemical system (remember that biological systems are also chemical systems):

EXAMPLE

$$Fe^{3+} + e^- = Fe^{2+}, \quad E^\circ = +0.77 \text{ V}$$

$$Eh = E^\circ - 0.059 \log \frac{[Fe^{2+}]}{[Fe^{3+}]} \tag{4.27}$$

or

$$Eh = E^\circ + 0.059 \log \frac{[Fe^{3+}]}{[Fe^{2+}]}$$

If activity $Fe^{3+} = 1$, then we have

$$Eh = E^\circ + 0.059 \, pFe^{2+} \tag{4.28}$$

Consider another reaction involving redox couple (H^+/H_2) at 25°C:

$$2H^+ + 2e^- = H_2$$

$$Eh = E^\circ - \frac{0.059}{2} \log \frac{[P_{H_2}]}{[H^+]^2} \tag{4.29}$$

If $P_{H_2} = 1$ atm, then we have

$$Eh = E^\circ - \frac{0.059}{2} \log \frac{1}{[H^+]^2} \tag{4.30}$$

$$Eh = E^\circ - \frac{0.059}{2} 2\,pH \tag{4.31}$$

If $E^\circ = 0$ V, then Equation 4.31 becomes

$$Eh = -0.059 \, pH \tag{4.32}$$

Thus for each increase of one pH unit, E° becomes more negative by 0.059 V or 59 mV. This is commonly known as the Nernst slope in the oxidation–reduction reactions.

4.2.1 E° vs. log K

Based on the first and the second laws of thermodynamics, the following relationships can be obtained for any given reaction:

$$\Delta G_r = \Delta G_r^\circ + RT \ln Q \tag{4.17}$$

$$\Delta G_r = -nFE \tag{4.19}$$

Electrochemical Properties

and at standard-state conditions:

$$\Delta G_r^\circ = -nFE^\circ \tag{4.18}$$

For a reaction under equilibrium conditions $K_{eq} = Q$:

$$\text{Oxidant} + e^- = \text{Reductant}$$

$$\Delta G = \Delta G_r^\circ + RT \ln \frac{[\text{Reductant}]}{[\text{Oxidant}][e^-]} \tag{4.15}$$

$$\Delta G_r = \Delta G_r^\circ + RT \ln K \tag{4.33}$$

Since ΔG_r at equilibrium $= 0$,

$$\Delta G_r^\circ = -RT \ln K \tag{4.12}$$

Since

$$\Delta G_r^\circ = -nFE^\circ \tag{4.18}$$

we have

$$-nFE^\circ = -RT \ln K \tag{4.34}$$

$$E^\circ = \frac{RT}{nF} \ln K_{eq} = \frac{RT}{nF} 2.303 \log K \tag{4.35}$$

Substituting the numerical values for R, T, and F, Equation 4.35 can be modified as

$$E^\circ = \frac{0.059}{n} \log K \tag{4.36}$$

$$\log K = 16.9 nE^\circ \tag{4.37}$$

These equations suggest that in oxidation–reduction reactions, the relationship of chemical reactions can be expressed as standard electrode potentials or equilibrium constants.

4.2.2 pe vs. Eh

The oxidation–reduction equilibrium in soils is usually expressed in terms of Eh, but just as pH that measures proton activity on the basis of moles per liter, Eh can be expressed as pe (negative logarithm of the electron activity).

Since many disciplines now use pe as much as Eh to express electron activity in a system, it is worthwhile to discuss the relationships between these two variables (Lindsay, 1979). Five decades ago the Swedish chemist Lars Gunnar Sillen suggested that the electrons (e^-) can be considered as any other reactant or product in chemical reactions. Sillen and Martell (1964) tabulated equilibrium constants for redox reactions in terms of both E° (standard electrode potentials) and log K (equilibrium activity constants), and encouraged the use of log K to calculate pe values for redox systems.

Like pH, the electron activity in a reaction can be defined as

$$\text{pe} = -\log [e^-] \tag{4.38}$$

Consider the following generalized oxidation–reduction reaction at 25°C:

$$\text{Oxidant} + ne^- = \text{Reductant}$$

$$K = \frac{[\text{Reductant}]}{[\text{Oxidant}][e^-]^n} \tag{4.39}$$

Rearranging, we obtain

$$[e^-]^n = \frac{1}{K}\frac{[\text{Reductant}]}{[\text{Oxidant}]} \tag{4.40}$$

Taking negative logarithms on both sides and dividing by (n), we obtain

$$-\frac{n}{n}\log[e^-] = \frac{1}{n}\log K - \frac{1}{n}\log\frac{[\text{Reductant}]}{[\text{Oxidant}]} \tag{4.41}$$

Since pe = $-\log[e^-]$, we have

$$-\log[e^-] = \frac{1}{n}\log K - \frac{1}{n}\log\frac{[\text{Reductant}]}{[\text{Oxidant}]} \tag{4.42}$$

$$\text{pe}^\circ = \frac{1}{n}\log K \tag{4.43}$$

The following relationship is obtained by combining Equations 4.39, 4.42, and 4.43:

$$\text{pe} = \text{pe}^\circ - \frac{1}{n}\log\frac{[\text{Reductant}]}{[\text{Oxidant}]} \tag{4.44}$$

Since

$$E^\circ = \frac{RT}{F}\frac{1}{n}\ln K \quad \text{or} \quad -nFE^\circ = -RT\ln K \tag{4.45}$$

or

$$\frac{1}{n}\log K = E^\circ\frac{F}{RT2.303} = \text{pe}^\circ \tag{4.46}$$

$$\text{pe}^\circ = 16.9E^\circ \text{ (at 25°C)} \tag{4.47}$$

$$\text{pe} = \text{Eh}\frac{F}{RT2.303} \tag{4.48}$$

$$\text{pe} = 16.9\text{Eh (at 25°C)} \tag{4.49}$$

$$\text{Eh} = 0.059\,\text{pe} \tag{4.50}$$

If we multiply both sides of Equation 4.44 by $RT2.303/F$, as follows:

$$\frac{RT2.303}{F}\text{pe} = \frac{RT2.303}{F}\text{pe}^\circ - \frac{RT2.303}{F}\frac{1}{n}\log\frac{[\text{Reductant}]}{[\text{Oxidant}]} \tag{4.51}$$

then Equation 4.44 reduces to the Nernst equation written in the form shown below:

$$\text{Eh} = E^\circ - \frac{2.303RT}{nF}\log\frac{[\text{Reductant}]}{[\text{Oxidant}]} \tag{4.23}$$

Electrochemical Properties

EXAMPLE

Consider the following simple half reaction:

$$Fe^{3+} + e^- = Fe^{2+} \quad E° = +0.77 \text{ V}$$

The forward reaction is a reduction process with Fe^{3+} accepting electrons and acting as an oxidizing agent while being reduced to Fe^{2+}. The half-cell voltage $E°$ is the reduction potential in reference to SHE.

Even though free electrons do not exist in the solution, the following equilibrium expression can be obtained

$$K = \frac{[Fe^{2+}]}{[Fe^{3+}][e^-]} \tag{4.52}$$

Rearranging, we obtain

$$[e^-] = \frac{1}{K} \frac{[Fe^{2+}]}{[Fe^{3+}]} \tag{4.53}$$

The transformed Nernst equation in terms of electron activity is written as

$$pe = pe° - \log \frac{[Fe^{2+}]}{[Fe^{3+}]} \tag{4.54}$$

Since

$$pe° = 16.90 E° \tag{4.55}$$

the Fe^{2+}/Fe^{3+} redox reaction can now be written as

$$pe = 13.0 - \log \frac{[Fe^{2+}]}{[Fe^{3+}]} \tag{4.56}$$

Equations 4.44 and 4.50 demonstrate that redox relationships can be expressed in terms of either Eh or pe.

Based on our earlier discussions, we know that Eh can be measured as cell potential in reference to SHE. The question is, can we measure e^- activity in a similar manner as the measurement of H^+ activity. In the absence of temperature, pe is a simple hypothetical calculation and represents transformation of Eh. However, pe values have much narrower range than Eh values, and can be easily combined with pH values (see discussion in the latter part of this chapter). Although there are advantages in using the pe concept in handling oxidation–reduction reactions, electrons do not occur as an independent species in the solution, and it is questionable to think of their activity. From a practical standpoint, a potential in volts is probably simpler and has a wider applicability.

4.3 MEASUREMENT OF Eh

Oxidation–reduction potential is measured with an electrode pair consisting of an inert electrode and a reference electrode. Saturated calomel is probably the most common reference electrode used in Eh measurement.

The inert electrode used the most is the bright platinum electrode. The construction and calibration of platinum electrodes are described in the latter part of this chapter. The inert electrode acts as an electron acceptor of or a donor to the ions in the measured solution. The half-cell reaction written as a reduction reaction is

$$Hg_2Cl_2 + 2e^- = 2Hg° + 2Cl^-, \quad E° = 0.268 \text{ V}$$

Nernst equation

$$Eh = E° - \frac{RT2.303}{nF} \log \frac{[\text{Reductant}]}{[\text{Oxidant}]} \quad (4.23)$$

becomes

$$Eh = 0.268 - \frac{0.059}{2} \log \frac{[\text{Hg°}]^2[\text{Cl}]^2}{[\text{Hg}_2\text{Cl}_2]} \quad (4.57)$$

[Hg] and [Hg_2Cl_2] are in their standard states and their activities are unity. Thus, the cell potential is influenced by Cl⁻ ions, and at 25°C, 1 atm, the activity of Cl⁻ in a saturated KCl solution is 2.86 M. Therefore,

$$Eh = 0.268 - \frac{0.059}{2} \log [\text{Cl}^-]^2 \quad (4.58)$$

$$Eh = 0.268 - \frac{0.059}{2} \log [2.86]^2 \quad (4.59)$$

$$Eh_{SCE} = 0.24 \text{ V or } 240 \text{ mV} \quad (4.60)$$

$$Eh = Eh_{SCE} + E_{SCE} \quad (4.61)$$

where Eh_{SCE} is the electrode potential of a standard calomel half cell measured in reference to SHE, E_{SCE} the electrode potential of a redox couple measured in reference to SCE, and Eh the electrode potential of redox couple measured in reference to SHE.

If the potential difference between a redox couple and SCE is measured, then the potential difference is 0.24 V less than the Eh for the redox couple:

$$Eh = E_{SCE} + 0.24 \quad (4.62)$$

where E_{SCE} is the electrode potential of the redox couple measured using SCE.

EXAMPLE

Consider an aqueous solution containing the redox couple Fe^{3+}/Fe^{2+}:

$$Fe^{3+} + e^- = Fe^{2+}, \quad E° = +0.77 \text{ V}$$

Assume the concentration of Fe^{2+} = 0.01 M and Fe^{3+} = 0.001 M.
Eh of the Fe^{3+}/Fe^{2+} redox couple can be calculated as

$$Eh = E° - \frac{RT2.303}{nF} \log \frac{[\text{Reductant}]}{[\text{Oxidant}]}$$

$$= 0.77 - 0.059 \log \frac{[\text{Fe}^{2+}]}{[\text{Fe}^{3+}]}$$

$$= 0.77 - 0.059 \log \frac{[0.01]}{[0.001]}$$

$$= 0.77 - 0.059 = 0.712 \text{ V} \quad (4.23)$$

$$E_{SCE} = Eh - Eh_{SCE} \quad (4.63)$$

where E_{SCE} is the electrode potential of a redox couple measured in reference to SCE.

$$E_{SCE} = 0.712 - 0.24$$
$$= 0.47 \text{ V}$$

The potential of Fe^{3+}/Fe^{2+} redox pair (at 0.001 M [Fe^{3+}] and 0.01 M [Fe^{2+}] concentrations) measured in reference to SCE will be 0.47 V.

4.4 Eh–pH RELATIONSHIPS

As previously discussed, for any redox couple, the ratio of reductant/oxidant is a function of the two variables Eh and pH of the system. We have already defined Eh in terms of free energy, equilibrium constant, and electron activity. pH is defined in terms of proton activity. These two master variables regulate many biogeochemical reactions and determine the stability of minerals and nutrient regeneration in soils and sediments. Thus, the activities or concentrations of reductant and oxidant can be depicted as

$$\frac{\text{Reductant}}{\text{Oxidant}} = f(\text{Eh, pH}) \qquad (4.64)$$

The role of Eh and pH in reactions of biogeochemical importance in natural systems is described by Baas Becking et al. (1960) as follows:

- Reactions involving hydrolysis with no protons and electrons exchanged, i.e.,

$$Fe_2O_3 + H_2O = 2FeOOH$$

- Reactions involving only protons, i.e.,

$$H_2CO_3 = H^+ + HCO_3^-$$
$$NH_4^+ = H^+ + NH_3$$

- Reactions involving only electrons, i.e.,

$$Fe^{3+} + e^- = Fe^{2+}$$

- Reactions involving both electrons and protons, i.e.,

$$Fe(OH)_3 + 3H^+ + e^- = Fe^{2+} + 3H_2O$$

The first type of reactions are not influenced by pH or Eh because the reaction involves hydrolysis of a compound. The second type of reactions involve the dissociation of acids; thus, pH has a direct effect on these reactions. The third type of reactions involve only loss or gain of electrons affecting the oxidation state of a substance. However, most of the reactions in the biosphere involve a fourth type of reactions, where both H^+ and e^- are transferred.

Under standard-state condition we have defined the following relationship between Eh and pH:

$$\text{Eh} = -0.059\text{pH} \qquad (4.34)$$

This equation assumes $E° = 0$ V, the ratio of reductant/oxidant $= 1$, absolute temperature $= 298.15$ K, and pressure $= 1$ atm.

We can write the following general chemical reaction including both protons and electrons:

$$m(\text{Oxidant}) + m\text{H}^+ + ne^- = m(\text{Reductant})$$

According to the Nernst equation:

$$\text{Eh} = E° - \frac{RT}{nF} \ln \frac{[\text{Reductant}]^m}{[\text{Oxidant}]^m[\text{H}]^m} \quad (4.65)$$

$$= E° - \frac{RT}{nF}\left(\ln \frac{[\text{Reductant}]^m}{[\text{Oxidant}]^m[\text{H}]^m} + \ln \frac{1}{[\text{H}]^m}\right) \quad (4.66)$$

$$= E° - \frac{RT}{nF} 2.3 \log \frac{[\text{Reductant}]^m}{[\text{Oxidant}]^m} - \frac{RT}{nF} 2.303 m\text{pH} \quad (4.67)$$

$$= E° - \frac{0.059}{n} \log \frac{[\text{Reductant}]^m}{[\text{Oxidant}]^m} - 0.059 \frac{m}{n} \text{pH} \quad (4.68)$$

If the ratio of reductant/oxidant $= 1$, the following basic relationship between Eh and pH is obtained:

$$\text{Eh} = E° - 0.059 \frac{m}{n} \text{pH} \quad (4.69)$$

In the previous discussions, we have shown that for each increase of one pH unit, there will be a shift of 0.059 V or 59 mV (Equation 4.32). This relationship is true only when $m\text{H}^+ = ne^-$ or $m/n = 1$, for a given oxidation–reduction reaction. Following examples indicate that m/n may vary from 1 to 4, depending on the compounds involved in the oxidation–reduction reaction (Table 4.2).

In wetland soils, more than one redox couple can function at any given time; thus, the effect of [H$^+$] on Eh will depend on the concentration of the oxidant undergoing reduction. In spite of this complexity, ΔEh/ΔpH for many wetland soils was found to be in the range of 0.035–0.130 V

TABLE 4.2
Examples of Redox Reactions Showing the Effect of Protons on Eh

Reduction Reaction	m/n	Eh/pH (V/pH Unit)
$O_2 + 4H^+ + 4e^- = 2H_2O$	1	0.059
$NO_3^- + 2H^+ + 2e^- = NO_2^- + H_2O$	1	0.059
$NO_3^- + 12H^+ + 10e^- = N_2 + 6H_2O$	1.2	0.078
$MnO_2 + 4H^+ + 2e^- = Mn^{2+} + 2H_2O$	2	0.118
$Fe(OH)_3 + 3H^+ + 1e^- = Fe^{2+} + 3H_2O$	3	0.177
$Fe_3(OH)_8 + 8H^+ + 2e^- = Fe^{2+} + 8H_2O$	4	0.236

Electrochemical Properties

(Redman and Patrick, 1965; Yu, 1985). Soils dominated by Fe^{3+} compounds will show greater shift in Eh for each unit of pH change.

The following two examples show the upper and lower limits for stability of water. Oxidation–reduction reactions of the biogeochemical reactions occur within these limits.

Upper limit for stability of water:

$$O_2(g) + 4e^- + 4H^+ = 2H_2O, \quad E° = +1.23 \text{ V}$$

From Equation 4.68, we obtain

$$Eh = E° - \frac{0.059}{4} \log \frac{[H_2O]^2}{P_{O_2}} - 0.059 \frac{4}{4} pH \quad (4.70)$$

If the activity of pure water is unity and $P_{O_2} = 1$ atm, then Equation 4.70 reduces to

$$Eh = 1.23 - 0.059 pH \quad (4.71)$$

From Equation 4.44, we obtain

$$pe = pe° - \frac{1}{4} \log \frac{[H_2O]^2}{P_{O_2}} - \frac{4}{4} pH \quad (4.72)$$

$$pe = 20.8 - pH \quad (4.73)$$

Lower limit for stability of water:

$$2H_2O + 2e^- = H_2 + 2OH^-, \quad E° = -0.83 \text{ V}$$

From Equation 4.68, we obtain

$$Eh = E° - \frac{0.059}{2} \log \frac{[P_{H_2}]^2}{[H_2O]^2} - \frac{0.059}{2} \log [OH]^2 \quad (4.74)$$

If the activity of pure water is unity and $P_{H_2} = 1$ atm, then Equation 4.74 reduces to

$$Eh = -0.83 + 0.059 \, pOH \quad (4.75)$$

or

$$Eh = -0.83 + 0.059(14 - pH) \quad (4.76)$$

From Equation 4.44, we obtain

$$pe = pe° - \frac{1}{2} \log \frac{P_{H_2}}{[H_2O]^2} + \frac{2}{2} pOH \quad (4.77)$$

$$pe = -14 + pOH \quad (4.78)$$

$$pe = -pH \quad (4.79)$$

These upper and lower limits for stability of water provide boundary conditions for the Eh–pH or pe–pH diagrams. Both redox potential and pH define the conditions under which important

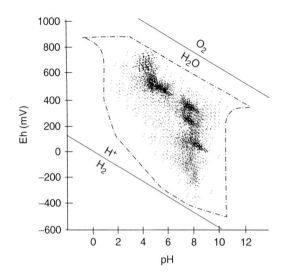

FIGURE 4.5 Distribution of Eh–pH values measured in aqueous environments. (Redrawn from Baas Becking et al., 1960.)

biogeochemical processes take place. The relationship between pH and Eh has been extensively developed for various minerals of geological interest (Krumbein and Garrels, 1952; Garrels and Christ, 1965; Truesdall, 1969; Pourbaix, 1966; Stumm and Morgan, 1981; Drever, 1982; Brookins, 1988). The distribution of Eh and pH values of natural aqueous environments is shown in Figure 4.5. The boundaries show limits of Eh–pH for natural systems with a few exceptions (acid mind drainage, acid sulfate soils, and black alkali soils); most of the earth's natural environments have Eh–pH values that lie within this parallelogram. In a classic paper, Baas Becking et al. (1960) used Eh–pH diagrams to delineate the Eh–pH field at which a number of important redox processes take place in sediment–water systems (Figure 4.5). Their delineations were based on published values of experimentally determined Eh and pH values. Using this approach, they were able to differentiate between normal (oxic) soils, wet (seasonally saturated) soils, and waterlogged (semipermanently saturated) soils. Their plot also showed that the pH range was narrower in reduced soils (more negative redox potentials). Similarly, chemical equilibrium and pe–pH relationships of several compounds of interest in soil science have been presented in detail by Lindsay (1979).

4.5 BUFFERING OF REDOX POTENTIAL (POISE)

The ability of a system to resist or retard the change in Eh upon addition of small quantities of oxidants or reductants is referred to as poising or the system at poise. This is analogous to the buffering capacity in pH measurements. Activities of the oxidants and reductants, and the number of electrons involved in the reactions are important in maintaining the poise of the system. It is the strongest when the ratio [reductant/oxidant] = 1. This will increase with the total concentration of the redox couple and the number of electrons involved in a reaction (Figure 4.6). As the ratio [reductant/oxidant] increases, Eh will decrease and vice versa. From Equation 4.26, for a redox couple with $n = 1$, the slope will be 0.059, and for a reaction with $n = 2$, the slope will decrease to 0.028. Addition of oxidants or reductants will lead to a decrease in Eh, with increase in the number of electrons involved in a reaction (Figure 4.6). In a system with mixed potential (such as a wetland soil system), the final potential is governed by a redox couple present in excess concentration compared to other redox couples. Thus, the dominant redox couple in the system will decrease the Eh value in permanently waterlogged soils. Mineral wetland soils and sediments are generally

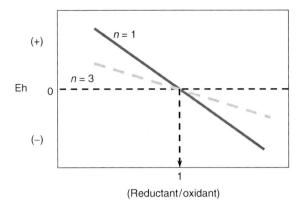

FIGURE 4.6 Influence of activity ratio (reductant/oxidant) and the number of electrons (n) on Eh.

highly poised because of the high concentration of reducible iron and manganese. Redox potentials in well-drained soils are poorly poised, because O_2/H_2O is an irreversible redox couple. In drained soils, the measurement of O_2 concentration in soil pore volume is the most reliable way of characterizing these soils.

4.6 MEASUREMENT OF REDOX POTENTIALS

4.6.1 Construction of Platinum Electrodes

The principles associated with the Nernst equation form the basis for developing electrodes to measure electron activity or electrical potential. The electrochemistry based on the electrode potential is related to ion activities, which results in the development of specific ion electrodes. There are several commercially available electrodes designed to measure Eh, pH, and specific ion activities. In this section we will present simple methods to construct redox electrodes for use in the laboratory and under field conditions. Many commercially available electrodes are bulky and are not suitable for use under field conditions. For the past three decades methods associated with the construction of redox electrodes were developed in our laboratories.

Field electrodes are typically made with platinum wire with a diameter of 1.024 mm cut into 1–1.5 cm segments. A smaller gauge platinum is not recommended because of poor reproducibility of potentials. These segments are soaked in a concentrated acid (1:1 HNO_3 and HCl mixture) for several hours (4 h) to remove surface contamination of metals and other impurities, followed by thorough washing in deionized distilled water. The platinum segments are soldered or fused to a copper lead (1.628–2.053 mm diameter) of desired length (Figure 4.7). Waterproof epoxy glue is used to cover the joint between copper and platinum. The copper wire is inserted into heat-shrinking tube to provide insulation. The platinum tip of the copper wire is sealed with epoxy as shown in Figure 4.7, while at the other end, about 3 cm of copper wire is exposed. This end is connected to an insulated cable connected to a pH/voltmeter. For seasonal measurements this type of electrode can be left installed in the field for a period of about 1 year. However, it is recommended that the electrodes are periodically (at least once every 3 months) checked and reinstalled in the field.

Laboratory electrodes are usually made with 20–30 cm of soft glass tubing (0.75 cm o.d.). Platinum segments (1.024 mm diameter) of 2 cm length are fused into one end of the glass tubing (Figure 4.8), in such a way that at least 1 cm of platinum extends out of the glass tube. The glass–platinum junction should be water tight. Small amounts of triple-distilled liquid mercury are poured into the glass tube to a depth of about 3 cm. This mercury will provide electrical contact between the copper wire, which

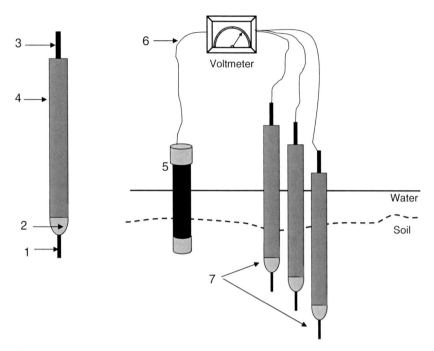

FIGURE 4.7 Schematic showing the construction and installation of field electrodes. (1) Platinum wire of 1.024 mm diameter, (2) waterproof epoxy, (3) copper wire, (4) heat-shrinking insulation, (5) reference electrode, (6) voltmeter, and (7) electrodes installed at different depths.

is inserted into the glass tube, and the platinum tip inside the tube. The Length of the copper wire is dependent on the use of electrodes. Other types of electrodes can be constructed by sealing 3–4 cm pieces of platinum wire in the side, near the bottom, of test tubes or other types of glass containers.

Reference electrodes (calomel electrodes) are available commercially through scientific catalogs or can be prepared in the laboratory as follows. A calomel reference electrode is prepared by fusing a 3–4 cm piece of 18 gauge platinum wire in the side, near the bottom, of a 20 × 150 mm glass test tube. Triple-distilled mercury is poured into the tubes until the protruding platinum wire is covered. A paste (about 1 cm thick) made from mercury, mercurous chloride, and saturated KCl is placed above the mercury in the reference tube. Saturated KCl is then added to the tube, and a salt bridge is used to connect the calomel half cell to the soil or sediment. The salt bridge is prepared by filling a glass tube with hot saturated KCl containing 4% agar. The liquid is solidified in the tube by cooling to room temperature. Calomel electrodes are checked by substituting a standard calomel electrode for the platinum electrode and checking for a zero potential difference between the two half cells. Commercially available reference electrodes are convenient for use under field conditions.

4.6.2 Standardization of Electrodes

Prior to use, all electrodes should be standardized using a solution containing redox couple of known potential at pH 7 and at pH approximating the soil pH. If soil pH varies between 4 and 7, calibration at pH 4 and 7 is adequate. The redox couples used to check electrodes are (i) buffer solutions containing saturated levels of quinhydrone and (ii) buffer solutions containing 0.1 M ferrous ammonium sulfate and 0.1 M ferric ammonium sulfate dissolved in 1 M H_2SO_4. The use of quinone/hydroquinone redox couple allows for the calibration of electrodes at pH values typical of wetland soils. The pH of solutions containing ferrous and ferric ammonium sulfate is zero, and Eh of the calomel reference electrode in this solution is 430 mV. The following example shows the use of quinone/hydroquinone redox couple to calibrate platinum electrodes.

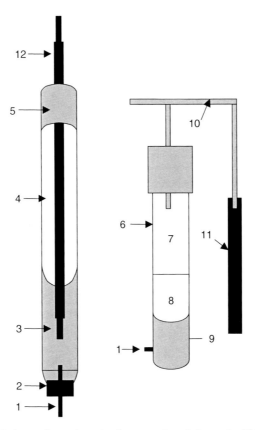

FIGURE 4.8 Laboratory platinum electrode and reference calomel electrode. (1) Platinum wire of 1.024 mm diameter, (2) waterproof epoxy, (3) triple-distilled mercury, (4) glass tubing, (5) silicon sealant covered with heat-shrink tubing, (6) test tube, (7) saturated KCl, (8) calomel–mercury–KCl paste, (9) mercury, (10) saturated KCl, (11) salt bridge, and (12) copper wire.

When quinone is added to the buffer solution, it is reduced to hydroquinone as shown below:

$$\text{Quinone} + 2\text{H}^+ + 2e^- = \text{Hydroquinone}, \quad E° = 0.699 \text{ V}$$

According to Equation 4.68

$$\text{Eh} = 0.699 - \frac{0.059}{2} \log \frac{[\text{Hydroquinone}]}{[\text{Quinone}]} - 0.059\,\text{pH} \tag{4.80}$$

In a saturated solution, if the ratio of quinone/Hydroquinone = 1, then Equation 4.80 reduces to

$$\text{Eh} = 0.699 - 0.059\,\text{pH} \tag{4.81}$$

$$\text{At pH 7, Eh} = 0.699 - 0.413 = 0.286 \text{ V} \tag{4.82}$$

$$\text{At pH 4, Eh} = 0.699 - 0.236 = 0.463 \text{ V} \tag{4.83}$$

Expected electrode potential with SCE can be calculated using the following equation:

$$E_{\text{SCE}} = \text{Eh} - \text{Eh}_{\text{SCE}} \tag{4.63}$$

$$\text{At pH 7.0, } E_{\text{SCE}} = 0.286 - 0.244$$
$$= 0.042 \text{ V or } 42 \text{ mV} \quad (4.84)$$

$$\text{At pH 4.0, } E_{\text{SCE}} = 0.463 - 0.244$$
$$= 0.219 \text{ V or } 219 \text{ mV} \quad (4.85)$$

Newly constructed platinum electrodes placed in quinone/hydroquinone solutions at pH 7.0 should give an electrode potential of 0.042 V or 42 mV. When placed in a pH 4.0 buffer solution, they should give a potential of 0.219 V or 219 mV. The amount of quinhydrone added to the pH buffer solution should be enough to obtain a final concentration of at least 1 g L^{-1}. Quinhydrone solutions should be made fresh every time electrodes are calibrated.

Eh measurements are sensitive to temperature. As shown in Table 4.3, Eh is inversely related to temperature. As such electrode potential with SCE will differ when electrodes are calibrated at different temperatures.

The surface condition of the platinum tip of an electrode directly influences the capacity of the platinum electrode to function as a source or sink for electrons. To maintain a proper surface condition, the platinum electrode surface is cleaned using (i) mechanical methods, (ii) chemical methods, and (iii) electrochemical methods. Mechanical cleaning involves polishing the platinum electrode surface with mild scouring powder or levigated alumina; for most electrodes, mechanical cleaning of the electrode surface is adequate. Chemical methods include soaking the platinum tips in a potassium dichromate–sulfuric acid solution or HNO_3, followed by rinsing with distilled water. Electrochemical methods involve anode stripping of the electrode, which removes the substance attached to the electrode surface. The electrodes with potentials deviating more than 10 mV from the standard potentials (see Equation 4.63) should be discarded or cleaned again until satisfactory results are obtained.

4.6.3 REDOX POTENTIALS IN SOILS

Redox potential measurements have been widely used to characterize wetland soils and sediments. The value of these measurements depends on their interpretation with due recognition to its theoretical and practical limitations. Redox potentials in soils are measured in (i) soil pore water, (ii) soil slurry, (iii) intact soil cores, or (iv) *in situ* under field condition. Under all these conditions, platinum electrode (custom made or purchased commercially) and reference electrode (usually calomel electrode) are

TABLE 4.3
Effect of Temperature on the Standard Potential of Calomel Half Cell and Platinum Electrode

| Temperature (°C) | Calomel Potential (mV) | Meter Reading | |
		pH 4 (mV)	pH 7 (mV)
5	257	236	67
10	254	232	60
15	251	227	54
20	247	223	47
25	244	218	41
30	241	213	34
35	238	209	28
40	234	204	21
45	231	198	14

Electrochemical Properties

placed in soil or soil slurry. Both electrodes are connected to a pH/millivolt (mV) meter. Within few minutes a stable reading can be obtained. Unstable readings under laboratory conditions suggest that the electrical circuit is not complete and all pertinent contact points should be checked. However, it is recommended that under field conditions, the electrodes are left for at least 24 h before measurements are taken. A longer equilibration period is necessary because of the heterogeneous nature of soil systems.

Under an *in situ* condition, a platinum electrode is inserted into the soil at a predetermined depth. Usually platinum electrodes are left in place for subsequent measurements. At the time of measurements, the reference electrode is inserted at a short distance into the wet soil. It is not necessary that the reference electrode be placed at the same depth as the platinum electrode, as long as there is enough soil moisture to ensure a good electrical contact. However, the electrical resistance of the measuring circuit can be affected by the distance between the platinum electrode and the reference electrode. This may not be a serious problem in saturated soils.

Stable electrode potential reading can be obtained if the electrode used is truly inert and reversible and a redox couple is established at the platinum surface. Rapid equilibration can be obtained in a soil system containing one redox couple at a relatively high concentration. A platinum electrode responds to all redox couples present in the soil, and provides an average mixed potential, which may not thermodynamically represent any one redox couple. In soils and sediments, redox equilibrium can never be reached due to continuous addition of electron donors and acceptors.

Two examples of Eh measurements under laboratory and field conditions are shown in Figures 4.9, 4.10, 4.11, and 4.12. Figure 4.9 shows Eh measurements made in a continuously stirred soil suspension

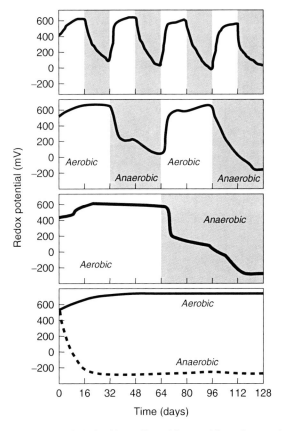

FIGURE 4.9 Changes in redox potential of soil as affected by aerobic and anaerobic conditions. (Redrawn from Reddy and Patrick, 1975.)

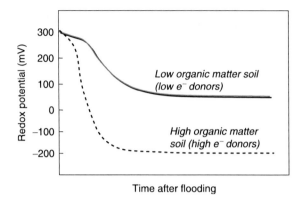

FIGURE 4.10 Influence of electron donor (soil organic matter) on soil redox potential.

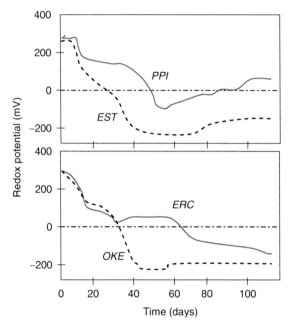

FIGURE 4.11 Laboratory redox potentials measured in flooded organic soils from the Everglades Agricultural Area in Florida. (Reddy, K. R., Unpublished Results, University of Florida.)

subjected to aerobic or anaerobic or alternate aerobic and anaerobic conditions. The Eh values readily respond to changes in oxidants in soil systems. Similarly, in Figure 4.10, Eh values are shown to respond to changes in hydrology under field conditions.

Redox potentials in soils are affected by a number of factors:

- Water table fluctuations (or hydrology) that introduce oxygen into the soil
- Loading of alternate electron acceptors
- Loading of organic matter (electron donor)
- Organic matter available in the soil (electron donor)
- Temperature
- Soil pH

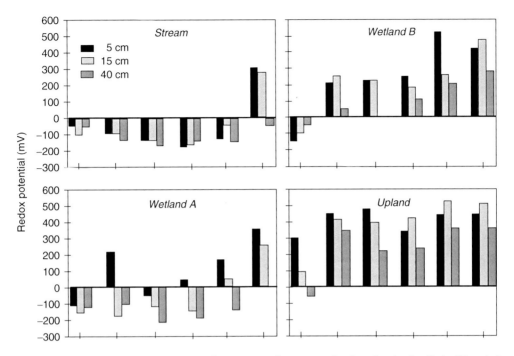

FIGURE 4.12 *In situ* potentials measured in stream sediments, wetland, and upland soils in Okeechobee Basin, Florida. (Redrawn from Scinto, 1990.)

The above-listed parameters are included in the Nernst equation to describe the Eh of a system. For example, pH and temperature have an inverse relationship with Eh. Similarly, an increase in electron donor supply (organic matter) will decrease Eh values, while an increase in electron acceptor supply (oxidants) will increase Eh values. Considering these factors, Eh can be used as an operational parameter to characterize soil anaerobiosis. It is a simple parameter, and because of its ease of measurement, it can be included in routine monitoring of soil quality and soil characterization. The Eh parameter by itself may have little value when it is related to the soil conditions at the time of measurement.

Soils vary spatially in Eh values, and thus a large number of measurements should be taken to adequately characterize a given area. This variability could be minimal in permanently waterlogged wetlands or areas with minimal hydrologic fluctuations. A platinum electrode inserted into the soil may be in direct contact with the organic matter or other solid phase, which may result in lower Eh values than those for the soils as a whole. For electrodes left in the field, microbial colonies can develop on the electrode surface, resulting in an alteration in Eh values.

Redox potential is the best available simple indicator of the oxidation–reduction status of the soil, for the following reasons. First, the range of Eh values in wetland soil is much wider, approximately 1000 mV as compared to a range of 300 mV in drained soils. Second, the higher concentrations of reductants that contribute to the mixed potential result in better poising and better reproducibility. Third, oxygen is easily reduced and, therefore, is usually present at negligible levels in wetland soils. The methods used to determine the oxygen status of drained soils cannot be used in wetland soils. Fourth, the direct measurement of reduced compounds (i.e., the presence of ferrous iron, manganous manganese, or sulfides or methane) is cumbersome and cannot be easily accomplished on a routine basis. Thus, redox potential remains a generally applicable, reasonably convenient way to identify the presence and intensity of reduction in wetland soils and sediments.

In addition to these practical problems, several thermodynamic limitations of redox potential measurements have been presented by Bohn et al. (1985). A brief summary of these comments is presented below:

- The quantitative capability of the Nernst equation to predict the activity of chemical species is valid only under equilibrium conditions. Most of the redox couples are not in equilibrium, except in highly reduced soils; steady-state condition may result in pseudo-equilibrium conditions. In soils, redox equilibrium is probably never reached because of the continuous addition of electron donors and acceptors. Biological systems add and remove electrons continuously. Thus, redox potential measurements cannot be used to accurately predict the activity of specific reductant and oxidant of the system.
- Platinum electrodes respond favorably to reversible redox couples than to irreversible couples. Under aerobic conditions, Eh deviates widely from the potential of O_2/H_2O couple. Under anaerobic conditions, Eh values may be related to activities of Fe^{2+} and Mn^{2+}, because these ions dominate the system.
- Redox potentials approximate closely the redox couples in the lower region of Eh–pH diagrams. Redox potentials do not approach O_2/H_2O potential, because of irreversibility.
- Redox potential and pH are closely related. The change in Eh per pH unit may range from 59 to 200 mV, depending on the redox couples and kinetics of reduction. Since Eh values represent mixed potentials, a simple correction of 59 mV per pH unit may not be adequate.
- A platinum electrode surface can be contaminated by coatings of oxides, sulfides, and other impurities.

In spite of a number of practical difficulties and thermodynamic limitations, Eh measurements provide a rapid, simple operational parameter to determine the intensity of reduction in wetland soils. Over the last half century, a large number of Eh measurements have been made in a wide range of natural systems, and in certain cases, Eh values have been related to various biogeochemical transformations. Thus, Eh values, at the minimum, provide a qualitative measure of the range of Eh values where selected processes affect the fate of many nutrients, trace metals, and toxic organic compounds.

4.7 pH

The significance of pH in natural systems is well known. The pH is defined as

$$\text{pH} = -\log [H^+] \tag{4.86}$$

where $[H^+]$ is the activity of H^+ ion in solution; more specifically H^+ is expressed as H_3O^+ (hydronium ion). Water undergoes a self-ionization, where it acts simultaneously as an acid and a base:

$$H_2O = H^+ + OH^- \tag{4.87}$$

or

$$2H_2O = H_3O^+ + OH^- \tag{4.88}$$

The equilibrium relationship can be written as

$$\frac{[H^+][OH^-]}{[H_2O]} = K_w \tag{4.89}$$

Electrochemical Properties

where K_w is the equilibrium constant, and the brackets are used to denote molar concentration. Because the concentration of [H_2O] is constant, Equation 4.89 can be written as

$$[H^+][OH^-] = K_w \qquad (4.90)$$

Since [H^+] = [OH^-], Equation 4.90 can be written as

$$[H^+]^2 = K_w \qquad (4.91)$$

Taking negative logarithm on both sides of Equation 4.91 and rearranging we obtain

$$-\log[H^+] = -\frac{1}{2}\log K_w \qquad (4.92)$$

Using chemical notations of $pK = -\log K$, Equation 4.92 can be written as

$$pH = -\log[H^+] \qquad (4.93)$$

If we assume for dilute solutions activity equals to concentration, Equation 4.86 will be the same as Equation 4.93.

At high pH, solutions have low H^+ activity; thus, compounds present in this system are not protonated. Similarly, low-pH systems are high in H^+ activity; thus, compounds are protonated.

4.7.1 Soil pH

The pH values of floodwater and underlying soils and sediments can be summarized as follows:

Floodwater	7.0–10.0
Flooded soils	6.5–7.5
Freshwater sediments	6.0–7.0
Ocean sediments	7.0–7.8
Marsh soils	5.0–7.0

Although the pH values of acid soils increase after flooding and those of alkaline soils decrease, soil properties and environmental variables such as temperature can markedly influence the pattern of changes. In Figure 4.13 influence of flooding on pH is demonstrated for several paddy soils of Philippines. Initial characteristics of these soils are given in Table 4.4.

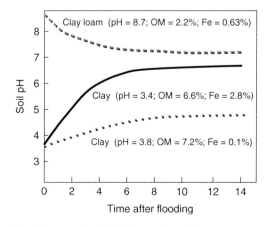

FIGURE 4.13 Influence of flooding on soil pH. (Redrawn from Ponnamperuma, 1972.)

TABLE 4.4
Selected Characteristics of Soil Used to Demonstrate the Effect of Flooding on Soil pH (see Figure 4.13)

Soil No.	Texture	pH	OM%	Fe%	Mn%
28	Clay	4.9	2.9	4.70	0.08
35	Clay	3.4	6.6	2.60	0.01
40	Clay	3.8	7.2	1.50	0.00
57	Clay loam	8.7	2.2	0.63	0.07
94	Clay	6.7	2.6	0.96	0.09
99	Clay loam	7.7	4.8	1.55	0.08

Source: Ponnamperuma (1972).

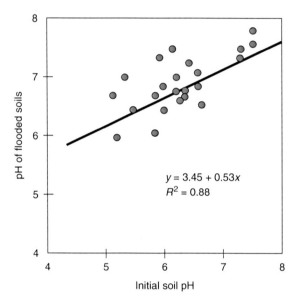

FIGURE 4.14 Relationship between pH before and after flooding soils from Louisiana. (Redrawn from Redman and Patrick, 1965.)

Soils high in carbon and reducible iron tend to attain a pH of about 6.5 within a few weeks after flooding. Increase in pH of acid soils low in carbon or reactive iron is much slower, and these soils may not attain a pH of 6.5. Acid sulfate soils especially those with low organic matter content may not attain a pH greater than 5 even after several months of flooding.

The pH of soils from the Mississippi River floodplain increased with flooding by as much as 1.6 pH units within 30 days (Figure 4.14). Regression analysis between pH measured before flooding and 30 days after flooding shows that for soils with original pH > 7.4, pH decreased after flooding, and for soils with original pH < 7.4, pH increased after flooding. The tendency of soils with low pH to decrease in acidity and soils with high pH to increase in acidity upon flooding suggests that the pH of flooded soils tends to be buffered around neutrality by substances consumed/produced during the oxidation–reduction reactions.

Addition of organic matter increases electron donor supply and favors the development of alkalinity through rapid reduction of oxidants. Effect was much greater in acid soils.

Flooding a soil results in the consumption of electrons and protons. Under many situations the ratio of proton to electron consumption may be greater than 1. Continuous consumption of protons during consumption of electrons results in an increase in pH. The increase in pH of flooded soils is largely due to the presence of iron and manganese in the form of hydroxides and carbonates:

$$Fe(OH)_3 + e^- + H^+ = Fe^{2+} + 3H_2O$$

$$MnO_2 + 2e^- + 4H^+ = Mn^{2+} + 2H_2O$$

$$SO_4^{2-} + 8e^- + 10H^+ = H_2S + 4H_2O$$

As shown in these reactions, reduction of iron and manganese results in the consumption of H^+ ions, thus resulting in a decrease in H^+ ion activity or an increase in pH. For example, in paddy soils of Philippines (Table 4.4) iron comprises 0.63–4.7% of the total soil mass, and manganese content can be as high as 0.1%. The proton to electron ratio is 3 for $Fe(OH)_3$ reduction, as compared to a ratio of 2 for MnO_2 reduction. Thus, iron reduction will have a greater effect on the pH increase in acid soils. Reduction of other electron acceptors such as nitrate and sulfate can also have a similar effect on the soil pH. In acid soils, an initial increase in pH can also occur due to rapid decomposition of soil organic matter and accumulation of carbon dioxide. However, acid soils eventually increase in pH due to reduction of oxides of Fe and Mn.

In alkaline soils, the following reactions tend to decrease pH:

$$CO_2 + H_2O = H_2CO_3$$

$$H_2CO_3 = HCO_3^- + H^+$$

$$HCO_3^- = H^+ + CO_3^{2-}$$

The pH of alkaline soils is highly sensitive to changes in the partial pressure of carbon dioxide. The carbonates of iron and manganese can also buffer the pH of soil to neutrality:

$$FeCO_3 = Fe^{2+} + CO_3^{2-}$$

$$CO_3^{2-} + 2H_2O = H_2CO_3 + 2OH^-$$

$$Fe^{2+} + 2H_2O = Fe(OH)_2 + 2H^+$$

Although the increase in pH of acid soils is obtained by soil reduction in alkaline soils, a relatively stable pH is obtained after a few weeks of flooding, primarily due to the buffering capacity of carbon dioxide. Ponnamperuma (1972, 1981) observed the following empirical relationship for flooded acid soils:

$$pH = 6.1 - 0.58 \log P_{CO_2}$$

Depletion of carbon dioxide from a reduced soil can result in an increased pH by as much as 2 units. The pH of most of the reduced soils equilibrated with carbon dioxide at 1 atm is 6.1.

In soils, the pH of soil pore water tends to decrease with depth. The pH values are usually above 7 at the soil–floodwater interface and decrease to native soil pH at a lower depth. Some examples of soil pore water pH as a function of depth in selected Florida's wetland soils are shown in Figure 4.15. High pH values at the soil–floodwater interface are due to the photosynthesis activity of algae.

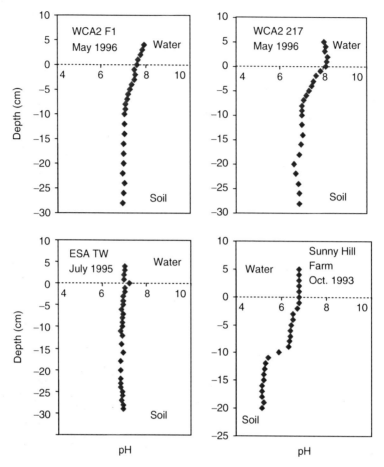

FIGURE 4.15 Soil pore water pH in selected wetlands of Florida. (From Reddy, K. R., unpublished results.)

4.7.2 FLOODWATER pH

Floodwater pH in wetlands is usually regulated by the photosynthetic activity of periphyton communities and submerged macrophytes. Depending on the alkalinity of water, the pH values can increase to about 9 during the day and decrease to near neutral levels during the night. These diel changes are directly linked to the absence or presence of carbon dioxide. The possible sources of carbon dioxide in floodwater are as follows:

- Respiration of periphyton and plankton
- Detrital tissue from macrophytes
- Dissolution of carbonated minerals
- Atmospheric carbon dioxide in floodwater

The possible sinks of carbon dioxide in floodwater are as follows:

- Photosynthetic activity of periphyton and plankton
- Alkalinity
- Loss to atmosphere

One example of diurnal pH fluctuations in the water column of a shallow reservoir is shown in Figure 4.16. The presence of submerged macrophytes and algae showed a significant change in pH

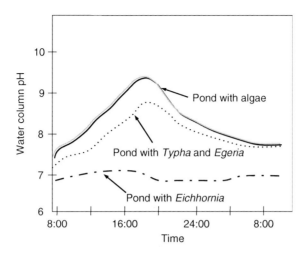

FIGURE 4.16 Diurnal variations in pH of water in selected aquatic systems containing water hyacinth (*Eichhornia crassipes* [Mart] Solms), cattails (*Typha latifolia* L.) and Egeria (*Egeria densa*), and control (no macrophytes but contained algae). (Reddy and Patrick, 1984.)

(as much as about 2 units). However, the presence of floating macrophytes such as *Eichhornia crassipes* or *Hydrocotyle umbellata* resulted in no such pH increase. Diel fluctuation in the pH of floodwater has a major influence on the cycling of nutrients, as discussed in the latter part of this book.

4.7.3 pH Effects

The pH changes in wetland soils significantly influence hydroxide, carbonate, sulfide, and silicate equilibria. These equilibria regulate the precipitation and dissolution of solids, the sorption and desorption of ions, and the concentrations of nutritionally significant ions or substances such as Al^{3+}, Fe^{2+}, H_2S, H_2CO_3, and undissociated organic acids.

When acid sulfate soils are flooded, reduced Fe^{2+} can combine with sulfide and form insoluble compounds, finally resulting in mineral formation (see Chapters 3 and 10 for details). Pyrite formation can also occur in swamp sediments. When oxygen is introduced into these systems by draining, sulfides are oxidized to sulfates biologically. This lowers the pH. The pH of these soils can drop as low as 3.0, where *Thiobacillus* may not function. These soils are also called *cat clays*—predominant in most of the coastal areas of the world.

The number of reactions that can be expressed with pH as a variable is remarkable, including reactions involving:

- Oxides
- Hydroxides
- Basic salts
- Carbonates
- Silicates
- Phosphates
- Sulfides

The following examples illustrate the relationship between iron oxide and pH:

$$Fe_2O_3 + 3H_2O = 2Fe^{3+} + 6OH^-$$

$$6OH^- + 6H^+ = 6H_2O$$

$$Fe_2O_3 + 6H^+ = 2Fe^{3+} + 3H_2O$$

$$K = \frac{[Fe^{3+}]^2 [H_2O]^3}{[H^+]^6 [Fe_2O_3]}$$

When activity of $Fe_2O_3 = 1$,

$$K = \frac{[Fe^{3+}]^2}{[H^+]^6}$$

Taking log on both sides, we obtain

$$\log K = 2 \log [Fe^{3+}] - 6 \log [H^+]$$

Since $pH = -\log [H^+]$, we have

$$\log K = 2 \log [Fe^{3+}] + 6\, pH$$

A similar analysis can be done for anaerobic soil systems containing the following redox couple:

$$Fe(OH)_3 + 3H^+ + e^- = Fe^{2+} + 3H_2O; \quad E° = 1.057$$

$$Fe^{3+} + e^- = Fe^{2+}$$

and the Nernst equation is written as

$$Eh = E° + \frac{RT}{nF} \ln \frac{[H^+]^3}{[Fe^{2+}]}$$

$$Eh = 1.057 - 0.059 \log [Fe^{2+}] - 0.177\, pH$$

4.8 REDOX COUPLES IN WETLANDS

A number of redox couples simultaneously function in wetlands, making it difficult to quantitatively use thermodynamic equilibrium relationships in predicting the activities of a redox pair (Rowell, 1981). However, if any of the redox pair is sufficiently high in concentration, it is possible to predict the behavior of that redox couple (assuming that the effect of other redox couples is minimal). The Eh values are good indicators of redox couples with reversible reactions at the electrode surface. For example, Eh may be useful in determining the concentrations of a redox system dominated by $Fe(OH)_3/Fe^{2+}$ and MnO_2/Mn^{2+}. But electrodes poorly respond to O_2/H_2O, NO_3^-/N_2, SO_4^{2-}/H_2S, and CO_2/CH_4 redox couples. Thus, caution should be exercised while interpreting the Eh values.

Although several redox couples function in wetland soils, the reactions shown in Table 4.5 are the most common reduction reactions involving a specific redox couple. A discussion on each of these redox couples is presented in respective chapters.

The reduction of inorganic redox systems in wetland soils can be described by intensity and capacity.

TABLE 4.5
Selected Redox Reactions in Wetland Soils

Electrochemical Reaction		pe_0	pe_7
I	$O_2 + 4H^+ + 4e^- = 2H_2O$	20.8	13.8
II	$2NO_3^- + 12H^+ + 10e^- = N_2 + 6H_2O$	21.0	12.7
III	$MnO_2 + 4H^+ + 2e^- = Mn^{2+} + 2H_2O$	20.8	6.8
IV	$Fe(OH)_3 + 3H^+ + e^- = Fe^{2+} + 3H_2O$	17.9	−3.1
V	$SO_4^{2-} + 10H^+ + 8e^- = H_2S + 4H_2O$	5.1	−3.6
VI	$CO_2 + 8H^+ + 8e^- = CH_4 + 2H_2O$	2.9	−4.1

4.8.1 Intensity

Intensity is the ease of reduction usually represented by the free energy of reduction or by the equivalent EMF of the reactions—oxidation–reduction potential or redox potential.

The reduction processes in soils are a result of oxidation of organic matter utilizing alternate electron acceptors such as nitrate, manganic manganese, ferric iron, sulfate, and carbon dioxide. After oxygen disappears, nitrate is utilized followed by manganic manganese, ferric iron, sulfate, and finally carbon dioxide. For this reason the soil must be considerably more reduced for sulfate to be reduced than for nitrate to be reduced. If nitrate, sulfate, and a suitable organic substrate are all present in a flooded oxygen-deficient soil, the denitrifiers will reduce all of the nitrate before any of the sulfate is reduced by the sulfate reducers. The intensity of reduction is different for the two reduction processes. The soil is not highly reduced (low reduction intensity or low electron pressure) during the reduction of nitrate whereas it is highly reduced during the reduction of sulfate (high reduction intensity or high electron pressure).

This intensity of reduction can be thought of in terms of the pressure of the electrons that the microorganisms need to dispense with as they carry out the oxidation of the energy source. It does not require much electron pressure for the nitrate to accept electrons (and even less electron pressure for oxygen, if it is present, to accept electrons). This is why a large energy yield can be obtained from the oxidation of the energy source since not so much of the possible energy from the oxidation process is utilized in forcing the electron acceptor to accept the electrons from the organic matter oxidation.

The situation is different for electron acceptors that are difficult to reduce (require more electron pressure to force the electrons to move to the electron acceptor). Such is the case for sulfate, for example. Sulfate reducers can oxidize organic substrates but a great deal of energy derived from the oxidation is used to force the sulfate ion to accept electrons and be converted to sulfide. This leaves less energy that can be used by the microorganism for its needs. A high intensity of reduction is required for sulfate to accept the electrons produced during the oxidation of the energy source.

Thus, there is a difference in the intensity of the reduction process. If you take measurements of the inorganic redox systems (electron acceptors) in a soil and find that oxygen is present, you know that the reduction intensity is very low and that all of the other electron acceptors present in the soil (nitrate, manganic, ferric, sulfate, and carbon dioxide) are in the oxidized form. If, for example, you find that oxygen, nitrate, and manganic manganese have been reduced but ferric iron, sulfate, and carbon dioxide are still present (not reduced) you can conclude that the soil is intermediately reduced or has a moderate reduction intensity. If oxygen and nitrate are used as electron acceptors and the soil contains manganous manganese, ferrous iron, and sulfide, you know that the soil is highly reduced or has a high reduction intensity.

In order to quantify this intensity of reduction in the soil, one approach would be to measure the above components and determine whether they are present in oxidized or reduced form.

This would be an excellent approach for determining the intensity of reduction. It is time-consuming and expensive extracting the soil solution under anaerobic condition (particularly when oxygen from the atmosphere readily contaminates the solution and reoxidizes the reduced components before they can be analyzed).

Another approach to measuring the intensity of reduction is to obtain some measure of the electron pressure in the soil. If soil microorganisms oxidize the energy sources and their enzyme systems transfer the electrons to a suitable acceptor, the soil solution will have a definite electron pressure that reflects the intensity at which these electrons are transferred to the acceptor. If a material is present in the soil that measures this electron pressure, an indication of the intensity of reduction can be obtained. This is done in practice by inserting an inert metal electrode (usually platinum) onto which the electrons readily flow. By exerting electron pressure on the electrode that corresponds to the electron pressure exerted by the soil solution on the platinum, the electron pressure on the platinum electrode can be measured. This is done by determining the voltage required to neutralize the charge on the platinum electrode with a suitable electrical arrangement (see Section 4.6). This approach to determine the intensity of reduction is called the oxidation–reduction or redox potential of the soil.

In most biological systems the intensity of reduction does not go much lower than the level at which carbon dioxide is reduced to methane and the intensity of oxidation does not become much higher than the level that exists in systems where oxygen is stable. The reason for this lower limit is the reduction of water to hydrogen gas under extremely reducing condition. Since there is an unlimited supply of water, the reduction of water to hydrogen and hydroxide ions keeps the system at this intensity until all of the water is reduced, which is an impossible situation. Likewise, if something is added to increase the oxidation intensity in biological systems, water will be oxidized to oxygen and hydrogen ions. Thus, the intensity of water reduction and the intensity of water oxidation provide the outside limits of the intensity of reduction and oxidation in biological systems (Patrick, 1981).

4.8.2 Capacity

Capacity is the amount of the redox systems undergoing reduction. The capacity factor can be best described in terms of oxygen equivalent.

The capacity factor of the various redox systems varies from one soil to another. The amount of oxygen present in the soil at the time of flooding is usually low. The amount of nitrate present in the soil is dependent on the soil organic matter and can vary. Reducible Mn can be present in soils at varying levels (<100–$1,000$ mg kg^{-1}), while reducible Fe can be in the range of 500–$3,000$ mg kg^{-1} of soil.

Although critical Eh values at which the inorganic redox systems become unstable provide valuable information, they do not provide any indication of the total capacity of the system to accept electrons, thereby supporting respiration. Capacity factor is equivalent to the total number of electrons accepted by oxidants in support of microbial respiratory activity. The redox systems present in the lowest amount generally have the smallest capacity for supporting microbial respiration. In mineral wetland soils, reducible manganese and iron support the organic matter decomposition. In coastal areas that receive seawater containing large amounts of sulfate, the reduction of sulfate to sulfide supports the microbial respiratory activities (see Chapter 5 for detailed discussion).

The decomposition of organic matter supported by nitrate, manganese, and iron systems is similar to the decomposition supported by oxygen since carbon dioxide and the reduced oxidant are the major products of this type of decomposition. Obligate anaerobes, however, produce organic acids, aldehydes, and organic sulfur compounds that under certain conditions are toxic to rice plants. The presence of these substances is one of the major reasons why intense reducing conditions, such as those characterized by redox potentials lower than that at which all of the reducible iron is converted to the ferrous form, are considered to be undesirable for the optimum growth of plants.

Electrochemical Properties

Another beneficial function of the large amount of inorganic oxidants in the soil is the nutritional effect of nitrogen, sulfur, and phosphorus in the decomposing organic matter. It is likely that much of the decomposition of organic matter in wetland soils is carried out by the facultative anaerobes that reduce nitrate, manganic manganese compounds, and especially ferric iron compounds.

In coastal areas that receive seawater containing large amounts of sulfate, the reduction of sulfate to sulfide provides a few species of true anaerobes with a respiratory system that can support considerable oxidative activity. In noncultivated coastal marshes where most of the active iron is in the reduced form, it has been estimated that much of the respiratory activity is due to sulfate. In these areas the soil often stays wet most of the year preventing the iron from oxidizing to the ferric form, while sulfate is continuously supplied from the sea. In inland areas where rice culture is more important the limited amount of sulfate in the soil does not support significant anaerobic respiration.

4.9 REDOX GRADIENTS IN SOILS

In wetlands, redox gradients in soil profile are created by

- Hydrologic fluctuations (alterations in water table depth)
- Addition of alternate electron acceptors (such as loading of nitrate and sulfate)
- Addition of electron donors (such as organic matter loading from external sources or internal production)

As stated earlier, in typical wetland soils Eh values vary from −300 to 700 mV, with a total range of about 1,000 mV. Aerobic soils where O_2/H_2O redox couple functions, the Eh range is between 300 and 700 mV. This narrow range of Eh values and due to lack of adequate poise and poor reproducibility makes Eh a less reliable parameter in aerobic soils. Selected redox couples functioning at various Eh levels are shown in Figure 4.17. At negligible oxygen levels (few hours after flooding a soil), Eh values (at pH = 7) are in the range of 320–340 mV, indicating that below these Eh values oxygen is absent in soils.

Nitrate reduction occurs at Eh values in the range of +200 to +300 mV. Nitrate can poise soil Eh for several days as observed in conditions when drained organic soils (usually containing high levels of nitrate) are flooded (Figure 4.18). In an organic soil, maintenance of nitrate levels between 25 and 50 mg L^{-1} resulted in steady Eh levels of soil slurry (Figure 4.19).

Depending on the hydrologic regime and input of electron donors (reductants) and acceptors (oxidants), steep Eh gradients are often observed:

- At the soil–floodwater interface
- At the root surface or in the rhizosphere
- Around saturated soil aggregate in a poorly drained soil

Some examples of Eh gradients at these interfaces are shown in Figures 4.20–4.23.

The authors and their associates have devoted considerable attention to studying the intensity aspects of these inorganic redox systems. The approach taken has been to determine the redox potential at which the oxidized component becomes unstable and accepts electrons from respiring microorganisms. The redox potential was chosen as an index of intensity for these studies because it covers the entire range over which these various inorganic redox systems function. As shown in Figure 4.17 and discussed earlier, the redox potential range encountered in waterlogged soils extends from approximately −300 mV on the reducing end to approximately +700 mV on the oxidizing end or about the entire range encountered in all biological systems. In well-drained soils that

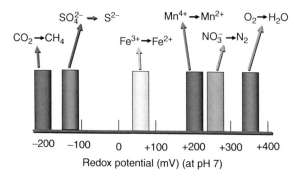

FIGURE 4.17 Critical redox potential at which oxidized inorganic redox systems begin to undergo reduction in wetland soils.

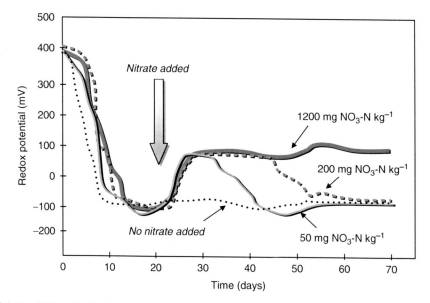

FIGURE 4.18 Effect of added nitrate on redox potential of a wetland soil.

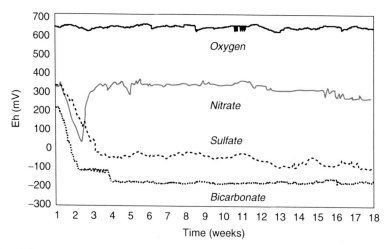

FIGURE 4.19 Influence of added electron acceptors on redox potential of a wetland soil. (McLatchey and Reddy, 1998.)

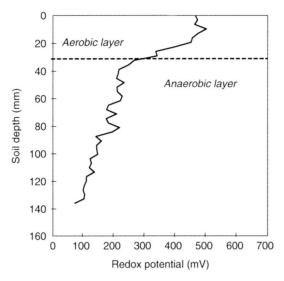

FIGURE 4.20 Redox potential as a function of depth in stream sediments. (Reddy, K. R., Unpublished Results, University of Florida.)

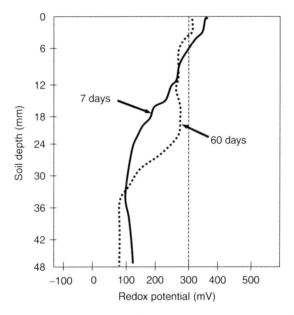

FIGURE 4.21 Redox potential as a function of depth in flooded rice soil. (Reddy, K. R., Unpublished Results, University of Florida.)

are permeated with oxygen from the atmosphere, the normal range of redox potential encountered is much less and occurs in a narrow range at the oxidizing end of the redox scale.

One experimental approach to determine the redox intensity at which each of these inorganic systems function has been to set up a system in which the redox potential is closely controlled and the reduction of the oxidized component of various inorganic redox systems studied. This technique involves the use of stirred soil suspensions that have their redox potential closely controlled at any point in the range indicated in Figure 4.17. A suspension with soil-to-water ratio of 1:4 for mineral soils or 1:10 for organic soils is incubated in a sealed chamber and the desired redox potential is obtained by automatically adding very small amounts of oxygen or appropriate amounts of alternate

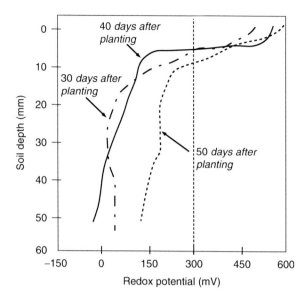

FIGURE 4.22 Redox potential of soil planted with rice, showing the influence of root oxygenation. (From Reddy and Patrick, 1984.)

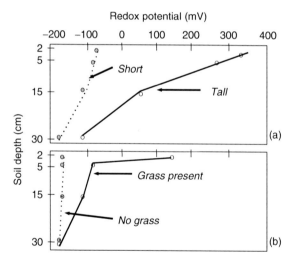

FIGURE 4.23 (a) Redox potential of soils with tall and short strands of *Spartina alterniflora*. (b) Profiles of Eh in an area devoid of vegetation and in nearby clumps of plants. (Adapted from Howes et al., 1981.)

electron acceptors. A controlled Eh–pH system was first developed by Patrick and coworkers (Patrick, 1960, 1966; Patrick et al., 1973), which has been used in several studies to determine the influence of pH and Eh on various biogeochemical processes (Figure 4.24). This method involves control of Eh by the addition of either air or inert gases such as nitrogen, argon, or helium. This method is suitable for studying chemical reactions, to some extent, microbial reactions functioning at the upper end of redox scale. In this system the effects of oxygen additions to control Eh are assumed to be minimal. Using this system, the stability of inorganic and organic compounds was studied as a function of Eh and pH.

A modified method uses addition of appropriate electron acceptors to buffer soil Eh (Mclatchey and Reddy, 1998). For example, Eh values of a soil suspension can be buffered at about 200 mV

Electrochemical Properties

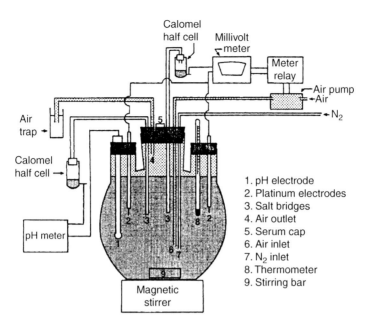

FIGURE 4.24 Schematic of a batch reactor used to control Eh and pH of soil suspensions.

by frequent additions of nitrate as an electron acceptor (Figure 4.19). Similarly, Eh values can be buffered at −100 mV using sulfate as an electron acceptor. This method provides ideal microbial communities functioning at distinct Eh values.

The pH as well as the redox potential can be closely controlled in this system, which allows the two major parameters involved in Nernst-type redox reactions to be utilized in the study of redox systems. Results of many studies in which the effects of both redox potential and pH on various soil systems were studied are discussed in various chapters in this book.

A summary of the results of a number of studies of the critical redox potential at which the various inorganic redox systems become unstable is shown in Figure 4.17. Going from the most easily reduced to the most difficultly reduced systems shows that oxygen is reduced first, followed by nitrate and oxidized manganese compounds, and then followed by ferric iron compounds. After the reduction of ferric iron the next system to become unstable is sulfate followed by the reduction of carbon dioxide to methane. The reduction of many of these oxidized redox systems is sequential; that is, one system is completely reduced before the next system begins to undergo reduction. An example of this sequential reduction or lack of overlap in the reductions of many of the systems is that no ferric iron is reduced to the ferrous form as long as any oxygen or nitrate is present in the soil. Likewise, sulfate and carbon dioxide will not be reduced if oxygen or nitrate is present. At the more reducing end of the scale, almost all of the sulfate must be reduced to sulfide before any methane appears.

The above results show that the soil inorganic redox systems differ considerably in their reduction intensities with the systems at the oxidized end of the redox scale accepting electrons from respiring microorganisms much more readily than the oxidants shown on the reducing end of the redox scale. On the basis of the energy made available for microbial respiration more energy is released per electron transferred where an easily reduced redox system is involved (oxygen, for example) as compared to reduction of the oxidized iron system. There are many uses of such information on reduction intensity; one example is that the knowledge of the redox of a soil will give an indication of the oxidation–reduction status of these various components. For example, Eh of 0 mV indicates that oxygen and nitrate are not likely to be present and that the bioreducible iron and manganese compounds are in a reduced state. At this same potential, however, sulfate is stable in the soil with no sulfide being formed and there also will be no methane produced at this potential.

4.10 MICROBIAL FUEL CELLS

As discussed in this chapter, it is well known that wetlands are not limited by electron donor, and anaerobic conditions in these systems provide high electron pressure. Some of the end products (ethanol, methane, and hydrogen) of organic matter decomposition are also known to produce electricity. However, capturing these electrons into fuel cells is a relatively new and novel idea. Lovley (2006) presented a review on how these electrons can be captured to produce electricity. Recent studies presented in this review show that during microbial respiration some of the microorganisms conserve energy to support growth by oxidizing organic compounds to carbon dioxide with direct quantitative electron transfer to electrodes. The idea of electricity production in microbial cultures was developed by Potter (1911, cited by Lovley, 2006) some 90 years ago. The microbial fuel cell essentially an electrochemical cell consists of an anode that accepts electrons from the microbial culture, and a cathode that transfers electrons to an electron acceptor, typically oxygen for most perceived practical applications. The anode is placed under anaerobic conditions, whereas the cathode is typically placed under aerobic conditions such as water column or air. Electrons flow from the anode to the cathode through an external electrical connection that assists in abiotic transfer of electrons to oxygen. This aids in harvesting some of the energy before microorganisms convert it to ATP by oxidative phosphorylation (Lovley, 2006). The review by Lovley (2006) provides a discussion on practical application of this novel concept in wetland soils and aquatic sediments to power electronic devices, such as monitoring equipment. Lovley calls this microbial fuel cell as a benthic unattended generator or BUG . The BUGs produce electricity from the organic matter stored in aquatic sediments.

4.11 SPECIFIC CONDUCTANCE

Specific conductance of most soils increases after O_2 depletion, attains a maximum, and declines to a fairly stable value, which varies with soil (Figure 4.25). In slightly acid and acid soils, reduction of insoluble Fe^{3+} and Mn^{4+} compounds to soluble forms accounts for an increase in cations in the soil solution. In neutral and alkaline soils Ca^{2+} and Mg^{2+} in the soil pore water make significant

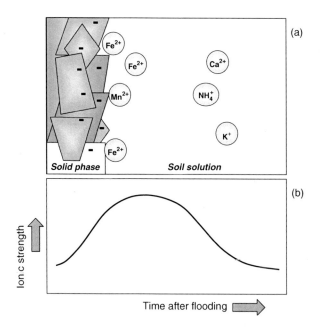

FIGURE 4.25 (a) Schematic of soil cation exchange showing the displacement of Ca^{2+} and NH_4^+ by Fe^{2+} and Mn^{2+} ions. (b) Change in the ionic strength of soil solution when influenced by flooding.

contributions to the ionic strength. Ferrous and manganous ions (Fe^{2+} and Mn^{2+}, respectively) produced through reduction displace other cations from the exchange complex to the soil pore water.

4.12 SUMMARY

Wetland soils undergo dynamic changes in physicochemical properties as a result of hydrologic fluctuations, loading of nutrients, and loading of contaminants. The physicochemical properties affected by flooding are Eh, pH, and ionic strength of soil solutions. Soil Eh is the most important parameter for characterizing the degree of oxidation or reduction of soil. Quantitative relationships between the oxidized and reduced components in soils have been derived by using the Nernst equation.

- Reductants (electron donors) in wetland soils are
  - Organic matter and various organic compounds
  - Reduced inorganic compounds such as NH_4^+, Fe^{2+}, Mn^{2+}, S^{2-}, CH_4, and H_2.
- Oxidants (electron acceptors) in wetland soils are
  - Inorganic compounds such as O_2, NO_3^-, MnO_2, FeOOH, SO_4^{2-}, and HCO_3^-
- Wetland soils are usually limited by electron acceptors and have abundant supply of electron donors. Upland soils are usually limited by electron donors and have abundant supply of electron acceptors (primarily oxygen):

$$\text{Oxidant} + e^- = \text{Reductant}$$

- Oxidant is the electron acceptor, and reductant is the electron donor.

$$\text{Eh} = E^\circ - \frac{RT}{nF} \ln \frac{[\text{Reductant}]}{[\text{Oxidant}]} \tag{4.22}$$

or

$$\text{Eh} = E^\circ + \frac{RT}{nF} \ln \frac{[\text{Oxidant}]}{[\text{Reductant}]}$$

$$\text{Eh} = E^\circ - \frac{0.059}{n} \log \frac{[\text{Reductant}]}{[\text{Oxidant}]}$$

$$\text{Eh} = E^\circ - \frac{0.059}{n} \log \frac{[\text{Reductant}]}{[\text{Oxidant}]}$$

- Under standard-state conditions, the relationship between Eh and pH is defined as follows (assuming $E^\circ = 0$ V and the ratio of reductant to oxidant as unity):

$$\text{Eh} = -0.059\,\text{pH} \tag{4.32}$$

- For many oxidation–reduction reactions:

$$m(\text{oxidant}) + mH^+ + ne^- = m(\text{reductant})$$

$$\text{Eh} = E^\circ - \frac{0.059}{n} \log \frac{[\text{Reductant}]}{[\text{Oxidant}]} - 0.059 \frac{m}{n} \text{pH} \tag{4.68}$$

$$\text{Eh} = E^\circ - \frac{RT}{nF} \ln \frac{[\text{Reductant}]^m}{[\text{Oxidant}]^m [H]^m} \tag{4.65}$$

- Redox potential of soils is affected by (i) the activities of electron donors (biodegradable organic matter and reduced inorganic substances), (ii) the activities of electron acceptors (oxidants such as O_2, NO_3^-, MnO_2, $FeOOH$, SO_4^{2-}, and HCO_3^-), and (iii) the temperature.
- Distinct Eh gradients are present (i) at the soil–floodwater interface, (ii) at the root zone of wetland plants, and (iii) around the soil aggregates in drained portions of wetlands during low water table depths.
- Redox potential can be used as an operational parameter for characterizing hydric soils for the presence or absence of reduced substances, and the degree of soil anaerobiosis. The Eh values obtained should be used in the context of related soil physicochemical properties such as soil pH and hydrologic conditions.
- Spatial variability in soils should be recognized in making Eh measurements. Equilibration period of at least 24 h should be used before making any measurements.

STUDY QUESTIONS

1. Briefly discuss the mechanisms involved in the development of the oxidized layer at the soil–water interface.
2. Briefly discuss the exchange of dissolved N, O_2, Fe, Mn, S, and C species between wetland soil and the overlying water column.
3. List the initial conditions assigned to standard hydrogen electrode (SHE).
4. List the electron donors and electron acceptors present in wetland soils.
5. Write the balanced reaction and the Nernst equation (at 25°C) for the following redox couples. Show all necessary steps in balancing the equations.

Redox Couple	$E°$ (V)
O_2/H_2O	+1.23
NO_3/N_2	+1.24
MnO_2/Mn^{2+}	+1.23
$Fe(OH)_3/Fe^{2+}$	+1.06
SO_4/S^{2-}	+0.34
CO_2/CH_4	+0.17

6. Using the Nernst equation, calculate the redox potential for the following systems at the concentrations and pH given:

Redox Couple	pH	Concentration
MnO_2–Mn^{2+}	5.0	Mn^{2+} = 88 mg L^{-1} in solution
$Fe(OH)_3$–Fe^{2+}	4.5	Fe^{2+} = 156 mg L^{-1} in solution
SO_4^{2-}–S^{2-}	6.0	SO_4^{2-} = 113 mg L^{-1} in solution
H^+–H_2	4.0	—

7. Discuss the significance of "poise" in redox systems of wetland soils.
8. Discuss briefly the limitations of redox potential measurements in wetland soils.
9. Why does soil pH approach neutrality when acid or alkaline soils are flooded? Discuss the mechanisms involved.
10. The $E°$ of the $2H^+ + 2e^- = H_2$ half reaction is arbitrarily set at zero. Show the relationship between Eh and pH (at 25°C) using the appropriate Nernst equation. Calculate the Eh for a solution at pH 8.
11. Derive the Nernst equation using thermodynamic relationship and by following the equation shown below:

$$\Delta G = -nFE$$

12. Calculate the change in free energy of formation for the following reaction:

$$H_2S = S° + 2H^+ + 2e^-, \quad E° = -0.141 \text{ V}$$

13. Calculate the Eh and pe for O_2/H_2O and $Fe(OH)_3/Fe^{2+}$ systems under the following conditions:

    O_2/H_2O system:

    $$O_2 = P_{O_2} = 1.0, 0.21, \text{ and } 0.021 \text{ atm } (E° = 1.23 \text{ V})$$

    $$\text{pH} = 4, 6, \text{ and } 8$$

    $Fe(OH)_3/Fe^{2+}$ system:

    $$Fe^{2+} = 1.0, 0.1, \text{ and } 0.01 \text{ mol L}^{-1} (E° = 1.06 \text{ V})$$

    $$\text{pH} = 4, 6, \text{ and } 8$$

    Show the appropriate steps in calculations. Calculate pe + pH for each concentration. Plot pe vs. pH. Write appropriate balanced equations, and briefly explain the significance of results obtained in your calculations.

14. Strengite ($FePO_4 \cdot 2H_2O$) is converted to Fe^{2+} and $H_2PO_4^-$ under reducing slightly acid conditions. At pH 6 and redox potential of –100 mV calculate the concentration of $H_2PO_4^-$ ppm P. What is the concentration of Fe^{2+}?

Reaction	$E°$ (V)	Reaction	$E°$ (V)
O_2 to H_2O	1.23	SO_4^{2-} to S^{2-}	0.16
NO_3 to N_2	1.246	CO_2 to CH_4	0.18
MnO_2 to Mn^{2+}	1.23	H^+ to H_2	0.00
$Fe(OH)_3$ to Fe^{2+}	1.057		
$FePO_4 \cdot 2H_2O$ to Fe^{2+}	0.316		

15. Make hypothetical plots for Eh vs. time under the following conditions:
    (a) Mineral soil and flooded
    (b) Organic soil and flooded
    (c) Mineral soil high in $Fe(OH)_3$ and flooded
    (d) Organic soil high in NO_3-N and flooded
    Briefly discuss the implications of each condition on soil reduction.

16. A quinone/hydroquinone redox couple is used to test the functioning of platinum electrode. The reduction reaction of this couple is quinone + $2H^+ + 2e^-$ = hydroquinone; $E° = 0.699$ V. Assume the ratio of the activities to be [quinone]/hydroquinone] = 1. The electrodes are tested in buffer solutions with pH of 7 or 4 containing the redox couple at 25°C.

FURTHER READINGS

Baas Becking, L. G. M., I. R. Kaplan, and D. Moore. 1960. Limits of the natural environment in terms of pH and oxidation–reduction potentials. *J. Geol.* 68:243–284.

Bartlett, R. J. and B. R. James. 1991. Redox chemistry of soils. *Adv. Agron.* 50:151–208.

Faulkner, S. P., W. H. Patrick, Jr., and R. P. Gambrell. 1989. Field techniques for measuring wetland soil parameters. *Soil Sci. Soc. Am. J.* 53:883–890.

Garrels, R. M. and C. L. Christ. 1965. *Solutions, Minerals, and Equilibria*. Harper & Row, New York. 450 pp.
Lindsay, W. L. 1979. *Chemical Equilibria in Soils*. Wiley, New York. 449 pp.
Lovley, D. R. 2006. Bug juice: harvesting electricity with microorganisms. *Nature Rev.* 4:497–508.
Ponnamperuma, F. N. 1972. The chemistry of submerged soils. *Adv. Agron.* 24:29–96.
Rowell, D. L. 1981. Oxidation and reduction. In D. J. Greenland and M. H. B. Hayes (eds.) *The Chemistry of Soil Processes*. Wiley, New York. pp. 401–461.
Snoeyink, V. L. and D. Jenkins. 1981. Oxidation–reduction. In *Water Chemistry*. Wiley, New York. pp. 316–430.
Stefansson, A., S. Arnorsson, and A. E. Sveinbjornsdottir. 2005. Redox reactions and potentials in natural waters at disequilibrium. *Chem. Geol.* 221:289–311.
Stumm, W. and J. Morgan. 1981. *Aquatic Chemistry*. Wiley, New York. 780 pp.

5 Carbon

5.1 INTRODUCTION

Carbon is the predominant constituent of all life forms. The structure, energetics, and functioning of life forms depend on the linkage of carbon with other major elements such as oxygen, hydrogen, nitrogen, phosphorus, and sulfur. This linkage is provided through long covalent chains and rings, and forms the foundation for many organic molecules associated with carbon in the biosphere. The biogeochemical cycle of carbon involves complex interactions between and within organic and inorganic carbon reservoirs. These interactions have been widely studied by scientists from various disciplines. The focus of research in recent years has been on the role of carbon cycle in regulating carbon dioxide and methane in the atmosphere and resulting "greenhouse effect."

Among the carbon reservoirs of the biosphere, a large proportion is stored in soil organic matter and marine sediments (Bolin, 1977). The accumulation of carbon in soils and sediments is a function of the organic carbon balance between net primary production (carbon fixation) and heterotrophic metabolism (decomposition). The fixation of atmospheric carbon through photosynthesis is the major source of carbon to terrestrial, wetland, and aquatic ecosystem.

$$6CO_2 + 6H_2O = C_6H_{12}O_6 + 6O_2 \text{ (Photosynthesis)}$$

$$C_6H_{12}O_6 + 6O_2 = 6CO_2 + 6H_2O \text{ (Respiration [Aerobic])}$$

Carbon atoms have various oxidation states ranging from +4 to −4 (Table 5.1). The highest oxidized forms of carbon are carbon dioxide, bicarbonate, and carbonate, which is the end product of decomposition processes. Carbon fixation during photosynthesis produces reduced carbon (CH_2O). Methane, another gaseous form of carbon, is the most reduced form and an end product of anaerobic decomposition process.

The two most common stable isotopes of carbon are ^{12}C (98.9% of the carbon) and ^{13}C (1.1% of the carbon). There are several radioactive isotopes including ^{10}C, ^{11}C, ^{14}C, ^{15}C, and ^{16}C, with half-lives ranging from 0.74 s for ^{16}C to 5726 years for ^{14}C. The ^{14}C isotope is routinely used in studies related to carbon cycling in ecosystems. Relatively abundant stable isotopes ^{12}C and ^{13}C are widely used in identifying the sources of carbon entering and leaving the reservoirs.

On a global scale, the major reservoirs of carbon are

- Atmosphere
- Hydrosphere
- Biosphere
- Lithosphere

Carbon is present in the atmosphere as CO_2, CH_4, CO, and volatile organic compounds (Table 5.2). Atmospheric CO_2 concentration increased from 270 ppm (v/v) in 1850 to 360 ppm in 2000 (Figure 5.1). During the same period, CH_4 concentration increased from 0.7 to 1.7 ppm. Much of the increase in CO_2 is attributed to burning of fossil fuels since the industrial revolution. However, a number of sources including natural wetlands, rice paddies, livestock, termites, and biomass burning have been attributed to an increase in CH_4 concentration of the atmosphere (Table 5.3). The total amount of CO_2 in the atmosphere is estimated at about 735 Pg C (1 Pg = 10^{15} g; Pg = petagram), as compared to 3 Pg C for CH_4. The amount of CO is estimated to be about 0.2 Pg. Recent studies have

TABLE 5.1
Selected Examples of Oxidation Number of Carbon

Compound	Oxidation Number for Carbon
CO_2	+4
HCO_3^-	+4
CO_3^{2-}	+4
CO	+2
CH_2O	0
CH_4	−4

TABLE 5.2
Reservoirs of Carbon

Reservoir	Amount of Carbon (Pg)
Atmosphere	
CO_2	732
CH_4	3
CO	0.2
Hydrosphere	
DIC	37,900
DOC	1,000
POC	30
Biota	3
Sediments	>10,000,000
Lithosphere	
Fossil fuels	5,000

Note: Pg = petagram = 10^{15} g.
Source: Bolin et al. (1979).

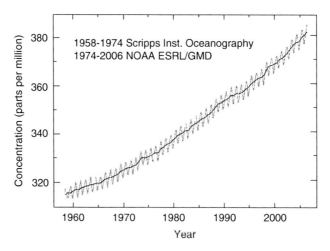

FIGURE 5.1 Atmospheric carbon dioxide at the Mauna Loa Laboratory. Data prior to May 1974 are from the Scripps Institution of Oceanography(SIO, blue), data since May 1974 are from the National Oceanic and Atmospheric Administration (NOAA, red). A long-term trend curve is fitted to the monthly mean values. (U.S. Department of Commerce, National Oceanic and Atmospheric Administration Earth System Research Laboratory. http://www.esrl.noaa.gov/media/2007/50YearCO2Record.html) (v/v) in 1850 to 360 ppm in 2000.

TABLE 5.3
Natural and Anthropogenic Sources of Methane in the Atmosphere

	Amount of Carbon (Tg CH_4–C $year^{-1}$)	
Source	Average	Range
Natural sources (Total)	150	100–300
Wetlands	110	70–170
Termites	20	10–50
Oceans	10	5–20
Freshwater	5	1–25
Gas hydrates	5	0–5
Anthropogenic sources (Total)	360	400–610
Coal mining natural gas, petroleum industries	100	70–120
Rice farming	60	20–150
Domesticated livestock	80	65–100
Livestock manure	25	10–20
Wastewater treatment	25	20–25
Land fills	30	20–70
Biomass burning	40	20–80

Note: Tg = teragram = 10^{12} g.
Source: Chynoweth (1996).

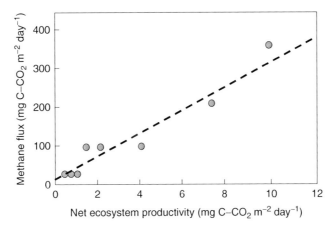

FIGURE 5.2 Relationship between net ecosystem productivity and net methane flux, (Redrawn from Whiting and Chanton, 1993).

shown a significant relationship between net primary productivity of wetland plants and CH_4 emission (Figure 5.2). This relationship indicates that approximately 3% of the net ecosystem production (carbon fixation) in wetlands is emitted back into the atmosphere, as methane (Whiting and Chanton, 1993).

In the hydrosphere, carbon is present as dissolved inorganic carbon (DIC), dissolved organic carbon (DOC), particulate organic carbon (POC), and the marine biota (Table 5.2). The amount of carbon present in each of these components is estimated to be 3 Pg in biota, 1,000 Pg as DOC, 30 Pg as POC, and 37,900 Pg as DIC.

The amount of carbon present in the biosphere is estimated to be 700 Pg in the living phytomass and about 150 Pg in the dead phytomass (Bolin et al., 1979) (Table 5.2). Net primary production of the land biota is estimated to be 63 Pg C $year^{-1}$. It is also estimated that 80% of the net primary productivity may contribute to annual litter fall. The total amount of carbon in the soil is estimated to be 1,672 Pg, with approximately 50% of this storage accounted for in peatlands (Bolin et al., 1979). Storage in swamps and marshes is estimated to be 27 Pg. Total estimated carbon in fossil fuels is estimated to be about 5,000 Pg, of which approximately 85% is in the form of coal (Table 5.2).

5.2 MAJOR COMPONENTS OF CARBON CYCLE IN WETLANDS

Organic carbon and inorganic carbon are the two major forms present in soil–water–plant components of wetland ecosystem. Carbon cycle in this system can be depicted as a storage of carbon in major reservoirs, which serve as either a source or a sink, and flux between reservoirs (Figure 5.3).

The reservoirs of carbon in a wetland can be grouped as follows:

- Plant biomass carbon (standing stock, live)
- POC (detritus, soil, and water)
- DOC (detritus, soil, and water)
- Microbial biomass carbon (MBC) (detritus, soil, and water)
- Gaseous end products (atmosphere, detritus, soil, and water)

5.2.1 Plant Biomass Carbon (Net Primary Productivity)

The vegetation reservoir (including macrophytic and algal species) represents transformers of inorganic carbon (CO_2) to organic carbon (primary production) through photosynthesis. Net primary productivity of wetland ecosystems is higher than that of many terrestrial ecosystems and is approximately of the same order of magnitude as that of tropical rain forests (Table 5.4). Although wetlands cover only a small portion of the earth's land surface (about 6%), their relative importance to the earth's biogeochemical cycle far exceeds their surface area, for they contain 68% of terrestrial soil carbon reserves and represent a major land-based carbon sink (Schlesinger, 1997). The primary productivity in wetlands is highly variable and is influenced by the type of vegetation, the geographic location, the nutrient status, and the methods used in determination of productivity (Table 5.5). The values reported in Table 5.5 are for aboveground biomass productivity. The ratio of belowground to aboveground biomass can range from 0.2 to 3.9, depending on the species and the geographic location (Gopal and Masing, 1990).

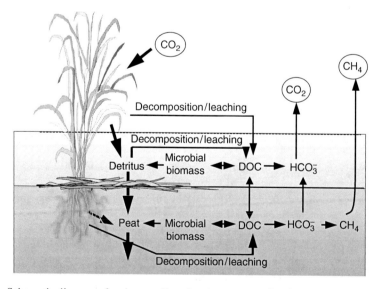

FIGURE 5.3 Schematic diagram of carbon cycling showing storage of carbon.

TABLE 5.4
Net Primary Productivity of Selected Ecosystems

Ecosystem	Net Primary Productivity ($gC\,m^{-2}\,year^{-1}$)
Desert	80
Boreal forest	430
Tropical forest	620–800
Temperate forest	65
Wetland	1,300
Cultivated land	760
Tundra	130

Source: Houghton and Skole (1990).

TABLE 5.5
Net Aboveground Primary Productivity of Selected Wetland Ecosystems

Wetland Type	Net Primary Productivity ($gC\,m^{-2}\,year^{-1}$)
Salt marsh	
Spartina	65–1,850
Juncus	2,938–4,043
Bogs and fens	
Sphagnum	30–1,660
Other mosses	10–507
Species excluding mosses	177–1,027
Marshes (freshwater)	
Submerged macrophytes	1–1,000
Floating macrophytes	10–2,067
Emergent macrophytes	155–6,180
Mangroves	1,000–4,599
Riparian	334–804
Southern deepwater swamps	103–770

Source: Gopal and Masing (1990).

Closed systems, such as the wetlands that receive most of their water from precipitation (e.g., bogs, pocosins, and some seasonal or ephemeral wetlands), are oligotrophic and typically have low primary productivity. Increased inputs from surface and ground water can increase primary productivity in fens and marshes. (Table 5.5) Wetlands that receive pulses of nutrients, such as river floodplains, littoral zones in lakes, and tidal marshes, are typically very productive (Sharitz and Pennings, 2006).

5.2.2 Particulate Organic Matter (Detrital and Soil)

As plants senesce, a portion of the aboveground biomass, and in some cases all of the aboveground biomass is returned to detrital pool, where it undergoes decomposition. In emergent marshes (such as freshwater and saltwater), most of the organic matter produced is converted to detrital pool. In forested wetlands, litter fall is the primary source of detrital pool. For example, litter fall accounts for approximately 50% of the net aboveground productivity in deepwater swamps in the southeastern United States (Mitsch and Gosselink, 2007). In addition, belowground biomass (but rarely quantified) is another major source of detrital inputs to the soil.

The heterotrophic microflora represents transformers of organic carbon back to inorganic carbon through catabolic activities. Organic carbon is stored in both living (vegetation, algae, and microbial biomass) and nonliving (dead plant detritus, native soil organic matter containing both humic and nonhumic substances) reservoirs. Nonliving reservoir of organic carbon is proportionally large in wetlands in relation to other ecosystems. This storage provides a substantial energy reserve to the ecosystem, which is slowly released through microbial catabolic activities.

The following are operation definitions of various types of particulate organic matter (POM) and dissolved organic matter (DOM):

Coarse POM (CPOM)	>1 mm
Fine POM (FPOM)	250 μm–1 mm
Ultrafine POM (UPOM)	0.45–250 μm
DOM	<0.45 μm
Colloidal DOM	0.2–0.45 μm

At the ecosystem scale, POM is generated through litter fall (in forested wetlands) and detrital production. This is the major reservoir of organic carbon above the soil surface and constitutes approximately 95% of the total carbon. The POM is broken down into simple readily utilizable organic forms, followed by simultaneous mineralization to inorganic carbon.

5.2.3 Microbial Biomass Carbon

The decomposition of organic matter is the primary ecological role of heterotrophic microflora in soils. Microbial biomass is a small fraction of plant detritus and soil organic matter; yet most of the net ecosystem production of wetland passes through the microbial loop at least once and typically several times. Microbial decomposers derive their energy and carbon for growth from detrital and soil organic matter and facilitate recycling of energy and carbon within and outside wetland ecosystem. Soil microbes exert a significant influence on ecosystem energy flow in the form of feedback because mineralization of organically bound nutrients is a regulator of nutrient availability for both primary production and decomposition.

During heterotrophic breakdown of POC and DOC, a portion of labile organic compounds hydrolyzed by enzymes is assimilated into microbial biomass. Under aerobic conditions, approximately 50% of the monomers formed can be assimilated into cell biomass, whereas the remaining is oxidized to carbon dioxide. The turnover of microbial biomass results in recycling of organic carbon into POC and DOC in the detrital layer and soil. As the decomposition proceeds, the refractory portion of POC and DOC increases, as more labile components are either mineralized to carbon dioxide or assimilated into cell biomass. Microbial biomass represents a small fraction of soils, accounting for about 3–5% of the total organic carbon (TOC) (Figures 5.4 and 5.5). Turnover of active biomass is rapid and may be in the order of days, as compared to several decades for soil organic matter.

Microbes and macrophytes often compete for available nutrients. In a nutrient-limited wetland, nutrients may be held tightly within the microbial biomass, reflecting efficient recycling of remineralized organic compounds. Environmental perturbations such as alterations in redox conditions may result in microbial mortality, resulting in a significant remineralization of nutrients.

Soil microbial biomass constitutes a significant carbon sink, in that it represents a significant portion of the "active" organic carbon pool. Turnover of active biomass is generally on the order of days, as compared to several years for the soil organic matter. Nutrients may be held tightly

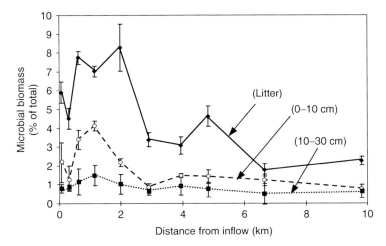

FIGURE 5.4 Microbial biomass composition in soils along the distance from inflow in WCA-2A area in Everglades (DeBusk and Reddy, 1998).

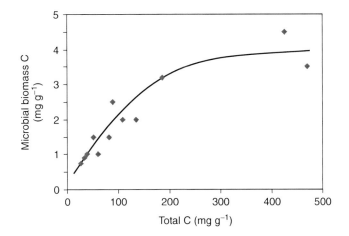

FIGURE 5.5 Microbial biomass carbon as a function of total carbon (DeBusk and Reddy, 1998).

within the microbial biomass component of low-nutrient wetlands reflecting efficient recycling of remineralized organic compounds.

5.2.4 Dissolved Organic Matter

DOM is operationally defined by many investigators and represents the material that passes through 0.45 μm glass fiber filter. In addition, DOM may also include a significant colloidal fraction. In this chapter, we focus on carbon component of DOM or DOC. In soils and sediments, DOC is extracted from pore waters, or in some cases soils are extracted with water and filtered through 0.45 μm filter. Some of the DOM in the detrital layer is also produced by leaching of materials from the detrital pool. The DOM represents approximately <1% of total soil organic matter. However, in surface waters, the DOC accounts for 90% of the TOC (Figure 5.6). The DOM represents a broad spectrum of organic compounds of varying environmental recalcitrance. Only a small fraction of the DOM extracted from stable organic matter may be available. The DOC pool is a relatively stable component of DOM in terms of both the size and the quality of the pool (Wetzel, 1993).

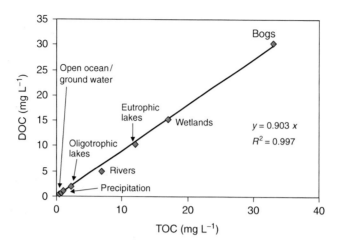

FIGURE 5.6 Relationship between total organic carbon and dissolved organic carbon of the water column. (Data from Wetzel, 2001.)

Decomposition of organic substrates involves both labile and recalcitrant organic matter, the latter comprising the bulk of the DOC. Turnover of labile pool is very rapid and often may approach a rate of 5–10 times per day; thus, actual concentration (storage) of labile DOC is very low (DeBusk and Reddy, 1998).

5.2.5 Gaseous Forms of Carbon

Carbon dioxide and methane are two gaseous end products of decomposition of organic matter under anaerobic conditions, whereas only carbon dioxide is produced under aerobic conditions. Carbon dioxide readily dissolves in water and is partitioned into H_2CO_3, HCO_3^-, and CO_3^- as follows:

$$CO_2 + H_2O = H_2CO_3$$
$$H_2CO_3 = H^+ + HCO_3^-$$
$$HCO_3^- = H^+ + CO_3^{2-}$$

The balance between dissolved and gaseous inorganic carbon depends on the pH of the water column and soil pore water. In wetlands, pH of soil may be buffered between 6 and 7, depending on the rate of reduction of various electron acceptors (see Chapter 4 for a detailed discussion). Water column pH may fluctuate between day and night depending on the photosynthetic activities of algae and submerged macrophytes.

At the ecosystem scale, the flux of carbon among various reservoirs is a continuous process and follows a steady "decay continuum." In a stepwise manner, carbon dioxide fixed during photosynthesis is returned to the atmosphere by the decomposition process, and the undecomposed organic matter is retained in the soil (Figure 5.7). The steps are as follows:

- Atmospheric carbon dioxide is fixed during photosynthesis, and organic carbon is produced.
- As plants age, older tissue is converted into detrital tissue and is subjected to fragmentation, leaching, and decomposition while attached to living plants. Some of this detrital matter may be above the floodwater surface.
- Detrital plant tissue is detached from original plant and deposited on the soil surface. Various macro- and microscale processes (as discussed in the latter part of this chapter) are involved in the breakdown of this material.

Carbon

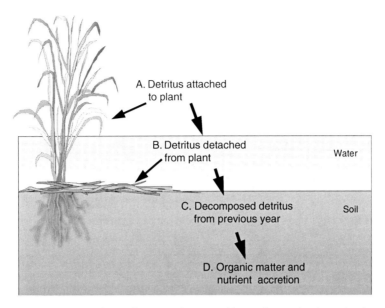

FIGURE 5.7 Decay continuum of organic matter decomposition and accretion in wetlands.

- Undecomposed older material is buried under new material.
- Finally, the undecomposed detrital material is integrated into soil organic matter by humification. At this stage, organic matter provides long-term storage for nutrients and contaminants.
- Gaseous end products (carbon dioxide and methane) produced during decomposition process are released into the atmosphere.

Both methane and carbon dioxide are greenhouse gases and contribute to global warming.

5.3 ORGANIC MATTER ACCUMULATION

Wetlands are characterized by aerobic and anaerobic interfaces in soil and water column and accumulation of organic matter. In wetland ecosystems, the primary productivity often exceeds the rate of decomposition processes, resulting in net accumulation of organic matter. The net accumulation of organic matter is regulated by the activity of various decomposers, including benthic invertebrates, fungi, and bacteria. In a simplistic way, decomposition may be viewed as a three-step process:

- Breakdown of POM by fragmentation by grazers
- Hydrolytic activity of extracellular enzymes involved in the conversion of POM (polymers such as polysaccharides).
- Microbial catabolic activities (conversion of monomers into carbon dioxide and methane).

As the detrital organic matter is decomposed and transformed, it is also subjected to burial, a process that generally results in a shift from aerobic to anaerobic conditions in the profile (Figure 5.8). At this stage, the decomposition rates are drastically reduced (see discussion on anaerobic processes in the latter part of this chapter). The decomposition process in wetlands differs from that in upland ecosystems. The predominance of aerobic conditions in upland soils generally results in a rapid decomposition of detrital organic matter. The net accumulation of organic matter is minimal in this case and involves accumulation of highly resistant compounds that are relatively stable. In wetland soils, the decomposition process occurs at significantly slower rates due to predominance of anaerobic conditions. Because of this, moderately decomposable organic matter accumulates along with lignin

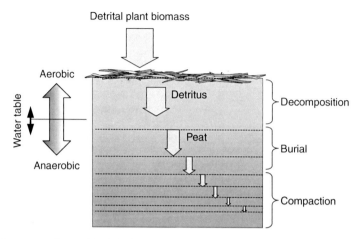

FIGURE 5.8 Decomposition and burial of organic matter and genesis of organic soil.

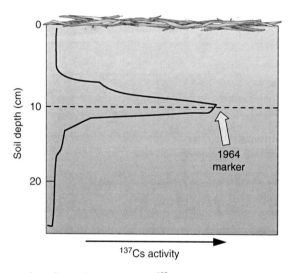

FIGURE 5.9 Vertical accretion of organic matter using ^{137}Cs as an indicator in a wetlatnd soil profile (Reddy et al., 1993).

and other recalcitrant fractions. Thus, the accumulation of organic matter in wetlands is typically characterized by a stratified buildup of partially decomposed plant remains, with a low degree of humification. The biodegradability of organic matter decreases with depth, as the material accreted deep into the soil ages and undergoes humification compared to the new material accumulating in the surface layers.

Long-term accumulation of organic matter in wetlands is determined by a balance between the inputs and the outputs. A number of approaches have been used to determine the long-term accumulation of organic matter in wetlands. Simple methods involved placing markers (a thin layer of feldspar) and determining net accumulation of organic matter above this marker for a defined period of time. Vertical accretion of organic matter is also determined using ^{137}Cs, a radioactive fallout product of nuclear bomb testing. The first significant fallout levels of ^{137}Cs were detected in 1954, whereas peak concentrations were measured in 1963 and 1964. Thus, organic matter accretion above this peak has been used as an indicator for accumulation since 1963–64 (Figure 5.9).

TABLE 5.6
Long-Term Accumulation of Organic Matter in Selected Wetlands

Wetland	Locations	Carbon Accumulation ($g\,m^{-2}\,year^{-1}$)	Reference
Peatland	Alaska	22–122	Billings (1987)
Everglades	Florida		
Typha sp.		163–387	Reddy et al. (1993)
Cladium sp.		86–140	
Salt marsh	Louisiana	200–300	Hatton et al. (1983)
Coastal marsh		50–500	Rabenhorst (1995)
Sphagnum	Wisconsin	34–75	Krazt and Dewitt (1986)
Taxodium	Georgia	45	Cohen (1974)
Bogs	Sweden	20–30	Armentano and Menges (1986)
Mangroves	Mexico	100	Twilley (1992)

Organic matter accretion was measured in various wetland ecosystems using this technique as shown in Table 5.6.

Organic matter accumulation regulates the following key functions in wetland ecosystems:

- Potential energy source to microbial communities
- Long-term storage of associated compounds (nutrients, heavy metals, and organics)
- Major component of the global carbon cycle and source/sink for greenhouse gases

5.4 CHARACTERISTICS OF DETRITUS AND SOIL ORGANIC MATTER

Organic matter is added to wetlands from both external and internal sources. External sources include particulate matter added through point and nonpoint sources. Point sources involve organic matter loads from municipal sewage effluents. Under most conditions, these waste effluents are added only when wetlands are used as a means of treating these waters. Wetlands are very effective in removing suspended solids from effluents. For this reason, wetlands are often used as buffers to remove suspended solids. For example, riparian wetlands are effective in removing sediment and other particulate matter discharged from upland portions of the watershed. In other cases, wetlands have been used as settling basins for suspended planktonic material from eutrophic lake water, which results in building up of organic matter in a short period of time (D'Angelo and Reddy, 1994a). The composition of the particulate material added to wetlands depends on the source of the effluent. The biodegradability of this material may range from easily decomposable to highly resistant.

Internal sources include detrital plant matter from macrophytes, litter fall in forested wetlands, and detrital matter from algal and microbial mats. The largest fraction of all organic matter entering a wetland is contributed by detrital organic matter from emergent macrophytes in marshes and trees in forested wetlands. In addition, belowground portion (roots and rhizomes) also contributes to detrital pool. The contribution of this pool is often integrated into the measurement of soil organic matter. Root exudation of organic compounds including ethanol, carbohydrates, and amino acids may also contribute to soil organic matter. Most of the root exudates are easily biodegradable under aerobic and anaerobic conditions. In addition, benthic invertebrates and soil microorganisms may also contribute significant amounts of organic carbon to wetland soils.

TABLE 5.7
Major Carbon Components of Plant Tissue, Peat, and Soil Organic Matter

Carbon Fraction	Monomer	Linkage	SOM (%)	Peat (%)	Plant Tissue* (%)
Water soluble fatty acids, sugars, amino acids	—	—		< 1	1–5
Ether extractable lipids, oils, and waxes	Fatty acids, sugars, phosphate	Esters and C-C		2–10	1–11
Proteins	Amino acids	Peptide	2–20	4–20	2–15
Cellulose	Glucose	β-1,4 ether	2–10	15–30	15–60
Hemicellulose	5C and 6C sugars	β-1, 4 and β-1, 3 ether	0–2	10–30	10–30
Lignin	Phenyl propane	Aryl ether and C-C	30–50	20–60	12–40
Mineral matter				5–30	3–20

Source: Clymo, 1983; Stevenson, 1994; DeBusk, 1996; and several other sources in the literature.

* Plant tissue represents newly dead tissue.

The following are the major constituents of detrital plant matter.

- Soluble substances such as sugars, fatty acids, and amino acids
- Cellulose
- Hemicellulose
- Lignin
- Fats, waxes, and resins
- Mineral matter

The relative proportion of these constituents varies with the type and source of detrital matter, the degree of decomposition, and the age of the material. Approximate measures of these constituents in plant material, peat, and soil organic matter are shown in Table 5.7.

The organic matter in waters and soils can be viewed as a complex of plant, microbial, and animal products in various stages of decomposition. The chemical constituents of organic matter can be grouped into the following (Stevenson, 1994):

- Nonhumic substances
  - Carbohydrates, proteins, and fats
- Phenolic substances
  - Lignins and tannins
- Humic substances
  - Heterogeneous mixtures of high-molecular-weight aromatic structures that result from secondary synthesis reactions.

5.4.1 Nonhumic Substances

Compounds in this group are well characterized in organic chemistry. This group includes carbohydrates in the form of simple sugars, hemicellulose, and cellulose, proteins, lipids, waxes, and oils.

5.4.1.1 Carbohydrates

Water-soluble carbohydrates include simple sugars, which are combined in a number of configurations in detritus and soil organic matter, and microorganisms. These simple sugars are referred to as

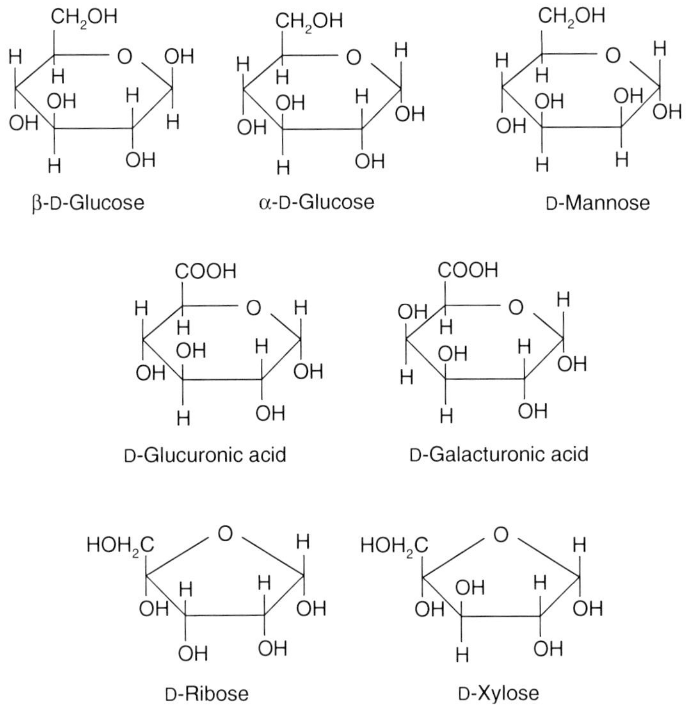

Carbon

FIGURE 5.10 Examples of monosaccharides commonly found in soils.

monosaccharides. These monomers are readily passed through microbial cells and rapidly degraded, and are thus considered labile sources of energy. Two characteristic functional groups of monosaccharides are (1) an aldehyde and (2) an alcohol. The most common monosaccharides having five and six carbons in their structure are called pentoses and hexoses, respectively (Figure 5.10).

Most of the carbohydrates found in detritus and soil organic matter occur as polysaccharides of high molecular weight. The most common polysaccharides are starch, cellulose, and hemicellulose. Starch is the most important food reserve of plant cells and is abundant in seeds, roots, and rhizomes of plants. Starch contains two types of glucose polymers: amylose and amylopectin. The amylose consists of a long, unbranched chain of glucose units connected by α(1, 4) linkages, whereas in amylopectin glucose units are connected by α(1, 6) linkages (Figure 5.11). These chains vary in molecular weight from a few thousand to 500,000.

Cellulose is a major constituent in vascular plants and is generally absent in algal and microbial tissue. Cellulose is fibrous, water insoluble, and usually present in protective cell walls of plants, such as stalks, stems, and all the woody portion of the plant. Cellulose constitutes 45–90% of woody tissue, and 15–30% of herbaceous plant tissue. Cellulose consists of a linear, unbranched polysaccharide of 10,000 or more glucose units connected by β(1, 4) linkages (Figure 5.12).

Hemicellulose is a rather poorly defined group of structural polysaccharides present in plant cell walls, but absent in algae and microbes. Hemicelluloses are diverse groups of polymers comprising

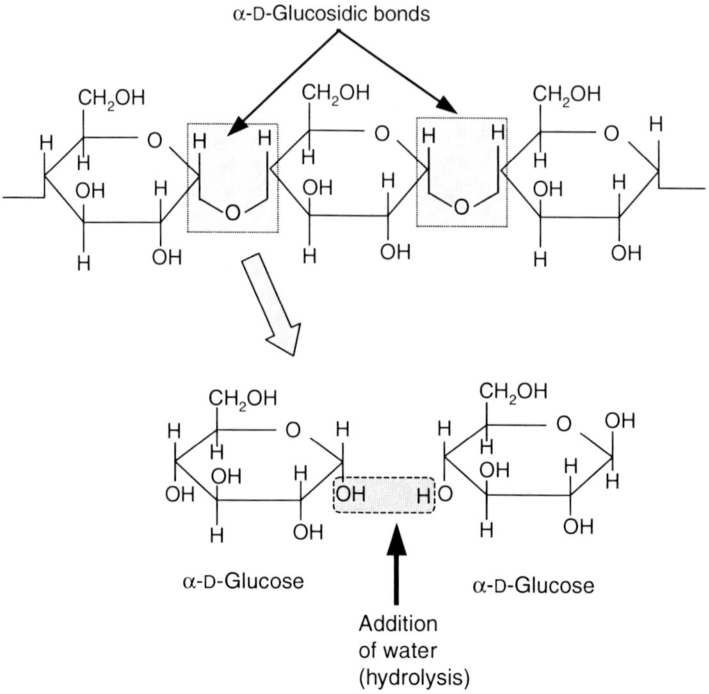

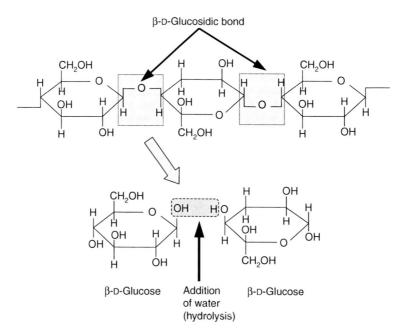

FIGURE 5.11 Hydrolysis of α-glucosidic molecules in a starch molecule.

FIGURE 5.12 Hydrolysis of β-glucosidic bonds in a cellulose molecule.

TABLE 5.8
Chemical Composition of a Bacterial Cell

Compound	Wet Weight Basis (%)	Dry Weight Basis (%)
Polysaccharide	3	5
Proteins	15	55
Lipid	2	9
DNA	1	3
RNA	8	20
Total polymers	26	96
Amino acids	0.5	0.5
Sugars	2	2
Nucleotides	0.5	0.5
Total monomers	3	3.5
Inorganic ions	1	1
Water content	70	—

Source: Madigan and Martinko (2006).

five and six carbon sugars, uronic acids, and other sugars. Most of the monomeric units are connected via β(1, 4) or (1, 3) linkages.

A number of methods are available for the determination of total carbohydrates in detritus and soil organic matter. For an additional discussion on characterization and associated methodologies, see Stevenson (1994).

Proteins are the most abundant macromolecules in cells and constitute more than 50% of the microbial biomass, 2–20% plant tissue, and soil organic matter (Table 5.7). Proteins consist of polymers of 20 different amino acids linked via peptide bonds (see additional discussion on proteins in Chapter 8). Proteins are a dominant component of bacterial cells (Table 5.8).

Lipids constitute a small portion of plant tissue. They constitute 5–10% of bacterial biomass and 10–25% of fungal biomass (Stevenson, 1994). Lipids are water insoluble, oily, or greasy organic substances that are extractable from plant tissue and soil organic matter by nonpolar solvents, such as chloroform or ether. Fatty acids are the characteristic building-block components of most of the lipids. Fatty acids are long-chain organic acids having 4–24 carbon atoms. Fatty acids are present in both saturated (no double bonds between carbons) and unsaturated (one or more double bonds between carbons) lipids. Lipids consist of a three-carbon glycerol backbone. Two of the glycerol carbons are ester linked with two long-chain monocarboxylic acids, and the third is ester linked with either a phosphate (e.g., phosphatides) or a sugar (glucolipid) moiety.

5.4.2 Phenolic Substances

Lignins and tannins are the common phenolic substances in detritus and soil organic matter. Lignin is present in secondary cell walls of plants and accounts for about 15–30% of the woody plant tissue and <10% of the herbaceous vegetation. Lignin is a highly branched random polymer of phenyl propanoid unit (which contains a basic unit of an aromatic ring, a phenyl group, with a three-carbon side chain, phenyl propanoid subunit) connected by C–C and C–O linkages with methoxy or hydroxy groups (Figure 5.13). Unlike many polysaccharides such as starch, cellulose, and hemicellulose, lignin does not have identical linkages of regular intervals.

Tannins are a heterogeneous group of phenolic compounds that precipitate proteins. Both tannins and lignins play an important role in the decomposition of labile plant constituents by complexing with proteins, exhibiting antibiotic activity, and forming an association with cellulose and hemicellulose (Boultin and Boon, 1991).

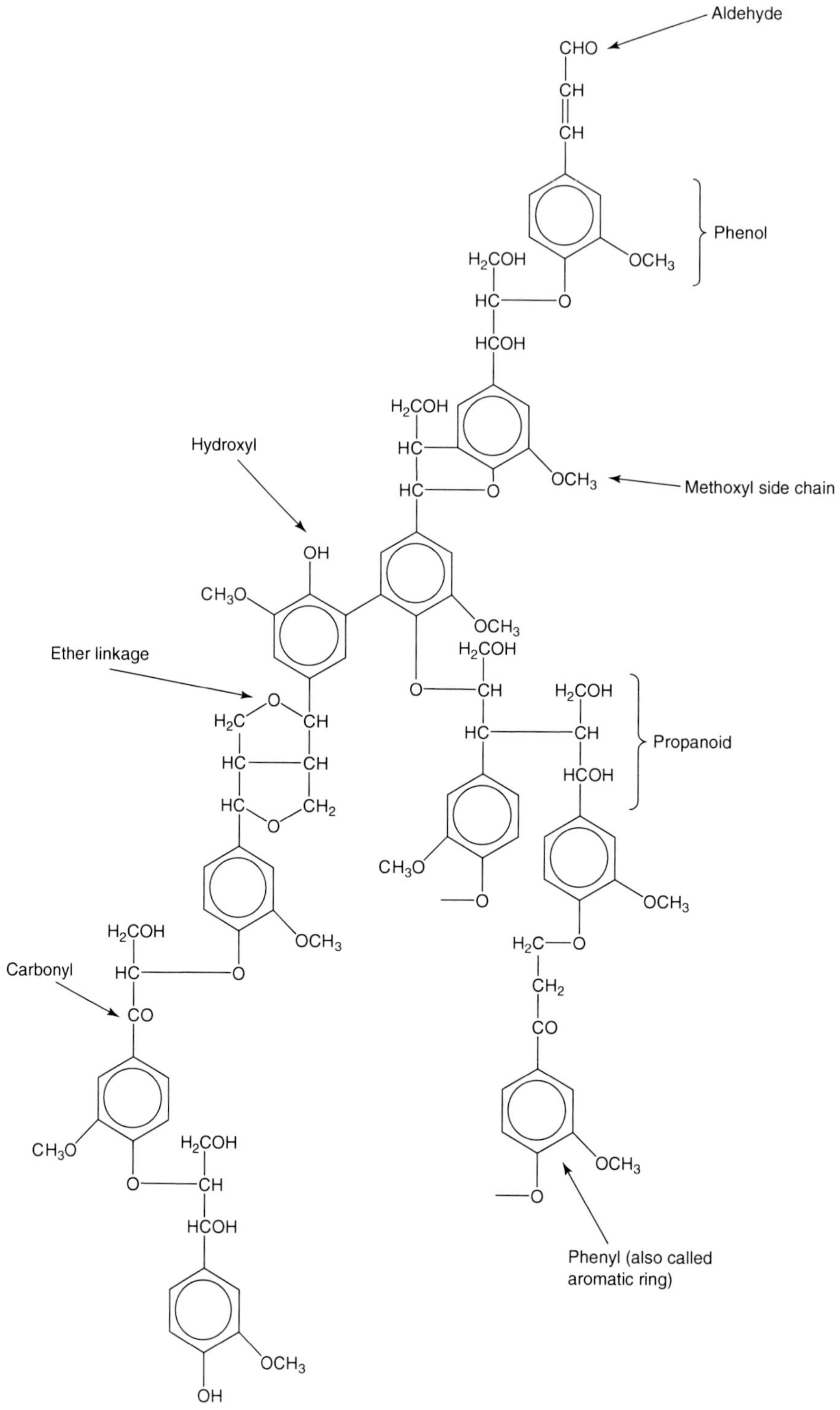

FIGURE 5.13 Generalized lignin structure, showing common functional groups. (Adapted from Paul and Clark, 1996.)

5.4.3 Humic Substances

Humic substances consist of heterogeneous mixtures of compounds with no single structural formula. These compounds are formed by dynamic alterations of resistant tannins and lignins by abiotic and biotic reactions, resulting in accumulation of humic substances. Humic substances are divided into three major groups on the basis of chemical characteristics and their solubility in acids and bases (Figure 5.14):

- Fulvic acid (acid and base soluble)
- Humic acid (acid insoluble and base soluble)
- Humin (acid and base insoluble)

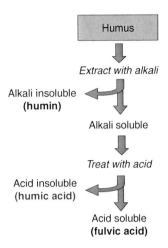

FIGURE 5.14 Three major groups of humic substances.

Humic substances contain aromatic, acid functional groups (total acidity 560–1,420 meq per 100 g, yellow-to-black coloration, high molecular weights (700–300,000), and high amounts of oxygen-containing functional groups, such as carboxyl, phenolic hydroxyl, and carbonyl structures (Figure 5.15a). The COOH group of humic substances contributes most of the acidity, compared to other functional groups.

A number of pathways have been proposed to describe the formation of humic substances from plant detritus (Stevenson, 1994). For many years, it was assumed that humic substances were derived primarily from the residual lignin, which remained after microbial attack, through a series of reactions involving demethylation, oxidation, and condensation of amino compounds. At present, the most accepted pathway involves transformation of residual lignin, cellulose, and other nonhumic substances into humic substances through a series of enzymatic reactions, sometimes referred to as "browning reactions" (Figure 5.15b). Phenolics, aldehydes, and acids released during microbial attack of lignin undergo enzymatic conversion into quinones.

Similarly, polyphenols are synthesized by microorganisms from nonhumic substances, which are subsequently converted to quinines through enzymatic reactions. The conversion of polyphenols formed from both sources to quinines is catalyzed by oxygen-requiring phenol oxidase enzymes. Quinones polymerize with or without amino compounds to from complex humic substances (Figure 5.15b).

A variety of functional groups are present in humic substances:

Carboxyl	O– –C– –OH
Phenolic OH	Aromatic ring
Alcoholic OH	–OH
Quinone and ketone	C=O
Amino	–NH_2
Sulfhydryl	–SH

Major elements in humic and fulvic acids are oxygen and carbon. The carbon content of fulvic acid ranges from 41 to 51% as compared to 54–59% for humic acid. In contrast, the oxygen content of fulvic acids is higher (40–50%) compared to 33–38% for humic acids (Table 5.9). These data demonstrate that distinct differences exist between fulvic and humic acids.

Fulvic acids tend to contain more oxygen and less carbon than humic acids. The molecular weight of fulvic acids is in the range of 1,000–30,000 as compared to 10,000–100,000 for humic acids. Fulvic acids are in less advanced stage of decomposition than humic acids. Functional acidity of fulvic acids is higher than that of humic acids. The reactivity of fulvic acids is due to their higher content of oxygen-containing functional groups, including COOH, phenolic OH, and alcoholic OH

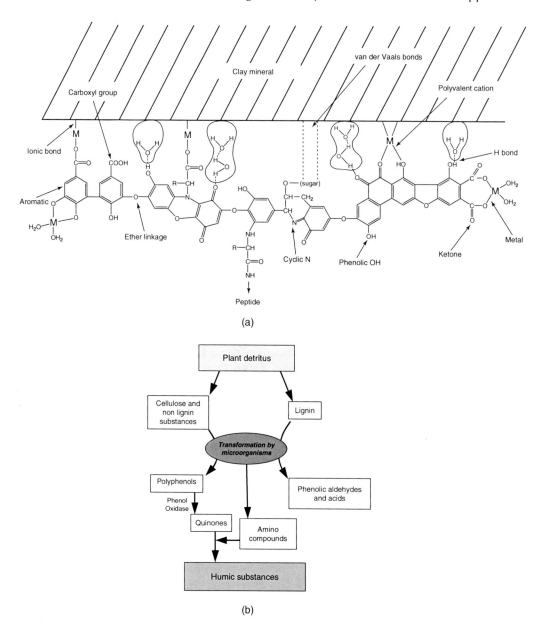

FIGURE 5.15 (a) Generalized structure of humic substances showing various functional groups. (b) Schematic diagram showing degradative pathways of detrital material (Stevenson, 1994).

TABLE 5.9
Elemental Content of Fulvic and Humic Acids

Element	Fulvic Acids (%)	Humic Acids (%)
Carbon	41–51	54–59
Oxygen	40–50	33–38
Hydrogen	4–7	3–6
Nitrogen	0.9–3.3	0.8–4.3
Sulfur	0.1–3.6	0.1–1.5

TABLE 5.10
Distribution of Oxygen-Containing Functional Groups in Humic and Fulvic Acids Isolated from Soils

Functional Group	Fulvic Acids (cmol kg^{-1})		Humic Acids (cmol kg^{-1})	
	Average	Range	Average	Range
Total acidity	1,030	640–1,420	670	560–890
COOH	820	520–1,120	360	150–570
Acidic OH	300	30–570	390	210–570
Weakly acidic + alcoholic OH	610	260–950	260	20–490
Quinone C|O and ketonic C|O	270	120–420	290	10–560
OCH$_3$	80	30–120	60	30–80

Note: Values are range and average for soils from Arctic temperature acid and neutral soils, and subtropical and tropical soils (Stevenson, 1994).

groups (Table 5.10). The oxygen content of these functional groups in fulvic acids is substantially higher than that in any other naturally occurring organic polymer (Stevenson, 1994).

Humin is the portion of soil organic matter that remains as a residue after extraction of the soil with dilute alkali. This fraction is highly resistant to biotic breakdown and insoluble in alkali. This material is highly humified with carbon content >60%. Recent studies using ^{13}C NMR showed that humin contains significant amounts of paraffinic substances that are insoluble in alkali (Stevenson, 1994). Organic solvents used to extract soil lipids can potentially dissolve some component of humin.

5.5 DECOMPOSITION

Decomposition of plant detritus involves stepwise conversion of complex organic molecules to simple organic constituents as a result of processes including (Figure 5.16)

- Physical leaching and fragmentation
- Extracellular enzyme hydrolysis
- Aerobic and anaerobic catabolic activities of heterotrophic microorganisms.

5.5.1 LEACHING AND FRAGMENTATION

As plants senesce, a portion of the aboveground biomass is returned to the detrital pool. Because of long residence of standing dead tissue of emergent macrophytes such as *Spartina* sp. and *Typha* sp., significant decomposition occurs by the time detrital material reaches the soil surface. Some of this areal decay is due to fungal activity resulting from high availability of oxygen. The initial step in this process is leaching of water-soluble organic and inorganic compounds as a result of cell autolysis. Water-soluble components include soluble carbohydrates, nucleotide bases, fatty acids, and amino acids. These components are readily bioavailable and serve as energy and nutrient sources to bacterial and fungal communities colonized on detrital plant tissue.

Physical leaching of soluble components is generally over within a few weeks of plant senescence, depending on structural components, temperature, and water flow. Leaching can be a significant factor in decomposition of nonlignocellulose materials such as periphyton mats. However, approximately 10–20% of detrital tissue containing lignocellulose materials (such as plants) is lost within 4 weeks after immersion into the water column (Benner et al., 1985).

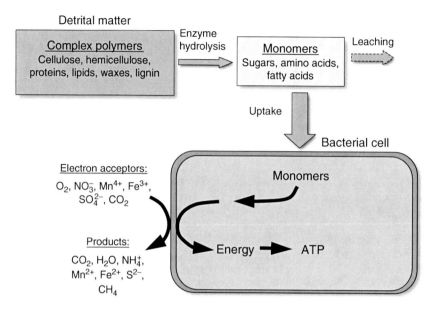

FIGURE 5.16 Schematic diagram of microbial degradation of soil organic matter.

Plant detritus (either attached or detached from the plant) undergoes physical fragmentation through the action of waves and currents, UV exposure, and through grazing activities of macro invertebrates. Fragmentation of plant detritus into small fractions (<1 mm) results in increased surface areas and accelerated microscale processes (such as enzymatic hydrolysis and catabolic activities).

5.5.2 Extracellular Enzyme Hydrolysis

After initial leaching and fragmentation, breakdown of plant detritus depends on the amount of major polymers, including lignin, cellulose, hemicellulose, lipids and waxes, and proteins. These structural components are mainly particulate matter and must be broken down into smaller units before they can be assimilated by microbes as energy and nutrient sources. Extracellular enzymes secreted by microbes (bacteria and fungi) aid in the hydrolysis of high-molecular-weight organic matter (HMW-OM) (>600 Da) associated with plant detritus and soil organic matter, which in its original form are too large to be transported into the periplasmic space of the cell (Weiss et al., 1991).

Extracellular enzymes are generally defined as enzymes that have crossed the cytoplasmic membranes of the microbial cell. These enzymes are divided into two groups.

- Enzymes released into outside environment and not attached to the producer. These are often referred to as exoenzymes or abiotic enzymes.
- Enzymes released that remain associated with producer cells such as within periplasmic space, or attached to the cell membrane or cell wall. These are often referred to as ectoenzymes.

A majority of the extracellular enzymes released by the microbes are bound to the solid surface (such as particulate detrital matter or soil organic matter in clays), but a small fraction remains in the soil pore water (Figure 5.17). Those free in soil pore water are most susceptible to microbial degradation and chemical alteration. Surface-bound enzymes may not be as effective as free enzymes because of a slow rate of substrate diffusion to the sites where enzymes are present.

Because extracellular enzyme activity is the rate-limiting step in microbially mediated decomposition of plant detritus in wetlands and aquatic systems, many researchers have studied enzyme

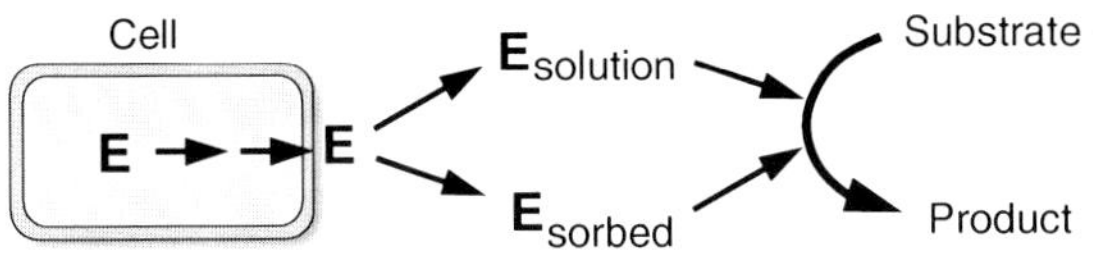

FIGURE 5.17 Fate of extracellular enzymes after they are released form microorganisms.

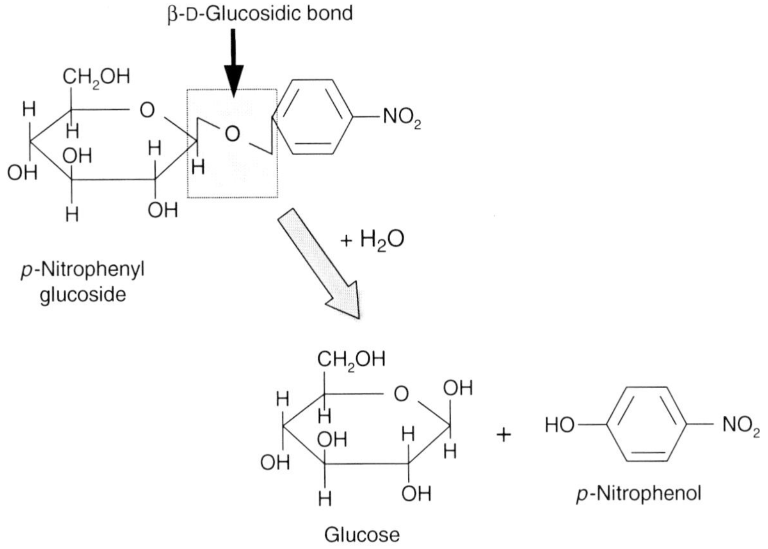

FIGURE 5.18 Substrate bonded to *p*-nitrophenyl glucoside is hydrolyzed by enzymes and released as *p*-nitrophenol, which determined by colorimetric methods.

activities over scales ranging from the molecular to the ecosystem levels (Sinsabaugh et al., 1993). Activities of β-glucosidase, endocellulase and exocellulase have been measured as indicators of decomposition of organic matter in wetlands and aquatic systems. These activities are determined by using artificial substrates as proxies for many high-molecular-weight substrates such as polysaccharides. For example, methylumbelliferyl (MUF) substrates consist of MUF fluorophore attached to the monosaccharide. Hydrolysis of MUF substrate leads to an increase in fluorescence signal from MUF flourophore (Arnosti, 1998). In soil research, *p*-nitrophenol bound to the monosaccharide is used as a proxy for HMW-OM (Figure 5.18). Colorimetric methods are used to determine the activity of *p*-nitrophenol released after hydrolysis (Eivazi and Tabatabai, 1988). Activity should be on a relative basis, when comparing the results from various ecosystems.

Most of the extracellular enzymes are hydrolytic, i.e., they involve addition of a water molecule across the enzyme-susceptible linkage (the most common hydrolytic extracellular enzymes are cellulases, proteases, and phosphatases, involved in mineralization of organic carbon, nitrogen, and phosphorus in wetlands). In this chapter, discussion on enzymes will be restricted to those involved in carbon cycling. For example, the hydrolytic enzymes catalyze the cleavage of covalent bonds such as C–O (esters and glycosides), C–N (proteins and peptides), and O–P (phosphates). Hydrolytic enzymes are not involved in the breakdown of aromatic ring structure of phenolic compounds, but can react with the functional groups (such as –OH) associated with these compounds (Wetzel, 1991). Nonhydrolytic enzymes such as oxygenases including peroxidase and phenol oxidase are involved in the breakdown of aromatic ring structure in phenolic compounds. Because of oxygen requirements, these reactions are restricted to aerobic environments. Thus, enzymatic breakdown of lignin and humic substances is not significant in the anaerobic portion of wetlands.

Microorganisms are the dominant producers of extracellular enzymes. Some examples of microbial groups include

Fungi
- Alternaria
- Cladosporium
- Fusarium
- Trichoderma

Bacteria
- Bacillus
- Pseudomonas
- Clostridium

Extracellular enzymes are generally a part of the rate-limiting step in the overall degradation of organic matter. Kinetic limitations to enzymatic breakdown include slow rate of diffusion of enzymes and substrates, adsorption of enzymes on solid phases, and complexation of enzymes to humic substances (Sinsabaugh et al., 1993; Wetzel, 1991). Selected examples of enzymes involved in degradation of cellulose, hemicellulose, and lignin are shown in Table 5.11. Breakdown of these structural polymers may require a multicomponent enzyme system comprising a wide range of microbial species. Rarely any single microorganism can produce all the enzymes required for the breakdown of organic substrates.

The hydrolysis of cellulose to glucose is mediated by the cellulase enzyme complex, which hydrolyzes the β-1-4-glycosidic bonds. This enzyme complex includes endocellulases, exocellulases, and β-glucosidases. These enzymes are responsible for producing soluble cellobiose or glucose. Similarly, exoxylanase and β-xylosidase are some examples of the enzymes involved in hydrolyzing polymeric hemicellulose into simple units such as xylan and xylobiose. The activity of enzymes involved in cleaving bonds is linked directly to the mineralization of organic matter in soils and sediments.

β-Glucosidase activity is influenced by soil redox conditions, with higher values under aerobic conditions than under anaerobic conditions (Figure 5.19). β-Glucosidase activity is influenced by

TABLE 5.11
Enzymes Involved in the Degradation of Lignocellulose

Process	Enzymes	EC Number	Substrates
Starch	X-Amylase		Amylose
Cellulose degradation	Exocellulase	3.2.1.4	Cellulose
	Endocellulase	3.2.1.9.1	Cellulose
	β-Glucosidase	3.2.1.2.1	Cellobiose
Hemicellulose	Exoxylanase		Xylan
degradation	Endoxylanase	3.2.1.8	Xylan
	β-Xylosidase	3.2.1.37	Xylobiose
Lignin degradation	Phenol oxidase	1.10.3.12	Lignin
	Peroxidase	1.11.1.7	Lignin

Note: EC number = Enzyme commission number. The first three numbers define major class, subclass, and sub–sub class, respectively. The last is a serial number in the sub–sub class, indicating the order in which each enzyme is added to the list (Mathews and van Holde, 1990).

Source: Sinsabaugh et al. (1991).

Carbon

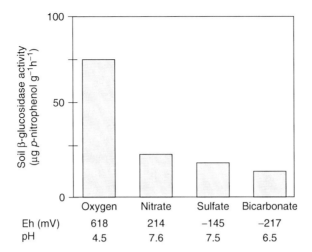

FIGURE 5.19 β-Glucosidase activity is higher under aerobic conditions. (Redrawn from McLatchey and Reddy, 1998.)

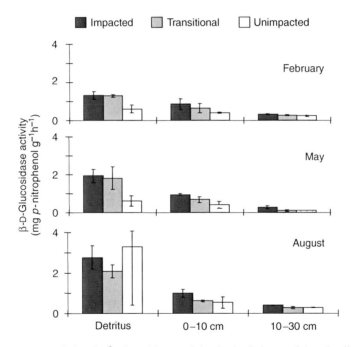

FIGURE 5.20 Seasonal variation in β-glucosidase activity in detrital material and soils (0–10 cm deep, 10–30 cm deep) in different trophic systems (Wright and Reddy, 2001a).

temperature (maximum activity at 30°C) and pH (maximum activity at 8.5) (King, 1986). The accumulation of labile organic compounds can repress the activity of glucosidase. Addition of complex polymeric organic compounds can induce high activity of enzymes. High activity of glucosidase is observed in the detrital layer of wetland soil and decreases with depth (Figure 5.20). Under most conditions, extracellular enzymes are secreted by microorganisms at the base level. These enzymes rapidly hydrolyze available substrates, and low-molecular-weight matter (LMW-OM) is released. High concentrations of products of enzyme hydrolysis may repress the activity of respective enzymes.

Phenol oxidase is active under aerobic conditions, and its activity is linked to the degradation of lignin and other phenolic compounds. Phenol oxidase activity is higher in the detrital layer and decreases with soil depth (Figure 5.21). In addition to being substrates, many phenolic compounds are also known to inhibit the activity of enzymes. Phenolic compounds are a major component of DOC in freshwater (Wetzel, 1991). These phenolic groups are known to induce precipitation of proteins (enzymes) by binding a monolayer to the protein surface (Wetzel, 1991). Under most conditions, pH of freshwater may range from 5 to 8 where carboxylic acid groups are ionized. Major cations (e.g., Ca^{2+}) react with these acids and can reduce the precipitation and inactivation enzymes. In soft water (low base cations), the interference of humic substances with enzyme activities may be greater than that in hard water with Ca^{2+} and Mg^{2+} concentration in the range of 40–60 and 15–25 mg L^{-1}, respectively (Wetzel, 1991) (Figure 5.22).

Phenol oxidase activity is highly sensitive to the availability of oxygen. The activity of this enzyme may be of great significance in soils with high organic matter content. Because of its oxygen

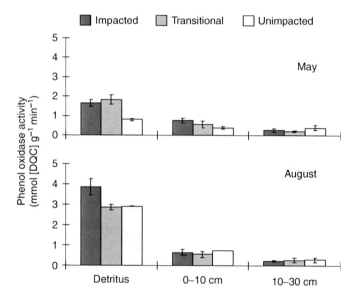

FIGURE 5.21 Seasonal variation in phenol oxidase activity in detrital material and soils (0–10 cm deep, 10–30 cm deep) in different trophic systems (Wright and Reddy, 2001a).

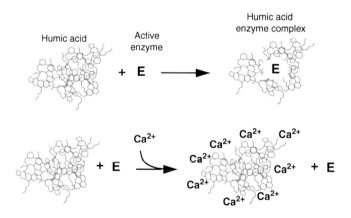

FIGURE 5.22 Presence of cations such as Ca^{2+} prevents soil enzymes inhibition by interacting with humic acids.

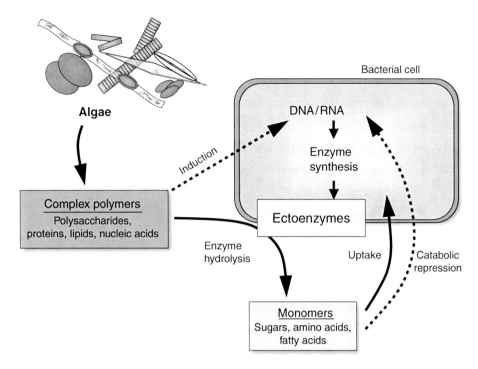

FIGURE 5.23 Algae and bacterial interaction.

requirements, the activity of phenol oxidase may be restricted to the aerobic portions of the wetland (water column with detrital material or aerobic root zone). Large exports of dissolved humic (phenolic) substances from organic-rich soils into adjacent aquatic systems are due to low phenolic degrading enzyme activities in these soils. A negative correlation has been found between phenol oxidase activity and the concentration of phenolic compounds (Benner et al., 1984). Phenol oxidase has been detected at a pH range of 2–10, with optima at 8 (Pind et al., 1994).

Enzyme assays are simple and serve as excellent indicators of microbial activities. They have also been used as indicators to evaluate nutrient/contaminant impacts in wetlands. However, there are several limitations to their use in quantitative evaluation of carbon and nutrient cycling in wetlands (Sinsabaugh et al., 1991). Some of the limitations are

- Lack of standard methodology
- Heterogeneity of detritus and soil samples
- Use of artificial substrates as proxies for HMW-OM.

Enzyme research in wetland ecosystems is not as extensively reported as it is for uplands and other aquatic ecosystems. On the basis of the results of various experiments under laboratory and field conditions, Chrost (1991) proposed a conceptual model for the role of microbial ectoenzyme synthesis, control, and activity in aquatic environments. The following general conclusions are drawn on the interaction of algae and microbial communities in transforming HMW-OM into readily utilizable LMW-OM (Figure 5.23):

- Algae excrete a variety of LMW-OM end products during metabolism and photosynthesis. These compounds support bacterial growth and metabolism.
- The LMW-OM (e.g., monosaccharides, amino acids, and organic acids) inhibits the activity (end-production inhibition) and represses the synthesis (catabolic repression) of ectoenzymes in heterotrophic bacteria.

- Upon senescence, algae release HMW-OM including polysaccharides, proteins, lipids, and nucleic acids
- Low concentration of LMW-OM and high concentration of polymeric HMW-OM (substrate for ectoenzymes) can derepress and induce enzyme synthesis in heterotrophic bacteria.

Similar interactions (described above) can occur in the water column of wetlands dominated by periphyton or microbial mat communities, and detrital matter produced from emergent macrophytes.

Extracellular enzymes play a pivotal role in catalyzing the rate-limiting steps during decomposition of plant detritus and soil organic matter, and release of bioavailable nutrients. The relationship between enzyme activity (such as β-glucosidase activity) and microbial respiration and labile carbon pools, such as MBC and DOC, suggests a role of these enzymes in carbon cycling.

5.5.3 Catabolic Activity

Catabolism refers to metabolic pathways that allow organisms to obtain energy from degradation of organic molecules; it is the last step in the overall decomposition of organic substrates (Figure 5.24). Some end products of decomposition (e.g., nutrients) are assimilated by microorganisms for their growth. This process is referred to as anabolism. In wetlands, diverse groups of heterotrophic microorganisms may be involved in the catabolic breakdown of organic matter, depending on the type of organic substrate, the availability of electron acceptors, and the environmental factors.

When organic compounds are oxidized, chemical energy stored in these compounds is released. Similarly, energy stored in inorganic reduced compounds is also released during their oxidation. This energy is expressed in calories, which is defined as the amount of heat energy necessary to raise the temperature of 1 g of water by 1°C (1 cal or 4.184 J).

The utilization of chemical energy by microorganisms involves oxidation–reduction reactions (see Chapter 4). The reductant is the source of energy and serves as the electron donor, whereas the oxidant accepts the electrons released by the donor, resulting in the production of energy. The larger the difference in E of two substances, the greater the energy yield, as shown in the following equation (see Equation 4.19, Chapter 4):

$$\Delta G = -nF\Delta E$$

where ΔG is the change in Gibbs free energy for the reaction (in kcal mol^{-1}), n the number of electrons transferred between redox couples, F the Faraday constant (23.061 kcal V^{-1}), and ΔE the electrical potential difference between two redox couples (in volts) (see Chapter 4 for additional discussion).

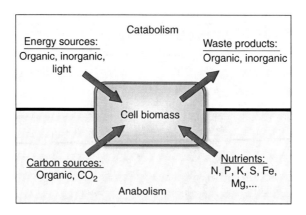

FIGURE 5.24 Schematic diagram depicting catabolic and anabolic reactions in a microbial cell.

The energy yield during aerobic respiration using glucose as energy source is shown in the following example:

$$C_6H_{12}O_6 + 6H_2O \rightarrow 6CO_2 + 24e^- + 24H^+ \ (E° = -0.43 \text{ V})$$
$$6O_2 + 24e^- + 24H^+ \rightarrow 12H_2O \ (E° = 0.82 \text{ V})$$
$$C_6H_{12}O_6 + 6O_2 \rightarrow 6CO_2 + 6H_2O \ (\Delta E° = 1.25 \text{ V})$$
$$\Delta G° = -nF\Delta E°$$
$$= -(24)(23.06 \text{ kcal V}^{-1})(1.25 \text{ V})$$
$$= -688 \text{ kcal mol}^{-1}$$
$$= -2{,}879 \text{ kJ mol}^{-1}$$

The enzymatic hydrolysis of polymeric organic compounds results in production of several monomers, including glucose, xylose, fatty acids, and amino acids. The chemical energy stored in these reduced compounds is released upon their oxidation, resulting in release of electrons.

In this example, oxygen serves as an electron acceptor (oxidant) for supporting biological oxidation (respiration) (Figure 5.25). As long as there is oxygen supply from the atmosphere, the other oxidants that microorganisms can use are not reduced. Oxygen is used preferentially because it takes electrons from the reductant material (organic matter) more readily than other oxidants. When oxygen becomes limiting, the other oxidants begin to accept electrons and keep the respiration of certain microorganisms going. During this process, nitrate is reduced by the electrons to nitrogen gas (N_2) and ammonium nitrogen, and Mn(IV) oxides are reduced to manganous compounds (Mn(II)). Ferric compounds are reduced to ferrous compounds, sulfate is reduced to sulfide if the demand for electron acceptors is high enough for the sulfate reducers to function, and under even more reducing conditions, carbon dioxide can even be reduced to methane.

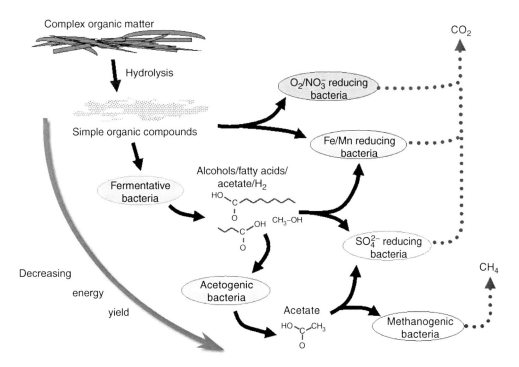

FIGURE 5.25 Heterotrophic microbial metabolism occurs at various trophic levels with decreasing energy yields.

5.5.3.1 Aerobic Catabolism

Aerobic microorganisms require oxygen for two purposes:

- As a terminal acceptor for electrons that are released during oxidation of organic substrates
- As a reactant during enzymatic attack of organic substrates

This dual role of oxygen allows aerobic microorganisms to outcompete other groups, by obtaining higher energy yield from oxidation of reduced organic and inorganic compounds. Higher energy yield during aerobic breakdown of organic substrates is due to the following reasons:

- Carbon atoms in the organic substrates are completely oxidized to CO_2.
- Oxygen redox couple has relatively high positive reduction potential. This leads to a large net difference in electrical potentials between electron donor (organic substrate) and terminal electron acceptor (oxygen).

The overall pathway of aerobic catabolism by which microorganisms oxidize organic carbon material is shown in Figure 5.26. Monomers (such as glucose) generated from extracellular enzyme hydrolysis are transported into microbial cells via diffusion or active transport. The generation of metabolic energy from glucose begins with glycolysis, which converts one molecule of glucose into two molecules of pyruvate and two molecules of ATP. This process involving 10 reactions can be grouped into two phases (Mathews and van Holde, 1990):

- *Energy investment phase*, in which sugar phosphates are synthesized at the expense of ATP conversion to ADP. Six-carbon substrate (glucose) is converted to two three-carbon sugar phosphates
- *Energy generation phase*, in which sugar phosphates are converted into energy-rich compounds, with phosphate transferred to ADP, resulting in the synthesis of ATP

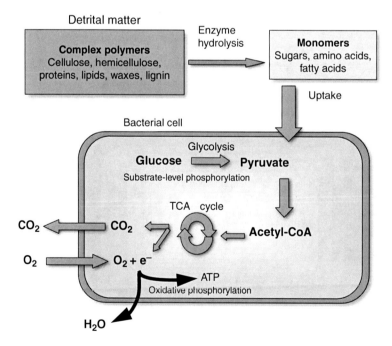

FIGURE 5.26 Complete oxidation of organic carbon compounds during microbial decomposition leads to production of carbon dioxide accompanied by reduction of oxygen to water.

The reaction in glycolysis is often referred to as the *Embden–Meyerhof pathway*.

$$\text{Glucose} \rightarrow 2 \text{ pyruvate}$$

$$2 \text{ ADP} \rightarrow 2 \text{ ATP}$$

$$2 \text{ NAD}^+ \rightarrow 2 \text{ NADH}$$

The energy investment phase results in expending 2 ATP, whereas the energy regeneration phase results in synthesizing 4 ATP, with overall glycolysis pathway yielding net 2 ATP. During this process, 2 moles of NAD^+ (nicotine adenine dinucleotide) are reduced to NADH, functioning as internal electron acceptor. The ATP generation in glycolysis does not involve a transfer of electrons to an inorganic electron acceptor (such as oxygen). This glycolysis process is often referred to as *substrate-level phosphorylation*. This process results in release of only a small fraction of chemical energy available in glucose.

The metabolic intermediate (e.g., pyruvate) undergoes complete oxidation to CO_2, through the pathway referred to as tricarboxylic acid cycle (TCA cycle). The following reactions are involved in TCA cycle. The first step involves conversion of pyruvate to acetyl-CoA through decarboxylation and production of NADH. The acetyl-CoA (2 carbon) combines with the four-carbon compound oxalacetate, leading to the formation of citric acid (6 carbon). The TCA cycle is also referred to as citric acid cycle. A series of reactions including dehydration, decarboxylation, and oxidation are involved in the conversion of citric acid to carbon dioxide. The electrons released are transferred to enzymes containing the coenzyme NAD^+.

$$NAD^+ + H^+ + e^- \rightarrow NADH$$

Reduced coenzymes are transported from the cytoplasm to the cell membrane (bacteria) or mitochondrial membrane (fungi), which are the sites of the ETS (Figure 5.27). This process is called oxidative phosphorylation. The electron transport in ETS can be viewed as follows:

- Transfer of primary donor (organic substrate) to coenzyme NAD^+
- Transfer of electrons through membrane-associated electron carriers (a series of coenzymes)
- Transfer of electrons to the terminal electron acceptor

As the NADH is oxidized, the electrons released are removed by specific carriers, and the protons are transported from cytoplasm to outside the cell. Removal of H^+ causes an increase in the number of OH^- ions inside the membrane. These conditions result in a proton gradient (pH gradient) across the membrane. This gradient of potential energy, termed as proton motive force, can be used to do useful work. This potential energy is captured by the cell by a series of complex membrane-bound enzymes, known as the ATPase in the process called oxidative phosphorylation. In 1961, the concept of proton gradient was first proposed as *chemiosmotic* theory by Peter Mitchell of England, who won the Nobel Prize for this scientific contribution.

Microorganisms capture about 30–50% of the energy released during the oxidation of glucose to carbon dioxide. Aerobic catabolism yields as many as 38 ATPs for each mole of glucose oxidized. Two ATPs are formed during *glycolysis* pathway, 30 ATPs in TCA cycle, and 6 ATPs in ETS. If we assume that each mole of ATP contains 30 kJ (7 kcal) of energy, then 38 ATPs formed during aerobic catabolism of glucose will conserve 1,140 kJ (273 kcal) of energy releases. This suggests that aerobic catabolism is about 40% efficient. It should be noted that catabolism results in the release of 100% of the energy stored in the glucose molecule, but approximately 40% of the energy released is conserved, whereas the remaining is lost as heat.

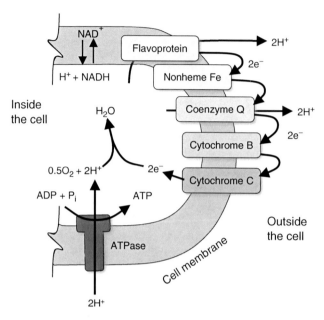

FIGURE 5.27 Energy in the aerobic organisms is generated by proton gradient generated across the membrane with the electron transport chain. (Redrawn from Madigan and Martinko, 2006.)

Oxygen reduction in ETS requires addition of four electrons. This reduction occurs in the following steps:

$$O_2 + e^- \rightarrow O_2^- \text{ (Superoxide)}$$

$$O_2^- + e^- + 2H^+ \rightarrow H_2O_2 \text{ (Hydrogen peroxide)}$$

$$H_2O_2 + e^- + H^+ \rightarrow H_2O + OH \text{ (Hydroxyl radical)}$$

$$OH + e^- + H^+ \rightarrow H_2O \text{ (Water)}$$

$$O_2 + 4e^- + 4H^+ \rightarrow 2H_2O$$

The intermediate compounds formed can be toxic for anaerobic organisms. Aerobic organisms have enzymes that are capable of decomposing these toxic intermediates. Enzymes catalase and peroxidase act on hydrogen peroxide, whereas superoxide dismutase acts on superoxide (Madigan and Martinko, 2006).

Utilization of oxygen as a reactant during aerobic breakdown of organic substrates is accomplished by the action of oxygenases. These enzymes are capable of catalyzing the incorporation of either one atom (monooxygenase) or both atoms (dioxygenase) of molecular oxygen into various organic chemicals. Production of oxygenases by microorganisms allows them to degrade two groups of organic substrates (Colberg, 1988):

- Aromatics and saturated alkanes with no functional groups
- Molecules that are synthesized only in the presence of oxygen, including lignin and humic compounds

For example, during degradation of benzene and polyaromatic hydrocarbon, activity of monooxygenase results in the formation of catechol or protocatechuate molecules, followed by ring opening,

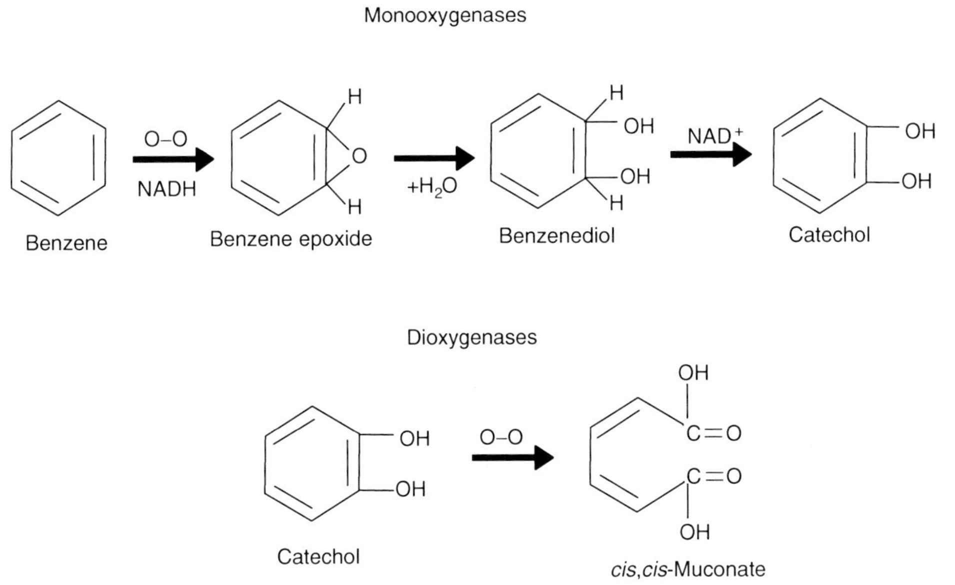

FIGURE 5.28 Production of oxygenases by microorganisms catalyzes the organic chemicals by incorporating molecular oxygen. (Madigan and Martinko, 2006.)

conversion to metabolic intermediate (e.g., acetyl-CoA, pyruvate), and finally oxidation to carbon dioxide (Madigan and Martinko, 2006) (Figure 5.28).

5.5.3.2 Anaerobic Catabolism

Oxygen supply in wetlands is restricted to the water column and to a thin layer of surface soil. Oxygen is also transported by wetland macrophytes to their root zone, resulting in the creation of aerobic conditions on root surfaces (see Chapters 3 and 6 for a detailed discussion on aerobic–anaerobic interfaces in wetlands).

In addition, seasonal fluctuations in hydrology and water table fluctuations can introduce oxygen into the soil profile. Thus, aerobic catabolism is restricted to small volumes of wetland soil, whereas in the remaining portion of the soil, the dominant microbial groups are anaerobes. These organisms gain energy from oxidation of various organic and inorganic reduced substrates, while using electron acceptors other than oxygen, including nitrate, Mn(IV) and Fe(III), sulfate, and carbon dioxide, and simple organic compounds (Table 5.12). Organisms conducting these processes obtain less energy from organic substrates compared to aerobic catabolism.

Anaerobic microbial groups are largely incapable of utilizing substrates such as lignin and humic substances that require initial oxygenation by oxygenases. Decomposition of organic substrates under anaerobic conditions results in the accumulation of reduced species including methane, sulfides, volatile fatty acids, ferrous iron, manganous manganese, ammonium nitrogen, and hydrogen. These reduced compounds store substantial amounts of energy released from organic substrates during anaerobic catabolism. This energy stored in inorganic reduced compounds is largely unavailable to heterotrophs under anaerobic conditions. Certain groups of organisms can obtain this energy through oxidation of reduced organic compounds and are called lithotrophs (see discussion in the latter part of this chapter).

Anaerobic decomposition of organic substrates is initiated by direct utilization of monomers (such as glucose) and fermentation of these monomers to simple compounds such as volatile fatty

TABLE 5.12
Dominant Processes Involved in Breakdown of Organic Substrates in Wetlands

Redox Potential (mV)	Electron Acceptor	Decomposition End Products	Microbial Groups
Aerobic			
>+300	O_2	CO_2, H_2O	Aerobic fungi and bacteria
Fermentation			
<−100 to +300	Organics	Organic acids, CO_2, H_2, alcohols, amino acids	Fermenting bacteria
Facultative			
+100 to +300	NO_3^-	N_2O, N_2, CO_2, H_2O	Denitrifying bacteria
	Mn^{4+}	Mn^{2+}, CO_2, H_2O	Mn^{4+} reducers
+100 to 100	Fe^{3+}	Fe^{2+}, CO_2, H_2O	Fe^{3+} reducers
Obligate anaerobic			
<−100	SO_3^{2-}	HS^-, CO_2, H_2O	Sulfate reducers
<−100	CO_2, acetate	CH_4, CO_2, H_2O	Methanogens
<−100	Organic acids	Acetate, CO_2, H_2	H_2-producing bacteria

acids, alcohols, hydrogen, and carbon dioxide. Utilization of the monomers or the fermentative products as energy sources depends upon

- Availability of electron acceptors
- Type and availability of organic substrates
- Competition between microbial groups for substrates

Monomers formed from hydrolysis of polymeric compounds can only be used by facultative anaerobic and sulfate reducers, whereas methanogens can only oxidize simple compounds such as acetate and H_2. Both facultative and obligate anaerobes can use fermentative products as their energy source. In general, metabolic activities that yield the highest amount of energy from oxidation of a common substrate will dominate if a given electron acceptor is present. This principle results in the observed sequential reduction of oxygen, nitrate, Mn(IV) and Fe(III) oxides, sulfate, and carbon dioxide following soil flooding and observed vertical profiles of dominant microbial processes in wetlands.

5.5.3.2.1 Fermentation

Water-soluble monomers released from hydrolysis of organic substrates are subsequently degraded to simpler products by fermenting bacteria. Fermenting bacteria can derive energy through internal oxidation–reduction reactions involving various organic substrates, through a process called fermentation. Under these conditions, selected organic compounds function as both electron donors and acceptors (Figure 5.29). Thus, no external supply of electron acceptors is required. Fermentation results in only partial oxidation of organic substrates; therefore, only a small amount of potential chemical energy is released. During fermentation, ATP is produced in a process called substrate-level phosphorylation.

In the following reaction, fermentation of glucose to ethanol and lactic acid is shown:

$$\underset{\text{(glucose)}}{C_6H_{12}O_6} \rightarrow 2\underset{\text{(ethanol)}}{CH_3CH_2OH} + 2CO_2 \quad \Delta G° = 239 \text{ kJ mol}^{-1}$$

$$\underset{\text{(glucose)}}{C_6H_{12}O_6} \rightarrow 2\underset{\text{(lactic acid)}}{C_3H_6O_3} \quad \Delta G° = 122 \text{ kJ mol}^{-1}$$

Carbon

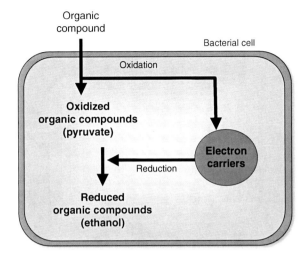

FIGURE 5.29 During fermentation, selected organic compounds present in the cell are used as electron acceptors.

The steps involved in the fermentation of glucose are as follows:

- Conversion of glucose to sugar phosphates. Two ATP are used in this reaction (energy investment phase).
- Oxidation of sugar phosphates to pyruvate. NAD^+ functions as electron acceptor. Two ATPs are synthesized in this reaction (energy generation phase).
- Two moles of pyruvate is oxidized to two moles of ethanol. NADH functions as electron donor. Two ATPs are synthesized in this reaction (energy generation phase).

The first two steps involve glycolysis (Embden–Meyerhof pathway), which generates pyruvate. This process occurs in aerobic, facultative anaerobic, sulfate reducer, and fermenting bacteria. The third step involves fermentation, specific to select groups of bacteria. Overall, conversion of glucose to ethanol results in synthesis of two ATPs. Assuming 30 kJ mol^{-1} of ATP, 60 kJ energy is conserved during fermentation. Total amount of energy released during fermentation of glucose to ethanol is 239 kJ mol^{-1}. This suggests that the efficiency of fermentative bacteria in energy conservation is about 25%, and the remaining energy is stored in ethanol or lost as heat.

Fermenting bacteria in wetland soils are largely obligate anaerobes of the following genera (Molongoski and Klug, 1976):

- *Clostridium*
- *Bacteroides*
- *Eubacterium*
- *Peptostreptococcus*

These organisms possess a multitude of fermentative pathways that allow them to utilize different organic substrates, including organic acids, sugars, alcohols, and amino acids. Typical end products of fermentation pathways include carbon dioxide, hydrogen, acetate, formate, butyrate, lactate, propionate, valerate, succinate, and ethanol. The type and amount of these end products depend largely on the type of organic substrate, the microbial species, and the environmental conditions. A second group of fermenting bacteria, referred to as hydrogen-producing acetogenic bacteria utilize volatile fatty acids to produce acetate, hydrogen, and carbon dioxide. Activities of these microbial groups are important in freshwater wetlands, where methanogenesis is a major terminal catabolic process.

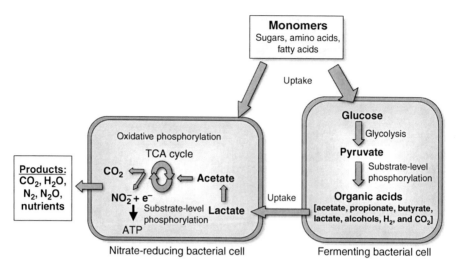

FIGURE 5.30 Biochemical pathways of nitrate-reducing bacteria are similar to aerobic microbial metabolism.

Fermentation of organic substrates is regulated by

- Oxygen concentration: Intermediate compounds of oxygen reduction are toxic to fermentative bacteria because they lack the enzymes superoxide dismutase, catalase, and peroxidase.
- Aerobes and facultative anaerobes: These outcompete most fermenting bacteria for electron donors.
- Buildup of fermentation reaction end products: These products typically include organic acids, alcohols, and hydrogen. In wetlands, fermentation end products are maintained at low levels by various catabolic pathways mediated by microorganisms, including denitrifiers, Mn and Fe reducers, sulfate reducers, and methanogens.

5.5.3.2.2 Nitrate Respiration
Nitrate reducers are one of the most widely studied microbial groups, largely because of their role in N cycling (see Chapter 8 for a detailed discussion on this process). Nitrate reduction in wetlands follows the following pathways:

Denitrification:

$$NO_3^- \to NO_2^- \to NO \to N_2O \to N_2$$

Dissimilatory nitrate reduction to ammonia (DNRA):

$$NO_3^- \to NO_2^- \to NH_4^-$$

Denitrifiers are heterotrophic bacteria (most of them facultative aerobes) that couple oxidation of organic substrates to reduction of NO_3^- to N_2 as the terminal electron acceptor in the ETS (Figure 5.30). This reaction occurs under moderately reduced conditions and in the absence of oxygen. Denitrifiers are capable of utilizing a wide range of substrates including the monomers formed as a result of extracellular enzyme hydrolysis of organic substrates, and the end products of fermentation. All $\Delta G°$ values for the reactions shown below are based on pH = 7. Following are the two examples of such reactions:

Glucose oxidation:

$$5C_6H_{12}O_6 + 30H_2O \to 30CO_2 + 120H^+ + 120e^-$$

$$24NO_3^- + 144H^+ + 120e^- \to 12N_2 + 72H_2O$$

$$5C_6H_{12}O_6 + 24NO_3^- + 24H^+ \rightarrow 30CO_2 + 12N_2 + 42H_2O$$

$$\Delta G° = 2{,}721 \text{ kJ mol}^{-1}$$

Acetate oxidation:

$$5CH_3COO^- + 8NO_3^- + 13H^+ \rightarrow 10CO_2 + 4N_2 + 14H_2O$$

$$\Delta G° = 907 \text{ kJ mol}^{-1}$$

Many different groups of bacteria, including *Bacillus, Pseudomonas*, and *Thiobacillus*, are capable of denitrification. The primary biochemical pathways for organic substrate oxidation by denitrifiers are similar to that described for aerobic catabolism. Because most of the denitrifiers are facultative anaerobes, they possess a functional TCA cycle that allows them to metabolize substrates completely to carbon dioxide and water. Many denitrifiers do not produce extracellular enzymes required for hydrolysis of polymers; thus, they generally rely on hydrolytic enzymes and fermenters to provide readily available substrates (Ljundahl and Erickson, 1985).

Dissimilatory reduction of nitrate to ammonia is performed by obligate and facultative anaerobes with fermentative metabolism, including *Clostridium* and *Bacillus* species (Tiedje, 1988). These organisms, in contrast to denitrifiers, usually do not rely on nitrate as electron acceptor. Therefore, DNRA involves $8e^-$ transfer as compared to $5e^-$ transfer for denitrification, suggesting that more organic substrate can be potentially degraded by DNRA. However, nitrate availability under DNRA conditions is usually very low because much of the nitrate formed during nitrification under aerobic conditions is rapidly consumed by denitrifiers in adjacent anaerobic environments.

The energy yield of glucose oxidation during nitrate respiration is slightly lower than the energy release using oxygen as terminal electron acceptor. Nitrate respiration accounts for approximately 95% of the potential energy released during aerobic catabolism of glucose (100% of potential energy stored in glucose is released during aerobic respiration). Although glucose oxidation during nitrate respiration results in the release of 100% of the energy stored in glucose, 95% of the energy is released for use in ATP synthesis, whereas the remaining 5% is conserved in the reduced end product N_2. This energy is not available, except through biological fixation of N_2 to ammonium and subsequent oxidation of ammonium to nitrate through nitrification. Because this energy availability is quite large, it is not surprising to discover that there are a number of denitrifying bacteria capable of exploiting the use of nitrate as terminal electron acceptor. In wetlands, nitrate does not normally support a significant amount of organic matter oxidation, due to low nitrate concentration in the soil pore water and the water column. In these systems, nitrification is restricted to zones where oxygen is present, which in turn limits nitrate supply for denitrification.

5.5.3.2.3 Manganese and Iron Respiration

As the demand for electron acceptors increases, facultative microbes can utilize oxidized forms of Mn(IV) and Fe(III) as electron acceptors during the catabolic breakdown of organic matter (Figure 5.31). Manganese and iron cycles in wetlands are discussed in detail in Chapter 10. The carbon flow during Mn and Fe reduction has not been studied as extensively as other electron acceptors (oxygen, nitrate, sulfate, and carbon dioxide).

Earlier studies did not differentiate between microbial and chemical reduction of metal oxides. However, in recent studies, microorganisms that can effectively couple degradation of organic matter have been isolated. These organisms were shown to oxidize short-chain fatty acids to carbon dioxide with Fe oxides as electron acceptor (Lovley et al., 1991).

A model for carbon flow through Mn and Fe reduction has been proposed by Lovley et al. (1991). In this model, organic matter degradation involves initial degradation of complex organic matter by hydrolysis, followed by fermentation, which results in the production of several readily

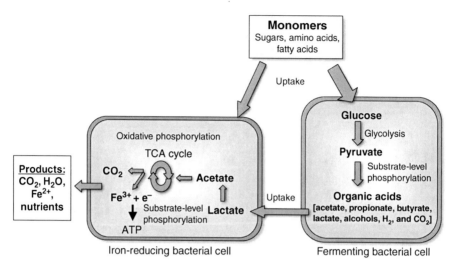

FIGURE 5.31 Microorganisms can couple organic matter oxidation with metal reduction.

utilizable simple organic compounds, which can be utilized by metal oxide–reducing microorganisms. The reduction of Mn and Fe oxides in anaerobic systems is supported by

- Stochiometric relationships between Mn and Fe reduction and carbon dioxide production during organic matter breakdown
- Direct microbial contact with metal oxides
- Accumulation of reduced Mn(II) and Fe(II)
- Inhibition of Mn and Fe reduction by electron acceptors with higher reduction potentials

Reduction of Mn and Fe oxides during oxidation of glucose is shown below:

$$C_6H_{12}O_6 + 12MnO_2 + 24H^+ = 6CO_2 + 12Mn^{2+} + 18H_2O$$

$$\Delta G_r^\circ = -2{,}027 \text{ kJ mol}^{-1}$$

$$C_6H_{12}O_6 + 24Fe(OH)_3 + 48H^+ = 6CO_2 + 24Fe^{2+} + 66H_2O$$

$$\Delta G_r^\circ = -441 \text{ kJ mol}^{-1} \text{ (pH = 6.0)}$$

$$\Delta G_r^\circ = -168 \text{ kJ mol}^{-1} \text{ (pH = 7.0)}$$

Similarly, reactions involving oxidation of acetate using Mn and Fe oxide as electron acceptors are shown in the following example:

$$CH_3COO^- + 4MnO_2 + 9H^+ = 2CO_2 + 4Mn^{2+} + 6H_2O$$

$$\Delta G_r^\circ = -572 \text{ kJ mol}^{-1}$$

$$CH_3COO^- + 8Fe(OH)_3 + 17H^+ = 2CO_2 + 8Fe^{2+} + 22H_2O$$

$$\Delta G_r^\circ = -48 \text{ kJ mol}^{-1}$$

Oxidation of organic substrates through Mn and Fe reduction results in complete breakdown to carbon dioxide and water, through the TCA cycle and ETS (Figure 5.31). For example, acetate is oxidized in the TCA cycle and the electrons are transferred to NAD^+, which is reduced to NADH.

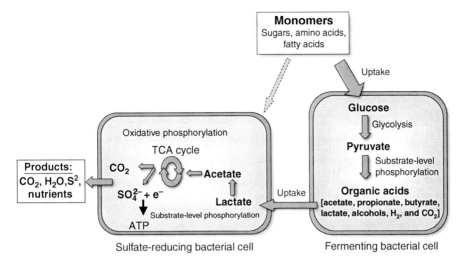

FIGURE 5.32 Organic matter decomposition involving sulfate respiration.

These electrons are subsequently transferred to cytochrome b, and then to the terminal electron acceptor, Fe oxide in contact with the cell membrane. This reaction is mediated by the enzyme Fe reductase. The net amount of energy released during glucose oxidation using Mn(IV) and Fe(III) is approximately 67 and 15%, respectively, of that released from glucose breakdown stored in the reduced end products such as Mn(II) and Fe(II). Reduction of Mn oxides occurs in wetland soils with Eh range of +200 to +300 mV, as compared to −100 to +100 mV for Fe oxide reduction.

5.5.3.2.4 Sulfate Respiration
Sulfate reducers are widely studied groups of microorganisms involved in the catabolic breakdown of organic matter, with special significance in coastal wetland ecosystems (for a detailed discussion on sulfur cycling in wetlands, see Chapter 11). Thermodynamically, sulfate is a much less favorable electron acceptor in the presence of electron acceptors with higher reduction potential including oxygen, nitrate, and Mn and Fe oxides. Sulfate reducers are obligate anaerobes that couple oxidation of organic substrates to carbon dioxide with the reduction of terminal electron acceptor sulfate to sulfides (Widdel, 1988). Sulfate-reducing bacteria cannot synthesize hydrolytic enzymes; thus, they cannot hydrolyze polymers such as polysaccharides and many groups cannot use monomers such as monosaccharides (e.g., glucose) as substrates for energy. As a result, sulfate reducers are dependent on fermenting bacteria to oxidize monomers to simple readily utilizable organic compounds (Figure 5.32).

Only select groups of sulfate reducers can oxidize simple organic compounds to carbon dioxide, whereas others can convert these compounds to acetate (Widdel, 1988).

- Group I sulfate reducers utilize lactate, pyruvate, malate, formate, and alcohols as their energy source but can only convert these compounds to acetate. They do not have TCA cycle and excrete acetate as waste products.

EXAMPLES

Desulfovibrio
Desulfomonas
Desulfotomaculum

- Group II sulfate reducers utilize all of the above-described organic compounds and acetate as their energy source and convert these compounds to carbon dioxide and water.

EXAMPLES

Desulfobacter
Desulfobacterium
Desulfonema

Oxidation of lactate to acetate during sulfate reduction is shown in the following reaction:

$$2\text{lactate} + SO_4^{2-} + 2H^+ = 2\text{acetate} + 2CO_2 + 2H_2O + H_2S$$

$$\Delta G_r^\circ = -88 \text{ kJ mol}^{-1}$$

Glucose and acetate oxidation to carbon dioxide during sulfate reduction is depicted in the following reactions. Glucose oxidation during sulfate reduction may be thermodynamically possible, but many known sulfate reducers cannot use this substrate directly.

$$C_6H_{12}O_6 + 3SO_4^{2-} + 6H^+ = 6CO_2 + 3H_2S + 6H_2O$$

$$\Delta G_r^\circ = -380 \text{ kJ mol}^{-1}$$

$$CH_3COO^- + SO_4^{2-} + 3H^+ = 2CO_2 + H_2S + H_2S + 2H_2O$$

$$\Delta G_r^\circ = -57 \text{ kJ mol}^{-1}$$

Sulfate reducers have a cytochrome-based ETS and ATP production that is linked to oxidative phosphorylation. Some of the ATPs synthesized are used for activation of sulfate to form APS (adenosine 5-phosphosulfate). In the absence of sulfate, these organisms are capable of utilizing other electron acceptors including sulfite, thiosulfate, nitrate (dissimilatory reduction to ammonia), and organic substrates. These reactions occur in soils with Eh values of <-100 mV. The net amount of energy release during glucose oxidation during sulfate reduction accounts for only 15% of the energy released during aerobic respiration. The remaining energy is stored in reduced end products, primarily in sulfides.

5.5.3.2.5 Methanogenesis

Carbon dioxide (including bicarbonate or carbonate) is the common electron acceptor used by a select group of microbes in anaerobic zones of wetlands, where the concentration of other inorganic electron acceptors with higher reduction potential is low. Methanogens are the only carbon dioxide–reducing bacteria in anaerobic environments, and also the major contributors of atmosphere methane. This group of organisms has been extensively studied in microbial ecology (Oremland, 1988). Methanogens are obligate anaerobes and can function in wetland soils with Eh values of <-200 mV.

Depending on substrate utilization, methanogens are divided into three groups (Madigan and Martinko, 2006):

- CO_2-type substrate utilizers: CO_2, CO, and formate
- Methyl substrate utilizers: methanol (CH_3OH); methylamine, $CH_3NH_3^+$; dimethylamine, $(CH_3)_2NH_2^+$, trimethylamine $(CH_3)_3NH^+$; methylmercaptan, CH_3SH; dimethylsulfide, $(CH_3)_2S$
- Acetate utilizers: acetate

Selected samples of methanogenic bacteria using these substrates have been isolated in aerobic soils:

- *Methanobacterium*: ($H_2 + CO_2$, formate)
- *Methanococcus*: ($H_2 + CO_2$, formate)
- *Methanothrix*: (acetate)
- *Methanosarcina*: ($H_2 + CO_2$, formate, methanol, methylamines, acetate)

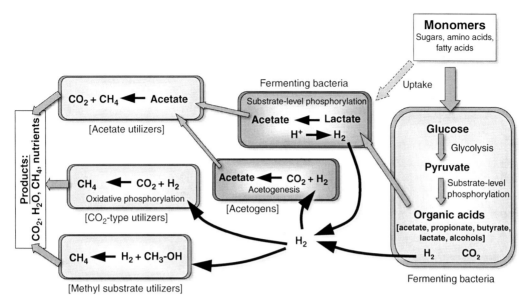

FIGURE 5.33 Pathways involved in organic matter decomposition during methanogenesis.

Methanogens are obligate anaerobes that grow autotrophically (use carbon dioxide as C source and as electron acceptor) and heterotrophically (use organic substrates as energy source). Like sulfate reducers, methanogens cannot directly utilize high-molecular-weight polymers, so methanogens must depend on at least three groups of microbes including hydrolytic, fermentative, and H_2-producing acetogenic bacteria (Figure 5.33) (Capone and Kiene, 1988).

The breakdown of glucose formed as a result of hydrolytic action on polysaccharides, under methanogenic environmental conditions, is described in the following sequence of reactions:

$$C_6H_{12}O_6 + 4H_2O \rightarrow 2CH_3COO^- + 2HCO_3^- + 4H_2 + 4H^+$$

$$\Delta G_r^\circ = -207 \text{ kJ mol}^{-1}$$

$$4H_2 + HCO_3^- + H^+ \rightarrow CH_4 + 3H_2O$$

$$\Delta G_r^\circ = -136 \text{ kJ mol}^{-1}$$

$$2CH_3COO^- + 2H_2O \rightarrow 2CH_4 + 2HCO_3^-$$

$$\Delta G_r^\circ = -31 \text{ kJ mol}^{-1}$$

$$\text{Sum: } C_6H_{12}O_6 + 3H_2O \rightarrow 3CH_4 + 3HCO_3^- + 3H^+$$

The biochemistry of methanogenesis varies among different methanogens, depending on substrate utilization. For example, autotrophic methanogens use carbon dioxide as electron acceptor and a carbon source with H_2 as primary electron donor. In this process, reduction of carbon dioxide involves sequential addition of hydrogens catalyzed by coenzymes unique to methanogens including methanofuran, methanopterin, and methyl coenzyme M, which is finally reduced to methane by the methyl reductase system. Production of methane through autotrophic reaction results in a net energy yield of -134 kJ mol^{-1}.

Heterotrophic methanogens use acetate directly for biosynthesis and as an energy source. During this process, methanogens convert acetate to carbon dioxide and methane by utilizing a different biochemical pathway than the autotrophic methanogens. The sequence of reactions involved are as follows (Madigan and Martinko, 2006): First, acetate is activated to acetyl-CoA, which then interacts with carbon monoxide dehydrogenase, following which the methyl group of acetate is transferred to vitamin B_{12}. Second, the methyl group is transferred to methanopterin, and then to coenzyme M to yield CH_3-CoM. Third, CH_3-CoM is then reduced to methane using electrons generated during oxidation of carbon monoxide to carbon dioxide, by the carbon monoxide dehydrogenase; conversion of acetate to CH_4 results in a net energy yield of -130 kJ mol^{-1}.

For both autotrophic and heterotrophic groups, electron and H^+ transfer processes result in a proton gradient across the microbial membrane, which drives the synthesis of ATP. Autotrophic and heterotrophic methane production results in a net yield of approximately 10% of the energy yield from aerobic respiration. As a result, methanogenesis is often the dominant terminal catabolic process involved in the breakdown of organic matter in freshwater wetlands. Organic matter turnover in freshwater wetlands may be limited by the supply of electron acceptors with higher reduction potentials such as oxygen, nitrate, iron and manganese oxides, and sulfate. Methanogens use carbon dioxide and acetate produced during decomposition of organic matter and thus may not be limited by electron acceptors.

Fermentation of simple sugars results in formation of simple organic compounds such as lactate, succinate, and ethanol. These compounds can be fermented further by other bacteria to acetate, carbon dioxide, methane, and hydrogen. For example, conversion of ethanol to hydrogen and acetate results in an unfavorable positive energy yield.

$$2CH_3CH_2OH + 2H_2O \rightarrow 4H_2 + 2CH_3COO^- + 2H^+$$
$$\Delta G_r^\circ = +42 \text{ kJ mol}^{-1}$$

However, hydrogen produced by this organism can be favorably used by methanogenic bacteria, with a net energy yield.

$$4H_2 + CO_2 \rightarrow CH_4 + 2H_2O$$
$$\Delta G_r^\circ = -143 \text{ kJ mol}^{-1}$$

On combining these two reactions we obtain:

$$2CH_3CH_2OH + CO_2 \rightarrow CH_4 + 2CH_3COO^- + 2H^+$$
$$\Delta G_r^\circ = -101 \text{ kJ mol}^{-1}$$

The process of coupled reaction is called *interspecies hydrogen transfer*, where hydrogen produced by a fermenter is used by methanogenic bacteria as an energy source. This relationship between two organisms is called syntrophy (Madigan and Martinko, 2006). Through this process, methanogens can aid in decreasing hydrogen levels allowing the fermenting bacteria to gain more energy from oxidation of organic substrates. The rate-limiting step in methanogenesis is the availability of hydrogen and acetate formed during fermentation. Once these substrates are formed, they are rapidly used by methanogens to produce methane.

5.5.3.3 Aerobic vs. Anaerobic Catabolism

The following is a brief summary of the comparison between aerobic and anaerobic catabolic processes:

- Many of the biochemical pathways involved in the breakdown of organic matter are similar for both aerobes and anaerobes because most pathways do not require oxygen.
- Both aerobes and anaerobes are capable of glycolysis (glucose degradation) and derive energy through substrate-level phosphorylation.

- Both aerobic and anaerobic microbes generate most of the ATP by oxidative phosphorylation. However, the terminal electron acceptor, energetics, and kinetics are different for aerobic and anaerobic microorganisms.
- Fermenting organisms generate most of their ATP by substrate-level phosphorylation.
- The ability of aerobes to incorporate oxygen into organic molecules via oxygenases allows these microbes to utilize aromatic and saturated alkanes with no functional groups (e.g., benzene).
- Aerobic catabolism results in complete conversion of organic substrates to carbon dioxide and water, resulting in the release of 100% of the energy stored in monomers. Anaerobic catabolism results in the storage of a large proportion of the energy released from organic substrates in reduced end products.

5.6 ORGANIC MATTER TURNOVER

Organic matter accumulation in wetlands is a function of the balance between net primary productivity and abiotic and biotic decomposition processes. Organic matter undergoes complex cycling in wetlands, and its fate depends on substrate quality and various environmental factors (see Section 5.7 for details). In general, easily degradable (labile) fractions are decomposed rapidly, whereas recalcitrant pools are accreted as new organic matter layers in soils. A range of laboratory and field techniques have been used to determine the rate of organic matter decomposition and turnover in wetlands. Studies may include laboratory batch incubation where a known amount of plant litter or soil organic matter is incubated under either aerobic or anaerobic conditions, and gaseous end products such as carbon dioxide and methane are measured as a function of time. In some of these studies, labeled carbon substrates with ^{13}C or ^{14}C are used to track the fate of the added compound. Field studies include the measurement of dry weight loss of known amount of plant litter placed in a litterbag. Loss in dry weight of detrital matter in these bags is assumed to be due to decomposition.

Accumulation of organic carbon in wetlands can be described in the simplest mathematical terms as follows:

$$\frac{dC}{dt} = P - kC$$

where C represents mass of organic carbon storage in the wetlands, k is a first-order decay rate constant (per time), and P is net primary production or organic carbon addition to the system. Under most conditions, we can assume that allochthonous input of organic carbon is negligible compared to primary productivity in wetland ecosystems. If the rate of organic matter production (net primary production) is approximately constant and k describes the overall rate of organic carbon decomposition in the system, then a steady-state level of carbon storage and the time required to attain that level may be predicted using the integrated form of the following equation:

$$C(t) = P/k(1 - e^{-kt})$$

It should be apparent that the model for organic carbon accumulation as shown by these two equations is not adequate to describe the complete process.

5.6.1 ABIOTIC DECOMPOSITION

Abiotic decomposition can be referred to as physical processes attacking the detrital matter, which may include

- Initial rapid loss due to leaching
- Mechanical fragmentation
- Degradation of DOM by ultraviolet radiation (UV 200–400 nm).

Rapid leaching of soluble inorganic and organic materials can take place when detrital matter is immersed in the water column (Webster and Benfield, 1986). Soluble organic compounds may include sugars, organic acids, proteins, phenolics, and others, whereas inorganic materials may include potassium, calcium, magnesium, and other ions). Leaching is rapid during the first few hours followed by a gradual decline for an extended period. Turbulence, wave action, and high flow rates can accelerate leaching losses. Leaching can occur in both detrital matter attached to the live plant and the detrital matter detached and deposited on the soil surface. Physical fragmentation can also occur by benthic invertebrates.

Photolysis refers to abiotic oxidation or mineralization of DOM where high-energy solar radiation (UV light) breaks down or alters the makeup of DOM in surface waters. This process is often referred to as photooxidation or photomineralization or photochemical degradation. The ultraviolet light is highly energetic and is defined in three groups:

- Ultraviolet A (UVA) (320–400 nm)
- Ultraviolet B (UVB) (280–320 nm)
- Ultraviolet C (UVC) (200–280 nm)

The types of UV radiation most pertinent for photolysis are UVA and UVB, whereas UVC has low transmittance through atmosphere. Only a portion of DOM may be subjected to UV breakdown. This pool is referred to as chromophoric dissolved organic matter (CDOM). The molecular components, chromophores, absorb UV radiation, and conjugation of chromophores results in absorption of solar radiation >290 nm (Wetzel, 2001). The color of the CDOM is primarily associated with humic and fulvic acids of the colloidal or high-molecular-weight fractions. As the CDOM is exposed to UVA and UVB light, the bonds between carbon molecules absorb energy and are electronically excited, which leads to modification of the bonding structure of the high-molecular-weight molecules. The bonds between molecules can be broken if the energy absorbed exceeds the energy of formation for these chemical bonds. The partial photodegradation of DOM can result in the formation of several smaller organic molecules and increase their bioavailability. Other reactive species such as hydroxyl radicals and hydrogen peroxide, and other reactive moieties can also be formed as a result of photolysis (Obernosterer et al., 2001a, 2001b). Photochemical reactions can convert DOM into carbon dioxide, carbon monoxide, ammonia, and phosphate, with loss in absorption of UV absorbance.

Complete photolysis:

$$CDOM \rightarrow CO_2$$

Incomplete photolysis:

$$CDOM \rightarrow \text{Lower-molecular-weight compounds}$$

The major portion of the DOM essentially consists of humic substances that are recalcitrant and may not be bioavailable for bacterial utilization. Photochemical reactions can potentially convert some of these humic substances to lower-molecular-weight compounds, which can be biologically available for microorganisms. For example, the DOM leachate from *Juncus* detrital matter exposed to UVB light resulted in photolytic decoupling of humic substances with the production of various bioavailable fatty acids such as acetate, citrate, formate, levulinate, pyruvate, and others (Wetzel, 2001). Large macromolecules of humic substances bound to inorganic materials such as calcium carbonate particles and other particles constitute colloidal fraction of DOM. Colloidal fraction may be high in bog waters and hard waters rich in DOM. These waters are chromophoric and tend to absorb UV light; however, their absorbing characteristics decrease with time as DOM undergoes photooxidation. At a global scale, the increase in UV radiation as a result of ozone depletion in the atmosphere can have a significant effect on the bioavailability of DOM and the productivity of biota in surface waters of wetlands and aquatic ecosystems. Photolytic oxidation in surface waters in wetlands can be inhibited by shading resulting from vegetation and trees, as compared to areas with open waters.

Carbon

5.6.2 BIOTIC DECOMPOSITION

During biotic decomposition of organic matter, microorganisms use organic substrates as energy, resulting in the emission of carbon dioxide and methane. In wetland ecosystem, this process occurs at different rates, with high rates observed under aerobic conditions and slow rates under anaerobic conditions. Depending on the hydroperiod, wetlands offer a range of redox conditions. The sequence of decomposition processes involved in converting complex polymers to simple monomers and eventually to carbon dioxide and methane have been discussed in earlier sections of this chapter. In this section, we will focus on selected laboratory and field studies to determine the turnover of organic matter under a range of conditions. Various approaches have been used to determine the decomposition processes under laboratory and field conditions.

Under laboratory conditions, typically soils are incubated at predetermined temperatures and headspace gases (carbon dioxide and methane) are monitored as a function of time (Moodie and Ingledew, 1990; Bridgham and Richardson, 1992). In some laboratory experiments, carbon dioxide produced is flushed continuously and trapped in alkaline solutions and solutions are subsequently analyzed for inorganic carbon. Soil oxygen demand (SOD) is also used as proxy for microbial respiration. However, it is important to distinguish the BOD from chemical oxygen demand. BOD includes not only heterotrophic respiration but also oxidation of reduced compounds such as ammonium and methane by autotrophic nitrifiers and methanotrophs, respectively (see Chapter 6). The chemical demand is due to the oxidation of reduced iron, manganese, and sulfides. Use of SOD can potentially overestimate microbial respiration rates in soils with high levels of inorganic reduced compounds. These laboratory incubation experiments are now routinely used to determine the microbial respiration under aerobic or anaerobic conditions.

Simple incubation techniques are described in various manuals and research papers for measuring microbial activity or respiration in soils and sediments. In the following example, Benner et al. (1985) measured rates of mineralization of the [^{14}C-lignin]lignocellulose and synthetic [^{14}C]lignin (dehydrogenative polymerizate [DHP]) in salt marsh and freshwater sediments (Figures 5.34 and 5.35). During 25-day incubation period at 28°C, decomposition rates were two to three times faster in salt marsh sediments than in freshwater sediments (Figure 5.34). In salt marsh soils, the lignin components of lignocelluloses from *Spartina alterniflora* and *Carex walteriana* were mineralized three and four times faster, respectively, than the lignin component of *Rhizophora mangle* and synthetic lignin. In freshwater soils

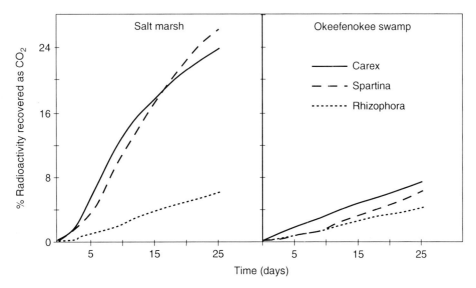

FIGURE 5.34 Rates of mineralization of lignocellulose in salt and fresh (Okefenokee) water system (Redrawn from Benner et al., 1985).

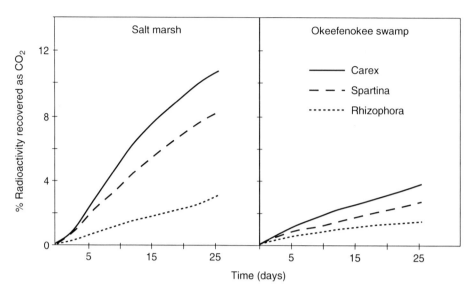

FIGURE 5.35 Rates of mineralization of synthetic lignin in salt and fresh (Okefenokee) water system (Adapted from Benner et al., 1985).

of Okefenokee swamp, the differences among rates were not significant (Figure 5.35). Similarly, each of the polysaccharide components of the lignocelluloses was mineralized more rapidly in salt marsh soils than in swamp soils (Figure 5.35). Low decomposition rates in swamp soils of Okefenokee were attributed to the low pH and low bioavailability of nutrients as compared to salt marsh soils.

The substrate-induced respiration (SIR) method was developed as a rapid means of estimating microbial activities in soils (Beare et al., 1990; Neely et al., 1991). The use of selective inhibitors such as streptomycin for bacteria and cycloheximide for fungi, in conjunction with substrate additions, has been used to quantify bacterial and fungal contributions to the total metabolism of microbial decomposers. The SIR procedure involves addition of a labile carbon source (e.g., glucose or acetate) to provide a carbon nonlimiting condition. The short-term increase in carbon dioxide production is proportional to the active microbial biomass and activity. The concept of addition of labile carbon to determine the kinetics of substrate utilization by microorganisms has been extensively studied in various ecosystems including wetlands and aquatic systems.

Under field conditions, the most common method used by biogeochemists and ecologists is the litterbag method. Litterbags with mesh sizes ranging from less than 1 mm to more than 10 mm have been used. Although there are several limitations to this method, investigators have consistently used this method because of its simplicity. Litterbags potentially alter the physical environment and exclude grazing benthic communities. Some of the DOC and fine POC can be lost from the system, thus resulting in overestimation of the decomposition rates. In field studies, the decomposition rates are expressed as mass loss rates using the first-order rate equation, with the basic assumption that the decomposition rate is in proportion to the amount remaining; they are expressed as follows:

$$C(t) = C_0 e^{-kt}$$

The linear form of the above equation is

$$\ln\left(\frac{C_t}{C_0}\right) = -kt$$

where C_t represents the amount of carbon remaining in the substrate at each decomposition period, C_0 is the initial carbon content of the substrate, k is the mass loss rate constant (expressed per time), and t is the time for the decomposition period (in days or months or years). For details, the reader is referred to a comprehensive review by Gopal and Masing (1990), Webster and Benfield (1986), Boultin and Boon (1991), and Chimney and Pietro (2006). The review by Boultin and Boon (1991) provides a critical review of field methods used in litter decomposition studies and offers some useful guidelines for using these techniques.

The k values presented in Tables 5.13a and 5.13b represent examples derived from various studies. The simple exponential decay does not consider the labile pools of carbon, which are either

TABLE 5.13a
Decomposition of Plant Matter as Determined by Mass Loss Using Litterbag Techniques

Plant	Decomposition Rate, k (day^{-1})
Chara	0.0041–0.1045
Azolla	0.0097
Eichhornia	0.0055–0.0382
Ceratophyllum	0.0192–0.0213
Hydrilla	0.02–0.0752
Elodea	0.026–0.0912
Najas	0.0296–0.0328
Potamogeton	0.0017–0.0963
Vallisneria	0.025–0.164
Lemna/Spirodela	0.011–0.35
Nymphoides	0.044–0.091
Phragmites	0.0012–0.0045
Panicum	0.0021–0.052
Polygonum	0.0039–0.0211
Typha	0.0008–0.0086
Scirpus	0.0018–0.0066
Spartina	0.0025
Juncus	0.0013–0.0097
Paspalum	0.0072–0.0187
Carex	0.0029–0.0046
Zizania	0.077

Source: Gopal and Masing (2006).

TABLE 5.13b
Decomposition of Plant Matter as Determined by Mass Loss Using Litterbag Techniques

Species	N	Mean (day^{-1})	Median (day^{-1})	CV (%)	t_{50} (days)
All emergent species	280	0.0083	0.0022	259	83
All floating species	80	0.0382	0.0243	88	18
All submerged species	107	0.0473	0.0280	111	15

Source: Chimney and Pietro (2006).

leached or decomposed rapidly. However, the decomposition of lignocellulose fractions is fairly well described by the first-order decay model. The decomposition of plant detrital matter occurs in two phases, with the first phase representing rapidly decomposable fraction and the second representing a slowly decomposable lignin fraction. The early stages of plant litter decomposition are characterized by a relatively rapid loss of soluble sugars and cellulose (Melillo et al., 1989) (Figures 5.36 and 5.37). In labile plant detrital matter, the time required for phase 1 decomposition can be shorter as shown for *Eichhornia* and *Pistia* in two riparian wetlands of the Okeechobee drainage basin in Florida (Reddy et al., 1995).

Because of the high residence time of standing dead tissue of emergent macrophytes such as *Typha* and *Cladium*, a significant degree of decomposition occurs by the time the leaf material enters the soil detrital layer. Leaching and mineralization of water-soluble, nonstructural components

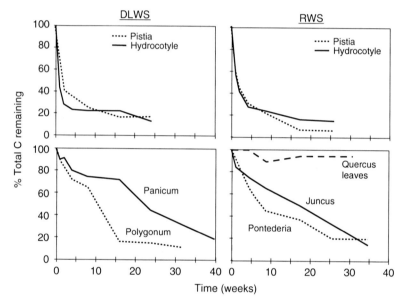

FIGURE 5.36 Plant litter decomposition in two wetlands of the Okeechobee drainage basin, Florida (Reddy, unpublished results).

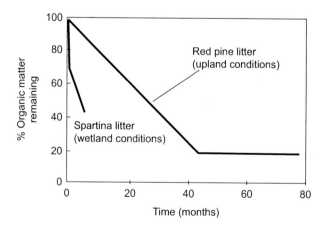

FIGURE 5.37 Difference in litter decomposition pattern represents the rapid loss of C by labile fractions and slowly decomposable lignin fraction (Melillo et al., 1989).

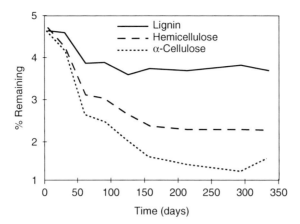

FIGURE 5.38 Rate of decomposition of different carbon fractions from emergent macrophytes. (Adapted from Moran et al., 1989.)

proceed rapidly after senescence. A significant loss of structural components (lignocellulose) may also occur in standing dead material. The recalcitrance of standing dead *Typha* and *Cladium* leaves in Everglades was demonstrated in a 2-year decomposition study using mass loss from litterbags placed on the soil surface (Davis, 1991). First-order rate constants calculated from these data were on the order of 0.001 day^{-1}, indicating that only the more recalcitrant components remained in the standing dead material.

Decomposition rate constants are negatively correlated to the total fiber content of the organic substrate but not well correlated to individual components such as cellulose, hemicellulose, and lignin. A field and laboratory study by Moran et al. (1989) measured the decomposition rates of whole litter and the lignocellulose components of the litter for the emergent macrophytes *S. alterniflora* and *C. walteriana*. Decomposition of individual carbon fractions varied with the rate of decomposition in the order of cellulose > hemicellulose > lignin (Figure 5.38, Moran et al., 1989).

5.7 REGULATORS OF ORGANIC MATTER DECOMPOSITION

The decomposition of organic matter in wetlands differs from that in adjacent upland ecosystems in a number of ways. The predominance of aerobic conditions in uplands generally results in rapid decomposition of organic matter. Net retention of organic matter is minimal and comprises accumulation of highly resistant and stable compounds. Wetlands provide a range of redox conditions, due to frequent-to-occasional anaerobic conditions resulting from flooding. As a result, significant accumulation of moderately decomposable organic matter occurs, in addition to lignin and other recalcitrant fractions (DeBusk et al., 2001).

Several factors control the decomposition of organic matter in wetlands (Figure 5.39). Although loading of anthropogenic nutrients stimulates the growth of vegetation in wetlands, a significant portion of nutrient requirements may be met through remineralization during decomposition of organic matter. The rate of organic matter turnover and nutrient regeneration is influenced by hydroperiod, characteristics of organic substrates, supply of electron acceptors, addition of limiting nutrients, temperature, and pH. Some of the important factors include

- Quality and quantity of organic matter
- Microbial communities and biomass
- Water table or soil aeration status
- Availability of electron acceptors with higher reduction potentials
- Temperature

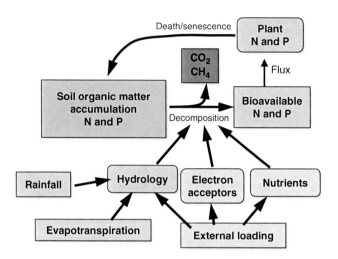

FIGURE 5.39 Schematic diagram of regulators of organic matter decomposition.

5.7.1 QUALITY AND QUANTITY OF ORGANIC MATTER

Quality of organic matter, also referred to as substrate quality, is determined by a combination of physical and chemical characteristics that determine the utilization by microorganisms. Substrate quality is not defined by any one of the characteristics; however, lignin and nitrogen content have been suggested as indicators of biodegradability. The influence of initial substrate quality on decomposition rate is evidenced by a wide variation in rates observed among plant species and among different parts of the same plant. Detrital plant matter is the major source of POM in wetlands. Approximately 50–80% of this material contains lignocellulose.

Plant materials high in cellulose, hemicellulose, and sugars and low in lignin content undergo decomposition at a faster rate than materials with high lignin content. The presence of lignin is the ultimate limiting factor in decomposition of plant matter. Lignin occurs in cells of conductive and supportive tissue and thus is not found in algae and mosses. Recently senesced labile organic matter and root exudates can have a significant impact on the decomposition of stable organic matter, a mechanism commonly known as "priming effect." This effect is more pronounced during the first few days of decomposition of labile material. This concept has been demonstrated in many agricultural soils receiving crop residues but not shown in many wetlands and other natural systems.

When describing the process of decomposition of plant detritus, it is not sufficient to consider only the disappearance of the original substrate, and under most conditions it is difficult to trace the fate of the original plant material. Decomposing plant detritus and soil organic matter represent a composite of original substrate including altered microbial communities and biomass. The relative proportion of living and nonliving components of the organic substrate changes continuously as the decomposition proceeds. As shown in Figure 5.40, the cellulose content decreased significantly in the order of standing dead plant tissue > litter layer > surface peat > subsurface peat (DeBusk and Reddy, 1998). The lignin content of samples increased significantly in the order of standing dead plant matter < litter layer < surface peat < subsurface peat. These trends reflect the age of organic matter. The cellulose content decreases whereas the lignin content increases with the age of organic matter.

The concept of a "decay continuum" proposed by Melillo et al. (1989) is reflected in the lignocellulose content of the substrate. The lignocellulose index (LCI) is now used as an indicator of substrate quality, which is defined as the ratio between the lignin content of the substrate and the total of lignin and cellulose content. During the initial stages of decomposition, the LCI of the substrate is low and increases as the material undergoes decomposition and becomes a part of the soil organic matter. For example, the LCI can range from approximately 0.2 in high-quality litter

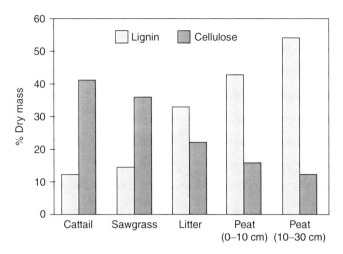

FIGURE 5.40 Age of the organic matter is reflected by the changes in the relative fractions of lignin and cellulose content. (Adapted from DeBusk and Reddy, 1998.)

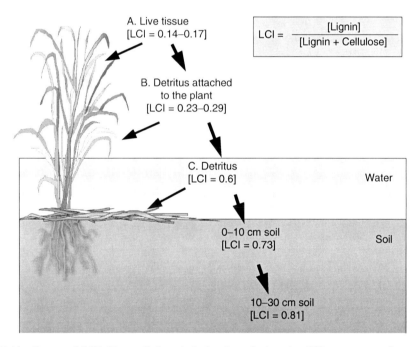

FIGURE 5.41 Range of LCI (lignocellulose index) values during the different stages of organic matter decomposition. (Data from DeBusk, 1996.)

to about 0.6 in low-quality litter (Figures 5.41 and 5.42). After an initial phase of decomposition in which most of the simple compounds including sugars and amino acids, and some of the cellulose and hemicellulose are lost, the LCI of the plant matter converges to a range of 0.7–0.8 found typically in stable soil organic matter (Figures 5.41 and 5.42). As the LCI increases, the plant detrital matter becomes increasingly resistant to microbial attack.

The estimated LCI for soil organic matter and litter in various forest and agricultural soils was reported to be in the range of 0.45–0.8 (Melillo et al., 1989). The implication of this is that after

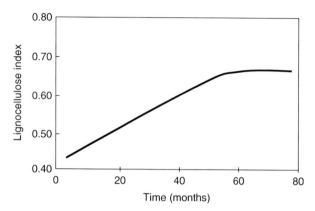

FIGURE 5.42 Increasing trend of LCI (lignocellulose index) with time is observed in decomposition of organic matter (Melillo et al., 1989).

the initial stages of decomposition, the rate of decomposition is regulated only by environmental conditions and is no longer a function of initial substrate composition. The concurrent decrease in labile or easily decomposable fractions and an increase in recalcitrant fractions of the organic matter lead to an overall decrease in decomposition rate. The decomposition process generally proceeds to a point where residual organic matter consists of only humic substances bearing little resemblance to the original material. The time required to reach this stage in wetlands depends on various environmental factors including the hydroperiod.

5.7.2 Microbial Communities and Biomass

Wetlands host a complex group of microbial communities in the soil and water column. These may include heterotrophic, chemotrophic, and phototrophic microbes. Much of the biogeochemical cycling that is critical to organic matter decomposition in wetlands is controlled by these groups. Considerable research has been conducted in recent years to define the energetic controls on carbon cycle in wetlands, and to gain a basic knowledge of the ecology and physiology of individual groups of many of the microorganisms responsible for organic matter turnover under a range of redox conditions. The diversity and activity of these microorganisms are dependent on the quality and quantity of electron donors and electron acceptors, and other related factors discussed as follows.

Microbial communities found in wetlands have been described as one-dimensional and respond to vertical gradients caused by the successive depletion of electron acceptors (see Chapter 3 for details and see Section 5.7.4). For example, depending on the oxygen availability, the surface layers (water column, detrital layers, surface soils) may support aerobic organisms including bacteria and fungi. Below the zone of oxygen depletion, other electron acceptors may support facultative and obligate anaerobes, whose activity is much lower than that of aerobic microorganisms.

Fungi have been characterized as the most important group of organisms responsible for plant litter decomposition in terrestrial and aquatic ecosystems. The relative importance of fungi in organic matter decomposition is not well documented for wetland environments. In acid peatlands and bogs, fungi play an important role in aerobic mineralization of soil organic matter. Decomposition of standing dead material and detrital matter in the water column may be dominated by fungi. Basidiomycetes (white-rot fungi) are the predominant decomposers in terrestrial ecosystems (Zeikus, 1981) and may occur in wetlands under favorable conditions such as lowered water table and exposure of surface soils and detrital matter to oxygenated conditions.

A hypothetical scheme of the development of a microbial community during cellulose decomposition in waterlogged soils is shown in Figure 5.43 (Saito et al., 1990). During cellulose decomposition,

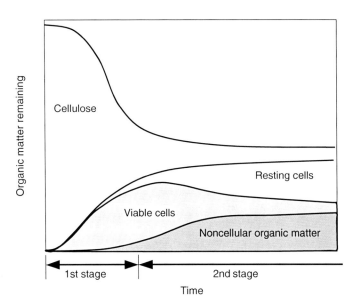

FIGURE 5.43 Conceptual model showing development of microbial community during cellulose decomposition in waterlogged soils (Adapted from Saito et al., 1990).

microbial succession can be represented in two stages. In the first stage, a few types of cellulolytic microorganisms predominate on the cellulose. Vigorous decomposition of the cellulose is accompanied by a rapid increase in microbial biomass, and H_2 is evolved from the microbial community on the cellulose. In the second stage, the rate of cellulose decomposition is typically very slow. The remaining cellulose is thickly covered with various types of microorganisms. The H_2 produced is consumed by microorganisms closely adhering to the remaining cellulose. In addition, noncellular organic nitrogen accumulates on the remaining cellulose. A large part of the microorganisms seems to be dormant at this stage.

Soil microbial biomass regulates the transformation and storage of nutrients. Because of its dynamic nature, it has the potential to be a sensitive indicator to detect changes resulting from anthropogenic impacts or disturbance to an ecosystem. The amount of MBC in detrital layer and soils depends on physicochemical characteristics and other associated environmental factors. Soils with high organic matter content typically have large amount of MBC. MBC was found to be higher in aerobic soils (2% of TOC) than in anaerobic soils (0.1–0.8% of TOC) (McLatchey and Reddy, 1998).

Indices based on microbial activity, microbial biomass, SIR, and organic carbon have been proposed to provide an operationally defined means for describing the response of soil microbial populations to substrate quality and environmental conditions (see Chapter 15 for details). The ratio of MBC to soil organic carbon (C_{mic}/C_{soc}) has been related to the soil organic carbon availability and the tendency of a soil to accumulate organic matter (Anderson and Domsch, 1989; Sparling, 1992). The ratio of basal microbial respiration rate per unit of MBC has been used as an index for specific respiration rate and is termed as the metabolic quotient (qCO_2-C). Microbial activity (expressed as aerobic respiration) was shown to be directly related to the microbial biomass in soils and detrital matter of Everglades soils (Figure 5.44). This represents the qCO_2-C of approximately 0.07 day^{-1} and a turnover period of approximately 15 days. The qCO_2-C has been used as a response variable to determine the effects of nutrient loading, temperature, soil management, ecosystem succession, and contaminant stress (Anderson and Domsch, 1993). Increased qCO_2-C is associated with a high resource availability, whereas lower qCO_2-C represents a low-nutrient matured system with low resource availability.

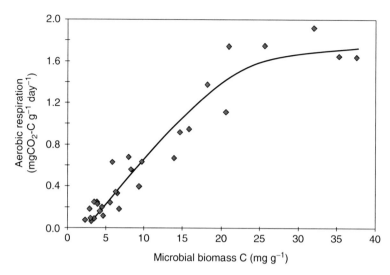

FIGURE 5.44 Relationship between the aerobic respiration in soils and microbial biomass carbon (DeBusk and Reddy, 1998).

5.7.3 Water Table or Soil Aeration Status

In wetlands, the presence of floodwater or saturated soil conditions limits the availability of oxygen in the soil profile; therefore, decomposition frequently proceeds at a much reduced rate with less favorable electron acceptors. Hydroperiod, which encompasses frequency, duration, and depth of flooding, is thus a primary factor governing the degree of organic matter accumulation in wetlands. Hydrologic fluctuations and hydroperiod can result in varying cycles of wetting and drying or flooding and draining. Decomposition of organic matter is affected by soil moisture or water table depth. This is primarily a reflection of the availability of oxygen. Under drained soil conditions, introduction of oxygen into the soil profile can result in rapid oxidation of organic matter. Drained soil conditions or lower water table depths promote the activity of aerobic microorganisms including bacteria and fungi.

Extracellular enzyme activities and microbial activities are affected by water level drawdown. Water level drawdown and introduction of oxygen into the soil profile were shown to increase the levels of phenol oxidase, an enzyme critical for degradation of phenolic compounds (Freeman et al., 2001a, 2001b). This dramatic effect on phenol oxidase activity and associated organic matter oxidation has been shown in peatlands that are subjected to water level drawdown. Low soil oxygen (high soil moisture) inhibits phenol oxidase activity, which potentially results in the accumulation of phenolic compounds (Freeman et al., 2001b). The phenolics inhibit the activity of other enzymes such as β-glucosidase, which is not oxygen limited, thus affecting overall organic matter decomposition. In Florida Everglades peatlands, decomposition rates of organic soils were significantly lowered when the water table depth was raised from 25 to 5 cm below soil surface (Volk, 1973). Similarly, carbon dioxide flux increased with decreasing water table in intact peat cores from Canadian wetlands (Moore and Dalva, 1993) and Florida's wetlands (Reddy et al., 2006) (Figure 5.45).

Decomposition of organic matter is rapid under aerobic conditions, which predominate in the water column and surface soil layers. These decomposition rates typically decrease with soil depth. Carbon dioxide fluxes from soil into the atmosphere increase with lowered water table depth or drained soil conditions. An enhanced rate of soil respiration in response to both drainage and nutrient enrichment has been previously reported for organic soils. For example, the interaction between hydrology and nutrient enrichment is graphically shown for the Everglades soils in Figure 5.46 (DeBusk and Reddy, 2003).

Carbon

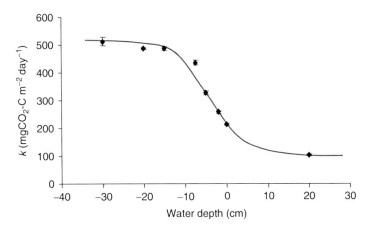

FIGURE 5.45 Effect of water table depth on organic matter decomposition and carbon dioxide evolution from peat soils (Reddy et al., 2006).

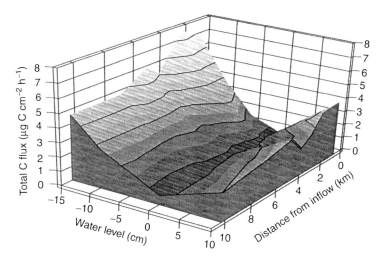

FIGURE 5.46 3-D surface plot showing interaction between nutrient enrichment and hydrology in the Everglades soils. (From DeBusk and Reddy, 2003.)

Introduction of oxygen influences a range of oxidation and reduction reactions related to cycling of various elements (as discussed in others chapters of this book). A few examples are shown below:

Introduction of oxygen as a result of lowering water table inhibits the following biogeochemical processes:
- Denitrification (reduction of nitrate)
- Dissimilatory reduction of iron and manganese compounds
- Sulfate reduction
- Methanogenesis (reduction of carbon oxide)

Introduction of oxygen promotes the following biogeochemical reactions (see Chapter 6):
- Nitrification (ammonium oxidation)
- Oxidation of reduced iron and manganese
- Sulfide oxidation
- Methane oxidation

5.7.4 Availability of Electron Acceptors with Higher Reduction Potentials

In wetland soils, organic matter decomposition is frequently limited by electron acceptor availability, rather than carbon availability as in upland ecosystems. The concentration and type of electron acceptors available in soils determine the types of microbial communities involved and the rate of decomposition process. Much of the detrital matter produced in wetlands is deposited on the soil surface. It is unlikely that there is enough oxygen in this matrix to decompose this material. Therefore, the decomposition of detrital matter is also dependent on the activity of anaerobic microorganisms using alternate electron acceptors. Similarly, the rate of organic matter decomposition in soils is dependent on the availability of electron acceptors (see for discussion in Chapters 3 and 4).

As discussed in earlier chapters, thermodynamic calculations indicated a progressively lower energy yield during microbial respiration coupled to sequential reduction of various inorganic electron acceptors in the order of oxygen > nitrate > manganese oxides > iron oxides > sulfate > carbon dioxide. Relative roles of these electron acceptors in organic matter decomposition depend on not only the quality of the organic substrate but also the capacity (or total amount) of inorganic electron acceptors. Under continuously flooded conditions, more energetically favored electron acceptors are consumed in surface soil layers of the profile (see Chapter 4 for details). Relative role of these electron acceptors in organic matter decomposition is presented in other chapters of this book.

Supply of electron acceptors is influenced by alternating wetting and drying cycles. Extent of organic matter decomposition decreased as the length of the anaerobic period increased (Reddy and Patrick, 1975). At the end of 4-month incubation, decomposition rates of organic matter were found to be the same as when a Mississippi River floodplain soil was incubated under either completely aerobic conditions (oxygen as the only source of electron acceptor) or 2–2 days or 4–4 days aerobic–anaerobic conditions (Figure 5.47). The most probable electron acceptors during short-term anaerobic conditions are probably nitrate and manganese and iron oxides.

It is evident that oxygen-, nitrate-, sulfate-reducing and methanogenic conditions have a profound effect on various biogeochemical properties regulating organic matter decomposition in wetland soils (Table 5.14). A review on the comparison of microbial dynamics in marine and freshwater system as influenced by the availability of electron acceptors is presented by Capone and Kiene (1988). Oxygen supply or availability is linked to hydrology and hydroperiod of a wetland. Under continuously flooded conditions, oxygen availability is restricted to water column and surface soil layers, and oxygen transport by plants into the root zone (Chapters 6 and 7). In the sequential reduction process, nitrate is the next favored electron acceptor. However, its role in organic matter

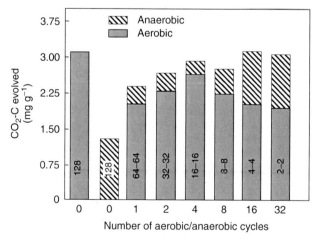

FIGURE 5.47 Effect of wet and dry cycles on decomposition of organic matter (Reddy and Patrick, 1975).

TABLE 5.14
Selected Biogeochemical Properties Related to Soil Organic Matter Decomposition in Wetland Soils Incubated under Various Redox Conditions

Biogeochemical Properties	Oxygen	Nitrate	Sulfate	Bicarbonate
Redox potential (mV)	620	310	−100	−220
pH	4.6	7.5	7.4	6.7
Total carbon (g kg^{-1})	392	388	402	414
Microbial biomass carbon (g kg^{-1})	10.1	3.4	0.7	0.35
β-Glucosidase (mg p-nitrophenol kg^{-1} h^{-1})	75	21	19	14
Phenol oxidase (mmol DQC kg^{-1} h^{-1})	79	ND	ND	ND
Dehydrogenase (mg formazan kg^{-1} h^{-1})	0.41	0.51	2.37	1.79
Basal respiration (mg C kg^{-1} day^{-1})	400	120	111	51
SIR (glucose added) (mg C kg^{-1} day^{-1})	1,238	280	229	146
SIR (acetate added) (mg C kg^{-1} day^{-1})	1,238	280	229	146

Note: Soil redox conditions were maintained by addition of nonlimiting supply of electron acceptors. DQC = dihydroindole quinone carboxylate; ND = not detected; SIR = substrate-induced respiration (McLatchey and Reddy, 1998).

decomposition is minimal due to low concentrations in soil pore water or overlying water column. In mineral wetland soils, iron and manganese oxides can provide large capacity for organic matter decomposition. Under normal pH conditions, oxides of manganese and iron are immobile and their availability to microbes is restricted because of solubility and mobility. Their role in organic matter decomposition can be more significant in soils that undergo frequent wetting and drying cycles. In permanently waterlogged conditions, most of the bioavailable iron and manganese oxides are already present in reduced forms. Oxidized forms may be restricted to soil–floodwater interface and to the root zone.

Decomposition of organic matter in freshwater and saltwater marshes is clearly distinguished by the relative rates of methanogenesis and sulfate reduction. There is ample evidence in the literature that methanogenesis is the significant pathway regulating organic matter decomposition in freshwater wetlands, whereas sulfate reduction is dominant in saltwater marshes. Electron acceptor (oxygen, nitrate, sulfate, and carbon dioxide) consumption rates in a range of wetland soils are significantly correlated to DOC (electron donor) and MBC (D'Angelo and Reddy, 1999). One example of oxygen consumption as a function of DOC in a range of wetland soils is shown in Figure 5.48.

The contribution of various processes to organic carbon mineralization in select aquatic systems is shown in Table 5.15. Similar data are not available for wetland soils at this time.

The examples presented in Table 5.15 suggest that in freshwater systems, methanogenesis dominates the decomposition of organic matter, whereas in salt marsh and marine ecosystems, sulfate reduction could be a significant process in regulating organic matter decomposition, reflecting in availability of respective electron acceptors. The influence of manganese and iron oxide reduction on organic matter decomposition is restricted to mineral soils dominated by these metals. Under drained conditions or during water table fluctuations, some of the reduced iron and manganese can be oxidized to amorphous forms of iron and manganese oxides, which is subsequently reduced during flooded conditions to support organic matter decomposition (see Chapter 10 for details). Decomposition using oxygen as an electron acceptor can occur across all ecosystems and depends on the hydroperiod and water table fluctuations (see Section 5.7.3). Typically, decomposition of organic matter is shown to occur at a rapid rate under aerobic conditions as compared to anaerobic conditions. Few examples of organic matter decomposition in soils from different ecosystems are shown in Figures 5.49 through 5.51).

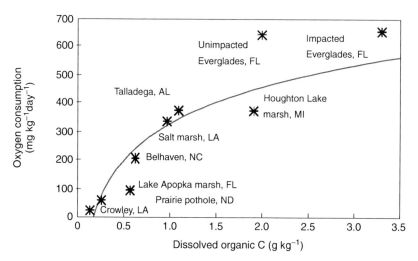

FIGURE 5.48 Oxygen consumption as a function of dissolved organic carbon in different aquatic systems (D'Angelo and Reddy, 1999).

TABLE 5.15
Relative Contribution of Select Electron Acceptors (Expressed as Percent of Organic Matter Oxidation) to Organic Matter Decomposition in Select Aquatic Systems

Site	Rate of Carbon Oxidation (mmol C m^{-2} day^{-1})	Oxygen	Sulfate	Methanogenesis
Marine				
Limfjorden	36	(47)	53	
Sippewissett salt marsh	458	10	90	
Sapelo I salt marsh	200	20	70	(10)
Sippewissett salt marsh	180	50	50	
Cape Lookout Bight	100		68	32
Lacustrine				
Blelham Tarn	—	42	2	25
Wintergreen	108	—	30	71
Lake Vechten	23	—		70
Lawrence Lake	3	—	30–81	19–70

Source: Capone and Kiene (1988).

In all cases, the ratio of aerobic to anaerobic decomposition ranges from 3 to 6. Anaerobic decomposition in these systems represents a combined effect of alternate electron acceptors. Anaerobic conditions here represent a combined effect of several alternate electron acceptors (other than oxygen) present on organic matter decomposition. As mentioned previously, typically all wetlands are limited in electron acceptors, whereas electron donors are nonlimiting in most wetlands. Annual carbon loss due to decomposition processes varied between 0.7 and 3.7% of total carbon in the top 24 cm soil from four freshwater wetlands in the southeastern United States (Schipper and Reddy, 1994). Approximately 70% of the carbon lost in these systems was due to methane production, suggesting the limitation of other electron acceptors.

Carbon

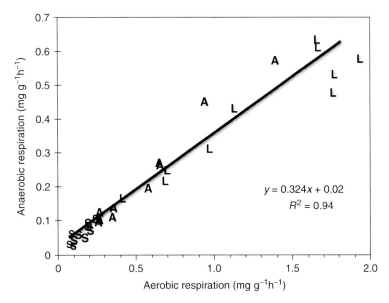

FIGURE 5.49 Aerobic and anaerobic decomposition of soil organic matter in Everglades wetland soils (L = litter; A = surface soil [0–10 cm]; and S = subsurface soil [10–30 cm]; [DeBusk and Reddy, 1998]).

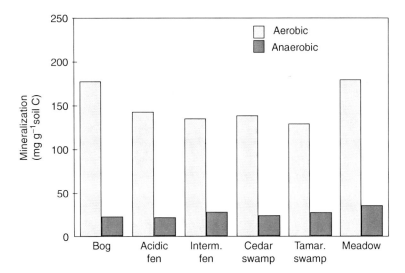

FIGURE 5.50 Aerobic and anaerobic decomposition of soil organic matter in different wetlands (Bridgham et al., 1998).

5.7.5 Nutrient Availability

Nutrient availability can affect decomposition rate by limiting the growth rate of microbial decomposers. Although nutrient loading is typically greater in wetlands than in uplands due to location within the landscape, nutrient availability may be low relative to the pool of available organic carbon in wetlands. Growth-limiting nutrients may be obtained by microorganisms from the organic substrates or from the water column or soil pore water. Both nitrogen and phosphorus have been identified as microbial growth–limiting nutrients in wetlands. Thus, the ratios of carbon to nitrogen (C:N) and carbon to phosphorus (C:P) in organic matter are often related to rate of decomposition.

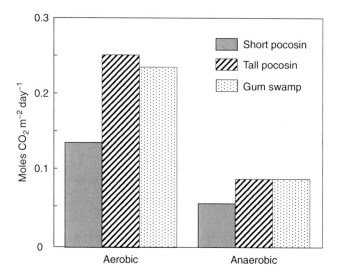

FIGURE 5.51 Aerobic and anaerobic decomposition of soil organic matter in different freshwater wetlands. (Redrawn from Bridgham and Richardson, 1992.)

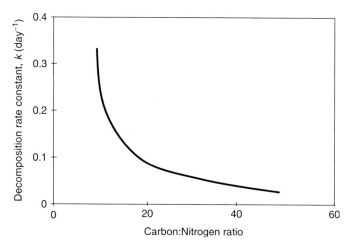

FIGURE 5.52 Relationship between rates of decomposition of soil organic matter and its C:N ratios. (Reddy, K. R., Unpublished Results, University of Florida.)

Microorganisms growing on low-nutrient substrates (e.g., high C:N ratio or high lignin:N ratio) tend to scavenge significant amounts of nutrients from the surrounding environment, resulting in net immobilization of growth-limiting nutrients. Plant matter with high C:N or C:P ratio decomposes slowly, as compared to that with low C:N or C:P ratio (Figures 5.52 and 5.53). In oligotrophic wetlands such as Everglades, high C:P ratio of organic matter resulting from phosphorus limitation was shown to influence microbial activity, thus affecting the overall decomposition. In fact, the decomposition of organic substrates was directly related to total phosphorus content of the substrate (Figures 5.54 and 5.55).

Several micronutrients such as molybdenum, nickel, boron, iron, zinc, vanadium, and cobalt have been shown to affect the microbial activities, especially of methanogens, but to a lesser extent of heterotrophs. The bioavailability of many of these trace metals is affected by the redox and pH of wetland soils. The significance of micronutrient deficiency in microbial processes in wetlands is not well understood. (Howarth, 1993; Megonigal et al., 2004).

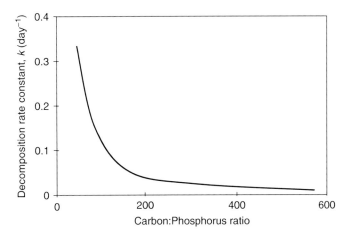

FIGURE 5.53 Relationship between rates of decomposition of soil organic matter and its C:P ratios. (Reddy, K. R., Unpublished Results, University of Florida.)

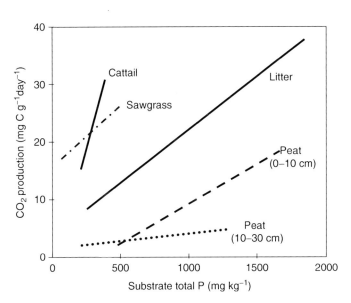

FIGURE 5.54 Effect of phosphorus concentrations on decomposition of organic matter with different nutrient ratios. (Adapted from DeBusk and Reddy, 1998.)

Nutrient loading to a wetland ecosystem can have a dramatic effect on the quality and quantity of plant matter produced and ultimately on the decomposition rates as demonstrated in the discussion presented in this section. As shown in Figure 5.56, nutrient loading can increase the plant litter production per unit area. Depending on the type of nutrient added, C:N or C:P ratio of the plant matter is altered. The relative proportion of refractory pool is higher in plant matter produced in low-nutrient system as compared to that produced in high-nutrient system. To some extent, slow decomposition rates in low-nutrient system are accounted for not only by high proportion of recalcitrant pool but also by nutrient limitation (high C:N or C:P ratio).

For terrestrial ecosystems, soil C:N ratio was significantly correlated to DOM export (Aitkenhead and McDowell, 2000). At a global scale, mean soil C:N ratio explained 99% of the variability in annual riverine DOC fluxes.

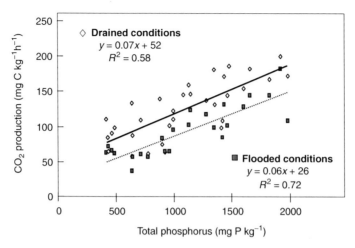

FIGURE 5.55 Relationship between total phosphorus and decomposition of organic matter in wetlands (Wright and Reddy, 2001b).

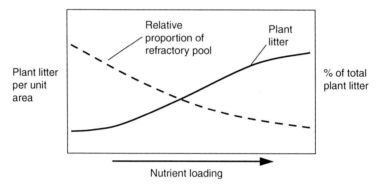

FIGURE 5.56 Relationship between nutrient loading and plant litter production.

5.7.6 TEMPERATURE

Temperature is one of the key regulators influencing biogeochemical processes in wetlands, by influencing the growth, activity, and survival of organisms. Chemical and enzymatic reactions regulating organic matter decomposition proceed at a faster rate as the temperature is increased. Microbial activity and organic matter decomposition are accelerated by an increase in temperature. Maximum activity may vary depending on the environmental conditions and microbial communities. Above certain temperatures, proteins, nucleic acids, and other cellular components may be irreversibly denatured: The plasma membrane may be collapsed, and thermal lysis of cells may occur. For each microbial community, there is a maximum temperature above which growth is inhibited, a minimum temperature below which growth no longer occurs, and an optimal temperature range in which growth is most rapid. Temperature response to biogeochemical processes is often expressed in terms of Q_{10} function:

$$Q_{10} = \left[\frac{k_2}{k_1}\right]^{(10/(T_2 - T_1))}$$

where k_1 and k_2 are rate constants for decomposition of organic matter at temperatures T_1 and T_2, respectively. For example, $Q_{10} = 2$ means that the decomposition would occur twice as fast at 20°C

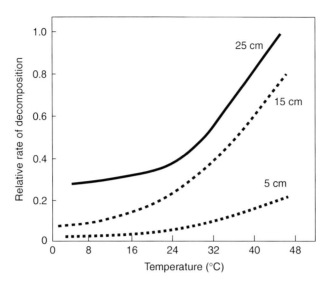

FIGURE 5.57 Effect of temperature on organic matter mineralization. (Redrawn from data by Volk, 1973.)

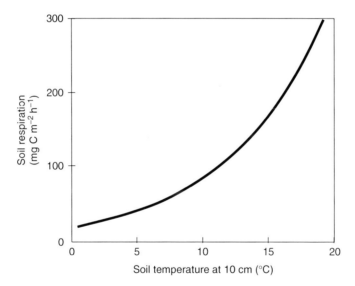

FIGURE 5.58 Effect of temperature on microbial respiration. (Redrawn from Davidson et al., 1998.)

as at 10°C (for details on basic concepts related to Arrhenius equation, see Chapter 2). Decomposition rate increases with temperature at 0°C with Q_{10} as high as 8, and the temperature sensitivity decreases with increasing temperature, as indicted by Q_{10} decreasing to 4.5 at 10°C and 2.5 at 20°C (Figures 5.57 through 5.59; Kirschbaum, 1995; Davidson et al., 1998).

The Q_{10} values for peatlands and bogs were reported to be in the range of 1.8–6.1 (Lafleur et al., 2005). Earlier studies have shown prolonged lag phase in carbon dioxide production at low temperatures (7°C) in soils amended with plant matter (Pal et al., 1975). When temperature was increased to 22°C, the decomposition rate increased by a Q_{10} value of 12 during the first 2 days, and a further increase in temperature to 37°C resulted in a Q_{10} value of 1.5. The Q_{10} values were highest during early stages of decomposition and decreased sharply and remained at a constant value for the remaining 4-month decomposition period (Pal et al., 1975). Surface soil temperatures were found to be better predictors of ecosystem respiration than temperatures in deeper soil depths (Lafleur et al., 2005).

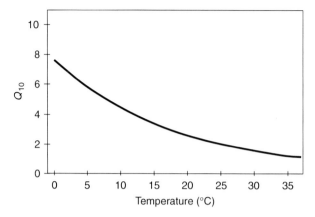

FIGURE 5.59 Effect of temperature on changes in the rate of decomposition. (Redrawn from Kirschbaum, 1995.)

TABLE 5.16
Temperature Dependence of Methane Production and Subprocesses Responsible for Methane Production

Sample Source, Process or Organism	Q_{10}
Methane production in soils	
Minerotrophic peat	1.5–6.4
Oligotrophic peat	2–28
Paddy soils	2.1–16
Methanogenesis of pure cultures	
Acetotrophic	2.9–9.0
Hydrogenotrophic	1.3–12.3
Anaerobic carbon mineralization processes	
Anaerobic carbon dioxide production: peat	1.5
Anaerobic carbon mineralization: paddy soil	0.9–1.8
Anaerobic hydrolysis of particulate organic matter	1.9
Acetate production from various substrates	1.7–3.6

Source: Segers (1998).

Temperature sensitivity to methanogenesis varied with Q_{10} values in the range of 1–28 (Table 5.16). This variable response to temperature may be due to the influence of compounding factors including the presence of electron acceptors of higher reduction potentials or intensity of soil anaerobiosis and availability of substrates. At lower temperatures, presence of electron acceptors with higher reduction potentials and their slow reduction can create a significant lag phase, thus influencing the rate of methanogenesis. At higher temperatures, reduction of these electron acceptors may be faster and lag phase for methanogenesis may be minimal.

The following list presents some of the general findings from the published literature on the effects of temperature on organic matter decomposition in terrestrial and wetland ecosystems:

- Temperature sensitivity of organic matter decomposition is much greater at a lower temperature and decreases with an increase in temperature. It is also suggested that temperature sensitivity is much greater for organic matter decomposition than for net primary productivity. This has important implications for organic matter storage in the ecosystem (Kirschbaum, 1995).

Carbon

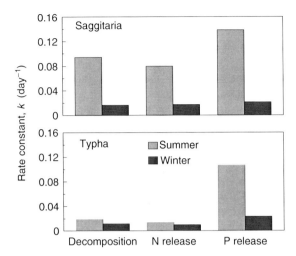

FIGURE 5.60 Seasonal variation in the soil organic matter decomposition in a Central Florida wetland. (Reddy and Fisher, Unpublished Results, University of Florida.)

- Decomposition of labile soil carbon is more sensitive to temperature than slowly degradable or resistant soil organic carbon. High levels of labile organic carbon pools are present in soils at low temperatures, whereas at higher temperatures relatively high levels of more recalcitrant organic matter are maintained (Dalias et al., 2001). However, recent findings suggest that temperature sensitivity for resistant organic matter does not differ significantly from that of labile pools, and that both types of soil organic matter will therefore respond similarly to global warming (Fang et al., 2005).
- Recalcitrant soil organic materials mineralize more efficiently at higher temperatures (Bol et al., 2003).
- Seasonal changes in temperature play a significant role in controlling the rates of biogeochemical processes regulating organic matter decomposition, including enzyme activities, DOM production, carbon dioxide, and methane emissions.
- Many studies using variety of plant matter and conducted in various types of freshwater systems have demonstrated seasonal variations in decomposition rates with faster breakdown during warmer periods (Webster and Benfield, 1986). One example of breakdown of detrital matter during summer and winter months is shown in Figure 5.60. Detrital matter from *Typha* was found to be less sensitive (1.5 times higher during summer months than winter months) to seasonal temperature than that from *Sagittaria* (5.5 times higher during summer months than winter months).
- Several studies have shown a strong interaction between temperature and soil moisture. Reddy et al. (1986) found that the effect of soil moisture was less significant at low temperature than at more optimum temperatures.

5.8 ENVIRONMENTAL AND ECOLOGICAL SIGNIFICANCE

Soil organic matter plays pivotal role in various ecological functions summarized as follows:

- It is a major component of the global carbon budget.
- It is a source of carbon dioxide and methane emissions into the atmosphere.
- It is a source of mineralizable nutrients for the growth of plants and other microorganisms.
- It is a source of energy for heterotrophic microorganisms.
- It is a source of DOC for downstream aquatic systems.
- It is a source of exchange capacity for cations in soils.

- It provides long-term storage for nutrients in soils.
- It complexes metals in soils.
- It is a strong adsorbing agent for toxic organic compounds in soils.

Wetlands are important components of the landscape (6% of the total land surface) and exert significant influence over global carbon budgets and climate change. Wetlands contain approximately 15–22% of the terrestrial carbon and are one of the major contributors to the global methane flux producing approximately 100–200 Tg CH_4 per year, which accounts for approximately 20–25% of global methane emissions into the atmosphere. Under drained soil conditions, methane emissions from wetlands are drastically reduced, whereas the accelerated organic matter decomposition in turn increases the rate of carbon dioxide emissions into the atmosphere. Under drained soil conditions, the ratio of net primary productivity to soil respiration decreases resulting in less accumulation of organic matter in wetlands. This potential feedback of carbon dioxide emissions resulting from accelerated decomposition has been evaluated for terrestrial ecosystems, but not for wetland ecosystem (Ise and Moorcroft, 2006). Emissions of carbon dioxide and methane from wetlands are controlled by:

- Production and partitioning of carbon dioxide and methane in anaerobic zones of soil
- Carbon dioxide production in aerobic zones of the soil profile
- Aerobic and anaerobic methane oxidation
- Transport processes regulating gaseous flux from soil to the water column and into the atmosphere

Methane and carbon dioxide produced in soils are transported into the atmosphere by diffusion and mass flow via two pathways: (1) the aerenchyma tissues of plant roots and stems and (2) flux from soil to the overlying water column (Figure 5.61). Gas exchange in plants is discussed in detail in Chapter 7. Carbon dioxide is highly soluble and undergoes various chemical reactions, and it may be difficult to estimate flux accurately without considering all associated reactions. Because of the potency (on molecule-to-molecule basis, methane absorbs 25 times as much infrared radiation as carbon dioxide) of methane as greenhouse gas, we will focus our discussion on methane emissions from wetlands.

Methane emissions are measured *in situ* by using closed chamber placed over a unit area of a wetland. The chamber may or may not include vegetation. A net increase in methane concentration

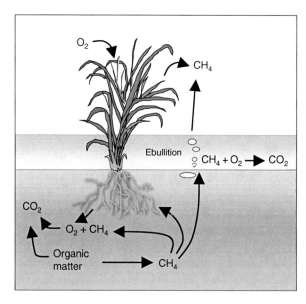

FIGURE 5.61 Pathways for emission of methane and carbon dioxide produced in wetlands.

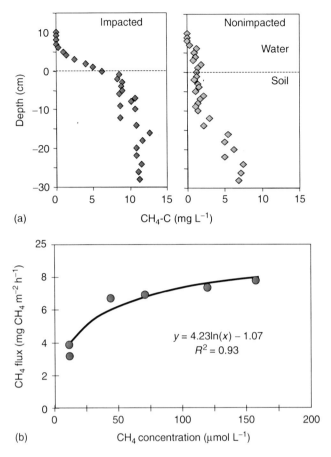

FIGURE 5.62 (a) Relationship between soil depth and methane concentrations (Reddy, K. R., Unpublished Results, University of Florida). (b) Relationship between methane concentration in soils and methane emissions from freshwater marsh (Ding et al., 2005).

in the chamber over a short period of time (less than 1 h) is measured and used to estimated methane fluxes. The second method commonly used is pore water equilibrators to determine the dissolved pore water methane concentrations in the soil profile. Dissolved methane concentration gradients are used to estimate methane flux from soil to the overlying water column (Figure 5.62a) (see Chapter 14 for details). The net methane flux measured by both methods accounts for the oxidation of methane either in the root zone or at the soil–floodwater interface and water column.

Dissolved methane profiles in soil pore water exhibit distinct gradients with low concentrations in surface layers and increase with depth (Figure 5.62a). The gradients are often regulated by the presence of other electron acceptors and the density of plant roots. Dissolved methane concentrations in the soil profile significantly relate to methane emissions in freshwater marsh dominated by *Carex* sp. (Figure 5.62b; Ding et al., 2005). As dissolved methane diffuses through the soil surface layers, up to 90% can be oxidized to carbon dioxide by methane-oxidizing bacteria, and thus only a small fraction of methane is released into the atmosphere (King, 1992). For methane diffusing through plants, approximately 50% of methane produced is oxidized in the root zone, whereas the remaining diffuses through aerenchyma tissue of wetland plants (Schipper and Reddy, 1996). In northern wetland soils, methane production was greatest in the beaver meadows and intermediate fens and lowest in the acidic fens and bogs (Figure 5.63). The proportion of anaerobic carbon mineralization as methane varied extensively among wetland sites: bogs, 0.5%; acidic fens, 2%; intermediate fens, 10%; cedar swamps, 5%; tamarack swamps, 5%; meadows, 12% (Bridgham et al., 1998).

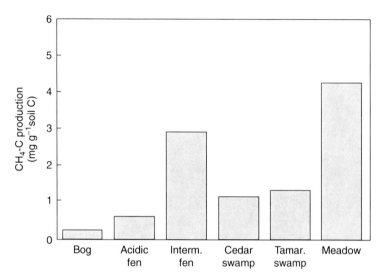

FIGURE 5.63 Methane production in select wetland soils from northern region in the United States (Bridgham et al., 1998).

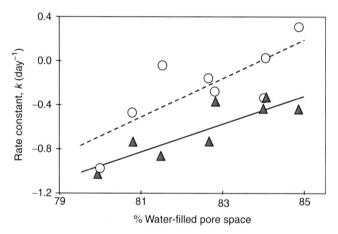

FIGURE 5.64 Rate of methane production in soils increases with increased water-filled pore space. (Adapted from Veldkamp et al., 2001.)

Globally, methane fluxes are dominated from (1) agricultural lands including rice paddies and (2) natural wetlands including forested wetlands, riparian wetlands, freshwater marshes, bogs, and fens (Milich, 1999). In subtropical marshes, approximately 3% of wetlands daily net ecosystem production is emitted into the atmosphere as methane (Whiting et al., 1993). Net methane fluxes vary, depending on location, hydroperiod, temperature, and other regulators (as discussed previously). Hydroperiod can significantly alter methane emission from wetlands. Lowering the water table can significantly decrease methane emissions. The methane produced can be readily oxidized in surface soil layers. Rate of methane production increases in soils with greater percent of soil pores filled with water (Figure 5.64).

Few examples of methane fluxes in wetlands are summarized in Table 5.17.

Globally, extensive areas of peatlands and wetlands have been drained and converted into agricultural lands. Drainage of organic matter–rich soils accelerated the decomposition process and emission of carbon dioxide. Many peatlands that have accumulated organic matter for centuries

TABLE 5.17
Potential Methane Production Rates from Select Wetlands

Location	Temperature	Flux (mg m^{-2} day^{-1})
Sudd (Sudan)	23.2–32.2	1,090
Papyrus swamps (Uganda)	21.1–22.2	463
Hardwood swamp (Michigan)	13–22	356
Rice paddy (Italy)	19–27	279
Pripett marshes (USSR)	−6.1 to 18.9	109
Temperate swamp (Virginia)	4–23	41–44
Boreal and tundra mires (Alaska)	−17.5 to 16	4–289
Everglades (Florida)	20–28	340–627
Sunny hill farm (Florida)	24–36	2,800
Apopka marsh (Florida)	19–27	510
Okefenokee swamp (Georgia)	15–34	1,000

Source: Westermann (1993) and Schipper and Reddy (1994).

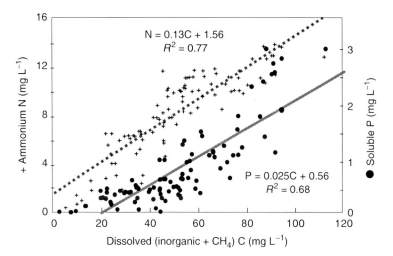

FIGURE 5.65 Relationship between decomposition product formation and nitrogen and phosphorus release (D'Angelo and Reddy, 1994a, 1994b).

have been oxidized, resulting in soil loss. Globally, the carbon dioxide production accounts for approximately 6% of that produced by combustion of fossil fuels (Maltby and Immirzi, 1993).

Accumulation of organic matter in wetlands results in storage of various nutrients including nitrogen, phosphorus, sulfur, and others (see Chapters 8 through 10). These nutrients are released during either microbial decomposition of organic matter or abiotic oxidation by fire. Nutrients released may be recycled within the ecosystem or exported from the system. In wetlands receiving minimal anthropogenic inputs of nutrients, breakdown of organic matter is the primary source of nutrients that supports the plant growth and ecosystem productivity.

Nutrient release into the environment is rapid in nutrient-rich wetlands (eutrophic wetlands) where nutrient supply meets the needs of microbial growth. Nitrogen and phosphorus release is directly related to the decomposition end products methane and carbon dioxide (Figure 5.65). In contrast, nutrients may be held tightly within the microbial biomass component of low-nutrient wetlands, reflecting efficient recycling of remineralized organic compounds. Microbes and plants often compete for nutrients in wetlands with limited supply of nutrients and tight nutrient cycles. Microbes outcompete plants for nutrients. The rate of organic matter accumulation is a critical determinant of how a wetland functions as an ecological unit within the landscape. The storage function is equally important for natural wetlands, especially those which represent an ecotone between terrestrial and aquatic systems, and constructed wetlands used for treatment of nutrient-enriched waters.

Wetlands are net exporters of POM and DOM to adjacent aquatic systems, when hydrologically connected to surface flow. TOC includes both particulate and DOC. In many natural systems, approximately 90% of the TOC is accounted for in DOC (Figure 5.6). The physicochemical characteristics of DOC and its concentrations in the water leaving a wetland are influenced by vegetation type, density, soil organic matter, and the extent of abiotic and biotic degradation.

Within wetlands DOC is produced (autochthonous) during decomposition and leaching of standing dead biomass, detritus, soil organic matter, and loading of allochthonous DOC from adjacent upland portion of the watershed. DOC provides several functions within the wetland including serving as an energy source to heterotrophic bacteria and source or sink for various elements and organic compounds through exchange, sorption, and complexation. DOC is the most mobile pool of organic matter in wetlands. DOC is hydrophobic and as a result can also transport some toxic elements and xenobiotics from wetlands to adjacent aquatic ecosystems. Its movement is controlled by water movement within the wetland. In addition, DOM can also flux from soil to the overlying water column.

As DOC passes through the wetland water column and through the submerged macrophytes and their epiphytic microbial communities, a selective removal of organic compounds occurs (Wetzel, 1993). In addition to biotic and abiotic degradation, certain compounds (such as amino acids) can adsorb inorganic particulate matter such as calcium carbonates and clays. As a result, the recalcitrance of DOC increases as it is exported to adjacent aquatic systems. Most of the labile pool of DOC is rapidly consumed by heterotrophic bacteria during their respiration or undergoes abiotic degradation, leaving a large portion of DOC refractory, and not readily available to microbes. Thus, a series of biogeochemical processes essentially increases the recalcitrance of DOC as it is exported to adjacent aquatic systems.

Long-term data on DOC exports from organic-rich watersheds in the United Kingdom has shown rising concentrations in adjacent streams. These increases are observed over timescales of up to 39 years (Freeman et al., 2001b; Worral et al., 2003), suggesting a shift in the carbon budget with decreasing storage of carbon. Further study (Worral and Burt, 2004) demonstrated that increase in DOC concentrations could not be explained by trends in stream flow, pH, conductivity, alkalinity, and turbidity. Other researchers in the region noted that increased DOC concentrations may be due to an increase in temperature, but this factor alone may not explain the DOC concentration.

Additional regulators controlling the increase in DOC may be due to increased enzymatic activity as a result of increased temperature. Anaerobic decomposition occurs very slowly in peat bogs, and the DOC production is restricted by repression of major biodegrading hydrolytic enzymes (Freeman et al., 2001a). In addition, lack of phenol oxidase activity under anaerobic conditions may result in accumulation of phenolic compounds. If the water table drops, introduction of oxygen into the system can result in activation of phenol oxidase, which may potentially degrade phenolic compounds. This process is referred to as "enzyme latch" mechanism (Freeman et al., 2001a). Thus, increased peat decomposition as a result of combined-effect enzyme latch and other factors such as hydrology and temperature can potentially explain elevated DOC exports from peatlands to adjacent streams (Worral et al., 2005).

5.9 FUNCTIONS OF ORGANIC MATTER IN SOILS

Nonliving organic matter (>95%) is the major constituent of soil organic matter and <5% is in living matter including microbial biomass and other soil fauna. Nonliving organic matter includes dead plant material and partially decomposed and resynthesized organic residues. Carbon compounds

Carbon

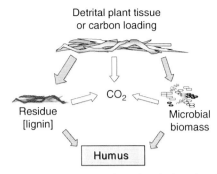

Humus: Total of the organic compounds in soil exclusive of undecayed plant and animal tissues, their "partial decomposition" products, and the soil microbial biomass

FIGURE 5.66 Undecomposed organic matter accumulated to form humic substances.

that are recalcitrant to aerobic and anaerobic decomposition tend to accumulate in soils as either undecomposed plant tissue (peat) or humic substances (Figure 5.66). The formation of humic substances takes place by condensation reactions between reactive phenolic groups of tannins and lignins with water-soluble nonhumic substances. This mechanism often termed as "humification" accounts for the high-molecular-weight and heterogeneous humic substances that contain significant amounts of nitrogen, phosphorus, and sulfur in their structure.

In the absence of oxygen, humic substances are resistant to decomposition and represent a significant carbon and nutrient storage in wetlands. Under drained conditions, humic substances are rapidly degraded, which releases nutrients to the bioavailable pool, thereby affecting downstream water quality. Humus is generally defined as the organic material that has been transformed by abiotic and biotic processes into stable form (see Section 5.4.3) and consists of two major types of compounds (Stevenson, 1986):

- Unhumified substances, which include carbohydrates, fats, waxes, and proteins
- Humified substances, which represent the most active fraction of humus and include a series of highly acidic, yellow-to-black colored HMW-OM commonly referred to as fulvic and humic acids

Approximately two-thirds of soil organic matter can be accounted for in the humified fraction, whereas the remaining exists in the nonhumified fraction. These two classes of compounds are not easily separated from one another, as some of the nonhumic materials such as carbohydrates may be covalently bound to the humic matter (Stevenson, 1986). Chemical characteristics of nonhumic and humic substances have been discussed in Section 5.4.3. In this section, the role and function of soil humus in biogeochemical cycling will be discussed.

Characteristics of soil humus in wetland soils are not known at this time. Much of the information presented relies on the literature from soils of terrestrial ecosystems. The following are the major characteristics of soil organic matter (Table 5.18).

- It is source of nutrients for plant growth.
- It is source of energy for soil microorganisms.
- It is source of exchange capacity for cations. The colloidal surfaces of humus are negatively charged. The sources of the charge are carboxylic or phenolic groups. The extent of negative charge is pH dependent with high values under alkaline conditions (Figures 5.67 and 5.68).
- It provides long-term storage for nutrients.
- It is a strong adsorbing agent for toxic organic compounds.

TABLE 5.18
General Properties of Humus in Soils

Property	Comments	Functions in Soil
Color	Dark color of many soil is caused by organic matter	May facilitate soil warming
Water retention	Organic matter can hold up to 20 times its weight in water	Improves moisture retention of soils. In wetlands during the period low water table conditions
Combination with clay minerals	Joins soil particles into structural units called aggregates	Permits gas exchange; stabilizes soil structure; increases permeability
Chelation	Forms stable complexes with copper, manganese, and zinc ions	Buffers the availability of trace metals to higher plants
Solubility in water	Insolubility of organic matter results partially from its association with clay; salts of divalent and trivalent cations with organic matter are also insoluble; source of dissolved organic matter	Exports relatively recalcitrant dissolved organic matter to adjacent aquatic systems
pH relations	Organic matter buffers soil pH in the slightly acid, neutral, and alkaline soils	Helps to maintain relatively constant pH
Cation exchange	Total acidities of isolated fractions of humus range from 3,000 to 14,000 mmol kg^{-1}	Increases the CEC of the soil; 20–70% of the CEC is due to organic matter
Mineralization	Decomposition of organic matter produces various gases and nutrients	Energy source for microbes; a source of nutrients for plant growth; source of greenhouse gases
Combination with organic molecules	Affects bioavailability, persistence, and biodegradability of toxic organic compounds	Decreases bioavailability of toxic organic compounds to biota

Source: Stevenson (1986).

- It inhibits enzyme activity.
- It complexes with metals (Figure 5.69):
  - Metal ions that would convert to insoluble precipitates are maintained in solution and enhance the bioavailability of metals.
  - Some organic complexes with metals may lower solubility and decrease the bioavailability of metals.
  - Soil organic matter plays a significant role in transporting metals from one ecosystem to another. Metals complexed with DOM can be transported from wetlands to adjacent aquatic ecosystems.

5.10 SUMMARY

- Carbon is important for living systems because it can exist in a variety of oxidation states (-4, 0, $+4$). Carbon is also a key component in cell structure and serves as a source of electrons for microbial processes.
- Soil and sediment organic matter sequesters the largest amount of carbon in the world (except for subterranean oil and natural gas reserves).
- Reservoirs of carbon in a wetland can be grouped as follows:
  - Plant biomass carbon (live)
  - POC (detritus and soil)

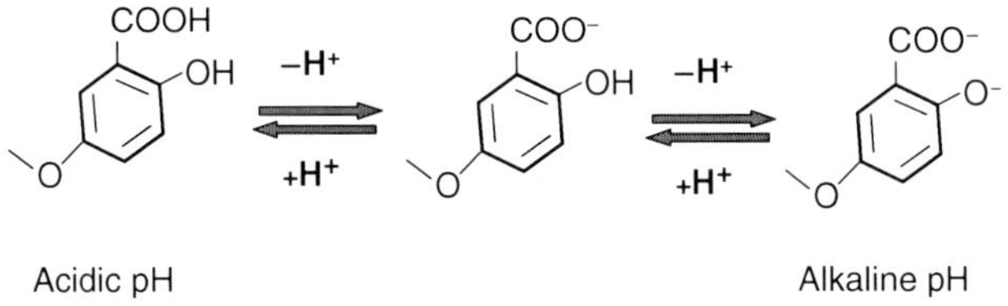

FIGURE 5.67 Extent of negative charge in soil organic matter as a function of pH.

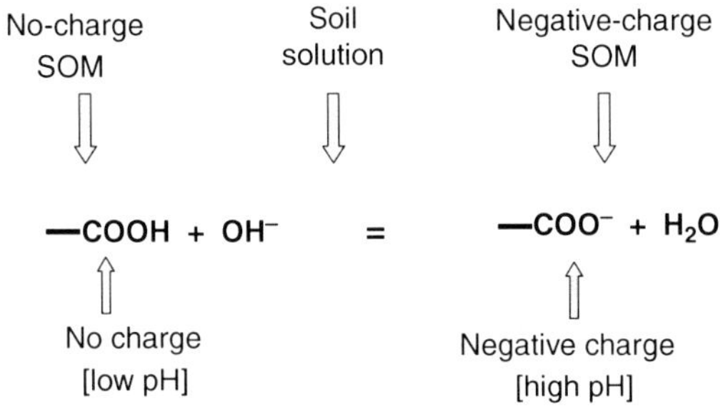

FIGURE 5.68 High pH conditions lead to increased negative charge in the soil organic matter.

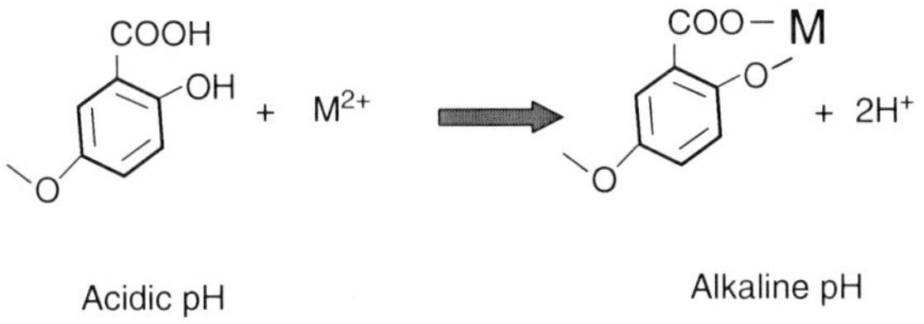

FIGURE 5.69 Soil organic matter complexation with metals (M = not labeled).

- DOC (detritus and soil)
- MBC (detritus and soil)
- Gaseous end products (atmosphere, detritus and soil)
• Organic carbon in the soil is an energy source because it acts as a source of electrons.
• Aerobic conditions readily decompose organic carbon such as detritus; however, as conditions become anaerobic, organic carbon is less readily decomposed and much is deposited as peat.
• Most of the decomposition of organic matter is driven by oxygen, but less efficient electron acceptors are used in anaerobic processes.
• Microbes
  - Microbes cannot use complex polymers directly and need to be broken down to monomers first.
  - Some extracellular enzymes are involved in catalyzing the breakdown polymers.
• Detrital matter is broken down into complex polymers (cellulose, proteins, lipids, lignin). Enzymes break these polymers into simple monomers (sugars, amino acids, fatty acids) that bacteria can consume.
• There is a direct relationship between microbial activity and enzyme activity; therefore, enzyme activity can be measured to determine the rate of organic matter decomposition by microorganisms.
• Some cells can excrete ectoenzymes that stay bound to the cell surface and aid the cell in polymer degradation.

- Anabolism is the building up of organic matter, and catabolism is the breaking down of organic matter.
- The end products of aerobic, nitrate, iron, and sulfate respiration include carbon dioxide, water, and nutrients. Hydrogen is a by-product of fermentation.
- Sulfate-reducing bacteria cannot directly assimilate monomers; they depend on fermenting bacteria cells for uptake of monomers.
- Lignin requires a lot of energy to decompose due to its aromatic structure; therefore, it accumulates in the humus portion of the soil.
- Organic mater is a source (short-term and long-term storage) of nutrients for plants and soil microbes. Organic matter often has a negative charge associated with the humus layer that attracts cations (which can be used as nutrients), metals, and pesticides.
- With increasing depth in the soil profile, the organic matter becomes more recalcitrant.
- Decomposition is regulated by substrate quality (substrate size, percentage of C, type of C, lignin cellulose ratio), electron acceptors (which, how many), limiting nutrients, and environmental factors such as temperature. Generally, an increase in temperature will increase the rate of organic matter decomposition; conversely, when a soil is saturated, anaerobic conditions slow down decomposition rates.
- Functions of organic matter
  - Source of nutrients for plant growth
  - Source of energy for soil microorganisms
  - Long-term storage for nutrients
  - Strong adsorbing agent for toxic organic compounds
  - Complexation of metals
- Ninety-five percent of the methane produced in the anaerobic zone is either taken up by aerenchyma of plants and released as gas or oxidized back to carbon dioxide in the water column and aerobic root zone.
- The following is a brief summary of the comparison between aerobic and anaerobic catabolic processes:
  - Many of the biochemical pathways involved in the breakdown of organic matter are similar for both aerobes and anaerobes because most pathways do not require oxygen.
  - Both aerobes and anaerobes are capable of glycolysis (glucose degradation) and derive energy by the substrate-level phosphorylation.
  - Both aerobic and anaerobic microbes generate most of the ATP by oxidative phosphorylation. However, the terminal electron acceptor, energetics, and kinetics are different for aerobic and anaerobic microorganisms.
  - Fermenting organisms generate most of their ATP by substrate-level phosphorylation.
  - The ability of aerobes to incorporate oxygen into organic molecules via oxygenases allows these microbes to utilize aromatics and saturated alkanes with no functional groups (e.g., benzene).
  - Aerobic catabolism results in complete conversion of organic substrates to CO_2 and H_2O, resulting in the release of 100% of the energy stored in monomers. Anaerobic catabolism results in accumulation of a large proportion of the energy released from organic substrates into reduced end products.
  - Aerobic decomposition results in the production of oxidized species (CO_2, H_2O, NO_3^-, SO_4^{2-}, and Mn^{4+} and Fe^{3+} oxides), whereas the anaerobic decomposition results in the production of reduced species (H_2, fatty acids, NH_4^+, N_2, N_2O, sulfides, CH_4, Fe^{2+}, and Mn^{2+}).
- Humic substances are divided into three major groups on the basis of chemical characteristics and their solubility in acids and bases.
  - Fulvic acid (acid and base soluble)
  - Humic acid (acid insoluble and base soluble)
  - Humin (acid and base insoluble)

Carbon

STUDY QUESTIONS

1. Why does organic matter accumulate in wetlands and poorly drained soils but not in upland (drained) soils?
2. Diagram and explain carbon and electron flow in aerobic, facultative anaerobic, and obligate anaerobic catabolic pathways. Where would you find each type of catabolism in nature?
3. Compare and contrast aerobic and anaerobic organic matter decomposition in terms of energy yield, rate, microbial biomass, efficiency, and end products.
4. List all inorganic and organic electron acceptors and show the relationship between redox potential and use of electron acceptors during catabolic activities of microorganisms.
5. What would be the ecological impact if a virus suddenly killed all methanogenic bacteria in wetlands and sediments?
6. What is the major difference in the methane fermentation in the rumen and anaerobic sediments?
7. List major carbon components of plant material. Graphically show their biodegradability under drained and wetland conditions.
8. Define extracellular enzymes and ectoenzymes. Discuss briefly their role in organic matter decomposition.
9. List and briefly discuss the factors influencing the decomposition of organic matter in wetlands.
10. Calculate moles of e^- acceptors needed to oxidize 1 mol acetic acid (CH_3COOH) under oxygen-, nitrate-, manganese (MnO_2)-, iron (FeOOH)-, and sulfate-reducing conditions. Write appropriate balanced equations.
11. How many moles of each of the electron acceptors (nitrate and sulfate) are required during oxidation of CH_3COOH to obtain same level of energy as would be generated during aerobic decomposition of 1 mol CH_3COOH?
12. Explain why decomposition of lignin is slower than that of carbohydrates in soils.
13. Can eutrophication of a wetland be driven by accelerated decomposition of organic matter following an increased loading of electron acceptors in the inflow water?
14. Define soil humus. How does humus differ from raw residues of plant material?
15. Discuss the importance of humus to soil fertility and environmental problems.
16. Briefly describe the classical chemical fractionation of soil humus.
17. Compare and contrast humic and fulvic acids. What is the significance of the higher oxygen content of fulvic acids compared to humic acids?
18. List important functional groups of soil humus. Indicate which group provides maximum acidity upon dissociation of protons.
19. Two wetland ecosystems: (a) highly eutrophic wetland receiving secondarily treated effluent containing nitrate and sulfate (nitrate nitrogen = 20 mg L^{-1} and sulfate sulfur = 50 mg L^{-1} and (b) oligotrophic wetland with water inputs only from rainfall. Which wetland potentially would produce more methane and why? Assume any initial conditions, but justify all your assumptions.
20. Name types of organic molecules (substrates) that may be utilized by aerobic, facultative anaerobic (i.e., nitrate-, iron-, and manganese-reducing bacteria), and obligate anaerobic (sulfate-reducing and methanogenic bacteria) in wetland environments.
21. Discuss, in terms of energy yields of the appropriate reactions, the importance of the syntrophic relationship between acetogens and methanogens, specifically the interspecies hydrogen transfer. If, in the absence of other electron acceptors, methanogenesis is inhibited, what impact would this have on the carbon cycle?
22. Soils in wetlands are often organic in nature, as a result of plant material that has slowly decayed and humified. How do you think that the soil characteristics in an herbaceous

marsh differ from those in a forested wetland? Do you think that the different plant litter quality (lignin-to-cellulose ratio, C:N ratio) will influence the soil characteristics?
23. Wetlands typically are described in the low-lying areas of the landscape (depressions), receiving and often storing water from the surrounding uplands. This is often not the case in acid peat bogs, which tend to be found in the higher areas of the landscape. Discuss how carbon is accumulated in these systems, i.e., how characteristics intrinsic to acid peat bogs affect the carbon cycle resulting in organic soils.

FURTHER READINGS

Boultin, A. J. and P. I. Boon. 1991. A review of methodology used to measure leaf decomposition in lotic environments: time to turn over an old leaf? *Aust. J. Marine Freshwater Res.* 42:1–43.

Capone, D. G. and R. P. Kiene. 1988. Comparison of microbial dynamics in marine and freshwater sediments: contrasts in anaerobic carbon metabolism. *Limnol. Oceanogr.* 33:725–749.

Cotner, J. B. and B. A. Biddanda. 2002. Small players, large role: microbial influence on biogeochemical processes in pelagic aquatic systems. *Ecosystems* 5:105–121.

Gopal, B. and V. Masing. 1990. Biology and ecology. In B. C. Pattan et al. (eds.) *Wetlands and Shallow Continental Water Bodies*. Vol. 1. SPB Academic Publishing, The Netherlands. pp. 91–239.

Howarth, R. W. 1993. Microbial processes in salt-marsh sediments. In T. E. Ford (ed.) *Aquatic Microbiology*. Blackwell Scientific Publications, Boston. pp. 239–259.

Lal, R., J. Kimble, F. Levine, and B. A. Stewart. 1995. *Soils and Climate Change*. CRC Press, Boca Raton, FL. 449 pp.

Madigan, M. T. and J. M. Martinko. 2006. *Brock Biology of Microorganisms*. 11th Edition. Pearson Prentice Hall, Upper Saddle River, NJ.

Megonigal, J. P., M. E. Hines, and P. T. Visscher. 2004. Anaerobic metabolism: linkages to trace gases and aerobic processes. In W. H. Schlesinger (ed.) *Biogeochemistry*. Elsevier-Pergamon, Oxford, UK. pp. 317–424.

Milich, L. 1999. The role of methane in global warming: where might mitigation strategies be focused? *Global Environ. Change* 9:179–201.

Pearl, H. W. and J. L. Pinckney. 1996. A min-review of microbial consortia: their role in aquatic production and biogeochemical cycling. *Microb. Ecol.* 31:225–247.

Schink, B. 1997. Energetics of syntrophic cooperation in methanogenic degradation. *Microbiol. Molec. Biol. Rev.* 61:262–280.

Segers, R. 1998. Methane production and methane consumption: a review of processes underlying wetland methane fluxes. *Biogeochemistry* 41:23–51.

Stevenson, F. J. 1986. *Cycles of Soil: Carbon, Nitrogen, Phosphorus, Sulfur, Micronutrients*. Wiley, New York. 380 pp.

Swift, R. S. 1996. Organic matter characterization. In *Methods of Soil Analysis, Part 3. Chemical Methods*. SSSA Book Series No. 5. Soil Science Society of America, Madison, WI. pp. 1011–1069.

Webster, J. R. and E. F. Benfield. 1986. Vascular plant breakdown in freshwater ecosystems. *Annu. Rev. Ecol. Syst.* 17:567–594.

Westermann, P. 1993. Wetland and swamp microbiology. In T. E. Ford (ed.) *Aquatic Microbiology*. Blackwell Scientific, Oxford.

Wetzel, R. G. 1992. Gradient-dominated ecosystems: sources and regulatory functions of dissolved organic matter in freshwater ecosystems. *Hydrobiologia* 229:181–198.

Wetzel, R. G. 2001. *Limnology: Lake and River Ecosystems*. Chapters 11 and 23. Academic Press, New York. 1006 pp.

Wigley, T. M. L. and D. S. Schimel. 2000. *The Carbon Cycle*. Cambridge University Press, Cambridge, UK. 292 pp.

6 Oxygen

6.1 INTRODUCTION

Wetlands are characterized by a range of conditions including both drained and flooded soil conditions. Depending on the hydrologic regime and hydroperiod, wetland soils often alternate between drained and flooded soil conditions. Gas exchange between atmosphere and soils is regulated by the hydrologic regime. The most important biogeochemical reactions involving gases include microbial and plant respiration that consume oxygen and produce carbon dioxide. During the day, algae and submerged aquatic vegetation in the water column consume carbon dioxide and produce oxygen by photosynthesis. Soil respiration consumes oxygen and produces carbon dioxide during both night and day. In many ecosystems, daytime oxygen production typically exceeds oxygen consumption.

In drained soils, most of the soil pores are filled with air and are interconnected to the atmosphere. This allows a rapid exchange of gases including oxygen between the atmosphere and soil pore spaces. The transport of oxygen in a well-drained soil is sufficient to supply the oxygen needed to support the growth of microbial and plant populations. Soils remain at high Eh levels (>300 mV) in the presence of oxygen. Gas exchange in these soils prevents the depletion of oxygen and excessive accumulation of carbon dioxide in the soil profile. Typically, oxygen flux is always from the atmosphere to the soil because the concentration of oxygen in soil pore spaces is lower than that in the atmosphere (Figure 6.1). The carbon dioxide flux is always from the soil to the atmosphere because the concentration of carbon dioxide in soil pores is typically higher than that in the atmosphere. The exchange of these gases is due to diffusion (in response to partial pressure gradients of individual gases) and mass flow (in response to total pressure gradients of all gases).

When soils are flooded, soil pores are filled with water and any dissolved oxygen present is consumed during microbial respiration. Under these conditions, oxygen is introduced into the wetland soil profile by diffusion and mass flow through (i) floodwater and (ii) plants (see Chapter 7). The oxygen concentration in floodwater varies both spatially and temporally and is often present at saturation levels during the photosynthetic period. The subsequent diffusion of oxygen into soil profile depends on the soil characteristics and the concentration of oxygen-demanding species. Wetlands are not always flooded. Depending on the hydroperiod, the water table can fluctuate leaving the upper portion of the soil unsaturated. After short-term flooding (due to heavy rainfall), saturated soil conditions can also exist in poorly drained soils (such as clay loam and silt loam), and for short periods of time, these soils exhibit the same biogeochemical characteristics as wetland soils. These conditions are typical of soils with short-term flooding or wetland habitats. In areas with deeper water columns, vertical stratification can occur in surface water, with low oxygen levels at the soil–floodwater interface. However, for wetlands with emergent macrophytes, shallow water depths of <1 m are common, and the water column in these areas is usually well mixed.

Oxygen plays two roles in biological systems: (i) as an electron acceptor during respiration of biota and (ii) as a reactant in certain biochemical reactions. In addition, oxygen is also involved in chemical oxidation of several reduced species. About 90% of the molecular oxygen present in microbial and plant cells is conserved and used for oxidative phosphorylation during respiration. The role of oxygen as a reactant in certain biochemical reactions is discussed in the latter part of this chapter. Oxygen is the most preferred electron acceptor because of its great affinity for electrons. In soil systems, oxygen is primarily used by microorganisms and plant roots: (i) microorganisms use organic materials as a source of energy for their metabolism, and during the process consume

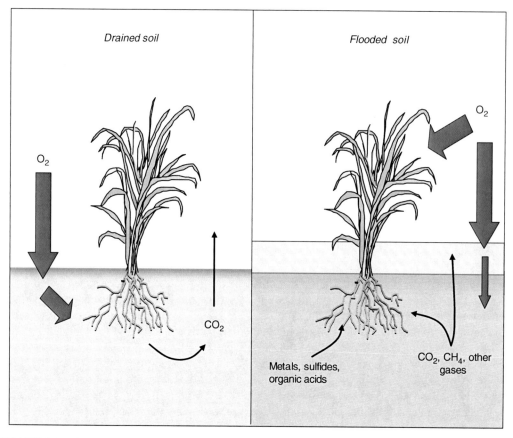

FIGURE 6.1 Gas exchange between drained and flooded soils.

oxygen, while oxidizing organic compounds to carbon dioxide; (ii) plant roots oxidize organic compounds synthesized in leaves and transported to roots. In both cases oxygen is consumed and carbon dioxide is produced.

6.2 OXYGEN–H$_2$O REDOX COUPLE

In natural systems, oxygen reduction occurs at the upper end of the Eh scale.

$$O_2 + 4e^- + 4H^+ = 2H_2O$$

The Nernst equation (Equation 4.65, Chapter 4) can be used to describe the relationship between oxygen content and Eh:

$$Eh = E° - \frac{RT}{nF} \ln \frac{[\text{reductant}]^m}{[\text{oxidant}]^m[H]^m} \qquad (4.65)$$

$$\text{For } O_2, \quad E° = 1.23 \text{ V}$$

then

$$Eh = 1.23 - \frac{0.059}{4} \log \frac{[H_2O]}{P_{O_2}} - \frac{0.059}{4} 4pH$$

Oxygen

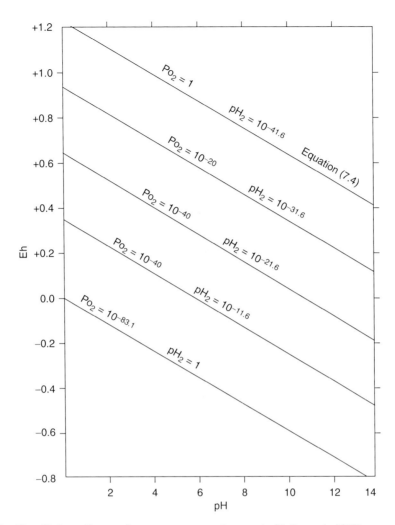

FIGURE 6.2 Eh–pH phase diagram for oxygen–water redox couple (Bolin et al., 1979).

When the activity of pure water is unity and $P_{O_2} = 1$ atm, then we obtain

$$Eh = 1.23 - 0.059 pH$$

At pH 7, Eh for oxygen reduction will be 0.817 V (Figure 6.2).

6.3 SOIL GASES

Depending on the hydrologic regime and intensity of soil reduction, wetlands accumulate various gases in soil pore water (Table 6.1). Many of these gases are end products of reduction reactions that occur in wetland soils. In addition to oxygen, these gases include carbon dioxide and methane (end products of respiration), nitrous oxide and nitrogen gas (end products of denitrification), hydrogen sulfide (end product of sulfate reduction), hydrogen (end product of fermentation), and ethylene. In alkaline soils, ammonia gas may also accumulate in soil pore water.

TABLE 6.1
Soil Gases Produced in Wetland Soils as a Result of Various Biogeochemical Processes

Biogeochemical Process	Soil Gases
Respiration	Carbon dioxide
Photosynthesis	Oxygen
Denitrification	Nitrous oxide, dinitrogen
Nitrification	Nitrous oxide
Ammonia volatilization	Ammonia
Sulfate reduction	Hydrogen sulfide
Fermentation	Hydrogen
Nitrogen fixation	Hydrogen
Methanogenesis	Methane
Phosphate reduction	Phosphine

TABLE 6.2
Henry's Law Constants for Select Gases at 1 bar Total Pressure and at Different Temperatures, Expressed in mol L^{-1} bar^{-1}

T (°C)	Oxygen	Carbon Dioxide	Nitrogen	Hydrogen Sulfide
0	2.18×10^{-3}	7.64×10^{-2}	10.5×10^{-4}	2.08×10^{-1}
5	1.91×10^{-3}	6.35×10^{-2}	9.31×10^{-4}	1.77×10^{-1}
10	1.70×10^{-3}	5.33×10^{-2}	8.30×10^{-4}	1.52×10^{-1}
15	1.52×10^{-3}	4.55×10^{-2}	7.52×10^{-4}	1.31×10^{-1}
20	1.38×10^{-3}	3.92×10^{-2}	6.89×10^{-4}	1.15×10^{-1}
25	1.26×10^{-3}	3.39×10^{-2}	6.40×10^{-4}	1.02×10^{-1}
30	1.16×10^{-3}	2.97×10^{-2}	5.99×10^{-4}	0.91×10^{-1}
35	1.09×10^{-3}	2.64×10^{-2}	5.60×10^{-4}	0.82×10^{-1}
40	1.03×10^{-3}	2.36×10^{-2}	5.28×10^{-4}	0.74×10^{-1}

Source: Pagenkopf (1978).

The concentration of gases in soil pore water and in the water column is related to their solubility in solution and their partial pressure in air, as described in the following equation:

$$[C_{gases}]_{aq} = K_H [P_{gases}] \quad (6.1)$$

where $[C_{gases}]_{aq}$ is the concentration of gaseous species in liquid medium in mol L^{-1}, K_H the Henry's law constant in mol L^{-1} bar^{-1}, and $[P_{gases}]$ the partial pressure of gaseous species in the air expressed in bars (1 atm = 1.01325 bars = 101.3 kPa). Henry's law constants at various temperatures for select gases are shown in Tables 6.2 and 6.3.

Dissolved oxygen concentration in the water column at 25°C:
Solubility of oxygen in water is proportional to the partial pressure of oxygen in the atmosphere and temperature, and can be expressed according to Henry's law as follows:

$$[O_2]_{aq} = K_H P_{O_2} \quad (6.2)$$

TABLE 6.3
Henry's Law Constants for Select Gases at 1 atm, Total Pressure and 25°C, Expressed in mol L⁻¹ atm⁻¹

Gas	K_H (mol L⁻¹ atm⁻¹ 25°C)
Methane	1.34×10^{-3}
Dinitrogen	5.2×10^{-4}
Nitrous oxide	2.6×10^{-2}
Ammonia	57.5
Hydrogen	7.90×10^{-4}

where $[O_2]_{aq}$ is the concentration of oxygen in liquid medium in mol L⁻¹, K_H the Henry's law constant in mol L⁻¹ atm⁻¹, and P_{O_2} the partial pressure of oxygen in the air expressed in atm.

In the atmosphere the oxygen partial pressure is 0.2095 atm. The vapor pressure of water at 25°C is 23.8 mm or 0.0313 atm.

$$P_{O_2} = \left[\frac{\text{(atmospheric pressure - vapor pressure of water)}}{\text{atmospheric pressure}} \right] \times \text{partial pressure of oxygen (volume fraction)}$$

$$= \left[\frac{(760 \text{ mm or } 1 \text{ atm} - 23.8 \text{ mm or } 0.0313 \text{ atm})}{760 \text{ mm or } 1 \text{ atm}} \right] \times 0.2095 \text{ atm}$$

$$= 0.209 \text{ atm}$$

Using Equation 6.2, at 25°C, dissolved oxygen concentration of the water column can be calculated as follows:

$$[O_2]_{aq} = (1.26 \times 10^{-3})(0.2029)$$

$$= 0.256 \times 10^{-3} \text{ mol L}^{-1}$$

$$= (0.256 \times 10^{-3})(32)$$

$$= 8.2 \times 10^{-3} \text{ g L}^{-1} \text{ or } 8.2 \text{ mg L}^{-1}$$

where 32 is the molecular weight of oxygen (g mol⁻¹).

The concentration of oxygen in soil pore waters is generally lower than that in the atmosphere above the soil, so that the net movement is downward, that is, from the atmosphere to soil. In wetlands, this net movement is prevented by the presence of floodwater within the pore space. However, when the water table is below the soil surface, the oxygen movement into soil is similar to uplands.

In well-drained mineral soils, the volume distribution is typically 50% solids, 25% water, and 25% air. The percent air-spaces are dictated by the water-holding capacity of soil. In organic soils, the solid mass may constitute only 15–20%, with the remaining volume consisting of water and air. Typical volume distribution of soil solids, air space, and water content of several Mississippi River floodplain soils of Louisiana is shown in Figure 6.3. The volume of solids remains constant for all soils, but the proportion of air and water volume changes. Well-drained coarse-textured soils generally have a larger volume of pore spaces filled with air, as compared to fine-textured soils. Similarly, organic soils have a high water-holding capacity and low solid mass, and a smaller volume of pore spaces filled with air.

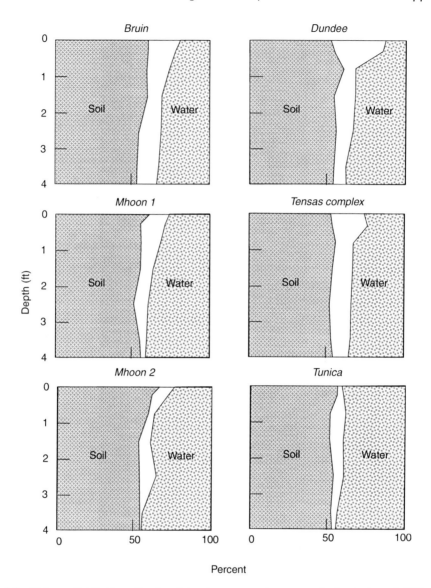

FIGURE 6.3 Volume distribution of soil solids, air space, and water at the beginning of 1970 growing season for Bruin, Mhoon 1, and Mhoon 2 soils at the Northeast Louisiana Experiment Station, and Dundee, Tensas Complex, and Tunica soils at Highland Plantation. (Redrawn from Patrick et al., 1973.)

Gas exchange is necessary to aerate plant roots. In general, oxygen enters the soil and carbon dioxide leaves the soil through the air-filled pores. Thus air-filled porosity directly controls oxygen diffusion and consequently soil oxygen content. Air-filled porosity varies with soil type and texture. In well-drained soils, most of the gas exchange occurs through the soil, whereas in flooded soils, the majority of gas exchange occurs through the plants. In upland soils, good soil aeration is defined as the gas exchange between the soil and the atmosphere, and this process is sufficiently rapid to prevent deficiency of oxygen or toxicity of carbon dioxide for normal functioning of plant roots and aerobic microbial populations. Optimum plant growth occurs in upland soils with air-filled porosities between 20 and 35%. In this case, aeration primarily refers to oxygen concentration. Under some soil conditions, carbon dioxide concentrations may reach toxic levels. The effective pore space for the movement of oxygen and other gases within a given soil is inversely related to soil moisture content. Therefore, the oxygen content of soils is controlled by (i) the percent of soil pores filled with water and (ii) the rate of oxygen consumption by the soil. It is probable that under certain soil conditions low concentrations of oxygen occur as a result of a temporary, but marked, increase in the rate of oxygen consumption by the soil, which cannot be readily offset by the normal process of diffusion.

The oxygen status of a soil is governed by the percent of soil pores filled with water, or air-filled porosity. Air-filled porosity is defined as the bulk volume of soil that is filled with air at any given time or under a given condition (such as specified moisture content), and can be described as follows:

$$S_a = S - \theta \tag{6.3}$$

where S_a is the fraction of soil pores filled with air, S the total porosity, and θ the volumetric water content.

Anaerobic soil volume is regulated by the soil type and water content. As the air-filled porosity increases (due to drainage or hydroperiod), the volume of anaerobic soil decreases (Figure 6.4). The time required to achieve complete aerobic conditions depends on the soil type and the concentration of oxygen-demanding species. In wetlands with saturated soil conditions, $S_a = \theta$; that is, all soil pores are filled with water. Under these conditions, oxygen is introduced into the soil profile through diffusion from overlying floodwater and through plants.

Diffusion of gases into soil can be described by Fick's first law:

$$j = -D\left(\frac{dC}{dX}\right) \tag{6.4}$$

where J is flux (mass diffusing across unit area per unit time, $ML^{-2}T^{-1}$), D the Fickian mass transport coefficient ($L^2 T^{-1}$), C the concentration (ML^{-3}), and X the distance over which a concentration

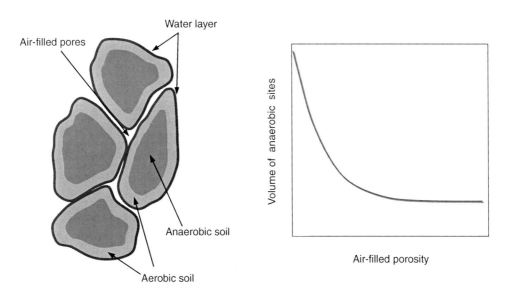

FIGURE 6.4 Relationship between air-filled porosity of soil and the volume of anaerobic sites.

change (gradient) is considered (L); M is mass in units of g or mg, L the distance in units of cm or m, T the time in units of s, min, or days. D is also called the diffusion coefficient when gradients are only due to molecular diffusion of chemical species (see Chapter 14 for details).

Relative diffusion coefficients for oxygen and carbon dioxide in air and water are shown below:
The diffusion of oxygen or carbon dioxide in water is approximately 10,000 times slower than that in still air. The diffusion of gases through soil pores is slower than that through still air, due to decreased area and increased distance. The size of the soil pores is relatively large for diffusion; thus, the slower diffusion is apparently not due to wall collision. Fine-textured soils have larger pore space (porosity) than coarse-textured soils. In field moist soils, the diffusion of gases in coarse-textured soils is typically faster than that in fine-textured soils, where a greater proportion of the pore spaces is filled with water. Since the cross-sectional area available for diffusion is influenced by the air-filled porosity of soil, the following linear relationship can be obtained to describe movement of oxygen in soils:

$$D_s = D_0 S_a T \quad (6.5)$$

where D_0 is the diffusion coefficient in air, D_s the diffusion coefficient in soil pores filled with air, and T the tortuosity factor. The T value is 0.66 for many mineral soils.

Soil oxygen or the aeration status of soils can be determined by various techniques, which can be used to characterize wetland soils undergoing seasonal hydrologic fluctuations and to delineate wetlands from uplands. These techniques are as follows:

- Redox potential
- Oxygen diffusion rate (ODR)
- Soil oxygen content

6.3.1 REDOX POTENTIAL

Redox potential that measures electron activity in soils is used as an indicator of soil aeration status in upland environments. Changes in Eh values are small if sufficient oxygen is present in the soil pores. Aerated soils have characteristic Eh values in the range of +300 to +500 mV and air-filled porosity of 0–60% (Figure 6.5). The low concentration of redox couples in aerated soils reduces

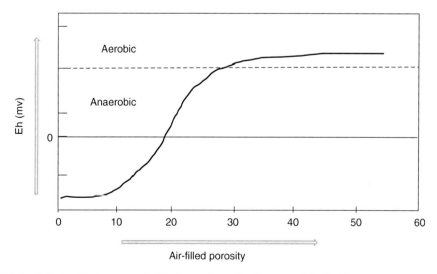

FIGURE 6.5 Relationship between air-filled porosity and redox potential of soil.

Oxygen

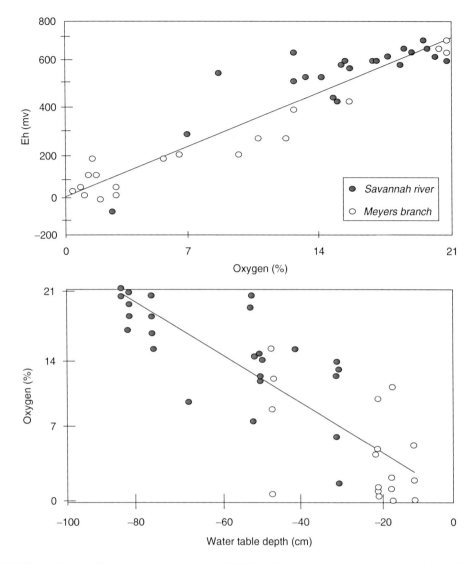

FIGURE 6.6 Relationships between redox potential (pH = 5), oxygen content, and water-table depth. Each point is an average of 22 monthly collections at each depth ($n = 4$) in each plot ($n = 10$) (Megonigal et al., 1993).

the stability, reproducibility, and general usefulness of Eh measurements (see Chapter 4 for detailed discussion on Eh). Soil Eh decreases rapidly when treated with a readily available energy source (e.g., glucose), especially at air-filled porosities of <20%. This is due to the higher rate of oxygen consumption than that of oxygen diffusion. However, at air-filled porosities >30%, oxygen diffusion in soils is adequate to counteract the oxygen consumption rate.

Field measurements of Eh are significantly correlated with soil oxygen (Figure 6.6), suggesting that Eh measurements can provide a reasonable indication of soil aeration status (Megonigal et al., 1993; Faulkner et al., 1989). Soil Eh and oxygen levels also respond to water table fluctuations in wetlands (Figure 6.7).

Relationships presented in Figure 6.7 indicate that oxygen was detected at Eh of 0 mV. However, laboratory studies using stirred soil suspensions repeatedly showed very little or no oxygen at Eh of <300 mV (pH = 7) (Figure 6.8). After short-term flooding of a silty clay loam soil, Eh decreased

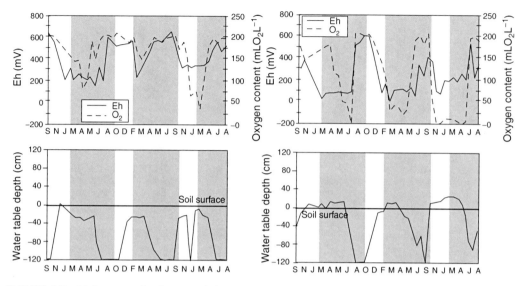

FIGURE 6.7 (a) Oxygen and redox potential (Eh) profile with water table depth for transitional plot (sampling period 1982–85). Shaded area denotes growing season. (b) Oxygen and redox potential (Eh) profile with water table depth for wetland plot. Shaded area denotes growing season. (Adapted from Faulkner and Patrick, 1992.)

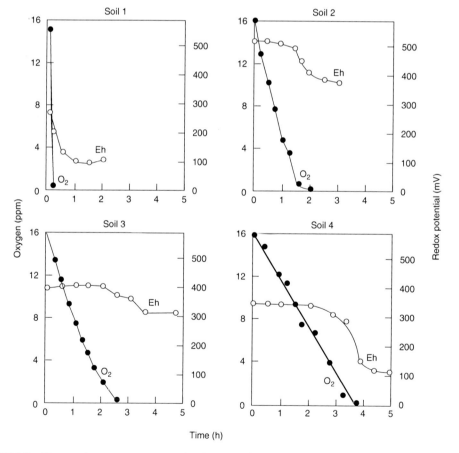

FIGURE 6.8 Changes in oxygen content and redox potential with time in stirred suspensions (Engler et al., 1976).

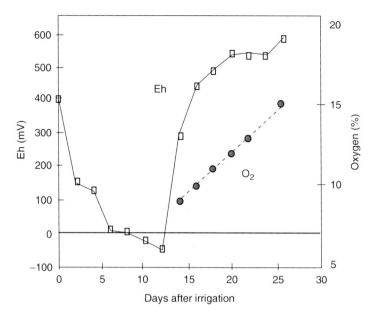

FIGURE 6.9 Redox potential and soil oxygen content after short-term flooding (Meek and Grass, 1975).

rapidly during the first 12 days (Figure 6.9), followed by a rapid increase when the soil was allowed to dry, indicating a rapid response in Eh to flooding and draining conditions (Meek and Grass, 1975). In spite of the limitations of Eh measurements, this technique provides a rapid evaluation of soil aeration status and can be measured on a larger number of sites with minimal expense.

6.3.2 OXYGEN DIFFUSION RATE

The ODR method relies on amperometric or polarographic techniques to measure the oxygen concentration of the soil pores. The amperometric techniques imply the measurement of current. The polarographic techniques imply that the electrode at which reduction of reactants occurs is in a polarized condition. Under polarized conditions, the concentration of reactants at the electrode surface is low, whereas the concentration of products is high. Under these conditions, the amount of current that the electrode will pass is directly proportional to the flux of reactant to the electrode surface. These principles are used in the ODR method.

Unlike Eh measurements, in ODR measurements external energy is used to make oxidation–reduction reactions take place. Reactions are generally nonspontaneous and are typically mediated by microbes. Depending on the pH, the following two reduction reactions have been proposed. Reduction of oxygen occurs in two steps, with H_2O_2 appearing as the intermediate product. During oxygen reduction to water, for one mole of oxygen a maximum of four electrons are used.

(a) $O_2 + 2H^+ + 2e^- = H_2O_2$

$H_2O_2 + 2H^+ + 2e^- = 2H_2O$

(b) pH = 5–12

$O_2 + 2H_2O_2 + 2e^- = 2OH^- + H_2O_2$

$H_2O_2 + 2e^- = 2OH^-$

The ODR method involves the chemical reduction of oxygen at the surface of a thin Pt electrode maintained at constant potential and negatively polarized with respect to a saturated calomel or

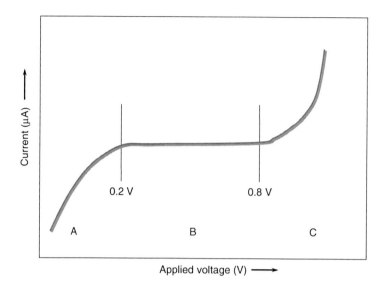

FIGURE 6.10 Current–voltage relationship where [A] = low applied voltage, where O_2 reduction rate is controlled by low activation energy limiting e^- transfer from electrode to O_2; [B] = plateau region where current becomes independent of applied voltage; [C] = increase in current flow at high voltage results in sufficient activation energy to reduce H^+ to H_2. (Redrawn from McIntyre, 1970.)

silver chloride electrode (Lemon and Erickson, 1955). A detailed review of the use of the ODR method to determine soil aeration is presented by McIntyre (1970). When electrical potential is applied between two electrodes placed in a soil system, current will flow (Figure 6.10). If a potential of the order of 0.1 V is applied, the current flow is negligible. When the potential is increased above 0.2 V, the current increases rapidly with each increment in voltage. This is because electrons on the Pt electrode surface acquire a potential sufficient to cause them to react with some of the dissolved oxygen at the surface of the electrode. As the potential increases, the reaction rate increases until a point is reached when the oxygen concentration at the electrode surface is zero. At voltages between 0.2 and 0.8 V, electrons on the Pt surface have sufficient activation energy to reduce the dissolved oxygen in the solution. The plateau region between 0.2 and 0.8 V indicates a point at which the current becomes independent of the applied voltage and is limited by the diffusion of oxygen to electrode surface.

The current intensity is proportional to oxygen flux at the electrode surface and is described as

$$i = nFAO_f \qquad (6.6)$$

where i is the current intensity (A), n the number of electrons per mole of O_2, A the electrode surface area (m^2), F the Faraday constant (96,500 C mol^{-1}), and O_f the oxygen flux (mol m^{-2} s^{-1}). (C = coulomb = A s). Oxygen flux (O_f) or ODR can be expressed as follows:

$$O_f\,(\mu g m^{-2} s^{-1}) = \frac{(MW)i}{nFA} = \left(\frac{32}{4 \times 96{,}500}\right)\frac{i}{A} \qquad (6.7)$$

where i is measured in ΦA and A in m^2, and MW is the molecular weight of O_2.

The ODR method was originally developed to determine soil aeration stress on plants. The diffusion of oxygen to the electrode surface was assumed to be similar to that to roots. Thus, the higher the ODRs, the greater the oxygen availability to roots. The ODR measurements are applicable in drained soils and have very little value in saturated soils where there is no oxygen. Nevertheless, the

ODR measurements were made in soils subjected to short-term flooding. As shown in Figure 6.11, the ODR change was minimal in the Eh range of −200 to +200 mV, as compared to rapid increase at Eh >200 mV. Although, both Eh and ODR respond to changes in the oxygen content of soils, ODR may be more sensitive at Eh >300 mV as compared to Eh measurement under oxygenated conditions.

For the ODR method, the Pt electrode surface is in direct contact with the soil, with a negatively polarized calomel reference electrode (Figure 6.12). The electrodes are similar to those of redox electrodes, except that the current is measured in the ODR method, whereas voltage is measured

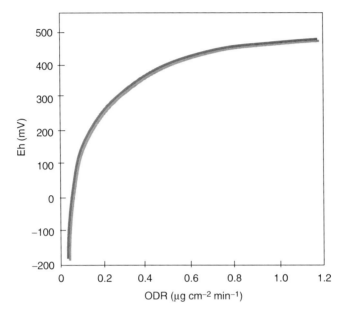

FIGURE 6.11 Oxygen diffusion rates (ODRs) and oxidation–reduction potential in North Yorkshire valley bog peat. (Redrawn from Armstrong, 1967.)

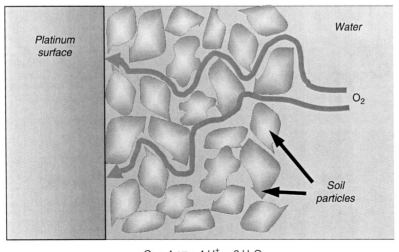

$$O_2 + 4\ e^- + 4\ H^+ = 2\ H_2O$$

FIGURE 6.12 Schematic showing the platinum tip of an electrode in direct contact with soil and water.

for Eh (see Chapter 4). A second type of electrode is the oxygen-permeable membrane–covered electrode, which is commercially available and is widely used in measuring the dissolved oxygen content of soil and water. The principle of operation in both electrodes is similar. The membrane is used as a filter and can prevent poisoning of the electrode. This electrode is bulky and cannot be installed directly in the field, and is restricted for use in water systems.

6.3.3 Soil Oxygen Content

Several approaches are used to determine the soil oxygen content, including obtaining gas samples from a soil profile and analyzing on a gas chromatograph (GC) equipped with a thermal conductivity detector. This approach is tedious and there is a potential of contaminating samples with atmospheric oxygen. Alternatively, a polarographic membrane–covered electrode can be used for direct measurement of oxygen in soil and water. A simple field method that adapts a polarographic oxygen electrode for this application was developed by Patrick (1977) and later modified by Faulkner et al. (1989). This method essentially involves withdrawing a small sample of air from permanently installed air reservoirs (Figure 6.13) in soils at predetermined depths into a cell containing a polarographic oxygen electrode that is connected to a portable oxygen meter. This method was modified for use in saturated

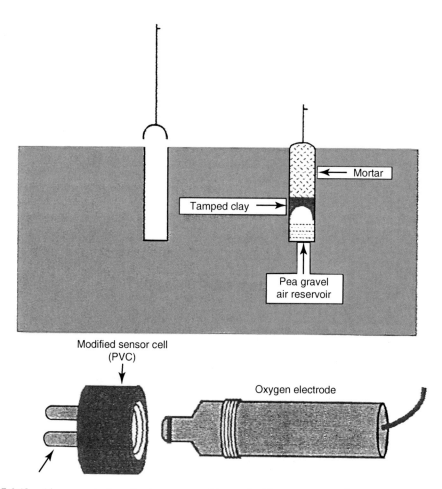

FIGURE 6.13 Air reservoirs in soil using copper tubing and rubber stoppers, and an oxygen analyzer system showing specially machined analyzer cell. Bottom view shows a specially machined cell on the left and an oxygen electrode on the right (Faulkner et al., 1989).

Oxygen

soils (for details of the method see Faulkner et al., 1989). Using this technique, the oxygen content of wetland soils was measured at predetermined depths (see Figure 6.6).

6.4 SOURCES OF OXYGEN

The sources of oxygen in wetlands are primarily the atmosphere (diffusion and mixing at the air–floodwater interface), rainfall, and oxygen production during photosynthesis. Typically, floodwater of wetlands contains near-saturation levels of oxygen during the photosynthetic period and often approach values of <20% saturation during the night. Thus, the oxygen content of floodwater can be highly variable within 24 h. Since the solubility of oxygen increases with a decrease in temperature, dissolved oxygen levels also exhibit a distinct seasonal trend.

Eutrophication of wetlands can influence the dissolved oxygen content of the water column as a result of an increase in organic matter production. The dissolved oxygen concentration of water and soil columns obtained from nutrient-impacted and -unimpacted sites in the northern Everglades wetland soils is shown in Figure 6.14. Oxygen profiles show distinct diel variations, with filamentous algae and periphyton playing a dominant role in the production and consumption of oxygen. Oxygen production during photosynthesis occurs as follows:

$$CO_2 + 4e^- + 4H^+ = CH_2O + H_2O$$
$$2H_2O = O_2 + 4e^- + 4H^+$$
$$CO_2 + H_2O = O_2 + CH_2O$$

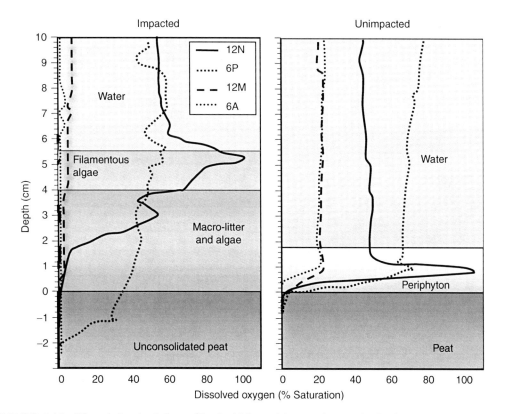

FIGURE 6.14 Diurnal dissolved O_2 profiles in high- and low-nutrient wetland microcosms (DeBusk and Reddy, 2003).

In the eutrophic soil cores, the oxygen concentration of the water column reached low levels during the night as a result of a high respiratory demand by microorganisms. During the day high oxygen levels are the result of active photosynthesis mediated by algae.

Peak oxygen production is observed during midafternoon in periphyton mats when photosynthesis is much greater than respiration. In the evening, oxygen production is slower, and oxygen diffuses into the litter layer. During the night, oxygen levels approach near zero levels in the soil columns obtained from areas impacted by nutrient loading, whereas oxygen levels decrease to only 20% of saturation in the soil columns obtained from unimpacted areas. Similarly, submerged macrophytes can also alter the dissolved oxygen level of the water column with supersaturation levels during the day and negligible levels during the night.

6.5 AEROBIC–ANAEROBIC INTERFACES

Flooding soils for either short-term (irrigation or rainfall) or long-term (wetlands or paddy fields) results in displacement of soil oxygen. Dissolved oxygen present in the pore water is rapidly consumed by aerobic bacteria during their respiratory activities, and depletion of oxygen results in

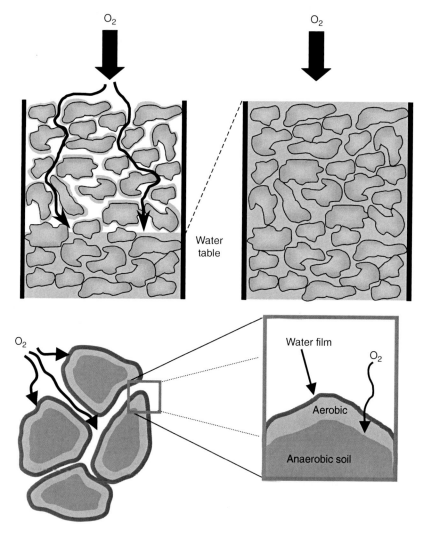

FIGURE 6.15 Schematic showing oxygen diffusion into a wetland soil: (a) saturated soil conditions and (b) water table below the soil surface.

switching of many aerobic bacteria to a facultative role and promotes the use of alternate electron acceptors. In soils flooded for a short period, renewal of oxygen is obtained after drainage of excess pore water, whereas in poorly drained soils or soils with high energy source (histosols or mineral soils with high organic matter), oxygen demand may remain much greater than oxygen supply.

When soils are flooded for a short time, as is the case in many poorly drained soils (such as soils with high clay content), displacement of soil oxygen increases the volume of anaerobic zones. As the soil is drained, air-filled porosity increases, resulting in decreased anaerobic sites. Anaerobic microsites within the soil aggregates, however, can persist for long periods, especially in fine-textured soils, creating aerobic–anaerobic interfaces within the soil aggregate (Figure 6.15). These anaerobic microsites play a significant role in nitrate losses through denitrification in upland soils. In addition to flooding, application of oxygen-demanding materials (such as organic wastes or ammoniacal fertilizers) can also result in rapid depletion of oxygen thus creating anaerobic zones. This is more prevalent in poorly drained soils.

In wetland soil, flooding creates two zones: (i) an oxygen-free soil zone and (ii) oxygen-containing floodwater. These conditions result in diffusion of oxygen to the anoxic sites and consumption of oxygen by the anoxic site. The oxygen concentration in floodwater varies both spatially and temporally and is often present at saturation levels during the photosynthetic period. Slow ODR in water and high oxygen demand by soil result in the consumption of oxygen at the soil surface and formation of thin oxidized or aerobic layer (Figure 6.16). Under poorly drained, low water table, or flooded soil conditions, two distinct soil zones are created: (i) an aerobic soil layer where aerobic microbes are involved in biogeochemical reactions and oxygen is used as an electron acceptor, and (ii) an anaerobic soil layer where facultative anaerobes and obligate anaerobes function. In this second layer where oxygen is absent, nitrate, oxides of iron and manganese, sulfate, and carbon dioxide are used as alternate electron acceptors. The aerobic layer at the soil–floodwater interface is not a fixed layer. It varies within a given day depending on the photosynthetic activity in the water column and the concentration of oxygen-consuming reduced compounds. Two examples of oxygen

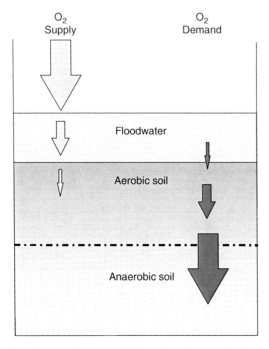

FIGURE 6.16 Schematic showing aerobic–anaerobic interfaces in a wetland soil.

profiles are (1) flooded soils with very little or no mixing in the water column and (2) saturated soil aggregate (see Figures 6.17–6.19). In flooded soil, oxygen concentration in the water column is relatively constant. Slow diffusion of oxygen and rapid consumption in the surface soil resulted in steep oxygen gradient, with concentrations approaching near zero levels at a depth of 10 mm (Figure 6.17). Similarly, oxygen concentrations within the soil aggregate also show steep gradients (Figures 6.18 and 6.19).

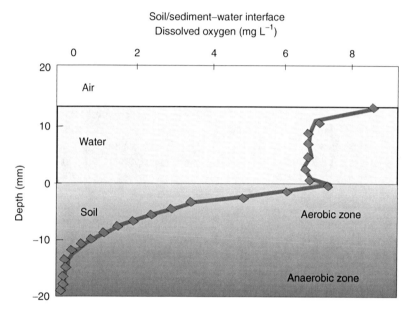

FIGURE 6.17 Aerobic–anaerobic interfaces in a wetland soil: dissolved oxygen concentration in soil and water column. (D'Angelo, E. M., and Reddy, K. R., unpublished results.)

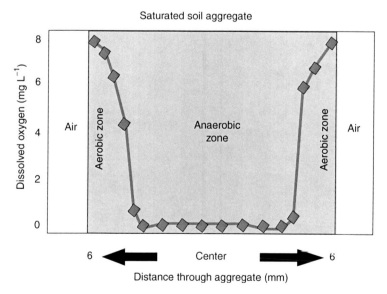

FIGURE 6.18 Aerobic–anaerobic interfaces in a wetland soil: dissolved oxygen concentration in a saturated soil aggregate. (D'Angelo, E. M., and Reddy, K. R., unpublished results.)

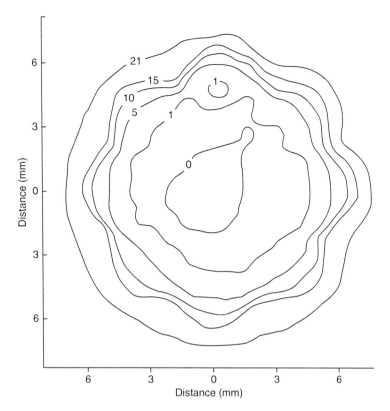

FIGURE 6.19 Spatial distribution of oxygen within a soil aggregate measured by oxygen microelectrode. Center of the aggregate showing anaerobic conditions (Sexstone et al., 1985).

Floodwater oxygen concentration in wetlands remains relatively high due to:

- Low density of oxygen-consuming organisms
- Photosynthetic oxygen production by algae and, possibly, by higher plants
- Mixing of water due to wind action and convection current

Photosynthesis influences oxygen level in the water column. Oxygen production during photosynthesis occurs as follows: In eutrophic wetlands, oxygen concentration of the water column typically reaches low levels or becomes anoxic during the night as a result of a high respiratory demand by microbes (Figure 6.14). During the day high oxygen levels in the water column can occur as a result of active photosynthesis mediated by algae. If the surface wetland soils and sediments receive light, even at very low intensities (<30 µmol quanta m^{-2} s^{-1}), photosynthesis of epipelic algal communities growing on the soils/sediments and particulate organic detritus can quickly (minutes) produce high, often markedly supersaturated (200–300% saturation), concentrations of oxygen within a community of less than 2 mm in thickness (Figure 6.20) (Carlton and Wetzel, 1987). This oxygen can diffuse several millimeters into the interstitial water of the supporting soils at rates greater than that of consumption by bacterial respiration and chemical oxidations. By this mechanism, diurnal changes occur in the oxidized microzones of soils from being fully oxidized with a 5- to 10-fold increase in oxidized depth during the day to being fully reduced at night.

In bacterially regulated (nonilluminated, nonphotosynthetic) organic sediments, rates of bacterial metabolism and oxygen consumption increased with increasing temperatures, particularly above 17°C (Kamp-Nielsen, 1975). Above this temperature, release of phosphorus to the overlying

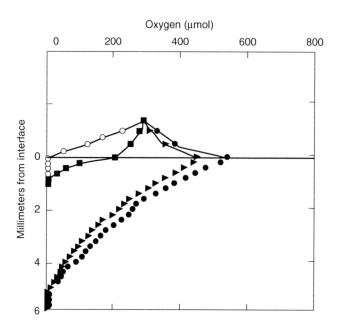

FIGURE 6.20 Oxygen microprofiles (right) during darkness (o—o), and after 1 h (■—■), 8 h (▲—▲), and 10 h (●—●) illumination with 10 μmol quanta m^{-2} s^{-1}. (Extracted from Carlton and Wetzel, 1987, 1988.)

water occurred as the oxidized microzone deteriorated even though dissolved oxygen occurred in the overlying water. Presumably diffusion of oxygen from the overlying water was insufficient to compensate for the microbial consumption. In contrast, Carlton and Wetzel (1988) found that when sediments were weakly illuminated, even less than 10 μmol quanta m^{-2} s^{-1}, and supported a modest epipelic algal community, oxygen production at the surface of the sediments and *within* the sediments quickly (minutes) supersaturated, oxidized the sediments, markedly increased the pH, and suppressed phosphate release into the overlying water. This process was reversed within minutes following darkness. Importantly, the effectiveness of the diurnal shifts between highly aerobic to totally anaerobic sediments was optimal at 17°C and above, and became weakly evident at 11°C or less (Carlton and Wetzel, 1988).

6.6 OXYGEN CONSUMPTION

6.6.1 Oxygen as Reactant

High levels of oxygen can be toxic to many microorganisms, especially to obligate anaerobes such as clostridia and methanogens. These bacteria are killed when exposed to air as a result of hydrologic fluctuations. Deleterious effects of oxygen on anaerobic bacteria were linked to the toxic intermediates formed during reduction of oxygen. Oxygen reduction to water occurs in four steps, with utilization of one electron during each step (Madigan et al., 2000). During reduction, introduction of external electrons into the orbital of oxygen results in the conversion of oxygen atom into:

- Superoxide (O_2^-): $O_2 + e^- = O_2^-$
- Peroxide (H_2O_2): $O_2^- + e^- + 2H^+ = H_2O_2$
- Hydroxyl radical (OH): $H_2O_2 + e^- + H^+ = H_2O + OH$
- Reduction of hydroxyl radical: $OH + e^- + H^+ = H_2O$

Oxygen

The formation of superoxide is the result of one electron transfer by several coenzymes in ETS, including flavins, flavoproteins, quinones, and iron sulfur proteins. This product has a longer half-life than other intermediates and is toxic to anaerobic bacteria. Peroxidase is formed by two electron transfers and mediated by flavoproteins. Peroxidase is further reduced to the hydroxyl radical with the addition of one electron followed by subsequent reduction to water by the addition of another electron. The oxidative effect of these intermediates can result in the destruction of cells. The aerobic bacteria have enzyme systems such as superoxidase dismutase, peroxidase, and catalase to reduce the toxic levels of these intermediates.

6.6.2 Oxygen as an Electron Acceptor

Oxygen reaching the soil surface from the water column is consumed by the following biochemical processes (Figure 6.21).

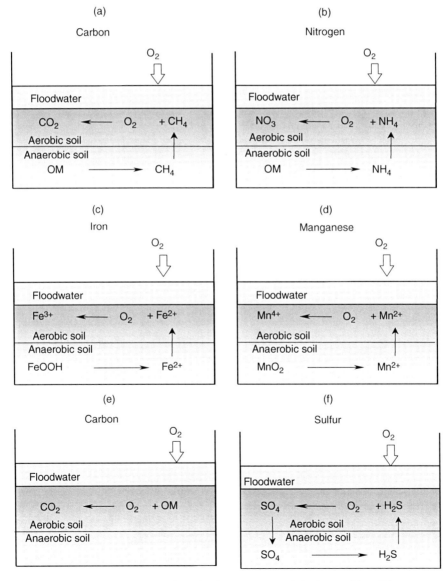

FIGURE 6.21 Schematic showing oxidation of reduced compounds in the aerobic soil layer.

1. Heterotrophic microbial respiration in aerobic soil layers, where oxygen is used as an electron acceptor:

$$C_6H_{12}O_6 + 6O_2 = 6CO_2 + 6H_2O$$

2. Chemical oxidation of reductants such as reduced Fe(II) and Mn(II) and sulfides that diffused from the anaerobic layer to the aerobic layer. These reduced compounds are the products of an alternate electron acceptor reduction during respiration by facultative anaerobes:

$$4Fe^{2+} + O_2 + 4H^+ = 4Fe^{3+} + 2H_2O$$

3. Lithotrophic oxidation of ammonium and methane by obligate aerobic organisms:

$$NH_4^+ + 2O_2 = NO_3^- + 5H_2O + 2H^+$$
$$CH_4 + 2O_2 = CO_2 + 2H_2O$$
$$H_2S + 2O_2 = SO_4^{2-} + 2H^+$$

Under all of these conditions, the aerobic layer serves as an effective sink for the reduced forms of several inorganic and organic compounds. A detailed discussion of oxidation and reduction of these compounds can be found in other chapters. For example, the role of oxygen in heterotrophic respiration is discussed in detail in Chapter 5.

The supply of oxygen to the soil surface and the consumption of oxygen in the soil are recognized as the major regulators determining the thickness of the aerobic soil layer.

Thickness of aerobic (oxidized) soil layer is influenced by

- Oxygen concentration in the overlying water
- Concentration of the reduced compounds (reductants) in the anaerobic soil zone
- Concentration of the energy source (organic matter)
- Photosynthetic activity of periphyton and submerged macrophytes
- Bioturbation by macroinvertebrates

The thickness of the aerobic layer varies from <1 mm to 3 cm. In relation to anaerobic soil volume, the aerobic soil volume at the soil–floodwater interface is small. However, this thin aerobic interface in the proximity of anaerobic soil is key to many unique biogeochemical processes functioning in wetlands. The differentiation of a wetland soil or sediment into two distinct zones as a result of limited oxygen penetration into the soil was first described by Pearsall and Mortimer (1939) and Mortimer (1941).

The thickness of aerobic layer can be determined by measuring

- The oxygen concentration of soil using microelectrodes
- The redox potential profile at the soil–floodwater interface
- The distribution of oxidized and reduced compounds

The oxygen concentration of the overlying water has a strong influence on the thickness of the aerobic soil layer. The increased levels of oxygen in the atmosphere above the floodwater can increase the flux of oxygen from the floodwater to underlying soil (Table 6.4). The thickness of the aerobic layer increased from 0.5 mm at 4% oxygen to 2.4 mm at 21% oxygen in a swamp soil with 6.2% organic matter (Table 6.5). Other studies have shown that the thickness of the aerobic soil layer can range from 2 to 15 mm, depending on the soil type and the amount of energy source available for heterotrophic respiration.

TABLE 6.4
Diffusion Coefficients for Oxygen and Carbon Dioxide in Air and Water

Medium	Oxygen (D cm² s⁻¹)	Carbon Dioxide (D cm² s⁻¹)
Air	0.226	0.181
Water	2.6×10^{-5}	2.04×10^{-5}

TABLE 6.5
Experimental Values for Oxygen Consumption and Thickness of the Aerobic Layer in Swamp Sediments

Oxygen Content (%)	Oxygen Consumption (mg m⁻² h⁻¹)	Thickness of Aerobic Layer (mm)[a]	
		A	B
4	10.4	0.5	0.8
8	21.6	1.3	1.8
12	20.5	1.6	2.1
16	22.7	1.7	2.0
20	26.3	2.4	2.2

[a] A = Estimated from Eh measurements and B = estimated from Fe^{2+} and Fe^{3+} distribution.
Source: Howeler (1972) and Howeler and Bouldin (1971).

TABLE 6.6
Influence of Added Carbon (as Rice Straw) to a Crowley Silt Loam Soil on the Thickness of Aerobic Soil Layer as Estimated by Redox Profile[a]

Organic Carbon Added (mg C g⁻¹)	Thickness of Aerobic Layer (mm)
0	13
0.45	10
2.25	2
9.0	0

[a] Aerobic conditions were assumed at Eh > 300 mV.
Source: Engler and Patrick (1974).

The thickness of the aerobic layer is also affected by the activity of heterotrophic organisms in the water column/soil–floodwater interface. Soils with high organic matter content typically have thin aerobic layers as compared to soils with low organic matter that have a thicker aerobic soil layer (Table 6.6). Dissolved oxygen profiles (Figures 6.17 and 6.18) show steep gradients in the oxygen profiles between soil and water column. The thickness of the aerobic layer during the day is higher (due to photosynthesis) and decreases to negligible levels during the night (due to high demand for oxygen during respiration and reduced photosynthetic activity in the water column).

Concentrations of reduced compounds (such as ferrous iron and manganous manganese, ammonium, sulfide, and methane) in the anaerobic layer can affect the thickness of the aerobic layer. The reduced compounds are in mobile and immobile forms. Mobile reductants are

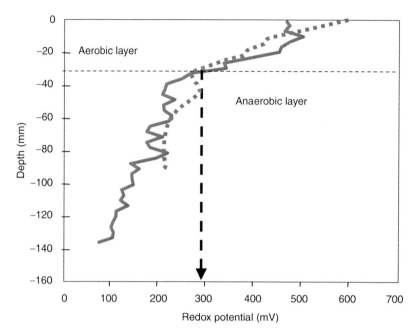

FIGURE 6.22 Aerobic–anaerobic interfaces in a wetland soil as determined by redox potential. (Fisher, M. M., and Reddy, K. R., unpublished results.)

produced in the underlying anaerobic layer and diffuse into the aerobic layer in response to a concentration gradient established across this surface. This source provides a steady supply of reduced compounds to the aerobic layer. In swamp soils, about 0.53 mg Fe^{2+} diffused into the aerobic layer in 260 h, and the amount of oxygen required to oxidize this reductant is 0.075 mg cm^{-2} (Howeler and Bouldin, 1971). The ferrous iron is oxidized to ferric iron and forms an insoluble precipitate. This form is immobile and resides in the aerobic layer. When oxygen depletion occurs in the aerobic layer due to either respiration during the night or the addition of organic reductants (such as litter and organic matter loading), the Fe(III) oxides are reduced to Fe(II) by microorganisms using this oxidant as an electron acceptor. The thickness of the aerobic layer predicted by Fe(III) and Fe(II) distribution agreed with the thickness measured using Eh profiles (Table 6.6).

The thickness of the aerobic layer can be determined by measuring Eh as a function of depth; redox potentials show sharp gradients at the soil–floodwater interface. Laboratory studies have indicated complete disappearance of O_2 at Eh <300 mV (pH = 7.0). This Eh value was used as a boundary between aerobic and anaerobic layers. A simple technique to determine redox profiles as a function of depth is described by Patrick and DeLaune (1972). This method involves a special motor-driven assembly that advances a platinum electrode at a rate of 2 mm h^{-1} through a soil profile. Redox potential is recorded continuously on a recorder or a data logger. Examples of Eh profiles are shown in Figure 6.22.

Benthic invertebrates can transport oxygen (and other electron acceptors) from water to the underlying soil during burrowing, processing sediments for food, or pumping oxygen through burrows for respiration. The resulting process of mixing at the soil–floodwater interface is known as bioturbation (Figure 6.23). A detailed discussion of bioturbation in sediments is presented by Aller (1982). The role of benthos in oxygen flux and nutrient cycling in lake sediments is widely studied, but no information is available for wetlands. Benthic animals in freshwater sediments are dominated by a few groups: oligochaetes, chironomids, amphipods, and bivalves. These animals play different roles, including feeding and defecation at the sediment surface, respiratory irrigation

Oxygen

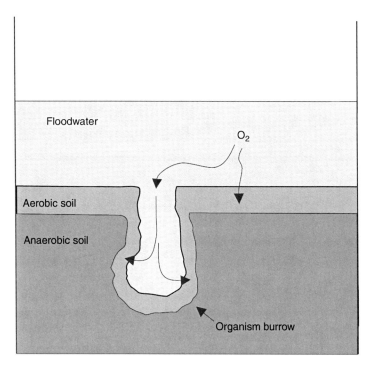

FIGURE 6.23 Schematic representation of burrowing by benthic invertebrates in sediments. Aerobic soil layer around irrigated burrow environment.

by constructing burrows in the sediments, and pumping of water and sediment at the interface. The movement of these animals at the sediment–water interface can result in the introduction of oxygen and increase in the thickness of the aerobic layer. For example, chironomid larvae may extend semipermanent tubes to depths more than 10 cm, resulting in the transport of oxygen along these tube channels.

The aerobic soil layer functions as an effective sink for reductants diffusing from the underlying anaerobic layer. Both chemical and biological processes regulate the consumption of oxygen in the aerobic layer. Oxygen consumption by anaerobic soils is best described by a two-phase first-order reaction (Figure 6.24; Reddy et al., 1980):

$$C = C_0 \exp[-(k_I + k_{II})t] \tag{6.8}$$

where C is the oxygen concentration (mg kg^{-1} soil), t the time (h), and k_I and k_{II} are, respectively, the first-order rate coefficients (h^{-1}) for chemical and biological oxidation processes. Oxygen consumption during phase I was described by a single rate constant of 0.15 h^{-1} (coefficient of variation = 20%). Oxygen consumption during Phase II was described by the following empirical relationship:

$$k_{II} = -0.0055 + 8.7 \times 10^{-6}[Fe^{2+}] + 9.9 \times 10^{-5}[NH_4\text{-}N]$$
$$r^2 = 0.82; n = 37 \tag{6.9}$$

Equation 6.9 is based on the oxygen consumption rates measured over a period of 70 h. Phase I reflects chemical oxidation and is primarily regulated by the concentration of reduced inorganic species such as Fe(II). Neither k_I nor k_{II} was correlated with the total or water extractable carbon, which suggests that the oxygen consumption was dominated by reducible Fe(II) and ammonium (which are much more sensitive to oxygen than organic carbon). It is likely that some of the water-soluble Fe(II)

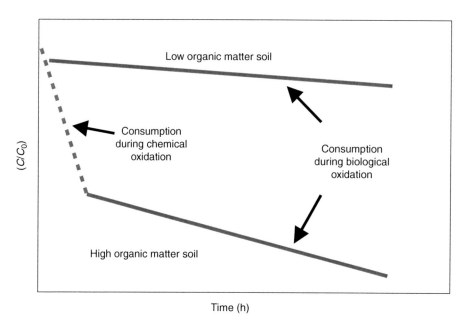

FIGURE 6.24 Rate of oxygen consumption by selected flooded soils from Mississippi river floodplain (Reddy et al., 1980).

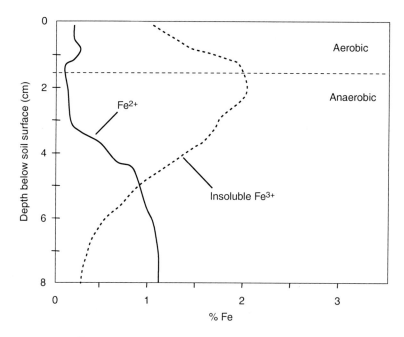

FIGURE 6.25 Mobile (Fe^{2+}) and immobile (Fe oxides) forms of iron in flooded soil profile (Howeler and Bouldin, 1971).

was chelated, thus preventing rapid oxidation during Phase I. However, this chelated Fe was probably oxidized along with the soluble organic carbon chelating this metal. Phase II reflects both chemical and biological oxidation processes. Howeler and Bouldin (1971) found that 50% of the total oxygen consumed by a swamp soil was used in oxidizing (i) water soluble Fe (diffusing upwards (from anaerobic layer to aerobic layer) and (ii) reduced Fe in the soil matrix (Figure 6.25).

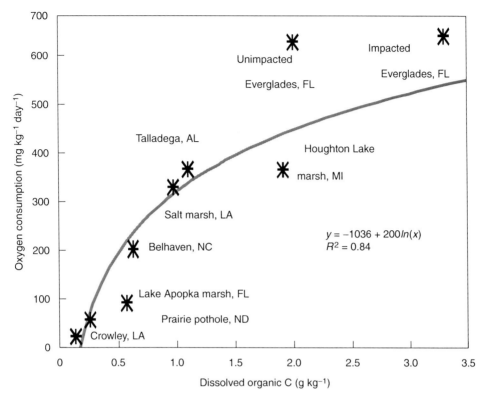

FIGURE 6.26 Relationship between soil oxygen demand (SOD) and heterotrophic microbial respiration by different wetland soils. Each data point represents soil from a different locations in the United States (D'Angelo and Reddy, 1999).

SOD is used as an indicator of oxygen consumption of wetland soils. In wastewater analysis, the SOD measurements are similar to BOD measurements, except that the consumption rates are measured over a 24 h period as compared to 5 days for BOD. The SOD is regulated by several factors, including the soil organic matter content, ammonium, ferrous iron, manganous manganese, sulfide, and methane. In wetland soils collected from various locations in the United States, oxygen consumption (expressed as SOD) was best described by heterotrophic respiration (Figure 6.26). The example in Figure 6.26 shows that oxygen consumption increased with increase in DOC in a range of wetland soils.

6.7 SUMMARY

- In biological systems, oxygen serves as an electron acceptor during respiration by bacteria and as a reactant in certain biochemical reactions. In addition, oxygen can be involved in chemical oxidation of reduced species in wetland soils.
- In drained soils, most of the soil pores are filled with air and are interconnected to the atmosphere, allowing for the rapid exchange of gases, including oxygen. Oxygen transport is sufficient to supply oxygen to support the microbial and plant populations.
- In flooded and waterlogged soils, the pores are filled with water and any dissolved oxygen is rapidly consumed. Under these conditions, oxygen is introduced into the wetland soil profiles by diffusion and mass flow through the floodwater and plants. The oxygen concentration in the soil pore space is lower than that in the atmosphere. In wetland soils, the net movement is restricted by the presence of water in the pore space. The diffusion of oxygen in water is 10,000 times slower than that in air.

- The diffusion of oxygen in air can be described by Fick's first law. Redox potential, which measures electron activity in soils, can be used as an indicator of soil aeration status. Aerated soils have Eh greater than +300 mV. Redox potential or Eh below +300 mV indicates little or no presence of oxygen. In addition to redox potential measurement, ODR and soil oxygen content can be used to determine the aeration status of wetland soils.
- The rate of oxygen consumption is related to the amount of organic matter present in soils. Eutrophication of wetlands can influence the dissolved oxygen content of the water column as a result of increases in the organic matter production. Oxygen demand is greater in the histosol systems that have high organic matter and energy sources.
- In wetland and flooded soils, the soil ODR in water and the high oxygen demand by soil result in the consumption of oxygen at the soil surface and formation of thin oxidized layers or aerobic layers overlying the reduced soils. Thin oxidized–reduced double layers are important in biogeochemical reactions and diffusion and exchange of gases and reduced substances between the water column and the atmosphere.
- Several factors influence the thickness of an oxidized surface layer. These include the oxygen concentration of the floodwater, the soil organic matter content, the amount of reduced compounds in the reduced or anaerobic soil zone, the photosynthetic activity of periphyton and aquatic macrophytes, and the bioturbation by macroorganisms.
- SOD is used as an indicator of oxygen in wetland soils. The SOD is regulated by soil organic matter content and reduced substances in the soil profile.

STUDY QUESTIONS

1. Discuss the factors influencing oxygen exchange in flooded or waterlogged soil. Explain the importance of water-filled porosity as compared to air-filled porosity on oxygen exchange.
2. What two roles does oxygen play in biological systems?
3. How can hydroperiod impact the oxygen status of wetland soils?
4. Write an equation showing oxygen reduction. At what soil Eh is oxygen present?
5. What is Henry's law? What factors are used in determining the solubility of oxygen in water?
6. Define Fick's first law for calculating the diffusion rates of gases in soil.
7. List the methods for determining oxygen or aeration status of soils.
8. What are the sources of oxygen in wetlands?
9. Describe the aerobic and anaerobic interface in wetland soils. List the factors that govern the thickness of the oxidized or aerobic surface layer.
10. List the general biochemical processes that consume oxygen in flooded soils and sediment.
11. What factors would support the rapid consumption of oxygen in soil?
12. What is soil oxygen demand (SOD)? How is SOD measured?
13. Calculate in grams the oxygen needed to oxidize 10 g of the following reductants. Write appropriate balanced equations and calculate the net free-energy yield for each reaction.
    1. $C_6H_{12}O_6$
    2. NH_4^+
    3. HS^-
    4. Mn^{2+}
    5. Fe^{2+}
    6. CH_4

FURTHER READINGS

Aller, R. C. 1982. The effects of macrobenthos on chemical properties of marine sediment and overlying water. In P. L. McCall and M. J. S. Tevesz (eds.) *Animal–Sediment Relations: The Biogenic Alteration of Sediments*. Plenum Press, New York. pp. 53–102.

Faulkner, S. P., W. H. Patrick, Jr., and R. P. Gambrell. 1989. Field techniques for measuring wetland soil parameters. *Soil Sci. Soc. Am. J.* 53:883–890.

McIntyre, D. S. 1970. The platinum microelectrode method for soil aeration measurement. *Adv. Agron.* 22:235–285.

Mortimer, C. H. 1941. The exchange of dissolved substances between mud and water in lakes. *J. Ecol.* 29:280–329.

Revsbech, N. P. 1989. An oxygen microelectrode with a guard cathode. *Limnol. Oceanogr.* 34:474–476.

Revsbech, N. P., B. B. Jørgensen, T. H. Blackburn, and Y. Cohen. 1983. Microelectrode studies of the photosynthesis and O_2, H_2S, and pH profiles of a microbial mat. *Limnol. Oceanogr.* 28:1062–1074.

Turner, F. T. and W. H. Patrick, Jr. 1968. Chemical changes in waterlogged soils as a result of oxygen depletion. *Trans. 9th Int. Congr. Soil Sci.* (Adelaide, Australia) IV:53–65.

7 Adaptation of Plants to Soil Anaerobiosis

7.1 INTRODUCTION

Plants are important in regulating biogeochemical cycles in wetlands. They are the primary source of organic matter in wetlands and play a major role in global carbon cycling. Organic matter is the primary energy source for various microbial communities regulating the biogeochemical cycling of nutrients in wetlands. There are a variety of conditions and processes that affect the type, distribution, and productivity of plants in wetlands. Many of these factors are known and described for plants in all systems (plant ecology); however, many are specific and found only in wetland environments.

Wetland hydrological cycles govern many abiotic and biotic factors including soil oxygen levels and nutrient availability, which in turn govern plant species composition and productivity. Soil reduction is an important process that influences the physiological functions, growth, and distribution of wetland plants. The specific impact of flooding on wetland plants depends on the following:

- Plant species
- Plant age (younger plants are more sensitive)
- Floodwater properties
  - Flowing water vs. stagnant water (flowing water provides more oxygen)
  - Nutrient status (high nutrient status; greater microbial activity)
- Soil properties (soils with high clay and organic matter content create greater oxygen deficiency in root environments)
- Level and duration of flooding
- Season of flooding (smaller impact during plant dormancy)

Saturated soil conditions in wetlands affect plant growth and productivity in several ways. The abundance of water seriously interferes with plant root metabolism, creating root oxygen deficiency. In addition, microbial processes in wetland soils can produce reduced substances potentially toxic to wetland plants. Saturated soil conditions in wetlands affect the reactivity of many inorganic redox-mediated processes, thus influencing adaptations of wetland plants.

The capacity of wetland plants to survive under anaerobic soil conditions is largely dependent on the rate of plant respiration and the rate of oxygen supply from photosynthetic tissue to belowground roots (Figure 7.1). If no molecular oxygen is available in the root tissue of plants, the oxygen deficit is typically termed as anoxia and root respiration is anaerobic. If molecular oxygen is present at low levels (less than atmospheric), the condition is called hypoxia. Oxygen is required for root respiration, which effectively reduces the concentration of soluble inorganic and organic reductants through chemical and biological oxidation (see Section 7.3.1 for various oxidation reactions in which oxygen is involved).

Plants use various morphological and physiological strategies to survive under anaerobic soil conditions (Crawford, 1989). These include the development of cortical intercellular airspaces that are continuous between shoots and roots and function as a conduit for transport of respiratory gases, especially oxygen for aerobic respiration (Figure 7.2). Under severe oxygen-deficient conditions, some plants are capable of respiring through anaerobic fermentation.

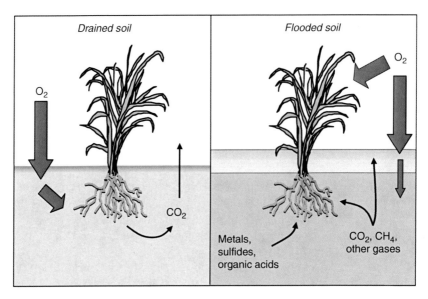

FIGURE 7.1 Schematic showing the exchange of gases in drained and flooded soils.

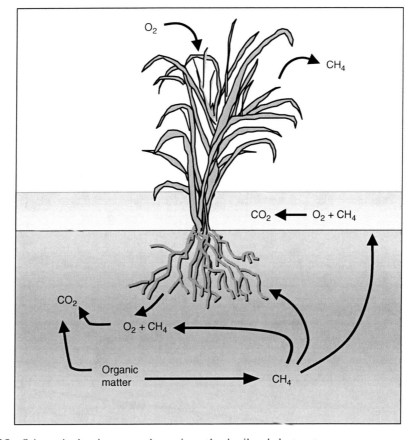

FIGURE 7.2 Schematic showing gas exchange in wetland soil and plant system.

The degree of soil reduction dictates the metabolic responses of the plant. Evaluation of the response of flood-tolerant plant species to oxygen-deficient conditions in wetland soil requires quantifying and characterizing soil substrate condition in which wetland plants grow. Simply quantifying the presence or absence of oxygen (aerobic/anaerobic conditions) in wetland soil is inadequate in evaluating the physiological response of flood-tolerant plant species to reducing soil environment.

Many terms are used to describe the wetland soil conditions including aerobic, anaerobic, waterlogged, flooded, oxidized, or reduced. Although descriptive, such terms refer only to excess water and low levels or absence of oxygen. In many cases, these terms do not truly depict or quantify wetland soil conditions that control plant growth. In contrast, soil redox potential can best reflect soil chemical properties. The redox processes of wetland soils have many uses in evaluating ecophysiology and growth responses of wetland plants. Redox potential values in a wetland soil range over 1,000 mV. Soils can be waterlogged and still have oxygen present, especially if little organic matter or energy source exists to consume soil oxygen. Likewise, a soil can be anaerobic, yet be only moderately reduced. For example, a redox potential of +400 mV indicates that oxygen may be present, even though there may be excess water. A redox potential of +200 mV would indicate the absence of oxygen or nitrate and that bioreducible iron and perhaps manganese compounds would be in oxidized state. A redox potential below −100 mV would indicate the presence of sulfide.

Anaerobic conditions represent Eh levels below +350 mV (DeLaune et al., 1990). Redox potentials below +100 mV are not uncommon in highly reduced soils. Since oxygen is absent at redox potential values of about +350 mV and lower, the absence of oxygen alone reveals little about the intensity of soil reduction. Evaluation of the physiological responses of plants grown at the upper end of the anaerobic range of the redox scale found in wetland soils may yield results that are not the same as for plants growing under more reducing soil environment with a high oxygen demand. Many flood-tolerant wetland species can grow in strong reducing soil conditions far below the redox value where oxygen disappears. Evaluation of the physiological responses of wetland plant species to such conditions should also distinguish differences between plant response to intensity and capacity of soil reduction and oxygen demand in the root environment. Even under strongly reducing conditions, it is unclear whether the plants are responding to stress caused by lack of oxygen in their roots or to toxins (e.g., sulfides) produced by the intense reduction.

7.2 DISTRIBUTION OF WETLAND PLANTS

Wetland plant species are commonly found along an environmental gradient (Reed, 1988). Species occurrence can be based on flooding and salinity regimes. Soil properties (organic matter, pH texture, salt, degree of anaerobiosis) also influence the distribution of wetland plant species. In addition to flooding regime and wetland soil properties, frequency of occurrence and diversity of wetland plant species may also depend on the relationship between intensity and capacity of reduction and root aeration capacity.

Wetland plants are classified by their frequency of occurrence in wetland conditions (Table 7.1). Such occurrence is strongly influenced by soil reduction and the ability of wetland plant species to maintain oxygenated root environment. This is supported by the observation that wetland vegetation can differ over a range of taxonomic soil series that exhibit similar flooding regimes or water-table fluctuations but differ in soil biochemical oxygen demand (BOD) as reflected in soil organic carbon content. Table 7.2 shows representative plants, which are classified as obligate, facultative wetland, facultative, and facultative upland (Reed, 1988). All the plants can tolerate some degree of hypoxia.

TABLE 7.1
Classification of Wetland Plants Based on Frequency of Occurrence in Wetland Environments

Obligate	Facultative Wetland	Facultative	Facultative Upland
Always found in wetlands (frequency 99%)	Usually found in wetlands (frequency 67–99%)	Sometimes found in wetlands (frequency 34–66%)	Seldom found in wetlands (frequency 3%)

Source: 1986 National Wetland Plant List U.S. Fish and Wildlife Services.

TABLE 7.2
Plant Species Representative of Categories Found Growing in Wetlands

Category	Species	Common Name
Obligate	*Panicum hemitomon*	Maiden cane
	Taxodium distichum	Baldcypress
	Nyssa aquatica	Tupelogum
	Cephalanthus occidentalis	Button bush
	Forestiera acuminata	Swamp privet
	Typha latifolia	Cattail
	Cladium jamaicense	Saw grass
Facultative wetland	*Quercus michauxii*	Cow oak
	Fraxinus pennsylvanica	Greenash
	Betula nigra	River birch
	Magnolia virginiana	Sweet bay
	Pinus elliottii	Slash pine
	Sabal minor	Palmetto
	Sorghum halepense	Johnson grass
	Spartina patens	Salt meadow cordgrass
Facultative	*Magnolia grandiflora*	Flowering magnolia
	Myrica cerifera	Wax myrtle
	Pinus taeda	Loblolly pine
	Quercus nigra	Water oak
Facultative upland	*Cornus florida*	Flowering dogwood
	Fraxinus americana	American ash
	Ilex opaca	American holly
	Sassafras albidum	Sassafras
	Carya ovata	Shagbark hickory

7.3 MECHANISMS OF FLOOD TOLERANCE

Most wetland plants are adapted to periods of soil oxygen deficiencies but may differ in ability to endure intense reducing conditions for extended periods (several days). A major stress for plants growing in flooded soil is the lack of external oxygen supply to support aerobic root respiration. Decrease in soil oxygen concentration results in increase in anaerobic root respiration and shift in ATP production via anaerobic pathway (Figure 7.3). Wetland plants use various strategies to cope with reduced oxygen levels in the root environment. These adaptations include the use of morphological, anatomical, and metabolic strategies.

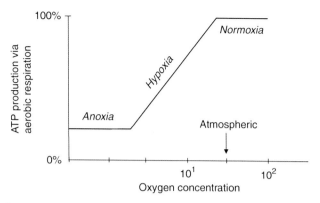

FIGURE 7.3 ATP production via aerobic respiration as a function of oxygen concentration (%).

7.3.1 METABOLIC ADAPTATIONS

Metabolic adaptations in addition to the morphological and anatomical adaptations are critical for wetland plant survival in anaerobic environments. The mechanisms involved are complex and include avoidance of accumulation of toxic compounds and maintenance of a continuous supply of carbohydrate. Under fully aerobic conditions, end products of root metabolism are carbon dioxide and water, where 1 mole of glucose produces 36 moles of ATP. Under anaerobic conditions, the glycolytic pathway produces pyruvic acid from glucose, which is converted into carbon dioxide and ethanol. The process is catalyzed by cytoplasmic enzymes with net conversion of two molecules of ATP from each glucose molecule, thus slowing down ATP synthesis and affecting energy-dependent processes.

Alcoholic fermentation is less efficient in producing ATP than aerobic root respiration; thus, net energy production by the anaerobic pathway is only a fraction of that produced by the aerobic respiration. Root energy metabolism under aerobic and anaerobic conditions is as follows:

- Aerobic conditions

$$C_6H_{12}O_6 + 6O_2 \rightarrow 6CO_2 + 6H_2O = 36 \text{ ATP}$$

  Therefore, 1 mole of glucose $\rightarrow$ 36 moles of ATP
- Anoxia (absence of oxygen)

$$C_6H_{12}O_6 \rightarrow 2C_2H_5OH + 2CO_2 = 2 \text{ ATPs}$$
$$\text{(ethanol)}$$

  1 mole of glucose $\rightarrow$ 2 moles of ATP

The initiation of anaerobic root metabolism reduces the amount of ATP required for the following processes (all of which are needed for plant growth):

- Biosynthesis processes
- Nutrient uptake
- Transport processes

Anaerobic root metabolism processes produce the following products, which can be toxic to plants:

- Ethanol—solubilizes membranes
- Acetaldehyde—solubilizes membranes
- Lactic acid—causes cytoplasmic acidosis (lower cytoplasmic pH)

An increase in anaerobic root respiration can reduce root carbohydrate deficiency by the following processes:

- Ethanol loss by diffusion from the root
- Increased root glucose consumption by some plants (keeps ATP rate stabilized)
- Reduced carbon transport from leaves due to inhibition of phloem loading
- Lower photosynthetic rates
- Reduced water uptake (first factor affecting flood-intolerant plants)

Flood tolerance in a given wetland plant is dependent on the ability to avoid ethanol production in the glycolytic pathway, thus maintaining a low level of ethanol biosynthesis (Figure 7.4). Anoxic or hypoxic conditions increase the activity of the fermentative enzyme, pyruvate decarboxylase (PDC). Alcohol dehydrogenase (ADH) catalyzes the terminal step in alcohol fermentation. Ethanol accumulation in the flood-sensitive species can cause death of the root cells. Plants use several mechanisms restricting the buildup of ethanol to toxic levels including transport from anaerobic tissue to aerated tissue, ethanol leakage to the surrounding areas by diffusion or by transpiration system, and rerouting of glycolytic intermediates to produce different end products such as malate and lactate.

Under prolonged flooded conditions or in wetland soils with high intensity of reduction, the amount of oxygen transported by plants may not be sufficient to support aerobic respiration. When this situation arises, certain wetland plants may survive by anaerobic metabolism. ADH activity, which is induced under anaerobic conditions, has been used as an indicator of root oxygen

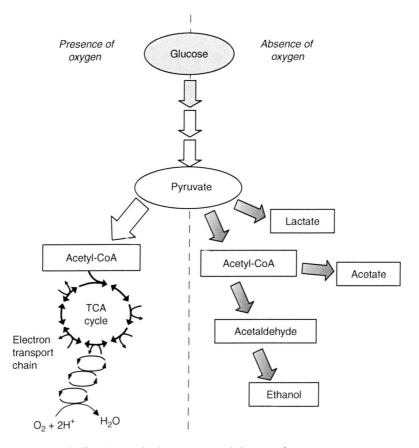

FIGURE 7.4 Plant metabolic pathways in the presence and absence of oxygen.

deficiency (Crawford, 1992; Mendelssohn et al., 1981). ADH is involved in catalyzing the reduction of acetaldehyde to ethanol, an end product during alcoholic fermentation, induced during anaerobic respiration. The ADH activity is inversely related to soil Eh, because of the possibility of anaerobic respiration. However, it is not clear whether the induction of ADH activity aids the survival of plants or merely signals deleterious effects of root oxygen deficiency (Crawford, 1992). Several studies have shown the relationship between intense anaerobic conditions induced by ADH activity and subsequent survival of plants (see Crawford, 1992, for references).

The adenylate energy charge (AEC) ratio (ATP/ATP+ADP+AMP) is also used as an indicator of root energy status. The AEC ratio is generally high in stream bank plants (examples: riparian wetlands or littoral zones of lakes), which reflects aerobic respiration. That is, AEC ratio is high under aerobic conditions and decreases with decrease in soil Eh. Increase in the AEC ratio of inland plants is due to consumption of glucose, resulting in ATP yield (process known as *Pasteur effect*) (Mendelssohn et al., 1981). Increased glucose consumption during alcoholic fermentation can create carbon deficits in plants and decrease productivity. In coastal marshes, *Spartina alterniflora* appears to develop this specific metabolic adaptation to combat the intensity of soil reduction, especially in inland areas. In their study, Mendelssohn et al. (1981) identified three zones of *S. alterniflora* vigor:

- Streamside zone is where plants are highly productive because of aerobic respiration and the production of ATP by oxidative phosphorylation.
- Inland zone is where loss in aerobic ATP production is compensated by anaerobic root respiration resulting in less productive plants.
- Continuously flooded inland zone is where hypoxia impacts root metabolism that restricts plant growth and in some cases causes dieback.

7.3.2 Morphological/Anatomical Adaptations

Morphological responses to flooding include epinasty and hypertrophy phenomena development of adventitious roots, lenticels on woody species, and pneumatophores on black mangrove (*Avicennia nitida*). In addition to the morphological changes, flood-tolerant plants are capable of developing specific anatomical structural characteristics to survive and function under anaerobic soil conditions. Such features include aerenchyma tissues and adventitious root development.

7.3.2.1 Roots

Three types of root regeneration occur in wetland plants (Hook, 1984). These are

- *Adventitious roots*: These occur on the stem above soil surfaces, usually within the flood zone. These are also called adventitious water roots. Root growth is stimulated by water movement, and these roots obtain most of their nutrients from water column.
- *Soil water roots*: These are formed by root regeneration within the soil. This occurs after die-off of the original roots on flooding and regeneration of secondary roots.
- *Altered roots*: These are produced upon flooding, and are morphologically different from original roots. For example, in sweetgum (*Liquidambar styraciflua*) the new roots produced are more succulent and clearer in appearance than original roots.

7.3.2.2 Pneumatophores

Black mangroves develop a unique root system of radial, spongy gas-filled roots running horizontally in the soil (Scholander et al., 1955; McKee and Mendelssohn, 1987). The root can send up

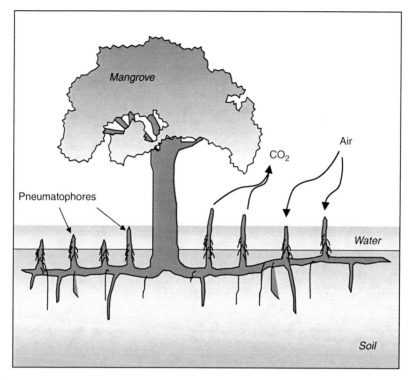

FIGURE 7.5 Schematic showing pneumatophores in mangrove trees.

numerous *air roots*, which provide direct connection to the atmosphere for gas exchange (Figure 7.5). These air roots are called pneumatophores.

7.3.2.3 Lenticels

In woody species such as mature trees, *lenticels* are produced in the periderm layer of impervious corky bark (tissue) of the stem to provide an opening for gas exchange between larger intercellular spaces and the atmosphere (Figure 7.6). Lenticels become hypertrophied on the stem, which consists of complementary cells with few closing layers at various distances from the phellogen (Hook, 1984). These closing layers usually have breaks that provide for gas exchange. Several studies provided conclusive evidence about the role of lenticels in gas exchange (see Hook, 1984). For example, in one study, when stem lenticels of *Nyssa aquatica* (swamp tupelo) were artificially covered, the rhizosphere was not as oxidized when compared to seedlings with lenticels exposed to air.

7.3.2.4 Intercellular Airspaces

Wetland plants can develop intercellular airspaces in different parts, including leaves, stems, rhizomes, and roots. These intercellular airspaces also known as aerenchyma tissue are present in cortical and xylem tissues. Because of increased cellular airspaces, flooding also increases air-filled porosity, which in turn reduces diffusion resistance in gas transport. In fully developed root systems, gas spaces occupy up to 12% of the total root cross-section area. Porosity increases with age, with older tissue having large proportions of root volume in air-filled pore spaces; thus, the porosity increases with distance from apex (Figure 7.7). Roots exhibit greatest demand for oxygen at the apex because of high respiratory activity.

Adaptation of Plants to Soil Anaerobiosis

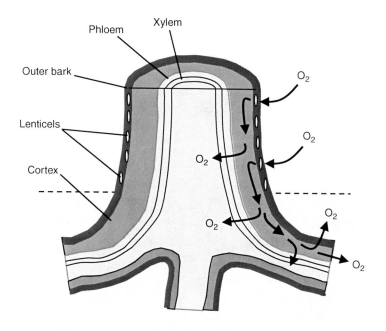

FIGURE 7.6 Swamp tupelo seedling showing lenticels and water roots. (Redrawn from Hook and Scholtens, 1978.)

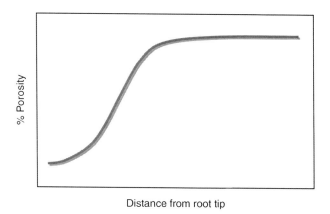

FIGURE 7.7 Porosity of rice roots as a function of distance from apex. (Redrawn from Armstrong, 1971.)

7.3.3 Aerenchyma Formation

Oxygen requirements of roots and rhizomes of wetland plants are largely met by transport of oxygen from the atmosphere through the gas-phase continuum of plant tissue. Development of aerenchyma and related internal pathways for oxygen diffusion to roots is a major adaptation mechanism in wetland plants. Aerenchyma is any tissue that contains large, air-filled intercellular space or lacunae. Roots develop aerenchyma that interconnects longitudinally and joins with the gas spaces of the stem base, thus providing a pathway for oxygen diffusion from air (Figures 7.8 and 7.9). Root porosity is a good predictor of root aerenchyma formation.

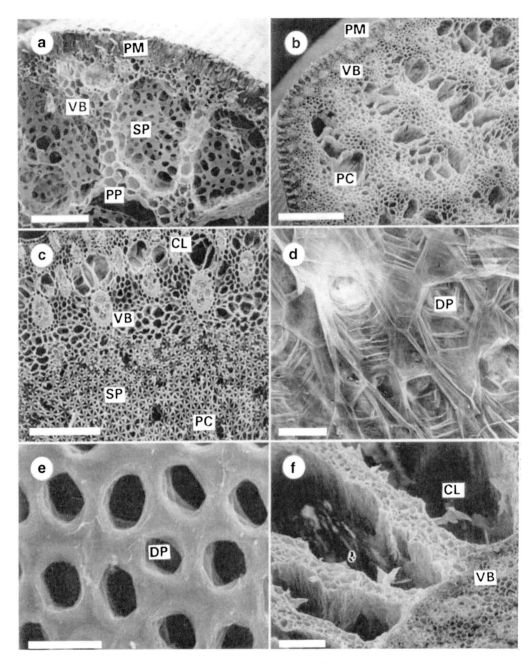

FIGURE 7.8 Micrographs of gas spaces in emergent plants. (a) TS *Schoenoplectus validus* culm with pith partitions and stretched stellate parenchyma (scale bar, 500 µm); (b) TS *Cyperus involucratus* stem (scale bar, 1 mm); (c) TS *Juncus ingens* culm, showing stellate pith parenchyma and cortical lacunae (scale bar, 500 µm); (d) surface airspaces (scale bar, 100 µm); (e) surface view of *Eleocharis sphacelata* pith diaphragm with simple pores (scale bar, 100 µm); (f) TS *Baumea articulata* rhizome showing large cortical lacunae and pith parenchyma (scale bar, 500 µm). SP—stellate parenchyma, PP—pith partition, PM—palisade mesophyll, VB—vascular bundle, PC—pith cavity, CL—cortical lacunae, DP—diaphragm pores. (From Brix, et al., 1992.)

Adaptation of Plants to Soil Anaerobiosis

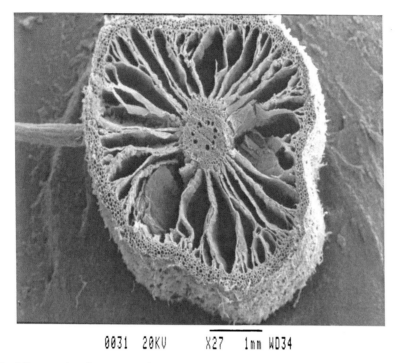

FIGURE 7.9 Micrographs of gas spaces in emergent plants. (Courtesy: Hans Brix, University of Aarhus, Denmark.)

Aerenchyma tissue develops in response to flooding. Increased ethylene-induced cellulase activity has been implicated in this process. Aerenchyma tissue development is a process resulting from cell separation or cortex breakdown producing gas-filled lacunae. In some wetland plants, low soil oxygen is needed for aerenchyma formation, whereas development of these systems in many species such as rice does not require oxygen stress; that is, it is controlled genetically (Smirnoff and Crawford, 1983). For instance in rice, aerenchyma develops under well-aerated conditions while hypoxic conditions further enhance the process (Jackson et al., 1985). Aerenchyma tissue enhances root oxygenation in wetland plants. It may also be found in the meristematic zone where there is a great need for oxygen.

Functions of aerenchyma in wetland plants and oxidized root zone are as follows:

- Serves as pathway for oxygen diffusion from atmosphere to roots
- Provides oxygen for aerobic respiration in root
- Enhances oxidation of rhizosphere around the root
- Buffers plant against soil toxins (ferrous iron and sulfide)
- Provides favorable habitat for microorganisms
- Reduces living cells, and hence reduces oxygen demand of root
- Serves as conduit for gaseous or volatile metabolites to move from the plant roots to the atmosphere (e.g., ethanol, acetaldehyde, methane, and carbon dioxide)

Physiological changes in plants prior to aerenchyma development following flooding include an increase in ADH activity and an increase in the production of ethylene. Ethylene production enhances aerenchyma production. There are two forms of aerenchyma: lysigenous aerenchyma and schizogenous aerenchyma. Lysigenous aerenchyma is formed by various degrees of cell wall separation and cell disintegration (lysis) of cortical cells. In lysigenous aerenchyma development, flooding causes hypoxia/anoxia leading to the synthesis of ethylene, which in turn increases cellulase activity, and subsequently cell lysis and aerenchyma formation. Schizogenous aerenchyma formation involves the enlargement of intercellular spaces without cell collapse.

Acceleration of airspace formation is attributed to production of ethylene and increased cellulase activity in the tissue (Kawase, 1981). The sequential processes in aerenchyma development are presented by McLeod et al. (1987). They suggest that flooding first results in soil oxygen depletion, followed by depletion of root oxygen. This results in ACC (1-aminocyclopropane-1-carboxylic acid) production that requires ATP. Ethylene is produced from ACC, and this process requires oxygen and is sensitive to temperature. Ethylene produced accelerates cellulase activity that softens tissue, resulting in the formation of aerenchyma tissue.

7.3.4 Intercellular Oxygen Concentration

Respiratory requirements of plant roots and shoots regulate oxygen concentration in intercellular plant spaces. Oxygen concentration in cells is generally greater during photosynthetic periods (during the day) and decreases significantly during the night. Respiratory activity also decreases with plant age. High respiratory activity (or demand for oxygen) is observed at the apex of the roots and the activity decreases with distance from apex (Figures 7.10 and 7.11). For a detailed discussion, see the review by Konings and Lambers (1991).

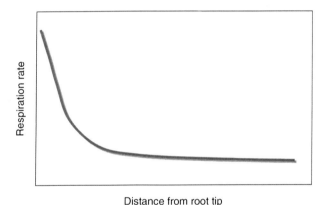

FIGURE 7.10 Root respiration as a function of distance from root tip (apex).

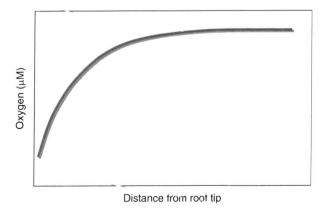

FIGURE 7.11 Oxygen leakage as a function of distance from root tip (apex).

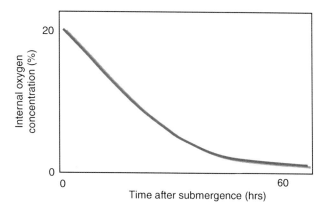

FIGURE 7.12 *Eriophorum angustifolium*. Internal oxygen concentration in the leaves following the submergence of the intact plant in anaerobic medium. (Redrawn from Armstrong and Gaynard, 1976.)

Studies reported by Armstrong (1971) suggest that the concentration of oxygen in the pore spaces of rice roots with lengths of 5–10 cm is about 12%. It is not clear what concentrations of oxygen provide adequate internal aeration for plant growth. Information on optimal oxygen concentrations to achieve root growth and to oxidize rhizosphere is limited. Root respiration of *Eriophorum angustifolium* increased with increase in gas-phase oxygen pressure until a point of critical pressure, where respiration rate was constant (Figure 7.12). The critical oxygen pressure (COP) for this wetland plant was 0.14 atm (14%), which is in the same range of those reported in the literature (mean = 0.1 atm) (Armstrong and Gaynard, 1976).

The COP values reported thus far are based on *in vitro* analysis. These values measured using intact plants are an order of magnitude lower. For example, for *E. angustifolium* the COP values obtained by intact plant method is 0.02 atm, as compared to 0.14 atm for *in vitro* method (Armstrong and Gaynard, 1976). Internal oxygen pressure is also influenced by diffusion and mass flow processes involved in moving gases in and out of wetland plants. Although internal airspaces function as a reservoir for oxygen, the amount of oxygen present only lasts for less than 1 h, if the oxygen supply is curtailed.

Based on the available literature, the following conclusions can be made on internal oxygen concentration of plants. The oxygen concentration in pore spaces of roots can be substantial, with abundant concentrations observed in the older portions of the root. The respiration requirements of older roots are minimal, and air-filled pore spaces are high compared to younger roots. Oxygen concentration increases with distance from apex, in accordance with air-filled porosity. Respiration rates are high in the region of the root tip (apex), and only small volume of root tip contains air-filled pore spaces.

7.4 MECHANISMS OF OXYGEN MOVEMENT IN WETLAND PLANTS

Oxygen entry into the plant varies with plant species. For many wetland plants, oxygen entry into the plant occurs through

- Stomata (herbaceous plants)
- Lenticels (woody species)

Transport of oxygen within the plant requires continuity of the intercellular airspaces to create a flow path with low resistance. Two major pathways are recognized for O_2 transport into and within the plant. These are

- Diffusion
- Mass flow

7.4.1 Diffusion

Net movement of gases by diffusion occurs when the partial pressures of individual gases in two neighboring systems are different, but the total pressure is the same in both systems (Figure 7.13). During this process, oxygen would diffuse from the atmosphere through the leaf (through stomata) to rhizomes and roots, where it would be respired. Similarly, the carbon dioxide produced during respiration would diffuse in the opposite direction, that is, from roots to shoots, where it is used during photosynthesis or escapes to the atmosphere. This process assumes that the gaseous phase in intercellular airspaces is stationary and individual gases move in response to partial pressure

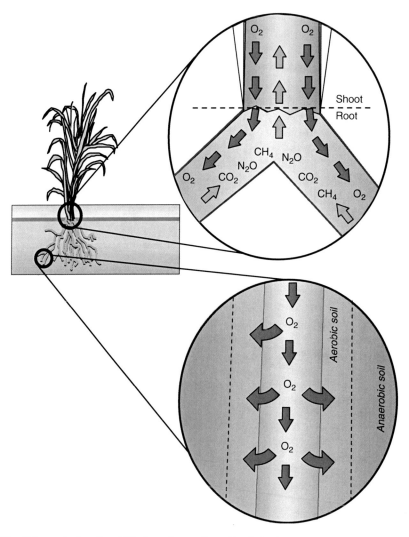

FIGURE 7.13 Schematic showing diffusion of gases in wetland plants.

gradients. A detailed discussion on root aeration by diffusion is presented by Armstrong et al. (1991a).

Oxygen diffusion in wetland plants is regulated primarily by the volume of intercellular airspaces and the degree of space continuity. Longitudinal pore spaces are continuous and relatively less tortuous for O_2 diffusion (Armstrong, 1967, 1968; Armstrong and Beckett, 1987; Armstrong et al., 1991a, 1991b). A second factor influencing oxygen transport is the diffusion path length. Assuming that the tortuosity in air-filled pore spaces is minimal, the path for longitudinal gas transfer can be approximately equal to root length (Armstrong and Armstrong, 1991). The length of root channel aerated due to diffusion is described as follows:

$$l = \frac{1}{2}\left(\frac{2D_o C_o}{Q}\right) \quad (7.1)$$

where l is the length of the channel aerated (cm), D_o the diffusion coefficient of oxygen in air (cm^2 s), C_o the oxygen concentration at the inflow section (atmosphere) (g cm^{-3}), and Q the uniform respiration rate (g cm^{-3} s^{-1}). The effective diffusion coefficient (D_e) can be obtained from the following expression:

$$D_e = D_o ET \quad (7.2)$$

where E is the fractional porosity and T is a tortuosity factor. The value of T is close to unity for oxygen moving through intercellular spaces (Armstrong, 1979).

7.4.2 Mass Flow

Mass flow refers to bulk flow of gases in response to the total pressure gradient without any regard to partial pressure of individual gases. Although this mechanism of gas transport was first reported about a century ago, diffusion was considered as the major mechanism of gas transport in many wetland plants. An early report by Merget (1874), as cited by Mevi-Schutz and Grosse (1988a), has showed that gas flow could be created by heating the surface of a leaf, which he claimed was caused by thermodiffusion. In reality, this gas flow was the result of temperature and humidity difference inside and outside the leaf (Mevi-Schutz and Grosse, 1988b).

These early observations, followed by recent reports by several researchers indicate that greater gas fluxes in intercellular airspaces of some wetland plants are due to pressurized convective flow (Dacey, 1981, 1987; Armstrong and Armstrong, 1991; Armstrong and Beckett, 1996; Brix et al., 1992). In a pressurized flow-through system, air enters the youngest leaves against a small gradient in total gas pressure and is convected through submerged parts (such as petioles) down to rhizomes and eventually to older leaves and dead culms (Figures 7.14 and 7.15). Mass flow measured in water lilies (Dacey, 1980, 1981, 1984; Dacey and Klug, 1979) and *Phragmites australis* (Armstrong and Armstrong, 1991) was shown to be several fold greater than the flow mediated by diffusion under stationary phase. A detailed discussion on convective gas flows in wetland plant aeration is presented by Armstrong et al. (1991a, 1991b).

Mass flow of gases in intercellular airspaces of wetland plants can be generated by several processes:

- Thermal transpiration
- Humidity-induced diffusion
- Solubilization of respiratory carbon dioxide
- Venturi effects across broken culms

Both thermal transpiration and humidity-induced diffusion require a porous partition within the plant tissue, with pore diameters of <0.1–3 µm (Armstrong et al., 1994), which is the molecular mean free path length for Knudsen diffusion. Armstrong et al. (1991a) defined Knudsen diffusion

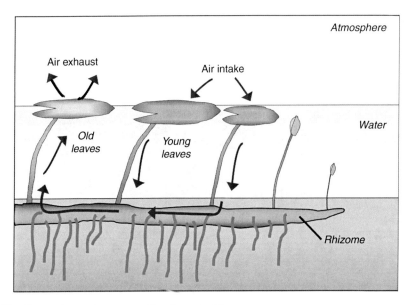

FIGURE 7.14 Gas exchange due to mass flow in water lily.

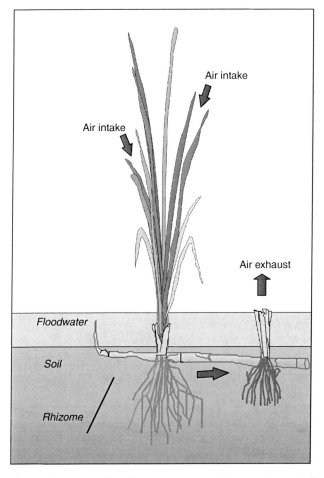

FIGURE 7.15 Gas exchange due to mass flow in emergent macrophytes such as cattails or reeds.

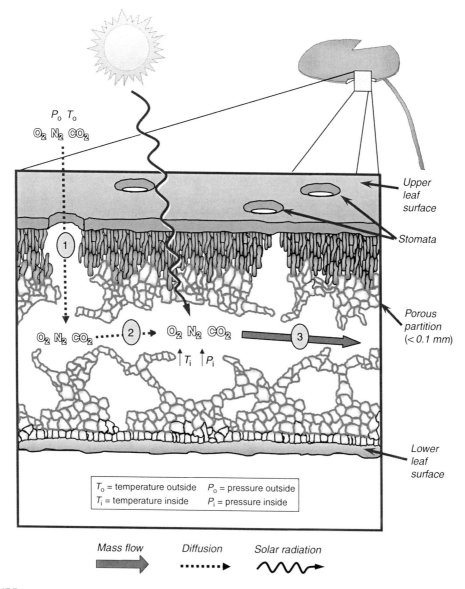

FIGURE 7.16 Convective air flow due to thermal transpiration and humidity-induced pressurization.

as "the gas movements through tubes (or porous media) where pressure is so low that collisions between molecules are so infrequent compared with collisions with the walls of the tube that they can be disregarded." As with normal diffusion, net movements of a gas species are determined by concentration gradients.

At atmospheric pressure, the Knudsen regime is realized only when tube/pore diameters are very small (molecular mean free path length <0.1 μm; Loeb, 1934). Knudsen diffusion coefficients are numerically smaller than those of normal diffusion in which molecule–molecule collisions dominate. Under this regime, gases enter through leaf tissue separating air-filled pore space from the palisade parenchyma that acts as porous partition (Figure 7.16). The pore size in these young leaves is <0.1 μm, which provides conditions for Knudsen diffusion and internal pressurization. Armstrong et al. (1994) reported that 3 μm may be sufficient to maintain adequate mass flow of gases.

Thermal transpiration refers to a temperature-induced diffusion across a porous partition where there is a temperature gradient. This results in a pressure gradient across the plant cell membrane and net movement of gases from cooler side to warmer side in an attempt to establish a stationary phase. Because of the mass flow in the direction of larger pore sizes (older leaves), the pressure on the warmer side is less than that under stationary phase. Gas enters the warmer compartment continuously through porous partition by Knudsen diffusion and vents through pore spaces with little or no resistance (Mevi-Schutz and Grosse, 1988a,b). Under ideal Knudsen conditions, the pressure difference induced by thermal transpiration can be related to temperature as follows (Brix et al., 1992):

$$P_i = P_a \left(\frac{T_i}{T_a}\right)^{1/2} \tag{7.3}$$

where P_i is the internal pressure in the leaf, P_a the atmospheric pressure outside the leaf, T_i the internal leaf temperature (K), and T_a the external temperature (K).

$$P_t = P_a \left[\left(\frac{T_i}{T_a}\right)^{1/2} - 1\right] \tag{7.4}$$

where P_t is the pressure differential induced by thermal transpiration.

Humidity-induced pressurization is the result of vapor pressure differential between the leaf and atmosphere separated by a porous partition (plant cell membrane) (Figure 7.16). The total pressure will be greater on the more humid side. Humidity-induced diffusion is more important than thermal transpiration because it can be increased with temperature and can function at a constant temperature, as well as across temperature gradient (Armstrong et al., 1991a, 1991b).

Under steady-state conditions, total internal pressure (P_i) approaches atmospheric pressure (P_a) plus the saturated water vapor pressure, as described by the following equation:

$$P_i = P_a + P_{wi} - P_{wa} \tag{7.5}$$

The pressure induced by humidity in the intercellular airspaces is the difference in water vapor pressure inside and outside the leaf (Brix et al., 1992):

$$P_w = P_{wi} - P_{wa} \tag{7.6}$$

where P_{wi} and P_{wa} are the water vapor pressure inside and outside the leaf, respectively, and P_w is the pressure differential ($P_i - P_a$).

The total pressure differential induced by thermal transpiration and humidity-induced pressurization is obtained by combining Equations 7.4 and 7.6:

$$P = P_a \left[\left(\frac{T_i}{T_a}\right)^{1/2} - 1\right] + P_{wi} - P_{wa} \tag{7.7}$$

Both processes operate simultaneously and independently (Brix et al., 1992). The theoretical basis for these processes is described in detail by Dacey (1987), Schroder et al. (1986), and Armstrong et al. (1991a,b).

Solubilization of respiratory carbon dioxide in solutions around the plant can mediate convective flows in several plants, including deepwater rice (Raskin and Kende, 1985), *Carex* (Koncalova et al., 1988), and young mangrove plants (Curran et al., 1986). Raskin and Kende (1985) provided evidence that convective flow of oxygen in deepwater rice is driven by solubilization of

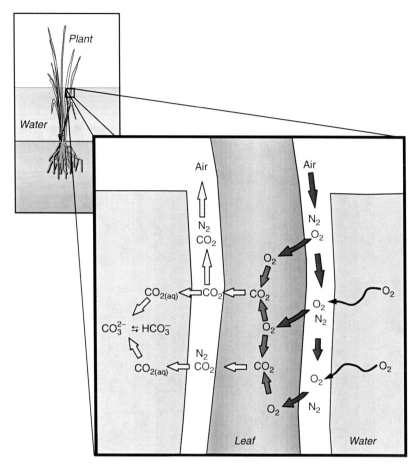

FIGURE 7.17 Pathways showing the influence of CO_2 solubilization on mass flow of O_2 in deepwater rice. (Redrawn from Raskin and Kende, 1985.)

respiratory carbon dioxide in water (Figure 7.17). The carbon dioxide produced during respiration readily solubilized in water, rather than diffusing upwards along the air film. At 25°C (pH = 7), carbon dioxide (as CO_2, H_2CO_3, and HCO_3^-) is 140 times more soluble than oxygen. This condition creates a pressure drop in the air film, resulting in convective flow of air from atmosphere into the air film. This flow of gases is continuous as long as the carbon dioxide activity gradient favors net movement of carbon dioxide from air film and intercellular airspaces into the surrounding water medium around the plant.

During the day, photosynthesis consumes carbon dioxide and releases oxygen. Since oxygen solubility in adjacent water is low, pressure gradients are not created in the air film, resulting in little or no intake of air. In fact, during the day flows can be reversed, that is, upward flux of oxygen from air films to the atmosphere. Using stimulation models, Beckett et al. (1988) demonstrated that convective flow induced by carbon dioxide solubilization is relatively small, compared to diffusion alone. The effect of carbon dioxide solubilization accounts for only 5% of the total oxygen intake as compared to 95% due to diffusion only (Armstrong et al., 1991a, 1991b).

Solubilization of oxygen can also play a significant role in mass flow of air in submerged plant portions (rhizomes and roots). For example, the potential for this process was measured in mangrove plants (Scholander et al., 1955; Curran et al., 1986), *Carex* roots (Koncalova et al., 1988), and

Cyperus papyrus (Li and Jones, 1995). These studies observed high rates of oxygen flux during the night, as a result of carbon dioxide pressure differentials resulting from solubilization in liquid medium. In *Carex* roots, when the pressure of carbon dioxide in liquid medium is higher than that in the roots, the oxygen supply due to mass flow decreases significantly (Koncalova et al., 1988). Under these conditions, oxygen supply due to diffusion can play a major role in root oxygenation.

Venturi-induced convective flow is driven by a pressure differential created by wind blowing across tall dead plant culms (Armstrong et al., 1992). This pressure differential results in mass flow of air into the underground system via broken culms closer to the water level (Figure 7.18). The Venturi effect of wind blowing across an open tube was described by Bernoulli's equation:

$$\Delta P = \frac{1}{2}\rho V^2 \tag{7.8}$$

where ΔP is the pressure differential developed (Pa), ρ the density of air (approximately 1.2–1.25 kg m^{-3}), and V the wind speed (m s^{-1}). The pressure differential is directly proportional to wind velocity and not affected by internal resistance or stem cross-section diameter. However, the convective flow rate is a function of wind speed and proportional to cross-sectional area of the culm (Armstrong et al., 1992).

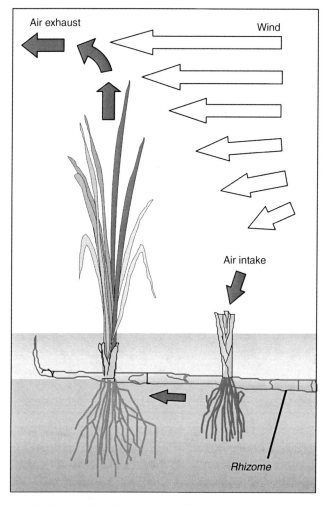

FIGURE 7.18 Influence of wind speed on the convective flow of gases as a result of Venturi effect.

The relative importance of each of the processes described above is difficult to ascertain because all the processes function simultaneously and independently, and various interactive factors regulate each process. Humidity-induced pressurization is usually the most dominant process, regulating convective flows in many wetland plants. Species with cylindrical culms and linear leaves usually have internal pressurization potential (Brix et al., 1992) and may be able to grow in deeper water than species dependent on root oxygenation due to diffusion only (Brix et al., 1992).

Several factors influence the convective flow of gases in wetland plants. These include

- Light intensity
- Leaf temperature
- Relative humidity
- Porosity in leaves and shoots
- Leaf surface area
- Wind velocity

Equations 7.4 and 7.6 demonstrate the effect of temperature and vapor pressure (mediated by humidity) on static pressure in the intercellular airspaces (Figures 7.19, 7.20a, and 7.20b). Light intensity, in the form of thermal radiation, exerts direct effect on pressurization in young leaves and shoots of wetland plants (Dacey, 1981; Armstrong and Armstrong, 1991). For example, in young leaves of water lily, static pressure increased from 80 Pa at a leaf temperature of 16°C to 220 Pa at 26°C (Dacey, 1981). Light intensity increases leaf temperature. In the dark, artificially increased leaf temperatures produced rapid changes in gas pressures of young leaves of water lily (Dacey, 1981).

Convective flows in young shoots of *Phragmites* increased by 37-fold when photosynthetically active radiation (PAR) increased from 0 to 1,400 µmol m^{-2} s^{-1}, as compared to 13-fold increase when PAR increased from 0 to 200 µmol m^{-2} s^{-1} (Figures 7.20a and 7.20b) (Armstrong et al., 1992;

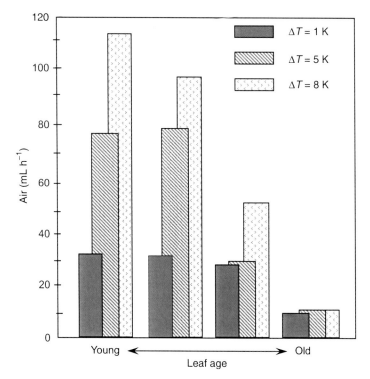

FIGURE 7.19 Influence of leaf temperature on pressurization of gases in intercellular airspaces. (Redrawn from Grosse, 1989.)

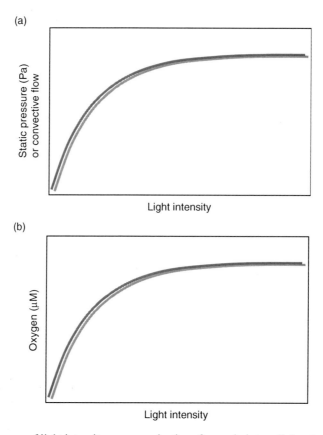

FIGURE 7.20 Influence of light intensity on pressurization of gases in intercellular airspaces.

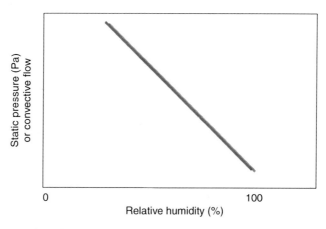

FIGURE 7.21 Influence of relative humidity on pressurization in the intercellular airspaces of wetland plants.

Christensen et al., 1994). Convective flows induced by humidity are inversely related to atmospheric relative humidity at constant PAR. When humidity decreased from 74 to 42%, convective flows increased approximately by threefold in *Phragmites* (Figure 7.21).

Mass flows mediated by thermal transpiration and humidity-induced pressurization require Knudsen's regime (i.e., pore diameter <0.1 μm). Static pressure in intercellular airspaces decreases

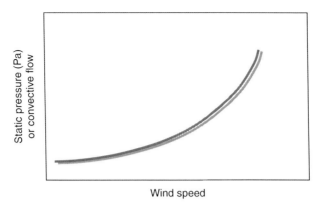

FIGURE 7.22 Influence of wind speed on intercellular pressurization.

with increase in pore diameter (Schroder et al., 1986). The effect is much greater with increased leaf temperature. Increased density of old culms in *Phragmites* stands decreases overall potential for mass flow and root oxygenation. This is probably due to increased porosity in older tissues. Mass flows also increased with the surface area of leaves (probably due to increased porosity in the tissues and increased number of stomata) (Armstrong and Armstrong, 1991). Wind speed was shown to increase convective flow of gases in *Phragmites* stands (Figure 7.22) (Armstrong et al., 1992).

7.5 OXYGEN RELEASE BY PLANTS

Perhaps the most significant long-term adaptation of wetland species to soil anaerobiosis is the development of the aerenchyma tissue in roots and shoots to assume the role of external gas diffusion path once provided by the soil atmosphere. Oxygen release by plants is an important adaptation that helps plants to overcome adverse anaerobic soil conditions. However, oxygen release is not crucial for the survival of many wetland plants under anaerobic soil conditions. Oxygen in aerenchyma tissue of wetland plants may be consumed by root respiration or be lost to the rhizosphere via radial oxygen diffusion from the root. Radial oxygen loss (ROL) decreases the amount of oxygen available to the apex of roots.

Rhizosphere oxygenation by ROL from roots occurs only after the respiratory requirements of the roots are met and conditions are ideal for leakage. Oxygen leakage from roots decreases rapidly with increasing distance from the root apex. This is due to high respiratory requirements for oxygen at the root apex. Porosity in the tissue increases with increasing distance from the root apex. Much of the oxygen leakage occurs through older tissue that has high porosity and offers least resistance to gas flows. However, the roots of many wetland plants contain a barrier to ROL along most of their length, especially at the basal portion of the roots, where older tissue (which may have high porosity) may contain suberin coating that may prevent ROL.

Processes regulating the movement of oxygen within the plant are discussed in Section 7.4. The primary question of ecological interest is: how much of the oxygen transported from shoots to roots is actually released into adjacent soil? As discussed earlier, (i) wetland plant roots contain significant volume in air-filled pore space through which oxygen can leak out, and (ii) wetland plant roots are adapted to use minimal amounts of oxygen for their respiration. There is ample evidence in the literature that these pore spaces contain oxygen and the root excretion of oxygen can oxidize (to some extent) the reduced compounds in the medium external to the root.

Although vast amounts of literature are available to explain movement of air through plants, there is very limited information on oxygen release capacity of wetland plants or the consumption of oxygen in the root zone by reduced compounds. Tables 7.3 and 7.4 show reported literatures on oxygen release by wetland plant species. Table 7.3 reports oxygen release rates in mg O_2 g^{-1} (dw) day^{-1}.

TABLE 7.3
Oxygen Release from Roots in Growing Media by Selected Wetland Plants

Plant Type	Oxygen Release (mg O_2 g^{-1} (dw) day^{-1})	Reference
Hydrocotyle umbellata	6.7	Reddy et al. (1990)
Eichhornia crassipes	4.3	Reddy et al. (1990)
Pistia stratiotes	1.8	Reddy et al. (1990)
Spartina alterniflora	0.5–1.7	Howes and Teal (1995)
Pontederia cordata	0.8	Reddy et al. (1990)
Typha latifolia	0.4	Reddy et al. (1990)
Carex lacustris	0.03	Bedford et al. (1991)
Canna flaccida	1.2	Reddy et al. (1990)
Scirpus pungens	2.3	Reddy et al. (1990)
Scirpus validus	1.2	Reddy et al. (1990)
Scirpus acutus	0.3	Bedford et al. (1991)
Taxodium distichum		
Drained	44.8	Kludze et al. (1994)
Flooded	147.2	Kludze et al. (1994)

TABLE 7.4
Methane Emission Rates through Plants and Open Water in Outdoor Wetland Microcosms ($n = 4$)

Plant Species	Plants (mg CH_4 m^{-2} h^{-1}) Mean	SD	Open Water (mg CH_4 m^{-2} h^{-1}) Mean	SD	% CH_4 Emitted Through Plants
Sagittaria latifolia	89.0	22.6	0.09	0.08	99
Colocasia esculenta	59.7	30.9	0.29	0.26	99
Canna flaccida	15.4	7.3	0.14	0.13	99
Thalia geniculata	7.5	2.3	0.07	0.05	99
Panicum hemitomon	2.6	2.6	0.21	0.26	93
Scirpus pungens	2.5	1.3	0.11	0.08	94
Cyperus articulatus	1.7	1.2	0.44	0.42	79
Sagittaria lancifolia	1.6	0.7	0.06	0.05	96
Scirpus validus	1.4	0.4	0.01	0.15	99
Phragmites australis	1.1	0.7	0.03	0.06	97
Typha latifolia	0.6	0.4	0.02	0.04	97
Pontederia cordata	0.5	0.2	0.11	0.01	81

Source: DeBusk, W. F., and Reddy, K. R., unpublished results.

On an areal basis, Armstrong et al. (1992) reported oxygen transport values of 1–7 g O_2 m^{-2} day^{-1} by *Phragmites*. However for same species, a much lower rate (0.02 g O_2 m^{-2} day^{-1}) was reported by Hower and Teal (1994). In addition to environmental conditions, the reported oxygen release measurements are all influenced by the various experimental procedures used in quantifying transport.

Root oxygenation in flooded plants is essential in the maintenance of aerobic respiration. In the absence of oxygen, the less energy-efficient anaerobic fermentation occurs. Decreasing oxygen pressure in plant roots results in depression of the root energy status as indicated by the AEC ratio (the ratio of phosphorylated adenine nucleotides to the adenine nucleotide pool). The other function of the aerenchyma-facilitated oxygen pathway to roots that has been related to wetland survival is the ability to provide an oxidizing atmosphere at the root surface for oxidation. Rhizosphere oxygenation by oxygen leakage from roots is of great importance because by this mechanism oxygen diffusing from the roots can oxidize soil toxins, such as hydrogen sulfide found in reduced soil environments. Hydrogen sulfide at levels of 1–2 mg L^{-1} can kill plant roots. Oxygen diffusion from aerial parts to the roots and the subsequent sulfide oxidation in the rhizosphere have been considered major mechanisms allowing plants to grow in soil with high level of sulfide.

7.6 MEASUREMENT OF RADIAL OXYGEN LOSS

Methods for measuring oxygen in soil solution include the use of platinum wire electrodes, paramagnetic oxygen analysis, gas chromatography, mass spectrometry, volumetric, polarographic, and colorimetric techniques. Some of these techniques have been used to study the oxidizing power of wetland plants or ROL. Oxygen release by roots in a deoxygenated liquid medium has been demonstrated using colorimeter and polarographic analyses.

A wide range of techniques are used to document and quantify oxygen release by wetland plants, including

- Oxygen microelectrodes (polarography)
- Redox potential
- Oxygen concentration in solution adjacent to roots
- Oxidation of reduced compounds

Oxygen microelectrodes, similar to those described by Revsbech (1989), have been used to determine the ROL and thickness of aerobic layers around the roots. Oxygen-depleted hydroponic solutions used in many ROL experiments are, however, a poor analog of wetland soil, because they do not mimic the high oxygen demand and low redox potentials found in wetland soil (DeLaune et al., 1990). Sediment oxygen demand acts as a sink for oxygen released by roots.

Redox potential measured in root zones can only indicate qualitative root oxygenation. Redox potential measured in soils with plant roots is generally greater than in soils without plants (see Chapter 4 for detailed discussion of Eh). High Eh values can be observed in soil cores obtained 90 days after planting of rice plants (Reddy and Patrick, 1984). As rice plants mature, root porosity and oxygen leakage increase, as compared to young plant roots, which have a high respiratory oxygen demand. In oligotrophic lakes, increased Eh values in the oxidized range were observed in sediments rooted with submerged macrophytes, as compared to eutrophic lake sediments (Jaynes and Carpenter, 1986). This is probably a result of high respiratory activity in eutrophic sediments as a result of organic matter accumulation. Other studies have shown that submerged macrophytes are capable of releasing oxygen, and increase in Eh values due to O_2 release results in an oxidized layer of 1 mm (Figure 7.23) (Flessa, 1994).

Laboratory methods have been used by creating an oxygen demand in the rhizosphere to quantify or evaluate plant physiological functions, including root oxygen release. One such method utilizes titanium citrate, a reducing agent, used to create a reducing medium to mimic wet-soil conditions (DeLaune et al., 1990; Kludze et al., 1994a; Lissner et al., 2003). This technique has been successful in screening the ability of various wetland plants to transport oxygen to rice roots. In this method, oxygen consumption by reductant is used to calculate oxygen transport capacity of wetland plants. Titanium (III) is a strong reducing agent and Ti^{3+}-citrate is purple blue in solution. Autooxidation of Ti^{3+} eliminates oxygen from a medium by first-order reaction with a rate constant, $k = 11.4 + (0.8 \times 10^{-5})$ s^{-1} at 25°C (Zehnder and Wuhrman, 1976; Kludze et al., 1994a, 1994b). The Ti^{3+}-citrate solution becomes colorless when oxidized to Ti^{4+}-citrate. Plants are placed in flasks

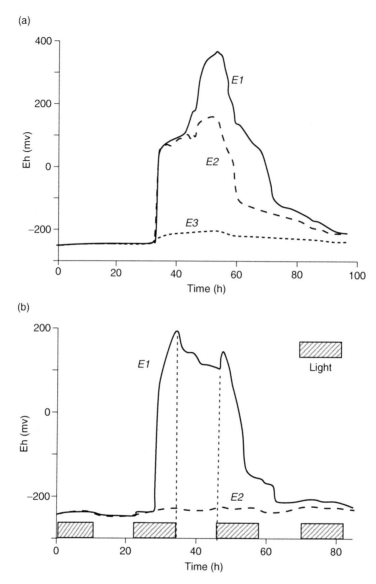

FIGURE 7.23 (a) Redox potential in the rhizosphere of *Ranunculus circinatus* at three electrode positions, E1 and E2 came into contact with the root tip after 32 h and distance of E3 to the root surface was 1 mm. Intensity of light: PAR 100 µE m^{-2} s^{-1}. (b) Redox potential in the rhizosphere of *Ranunculus circinatus* as influenced by light intensity. E1 came into contact with the root tip after 27 h and distance of E3 to the root surface was 1 mm. Intensity of light: PAR 100 µE m^{-2} s^{-1} (hatched boxes) and 15 µE m^{-2} s^{-1}. (From Flessa, 1994.)

containing titanium citrate (Figure 7.24). A sample of the solution is removed and oxidation of Ti^{3+} by oxygen diffusion from plant roots is determined colorimetrically.

Absorbance of the partly oxidized Ti^{4+}-citrate solution is measured at 527 nm on a UV/VIS spectrophotometer. Released oxygen on a whole-plant basis is determined by extrapolation of the measured absorbance to a standard calibration curve. These hydroponic studies using artificial redox buffer may be useful in screening related response of wetland plants to reduced soil conditions. However, the conditions do not truly mimic the conditions in wetland soils. Oxidation of other reductants such as DOC, ferrous iron, and ammonium can also be used to estimate oxygen transport capacity of wetland plants (Reddy et al., 1990; Burgoon and Reddy, 1996).

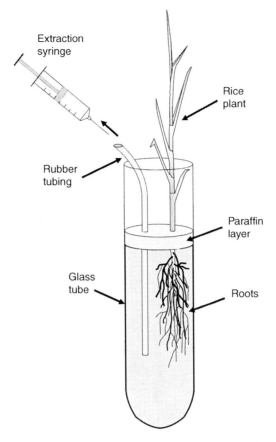

FIGURE 7.24 Experimental apparatus used to estimate radial oxygen loss from wetland plant roots into titanium citrate buffer solution. (Redrawn from Kludze et al., 1994a.)

7.7 SOIL PHYTOTOXIC ACCUMULATION EFFECTS ON PLANT GROWTH

Toxic substances produced in soil under anaerobic conditions at low redox potentials can influence plant growth. Potential phytotoxins in reduced soil include

- Manganous manganese, Mn^{2+}
- Ferrous iron, Fe^{2+}
- Hydrogen sulfide, H_2S
- Organic acids

Such toxins can have impact on roots not affected by oxygen depletion itself. For example, after prolonged periods of flooding, hydrogen sulfide can be produced by the bacterial reduction of sulfates and by dissimilation of sulfur-containing amino acids. *Desulfovibrio* and *Desulfotomaculum* and some species of *Bacillus*, *Pseudomonas*, and *Saccharomyces* are capable of utilizing sulfate as the terminal electron acceptor in their respiration while oxidizing organic acids and alcohols released under these anaerobic conditions. At only 2.5 mg L^{-1}, hydrogen sulfide can kill roots. Organic soils typically have low bulk density and limited mineral matter and can have high levels of sulfide. Some organic soils may not have enough reduced iron to precipitate sulfide to reduce toxicity to roots.

Oxygen diffusion from aerial parts of wetland plants to roots and the subsequent sulfide oxidation in the rhizosphere have been considered major mechanisms allowing a high level of sulfide tolerance in some species (Joshi and Hollis, 1976). Oxygen release from young seedlings

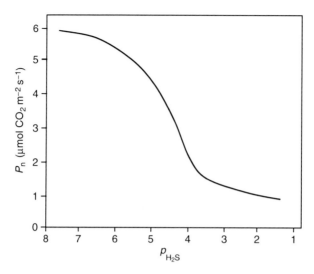

FIGURE 7.25 Relationship between net photosynthetic rate (P_n) and hydrogen sulfide concentration for *Spartina alterniflora*. (Redrawn from Pezeshki et al., 1988.)

and aerenchyma tissue development is inhibited when the plants are exposed to hydrogen sulfide concentration of 0.2 mg L^{-1}. Lack of sufficient aerenchyma development in immature plants limits oxygen transport ability of roots to rhizosphere.

Sulfide is a major factor contributing to the slow growth and low productivity of salt marsh. Salt marshes generate high sulfide as a result of sulfate in seawater or tidal water diffusing into the soil and being reduced to sulfide. Elevated soil sulfide causes dieback of *S. townsendii* in Great Britain and has been associated with low productivity for the short form of *S. alterniflora* stands in Louisiana salt marshes. Salt marsh soils low in mineral sediment and high in organic content may contain sulfide levels toxic to salt marsh vegetation. Pezeshki et al. (1988) reported that *S. alterniflora* growing in Barataria Basin salt marsh is sensitive to a H$_2$S level of 0.34 mg L^{-1} (Figure 7.25).

Under anaerobic soil conditions, organic substrates are often not decomposed completely to carbon dioxide. Incompletely oxidized intermediates and end products toxic to plants often accumulate in waterlogged soils. These intermediates and end products include lactic acid, ethanol, acetaldehyde, and aliphatic acids such as formic, acetic, or butyric acid. Ethylene is also sometimes present in abnormally high concentrations in waterlogged or anaerobic soil.

Oxygen release capacity of wetland plant varies with species, age, photosynthetic activity, and root zone environment. Oxygen consumption by reductants at the soil–floodwater interface is briefly discussed in Section 6.6. The concepts presented in Section 6.6 are also applicable to the oxygen consumption at the soil–root interface (Figure 7.26). The most striking observation of oxidizing power of plant roots is seen in mineral wetland soils containing iron. Oxidized root channel generally appears as reddish brown areas on the surface and around the roots, as a result of oxidation of Fe^{2+} to Fe^{3+} and precipitation of FeOOH (Figure 7.27). This plaque formation results from diffusion of Fe^{2+} toward the root zone in response to concentration gradients at the interface (similar to those observed at the soil–floodwater interface). In this manner, oxidized rhizosphere functions as a sink for Fe^{2+} and other reduced substances.

Oxidized root channels have been observed for few species, including rice (*Oryza sativa*), cattails (*Typha* sp.), reeds (*Phragmites*), *Spartina* sp., *Carex* sp., and *Potomogeton* sp. (see review of Mendelssohn et al., 1995). The iron-enriched plaques essentially consist of FeOOH minerals (see Section 7.8.1). Excessive ferric iron precipitation can block the uptake of nutrients. Gas exchange within the root can also be decreased by dense plaque formation.

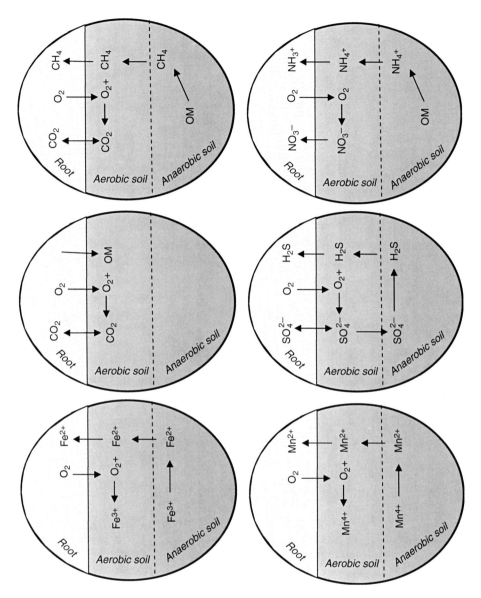

FIGURE 7.26 Schematic showing the oxidation of CH_4, NH_4, Fe^{2+}, Mn^{2+}, and sulfides in the root zone of wetland plants.

Sulfides are produced as a result of sulfate reduction and dissimilation of sulfur-containing amino acids (see Chapter 11 on sulfur). Sulfide has been implicated as one of the key regulators of plant growth in coastal marshes, especially in soil with low mineral matter (such as organic soils) (DeLaune et al., 1983). Oxidation of sulfides in the rhizosphere has been considered as the major mechanism allowing plants to tolerate relatively high levels of sulfides. At sulfide concentrations of 0.2 mg L^{-1}, oxygen release by young seedling of rice was inhibited (Joshi and Hollis, 1976).

Sulfide is oxidized in the rhizosphere and the resulting sulfate is taken up by the plants. Alternatively, sulfide can be taken by the plant and oxidized in the intercellular airspaces of roots (Carlson and Forrest, 1982). Sulfide precipitation with metals as metal sulfides (insoluble precipitates under anaerobic conditions) decreases pore water concentrations. However, metal sulfides formed in

FIGURE 7.27 (See color insert following page 392.) Photographs showing oxidized roots of rice plants.

the vicinity of roots can be oxidized in the rhizosphere. For example, labeled sulfide from Na_2S, MnS, FeS, ZnS, and CuS under low Eh conditions was oxidized in the root zone of rice plants, and the resulting sulfate formed was taken up (Engler and Patrick, 1975). The degree of oxidation of metal sulfides and uptake of ^{35}S by rice plants were directly related to solubility of sulfides (Figure 7.28).

Gas exchange by wetland plants has important ecological and environmental significance because of its role in

- Methane emissions and oxidation
- Nitrification–denitrification
- BOD (organic carbon) removal in constructed wetlands

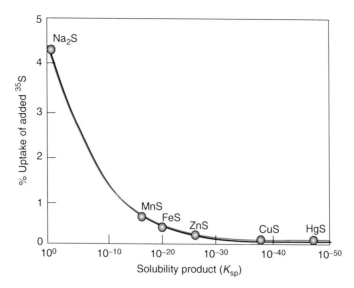

FIGURE 7.28 Percent uptake of ^{35}S by the rice from a flooded Crowley silt loam amended with selected metal sulfides such as Na_2S, MnS, FeS, ZnS, CuS, and HgS. (Adapted from Engler and Patrick, 1975.)

Methane is a key radiative gas contributing to the global warming, and wetlands have been shown to be important sources of atmospheric methane on a global scale (Bartlett and Harriss, 1993; see Chapter 5 for a detailed discussion). Plants transport up to 79–99% (Table 7.4) of methane released from rice fields and wetlands, as a result of diffusion and mass flow mechanisms discussed earlier. Plant species vary in their capability to transport methane because of differences in pressurization capacities. Plants can also retard the release of methane by high rates of oxygen transport and oxidation in the rhizosphere. The relationship between methane emissions and oxygen transport is shown in Figure 7.29, where generally, methane emission and oxygen transport by wetland plants are inversely related. Rhizosphere oxidation by *Sagittaria latifolia* averaged about 65% of the total methane produced in the soil (Schipper and Reddy, 1996).

Oxygen transport by plants and subsequent oxidation in the rhizosphere support oxidation of ammonium to nitrate (Figure 7.26). Nitrate formed diffuses into the adjacent anaerobic soil layer and undergoes reduction to nitrous oxide and nitrogen gas. Resulting gases are transported through plants by diffusion and mass flow to the atmosphere. These sequential processes were demonstrated by Reddy et al. (1989) for rice, pickerel weed (*Pontederia cordata*), and soft rush (*Juncus effusus*).

Evidence of BOD removal due to oxygen transport by floating and emergent macrophytes is presented by Reddy et al. (1990). Oxygen transport by plants resulted in BOD removal rates of 0.007–0.008 h^{-1}, as compared to 0.004–0.005 h^{-1} in systems with no plants but receiving mechanical aeration.

7.8 OXIDIZING POWER OF PLANT ROOTS

Although we presented evidence for oxygen release by roots and subsequent oxidation of reductants in the root zone, a fundamental question is whether the amount of oxygen released by roots is adequate to maintain oxidized rhizosphere to support aerobic respiration. As discussed earlier, oxygen released from roots will be first consumed during chemical oxidation of Fe^{2+}, Mn^{2+}, and S^{2-}, followed by heterotrophic and autotrophic oxidation by bacteria. The kinetics of chemical oxidation are much more rapid than biological oxidation.

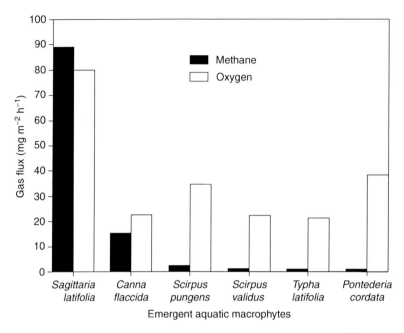

FIGURE 7.29 Methane and O$_2$ flux in selected emergent macrophytes cultured in flooded organic soil in Florida. (Reddy, K. R., Unpublished Results, University of Florida.)

Once the reductants present in the oxidized root zone are depleted, subsequent oxidation of these compounds will depend on the diffusive resupply across the aerobic–anaerobic interface. The ability of wetland plants to transport oxygen has attracted biologists and engineers to include this process into designing constructed wetlands for wastewater treatment.

Oxidizing power of wetland plant roots depends on

- Oxygen release capacity of plant roots
- Concentration of reductants in the root zone

As presented in Figure 7.26, various oxidation reactions mediated by oxygen function in the root zone, which aid in abating high levels of reductants in the root zone.

7.8.1 Root Iron Plaque Formation

The dominant and most reported component of root plaques is various oxidized compounds of iron. Microscopic observations of root plaques show a highly heterogenous morphology composed mostly of an amorphous material dispersed throughout nodules (50–300 nm in diameter), needles (50–100 nm in length), and filaments with variable lengths. This iron plaque formation on roots results from diffusion of Fe^{2+} toward the root zone in response to concentration gradients at the interface (similar to those observed at the soil–floodwater interface). The oxidized rhizosphere functions as a sink for Fe^{2+} and other reduced substances.

Oxidized root channels have been observed for few species, including rice (*O. sativa*), cattails, reeds, *Spartina* sp., *Carex* sp., and *Potomogeton* sp. (see review by Mendelssohn et al., 1995). The iron-enriched plaques essentially consist of FeOOH minerals (Bacha and Hossner, 1977). Iron plaque may be amorphous or crystalline, in the forms of iron such as ferric hydroxides, goethite, lepidocrocite, and siderite. Iron oxides or hydroxides in rhizosphere have high affinity for metals and metalloids.

A primary factor controlling the formation of oxidized root deposits is the ability of wetland plants to transport oxygen from the atmosphere through the plants into the roots and out into the surrounding soil. The amount of ROL increases with increase in oxygen demand in reduced soil environments. Many flood-tolerant plants are capable of oxygen movement along a concentration gradient through plant into anaerobic soil. Also, some plants exhibit root enzymatic oxidation, which may also cause the formation of an oxidized rhizosphere around the root.

Plaques accumulate because soluble Fe^{2+} moves along concentration gradients in reduced form toward the oxidized rhizosphere where Fe^{3+} compounds can precipitate. Iron plaque formation occurs both on root surface and within the root. The interior root penetration of iron deposition is variable. Differences in plaque penetration of the root may result from variance in the extent of the oxidized rhizosphere because of differential root ROL among wetland plant species. Iron plaques do not form on and around roots under aerobic conditions because soil iron is in the ferric form and is not affected by root oxygen loss and as a result, no root-induced iron precipitation occurs. Armstrong and Boatman (1967) showed that ferric iron functioned as a sink for oxygen, as evidenced by thick sheath of FeOOH around roots. Under intense reducing conditions, ferrous iron diffused into the root and oxidized on the cell walls bordering the intercellular spaces of wetland plants.

The amount of plaque formation is also governed by the amount of reducible iron in the root environment. The greatest degree of iron plaque accumulation occurs when the iron in solution is in the ferrous form. Plaque formation is dependent on the oxidation of Fe^{2+} to Fe^{3+}. Since, iron solubility or reduction increases with decreasing soil redox potential (Eh), greater plaque formation occurs at reduced potential below which soil iron is reduced (+150 mV). A highly negative Eh, however, may create such an intense SOD in the rhizosphere that oxidizing capacity of the root is overwhelmed, resulting in the absence of plaques.

Soil pH affects iron plaque directly by controlling the concentration of iron in solution and the resolubilization of precipitated root iron, and indirectly by influencing the net oxidizing capacity of the root. Soil texture and organic matter content are also important factors in determining the extent of iron plaque formation. Clay soils, for example, contain more reducible soil iron than sandy soils, and in general, more root plaque formation occurs if reduction takes place in fine-textured mineral soils that contain high iron content and sufficient organic matter for iron reduction. In summary, the presence of soluble soil iron and wetland plant species capable of creating an oxidized rhizosphere are the most important factors controlling plaque formation.

It is generally believed that root iron plaques will potentially reduce the ability of a plant to assimilate nutrients. This observation is based on the precipitation of phosphorus with ferric iron oxides. In alkaline soils, severe iron deficiency can be observed in wetland plants because of low solubility of Fe oxides at near neutral and alkaline pH values. Oxidized iron has a high binding capacity for phosphorus both through the formation of ferric phosphate minerals and through adsorption reactions (Figure 7.30a). Excessive ferric iron precipitation can block the uptake of nutrients, and gas exchange within the root can also be decreased by dense plaque formation. These conditions can potentially reduce uptake of phosphorus by plants. However, lowering of rhizosphere pH can result in solubilization of ferric phosphates and potential release of phosphorus for plant uptake (Figure 7.30b). Lowering of rhizosphere pH can occur as a result of (1) the release of protons during oxidation of reduced species, (2) the root release of protons to balance excess uptake of cations over anions (where dominant form of nitrogen is ammonium ions), and (3) the accumulation of carbon dioxide derived from respiration (Begg et al., 1994).

7.9 EFFECT OF INTENSITY AND CAPACITY OF SOIL REDUCTION ON WETLAND PLANT FUNCTIONS

Most hydroponic studies of the response of wetland plants to anaerobic root environments have been based on the purging of nitrogen to remove oxygen from solution. Such treatments lead to a redox potential of 330–400 mV (DeLaune et al., 1990). This value is much higher than redox values found in flooded soils that may reach as low as −250 mV. Artificial redox buffer (titanium citrate) used to

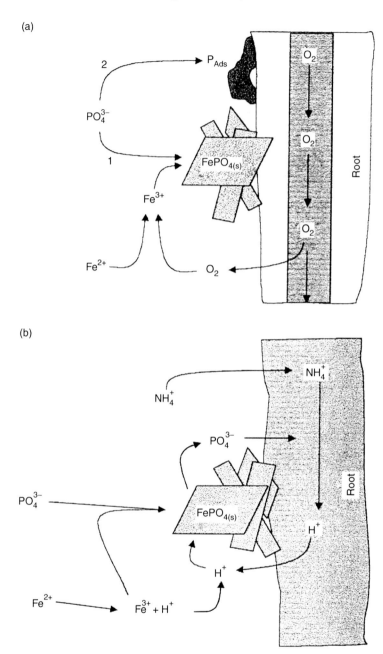

FIGURE 7.30 (a) Conceptual diagram showing possible root plaque interference with phosphorous uptake through ferric phosphate mineral formation (1) and phosphate adsorption reactions (Inglett, P. W., unpublished). (b) Conceptual diagram demonstrating the possible beneficial mechanism of root plaques for nutrient uptake by rice. Based on the hypothesis of Saleque and Kirk (1995). (From Inglett, P. W., unpublished.)

create a strong oxygen demand in hydroponic solution clearly demonstrated that plants grown under highly reduced condition respond differently than plants grown at moderately reducing conditions.

Both intensity and capacity of soil reduction appear to influence plant functioning in wetland ecosystems. In wetland soils, plants are faced with a substantial demand for oxygen in the rhizosphere and the potential for loss of oxygen to soil, and thus the plants must deal with additional root stress. As soil reduction continues and intensifies, a progressively greater demand is imposed on roots for oxygen supply, and thus a corresponding greater potential for loss of oxygen to the rhizosphere. The severity of oxygen loss and the effects of reduction intensity and capacity on plant functioning appear to be broad across wetland species. Under some reducing soil conditions, plant roots may not be able to create an oxidized rhizosphere, and thus fail to neutralize sulfide and other plant toxicants.

Soil reduction can be described in both intensity and capacity terms. The intensity factor determines relative ease of reduction and is normally denoted by the redox potential (Eh). The capacity factor, however, denotes the amount of the redox system undergoing reduction and is equivalent to total amounts of labile carbon compounds or total energy sources that are utilized during microbial activity (best described in terms of its oxygen equivalent). In effect, two different soils with the same reduction intensity may differ with respect to the capacity factor because of variations in oxygen demand.

The soil redox capacity factor is also important although much less is known about its effects on wetland plants than is known about the effects of the intensity factor. Soil reduction capacity can be determined using measurements of soil respiration carbon dioxide and calculating oxygen equivalent by stoichiometry (Kludze and DeLaune, 1995a). Experimental reduction capacity may be controlled by adding different amounts of granular D-glucose to the root medium that is also maintained under preset level of soil redox intensity or redox level (Eh < +350 mV; Kludze and DeLaune, 1995a,b).

7.9.1 Effect of Soil Reduction Intensity

Soil anaerobiosis results in plant stress symptoms similar to those caused by drought stress. Among these are stomatal closure and reduction in net carbon assimilation. Photosynthesis in plants susceptible to flooding rapidly declines under anaerobiosis. Reduction in photosynthesis in response to flooding has also been reported for flood-tolerant species.

In controlled laboratory mesocosms, *Spartina patens*, a dominant brackish marsh species found along the U.S. Gulf coast, and rice (*O. sativa*) showed a decrease in net photosynthesis in response to reduced soil redox potentials. Net photosynthesis decreased when soil redox potential or Eh was below −100 mV (Kludze and DeLaune, 1995b). A similar reduction in photosynthetic rates was observed in *O. sativa* with increase in intensity of reduction (Figure 7.31). However, wetland plants

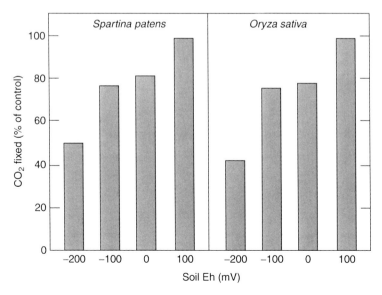

FIGURE 7.31 Carbon dioxide fixation in salt meadow cordgrass and rice as a function of soil redox intensity (Eh) along the portion of the redox scale. (Modified from Kludze and DeLaune, 1995a.)

are often capable of regaining their stomatal and photosynthetic functioning despite continuation of flooding and low soil Eh conditions (Brown and Pezeshki, 2000; Pezeshki, 2001).

Oxygen supply is essential for root elongation in both flood-tolerant and flood-sensitive plants. Low oxygen and high carbon dioxide levels can decrease root permeability to water by over 50%. When respiration is inhibited, water uptake decreases. Status of soil aeration influences root oxygen supply because of changes in oxygen concentration gradients between aerial parts and rhizosphere. Two types of root growth modification include increased branching of roots and adventitious root formation. Adventitious roots are important because they enhance oxygen transport. Normal growth and functioning of roots require more oxygen than do the root respiration processes. Under flooded conditions, oxygen reaches the roots through airspaces within soils (Kludze and DeLaune, 1995b). Low sediment Eh inhibits root growth and development. Restricted root elongation reduces ability of wetland plants to reach various soil horizons. Data of Pezeshki and DeLaune (1990) showed that in *S. patens*, root elongation was severely inhibited shortly after initiation of and during low soil redox potential treatment in the range of −50 to +70 mV (Figures 7.32–7.34). Inhibition of root growth in response to flooding has been reported for many vascular plants.

7.9.2 Relationship of Reduction Intensity with Root Porosity and Radial Oxygen Loss

Intensity of soil reduction (low soil Eh) promotes increase in both root porosity and ROL from root to the rhizosphere (Figure 7.35). There is an increasingly higher oxygen loss rate as soil Eh becomes more reduced. Root porosity in many wetland species is much higher, particularly if grown under anaerobic soil conditions. In *S. patens*, root porosity increased as soil Eh decreased, resulting in root porosity of 22% in plants grown at +200 mV, whereas porosity was 45% in plants grown at −300 mV. Also, ROL was significantly greater for plants in −300 mV Eh treatment as compared to +200 mV Eh (Kludze and DeLaune, 1994).

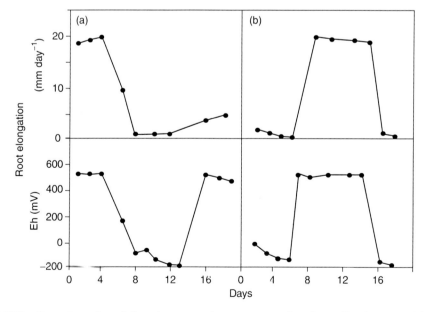

FIGURE 7.32 Root elongation of *Spartina patens* in response to changing soil redox potential (Eh). (a) Response to high Eh (aerobic) followed by low Eh (anaerobic) conditions. (b) Reverse procedure. (Modified from Pezeshki and DeLaune, 1990.)

Adaptation of Plants to Soil Anaerobiosis 251

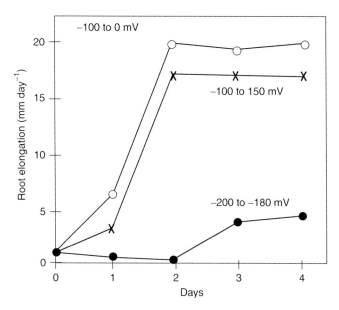

FIGURE 7.33 Recovery of root elongation in *Spartina patens* after an increase in soil redox potential (Eh) to +500 mV. The increase in Eh followed anaerobic conditions in which soil Eh was reduced for 5 days to a range of –180 to –200 mV (closed circles), –100 to 150 mV (x's), and 0 to –100 mV (open circles). (From Pezeshki and DeLaune, 1990.)

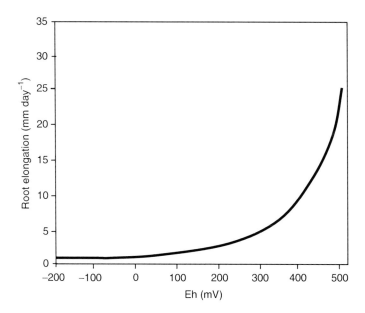

FIGURE 7.34 Relationship between root elongation of *Spartina patens* and soil redox potential (Eh). (Redrawn from Pezeshki and DeLaune, 1990.)

In *Taxodium distichum*, a significant increase in root porosity and ROL was noted at Eh intensity of −240 to −260 mV in root medium (Kludze et al., 1994a, 1994b). ROL increased from 12.7 in control treatments to 42.3 mmol O_2 g^{-1} day^{-1}. Similarly, root airspace increased from 13.3 to 41.4% in response to the intensity of reduction. In *O. sativa* (rice), ROL increased in response to a drop

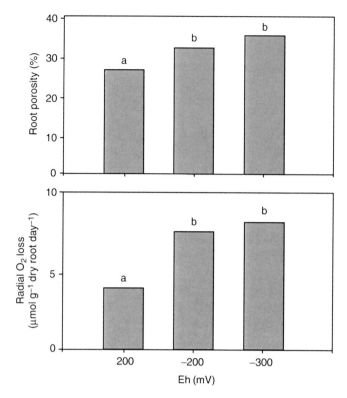

FIGURE 7.35 Root porosity and radial oxygen loss in *Spartina patens* grown under various soil redox intensity for 50 days. (Modified from Kludze and DeLaune, 1994.)

in soil Eh concomitant with root porosity that increased from 26.8 to 35% when Eh dropped from +200 to −300 mV (Kludze et al., 1993).

Increase in ROL that parallels the increase in root porosity in response to anaerobic soil conditions may not be sufficient to satisfy all the root respiratory needs. Despite a substantial enhancement of aerenchyma tissue formation in *S. patens*, ADH activity continued to be higher in flooded than control plants indicating continued oxygen stress in the roots of flooded plants. In addition, increase in ROL reported under intense soil Eh may explain the reported reductions in root growth of several wetland species under low soil Eh conditions. For instance, in *S. patens*, carbon assimilation decreased when soil Eh dropped from +100 to −200 mV. These results clearly indicate the influence of soil reduction intensity on plant growth. It has also been demonstrated that roots are more sensitive to Eh intensity than shoots (Kludze and DeLaune, 1994). Pezeshki and DeLaune (1990) reported cessation of root growth in *S. patens* under reducing conditions and concluded that such reduction may, in part, be responsible for a negative feedback inhibition of photosynthesis resulting in reduction in productivity of this species. The intensity of reduction in root zone and the resulting demand for oxygen in the root zone exert significant influence on plant physiological functioning (DeLaune et al., 1990).

7.9.3 Effect of Soil Reduction Intensity on Nutrient Uptake

Soil anaerobiosis also affects plant nutrient uptake in wetland environments. Anaerobiosis in the rhizosphere, a dominant factor in wetland areas, causes physiological stresses that can limit active uptake of essential elements such as nitrogen. The nitrogen status, in turn, can affect photosynthetic activity in plants. In addition to effects on plant growth, both intensity and capacity of reduction may govern nutrient uptake in wetland plants. In a study of seedlings of two bottomland woody species grown in soils maintained at three Eh levels, +560, +340, and +175 mV, fertilizer ^{15}N uptake decreased with decrease in soil redox potential, a response to the intensity of reduction (DeLaune et al., 1998) (Figure 7.36).

Phosphorous uptake decreased with decrease in redox potential or reduction intensity in the root medium. A study by DeLaune et al. (1999) indicated that the greatest uptake was measured under the oxidized treatment (+565 mV). Phosphorous uptake was less under two reducing treatments, and considerably less at −200 mV, in which a high oxygen demand was created using titanium (Ti^{3+}) citrate (Figure 7.37). These results suggest that nutrient uptake by wetland plants is governed by soil reduction intensity and capacity. The measured physiological responses suggest that responses of wetland plants may not be entirely or directly associated with flooding effects on plant function, but may also be associated with secondary effects such as changes in nutrient uptake. Increasing capacity of reduction (using a titanium citrate solution) resulted in a further decrease in phosphorous uptake (DeLaune et al., 1999), further confirming the effects of reduction capacity on nutrient uptake.

7.9.4 Soil Reduction Capacity Effects on Carbon Assimilation and Radial Oxygen Loss

Increasing the reduction capacity among wetland soils influences plants in many ways, including oxygen transport, rhizosphere oxygenation, and photosynthetic rates (Figure 7.38) (Kludze and DeLaune, 1995b). Increased capacity of reduction under a constant Eh intensity of −200 mV

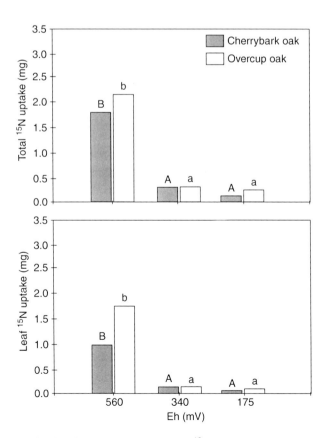

FIGURE 7.36 Relationship of soil redox potential to total ^{15}N uptake (whole plant) and leaf ^{15}N uptake in cherrybark oak and overcup oak. (From DeLaune et al., 1998.)

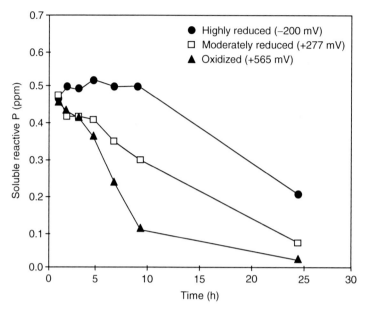

FIGURE 7.37 Phosphorus uptake by *Typha*. Change in phosphorus concentration in nutrient solution maintained under oxidized, moderately reduced, and highly reduced conditions. (From DeLaune et al., 1990.)

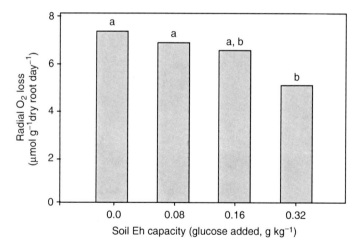

FIGURE 7.38 Radial oxygen loss (ROL) in *Spartina patens* grown under various soil reduction capacities while the reduction intensity was maintained at −200 mV. Values followed by the same letter are not significantly different at the 0.05 level. (From Kludze and DeLaune, 1995b.)

did not have any significant effect on root porosity in *S. patens*, but oxygen release increased in response to the increasing Eh captivity (Kludze and DeLaune, 1995b). However, the authors reported that there was a threshold of Eh capacity beyond which oxygen release remained constant or decreased, or both, in this species. The response was attributed to the potential effects of several factors such as soil phytotoxins as well as plant physiological responses including stomatal closure. However, the reasons for such a response remain unknown. Plant carbon fixation, root growth, and shoot growth were significantly inhibited in *S. patens* by increasing soil reduction capacity.

Both intensity and capacity of reduction influence plant growth functioning in wetland ecosystems. In wetland soils, plants are faced with a substantial demand for oxygen in the rhizosphere and the potential for loss of oxygen to the soil resulting in additional root stress. As soil reduction continues and intensifies, a progressively greater demand is imposed on roots for oxygen, and thus a greater potential for loss of oxygen to the rhizosphere. The severity of oxygen loss and effects of reduction intensity and capacity on plant functioning appear to be broad across wetland species. Studies that focus on plant response to a reducing environment should consider the influence of capacity of soil reduction, in addition to the intensity of soil reduction on wetland plant growth and functioning. In summary, the oxygen demand in reducing soil environment governs the following plant growth parameters:

- ROL from root
- Aerobic/anaerobic root metabolism
- Root development
- Nutrient uptake
- Photosynthesis

7.10 SUMMARY

Wetland plants are important in regulating the biogeochemical cycles in wetlands. The distribution and productivity of plants in wetlands are governed by both soil condition and individual plant adaptation for growing in wetland environment.

- Excess water and reducing conditions in wetlands affect plant growth and productivity in several ways. The abundance of water and reduced soil oxygen supply seriously interfere with plant root metabolism, creating root oxygen deficiency.
- Microbial processes in wetland soil can produce reduced substances such as sulfide, which are potentially toxic to wetland plants.
- Wetland plants have unique physiological strategies or adaptation mechanism for growing in soil during periods of soil oxygen deficiencies and for enduring intense soil reduction for extended periods.
- The intensity and capacity of soil reduction also influence plant functions in wetland ecosystems. As soil reduction intensifies, demand for oxygen in the root rhizosphere also increases. Under some reducing conditions, with high soil reduction capacity, plant roots may not be able to create an oxidized rhizosphere, which helps in neutralizing sulfide and other plant toxicants.
- Intensity of soil reduction promotes increase in root porosity and ROL from root to rhizosphere. There is an increasingly higher rate of oxygen loss as soil redox potential becomes more reduced.
- Increase in aerenchyma formation that parallels the increase in porosity in response to reducing soil conditions may not be sufficient to satisfy respiratory needs of roots for oxygen. Such conditions impact nutrient uptake and carbon assimilation in wetland plants including morphological, anatomical, and metabolic characteristics.
- Oxygen transport and gas exchange in wetland plants have important ecological and environmental significance because of their role in emission of greenhouse gases (methane and nitrous oxide) and nitrification–denitrification process in the rhizosphere.
- Oxygen transport and subsequent oxidation in the rhizosphere support oxidation of ammonium to nitrate. The nitrate diffuses into the adjacent anaerobic soil layers and is denitrified into nitrous oxide and nitrogenous gases. The resulting nitrogen gases including any methane produced in reduced soil environment are transported through plant stems by diffusion and mass flow to the atmosphere.

STUDY QUESTIONS

1. Discuss the various morphological, metabolic, and anatomical adaptations used by wetland plants for growing in wetland soils.
2. Discuss the mechanisms of oxygen transport in wetland plants.
3. What is the difference between diffusion and mass flow gas transport mechanism?
4. How is oxygen transport in wetland plants measured?
5. List potential phytotoxins in the reduced soils.
6. Explain differences between aerobic and anaerobic root metabolisms.
7. How do anatomical adaptations influence emissions of greenhouse gases from wetland?
8. What is the relationship of soil reduction intensity to the root porosity and radial oxygen loss (ROL)?
9. Discuss root iron plaque formation.
10. How can the intensity and the capacity of soil reduction affect nutrient uptake and carbon assimilation?
11. List ways in which excess water in the root zone influences plant physiological function.

FURTHER READINGS

Aller, R. C. 1982. The effects of macrobenthos on chemical properties of marine sediment and overlying water. In P. L. McCall and M. J. S. Tevesz (eds.) *Animal–Sediment Relations: The Biogenic Alteration of Sediments*. Plenum Press, New York. pp. 53–102.

Armstrong, W. 1967. The oxidizing activity of roots in waterlogged soils. *Physiol. Plant.* 20:920–926.

Armstrong, W. 1968. Oxygen diffusion from roots of woody species. *Physiol. Plant.* 21:539–543.

Armstrong, W., J. Armstrong, P. M. Beckett, and S. H. F. W. Justin. 1991. Convective gas flows in wetland plant aeration. In M. B. Jackson, D. D. Davies, and H. Lambers (eds.) *Plant Life under Oxygen Deprivation*. SPB Academic Publishing, The Hague, The Netherlands. pp. 283–302.

Armstrong, W. and P. M. Beckett. 1987. Internal aeration and the development of stellar anoxia in submerged roots. *New Phytol.* 105:221–245.

Armstrong, W. and P. M. Beckett. 1996. Pressurized aeration in wetland macrophytes: some theoretical aspects of humidity-induced convection and thermal transpiration. *Folia Geobot. Phytotaxinom.* 31:25–36.

Brix, H., B. K. Sorrell, P. T., Orr. 1992. Internal pressurization and convective gas flow in some emergent freshwater macrophytes. *Limnol. Oceanogr.* 37:1420–1433.

Crawford, R. M. M. 1992. Oxygen availability as an ecological limit to plant distribution. *Adv. Ecol. Res.* 23:93–185.

Dacey, J. W. H. 1980. Internal winds in water lilies: an adaptation for life in anaerobic sediments. *Science* 210:1017–1019.

Dacey, J. W. H. 1984. Water uptake by roots control water table movement and sediment oxidation in short *Spartina* marsh. *Science* 227:487–489.

Dacey, J. W. H. and M. J. Klug. 1979. Methane efflux from lake sediments through water lilies. *Science* 203:1253–1254.

Hook, D. D. and R. M. M. Crawford. 1978. *Plant Life in Anaerobic Environments*. Ann Arbor Science, Ann Arbor, MI. Chapters 9, 10, and 12. 549 pp.

Jackson, M. B., D. D. Davies, and H. Lambers. 1991. *Plant Life Under Oxygen Deprivation*. SPB Academic Publishing, The Hague, The Netherlands. Chapters 17, 18 and 19. 316 pp.

Joshi, M. M. and J. P. Hollis. 1976. Interaction of *Beggiatoa* and rice plant: detoxification of hydrogen sulfide in the rice rhizosphere. *Science* 195:179–180.

Kludze, H. K., S. R. Pezeshki, and J. H. Pardue. 1990. An oxidation–reduction buffer for evaluating the physiological responses of plants to root oxygen stress. *Environ. Exp. Bot.* 30:243–247.

Kozlowski, T. T. 1984. *Flooding and Plant Growth*. Academic Press, Orlando, FL. Chapter 8. 345 pp.

8 Nitrogen

8.1 INTRODUCTION

Nitrogen is one of the most limiting nutrients regulating the productivity in terrestrial, wetland, and aquatic ecosystems. These ecosystems contain a complex mixture of nitrogen compounds existing in both organic and inorganic forms. The relative proportion of each form depends on the sources of nitrogen entering these systems, and the relative rates and turnover times of these compounds. Organic forms are present in dissolved and particulate forms, whereas inorganic nitrogen (ammonium N, nitrite N, and nitrate N) is present in dissolved forms. Particulate forms are removed through settling and burial, whereas the removal of dissolved forms is regulated by various biogeochemical reactions functioning in soil and overlying water column. The relative rates of these processes are affected by physicochemical and biological characteristics of the soil and water column and the organic substrates present. Although the basic nitrogen transformations in terrestrial, wetland, and aquatic ecosystems are the same, relative rates and storages are different in each of these ecosystems.

8.2 FORMS OF NITROGEN

Nitrogen with its symbol as "N" has an atomic number of 7. The isotopes of nitrogen have mass numbers ranging from 12 to 18. Nitrogen isotopes 14 and 15 are stable with natural abundance of 99.64 and 0.366%, respectively. However, there are five other isotopes (N-12, N-13, N-16, N-17, and N-18), which are unstable and radioactive. Nitrogen, which is required by all organisms for basic processes of life, is found in nature in two basic forms: inorganic and organic nitrogen.

8.2.1 INORGANIC NITROGEN

Dinitrogen (N_2): It is the most common form of nitrogen and makes up to 78% of the atmosphere. It is colorless at room temperature, very stable, and slightly soluble in water. High activation energy is required to break the N_2 triple bond. Biologically this triple bond is broken by organisms capable of fixing N_2. Requirement of high energy to break the triple bond and inability of many biotic communities to directly utilize N_2 places heavy demand on other forms of nitrogen. Industrial fixation of N_2 through the Haber–Bosch process, hence forming nitrogen fertilizers, is largely responsible for meeting plant nitrogen demand in an agricultural ecosystem. This process altered the global N balance. Denitrification is the process that returns dinitrogen gas to the atmosphere through the reduction of nitrogen oxides in anaerobic environment.

Ammonia/ammonium N (NH_3/NH_4^+): Ammonia is colorless and readily soluble in water. The Ionic form is present under acidic soil conditions, whereas unionized form is present under alkaline conditions. Ammonia is also absorbed on soil cation exchange complex or fixed in the crystal lattice of clay minerals. Industrially, ammonia is produced through the "Haber process" at 400°C and 250 atm ($N_2 + 3H_2 \rightarrow NH_3$). Ammonia is a common source of fertilizers as (1) anhydrous ammonia directly injected into the soil and (2) ammonium salts such as ammonium nitrate or ammonium sulfate. Ammonium fertilizer applied to upland or aerobic soils is rapidly oxidized to nitrate. Intensive use of these fertilizers and excessive rainfall or irrigation have resulted in elevated levels of nitrate in groundwaters.

Nitrite (NO_2^-): It is a relatively short-lived form of nitrogen. It is converted to nitrate by microorganisms. It is readily soluble in water and highly mobile in soil. Nitrite is rapidly oxidized to nitrate in aerobic environments or reduced to nitrous oxide in anaerobic soil environments.

Nitrate (NO_3^-): Nitrate is readily soluble in water and highly mobile in soils. Together with ammonium, it is the nitrogen form used by plants, microbes, and other biota as a nutrient. It is also used as an electron acceptor by microorganisms. Nitrification of ammonia N in soil is a source of nitrate. In addition, nonpoint and point source discharges from terrestrial ecosystems can be a major source of nitrate to wetlands and aquatic systems. In the atmosphere, nitrogen oxides can react with photochemically produced free radicals and produce nitric acid, which contributes to acid rain. Nitrate is also a common fertilizer used in agricultural ecosystems.

Nitric oxide (NO): Nitric oxide is a colorless gas at room temperature and is slightly soluble in water. It is produced as a result of industrial emissions. In urban environment, automobile emissions can be a source of nitric oxide. In acid soils, nitrate can be reduced to nitric oxide.

Nitrous oxide (N_2O): Nitrous oxide is a colorless gas, commonly known as "laughing gas," and is slightly soluble in water. It is an intermediate in the nitrogen cycle, and it resembles oxygen in its behavior when heated with combustible substances. It is an end product of denitrification and is also produced as a result of chemical reactions in the atmosphere.

8.2.2 ORGANIC NITROGEN

Proteins are polymers of amino acids. They are the building blocks of life. They are essentially composed of long chains of polypeptides, which are amino acids in a covalent linkage called a peptide bond. Each of the various 20 amino acids has a carboxyl group (–COOH) and an amino group (–NH_2). This form of nitrogen accounts for approximately 50% of the organic nitrogen.

Nucleic acids are polymers of mononucleotides. Each mononucleotide is made up of a base–sugar–phosphate unit. These units are linked to each other by a phosphoric acid ester linkage to form nucleic acids. This form of nitrogen accounts for less than 1% of the total soil nitrogen.

Amino sugars are structural components of a broad group of substances such as mucopolysaccharides and chitin. Approximately 5–10% of organic N in the surface layer of most soils can be accounted for as nitrogen-containing carbohydrate or amino sugars.

Urea is another form of nitrogen, which is a waste product from animals and humans. It is also industrially produced and is commonly used as a fertilizer in agricultural ecosystems including rice paddies throughout the world.

8.3 MAJOR STORAGE COMPARTMENTS

Nitrogen, which is the most abundant gas in the atmosphere ($3,900 \times 10^6$ Tg), is also abundant in other spheres of the earth (Figure 8.1). Here we should note that 1 Tg is equal to 10^{12} g. The lithosphere, which is the rock and soil material portion of our planet, has a nitrogen content of $163,600 \times 10^6$ Tg. Two-third of the earth is covered by water and is termed as the hydrosphere, with a N storage of 23×10^6 Tg. The biosphere, which is the realm of life on earth, has the least nitrogen mass of 0.28×10^6 Tg. There is a constant exchange of nitrogen among these storage compartments of the earth. The nitrogen cycle maintains the balance among various layers of the earth. Nitrogen reactions within the biosphere are the key regulators of ecosystem productivity and functions. Wetlands account for approximately 6% of the biosphere.

The inputs of nitrogen to wetlands include biological N_2 fixation, and point and nonpoint loads from external sources (Figure 8.2). Examples include atmospheric nitrogen deposition, agricultural and urban runoff, application of fertilizers to rice paddies, stormwater runoff carrying nutrients and

Nitrogen

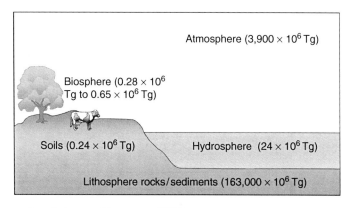

FIGURE 8.1 Reservoirs of nitrogen (Stevenson, 1986).

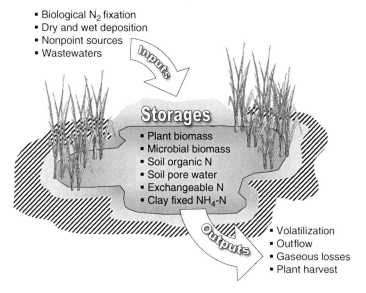

FIGURE 8.2 Major inputs and outputs of nitrogen in wetlands.

fertilizers, and nutrient-enriched water in the case of treatment wetlands. Measurable quantities of nitrogen also enter from precipitation and particulate matter (organic dust containing nitrogen or adsorption of volatized ammonia) in the atmosphere. A typical range of total nitrogen concentrations associated with rainfall is 0.5–2 mg L^{-1}. Typical dry fall nitrogen inputs are 0.07–0.25 g TN m^{-2} year^{-1}.

Organic and inorganic nitrogen are the two major forms present in soil–water–plant components of wetland ecosystem. The Nitrogen cycle in this system can be depicted as the storage of nitrogen in major reservoirs, which serve as either a source or a sink, and a flux between those reservoirs. Various nitrogen forms and associated transport and transformations are shown in Figure 8.3.

Reservoirs of nitrogen in a wetland can be grouped as follows:

- Plant and algal biomass nitrogen (live)
- Particulate organic nitrogen (detritus, soil, and water)
- Microbial biomass nitrogen (MBN) (detritus, soil, and water)
- Dissolved organic nitrogen (DON) (detritus, soil, and water)
- Inorganic forms of nitrogen
- Gaseous end products (atmosphere, detritus, soil, and water)

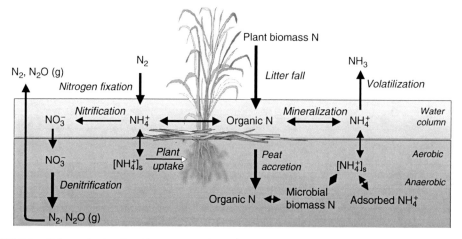

FIGURE 8.3 Schematic showing nitrogen transformations in wetlands.

8.3.1 PLANT BIOMASS NITROGEN

Nitrogen is an essential nutrient for macrophytes and algae, and is most frequently limiting the productivity in wetlands and aquatic systems. Next to carbon, nitrogen is by far the largest component of the plant biomass. Total amount present in this storage pool depends on the standing crop of aboveground and belowground biomass. Sources of nitrogen to macrophytes include external sources, mineralization of organic nitrogen in water and soil, and flux of ammonium from the soil to water column. In addition, algae may also obtain nitrogen through biological fixation. Nitrogen limitation can affect the photosynthetic activity in plants and algae, thus controlling overall productivity. Tissue nitrogen concentration is directly related to the amount of available nitrogen in water and soil. Plants and algae can assimilate both ammonium and nitrate nitrogen, and in some cases they can directly assimilate some soluble organic compounds. Nitrate assimilated is reduced to ammonia in roots and leaves. This process is catalyzed by enzymes; nitrate and nitrite reductases. Ammonia produced in these reactions is assimilated into several amino acids, and subsequently converted into proteins.

8.3.2 PARTICULATE ORGANIC NITROGEN

As plants senesce, a portion, and in some cases all, of the aboveground biomass is returned to detrital pool, where it undergoes decomposition and mineralization of organic nitrogen. In emergent marshes (such as freshwater and saltwater), most of the organic nitrogen produced is converted to the detrital pool. In forested wetlands, litter fall and associated organic nitrogen is the primary source for the detrital pool. For example, litter fall accounts for approximately 50% of the net aboveground productivity in deepwater swamps in the southeastern United States (Mitsch and Gosselink, 2000). In addition, belowground biomass is another major source of detrital nitrogen inputs to the soil.

Relative pool sizes of various forms of nitrogen in wetland soils are shown in Figure 8.4. The largest storage of nitrogen is in organic forms present in soil organic matter. Living forms of organic matter include submerged and emergent vegetation, fauna, and microbial biomass, whereas nonliving forms are detrital plant matter, peat, standing dead, and soil organic matter. Soil organic nitrogen is the largest storage of nitrogen in wetlands and is comprised primarily of humic compounds and complex proteins with a small percentage of amino acids and amines. The nitrogen present in soil organic nitrogen may slowly accrete over time but it is not readily bioavailable. There are a number of factors that can influence the bioavailability of nitrogen in wetland soils such as temperature, hydrologic fluctuations, water depth, electron acceptor availability, and microbial activity.

Nitrogen

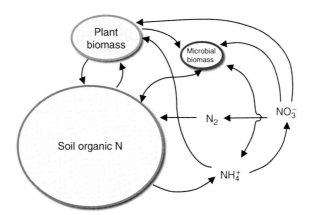

FIGURE 8.4 Relative pool sizes of various forms of nitrogen in wetland soils.

8.3.3 Microbial Biomass Nitrogen

Microbial biomass usually represents 0.5–3.0% of total nitrogen and is a key component of the nitrogen cycle. Microbes are responsible for a significant amount of the work in the system by utilizing organic and inorganic forms of nitrogen for cell growth. The nitrogen stored in the microbial biomass is the most active pool and regulates the amount of bioavailable nitrogen. Microbial consortia use detrital organic matter (produced from dead plant biomass) as an energy source and in the process break down organic nitrogen to ammonium nitrogen.

8.3.4 Dissolved Organic Nitrogen

The DON is defined arbitrarily by many investigators, and represents the material that passes through a 0.45 μm membrane filter. In soils and sediments, DON is extracted from pore waters or in some cases soils are extracted with water and filtered through 0.45 μm filter. Some of the DON in the detrital layer is also produced through leaching of materials from the detrital pool. The DON represents approximately <1% of total soil organic matter. DON can be very important in oligotrophic wetlands and aquatic systems, where most of the nitrogen is in the dissolved organic form. A significant portion of DON may be remineralized depending on ambient microbial and enzyme activity, nutrient status, C:N:P ratio, and chemical characteristics of the DON source material.

8.3.5 Inorganic Forms of Nitrogen

Inorganic forms of nitrogen stored in wetlands include ammonium, nitrate, and nitrite. They can be found in soil pore water. Ammonium may be present on the soil cation exchange complex or permanently trapped within the clay layer. Inorganic N forms are not typically very stable, comprising <1% of the total N within wetlands.

8.3.6 Gaseous End Products

Gaseous forms of nitrogen include ammonia, nitrous oxide, and dinitrogen, which are readily lost to the atmosphere, and comprise approximately <1% of total nitrogen within a wetland.

Nitrogen is exported from wetland systems via ammonia volatilization under alkaline conditions, where unionized ammonia in its gaseous form can be transported from soil and water column to the atmosphere. Another gaseous loss of nitrogen includes nitrous oxide and nitrogen produced during denitrification and emitted into the atmosphere.

Other outputs include harvested plant biomass and detrital export (removing the nitrogen stored in the biomass, or simply wetland outflows where nutrients are exported in water and POM leaving wetlands such as in tidal marshes or riparian wetlands).

8.4 REDOX TRANSFORMATIONS OF NITROGEN

Nitrogen occurs in various oxidation states, ranging from +5 in the most oxidized forms to −3 in the most reduced forms (Figure 8.5). Most of the nitrogen in wetland nitrogen cycle is bonded to either C, H, or O. Nitrogen bonded to C and H has an oxidation number of −3, whereas those bonded to O have positive oxidation numbers ranging from +2 to +5. Several nitrogen cycling processes are governed by oxidation–reduction reactions.

Pathway 1 shows ammonification, which is the first step in mineralization of organic nitrogen. *Nitrogen mineralization* refers to the biological transformation of organically combined nitrogen to ammonium nitrogen during the degradation of organic matter. *Ammonification* is defined as the biological transformation of organic nitrogen to ammonium, which can occur in either aerobic or anaerobic conditions, but is slower under anaerobic conditions due to less efficient decomposition. Ammonium nitrogen is the most reduced form with an oxidation state of −3.

Pathway 2 shows *immobilization*, whereby ammonium is assimilated into the biomass of plants and microbes. The amount and type of plant residue or organic matter in the soil can influence amount of nitrogen immobilized in microbial and plant biomass. A high amount of organic matter in the soil will result in immobilization. Microbial immobilization of ammonium nitrogen depends on the carbon-to-nitrogen ratio of organic residues undergoing decomposition (as discussed later in this chapter).

Pathway 3 shows *nitrification*, which is defined as the biological oxidation of ammonium to nitrate under aerobic soil conditions. The chemoautotrophic bacteria *Nitrosomonas* sp. converts $2NH_4^+ + 3O_2 \rightarrow 2NO_2^- + 2H_2O + 4H^+$ + energy and then *Nitrobacter* sp. converts $2NO_2^- + O_2 \rightarrow 2NO_3^-$ + energy. The oxidation state of nitrogen increases from −3 (NH_4^+) to +3 (NO_2^-) or +5 (NO_3^-). The autotrophic bacteria use ammonium ions as their primary source of energy.

Pathway 4 shows *denitrification*. This process is linked to microbial respiration where electrons are added to nitrate or nitrite, resulting in the production of nitrous oxide or nitrogen gas. The

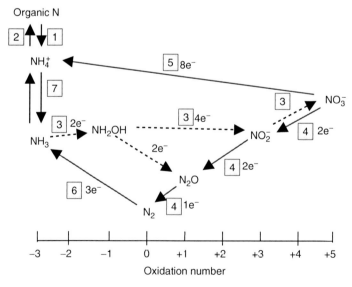

FIGURE 8.5 Oxidation and reduction reactions of nitrogen in wetlands. Numbers 1–7 refer to pathways of nitrogen reactions. 1 = ammonification; 2 = immobilization; 3 = nitrification; 4 = denitrification; 5 = dissimilatory nitrate reduction to ammonia; 6 = dinitrogen fixation; and 7 = ammonia volatilization.

oxidation state of N decreases from +5 to 0 going from nitrate to nitrogen gas. It is microbial mediated, where facultative bacteria such as *Bacillus* and *Pseudomonas* possess the enzymes that allow them to use nitrate, nitrite, and nitrous oxide as the terminal electron acceptor during oxidation of organic C in anaerobic environments.

Pathway 5 indicates two contrasting pathways. *Dissimilatory nitrate reduction to ammonia (DNRA)* occurs in anaerobic soil and involves the reduction of nitrate to ammonium rather than to nitrogen gas. It is carried out by obligate anaerobes such as *Clostridium* and *Streptococcus* under highly reduced conditions. The reaction ($NO_3^- + 10H^+ + 8e^- \rightarrow 3H_2O + NH_4^+$) consumes eight electrons per NO_3^-, whereas denitrification consumes only five electrons. It also consumes 10 protons, driving the pH towards alkaline conditions. The *assimilatory nitrate reduction to ammonia* (ANRA) pathway occurs in both aerobic and anaerobic soil conditions and involves the reduction of nitrate to ammonia or amino nitrogen as a cell constituent. This process is common to many microorganisms and most plants.

Pathway 6 shows *dinitrogen fixation*, which is the reduction of atmospheric, inert nitrogen to ammonia that plants and microbes can then use. It occurs in the water column, soil–floodwater interface, root zone, and anaerobic soil in wetlands. It is carried out by free-living bacteria (*Clostridium*), bacteria living in symbiotic association with plants (*Rhizobium*), cyanobacteria (*Anabaena*), and periphyton.

Pathway 7 shows *ammonia volatilization*, which is a physicochemical process controlled by the pH of the environment. There are many biotic processes that can alter the pH of a wetland including photosynthesis and denitrification. An alkaline pH favors the presence of unionized ammonia, whereas acidic or neutral pH favors that of ionized ammonia. Loss of nitrogen due to volatilization is insignificant at pH <7.5, but it dramatically increases at pH >7.5.

Nitrogen fixation (input to biosphere) and denitrification (output from biosphere) balance the nitrogen budget in the biosphere (Figure 8.6). Nitrogen fixation and other inputs account for 92×10^6 metric tons of nitrogen in the biosphere every year. Some of these other inputs include industrial fixation and combustion lightening fixation. Biological fixation contributes to approximately 78% of all nitrogen fixation for nonocean areas, whereas industrial and combustion lightening fixation accounts for 20 and 2%, respectively. Approximately 9 metric tons year^{-1} are fixed permanently into the biosphere every year. This means that 83×10^6 metric tons year^{-1} are released back into the atmosphere primarily through denitrification. Nitrogen gas is the most stable form of nitrogen, which may explain why the atmosphere is a major store of nitrogen. Nitrogen fixation is an energy-intensive process that

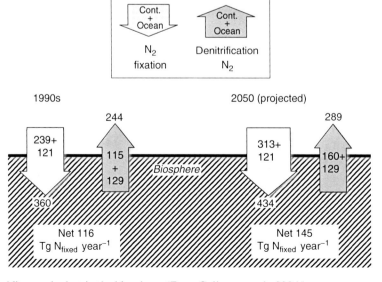

FIGURE 8.6 Nitrogen budget in the biosphere. (From Galloway et al., 2004.)

very few organisms can perform. Only about 3% of net primary production of organic matter involves nitrogen fixation from N_2.

What will happen to the atmospheric nitrogen gas if organisms responsible for denitrification are removed from the biosphere? If denitrifying organisms are removed from the biosphere, there will be depletion of nitrogen from the atmosphere as a result of biological fixation of dinitrogen. This depletion will result in enrichment of atmospheric oxygen and carbon dioxide, thus creating imbalance in atmospheric gases. One estimate suggests that in approximately 100 million years all nitrogen gas will be removed from the atmosphere if the process that returns dinitrogen to atmosphere is eliminated.

8.5 MINERALIZATION OF ORGANIC NITROGEN

Microbial communities in wetland soils play a key role in the oxidation of complex organic compounds and regeneration of available nitrogen for supporting primary productivity of biotic communities in wetlands. The processes regulating the breakdown of organic matter are related to the cycling of carbon and nitrogen. These two elements are key constituents of organic matter, and it is not surprising that the processes and factors controlling the turnover are the same (for details on carbon cycle, see Chapter 5). Of the various transformations nitrogen undergoes in soils, ammonification–immobilization cycle is the continuous one, the two individual steps often achieving opposite goals.

Ammonification refers to the breakdown of organic forms of N to ammonium (Figure 8.7). Ammonification is an oxidation process linked to microbial catabolic activity of general-purpose heterotrophs that utilize nitrogenous organic compounds as an energy source. This process is strongly related to organic C decomposition because a major portion of N in soil organic matter and plant detritus is bonded to C. As organic matter is decomposed, organic N is mineralized and released as ammonium. Depending on the microbial N requirements, mineralized or released ammonium is rapidly assimilated and utilized. Ammonification process is also referred to as *gross mineralization*. *Immobilization* is the conversion of inorganic N to organic forms (Figure 8.7). Inorganic forms of N (ammonium and nitrate nitrogen) are used as a nutrient source for cell synthesis and the building of biomass. Immobilization is an assimilative process linked to microbial anabolic activity. The inorganic N assimilated by microorganisms is rapidly metabolized into nitrogenous constituents within cells.

Because of this microbial control, ammonification is generally accompanied by immobilization: the two processes tend to counteract each other so far as the production of inorganic N is concerned.

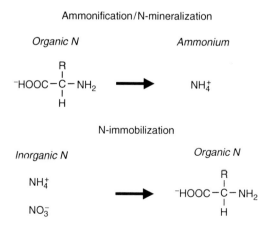

FIGURE 8.7 Processes involved in the transformation and conversion of ammonia in soils.

The difference between ammonification and immobilization is usually positive in wetlands resulting in a net release of ammonium N, which is often referred to as *net nitrogen mineralization*. Net nitrogen mineralization depends on the microbial nitrogen requirements. In wetlands, lower nitrogen requirements of anaerobic microbial communities favor net mineralization. Both ammonification and immobilization processes repeat continuously again and again as a result of a "continuous internal cycle," leading to exhaustion of labile organic matter introduced as plant detritus material. If no additional labile organic nitrogen is introduced into the wetland system, ammonification–immobilization cycle reaches a steady state, resulting in a stabilized C:N ratio of the organic matter. Net immobilization occurs in systems limited by nitrogen, whereas net mineralization occurs in systems limited by carbon. The C:N ratio concept can be used to determine the microbial nitrogen requirement and potential net mineralization, as discussed in the following section.

8.5.1 C:N Ratio Concept

Here we show nitrogen requirements of microorganisms involved in decomposing detrital matter in wetlands. The carbon-to-nitrogen (mass) ratio of microbial biomass is typically around 10. Microbes assimilate both carbon and nitrogen during cell synthesis to maintain this ratio. Efficiency of carbon assimilation by aerobic microorganism is approximately 20–60% (dry weight basis), with fungi more efficient in assimilating carbon than bacteria. The efficiency of carbon assimilation by anaerobic bacteria is in the range of 5–10%. Using these conditions and assumptions let us estimate the nitrogen requirements of microorganisms during decomposition of plant detritus. Carbon content of plant detritus is typically constant at approximately 40% (dry weight basis), whereas the carbon content of photosynthetic tissue can be approximately 45–50% (dry weight basis). The nitrogen demand by microbial biomass can be calculated as follows.

Let us assume that 100 units (dry weight basis) of plant detritus is undergoing decomposition in a wetland. The carbon content of detrital matter is 40 units assuming 40% carbon content (dry weight basis). Carbon and nitrogen use efficiency is set at approximately 40% for aerobes and 10% for anaerobes. Based on the above assumptions, 16 units of carbon are assimilated by aerobes and 4 units of carbon by anaerobes. To maintain C:N ratio of 10 in their biomass, aerobic microbes and anaerobes would require 1.6 units and 0.4 units of nitrogen, respectively.

The critical plant detritus nitrogen content can be estimated by dividing the units of nitrogen demand of microbes by the total units of the detrital matter (in this example it is 100 units). For aerobic decomposition, this represents 1.6 units of nitrogen of 100 total units, which equals 1.6%. For anaerobic decomposition this would represent 0.4 units of nitrogen of 100 total units, an amount equivalent to 0.4%. Finally, to determine the critical C:N mass ratio of the detrital matter, we can divide the percentage of carbon in the detritus by the percentage of critical plant detritus nitrogen. For aerobic decomposition of plant detritus, there is a 40% carbon and 1.6% critical nitrogen requirement, which equals to a C:N ratio of 25. For anaerobic decomposition, there is 40% carbon and 0.4% critical nitrogen requirement, which equals to a C:N ratio of 100. Using these critical C:N ratios, we can predict aerobic and anaerobic decomposition rates of plant detritus in relation to ammonification and immobilization.

Ammonification (A), as discussed previously, is the breakdown of organic nitrogen to ammonium (NH_4^+). Immobilization (I) is the buildup of organic matter or assimilation of inorganic nitrogen into microbial biomass. Therefore, using plant detritus as an example, under aerobic conditions, if the C:N ratio of plant detritus is greater than 25, net immobilization of inorganic nitrogen will occur as a result of nitrogen assimilation by microbes ($I > A$) during decomposition. If the C:N ratio is less than 25, then net ammonification will result in the release of inorganic nitrogen ($A > I$) as the nitrogen demand by microbes is adequately met. Under anaerobic conditions, if the C:N ratio is greater than 100, then a buildup of organic matter occurs ($I > A$) because of the low nitrogen requirements of microbes under these conditions. If the C:N ratio is less than 100, then ammonium will be released ($A > I$) as the microbes decompose the plant detritus. Differences in microbial

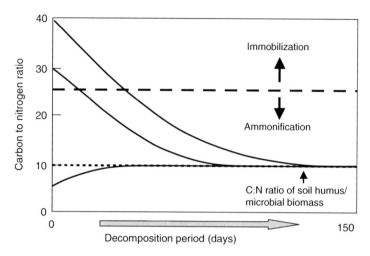

FIGURE 8.8 Relationship of decomposition of organic matter and detrital C:N ratio.

nitrogen requirements explain why ammonium release is higher under anaerobic conditions than under aerobic conditions.

The importance of C:N ratio of detritus in the realm of nitrogen immobilization–ammonification can be summarized as follows (Figure 8.8): C:N ratio of detritus reflects the ammonification–immobilization patterns of nitrogen. Organic substrates with wider C:N ratio stimulate immobilization, whereas those with narrower C:N ratio favor ammonification. During decomposition of plant detritus with high C:N ratio, a major portion of carbon is lost as carbon dioxide into the atmosphere during the consecutive cycles through turnover, whereas nitrogen is retained in the organic form (primarily microbial biomass) to maintain their critical C:N ratio of 10.

Repeated cycles of ammonification bring the C:N ratio of detritus below critical level, where net ammonification occurs. Now the material is at a stage when the "net mineralization" of nitrogen can be realized. As the reactions of mineralization and immobilization are simultaneous and opposite in direction, it is the net mineralization and immobilization that controls the amount of available inorganic nitrogen. This net effect depends on the type of carbonaceous material and its nitrogen content.

Hypothetical examples of C:N ratio changes in detritus during decomposition under aerobic conditions are shown in Figure 8.8. Two critical C:N ratios are represented by dotted lines. The first corresponds to plant detritus with C:N ratio of 5. For this residue, rapid ammonification and an increase in microbial biomass increased the overall C:N ratio close to 10. For the residues with C:N ratio of 30 and 40, as decomposition proceeds, the C:N ratio decreases steadily until the overall ratio reaches a steady-state value of approximately 10. Within the area above this value, inorganic nitrogen produced by decomposition is rapidly consumed by increasing populations of microorganisms responsible for decomposition. As a result, nitrogen is immobilized. It is easily noticed that the lower the initial C:N ratio, the sooner this phase ends. Net mineralization of organic nitrogen occurs as soon as the C:N ratio decreases below this critical value. As the decomposition continues, both organic carbon and nitrogen are converted into inorganic forms. Much of the carbon is lost as carbon dioxide and the nitrogen is released as ammonium nitrogen. As the decomposition proceeds, the C:N ratio of organic residue reaches a ratio of approximately 10. At this point, most of the labile portion of the organic residue is decomposed, and the remaining relatively nonlabile residue accumulates in the soil. The time required for decomposition of this phase depends on the initial composition of the residue and the environmental conditions. The residue essentially

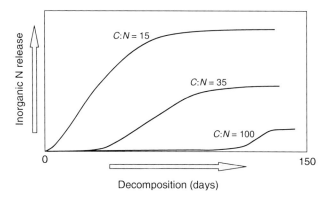

FIGURE 8.9 Mineralization of organic nitrogen plant residues with different C:N ratios. (Data from Tusneem and Patrick, 1971.)

consists of humic and fulvic substances, with an approximate C:N ratio of 10. The importance of C:N ratio can be summarized as follows:

- The C:N ratio reflects the mineralization–immobilization pattern of nitrogen. A wider C:N ratio promotes immobilization, whereas the narrower ratio favors mineralization.
- The C:N ratio is important because during the decomposition of plant detrital matter, carbonaceous substances provide energy source for microbes and nitrogen is assimilated and metabolized as proteins in the synthesis of cellular materials.

Mineralization of organic nitrogen plant residues with a range of C:N ratios is shown in Figure 8.9. Decomposition of plant residue with C:N ratio of 15 shows a rapid release of inorganic nitrogen, suggesting net mineralization. Decomposition of plant residue with C:N ratio of 35 showed no net release of inorganic nitrogen during first few days of decomposition, followed by net release of inorganic nitrogen. This suggests that the initial release of inorganic nitrogen is rapidly assimilated into microbial biomass, and the net release occurs after the initial lag phase. Decomposition of plant residue with C:N ratio of 100 showed a much longer lag phase before inorganic nitrogen release occurred. As decomposition occurs, the C:N ratio of the remaining organic matter decreases because carbon is being lost and nitrogen conserved. When C:N ratio of decomposing residue approaches <25, net ammonification occurs under aerobic conditions. As the residue undergoes decomposition, a portion of nitrogen released is assimilated into microbial biomass. Interestingly, the C:N ratio of microbial biomass is approximately 10. The residual organic residues are converted into newly formed humic substances and ultimately stable humus. The mean residence time of nitrogen in any given pool can range from a few days to several hundred years.

Net ammonification depends not only on C:N ratio, but also on the quality of the detrital matter. As discussed in Chapter 5, various organic substrates decompose at different rates, in the order of proteins > carbohydrates > cellulose and hemicellulose > lignin. The time required for microorganisms to lower the C:N ratio of carbonaceous plant residues to the level where inorganic nitrogen accumulate will depend upon such factors as climate and especially temperature, hydrology, lignocellulose ratio, and level of activity of the soil microorganisms. As a general rule of thumb, organic substrates rich in nitrogen favor net ammonification and those poor in nitrogen favor net immobilization.

8.5.2 Chemical Composition of Organic Nitrogen

Soil nitrogen is predominantly in organic forms with less than 5% present in inorganic forms (Figure 8.10a). The nature of nitrogen-containing organic compounds occurring in soils is not well documented. Several operationally defined chemical fractionation schemes have been used

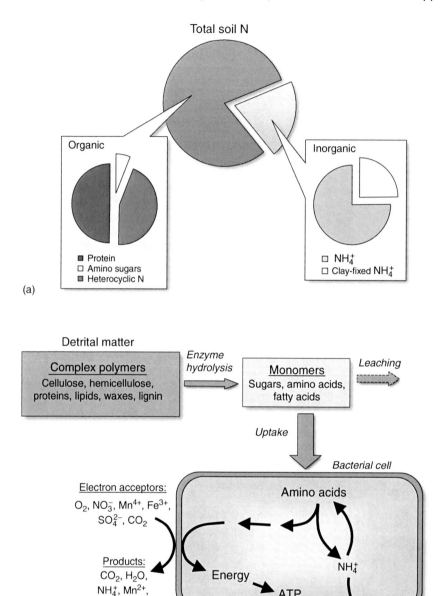

FIGURE 8.10 (a) Relative proportion of organic and inorganic forms in wetland mineral soils. (b) Microbial degradation of organic nitrogen to ammonia.

to characterize soil organic nitrogen (Kelley and Stevenson, 1995; Schulten and Schnitzer, 1998). Organic nitrogen in detritus and soil organic matter includes a broad range of nitrogenous organic compounds such as proteins, cell wall constituents (amino sugars and their polymers, chitin and peptidoglycan) and nucleic acids. The bulk of organic nitrogen includes proteins; other components account for a small fraction of total nitrogen. The various components of organic matter decompose at different rates, the order for plant residue components being proteins > carbohydrates > cellulose and hemicellulose > lignin. The initial substrate is often a macromolecule or *polymers* (such as protein, nucleic acid, and aminopolysaccharide), from which simpler N-containing

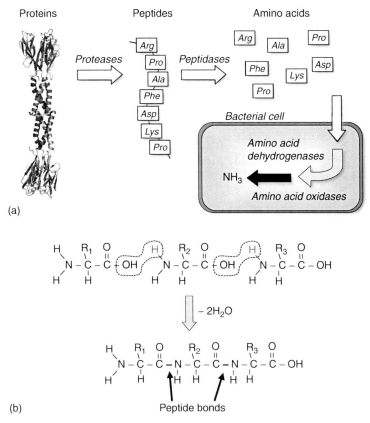

FIGURE 8.11 (a) Microbial degradation of proteins to ammonia. (b) Structure showing peptide bonds in proteins (polymers of amino acids).

monomers are formed (e.g., amino acids, pyrine and pyrimidine bases, amino sugars). The simpler compounds are usually taken up by microbial cells and subsequently degraded with the formation of ammonia (Figure 8.10b). The conversion of organic nitrogen to ammonia by heterotrophic bacteria occurs in two steps: (1) enzymatic hydrolysis of proteins and other complex polymers by aerobic and anaerobic organisms, releasing amino acids and other simple compounds; and (2) subsequent deamination of these compounds results in the formation of ammonia nitrogen (Figure 8.10b).

Amino acids are monomeric units of proteins. Two types of proteins are present in the cells: structural proteins and catalytic proteins (enzymes). Proteins are polymers of various lengths of defined sequence of amino acids that are covalently bonded by means of peptide linkages. Degradation of proteins and peptides involves an initial breaking of peptide bonds by proteases and peptidases to form amino acids, from which ammonia is released through the action of such enzymes as amino acid dehydrogenases and oxidases (Figures 8.11a, 8.11b, 8.12, and 8.13). Amino acids contain two key functional groups, an amino group ($-NH_2$) and a carboxylic acid group ($-COOH$). Approximately 35–40% of the total organic nitrogen can be accounted for inproteinaceous fractions (proteins, peptides, and amino acids) (Schulten and Schnitzer, 1998). The amino acid composition of soils is similar to those found in bacterial cells, suggesting the role of soil microbes in synthesizing proteins, peptides, and amino acids from plant residues and soil organic matter (Sowden et al., 1977) (Table 8.1).

The *nucleic acids* consist of individual mononucleotide units (pyrine or pyrimidine base–sugar–phosphate) joined by a phosphoric acid ester linkage through the sugar. During decomposition, nucleic acids are converted to mononucleotides by the action of nucleases, which catalyze the hydrolysis of

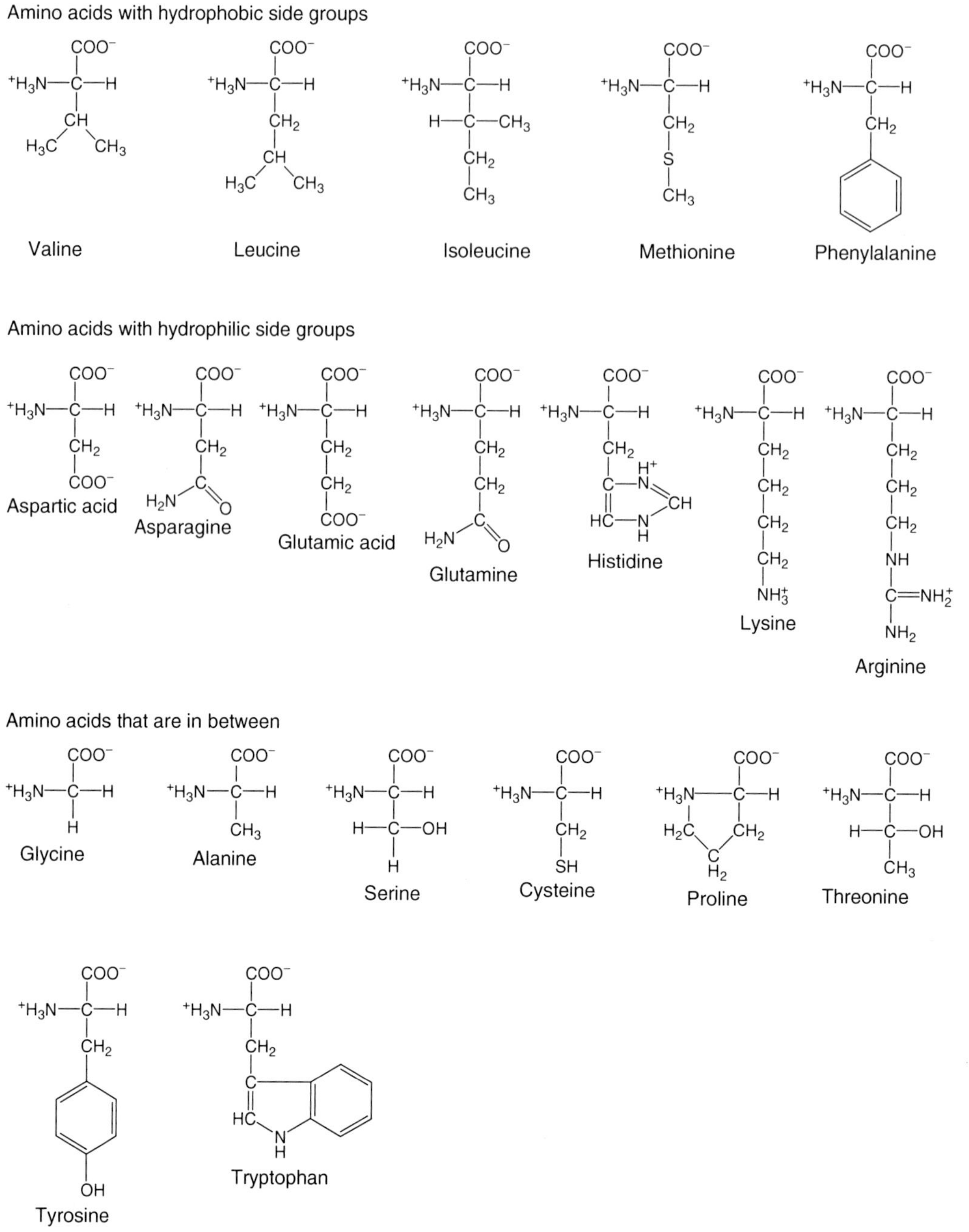

FIGURE 8.12 Chemical structure of amino acids that occur in soils.

ester bonds between phosphate groups and pentose units (Figure 8.14). The mononucleotides, in turn, are converted to nucleosides and inorganic phosphate by nucleotidases. The nucleosides are then hydrolyzed to purine or pyrimidine bases and the pentose component of the nucleosides, which are subsequently converted to ammonia by reactions catalyzed by amidohydrolases and amidinohydrolyases. Soils contain a small fraction in this pool with values in the range of 0.3% of total nitrogen in organic soils to 3% in mineral agricultural soils (Schulten and Schnitzer, 1998).

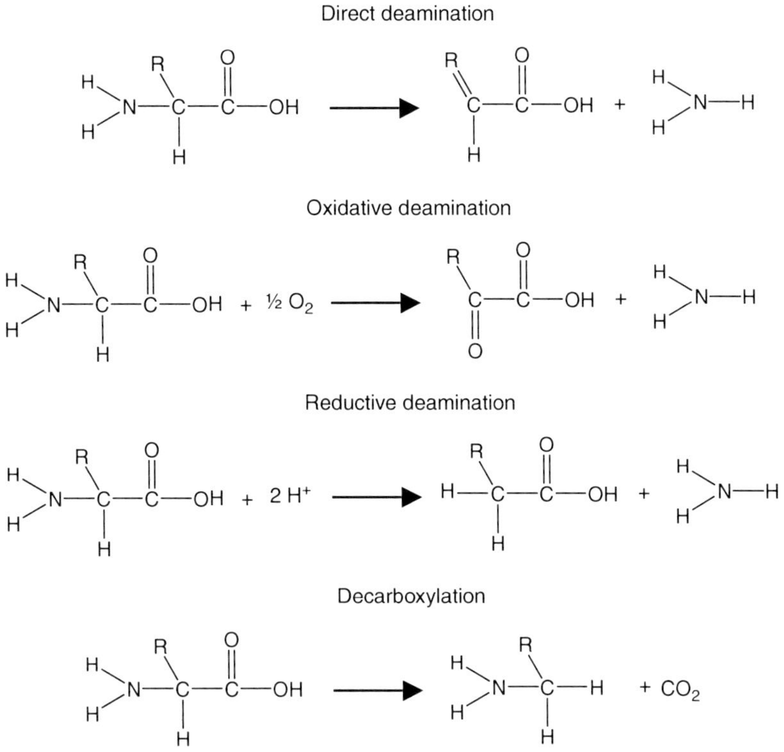

FIGURE 8.13 Deamination processes involved in ammonification of organic nitrogen.

TABLE 8.1
Common Amino Acids Found in Soils

Aliphatic amino acids	Glycine, alanine, valine, leucine, isoleucine
Amino acids with hydroxyl or sulfur-containing side chains	Serine, cysteine, threonine, methionine
Aromatic amino acids	Phenylalanine, tyrosine, tryptophan
Cyclic amino acids	Proline
Basic amino acids	Histidine, lysine, arginine
Acidic amino acids and their amides	Aspartic acid, glutamic acid, asparagine, glutamine

Source: Stevenson (1994).

Amino sugars are common as structural components of cell walls of microorganisms in combination with mucopeptides and mucoproteins. Formation of ammonia from glucosamine (a common soil amino sugar) is catalyzed by the combined action of glucosamine kinase and glucosamine-6-phosphate isomerase; the products are ammonia and fructose-6-phosphate (Figure 8.15). The most prominent amino sugars measured in soils are D-glucosamine and D-galactosamine, with the former present in greater amounts (Stevenson, 1994). Amino sugars account for approximately 5–10% of the organic nitrogen.

In addition to the nitrogenous components of plants or animal tissues, soils may contain organic nitrogen as *urea*, which is a constituent of the urine of grazing animals and is often added to the soil as fertilizer. Urea is rapidly hydrolyzed to ammonia (Figure 8.16). This reaction is catalyzed by urease, an enzyme found practically in all soils.

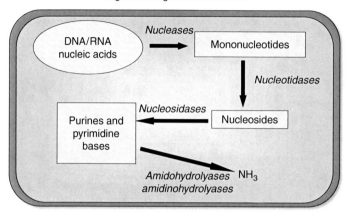

FIGURE 8.14 Enzymatic degradation of nucleic acids leading to ammonia formation.

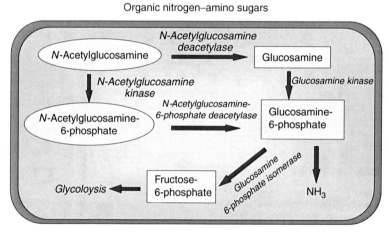

FIGURE 8.15 Enzymatic degradation of amino sugars to release ammonia.

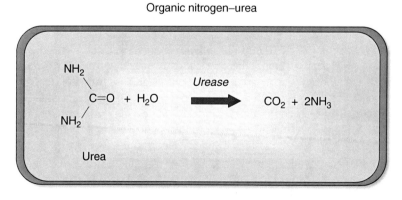

FIGURE 8.16 Enzymatic hydrolysis of urea to ammonia.

Mineralization is basically a sequence of enzymatic reactions. Ammonia is stable under anaerobic conditions; therefore, it is the primary form of nitrogen in wetland soils. Globally, ammonia represents only about 15% of the nitrogen released into the atmosphere. Although ammonification can occur in both aerobic and anaerobic environments, a higher concentration of ammonia is seen in anaerobic conditions due to the lower demand for nitrogen by anaerobic organisms.

A large proportion of organic nitrogen (approximately 50%) remains unidentified. In recent years, several approaches have been used to determine organic nitrogen forms. These include optical properties (UV and fluorescence), ^{15}N cross-polarization magic angle spinning nuclear magnetic resonance (^{15}N CPMAS NMR) spectroscopy, and x-ray photoelectron spectroscopy (XPS) (Maie et al., 2006). Acid-hydrolyzable amino acids (HAAs) are also used as an indicator of bioavailable organic nitrogen. Few studies have attempted to use ^{15}N NMR (solution and solid-state NMR) to characterize organic nitrogen forms in peats, plant residues, soils, and humic fractions. The conclusions drawn from these studies have shown considerable consistency, whether dealing with native soil N (natural-abundance studies) or the result of incubation with ^{15}N sources; in all cases, the N observed in the soil, humic fraction, or compost was largely present in amide structures, with minor amounts of indole, pyrrole, and amino N (Preston, 1996). The solid-state ^{15}N NMR spectrum of ^{15}N-labeled algae is shown in Figure 8.17 (Zang et al., 2000). The spectrum shows a predominance of amide signal corresponding to peptide nitrogen. This is not surprising because the bulk of algal biomass typically constitutes proteinaceous materials. Chemical fractionation methods have shown that a large part of the soil organic nitrogen is in "unknown" forms that is, not accounted for by hydrolysis and analysis as ammonium, amino acids, or amino sugars. This pool still remains unidentified.

The distribution of various organic nitrogen fractions in flooded Crowley silt loam soil (paddy soil in Louisiana) as determined by chemical fractionation scheme is shown in Figure 8.18 (Cheng and Kurtz, 1963; Tusneem and Patrick, 1971). Amino acid nitrogen accounted for approximately 45% of total nitrogen, whereas hydrolyzed ammonium nitrogen comprised of 28% of total nitrogen. The amount of nitrogen found in amino sugars was 5% of the total nitrogen. Humin nitrogen fraction (both soluble and insoluble fractions) comprised of 20% of the total nitrogen.

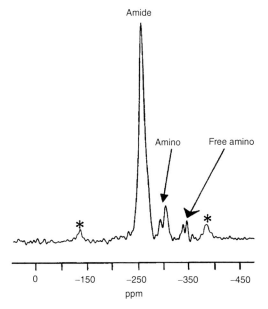

FIGURE 8.17 Solid-state ^{15}N NMR spectrum of freeze-dried ^{15}N-labeled algae. Spinning sidebands (SSB) are indicated by asterisks (Zang et al., 2000).

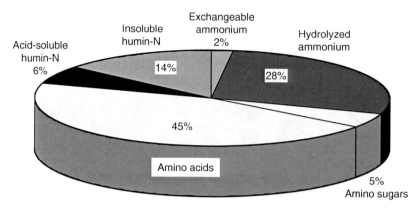

FIGURE 8.18 The distribution of organic nitrogen fractions in a flooded Crowley silt loam soil (paddy soil of Louisiana). Total nitrogen = 765 mg kg^{-1}. (Redrawn from data by Tusneem and Patrick, 1971.)

TABLE 8.2
Fractionation of Soil Organic Nitrogen by Acid Hydrolysis

Form	Method	Percentage of Soil Nitrogen
Acid-insoluble nitrogen	Nitrogen remaining in soil residue following acid hydrolysis. Usually obtained by difference (total soil nitrogen–hydrolyzable nitrogen)	20–35
Ammonia nitrogen	Ammonia recovered from hydrolysate by steam distillation with MgO	20–35
Amino acid nitrogen	Usually determined by the ninhydrin-CO$_2$ or ninhydrin ammonia methods	30–45
Amino sugar nitrogen	Steam distillation with phosphate–borate buffer at pH 11.2 and correction for ammonia nitrogen. Colorimetric methods are also used	5–10
Hydrolyzable unknown nitrogen (HUN) fraction	Hydrolyzable nitrogen not accounted for as ammonia, amino acids, or amino sugars	10–20

Source: Stevenson (1986).

An operationally defined chemical fractionation scheme was described by Stevenson (1986) (see Table 8.2). This method involves extracting soil organic nitrogen using hot 6N HCl, after which the nitrogen is separated into various fractions.

8.5.3 Microbial Degradation of Organic Nitrogen

Detrital matter consists of complex polymers including proteins, lipids and waxes, cellulose, hemicellulose, and lignin. Pathways of organic nitrogen mineralization are similar to those discussed for organic carbon breakdown (see Chapter 5). Organic nitrogen compounds are sequentially hydrolyzed into simpler compounds by the activity of hydrolytic enzymes. The extracellular enzyme hydrolysis of organic matter is considered to be the rate-limiting step in the overall decomposition process. Therefore, the activity of enzymes may provide an indication of mineralization rates of organic compounds in wetland soils. As discussed earlier, several substrate-specific extracellular enzymes are involved in hydrolyzing organic nitrogen polymers into monomers. In this

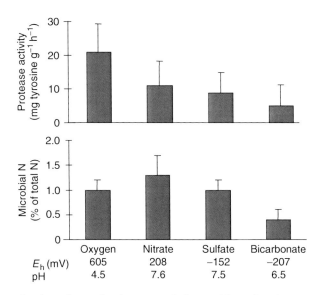

FIGURE 8.19 Mineralization of organic nitrogen as influenced by soil redox potential. (Redrawn from McLatchy and Reddy, 1998.)

example, we show the effect of soil redox conditions on protease activity. Microbes catabolize monomers efficiently with oxygen as an electron acceptor (Figure 8.19). Higher protease activity indicates greater mineralization of organic nitrogen. As expected, greater protease activity was noted under aerobic conditions than under anaerobic conditions. The protease activity (measured by hydrolysis of casein) decreased as the soil became more reduced. It reached the highest level in the aerobic soil, which would indicate a higher capability of degrading protein compared with anaerobic soils.

The breakdown of soil organic matter and detrital matter is typically slower under flooded soil conditions than under drained soil conditions. Hence, a lower gross nitrogen mineralization rate would be expected under flooded soil conditions than under drained soil conditions. Gross nitrogen immobilization is lower under flooded soil conditions because of low metabolic nitrogen requirements of anaerobic bacteria. Thus, the net effect of mineralization and immobilization is higher under flooded soil conditions than under drained soil conditions, leading to higher rates of inorganic nitrogen accumulation under flooded soil conditions. The characteristic features of anaerobic bacterial degradation of organic compounds are (1) incomplete breakdown of complex polymers into volatile fatty acids, methane, hydrogen, and carbon dioxide, with subsequent low energy yield; (2) low-energy fermentation results in the synthesis of fewer microbial cells per unit of organic matter degraded; (3) low nitrogen requirements during anaerobic metabolism; and (4) net release of ammonium ions during decomposition of plant detritus with wide C:N ratio.

In the absence of any one standardized method, many investigators have used laboratory batch incubation of wetland soils and aquatic sediments and measured ammonium production as a measure of nitrogen mineralization rates (Wang et al., 2001; White and Reddy, 2001) (Figure 8.20). These methods range from simple batch incubation experiments in which ammonium accumulation is measured over time to stable isotope dilution technique in which soil cores or anaerobic soil slurries are injected with ^{15}N-labeled ammonium nitrogen. This method allows to estimate both gross and net mineralization rates (Blackburn, 1979).

Under field conditions, the most common method used is the litter bag method. Litter bags with mesh sizes ranging from less than 1 mm to more than 10 mm have been used. Although there are several limitations to this method, investigators have consistently used this method because of its

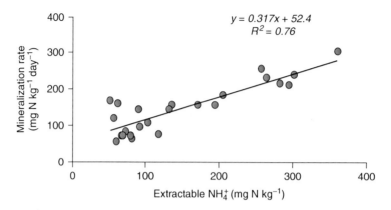

FIGURE 8.20 Relationship between organic mineralization rate and ambient ammonia levels in soils (White and Reddy, 2001).

simplicity. One drawback is that litter bags potentially alter the physical environment and exclude grazing benthic communities. Also some of the DON and fine particulate organic nitrogen can be lost from the system, thus resulting in overestimation of the mineralization rates.

A wide range of mineralization rates in wetlands are reported in the literature, with values ranging between 4 and 357 mg N m^{-2} day^{-1} (mean = 111 ± 124, n = 13; Martin and Reddy, 1997). Similarly, Herbert (1999) reported organic mineralization rates in the range of 7–430 mg N m^{-2} day^{-1} for unvegetated coastal sediments and 50–1,125 mg N m^{-2} day^{-1} for vegetated coastal sediments. Lower organic nitrogen mineralization rates in unvegetated sediments are due to low organic inputs as compared to vegetated sediments. Organic nitrogen mineralization rates are generally lower in ombrotrophic wetlands (low nutrient and closed systems) as compared to minerotrophic (relatively high nutrient and open systems), as measured in batch incubation experiments conducted in soils from various community types in northern wetlands of the United States (Bridgham et al., 1998). In this study, aerobic mineralization rates were highest in bogs and acidic fens and lowest in cedar swamps and intermediate fens (Figure 8.21). By contrast, anaerobic mineralization was highest in beaver meadows and lowest in intermediate fens. Differences in organic nitrogen mineralization rates in these communities are due to variability in the quality of organic matter inputs. Aerobic mineralization rates were approximately 2.6 times higher than anaerobic mineralization rates (Bridgham et al., 1998).

Net release of ammonium is determined by the balance between ammonification and immobilization, which is controlled by the nitrogen requirements of microorganisms involved, nature of organic nitrogen, and other soil and environmental factors. Because of low nitrogen requirements of anaerobic microorganisms and low oxidation rates of ammonium, wetland soils usually accumulate high levels of ammonium nitrogen, as compared to upland soils. Organic nitrogen mineralization can be described as a function of C:N ratio, extracellular enzyme (such as protease), microbial biomass, and soil redox conditions. Organic nitrogen mineralization was highly correlated with the biomass nitrogen of wetland soil maintained under different redox conditions (McLatchey and Reddy, 1998). MBN is directly correlated with MBC, with an approximate ratio of 10 (Figure 8.22). Measurement of these parameters can indicate the extent of organic nitrogen mineralization in soil and water column of a wetland.

Organic nitrogen mineralization is the primary source of nitrogen for wetland vegetation. Rate of ammonium nitrogen accumulation in wetland soils has considerable agronomic and ecologic significance. In the example shown in Figure 8.23, fertilizer nitrogen added to the soil was the primary source of nitrogen for the plant during early growing season. As the initial added fertilizer nitrogen was depleted either by plant uptake or through losses, mineralization of soil organic nitrogen became the primary source for the rice plant (Figure 8.23). Approximately 60% of the nitrogen

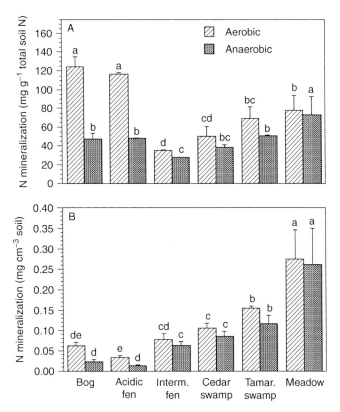

FIGURE 8.21 Cumulative organic nitrogen mineralization (59 weeks) in select temperate wetlands. (From Bridgham et al., 1998.)

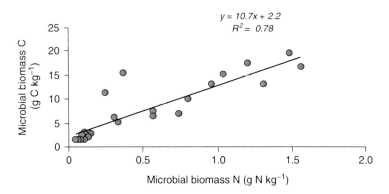

FIGURE 8.22 Relationship between microbial biomass carbon and microbial biomass nitrogen in organic soils (White and Reddy, 2001).

requirements of rice are supported by the mineralization of organic nitrogen. Ammonium nitrogen mineralized under anaerobic conditions can be used as an estimate of the amount of nitrogen available for wetland vegetation. Ecologically, ammonium nitrogen mineralized in soils supports the productivity of wetland vegetation.

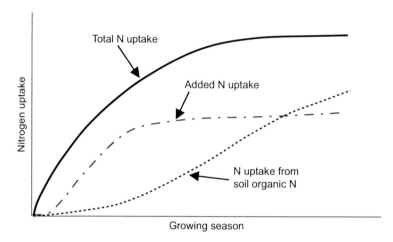

FIGURE 8.23 Relative contribution of added nitrogen and native soil organic nitrogen to nitrogen uptake by rice plant. (Data from Reddy and Patrick, 1976.)

8.5.4 REGULATORS OF ORGANIC NITROGEN MINERALIZATION

Inorganic nitrogen regeneration and metabolism are regulated by the quantity of organic matter accretion, organic matter quality including initial nitrogen content, microbial biomass, carbon-to-nitrogen ratio, pH and redox potential, availability of electron acceptors, temperature, and soil type.

Substrate quality: Organic nitrogen mineralization measured in a range of paddy soils was found to be influenced by the initial nitrogen content of organic matter. Ammonium accumulation rates are high in soils with higher nitrogen content (Figure 8.24), suggesting that the mineralization rates are influenced by the available ammonium nitrogen. At low soil ammonium levels net mineralization rate is slower, primarily due to assimilation of inorganic nitrogen into microbial biomass. As ammonia levels increase, the rate of mineralization also increases, as indicated in soils from wetlands of northern Everglades (Figure 8.20).

Carbon to nitrogen ratio: As discussed in previous sections, mineralization of organic nitrogen requires microbial breakdown of organic matter through various steps including enzymatic hydrolysis of complex organic nitrogen polymers. When the initial C:N ratio of a substrate is higher than that of the microbes, the fraction of organic nitrogen that is mineralized is assimilated into microbes to satisfy their nitrogen requirements. This results in overall decrease of C:N ratio of the remaining substrate. Once the C:N ratio of the substrate–microbe complex decreases below a critical value, net mineralization of nitrogen occurs in the system (Janssen, 1996). For all substrates, mineralization of organic nitrogen is inversely related to C:N ratio of organic substrates undergoing decomposition. Other studies have shown that the chemical nature of carbon in the litter material such as the amount of lignin (Herman et al., 1977; Melillo et al., 1982) and the content of tannins and other polyphenols (Palm and Sanchez, 1991), in combination with the C:N ratio may be better predictors of organic nitrogen mineralization (Gillon et al., 1999; Bruun et al., 2005, 2006).

Microbial biomass and enzyme activity: Organic nitrogen mineralization is directly proportional to microbial biomass and enzyme activity. Organic nitrogen mineralization rates were significantly related to microbial biomass nitrogen of Everglades wetland soils (White and Reddy, 2002). In upland soils, significant correlations between gross N mineralization rate, microbial biomass C and N, and enzyme activities (protease, deaminase,

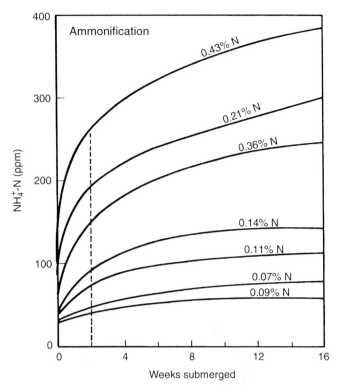

FIGURE 8.24 Relationship between carbon-to-nitrogen ratio of plant litter and mineralization of organic nitrogen (Ponnamperuma, 1981).

and urease), protease was the variable that most frequently described the gross N mineralization rate when included in the stepwise regression equation (Zaman et al., 1999). Similarly, McLatchy and Reddy (1996) showed a significant relationship between protease and gross organic nitrogen mineralization in a wetland soil maintained under different redox conditions.

Temperature: The rate of ammonification increases with an increase in soil temperature. In contrast to most microbial processes, optimum temperature for conversion of organic nitrogen to ammonium nitrogen is between 40 and 60°C. These temperatures are seldom encountered under field conditions. In general, several studies reported that the rate of ammonification doubles with a temperature increase of 10°C, especially in the temperature range of 15–40°C (Kadlec and Reddy, 2001).

Redox potential and hydrology: Redox status of wetlands is regulated by both water-table fluctuations and loading of alternate electron acceptors. These conditions can have a substantial control over organic nitrogen mineralization. Mineralization of organic nitrogen can proceed under both aerobic and anaerobic conditions. However, due to the restricted supply of oxygen, the influence of alternate electron acceptors on microbial respiratory pathways can regulate the rate at which organic nitrogen is mineralized. The data shown in Table 8.3 show the influence of different redox conditions on potentially mineralizable organic nitrogen. For example, aerobic mineralization of organic nitrogen in detrital layer of a wetland soil is approximately 4, 7, and 12 times greater than that under nitrate-reducing, sulfate-reducing, and methanogenic conditions, respectively (Table 8.3).

Soil pH: In most wetlands, pH is buffered around neutrality, whereas under drained soil conditions, pH of the soil decreases as a result of nitrate accumulation during mineralization. The optimum pH range for ammonification is between 6.5 and 8.5.

TABLE 8.3
Mean Organic N Mineralization Rates (mg N kg^{-1} day^{-1}) for Detritus and Soil from Northern Everglades Wetlands Soils (One Standard Error in Parentheses) ($n = 8$)

Redox Status	Detritus	Soil (0–10 cm)	Soil (10–30 cm)
Oxygen (aerobic)	237 (26.1)	143 (8.4)	74.9 (5.6)
Nitrate reducing	59.5 (12.0)	17.1 (6.3)	3.95 (1.1)
Sulfate reducing	36.0 (7.9)	8.94 (3.7)	3.05 (1.5)
Methanogenic	19.3 (2.0)	4.54 (1.3)	1.24 (0.2)

Source: White and Reddy (2001).

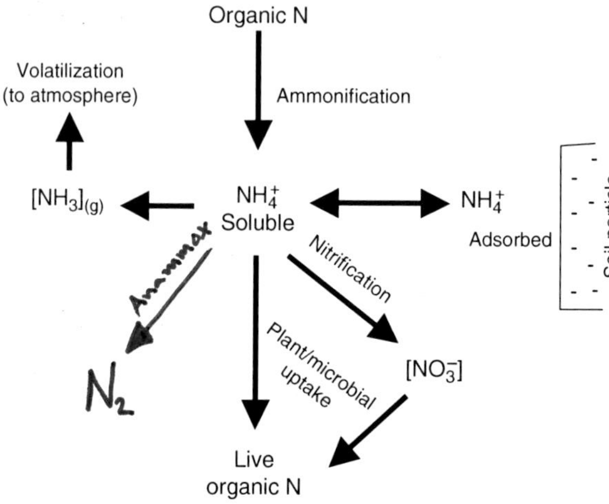

FIGURE 8.25 Schematic showing the fate of ammonia in wetlands.

The fate of ammonium nitrogen in wetland soils is summarized in Figure 8.25. Ammonification of organic nitrogen results in the release of ammonium into the soil solution, which is readily partitioned into dissolved phase and an adsorbed phase, maintaining a certain level of equilibrium between these two pools. Depending on the conditions found in wetlands, the fate of ammonium in soil solution includes (1) loss via volatilization as ammonia, (2) oxidation of ammonia by nitrifiers, (3) uptake by wetland vegetation and periphyton, (4) adsorption as ammonium ions on the cation exchange complex of soil, and (5) fixation into clay.

8.6 AMMONIA ADSORPTION–DESORPTION

Ammonium is adsorbed on the soil cation exchange complex. Soils with high CEC typically adsorb more ammonium and maintain low ammonia levels in the soil solution. The ammonium adsorbed on the exchange complex is expressed as mass (milligrams) of ammonium nitrogen per kilogram of dry soil, whereas the ammonium in soil solution or soil pore water is expressed as mass (milligrams) of ammonium per liter of pore water. The ratio of ammonium adsorbed (mg kg^{-1}) to ammonium in soil pore water (mg L^{-1}) is commonly referred to as partition coefficient or adsorption coefficient with units of L kg^{-1}. A typical ammonium adsorption isotherm is shown in Figure 8.26. At ammonia concentrations typically found in wetland soil pore water, the relationship between ammonium in solution (C) and ammonium in adsorbed phase (S) can be described by a simple linear equation.

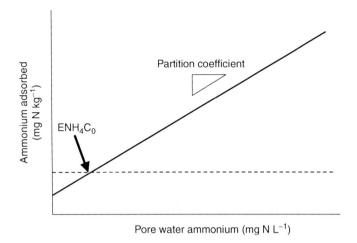

FIGURE 8.26 Ammonium adsorption isotherm. Relationship between ammonium adsorbed on cation exchange complex and pore water ammonium concentration.

At concentrations found in many wetlands, the relationship between S and C is typically linear, and can be described as follows:

$$S = K_d C - S_0 \tag{8.1}$$

where S_0 is the y-axis intercept representing the initial soil ammonium present in the adsorbed phase (mg kg^{-1}) and K_d is the linear adsorption coefficient or partition coefficient representing adsorption strength (L kg^{-1}). High K_d values represent strong ammonium adsorbing strength of soil cation exchange complex. Soils with high K_d values maintain low ammonium concentrations in soil solution (or soil pore water).

ENH$_4$C$_0$ (equilibrium ammonium concentration) is defined as the ammonium concentration in solution at which adsorption equals to desorption ($S = 0$). The ENH$_4$C$_0$ can be estimated as follows:

$$\text{ENH}_4\text{C}_0 = \frac{S_0}{K_d} \tag{8.2}$$

Low ENH$_4$C$_0$ values are observed in soils with high CEC such as clay or organic soils. Soils with low CEC or soils where exchange sites are occupied with other cations have high ENH$_4$C$_0$.

Ammonium equilibrium relationships between adsorbed and solution phases are schematically presented in Figures 8.27 and 8.28. The CEC of a soil refers to the amount of positively charged ions a soil can hold and is expressed as moles of positive charge per unit mass. In soil science literature, the CEC is typically expressed as centimoles of positive charge per kilogram of soil (dry weight basis). For example, soils with 10 cmol kg^{-1} of positive charge would adsorb 10 cmol of ammonium per kilogram of soil or 5 cmol of calcium or magnesium per kilogram of soil. Soils have slightly excess of negative charge sites on the silicate clay and organic matter. Because cations are held to the soil by their positive charge, they cannot be leached downward into the soil when water moves through the soil profile. Thus, the higher the clay content and the organic matter content, the higher the CEC of the soil. Soils with a high CEC will tend to hold onto the positively charged nutrients better than soils with a low CEC. The CEC of a soil can be increased somewhat by increasing the soil's organic matter content. Dissolved cations present in the pore water are attracted to the negatively charged soil particles. In soils, several positively charged cations compete with ammonium for the soil surface exchange sites because both substances are of about the same size.

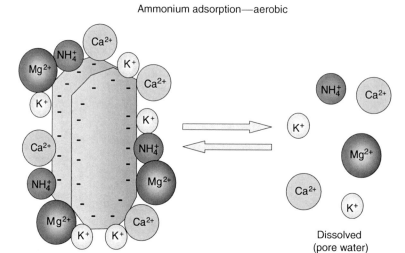

FIGURE 8.27 Schematic showing partitioning of ammonium between soil cation exchange complex and pore water under aerobic conditions.

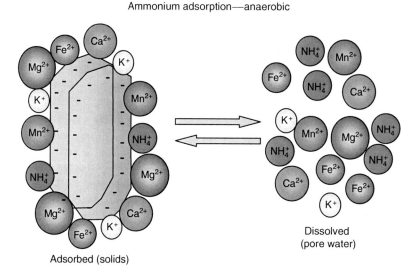

FIGURE 8.28 Schematic showing partitioning of ammonium between soil cation exchange complex and pore water under anaerobic conditions.

Under drained (aerobic) conditions, the cation exchange complex is typically occupied by various cations such as potassium, calcium, magnesium, and ammonium (Figure 8.27). Nitrification and adsorption of ammonium on exchange complex maintains low ammonium levels in soil pore water. The partition coefficients for ammonium are higher in drained soils than in wetland soils. Data from the literature suggest that exchangeable ammonium concentrations in freshwater wetlands are generally higher than in coastal wetlands. The ratio of exchangeable ammonium to soil pore water ammonium concentrations (also termed as partition coefficient) is also generally higher in freshwater wetlands than in coastal wetlands. This is primarily due to high concentration of sodium ions in coastal wetland soils, which can potentially displace ammonium from exchange complex.

A number of factors can influence the amount of exchangeable ammonium in soils. Soil porosity, clay mineral structure, presence of other cations, and organic matter content are a few examples. The cation concentration of freshwater wetlands is generally lower than that of coastal wetlands and would result in higher partition coefficients in freshwater wetlands relative to coastal wetlands.

Under wet soil (anaerobic) conditions, accumulation of reduced cations such as iron (II) and manganese (II) in mineral soils can potentially occupy much of the cation exchange sites (Figure 8.28). This can result in displacement of ammonium from the exchange complex. The partition coefficients for ammonium are generally lower in anaerobic soils than in drained or aerobic soils.

8.7 AMMONIA FIXATION

Soils dominated with 2:1 type clay minerals have the capacity to fix ammonium. Each unit cell of these minerals consists of an octahedral Al–O–OH layer sandwiched between two tetrahedral Si–O layers (Figure 8.29). The octahedral and tetrahedral layers are fundamental units of silicate clays. Replacement of ions by another similar-sized ions during mineral formation is known as isomorphic substitution. If the replacement ions have lower positive charge than the ions they replace, the mineral lattice is left with a negative charge. For example, the aluminum ions in octahedral sheets and silica in tetrahedral sheets are subject to isomorphic substitution or replacement by lower-valance cations such as potassium, ammonium, calcium, magnesium, and sodium. This can result in net negative charges. About 20% of the silicon sites in the tetrahedral sheet are occupied by aluminum atoms; therefore, a high net negative charge is created in the tetrahedral layer. To satisfy this charge, ammonium and potassium ions are trapped in spaces in adjoining tetrahedral layers. Once ammonium is entrapped (fixed) between these layers, it may be slowly available for plant uptake. Vermiculite has the greatest capacity to fix both ammonium and potassium ions, followed by fine-grained micas. Ammonium fixation can be significant in wetlands dominated by these minerals.

A buildup of exchangeable ammonium from the mineralization of organic nitrogen in wetlands could lead to a concentration gradient favoring the diffusion of ammonium into the interlayers of clay minerals. Anaerobic environment in wetlands can also increase ammonium fixation through the reduction and dissolution of iron oxide coatings on the surface of clay minerals thereby reducing obstacles for ammonium movement in and out of the interlayers of the clay minerals. In mineral

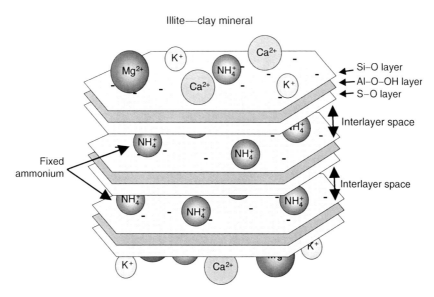

FIGURE 8.29 Schematic showing ammonium fixation in clay lattice.

wetland soils, ammonium fixation can also be enhanced by increasing the negative charge of interlayers of the clay minerals through the reduction of ferric iron associated with clay minerals. The temporary fixation of ammonium could protect nitrogen from losses, while still enabling a timely release of ammonium to plants. Nonexchangeable ammonium can be an important source of nitrogen to rice on submerged soils rich in vermiculite. Although the clay fixation of ammonium is not documented in wetlands, the significance of this process is reported for paddy soils. In paddy soils dominated by these minerals, up to 36% of ammonium was nonexchangeable and permanently fixed in clay lattice (Savant and DeDatta, 1982).

8.8 AMMONIA VOLATILIZATION

Ammonia loss through volatilization to the atmosphere is a complex process mediated by a combination of physical, chemical, and biological factors. The exchange of ammonia between water column, soils, and the atmosphere plays an important role in wetland nitrogen cycle. However, the significance of this process is not well established.

8.8.1 Physicochemical Reaction

Wetlands provide an ideal environment for ammonia (NH_3) volatilization. This process involves the balance between ionized form (NH_4^+) and unionized form (NH_3) (Figure 8.30).

The equilibrium between these two species is regulated by the pH of the solution they are present in. The ammonium nitrogen, as described earlier, is derived from several sources such as ammonification of organic nitrogen and inputs from external sources (fertilizers, wastewater, and rainfall). Ammonium ions present on the exchangeable sites are in equilibrium with those in the soil solution or pore water. Ammonium ions in solution are converted to ammonia gas through Equations 8.4 and 8.5:

$$OH^- + H^+ = H_2O \tag{8.3}$$

$$NH_4^+ = NH_3 + H^+ \tag{8.4}$$

$$NH_4^+ + OH^- = H_2O + NH_3 \tag{8.5}$$

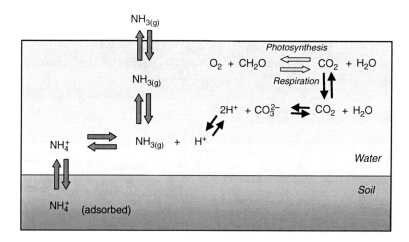

FIGURE 8.30 Schematic showing the reactions involved in ammonia volatilization.

Water is the source of OH⁻ ions. Under alkaline conditions, OH⁻ reacts with NH_4^+ resulting in the formation of NH_3 as NH_4^+ dissociates to form H^+, as shown in the preceding reaction. The H^+ thus formed will consume alkalinity of the water column, if present, or will otherwise lower pH. Ammonia volatilization is a physicochemical process and is directly related to the content of aqueous NH_3 or partial pressure of ammonia (p_{NH_3}) in water at the interface with the atmosphere. Aqueous NH_3 as a fraction of total ammoniacal N is directly influenced by water pH and temperature.

For additional information on equilibrium relationships see below:

$$NH_3 + H_2O = NH_4^+ + OH^- \quad (8.6)$$

For ammonium–ammonia equilibrium, the dissociation constant is given as follows:

$$\frac{[NH_4^+][OH^-]}{[NH_3][H_2O]} = K_1 = 1.82 \times 10^{-5} \text{ at } 25°C \quad (8.7)$$

The dissociation constant for water is given as follows:

$$H_2O = H^+ + OH^- \quad (8.8)$$

$$\frac{[H^+][OH^-]}{[H_2O]} = K_w = 10^{-14} \quad (8.9)$$

Total amount of inorganic nitrogen (NH_4^+) present in solution is expressed as the total ammoniacal nitrogen (TAN), which is defined as

$$TAN = [NH_3]_{(aq)} + [NH_4^+] \quad (8.10)$$

Combining Equations 8.6 and 8.10, we obtain

$$K_1 = \frac{[NH_4^+][OH^-]}{[NH_3]_{(aq)}}$$

or

$$[NH_4^+] = \frac{K_1[NH_3]_{(aq)}}{[OH^-]} \quad (8.11)$$

Combining Equations 8.10 and 8.11, we obtain

$$TAN = [NH_3]_{(aq)} + \frac{K_1[NH_3]_{(aq)}}{[OH^-]} \quad (8.12)$$

As $[OH^-] = K_w/[H^+]$, we can modify Equation 8.12 to

$$TAN = [NH_3]_{(aq)} + \frac{K_1[NH_3]_{(aq)}[H^+]}{K_w} \quad (8.13)$$

Dividing both sides of Equation 8.13 by $[NH_3]$, we obtain

$$\frac{TAN}{[NH_3]_{(aq)}} = 1 + \frac{K_1[H^+]}{K_w} \quad (8.14)$$

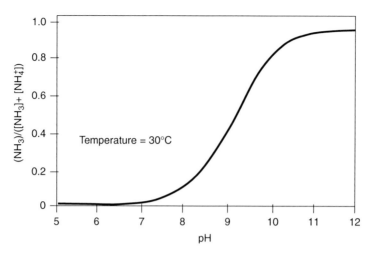

FIGURE 8.31 Fraction of unionized ammonia as function of pH.

Fraction of unionized NH_3 (F) can be calculated by modifying Equation 8.14 as follows:

$$\frac{[NH_3]_{(aq)}}{TAN} = \frac{1}{1 + (K_1[H^+]/K_w)} \qquad (8.15)$$

K_1 and K_w are constants and $[H^+]$ concentration (activity) can be obtained by measuring the pH of the water column or soil pore water. The fraction of unionized NH_3 as a function of pH is estimated using Equation 8.15 (plotted in Figure 8.31). Thus, NH_3 formation is directly related to pH of the water column, with up to 85% of the TAN present as NH_3 at pH of 10, whereas only 35% is present at pH 9. The rate of TAN loss is directly proportional to the difference in activity of NH_3 in water and in the atmosphere.

At equilibrium, NH_3 in the water column can be related to NH_3 in the atmosphere as follows:

$$[NH_3]_{(aq)} = K_H P_{NH_3} \qquad (8.16)$$

where K_H is Henry's law gas constant (mol L^{-1} atm^{-1}), P_{NH_3} is the partial pressure of ammonia in the air (atm), and $[NH_3]_{(aq)}$ is ammonia concentration in the water column (mol L^{-1}).

Ammonia volatilization primarily occurs in the water column and in benthic layers of periphyton. Typically, soil pH is not high enough in many wetlands to promote significant ammonia volatilization process. Two primary factors regulating this process are pH and ammonia concentration of the water column. Photosynthetic activity in the water column, temperature, vegetation density, air movement above the water surface, and mixing in the water column are among other factors that also regulate the rate of ammonia volatilization loss. This process can play a significant role if the influent water contains high levels of ammonium nitrogen. Two dominant factors, pH and ammonium nitrogen concentration, can be used as indicators to determine the extent of nitrogen removal through this process.

8.8.2 Regulators of Ammonia Volatilization

Photosynthetic activity in the water column and pH: Submerged aquatic vegetation and periphyton communities provide an ideal environment for pH alternation in the water column.

$$nCO_2 + nH_2O = (CH_2O)_n + nO_2 \qquad (8.17)$$

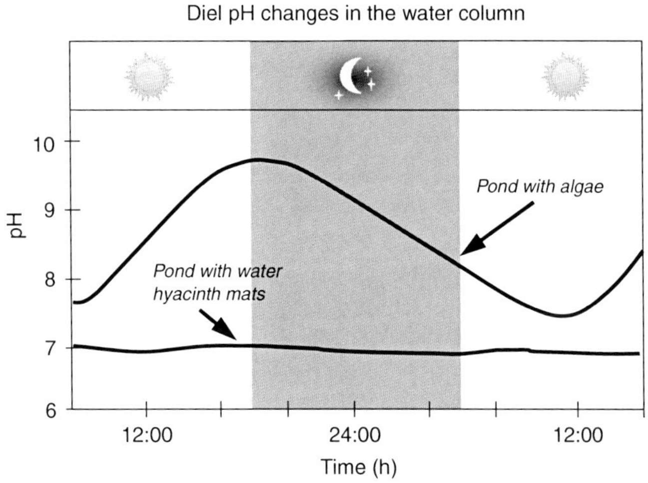

FIGURE 8.32 Diel fluctuations in pH of the water column of reservoirs: (a) dominated algae and (b) covered with floating mats of water hyacinths.

The above reaction results in a rise in pH. In low alkalinity systems, the pH of the water column can reach as high as 10 during high photosynthetic activity. As mentioned earlier, an increase in pH of the water column is directly related to the ammonia loss (Figure 8.32). When algae photosynthesize during the daytime, they consume dissolved carbon dioxide from the water. As a result, the pH is increased and the environment becomes favorable to ammonia volatilization (Figure 8.31). When the floodwater and soil are alkaline (more OH^-), more ammonium ions react with OH^-, resulting in the production of unionized ammonia. The fraction of unionized ammonia present in the water column or soil pore water is regulated by pH (Figure 8.31). Thus, for volatilization process to occur, ammonia should be in aqueous phase in the water column.

Alkalinity and buffer capacity: Dissociation of ammonium ions to ammonia results in the production of hydrogen ions. Unless the water column or soils are well buffered, the medium can be acidified and the rate of ammonia volatilization can decrease. Thus, a water column with high alkalinity and calcium carbonate content can buffer the system and maintain high-pH conditions. Alkalinity is affected by the balance between photosynthesis and respiration by algae and submersed macrophytes in the water column. Ammonia volatilization losses are directly proportional to the alkalinity of the system.

Ammonium concentration (Figure 8.33a): Under alkaline conditions, the rate of ammonia loss through volatilization is directly proportional to the total ammonium concentration in the water column. In most wetlands, ammonium concentrations in the water column are typically low. In constructed wetlands used for the treatment of ammonia-rich waters (primary sewage effluents, dairy effluents, poultry and swine lagoon effluents, and other industrial effluents), volatilization losses can be significant, if ideal environmental conditions are created in the water column. In addition to ammonia volatilization, nitrification and biotic uptake also regulate ammonia concentrations in the water column. Thus, any other factor that controls this concentration also controls volatilization losses. There are three primary sources of ammonium to the water column: decomposition of organic matter in the water column, external loading (nonpoint or point sources and precipitation), and flux of ammonium from soil to the overlying water column.

Floodwater depth (Figure 8.33b): Periphyton activity is higher in wetlands with shallow water depth and decreases with an increase in water depth. In shallow water depths, mixing and currents increase the rate of exchange of ammonia between the water column and the

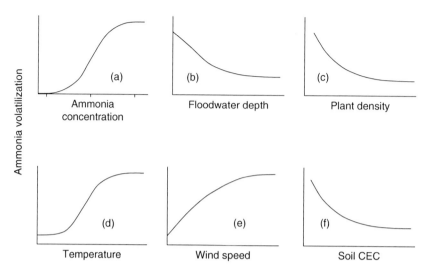

FIGURE 8.33 Influence of various factors controlling ammonia volatilization in wetlands.

atmosphere. Thus, high photosynthetic activity and greater mixing of the water column increase the rate of ammonia volatilization losses in wetlands with shallow water depths.

Plant density (Figure 8.33c): If the plant density is low, air movement above the water surface is high, which decreases with an increase in plant density; this further decreases the rate of ammonia exchange between water column and the atmosphere. However, high plant density can decrease light penetration and potentially reduce algal activity, and thus maintain the pH near neutral conditions.

Cation exchange capacity of soil (Figure 8.33d): Lower CEC soils (such as in sandy soils) cause more volatilization because less ammonium ions are held by the soils. Soils with high CEC tend to retain ammonium ions on the exchange complex and maintain low levels of ammonium ions in pore water. High partition coefficients for ammonium in high-CEC soils decrease ammonium mobility from soil to the overlying water column, thus influencing overall loss due to volatilization process.

Soil and water temperature (Figure 8.33e): Assuming relatively high ammonia concentrations and alkaline pH conditions, an increase in temperature increases the conversion of ammonium ions to ammonia. The dissociation constants for ammonium–ammonia equilibrium and water ionization constants are influenced by the temperature (Table 8.4). Thus, the fraction of undissociated ammonia present in the water column or soil pore water is regulated by temperature. In addition, temperature also affects the solubility of ammonia in the water. Thus, at higher temperature, solubility of ammonia in water is less and diffusion is greater, which promotes increased rate of ammonia volatilization. The Ammonia volatilization rate was shown to increase approximately from 1.3 to 3.5 times for each 10°C rise in temperature.

Wind speed (Figure 8.33f): High wind speed increases the transfer of ammonia from soil–water column to the atmosphere. This effect is more pronounced in wetlands with low plant density and shallow water depths.

Ammonia volatilization is recognized as the major process by which nitrogen fertilizers are lost from paddy fields, especially when urea or other ammonium-based fertilizers are surface applied to the water column. In tropical climates, high water temperatures associated with algal activity in the water column are conducive to enhance volatilization of added fertilizers. Urea, a common fertilizer for rice in Asia, is rapidly hydrolyzed within the week after application to submerged soils. Ammonia originating from the hydrolyzed urea accumulates in floodwater, and the peak concentration of ammonia in the floodwater of tropical rice fields typically occurs within 1–5 days after urea

TABLE 8.4
Effect of Temperature on Solubility and Dissociation Constants for Ammonia and Ionization Constants for Water

Temperature (°C)	K_w	pK_w	K_1	pK_1	K_H
0	0.114×10^{-14}	14.944	1.374×10^{-5}	4.862	52.8
5	0.185×10^{-14}	14.734	1.479×10^{-5}	4.830	45.6
10	0.292×10^{-14}	14.535	1.570×10^{-5}	4.804	40.2
15	0.451×10^{-14}	14.346	1.652×10^{-5}	4.782	35.7
20	0.681×10^{-14}	14.167	1.710×10^{-5}	4.767	30.5
25	1.007×10^{-14}	13.997	1.774×10^{-5}	4.751	27.8
30	1.469×10^{-14}	13.833	1.820×10^{-5}	4.740	24.0
35	2.089×10^{-14}	13.680	1.849×10^{-5}	4.733	21.7
40	2.917×10^{-14}	13.535	1.862×10^{-5}	4.730	19.9

Source: Freney et al. (1981).

application. High concentrations of ammonia together with high floodwater pH and temperature favor loss of added fertilizer N by ammonia volatilization. Up to 70% of the applied nitrogen can be lost through ammonia volatilization in rice fields (Buresh et al., 2008).

8.9 NITRIFICATION

Nitrification in wetlands is restricted to aerobic zones of soil and water column or under drained soils conditions, where ammonium is oxidized to nitrate. Nitrification reaction supports denitrification by supplying heterotrophs with nitrate as their electron acceptor. In a broader sense, nitrification is defined as the conversion of organic or inorganic compounds from reduced state to a more oxidized state. Three groups of microorganisms are capable of oxidizing ammonium under aerobic conditions:

- Chemoautotrophic bacteria
- Methane-oxidizing bacteria
- Heterotrophic bacteria and fungi

Because of limited oxygen availability in wetland environments, nitrification is restricted to the (1) aerobic water column, (2) aerobic soil–floodwater interface, and (3) aerobic root zone (Figure 8.34). In all these zones, nitrification is supported predominantly by chemoautotrophic bacteria, which use oxygen as their electron acceptor and ammonium as their energy source. Nitrification in these zones is often limited by the availability of ammonium.

8.9.1 CHEMOAUTOTROPHIC BACTERIA

Nitrification (often referred as aerobic ammonium oxidation) is defined as the biological oxidation of ammonium to nitrite and then to nitrate. This process is known to take place in two stages as a result of obligate chemoautotrophic bacteria of the genera "Nitroso" (e.g., *Nitrosomonas*, *Nitrosococcus*, *Nitrospira*, and *Nitrosovibrio*) involved in oxidizing ammonium to nitrite. The second-state oxidation is mediated by obligate chemoautotrophic bacteria of the genera "Nitro" (e.g., *Nitrobacter*, *Nitrococcus*, *Nitrospina*, and *Nitrospira*) involved in oxidizing nitrite to nitrate. These bacteria use ammonium nitrogen as an energy source or electron donor to support their metabolic activities. Nitrifiers can utilize carbon dioxide and build their cell constituents through the Calvin reductive pentose phosphate cycle (Schmidt, 1982). Two-step oxidation reaction is shown below.

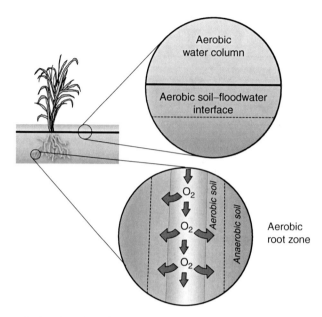

FIGURE 8.34 Schematic showing sites of nitrification in wetlands.

Nitrosomonas oxidizes ammonium to nitrite as follows:

$$NH_4^+ + 2H_2O = NO_2^- + 6e^- + 8H^+ \qquad (8.18)$$

$$1.5O_2 + 6e^- + 6H^+ = 3H_2O \qquad (8.19)$$

$$NH_4^+ + 1.5O_2 = NO_2^- + H_2O + 2H^+ \qquad G° = -272 \text{ kJ mol}^{-1} \qquad (8.20)$$

Nitrobacter oxidizes nitrite to nitrate as follows:

$$2NO_2^- + 2H_2O = 2NO_3^- + 4e^- + 4H^+ \qquad (8.21)$$

$$O_2 + 4e^- + 4H^+ = 2H_2O \qquad (8.22)$$

$$2NO_2^- + O_2 = 2NO_3^- \qquad G° = -75 \text{ kJ mol}^{-1} \qquad (8.23)$$

These reactions involve a change in oxidation number of N from −3 to +5, a span involving eight electrons. The electrons produced from nitrogenous compounds enter the electron transport chain, and the electron flow establishes membrane potential and proton gradient and ATP synthesis via electron-transport phosphorylation through cytochrome system. Oxygen is essential for nitrification. It takes 2 mol of oxygen to oxidize 1 mol of ammonium to nitrate. Stoichiometrically, these organisms require 4.57 g of oxygen per gram ammonium oxidized. These reactions therefore potentially use large quantities of oxygen. Nitrification can at times produce oxygen-depleted water in lakes, streams, and estuaries subject to high loadings of nitrogenous waste materials.

Nitrifying bacteria use stepwise enzymatically catalyzed reactions for oxidation of ammonium to nitrate (Figure 8.35). The intermediates, hydroxylamine (NH_2OH) and nitroxyl (NOH), are thought to be enzyme bound. Driven by oxygenases, the oxidation of ammonium to hydroxylamine (NH_2OH) is the first intermediate form, which creates the −1 oxidation state of nitrogen. This step utilizes two electrons. In the next step, hydroxylamine is oxidized to nitrite, where additional four electrons are utilized. However, the oxygen stress associated with this reaction in wetlands has at least two effects. The formation of NH_2OH is more likely to be mediated by microbial enzymatic

Nitrogen

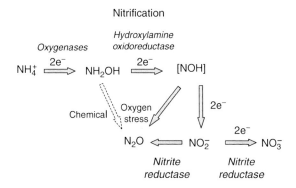

FIGURE 8.35 Pathways and intermediate products of nitrification of ammonium to nitrate.

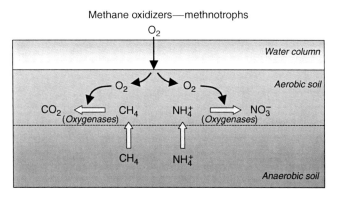

FIGURE 8.36 Schematic showing the role of methane oxidizers in nitrification.

activity involving oxidoreductase, forming another intermediate compound known as nitroxyl (NOH) with a +1 oxidation state of nitrogen. At high ammonium levels, nitrification can result in a greater oxygen demand. This creates oxygen depletion, resulting in reduction (NOH) to nitrite and then to nitrous oxide. Nitrite oxidation to nitrate is mediated by nitrite oxidase system. This reaction occurs in a single step and uses two electrons. The electrons are transported to oxygen using ETS, with synthesis of ATP.

The nitrifying chemoautotrophs involved derive carbon for cell synthesis mostly from carbon dioxide and obtain energy for reduction of carbon dioxide through the oxidation of nitrogenous compounds. These organisms cannot use organic carbon as the sole carbon source for growth or obtain energy by oxidizing substrates other than those containing nitrogen.

8.9.2 Methane-Oxidizing Bacteria

In addition to chemoautotrophs, methanotrophic bacteria are also capable of oxidizing ammonium to nitrate. Several similarities exist between methane oxidizers and autotrophic ammonium oxidizers (Figure 8.36). Numerous studies have shown that methane (CH_4) oxidizers can cooxidize ammonium to nitrite, nitrate or both. Collectively, methylotrophs are bacteria that are capable of growth on one-carbon compounds as their sole carbon and energy source. Thus, methanotrophs are a group of methylotrophic bacteria that utilize methane as their sole source of carbon and energy.

Ammonium N has been shown to be a competitive inhibitor of methane oxidation, suggesting that ammonia and ammonium are oxidized by the methane monooxygenase system. There are striking similarities between ammonium oxidation by *Nitrosomonas* and methane oxidation by methanotrophs. The methanotrophic bacteria are obligate aerobes and closely associated with the aerobic interfaces in wetlands. Both methane and ammonium diffuse from anaerobic zones into aerobic zones where they can be oxidized by these methylotrophs. Oxygen that has diffused down from the water column or leaked from plant roots is potentially available as an electron acceptor. The carbon dioxide produced escapes into the atmosphere and the nitrate produced would likely diffuse down into the soil to be taken up by plants or used as an electron acceptor by microbes.

The following conditions are associated with methylotrophs:

- Nitrification is not essential for their growth.
- They grow best with nitrate as the nitrogen source.
- High ammonia levels (>200 mg L^{-1}) inhibit their growth.
- The rates of ammonia oxidation by methylotrophs are lower than those by autotrophic nitrifiers.

8.9.3 Heterotrophic Bacteria and Fungi

Heterotrophic nitrifiers use organic substrates as the energy source and gain no energy from the oxidation of ammonium. Earlier studies have identified that *Achromobacter* (bacteria) and *Aspergillus* (fungi) are capable of oxidizing ammonium. Heterotrophic nitrifiers do not accumulate large amounts of end products, as compared to autotrophic nitrifiers. Under oxygen-limited conditions, some of these heterotrophic oxidizers can also reduce nitrite and nitrate to gaseous end products. In cocultures of *Thiosphaera pantotropha* and *Nitrosomonas europaea*, Kuenen and Robertson (1994) suggested that at low C:N ratios autotrophic nitrification dominates the oxidation of ammonia, whereas at high C:N ratios heterotrophic nitrification is higher than autotrophic nitrification. Overall, however, the relative importance of this group in the oxidation of ammonium is not clearly understood and the significance of this process in wetlands is not documented.

8.9.4 Regulators of Ammonium Oxidation

Ammonium oxidation (nitrification) in wetlands is restricted to the (i) water column, (ii) surface aerobic soil layer, and (iii) aerobic root zone. In these zones, nitrification is often regulated by the availability of ammonium and the supply of oxygen. Ammonium oxidation results in the production of protons, leading to acidification of soils. *Nitrobacter* is more sensitive to pH than *Nitrosomonas*. As a result, further oxidation of ammonia can be inhibited due to nitrous acid formation (HNO_2). This is analogous to oxidation of sulfide and the formation of sulfuric acid in coastal wetlands. This decrease in pH can subsequently inhibit ammonium oxidation. Under optimal conditions, ammonium oxidation can follow with initial accumulation of nitrite, followed by accumulation of nitrate (Figure 8.37). Thus, the magnitude of nitrification is regulated by the fraction of soil volume

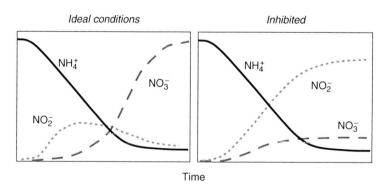

FIGURE 8.37 Ammonium oxidation under ideal conditions where very little nitrite accumulates as an intermediate product. Second step of nitrification is inhibited resulting in accumulation of nitrite.

with oxygen and the supply of ammonium to these zones. Other factors regulating nitrification are alkalinity, inorganic carbon source, nitrifying population, and pH. Ammonification provides the starting substrate for nitrification. Oxygen supply to aerobic soil–floodwater interface depends on the diffusion through the water column or photosynthetic production by periphyton communities or submersed aquatic vegetation. In addition, oxygen flux through wetland plants into the root zone can maintain aerobic zone in the root zone, which can also support nitrification.

Ammonium concentration: High concentration of ammonium nitrogen has been found to inhibit the activity of *Nitrobacter*, particularly at high pH values, probably due to production of free ammonia. However, in wetlands, ammonium levels are never high enough to affect nitrifiers in an adverse manner.

Oxygen availability: Oxygen is essential for the obligate aerobic requirements of nitrification. However, the actual limit is more dependent on the oxygen flux into the system rather than on the oxygen level at a given time.

pH, alkalinity, and carbon dioxide: Acidity is one of the major factors controlling nitrification in soils. Generally, there are few or no autotrophic nitrifying bacteria present under acid soil conditions. Relatively alkaline pH conditions favor nitrification. Removal of carbon dioxide during photosynthesis results in the consumption of H^+ and accordingly an increase in pH. Therefore, high carbon dioxide concentration in soils will have an inverse effect. Under optimal conditions, oxidation of nitrite to nitrate proceeds at a faster rate than the oxidation of ammonium, thus resulting in very little or no accumulation of nitrite in soils (Figure 8.37). However, nitrite may accumulate to some extent under high- or low-pH conditions. *Nitrobacter* is sensitive to unionized ammonia under alkaline conditions, resulting in accumulation of nitrite. For example, in calcareous soils, unionized ammonia accumulation can inhibit the activity of *Nitrobacter*, thus influencing the accumulation of nitrite. Optimal pH for maximum nitrification rate is in the range of 6–8. Nitrification is slower in a pH range of 4–6, and can be inhibited at pH values below 4 (Schmidt, 1982).

Temperature: The optimum temperature for nitrification in pure cultures ranges from 25 to 35°C and in soils from 30 to 40°C. Lower temperatures (below 15°C) have a much more drastic effect on nitrification rate compared to the temperatures between 15 and 35°C. (Kadlec and Reddy, 2001).

Nitrifying population: The process of nitrification depends on the metabolism of nitrifying organisms; thus, it is imperative that they be present in adequate numbers to achieve the oxidation of ammonium nitrogen. Generally, fertilized soils have larger populations of nitrifiers compared to unfertilized soils. It should be recognized that nitrifiers use ammonia as an energy source. Thus, the activity of nitrifiers depends on the availability of this energy source.

Cation exchange capacity: The CEC plays an important role in buffering rapid changes in ammonium concentration in soil solution; therefore, nitrification rates are slower in soils with high CEC such as clay loam soils. Again, diffusion of ammonium to microbes may be slower in soils with high CEC as these soils maintain low ammonium levels in soil pore water.

Redox potential: It is usually considered that no autotrophic nitrification will occur when the redox potential is below 250 mV. Under mildly aerobic conditions, the oxidation of ammonium to nitrate is carried out jointly by autotrophic and heterotrophic bacteria, whereas under anaerobic conditions with the redox potential below –85 mV, only heterotrophic nitrification takes place.

The relative importance of aerobic zones in overall nitrification depends on oxygen availability and ammonium concentration (Reddy and Patrick, 1984; Henriksen and Kemp, 1988). Nitrification rates were reported to be in the range of 0.01–0.161 g N m^{-2} day^{-1} (mean = 0.048 ± 0.044; $n = 9$;

Martin and Reddy, 1997). These values are lower than the values reported for mineralization, suggesting that oxygen and ammonium availability limits nitrification. Measurement of oxygen concentration in the water column or Eh values in soils can provide a reliable indication of potential nitrification. Measurement of nitrate in aerobic zones of wetlands may underestimate overall nitrification, as some of the nitrate can diffuse into anaerobic zones and is lost through denitrification. However, in well-aerated soils, nitrification potentials are determined by measuring accumulation of nitrate over time.

8.10 ANAEROBIC AMMONIUM OXIDATION

In the classical view of nitrogen transformations in wetland soils and aquatic sediments, ammonia is oxidized aerobically to nitrate by nitrifying bacteria (nitrification) (see Section 8.9) and nitrate is reduced anaerobically to dinitrogen by denitrifying bacteria (denitrification) (see Section 8.11). However, recent studies have demonstrated that under anaerobic conditions, thermodynamically it is possible that several other alternate electron acceptors potentially can oxidize ammonium nitrogen to nitrogen gas (Figures 8.38). However, this pathway assumes that some of the anaerobic bacteria are capable of using ammonium as their electron donor and derive energy through oxidation. The significance of this process has been documented recently in wastewater lagoons and marine and estuarine sediments (Mulder et al., 1995; Kuypers et al., 2002; Thamdrup and Dalsgaard, 2002; Schmidt et al., 2004). Ammonium oxidation to dinitrogen using nitrite as an electron acceptor is now commonly known as anammox (Mulder et al., 1995). The proposed reactions for this process are

$$5NH_4^+ + 3NO_3^- = 4N_2 + 9H_2O + 2H^+ \tag{8.24}$$

$$NH_4^+ + NO_2^- = N_2 + 2H_2O \tag{8.25}$$

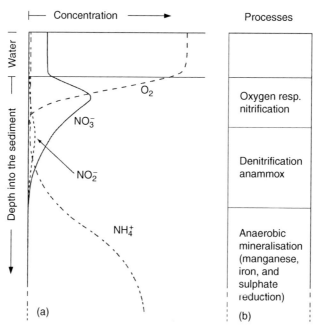

FIGURE 8.38 Schematic representation of selected dissolved species in the soil pore water and overlying water column and indication of depth intervals in which selected processes occur. Concentrations not drawn to scale. (Drawn from Dalsgaard et al., 2005.)

In continental shelf sediments, up to 67% of the N_2 formation was found to be due to anaerobic ammonium oxidation with nitrate (or possibly nitrite) and only 33% of the N_2 production was due to denitrification (Thamdrup and Dalsgaard, 2002). At present this reaction is not reported in wetland soils. Anaerobic ammonium oxidation is regulated by

- Availability of ammonium, nitrate, and nitrite
- Organic matter content
- Temperature
- Presence of other oxidants (such as iron and manganese oxides)

In continental shelf sites, contribution of anaerobic ammonium oxidation to N_2 was observed to be greatest in water with low organic loading (67%); and closer to shore where organic matter loading to water was high, the contribution of this process decreased to 24%. In a eutrophic coastal bay, the contribution decreased to 2% (Thamdrup and Dalsgaard, 2002). However, the significance of anaerobic ammonium oxidation to N_2 production decreased in sediments along an estuarine gradient with decreasing organic matter content to <1% (Trimmer et al., 2003). These observations are in sharp contrast to other studies, where anaerobic ammonium oxidation increased with an increase in organic matter loading to the water column. These results suggest that it is likely that the effect of organic matter loading may be compounded by salinity gradients and availability of electron donors and acceptors. Anaerobic ammonium oxidation was highest at 15°C and decreased sharply above 25°C, reaching zero around 37°C (Figure 8.39) (Dalsgaard and Thamdrup, 2002).

The contribution of anaerobic ammonium oxidation reaction to overall nitrogen loss in wetlands and aquatic systems has been a subject of great speculation. The significance of this process has been demonstrated in marine systems, but to our knowledge there is no reported documentation in freshwater systems. Anaerobic ammonium oxidizers are obligate anaerobes and it is likely their activity can be promoted by other inorganic electron acceptors such as iron and manganese oxides and sulfates. However, there is no reported evidence that these reactions occur in wetland soils and aquatic sediments.

Major limiting factors controlling anaerobic ammonium oxidation reaction are the availability of nitrite and the competition for electron acceptors by heterotrophs. Soils and sediments with high organic matter content can create higher demand for electron acceptors (nitrite and nitrate). Under these conditions, anammox may not be able to keep up with denitrification when electron donor availability is very high. It is likely under available carbon-limiting conditions that some autotrophs

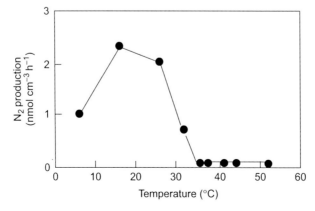

FIGURE 8.39 Rates of dinitrogen production by anaerobic oxidation of ammonium with nitrite as a function of temperature. (Drawn from Dalsgaard and Thamdrup, 2002.)

may use ammonium as an energy source and potentially oxidize it to dinitrogen gas. Typically, many wetland soils are not limited by carbon, and if nitrate and nitrite are present in these systems, they will be rapidly used by heterotrophic denitrifiers. So, under what conditions could one expect anaerobic ammonium oxidation as a significant contributor to N_2 production? Here are some possible scenarios:

1. In soils and sediments with high levels of ammonium nitrogen and low levels of bioavailable organic carbon, anaerobic ammonium oxidizers can compete with denitrifiers for available nitrite.
2. In calcareous soils, presence of unionized ammonia can potentially inhibit the activity of *Nitrobacter*, thereby reducing the rate of conversion of nitrite to nitrate. This may result in accumulation of nitrite. Migration of nitrite from aerobic zones to anaerobic sites can potentially promote ammonium oxidation.
3. High rates of nitrification in aerobic zones can result in accumulation of nitrate, which can diffuse into anaerobic zones. In this zone if denitrification is nitrate saturated, some of the nitrite formed may be available for anammox process (Dalsgaard et al., 2005).
4. It has been found that the temperature optima for anaerobic ammonium oxidizers is around 15°C, as compared to a much higher temperature for denitrifiers. This suggests that anaerobic ammonium oxidizers may be more active at low temperatures where the activity of denitrifiers slows down.
5. Anaerobic ammonium oxidation is promoted by the reduction of other oxidants such as iron and manganese oxides and sulfates.

8.11 NITRATE REDUCTION

In wetlands, nitrate is used by microbes and plants as a nitrogen source or as an electron acceptor to support catabolic activities of select heterotrophic bacteria. As early as 1882, Maquenne reported the following observations on nitrate reduction in soils:

- Nitrate is reduced under certain conditions in arable land releasing nitrogen gas.
- Nitrate reduction occurs in arable soil that contains high organic matter.
- Nitrate reduction occurs when the soil atmosphere is completely stripped of oxygen.

The conclusions presented reflect the use of nitrate as an electron acceptor by facultative bacteria.
In soils, nitrate is reduced as follows:

- Catabolic nitrate reduction:
  – Denitrification (mediated by aerobic bacteria capable of anaerobic growth with oxidized nitrogenous compounds, also known as facultative bacteria)
  – Aerobic denitrification—under aerobic but low-oxygen conditions, the nitrite formed can be reduced to nitrous oxide by select chemoautotrophic and heterotrophic nitrifier
  – DNRA (mediated by anaerobic bacteria)
- Anabolic nitrate reduction:
  – ANRA (mediated by bacteria, fungi, algae, and plants)

Catabolic nitrate reduction involves the reduction of nitrate by select bacterial groups to gaseous products such as nitrous oxide and dinitrogen or ammonia. In this process, nitrate is used as an alternate electron acceptor in the energy-producing oxidation of organic substrates (sometimes reduced inorganic substrates) during cellular respiration. This process is carried out by dissimilative nitrate reductase enzymes. These enzymes are membrane-bound proteins whose synthesis is repressed by molecular oxygen; thus, they are only synthesized under anoxic conditions.

Nitrogen

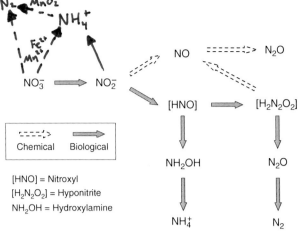

FIGURE 8.40 Pathways and intermediate products during nitrate reduction reaction.

Anabolic nitrate reduction or assimilatory nitrate reduction involves reduction of nitrate to ammonia, which becomes a nitrogen source for cell growth or biosynthesis. It occurs in all plants, most fungi, and many prokaryotes, where nitrogen from nitrate is incorporated into cellular organic nitrogen compounds. In this process, assimilative nitrate and nitrite reductases reduce nitrate nitrogen from an oxidation state of +5 to ammonia that has a nitrogen oxidation state of –3. This process occurs primarily in aerobic soils where nitrate is stable. Many wetland plants, fungi, and bacteria prefer ammonium over nitrate and only carry out assimilatory nitrate reduction when ammonium concentrations are relatively low. Ammonia, in fact, represses the function of assimilative nitrate reductase. The concentration of ammonium in aerobic soil layers is largely controlled by diffusive flux from anaerobic layers, and decomposition of organic matter in the water column and the aerobic soil layer. The assimilatory nitrate reduction is not sensitive to oxygen.

Nitrate added to wetlands or derived through nitrification rapidly diffuses into anaerobic soil layers, where it is used as an electron acceptor and reduced to gaseous end products (nitrous oxide and dinitrogen) or ammonium (Figure 8.40). During bacterial metabolic activity, ATP is generated by transport of electrons from organic or inorganic compounds to nitrogen oxides. During this process nitrate serves as the terminal electron acceptor of electron transport chain. The former pathway known as denitrification occurs at higher Eh levels (200–300 mV), whereas the latter process known as dissimilatory nitrate reduction to ammonium (DNRA) occurs at low Eh levels (<0 mV).

Several reviews have addressed the biochemistry, physiology, and environmental regulators of catabolic nitrate reduction (Payne, 1973; Delwiche, 1981; Firestone, 1982; Seitzinger, 1988; Zehr and Ward, 2002). The microbial groups involved in catabolic reduction of nitrate either via denitrification or DNRA occur in similar habitats. Denitrification is the dominant process in moderately reduced soil conditions in wetlands undergoing seasonal hydrologic fluctuations. In long-term anaerobic habitats such as permanently waterlogged wetlands, lake sediments, paddy soils, animal rumen, and anaerobic digesters, DNRA can be a significant process. However, in these habitats, nitrate production and availability often limits the overall DNRA.

8.11.1 Denitrification

Denitrification is defined as the biological reduction of nitrate or nitrite to gaseous end products such as nitrous oxide and nitrogen gas. Denitrification is mediated by facultative bacteria and in the absence of oxygen these bacteria have the ability to use both molecular oxygen and nitrogen oxides as alternate electron acceptors during cellular respiration. These bacteria tend to live in conditions with varying degrees of oxygen availability.

The actual process of denitrification, however, occurs only in anaerobic regions of the soil profile. During denitrification, nitrate nitrogen that has an oxidation state of +5 is reduced to molecular nitrogen that has an oxidation state of 0, with transfer of five electrons. Facultative bacterial groups do not require high availability of electrons to carry out denitrification. They can accomplish it with a low electron pressure, which corresponds to moderately reduced soil conditions. Several intermediate compounds are formed during nitrate reduction to N_2 (Figure 8.40). Nitroxyl (HNO) and hyponitrous acid ($H_2N_2O_2$) are some of the possible intermediates during denitrification. These compounds are unstable and are rapidly reduced to nitrous oxide.

Each step of the denitrification pathway is catalyzed by a distinct enzyme, nitrogen oxide reductase (nitrate reductase, nitrite reductase, nitric oxide reductase, and nitrous oxide reductase), that transfers electrons from the chain to the particular intermediate. Thermodynamically, in the absence of oxygen, nitrogen oxides are the most preferred electron acceptors by facultative bacterial groups. The role of nitrogen oxides in regulating organic matter decomposition has been discussed in earlier chapters (see Chapter 5).

$$C_6H_{12}O_6 + 6NO_3^- + 6H^+ = 6CO_2 + 3N_2O + 9H_2O \quad G^\circ_{298} = -2{,}369 \text{ kJ mol}^{-1} \quad (8.26)$$

$$C_6H_{12}O_6 + 12NO_2^- + 12H^+ = 6CO_2 + 6N_2O + 12H_2O \quad G^\circ_{298} = -2{,}571 \text{ kJ mol}^{-1} \quad (8.27)$$

$$C_6H_{12}O_6 + 24NO = 6CO_2 + 12N_2O + 6H_2O \quad G^\circ_{298} = -3{,}795 \text{ kJ mol}^{-1} \quad (8.28)$$

$$5C_6H_{12}O_6 + 24NO_3^- + 24H^+ = 30CO_2 + 12N_2 + 42H_2O \quad G^\circ_{298} = -2{,}202 \text{ kJ mol}^{-1} \quad (8.29)$$

$$C_6H_{12}O_6 + 8NO_2^- + 8H^+ = 6CO_2 + 4N_2 + 10H_2O \quad G^\circ_{298} = -3{,}210 \text{ kJ mol}^{-1} \quad (8.30)$$

$$C_6H_{12}O_6 + 12NO = 6CO_2 + 6N_2 + 6H_2O \quad G^\circ_{298} = -3{,}918 \text{ kJ mol}^{-1} \quad (8.31)$$

$$C_6H_{12}O_6 + 12N_2O = 6CO_2 + 12N_2 + 6H_2O \quad G^\circ_{298} = -4{,}129 \text{ kJ mol}^{-1} \quad (8.32)$$

A wide range of bacteria are capable of using nitrogen oxides as electron acceptors. Some organisms are capable of following through the entire reduction pathway, whereas others are capable of catalyzing only one or two steps of the pathway. Ecologically, any organism either involved completely or partially can be considered a denitrifier. Partial denitrification by certain groups of bacteria may be due to (1) unavailability of nitrate, (2) inability to synthesize nitrogen oxide reductases, or (3) environmental factors such as pH, oxygen concentration, or concentration of intermediate compounds (Table 8.5).

8.11.2 Nitrifier Denitrification

Nitrifier denitrification refers to reduction of nitrite to nitrous oxide in aerobic cultures formed during nitrification by ammonium-oxidizing bacteria. The ammonium-oxidizing bacteria includes chemoautotrophs, methanogens, and some heterotrophs. All these groups are capable of reducing some of the intermediate compounds such as hydroxylamine (NH_2OH) and nitrite to nitrous oxide, especially under low oxygen conditions (Zehr and Ward, 2002; Wrage et al., 2001; Sutka et al., 2006). Nitrifier denitrification occurs in two steps: First step involves oxidation of ammonia to nitrite, and the second involves conversion of nitrite to nitrous oxide (Figure 8.41). The nitrous oxide can also be produced and consumed by heterotrophic denitrifying organisms. Both nitrifiers and denitrifiers use similar pathways to reduce nitrite and produce nitrous oxide. Seasonal variations in hydroperiod and water-table fluctuations in wetlands result in uneven distributions of water flow, nutrients, and microbial populations, and create gradients in the soil profile with respect to aerobic and anaerobic conditions. These conditions provide zones of low oxygen concentrations,

TABLE 8.5
Genera of Bacteria Reported to Contain Strains that Carry Out Nitrate Respiration

Actinobacillus	Escherichia	Peptococcus
Actinomyces	Eubacterium	Photobacterium
Aeromonas	Ferrobacillus	Planobiospora
Agrobacterium	Flavobacterium	Planomonospora
Alcaligenes	Fusobacterium	Plesiomonas
Arachnia	Geodermatophilus	Propionibacterium
Arthrobacter	Haemophilus	Proteus
Bacillus	Halobacterium	Pseudomonas
Bacterionema	Halococcus	Rhizobium
Bacteroides	Hyphomicrobium	Rothia
Beneckea	Hyphomonas	Salmonella
Bordetella	Klebsiella	Selenomonas
Branhamella	Lactobacillus	Serratia
Brucella	Leptothrix	Shigella
Campylobacter	Listeria	Simonsiella
Cellulomonas	Lucibacterium	Spirillum
Chromobacterium	Microbispora	Sporosarcina
Citrobacter	Moraxella	Streptonmyces
Corynebacterium	Mycobacterium	Streptosporangium
Cytophaga	Neisseria	Thiobacillus
Dactylosporangium	Nocardia	Thiomicrospira
Enterobacter	Paracoccus	Veillionella
Erwina	Pasteurella	Vibrio

Source: Ingraham (1981).

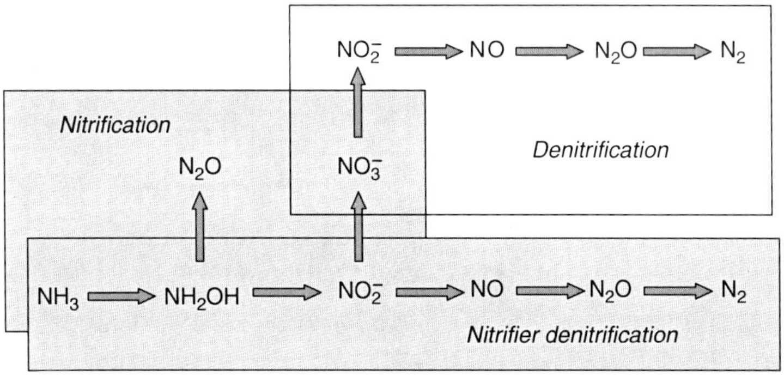

FIGURE 8.41 Schematic of pathways showing denitrification during nitrification reaction (Wrage et al., 2001).

creating demand for electron acceptors. One of the first studies suggesting that nitrifying bacteria may produce nitrous oxide was reported by Corbet (1935). This study demonstrated that certain bacteria convert ammonium or hydroxylamine to nitrite and suggested that nitrous oxide may be produced during this reaction. However, at that time it was generally regarded that nitrous oxide production during nitrification was insignificant. Subsequently, several researchers have demonstrated significant nitrous oxide evolution in ammonium-amended soils. The key findings of studies conducted during 1970s and early 1980s are summarized by Blackmer and Bremner (1981) as follows:

1. Nitrous oxide production was observed in well-aerated soils with moisture content of less than 5% (Table 8.6).
2. Nitrous oxide emissions from well-aerated soils are not significantly correlated to soil nitrate content, but are very significantly correlated to nitrifiable nitrogen content (or their capacity to produce nitrate under aerobic conditions).
3. Nitrous oxide emission from well-aerated soils is stimulated by the addition of nitrifiable forms of nitrogen (ammonium, urea, alanine, and other compounds), but is not significantly affected by nitrate and glucose (Table 8.7).
4. Inhibition of nitrification by amending soils with nitrapyrin (N-Serve or nitrification inhibitor) greatly reduced nitrous oxide emissions (Table 8.8).

TABLE 8.6
Nitrous Oxide Emissions from Well-Aerated Soils

Soil	pH	Organic Matter (%)	Nitrous Oxide Production (ng g^{-1})
Dickinson	6.6	1.2	9
Chelsea	6.4	1.4	5
Ida	8.2	1.5	8
Lindley	5.3	2.9	10
Monona	5.9	3.4	6
Marshall	5.9	4.1	11
Harps	7.9	7.1	5
Webster	6.8	10.2	10
Okoboji	4.8	11.2	22
Muck	5.9	28.4	30
Peat	6.0	40.8	62

Source: Blackmer and Bremner (1981).

TABLE 8.7
Effects of Various Amendments on Emissions of Nitrous Oxide (ng g^{-1} in 7 Days) from Well-Aerated Soils

Amendment	Harps Soil	Webster Soil
None	4	6
Ammonium nitrogen	247	50
Urea	292	75
Alanine	218	81
Nitrate nitrogen	4	7
Glucose	1	5
Nitrate + glucose	4	8

Note: Field moist soils incubated at 60% water-holding capacity.
Source: Blackmer and Bremner (1981).

TABLE 8.8
Effects of Acetylene and Nitrapyrin on Production of Nitrous Oxide in Samples of Nonsterile Storden Soil Incubated at Different Moisture Contents after Treatment with N as Ammonium Sulfate or Potassium Nitrate (200 μg N g^{-1} of Soil)

Form of Nitrogen Added	Soil Moisture (cm^3 g^{-1} of Soil)	N_2O Evolved in 14 Days (ng g^{-1} of Soil)		
		Control	Acetylene Added	Nitrapyrin Added
Ammonium N	0.23	419	<1	<1
	0.47	442	3	2
	0.61	3,160	1	1
Nitrate N	0.23	<1	1	<1
	0.47	2	1	1
	0.61	4	4	2

Source: Blackmer et al. (1980).

8.11.3 Aerobic Denitrification

In recent years, availability of modern tools including oxygen and ion-specific electrodes, culture vessels, and analytical methods has made it possible to determine the environmental conditions that support denitrification. These studies have shown that it is possible that a group of bacteria (mostly heterotrophic nitrifiers) are able to simultaneously utilize oxygen, nitrite, or nitrate as electron acceptors, even when oxygen levels in cultures approach near saturation levels (Zehr and Ward, 2002). Aerobic denitrification was described in *Paracoccus pantotrophus* (Rainey et al., 1999). The extent to which this process regulates overall nitrate loss is not known at this time. In wetlands, potentially this process can occur in aerobic portions of the soil profile. To date there is no evidence that the aerobic denitrification occurs in wetland environments.

8.11.4 Chemodenitrification

Chemodenitrification refers to abiotic conversion of ammonium to nitrite or reaction of nitrite itself with organics such as amines and inorganics such as metals, resulting in conversion of nitrite to gaseous end products (Figure 8.40). This alternative pathway is documented in agricultural soils and marine sediments (Nelson, 1982; van Cleemput and Baert, 1984). Substantial nitrogen deficits were observed in agricultural soils amended with urea or ammoniacal fertilizers. These losses could not be accounted for either ammonia volatilization or biological denitrification. These soils also exhibited significant accumulation of nitrite, suggesting that the second step in nitrification is inhibited in soils with high ammonium levels. The evidence of an alternate pathway of nitrogen loss is summarized as follows (Nelson, 1982):

1. Substantial loss of nitrogen has been observed in soil environment that is not favorable for ammonia volatilization and biological denitrification.
2. Significant gaseous loss of fertilizer nitrogen has been observed under conditions that lead to accumulation of nitrite.
3. Nitrite added to sterilized acid soils is rapidly converted to gaseous end products such as nitric oxide and nitrous oxide.
4. Larger proportion of nitrite is present as nitrous acid under acidic conditions, with 1.6, 14, and 63% present as nitrous acid at pH 5, 4, and 3, respectively, with the remaining present as nitrite.

The nitrite and nitrous acid equilibrium in solution is given as follows:

$$HNO_2 = NO_2^- + H^+ \tag{8.33}$$

$$\frac{[H^+][NO_2^-]}{[HNO_2]} = 6.0 \times 10^{-4} \tag{8.34}$$

Although significance of these losses is not documented in wetlands, it is important to recognize that wetlands receiving high nitrogen loading (such as treatment wetlands) can provide conditions to promote these reactions. The following mechanisms have been proposed (Nelson, 1982).

Self-decomposition of nitrous acid: Nitrous acid is produced when nitrite is added or formed in acidic conditions. The rate of reaction increases with a decrease in soil pH because larger proportion of nitrite is present as nitrous acid.

$$2\,HNO_2 = NO + NO_2 + H_2O \tag{8.35}$$

Reactions with metal cations: Thermodynamically, several reactions involving metals can abiotically convert nitrite to nitric oxide as shown below:

$$Mn^{2+} + HNO_2 + H^+ = Mn^{3+} + NO + H_2O \tag{8.36}$$

$$Fe^{2+} + HNO_2 + H^+ = Fe^{3+} + NO + H_2O \tag{8.37}$$

Reaction with amino groups: It has been shown that organic compounds containing free amino groups can react with nitrous acid under acid conditions and produce dinitrogen.

$$R\text{–}NH_2 + HNO_2 = R\text{–}OH + N_2 + H_2O \tag{8.38}$$

8.11.5 Dissimilatory Nitrate Reduction to Ammonia (DNRA)

DNRA is a process that reduces nitrate to ammonia (Figure 8.43). This process is carried out by dissimilative nitrate reductase enzymes. Dissimilative nitrate reductases are membrane-bound proteins, whose synthesis is repressed by molecular oxygen and thus are only synthesized under anoxic conditions. For this reason, DNRA is mediated by obligate anaerobes that use nitrate as alternate electron acceptor in the process of cellular respiration.

Obligate anaerobes that carry out DNRA tend to live in anoxic lake sediments or permanently waterlogged wetlands. In this process, nitrate nitrogen that has an oxidation state of +5, is reduced to ammonia that has an oxidation state of −3, with transfer of eight electrons. Obligate anaerobes require a high availability of electrons to carry out this process. A large supply of electrons provides the bacteria with the force they need to carry out this process to completion. The high electron pressure needed corresponds to a redox potential (Eh) of less than 0 mV.

$$C_6H_{12}O_6 + 3\,NO_3^- + 3\,H^+ = 6\,CO_2 + 3\,NH_3 + 3\,H_2O \tag{8.39}$$

The DNRA process requires high electron pressure and low redox potentials. In other words, the DNRA is the highest in environments where electron donor (carbon)-to-electron acceptor (nitrate) ratio is high, and decreases with a decrease in this ratio. For example, environments that are highly reduced, such as anaerobic digesters, rumen, and lake sediments, support high activity of obligate anaerobes that are capable of DNRA (Figure 8.42).

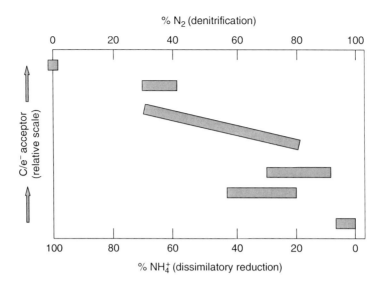

FIGURE 8.42 Relationship between electron donor-to-electron acceptor ratio and denitrification and dissimilatory nitrate reduction to ammonia (Tiedje, 1988).

8.11.6 Regulators of Nitrate Reduction

Catabolic nitrate reduction in wetlands is regulated by various external and internal factors including oxygen content of soil, presence of electron acceptors, supply of electron donors, plant root density, enzyme activity, temperature, pH, and nitrate flux from aerobic zones to anaerobic sites. As discussed earlier, catabolic nitrate reduction includes both denitrification and DNRA. Much of the research reported on the influence of environmental regulation on this process is based on denitrification, with very little or no information available on DNRA. Unfortunately, many of these studies were conducted under laboratory conditions by varying one regulator, while keeping other regulators at a constant level. These experiments provide the first approximation of the effect, but do not reflect the ability of bacterial communities to adapt to environmental changes and function accordingly.

Oxygen content of the soil: Depending on hydroperiod and water-table fluctuations in wetlands, a significant portion of the soil profile can be aerated. Saturated soil conditions in wetlands soils restrict oxygen supply to the soil–floodwater interface and water column, and to the rhizosphere. Thermodynamically, oxygen is the most preferred electron acceptor for facultative bacteria capable of reducing nitrate. Thus, catabolic nitrate reduction rates are inversely related to soil oxygen content (Figure 8.43). However, in soils with high water-holding capacity (soils with high clay and organic matter content), a significant portion of the soil profile can have anaerobic sites. Nitrate reductase is a molybdenum-containing enzyme involved in catabolic nitrate reduction. This enzyme is a membrane-bound protein and is repressed by oxygen. Thus, it is synthesized or activated only under oxygen-free (anaerobic) environments.

Oxygen content of the soil has no effect on anabolic nitrate reduction, as aerobes use nitrate as nutrient source and reduce it to ammonia during cell synthesis. The assimilative nitrate reductases are soluble proteins and are repressed by high ammonia levels.

Presence of nitrogen oxides: Nitrate reduction in wetlands is often limited by the availability of nitrogen oxides including nitrate. Sources of nitrate in wetlands are point and nonpoint sources from adjacent watersheds, groundwater inputs, atmospheric deposition, nitrification of ammonium in aerobic zones, and nitrification of ammonium released from anaerobic soil layers either by diffusion, advection, or bioturbation. Nitrate loading to most wetlands

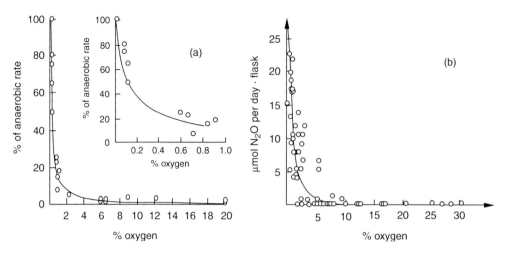

FIGURE 8.43 Effect of oxygen concentration on denitrification rates. (Modified from Tiedje, 1988.)

is lower than their capacity to process through denitrification or DNRA; therefore, nitrate added to wetlands is typically removed in areas closer to points of inflow. In this area, denitrification rates are typically higher than those measured in the marsh interior.

In interior wetland areas, the source of nitrate is primarily nitrification, and denitrification rates in these areas are limited by nitrate availability. Several studies have shown that denitrification rates increase with an increase in nitrate concentration. Half-saturation constants (one of the Michaelis–Menton kinetics parameter) reported for denitrification range from 27 to 344 µM for lake and marine sediments (Seitzinger, 1988) and 130 to 1,200 µM for soils (Firestone, 1982). Low half-saturation constants probably reflect carbon limitation in soil or diffusion of nitrate from aerobic sites to anaerobic sites.

Supply of electron donors: Facultative bacteria that utilize nitrate as electron acceptor have the capability to use various organic and inorganic electron donors. Wetlands provide abundant supply of electron donors that can be used during catabolic nitrate reduction. Complex polymers of plant detrital matter and soil organic matter must undergo enzymatic hydrolysis before nitrate reducers can use this material as an energy source (see Chapter 5). Many extracellular enzymes are involved in hydrolyzing complex polymeric compounds into simple organic compounds, which are subsequently used by microbes. In many soils, this process can limit the availability of organic carbon. Organic donors include carbohydrates, fatty acids, amino acids, alcohols, and other labile organic compounds. Both aboveground and belowground biomass can contribute to the detrital pool and eventual use as an energy source by nitrate reducers. Plant roots can influence nitrate reduction in a number of ways: (1) they provide a labile carbon source through root exudates; (2) detrital matter from fine roots provides labile organic carbon; (3) high root density can effect oxygen consumption through root respiration, creating anaerobic conditions in the rhizosphere; and (4) the roots remove nitrate through plant assimilation. Reduced inorganic compounds that are used as electron donors include ferrous iron, manganous manganese, sulfides, ammonium, and others.

Nitrate reduction rates have been shown to be highly correlated to soil organic matter and soluble or available organic carbon (determined as extractable organic carbon) in soils. Several studies have shown a strong relationship between nitrate reduction and available carbon in soils (Buford and Bremner, 1975; Reddy et al., 1982). Nitrate reduction in wetland soils can be coupled to organic matter mineralization (see Chapter 5), as facultative bacteria use nitrate as an electron acceptor during oxidation of organic substrate. The relationship between DOC and nitrate reduction for several wetland soils is shown in Figures 8.44 and 8.45.

Denitrification enzyme activity: Catabolic nitrate reduction involves several enzymes including dissimilatory nitrate reductase (nitrate to nitrite), nitrite reductase (nitrite to nitric oxide), nitric oxide reductase (nitric oxide to nitrous oxide), and nitrous oxide reductase (nitrous oxide to nitrogen gas). The combined effect of all these enzymes is operationally defined as denitrification enzyme activity (DEA), which is now routinely measured under

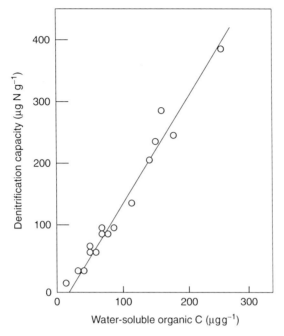

FIGURE 8.44 Relationship between water extractable organic carbon and denitrification capacity. (From Burford and Bremner, 1975.)

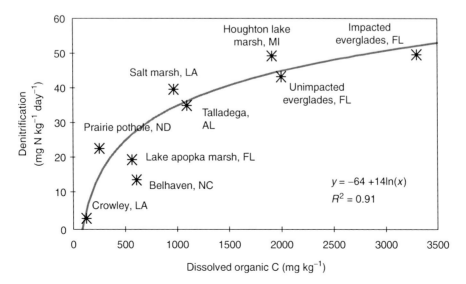

FIGURE 8.45 Relationship between dissolved organic carbon and denitrification rates in select wetland soils. (From D'Angelo and Reddy, 1999.)

TABLE 8.9
Denitrification Enzyme Activity in Select Wetlands

Wetland	Denitrification Rate (mg N kg^{-1} day^{-1})
Houghton Lake wetland, Michigan	
Near inflow	74
Interior	11
Orange County wetland, Florida	
Near inflow	9
Interior	1
Everglades WCA-2A, Florida	
Near inflow	6
Interior	1
Constructed wetlands, Massachusetts	<2–22
Natural wetlands, Rhode Island	4–9

Source: Duncan and Groffman (1994), Gale et al. (1993), and D'Angelo and Reddy (1999).

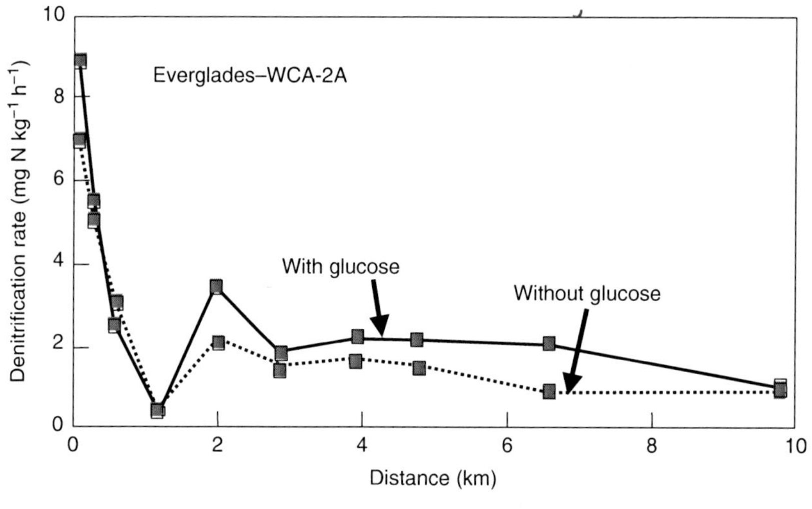

FIGURE 8.46 Denitrification enzyme activity as a function of distance from inflow of a wetland. (From White and Reddy, 1999.)

laboratory conditions. The nitrous oxide reductase activity is inhibited by sulfides and acetylene. This characteristic was used in the measurement of DEA, as described by Tiedje (1994) (Table 8.9 and Figure 8.46).

Temperature: Catabolic nitrate reduction rates are influenced by temperature, similar to any biological reactions. The increase in biological activity as soil temperature increases results in an exponential increase of DNRA and denitrification rates. From the results reported in several studies, it can be concluded that the catabolic nitrate reduction rates increased 1.5–2-fold with 10°C rise in temperature (Figure 8.47).

pH: Optimal pH for bacterial groups involved in catabolic nitrate reduction is in the range of 6–8. This is a typical range for many wetland soils. Rates are lower in acid soils with pH of <5. The proportion of nitrous oxide occurring as the end product of denitrification

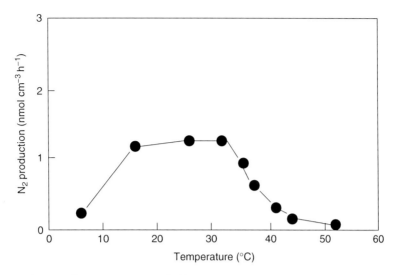

FIGURE 8.47 Influence of temperature on denitrification rates in marine sediments. (Drawn from Dalsgaard and Thamdrup, 2002.)

increase with a decrease in soil pH, suggesting that soil acidity influences gaseous end products. In acid soils, nitrite can be reduced chemically to nitric oxide through a process not related to catabolic nitrate reduction.

Nitrate flux from aerobic zones to anaerobic sites: For nitrate reduction to occur in wetlands, nitrate must be present in anaerobic zones. Thus, nitrate reduction rates are regulated by transport of nitrate either by diffusion or by mass flow from aerobic zones to anaerobic portions of the soil. Similarly, the rate of nitrification and the oxygen availability in the soil regulate nitrate concentrations in aerobic zones of the soil. In wetlands with limited inputs of nitrate from external sources, nitrification and atmospheric deposition are the primary sources of nitrate. In these systems, denitrification rates are tightly coupled to nitrification rates.

8.11.7 Nitrate Reduction Rates in Wetlands and Aquatic Systems

A wide range of techniques are used to determine catabolic nitrate reduction rates. These include mass balance methods using input–outputs, acetylene inhibition techniques, dinitrogen production rates, nitrate consumption rates, nitrate pore water profiles, and stable isotope ^{15}N tracer techniques. The limitations and advantages of these methods are discussed by Seitzinger (1988) and Herbert (1999).

Denitrification is typically the dominant pathway of nitrate removal in wetlands. Thus, the nitrate reduction rates presented in Tables 8.10 and 8.11 represent denitrification rates. Rates are presented on areal basis (Table 8.10) and as first-order rate constants (Table 8.11). Nitrate reduction rates reported for constructed wetlands are in the range of 3–1,020 mg N m^{-2} day^{-1}. In most wetlands, denitrification rates are limited not only by nitrate concentration but also by hydraulic retention time (or contact time of nitrate with anaerobic zones) and diffusion/mass flow of nitrate from aerobic zones to anaerobic sites (Martin and Reddy, 1997). Denitrification rates are usually higher in soils receiving steady loading of nitrate than in soils receiving low or negligible nitrate levels (Cooper, 1990; Gale et al., 1993).

In a system with active denitrification, nitrate levels are usually low, thus, measurement of nitrate in soil and water column does not provide a reliable indication of this process. As denitrification is mediated by heterotrophic microorganisms, its rate may be regulated by nitrate concentration (electron acceptor) and available C (electron donor). Significant correlations were observed between denitrification rates and available organic C (mineralizable organic C) (Gale et al., 1993; D'Angelo and Reddy, 1999). Similarly, MBC and denitrification enzyme assay (DEA) can also serve as potential indicators of denitrification rates (White and Reddy, 1999).

TABLE 8.10
Nitrate Reduction Rates Reported for Various Wetlands and Aquatic Systems

Ecosystem	Denitrification Rate (mg N m^{-2} day^{-1})	Reference
Freshwater wetlands	7–5,914	Buresh et al. (2008)
Constructed wetlands	3–1,020	Martin and Reddy (1997)
River and stream sediments	18–116	Seitzinger (1988)
Lakes	34–57	Seitzinger (1988)
Coastal and marine sediments	<0.1–359	Seitzinger (1988)

TABLE 8.11
Nitrate Reduction Rates Reported for Flooded Soils and Sediments

Ecosystem	Denitrification Rate, k (day^{-1})
Mineral soils	0.03–0.9
Organic soils	0.1–1.5
Lake sediments	0.04–0.76

Note: Rates presented for a range of soils and sediments.
Source: Summarized by Reddy and Patrick, (1984).

TABLE 8.12
Nitrate Reduction in Estuarine Sediments

E_h (mV)	Nitrate N	Ammonium N	Organic N	Unaccounted for
300	14.0	<1.0	4.3	81.7
0	0	13.0	5.1	81.9
−200	0	26.0	8.7	65.5

Note: Stable isotopic nitrate ^{15}N was added to sediments maintained at three different Eh levels. At the end of 210 h, ^{15}N tracer was determined in various nitrogen pools. The data are expressed as percentage of added nitrate ^{15}N. Unaccounted pool of nitrogen is assumed to be lost as nitrous oxide and dinitrogen gas.
Source: Buresh, and Patrick (1981).

Although denitrification is the dominant process in regulating nitrate loss in wetlands and aquatic systems, DNRA can be significant under certain conditions. Examples shown in Table 8.12 and Figure 8.48 show the significance of DNRA in sediments. Obligate anaerobes are known to respire through DNRA process. The DNRA increases with a decrease in Eh of sediments. In highly reduced estuarine sediments (Eh = −200 mV), up to 26% of the added nitrate was reduced to ammonium. This is primarily due to high electron pressure in highly reduced sediments. Similarly, in freshwater lake sediments, approximately 20–30% of the added nitrate was reduced through DNRA (Figure 8.48).

Nitrogen

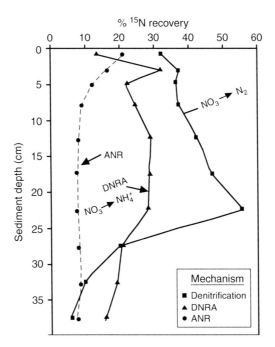

FIGURE 8.48 Effect of sediment depth on partitioning of nitrate reductive processes of denitrification, dissimilatory nitrate reduction to ammonia (DNRA), and assimilatory nitrate reduction (ANR). Each value represents the mean of six relications (D'Angelo and Reddy, 1993).

8.12 NITROGEN FIXATION

Nitrogen fixation is a significant process in wetlands. In wetlands and aquatic systems with limited nitrogen loading, biological nitrogen can play a significant role in supplying nitrogen and maintaining ecosystem production and soil organic matter. Biological nitrogen fixation can occur in the following zones of wetlands:

- Periphyton mats in the water column
- Rhizosphere of wetland plants
- Aerobic or anaerobic soil layer

Biological dinitrogen fixation involves the enzymatically catalyzed reduction of atmospheric dinitrogen to two molecules of ammonia. This reaction requires cleaving the triple bond of the N_2 molecule. The triple bond between two nitrogen atoms is extremely stable and biological reduction of dinitrogen to ammonia is an energy-demanding process with approximately 16 ATPs consumed per molecule of dinitrogen fixed (Postgate et al., 1987). Once this triple bond is broken, the single atoms of N combine with H^+ to form NH_3 with transfer of electrons, as shown below:

$$N_2 + 6e^- + 6H^+ = 2NH_3 \quad (8.40)$$

$$2H^+ + 2e^- = H_2 \quad (8.41)$$

$$N_2 + 8e^- + 8H^+ = 2NH_3 + H_2 \quad (8.42)$$

Reduction of dinitrogen to ammonia requires only six electrons, while two electrons are used to produce hydrogen. Because of the high energy input required to break the triple bond, the enzyme nitrogenase is required to catalyze the reaction. It is interesting to note that the enzyme nitrogenase (synthesized by only a few prokaryotic organisms) catalyzes this reaction at ambient temperatures and normal atmospheric pressure. In contrast, the commercial Haber–Bosch process requires high temperatures and pressures to perform the same reaction to produce ammonia.

Nitrogen-fixing prokaryotic organisms are also capable of synthesizing nitrogenase. The nitrogenase enzyme complex consists of two proteins, the MoFe protein and Fe protein; is sensitive to oxygen; contains iron and molybdenum; needs magnesium to be active; converts ATP into ADP when functioning, is inhibited by ADP; reduces dinitrogen and other small triple-bonded molecules (e.g., acetylene, cyanide, carbon monoxide); needs up to 30 molecules of ATP per mole of dinitrogen; and reduces some H^+ to H_2 (Zuberer, 1999). Hydrogen gas produced from this reaction is used by some bacterial groups (such as methanogens) as an energy source.

The Nitrogenase enzyme complex is not only specific to reduction of dinitrogen, but also reduces acetylene (C_2H_2) to ethylene (C_2H_4). This reaction requires the transfer of two electrons.

$$C_2H_2 + 2e^- + 2H^+ = C_2H_4 \qquad (8.43)$$

Although this reaction has very little use for cell growth, it was found to be very useful for indirect measurement of nitrogenase activity (Stal, 1988). The method essentially consists of incubating substrate in a gas-tight glass vial with 10% headspace containing acetylene. At the end of 2–4 h incubation at constant temperature, the ethylene produced is analyzed using GC. Acetylene reduction values are used to estimate the actual rates of dinitrogen fixation using a theoretical conversion ratio of 3 mol of acetylene to 1 mol of dinitrogen. Several studies have used this technique to estimate dinitrogen fixation in a wide range of habitats, including wetlands (Inglett et al., 2004).

The microorganisms that are capable of fixing atmospheric dinitrogen include free-living autotrophs such as blue-green algae (cyanobacteria); free-living heterotrophs such as aerobes, facultative, and obligate anaerobes; and symbiotic microorganisms. These groups of organisms are also known as *diazotrophs* (organisms that can use atmospheric dinitrogen as their sole source of nitrogen for growth). In wetland environments, autochthonous biological nitrogen fixation is mediated by cyanobacteria and phototrophic bacteria inhabiting the water column, plant detrital matter, and the soil surface, and the heterotrophic bacteria throughout the soil profile including rhizosphere. Allochthonous biological nitrogen fixation comprises diazotrophs such as nitrogen-fixing cyanobacteria living in symbiosis with *Azolla* and rhizobia associated with wetland legumes.

8.12.1 Regulators of Dinitrogen Fixation

Various biogeochemical factors regulating nitrogen fixation are summarized in Figure 8.49 (Howarth et al., 1988). The influence of macroelements such as nitrogen and phosphorus is well established, whereas the role of micronutrients is not well understood.

> *Microbial communities*: Cyanobacterial communities are one of the major microbial groups involved in biological fixation of dinitrogen. This group of diazotrophs was once referred to as the "blue-green algae" and consists of heterogeneous group of phototrophic organisms. The activity of these organisms is restricted to the water column and benthic surfaces. These mats consist of not only phototrophic organisms but also a wide range of heterotrophic bacteria. In soils and sediments of these systems, there exists a wide range of nitrogen-fixing community including a diverse array of chemoautotrophic and heterotrophic bacteria ranging from obligate aerobes, facultative anaerobes, and Archaea. These may include *Azotobacter, Azospirillum, Beggiatoa, Enterobacter, Klebsiella, Desulfobacter, Desulfovibrio, Clostridium, Methanococcus, Methanosarcina*, and others (Herbert, 1999).
>
> *Energy source*: Nitrogen fixation is an energy-intensive process and the factors controlling rates of photosynthesis or the distribution of photosynthate within the nitrogen-fixing organisms

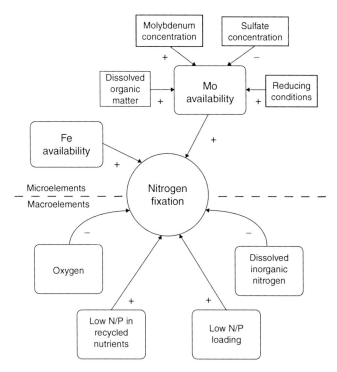

FIGURE 8.49 Schematic showing biogeochemical controls regulating nitrogen fixation in wetlands and aquatic systems. Symbols represent positive (+) and negative (–) effects. (Adapted from Howarth et al., 1988.)

would affect overall dinitrogen fixation rates. In higher plants, high rate of photosynthate flow to roots was shown to increase nitrogen fixation in nodules of legumes. Similarly, increased photosynthetic activity within the cyanobacterial mats can also increase rates of nitrogen fixation. Addition of high C:N ratio by organic substrates can immobilize available inorganic nitrogen, creating nitrogen limitation. Addition of glucose to estuarine sediments was shown to promote heterotrophic nitrogen fixation by approximately three times, under both aerobic and anaerobic conditions (Herbert, 1999). Others have demonstrated a large increase in heterotrophic nitrogen fixation when sediments are amended with plant storage polysaccharides such as xylan, alginate, laminarin, and glycogen (Herbert, 1999). These conditions can promote nitrogen fixation by certain groups of heterotrophic bacteria. This is due to increased activity of heterotrophic organisms from substrate carbon additions, which can create increased demand on nitrogen. Release of root exudates such as organic acids can also promote heterotrophic nitrogen fixation. Photosynthetic nitrogen-fixing organisms do rely on organic substrate as an energy source.

Nutrients: Availability of inorganic nitrogen, either ammonium or nitrate, can inhibit biological nitrogen fixation. Availability of inorganic nitrogen represses nitrogenase activity by organisms, which results in avoiding energy expense in synthesizing this enzyme. At high levels of ammonium nitrogen, there is clear evidence that nitrogenase activity is inhibited. However, in nitrogen-limited wetlands, addition of inorganic nitrogen can stimulate growth of vegetation, which can result in increased release of root exudates. This can cause indirect stimulation of nitrogenase activity in the root zone.

Cyanobacteria are commonly found to dominate wetlands and aquatic systems under conditions of nitrogen limitation or when levels of other potentially limiting elements (e.g., phosphorus) are

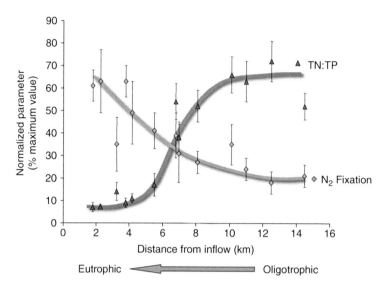

FIGURE 8.50 Biological nitrogen fixation in periphyton mats along eutrophic gradient in northern Everglades wetland. (From Inglett et al., 2004.)

high compared to that of N (Paerl, 2000). This observation is commonly explained by the ability of many cyanobacteria to convert atmospheric dinitrogen gas into the biologically available ammonium form via the nitrogenase enzyme complex (Inglett et al., 2004). In the oligotrophic wetlands and aquatic systems, nitrogen-fixing organisms are limited by phosphorus. However, in eutrophic systems, phosphorus loading results in a decrease of nitrogen to phosphorus ratios in algal mats, resulting in nitrogen limitation (Figure 8.50). This may result in increased rates of biological nitrogen fixation. Similarly, applied phosphorus has been shown to simulate nitrogen fixation by blue-green algae in paddy soils (Buresh et al., 2000).

Both iron and molybdenum are known to be essential components of nitrogenase required for dinitrogen fixation (Howarth et al., 1988). The nitrogenase complex has approximately 30 atoms of iron compared to 2 atoms of molybdenum. The availability of iron and molybdenum may vary among wetlands and aquatic systems, which to some extent can explain the variability observed in dinitrogen fixation rates. The importance of these elements in dinitrogen fixation in wetland environments is not established. High levels of sulfate are known to inhibit molybdenum availability, thereby affecting dinitrogen fixation rates.

Light: Photosynthetic nitrogen fixers are confined to open areas of wetlands. Shading by wetland vegetation inhibits the growth of cyanobacterial mats, thus affecting overall contribution of nitrogen via biological fixation. In general, nitrogenase activity is stimulated by light in photosynthetic organisms. However, in the presence of photosynthetically derived oxygen, heterotrophic nitrogenase activity is decreased. In addition, diurnal variations in nitrogenase activity by cyanobacterial mats are influenced by light intensity, with high values observed during daytime. Light intensity can have a direct effect on photosynthetic activity, which may decrease the supply of organic substrates and reduce nitrogenase activity.

Redox potential and pH: Nitrogen fixation rates were found to increase with a decrease in redox potential. Nitrogen fixation rates were highest in soils at Eh levels of −200 to −250 mV. This suggests that obligate anaerobes are involved in nitrogen fixation. Nitrogenase activity is highest at soil pH close to 7.

Temperature: Like any other biological reactions, nitrogen fixation rates are influenced by temperature with Q_{10} values around 2. Contrasting differences were noted in nitrogen fixation rates of seagrass meadows in temperate and tropical climates (summarized by Herbert, 1999). In temperate climates, nitrogen fixation rates were in the range of 0.1–7 mg N m^{-2} day^{-1}, as compared to 5–140 mg N m^{-2} day^{-1} in tropical climates.

8.12.2 Nitrogen Fixation Rates

Nitrogen fixation rates (expressed as nitrogenase activity) are measured using the acetylene (C_2H_2) reduction assay. The nitrogenase enzyme system in microbial cells catalyzes the dinitrogen reduction to ammonia and subsequent synthesis of amino acids. Nitrogen fixation rates can also be measured directly using ^{15}N-labeled dinitrogen gas.

A wide range in nitrogen fixation rates is reported in wetlands and aquatic systems (Table 8.13; Buresh et al., 1980; Howarth et al., 1988; Herbert, 1999). Rice paddies and coastal wetlands showed higher nitrogen fixation rates than freshwater wetlands, peat bogs, and lakes.

High rates of nitrogen fixation are typically confined to open areas of wetlands colonized by cyanobacterial mats (Table 8.14). However, these regions occupy a small fraction of the total area of wetlands. Shading by macrophytes restricts the growth of cyanobacterial mats, thus decreasing overall nitrogen fixation rates. However, nitrogen fixation can be a major contributor in unvegetated areas of wetlands and other shallow water bodies. Reported nitrogen fixation rates by cyanobacterial mats are in the range of 4–208 mg N m^{-2} day^{-1}. Nitrogen fixation rates in the root zone of wetland plants depend on plant type and microbial community structure in the root zone (Table 8.15). Reported nitrogen fixation rates are in the range of 1–140 mg N m^{-2} day^{-1}.

TABLE 8.13
Nitrogen Fixation Rates (Estimated Acetylene Reduction Assay) in Select Wetland Ecosystems

Ecosystem	Dinitrogen Fixation Rate (mg N m^{-2} day^{-1})	Reference
Rice paddies	2–48	Buresh et al. (1980)
Coastal wetlands	1–126	Buresh et al. (1980)
Freshwater marshes	0.1–16	Howarth et al. (1988)
Cypress swamps	1–8	Howarth et al. (1988)
Peat bogs	0.1–6	Howarth et al. (1988)
Flax Pond mud flats	1.8	Herbert (1999)
Estuaries	0.2–5	Howarth et al. (1988)
Oligotrophic lakes	0–5	Howarth et al. (1988)
Mesotrophic lakes	0.03–0.25	Howarth et al. (1988)
Eutrophic lakes	0.5–25	Howarth et al. (1988)

TABLE 8.14
Nitrogen Fixation Rates in Cyanobacterial Mats

System	Dinitrogen Fixation Rate (mg N m^{-2} day^{-1})	Reference
Cyanobacterial mats	4–208	Howarth et al. (1988)
Cyanobacterial mats, Everglades	5–48	Inglett et al. (2004)
Salt marsh, Massachusetts	3.9	Herbert (1999)
Colne Point marsh, United Kingdom	16.4	Herbert (1999)
Flax Pond salt marsh, New York	36.8	Herbert (1999)

TABLE 8.15
Estimated Dinitrogen Fixation Rates in the Root Zone of Wetland Plants

Plant	Dinitrogen Fixation Rate (mg N m^{-2} day^{-1})
Thalassia testudinum	3–137
Spartina alterniflora	0.7–141
Glyceria borealis	16
Typha sp	18
Juncus balticus	6.5–80
Carex scoparia	18
Scirpus polyphyllus	10
Zostera marina	1–6

Sources: Buresh et al. (1999).

Nitrogen fixation is probably a minor component of the total nitrogen input to many wetlands. The contribution of nitrogen fixation to the total nitrogen budget in closed systems is higher because of low external inputs. For example, total nitrogen input from nitrogen fixation in bogs was estimated to be 59% of the total nitrogen while in Florida's cypress domes it was reported to be in the range of 46–76% (see review by Howarth et al., 1988). Caution should be exercised in using these values because these estimates are based on short-term incubations of limited scale. Other estimates based on short-term incubations suggest that biological nitrogen fixation contributes approximately 1–2% of the total nitrogen budget in salt marshes and up to 10% in rice paddies. In tropical seagrass meadows, nitrogen fixation can supply up to 50% of the nitrogen requirements of plant communities.

Nitrogen fixation in lake sediments contributes less than 2% of the total nitrogen input (Howarth et al., 1988). However, in lakes with low external nitrogen inputs, nitrogen fixation rates accounted for 32% of the total nitrogen input in Lake Tahoe, and 6% of the total nitrogen input in Mirror Lake (see review by Howarth et al., 1988).

8.13 NITROGEN ASSIMILATION BY VEGETATION

Vegetation plays several roles in wetlands including (i) assimilating nitrogen into plant tissue for growth, (ii) supplying carbon substrates to microbes, (iii) providing an optimum environment in the root zone for nitrification–denitrification to occur, and (iv) building organic matter in soils. Nitrogen uptake by plants is defined as the amount of nitrogen that is assimilated in the aboveground and belowground portions of a plant. Efficiency of nitrogen utilization (defined as an increase in plant nitrogen per unit mass of available nitrogen) by aquatic vegetation is highly variable, depending on the type of wetlands (forested vs. herbaceous), soil type, nutrient status, and climate. Ecologists have used agronomic principles developed in crop production to determine the nutrient use efficiency of vegetation in natural systems. Nutrient (such as nitrogen and phosphorus) availability to vegetation is determined by external loading and nutrient use efficiency (Shaver and Melillo, 1984; Killingbeck, 1996; Whigham, 1999). Nutrient use efficiency provides an indication of the effectiveness of vegetation to produce biomass per unit of the available nutrient. Under nitrogen-limiting conditions, some of the plant nitrogen from senescing tissue may be translocated and stored in the belowground tissue, with very little of the nutrient in the detrital matter. Plants grown in these environments typically exhibit high nutrient use efficiency. In areas with high nitrogen availability, plants are inefficient in using available nutrients and exhibit low nutrient use efficiency.

Nitrogen

Plant tissue nitrogen concentrations have been widely used to assess the availability for plant growth and degree to which nitrogen is limiting plant growth. Nitrogen content of plant tissue is highly variable and is dependent on age of the plant, nitrogen availability, genetic ability of plant to assimilate nitrogen, soil type, and environmental conditions (Gusewell and Koerselman, 2002). Significant differences in nitrogen content among wetland plants were observed when these plants were grown under identical soil and nutrient conditions (Figure 8.51) (McJannet et al., 1995). Nitrogen content of plant tissue is inversely related to biomass (Figure 8.52). It is expected for plants with high biomass production that concentration of nitrogen may be lower as a result of dilution and distribution within the tissue. Similarly, younger plants have higher nitrogen content (>25 g kg^{-1}) and as the plant approaches maturity, nitrogen content decreases (Figure 8.53). During the same time period, total nitrogen storage in the plant tissue increases (Table 8.16).

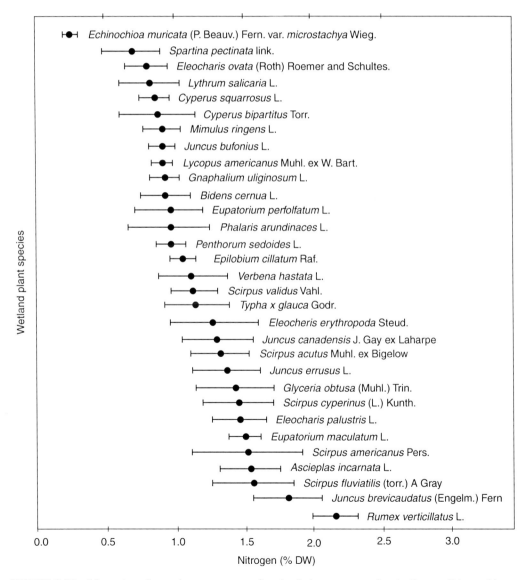

FIGURE 8.51 Mean plant tissue nitrogen content of wetland plant grown under similar conditions with sustained fertilization. Error bars denote one standard deviation ($n = 4$–5) (From McJannet et al., 1995).

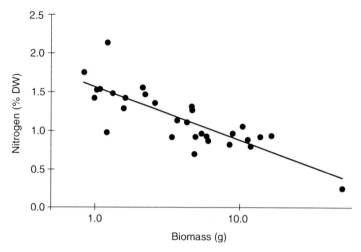

FIGURE 8.52 Relationship between plant tissue nitrogen and the aboveground biomass of wetland plants. The solid line represents the linear best-fit to the data ($r^2 = 0.67$, $p < .001$, $n = 31$) (McJannet et al., 1995).

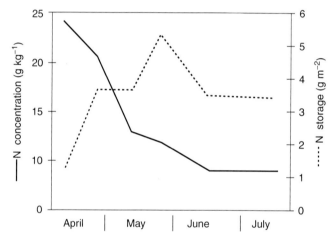

FIGURE 8.53 Temporal changes in tissue nitrogen content and nitrogen storage of cattails (*Typha latifolia*). (Adapted from Reddy and DeBusk, 1987.)

Leaf N:P ratio also has been used to determine whether a wetland is N-limited or P-limited (Verhoeven et al., 1996). It has been suggested that N:P < 15 (mass basis) indicates nitrogen limitation, whereas N:P > 15 indicates phosphorus limitation (Verhoeven et al., 1996). The N:P ratios show a closer relationship with the nutrient availability than nitrogen and phosphorus concentrations (Gusewell and Koerselman, 2002). As explained by Gusewell and Koerselman (2002), this is due to the following reasons: (1) the N:P ratios varied relatively less among species than individual nitrogen and phosphorus concentrations, (2) the N:P ratios varied less in response to the total supply of nitrogen and phosphorus, and (3) the N:P ratio varied more in response to differences in the relative availability of nitrogen and phosphorus.

The nutrient use efficiency of vegetation and the C:N ratio of the plant litter decreases with an increase in nutrient loading (Shaver and Melillo, 1984; Koch and Reddy, 1992). Plants derive most of their nitrogen from soil pore water with only a small amount of the floodwater nitrogen being directly utilized. Floodwater nitrogen is rapidly utilized by algae and microbial communities growing on plant litter substrates or lost through ammonia volatilization and nitrification–denitrification reactions

TABLE 8.16
Range of Nitrogen Concentrations in Wetland Vegetation

Plant	Nitrogen (g kg^{-1})	
Herbaceous vegetation		
Typha (cattail)	5–24	
Juncus (rush)	15	
Scirpus (bulrush)	8–27	
Phragmites (reed)	18–21	
Eleocharis (spike rush)	9–18	
Saururus	15–25	
Forested vegetation	Leaves	Wood
Taxodium	15–18	
Nyssa	19–24	1.4
Acer	10–22	1.7
Magnolia	19–25	4.1

Source: Reddy and DeBusk (1987).

(Reddy and Patrick, 1984; DeBusk and Reddy, 1987). The extent of these processes also increases with nitrogen loading. Measurement of plant tissue nitrogen and biomass can provide an indication of nitrogen removal efficiency by wetlands.

Nitrogen assimilation by herbaceous vegetation is usually short-term and usually cycled rapidly within the systems. Elevation of temperature from 10 to 25°C was shown to increase nitrogen use efficiency by *Typha* spp. from 5 to 38% (Reddy and Portier, 1987). In temperate climates much of the nitrogen assimilation occurs during growing season. During winter months, aboveground biomass is killed and accumulates as detrital tissue. At the same time, a significant portion of nitrogen is translocated to belowground portions. Nitrogen release during decomposition of detrital tissue *Typha* spp. was more rapid during summer months than during winter. Unlike herbaceous macrophytes, forested wetlands provide long-term storage in the form of woody biomass. However, a significant portion of nutrients stored in leaves are returned to the forest floor, and eventually incorporated into the soil (Reddy and DeBusk, 1987).

Several studies have used mass balance methods to estimate nitrogen retention within the wetlands (DeLaune et al., 1989; White and Howes, 1994). These studies showed that nitrogen demand of aquatic vegetation and microbial and periphyton communities is not met by external inputs alone, which suggests the role of internal cycling and turnover. Only a few studies have quantified the role of remineralization of organic nitrogen during plant litter decomposition and belowground biomass turnover, and several studies have reported values of 54–95% of plant uptake demand being met through this process (Hopkin'son and Schubauer, 1984; DeLaune et al., 1989; White and Howes, 1994).

8.14 NITROGEN PROCESSING BY WETLANDS

Nitrogen reactions in wetlands effectively process inorganic N through nitrification and denitrification, ammonia volatilization, and plant uptake (Figure 8.3). These processes aid in maintaining low levels of inorganic N in the water column. A significant portion of DON is returned to the water column during breakdown of detrital plant tissue or soil organic matter, and majority of this DON is resistant to decomposition. Under these conditions, surface water leaving wetlands may contain elevated levels of nitrogen in the dissolved organic form. Relative rates of these reactions will, however, depend on the optimal environmental conditions present in soil and water column.

Exchange of dissolved nitrogen species between the soil and water column support several nitrogen reactions. For example, nitrification in aerobic soil layer is supported by ammonium flux from the anaerobic soil layer. Similarly, denitrification in anaerobic soil layer is supported by nitrate flux from the aerobic soil layer and water column (see Chapter 14 for discussion on transport processes).

8.14.1 Ammonium Flux

Ammonium in the aerobic soil layer and the water column is derived from (1) decomposition of organic matter in the water column, (2) mineralization of organic nitrogen in the aerobic soil layer, and (3) diffusion and mass flow of ammonium from anaerobic soil layer to aerobic soil layer and water column. The later process contributes a major portion of ammonium to the overlying aerobic soil layer and water column. Flux of ammonium is accomplished by diffusion, advection, bioturbation, and mixing at and near soil–floodwater interface. High concentrations of dissolved ammonium and soluble organic nitrogen in anaerobic soil layer establish steep gradients between the soil and overlying water column. As a result, flux is always from anaerobic soil layer to the overlying water column, suggesting that wetland soils function as a source of ammonium and DON to the water column (Figure 8.54).

Ammonium flux from anaerobic soil layer is governed by the (1) concentration gradient established as a result of ammonium consumption in the aerobic zone due to nitrification and ammonia volatilization, (2) ammonium regeneration rate in the anaerobic soil layer, (3) adsorption coefficient for ammonium, (4) soil CEC, (5) intensity of soil reduction and accumulation of reduced cations, (6) bioturbation at the soil–floodwater interface, and (7) soil porosity.

Soil pore water ammonium profiles in several wetlands and aquatic systems are shown in Figures 8.55 and 8.56. In all these ecosystems, ammonium concentrations in soil pore water are higher in subsurface layers and lower in surface layers. Steep concentration gradients are noted between the soil and overlying water column. These data suggest that ammonium removal from these soil layers was due to upward flux and nitrification and ammonia volatilization in the aerobic soil layer and water column. Ammonium consumption rates in surface layers can increase overall flux from anaerobic soil layers. In mineral wetland soils, reduced Fe and Mn occupy most of the cation exchange complex, enabling ammonium ions in soil pore water, thus resulting in high flux rates. Ammonium flux rates

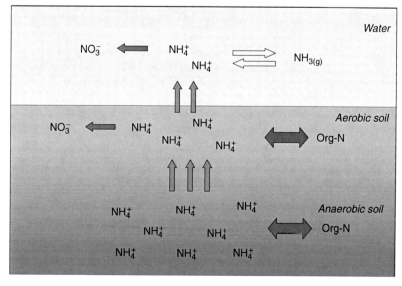

FIGURE 8.54 Schematic showing the flux of ammonium from anaerobic soil layer to aerobic soil layer and overlying water column, and nitrification of ammonium in aerobic soil layer.

Nitrogen

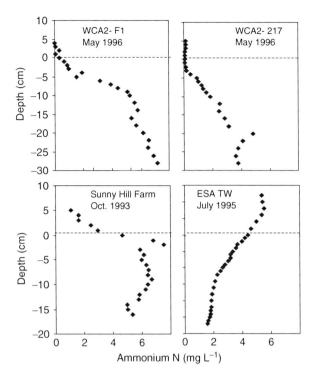

FIGURE 8.55 Pore water ammonium profiles of select wetlands in Florida. WCA-2-F1 = water conservation area-phosphorus-enriched site; WCA-2-217 = water conservation area-phosphorus-limited site; Sunny Hill Farm = agricultural land converted to wetland; ESA-TW = natural wetland used for wastewater treatment.

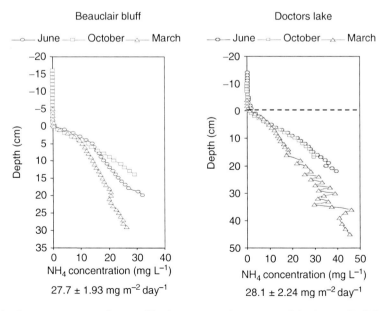

FIGURE 8.56 Pore water ammonium profiles in two aquatic systems of the Lower St. Johns River Basin (Malecki et al., 2004).

can be inversely related to soil CEC. Soils with low CEC (sandy soils) tend to maintain high levels of ammonium in soil pore water than soils with high CEC (clay loams or organic soils).

8.14.2 Nitrate Flux

Nitrate in the floodwater and aerobic soil layer is derived from (1) nitrification of ammonium in these zones and (2) external loading to the system. In wetlands with low external nutrient loading, nitrification is the main source of nitrate in the aerobic soil and water column (Figure 8.57). Nitrate present in these layers can potentially be assimilated by periphyton and vegetation and readily diffused into anaerobic soil layers where it can be used as electron acceptor by facultative bacteria. Unlike ammonium, nitrate being an anion is highly mobile in the soil profile. Nitrate nitrogen present in the water column readily diffuses into the anaerobic layer in response to sharp concentration gradients (Figure 8.58). Under most conditions, very little or no denitrification occurs in the water column (Figure 8.59).

Nitrate flux from the aerobic portion of the soil is controlled by (1) labile organic carbon supply in anaerobic portion of the soil, (2) thickness of aerobic soil layer, (3) water column depth, (4) mixing and aeration in the water column, (5) nitrate concentration, and (6) temperature. The flux of nitrate from the floodwater to underlying soil increases with an increase in temperature (Figure 8.59). At low temperatures, nitrate can diffuse to deeper layers into anaerobic zones. Under these conditions, it is likely that nitrate may play significant role in ANAMOX reactions as temperature optima for this reaction is between 10 and 15°C, as compared to denitrification that has temperature optima around 30°C (see Figures 8.39 and 8.47).

The importance of coupled nitrification–denitrification in wetlands and rice paddies has been recognized by several researchers around the world. These reactions coupled with other nitrogen loss mechanisms such as ammonia volatilization and anammox control overall nitrogen loss from wetlands and aquatic systems. Ecologically these reactions maintain the nitrogen balance in the system and regulate excessive accumulation of nitrogen, especially in wetlands receiving large external inputs. In paddy soils, these processes are not desirable because they tend to decrease overall nitrogen availability to plants. Hydroperiod and water-table fluctuations can have major influence on nitrification–denitrification reactions. For example, during high water table and anaerobic conditions, soils accumulate ammonium resulting from mineralization of organic matter. During a low water table and drained soil conditions, ammonium is rapidly oxidized to nitrate. Subsequent flooding

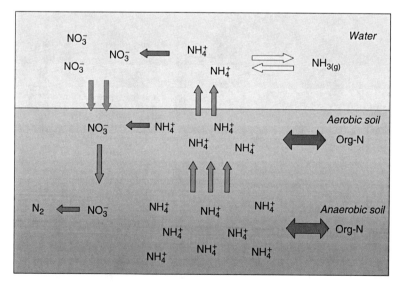

FIGURE 8.57 Schematic showing the flux of nitrate nitrogen from aerobic soil layer to anaerobic soil layer and overlying water column, and denitrification of nitrate in anaerobic soil layer.

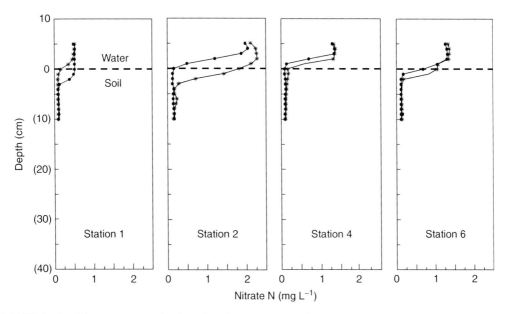

FIGURE 8.58 Nitrate concentration in soil and water column of a wetland.

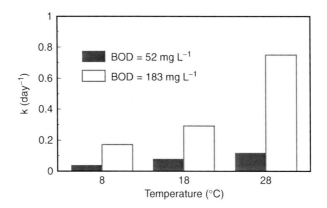

FIGURE 8.59 Influence of temperature on nitrate removal in wetlands.

of these soils can result in the reduction of nitrate via catabolic processes. These alternate flooding and drained soil conditions can result in substantial loss of nitrogen in wetlands. Similarly, wetland plants create aerobic conditions in the root zone, which provide environment for nitrification. Nitrate formed in the root zone can diffuse into adjacent anaerobic zones and be reduced to nitrogen gas.

The significance of wetlands as a source of nitrous oxide is not well established. As discussed in previous chapters (Chapters 3–5), wetlands are limited by electron acceptors. Thus, nitrate when used as an electron acceptor undergoes reduction to nitrite, nitrous oxide, and then nitrogen gas. Under most wetland conditions, nitrate is reduced to nitrogen gas with very little or no accumulation of nitrous oxide. However, in wetlands recently flooded after long dry period or with high nitrate loading, there is likelihood of producing significant amounts of nitrous oxide. The amount of nitrous oxide evolution during denitrification varies greatly with nitrogen-to-nitrous oxide ratio greater than 100:1 to less than 10:1. Wider ratios are observed in soils with high carbon content or greater demand for electron acceptors, whereas lower ratios are observed in soils with low carbon and high nitrate levels, and demand for electron acceptors is not as high.

Wetlands provide ideal conditions for nitrogen processing through various biogeochemical reactions (Figure 8.3). In natural wetlands with low external nitrogen loading, internal cycling mediated by mineralization, nitrification, and nitrate reduction couples with exchange processes between aerobic and anaerobic soil layers.

To evaluate the overall nitrogen loss from wetlands and aquatic systems through biogeochemical processes, it is important to determine which of the processes are limiting the overall loss mechanisms. For nitrification–denitrification, the processes that regulate overall nitrogen loss are (1) mineralization of organic nitrogen, (2) flux of ammonium from anaerobic zone to aerobic zones, (3) nitrification of ammonium in the aerobic zone (which is controlled by oxygen flux from water column to soil), (4) nitrate flux from aerobic zone to anaerobic zones, and (5) nitrate reduction in the anaerobic zones. Several studies have indicated that in most wetlands, mineralization of organic nitrogen, nitrate flux from aerobic to anaerobic zones, and nitrate reduction in anaerobic zones are not the limiting processes regulating overall nitrogen loss. This loss is controlled to a large extent by ammonium flux from anaerobic zones to aerobic zones and nitrification in aerobic zones.

8.15 SUMMARY

- All living organisms require nitrogen for the basic metabolic processes.
- Much of the N is associated with C, P, and S in the living matter; hence, the nitrogen cycle plays a critical role in the biosphere.
- Point and nonpoint sources, precipitation, and biological N fixation are the major inputs to wetlands, whereas N losses from wetlands include biotic (denitrification) and abiotic (volatilization) reactions and surface/subsurface flows. Wetlands often function as effective reservoirs of N.
- Many of the nitrogen cycle processes are governed by oxidation–reduction reactions.
- Nitrogen is usually the limiting nutrient in wetlands. Organic nitrogen is the major storage pool in wetlands.
- The bioavailability of nitrogen in a wetland is influenced by temperature, hydrologic fluctuations, water depth, electron acceptors availability, and microbial activity.
- The carbon-to-nitrogen ratio, enzyme activity, and potentially mineralizable nitrogen (PMP) are useful indicators that can be used to determine the soil's capacity to release ammonium N.
- Coupled processes of nitrification and denitrification occur simultaneously in most wetlands.
- The ammonium flux from the anaerobic zone (where NH_4^+ is in high concentrations) to aerobic zone (where the concentration is low) is regulated by many factors: ammonium concentration gradient, ammonium production rate, CEC, adsorption–desorption, intensity of soil reduction, temperature, bioturbation and mixing, and ammonium consumption rate.
- The nitrate flux from the aerobic zone (where the concentration is high) to anaerobic zone (where the concentration is low and the demand for electron acceptors is high) is regulated by many factors: nitrate concentration gradient, nitrification rate, reduction or denitrification rate, and bioturbation and mixing.

STUDY QUESTIONS

1. Draw a nitrogen cycle and list all pathways regulating nitrogen transformations and fluxes in a wetland soil. Show which pathways involve heterotrophic and chemoautotrophic organisms. Indicate how many electrons are involved per mole of oxidant or reductant and by using arrows show the direction of each pathway. Note that some pathways may go in both directions. Explain the factors regulating diffusive flux of ammonium from soil to overlying water column. What process controls overall nitrogen loss from wetland ecosystem?

2. A constructed wetland is receiving ammonium nitrogen loading at a rate of 10 g m^{-2} day^{-1}. Assume all of this ammonium is in floodwater and will be available for nitrification. Calculate how much oxygen should be present in the floodwater to oxidize ammonium to nitrate. Express oxygen requirements in g m^{-2} day^{-1}.

3. You are asked to design a constructed wetland for nitrate nitrogen removal. Your system is limited by the availability of carbon. You know from this course on nitrogen transformations that you can use glucose ($C_6H_{12}O_6$) as an energy source to promote NO_3^- removal through denitrification. Calculate how much glucose carbon is needed to oxidize 10 g NO_3-N m^{-2}.

4. Calculate diffusive flux of ammonium nitrogen from soil to the overlying water column using Fick's first law. The following measurements are available for your use: soil porosity = 0.8 cm^3 cm^{-3}; diffusion coefficient = 17.1 cm^2 day^{-1}; ammonium N concentration at the soil–floodwater interface = 0.2 mg L^{-1} (μg cm^{-3}); ammonium nitrogen concentration at the soil depth of 5 cm = 5.0 mg L^{-1} (μg cm^{-3}); assume overlying water column is well mixed with gradient. Also, assume linear gradient between soil depth of 0–5 cm.

5. Several chemotrophs can use select reductants (such as ferrous iron and hydrogen sulfide) as their energy source, while using nitrate as their electron acceptor. Determine the moles of nitrate needed to oxidize one mole of ferrous iron and hydrogen sulfide. Write appropriate balanced reactions for oxidation of each of these reductants using nitrate.

6. Suppose you have applied 286 kg of ammonium nitrate per hectare to the water column (containing algae) of a wetland soil containing no macrophytes. Discuss which nitrogen reactions are involved in removing added nitrogen from the system. Write appropriate balanced equations. How much of the added ammonium nitrate still remains in the system if 50% of the added ammonium nitrogen and 90% of the added nitrate nitrogen are lost from the system?

FURTHER READINGS

Dalsgaard, T., B. Thamdrup, and D. E. Canfield. 2005. Anaerobic ammonium oxidation (anammox) in the marine environment. *Res. Microbiol.* 156:457–464.

Galloway, J. N., F. J. Dentener, D. G. Capone, E. W. Boyer, R. W. Howarth, S. P. Seitzinger, G. P. Asner, C. C. Cleveland, P. A. Green, E. A. Holland, D. M. Karl, A. F. Michaels, J. H. Porter, A. R. Townsend, and C. J. Vorosmarty. 2004. Nitrogen cycles: past, present, and future. *Biogeochemistry.* 70:153–226.

Groffman, P. M., M. A. Altabet, J. K. Bohlke, K. Butterbach-Bahl, M. B. David, M. K. Firestone, A. E. Giblin, T. M. Kana, L. P. Nielsen, and M. A. Voytek. 2006. Methods for measuring denitrification: diverse approaches to a difficult problem. *Ecol. Appl.* 16:2091–2122.

Gusewell, S. and W. Koerselman. 2002. Variation in nitrogen and phosphorus concentrations of wetland plants. *Perspect. Plant Ecol. Evol. Systemat.* 5:37–61.

Herbert, R. A. 1999. Nitrogen cycling in coastal marine sediments. *FEMS Microbiol. Rev.* 23:563–590.

Howarth, R. W., R. Marino, J. Lane, J. J. Cole. 1988. Nitrogen fixation in freshwater, estuarine, and marine ecosystems, 1: rates and importance. *Limnol. Oceanogr.* 33:669–687.

Schulten, H. R. and M. Schnitzer. 1998. The chemistry of soil organic nitrogen: a review. *Biol. Fertil. Soils* 26:1–15.

Seitzinger, S. P. 1988. Denitrification in freshwater and coastal marine ecosystems: ecological and geochemical significance. *Limnol. Oceanogr.* 33:702–724.

Seitzinger, S., J. A. Harrison, J. K. Bohlke, A. F. Bouwman, R. Lowrance, B. Peterson, C. Tobias, and G. Van Drecht. 2006. Denitrification across landscapes and waterscapes: a synthesis. *Ecol. Appl.* 16:2064–2090.

Stevenson, F. J. 1994. *Humus Chemistry.* 2nd Edition. Wiley, New York. 496 pp.

Zehr, J. P. and B. B. Ward. 2002. Nitrogen cycling in the ocean: new perspectives on processes and paradigms. *Appl. Environ. Microbiol.* 68:1015–1024.

9 Phosphorus

9.1 INTRODUCTION

Phosphorus (atomic weight 30.974) is the 12th most abundant element in the lithosphere. Phosphorus concentration in the earth's crust is reported to be approximately 0.1%. Under natural conditions, weathering of minerals releases phosphorus into the environment. The phosphorus cycle is dynamic and involves interaction or exchange between biotic and abiotic pools. Phosphorus inputs through atmospheric deposition and weathering of natural minerals can maintain wetlands and aquatic systems enriched with phosphorus and can significantly impact trophic conditions. On any given landscape, phosphorus transfer is typically from uplands to wetlands, and then to the aquatic environment (Figure 9.1). Phosphorus loads from uplands to many aquatic systems rapidly increased during the industrial and green revolution as a result of heavy fertilizer use and the demand to produce more food to meet the demand of the population explosion. Converting wetlands to agricultural and urban lands decreased the capacity of existing wetlands to retain phosphorus. This has resulted in phosphorus enrichment of many lakes, rivers, estuaries, and coastal waters. At the landscape level, wetlands can function as buffers for phosphorus retention between uplands and adjacent aquatic systems such as lakes, streams, and estuaries. Thus, it is important to understand the transfer of phosphorus from uplands to wetlands, and the biogeochemical processes regulating its availability and retention in wetlands.

Phosphorus can be a major limiting nutrient in many freshwater aquatic ecosystems such as lakes and streams. During the last five decades, many studies have been conducted to determine the fate of phosphorus in aquatic ecosystems. At the landscape level, streams and wetlands form a critical interface between uplands and adjacent water bodies, as all of these ecosystems are hydrologically linked. Water and associated contaminants (such as phosphorus) are transported from uplands by either surface or subsurface flow. Surface flow can include first-, second-, and third-order streams as well as the associated riparian floodplains, marshes, and swamps. In low-gradient systems, streams are largely composed of interconnected marshes and swamps. To improve drainage in many agricultural areas, ditches are often cut to lower regional water tables for increased agronomic production; ditches often connect wetlands that otherwise would have been isolated from surface flows. The resulting flow of water in such areas follows a complex path through wetlands, ditches, and streams. Thus, phosphorus loading to the receiving aquatic system depends on the retention capacity of these landscape components of the basin.

In the biosphere, soils and sediments are the major reservoirs for phosphorus. Biota on the land contains more phosphorus than the biota in marine systems. Atmospheric deposition can range from 20 to 80 mg P m^{-2} $year^{-1}$, and in some oligotrophic ecosystems, this source can be significant in regulating trophic status. Global flux of phosphorus from rivers to the sea is about 21×10^{12} g P $year^{-1}$ with approximately 10% of this being bioavailable to marine biota, and the remaining associated with sediment particles is deposited (Schlesinger, 1997) (Table 9.1).

Many wetlands are also impacted by phosphorus loading and are considered eutrophic. Phosphorus is not a limiting factor in many wetlands, although there are a few exceptions. Several oligotrophic wetlands exist in the biosphere. A classic example is the Florida Everglades (see Chapter 17 for details). In comparison to wetlands, many freshwater aquatic systems such as lakes and streams are phosphorus limited; thus, phosphorus availability in these systems regulates primary production. The significance of phosphorus limitation is typically documented by a strong correlation

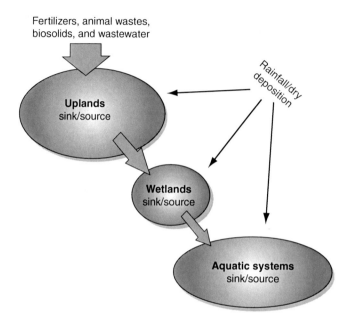

FIGURE 9.1 Schematic showing phosphorus loads and transport in a watershed.

TABLE 9.1
Major Reservoirs of Phosphorus in the Biosphere

Reservoir	Total Phosphorus × 10^{12} kg
Land	
Soils	96–160
Minable rock	19
Biota	2.6
Freshwater (dissolved)	0.09
Ocean	
Sediments	840,000
Dissolved (inorganic)	80
Detritus (particulates)	0.65
Biota	0.05–0.12

Source: Stevenson (1986).

between water column total phosphorus and chlorophyll a concentrations. This leads to setting up of boundaries for trophic status of water bodies using total phosphorus concentrations of the water column, as described below:

- Oligotrophic systems: <10 µg L^{-1}
- Mesotrophic systems: 10–35 µg L^{-1}
- Eutrophic systems: 35–100 µg L^{-1}
- Hypereutrophic systems: >100 µg L^{-1}

In these phosphorus-limited wetlands and aquatic systems, phosphorus availability is tightly controlled and cycled to maintain ecosystem stability. Phosphorus requirements of biotic communities are usually much lower than that of nitrogen (typical atomic N:P ratio of 16:1), as compared to available nitrogen in these systems. Thus, many biotic communities can assimilate phosphorus beyond their needs through a process called "luxury uptake" and store phosphorus in their tissues as polyphosphates. Depending on phosphorus accumulation, wetlands can function as both a source and sink for phosphorus. For example, eutrophic marshes can be the sources of phosphorus to adjacent estuaries and coastal waters.

Typically, phosphorus is added in various forms to a watershed (Figure 9.1). These include fertilizers, nonhazardous wastes (animal manures and biosolids), and nutrient-enriched waters. Historically, organic wastes such as animal manures were applied to agronomic crops and pastures on the basis of the nitrogen availability, which has resulted in excessive application of phosphorus. As a result, many uplands used for land application of wastes have accumulated phosphorus in excess amounts. A major portion of the phosphorus added to uplands is retained within the soil. However, as upland soils become saturated or overloaded with phosphorus, a significant portion of the stored phosphorus can be released and transported with water during runoff events.

One example of phosphorus transfer in a watershed dominated by agriculture is the Lake Okeechobee drainage basin of south Florida. Major land uses in this basin are pasture, dairy, citrus, and wetlands. Although the dairy industry constitutes a small portion of the overall land use in the basin, it is a major contributor of phosphorus (approximately 50% of the load) to the lake. To reduce phosphorus loads, the state of Florida bought several dairies (Dairy Rule) and provided resources to implement best management practices to improve the nutrient management at the remaining dairies. Approximately 82% of the phosphorus added to the Okeechobee drainage basin remains in upland soils (Figure 9.2). Eight percent of the phosphorus is transferred into the wetland environment, and the remaining 10% is transferred through the wetlands downstream into the lake (per year). The buyout of several dairies was not very effective in reducing inflow phosphorus concentrations into wetlands and ultimately to Lake Okeechobee. Further, due to historical phosphorus loading of soils in the landscape, phosphorus retained in soils was subsequently released into surface water. Under these conditions, wetlands may not function as effective sinks for phosphorus, as the phosphorus gradient is from the soil to the water column. The possible explanation for the ineffectiveness of these

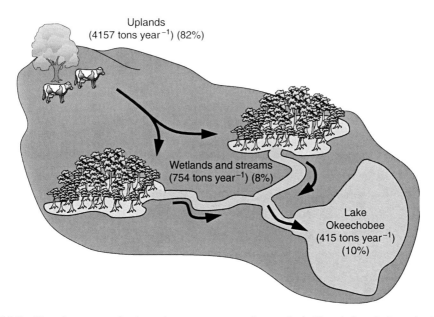

FIGURE 9.2 Phosphorus retention in various components of watershed: Okeechobee drainage basin as case example.

wetlands could be any of the the preceding factors but most likely it is caused by a saturated system due to years of phosphorus loading. This residual phosphorus is termed as "phosphorus memory" in the system, which may last a few years to several decades (a detailed discussion is presented later in this chapter). Similar situations can exist in other watersheds with intensive land use practices.

9.2 PHOSPHORUS ACCUMULATION IN WETLANDS

9.2.1 WHY DOES PHOSPHORUS ADDED TO WETLANDS ACCUMULATE IN SOILS?

Phosphorus accumulation in wetlands is regulated by vegetation, periphyton and plankton, plant litter and detrital accumulation, soil physicochemical properties, water flow velocity, water depth, hydraulic retention time, length-to-width ratio of the wetland, phosphorus loading, and hydrologic fluctuations. In evaluating wetlands for phosphorus accumulation, it is necessary to consider the following:

- Short-term storage mediated by assimilation into vegetation and periphyton and incorporation into detrital tissue
- Long-term storage mediated by soil assimilation, and accretion of organic and mineral matter

Short-term storage and associated phosphorus dynamics significantly impact long-term storage and discharge water quality.

Unlike lakes and reservoirs, wetlands can function under steady state as "plug flow" system (Figure 9.3). Water added to wetlands is not well mixed, resulting in typically high concentrations near inflow points, which decrease with distance from the source. Natural or constructed wetlands receiving inflows of water containing phosphorus all exhibit similar gradients. The concentration of phosphorus is highest at the inflow point and is diluted as the phosphorus is carried in a plug flow fashion through the wetland. Eventually, the phosphorus reaches the outflow as all wetlands have a finite capacity for accumulation. The reason for this finite capacity is that there is no gaseous escape from the system; so all phosphorus brought into the system must be transformed or retained by the system in some fashion. The distribution of phosphorus in soils is mainly controlled by hydrologic factors, substrate and soil composition, and redox conditions.

Following are the examples of wetlands, where long-term phosphorus loading has resulted in soil phosphorus gradients:

- Everglades Water Conservation Area 2A (WCA-2A) in the Everglades receiving agricultural drainage effluents
- Houghton Lake wetland in Michigan receiving secondarily treated sewage effluents
- Orlando Easterly wetland receiving secondarily treated sewage effluents

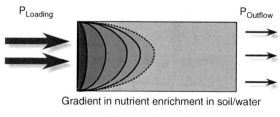

FIGURE 9.3 Soil and water column phosphorus gradients resulting from point sources of external loading to a wetland.

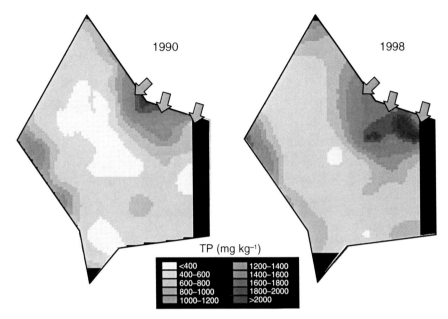

FIGURE 9.4 Soil phosphorus enrichment in the northern Everglades resulting from point sources of external loading to a wetland. (From DeBusk et al., 2004.)

A classic example of phosphorus accumulation in wetland systems can be seen in the WCA-2A (see Figure 9.4). Surface water inflow to WCA-2A occurs through control structures located on the Hillsboro Canal. Phosphorus accumulation in this wetland represents almost 40 years of loading from the adjacent agricultural area. High total phosphorus concentrations are in areas impacted by long-term phosphorus loading, and low total phosphorus concentrations are from soils collected from unimpacted sites of the Everglades. The dark area shown in Figure 9.4 represents phosphorus concentrations in excess of 2,000 mg kg^{-1}. These areas correlate to the water column phosphorus concentration and tissue phosphorus concentration in the vegetation. If loading rates continue at current levels, the wetland will eventually accumulate the maximum amount of phosphorus possible and the concentration of phosphorus in the outflow will begin to increase. This can be avoided or controlled by proper management of the area, e.g., decreasing inflow phosphorus concentrations by reducing external phosphorus loads.

As discussed earlier, most of the phosphorus entering wetlands accumulates within the system. Surface soils in nutrient-impacted wetlands are often enriched as a result of recent accumulation, decomposition processes, and remobilization of phosphorus from subsurface soils to surface through plant uptake and deposition as detritus material. Thus, total phosphorus content of surface soils is higher than that of subsurface soils. Similar total phosphorus profiles have been seen for many wetlands and aquatic systems. In the impacted site, subsurface total phosphorus content can also represent the background levels of phosphorus for these soils, assuming that the surface material is the result of recent accumulation. Much of this phosphorus accumulation is due to organic matter accretion (detrital matter deposition) associated with phosphorus sorption to particulate matter.

A long-term database on a northern peat wetland used for wastewater treatment indicates the reliability of wetlands in removing 97% of the added phosphorus consistently over a 15-year study period (Kadlec, 1993). In this wetland, peat accreted at 2–3 mm year^{-1} during the 15 years of operation (Kadlec, 1993). Sedimentation (mineral and organic sediments derived from external sources) and peat accretion (recalcitrant organic matter derived from aquatic biota) result in long-term phosphorus storage in wetlands. Johnston (1991) summarized the annual phosphorus accumulation rates in selected wetlands with mineral and organic soils. Wetlands with mineral

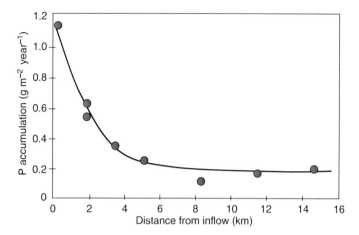

FIGURE 9.5 Long-term phosphorus accumulation rates in the northern Everglades. (From Reddy et al., 1993.)

soils accumulated 0.1–8.2 g P m^{-2} year^{-1} (average = 1.46 g m^{-2} year^{-1}), as compared to 0.04–1.1 g P m^{-2} year^{-1} (average = 0.26 g P m^{-2} year^{-1}) in wetlands with organic soils (Johnston, 1991). In the Everglades WCA-2A, Craft and Richardson (1993) and Reddy et al. (1993) measured phosphorus accumulation rates in the range of 0.04–1.12 g P m^{-2} year^{-1} (Figure 9.5). After a decade of sewage effluent loading to a wetland, Cooke (1992) reported phosphorus retention of approximately 30 g P m^{-2} year^{-1}.

Although wetlands accumulate large quantities of phosphorus, the outflow phosphorus concentration of wetland increases with phosphorus loading. The phosphorus concentration of surface water may still remain high even after external phosphorus loads are curtailed. The time required for these systems to recover depends on the amount of phosphorus stored in the system and its bioavailability. This will be discussed later in this chapter.

9.3 PHOSPHORUS FORMS IN WATER COLUMN AND SOIL

The speciation of phosphate ions in solution is dependent on pH and is characterized by three equilibrium constants:

$$[H_3PO_4] \leftrightarrow [H_2PO_4^-] + [H^+] \quad \log K^0 = -2.15 \text{ (primary ionization)} \tag{9.1}$$

$$[H_2PO_4^-] \leftrightarrow [HPO_4^{2-}] + [H^+] \quad \log K^0 = -7.20 \text{ (secondary ionization)} \tag{9.2}$$

$$[HPO_4^{2-}] \leftrightarrow [PO_4^{3-}] + [H^+] \quad \log K^0 = -12.35 \text{ (tertiary ionization)} \tag{9.3}$$

Under acid soil conditions, the dominant phosphate species is orthophosphoric acid (H_3PO_4), which is a weak acid, colorless, and freely soluble in water. The dominant species under alkaline soil conditions is PO_4^{3-}. Under most natural conditions, dominant phosphate species are $H_2PO_4^-$ and HPO_4^-. The mole fraction of each phosphate species varies with pH as shown in Figure 9.6. For example, at the first intersection point at pH 2.2, both $[H_3PO_4]$ and $[H_2PO_4^-]$ are present in equal amounts or at mole fraction of 0.5, or 50% of total phosphate. Similarly, at pH 7.2, there are approximately equal amounts of $H_2PO_4^-$ and HPO_4^{2-}. At extremely low pH, H_3PO_4 predominates. In the pH range of most soils (4–6.5), $H_2PO_4^-$ is the dominant form of orthophosphate. HPO_4^{2-} is also present in alkaline soils. The relative deprotonation and protonation determine their reactivity as inorganic ligands

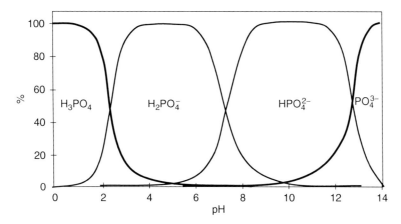

FIGURE 9.6 Relative distribution of phosphate species as a function of pH.

or ions pairs, particularly with iron and aluminum under acidic conditions and with calcium and magnesium under alkaline conditions.

Oxidation state of phosphorus in the environment: Phosphorus is known to occur in several oxidation states including phosphides (−3), diphosphides, elemental phosphorus (0), hypophosphite (+1), phosphite (+3), and phosphate (+5). In biogeochemical systems such as wetlands, phosphate is the most common form (Morton and Edwards, 2005). Oxidation state of P in all phosphate ions (H_3PO_4, $H_2PO_4^-$, HPO_4^{2-}; and PO_4^{3-}) is +5. Under most conditions, phosphate reactions do not involve the transfer of electrons; therefore, the redox potential does not directly affect the inorganic phosphorus speciation in most ecosystems. However, in some highly reduced (systems with high electron pressure) acidic environments, phosphate can be reduced to phosphine gas, PH_3. Phosphine is analogous to ammonia in structure and formula. The formation of PH_3 constitutes the only redox reaction that directly involves inorganic phosphorus. The significance of this process in wetlands is not well documented. Ideal conditions for reduction of phosphate to phosphine are

- High electron pressure or highly reduced conditions or extremely low redox potentials
- Low pH or acidic conditions (pH < 3)
- Reduction of phosphate to PH_3, which proceeds through several steps of extremely low redox potential
- HPO_4^{2-}/HPO_3^{2-}, −690 mV
- $HPO_3^{2-}/H_2PO_2^-$, −910 mV
- $H_2PO_2^-/P$, −922 mV
- P/PH_3, −525 mV

Phosphine is not stable under aerobic environments. The aerobic soil layer can function as the sink for PH_3 through abiotic oxidation $PH_3 + 2O_2 = H_3PO_4$. Trace amounts of PH_3 have been detected in wetlands, paddy fields, and sewage treatment systems (Dévai et al., 1988; Dévai and DeLaune, 1995; Morton and Edwards, 2005).

Phosphorus solubility is also directly affected by the changes in redox potential (Patrick, 1964). In well-drained mineral soils, some of the inorganic P is bound to oxidized forms of iron such as iron oxyhydroxides. Ferric phosphate is a common phosphate compound in these systems. Similarly, in oxidized portions of mineral wetland soils, some of the inorganic phosphorus is also present as ferric phosphate. The mineral form of $FePO_4$ is called strengite.

We know that under anaerobic soil conditions, oxidized forms of Fe function as electron acceptor and are reduced to ferrous iron.

$$Fe^{3+} + e^- = Fe^{2+} \tag{9.4}$$

$$FePO_4 + H^+ + e^- = Fe^{2+} + PO_4^{3-} \tag{9.5}$$

As Fe^{3+} in $FePO_4$ is reduced to Fe^{2+}, phosphate is released. These reactions will be discussed in detail later in this chapter.

9.3.1 WATER COLUMN

Phosphorus entering a wetland water column is typically present in both organic and inorganic forms. The relative proportion of each form in wetlands depends on the soil, vegetation, and land use characteristics of the drainage basin. Accurate speciation of phosphorus forms is often difficult with the currently available analytical methodologies. To trace the transport and transformations of phosphorus within wetlands, it is convenient to classify forms of phosphorus entering into these systems as (i) dissolved inorganic phosphorus (DIP), (ii) dissolved organic phosphorus (DOP), (iii) particulate inorganic phosphorus (PIP), and (iv) particulate organic phosphorus (POP) (Figure 9.7). The particulate and soluble organic fractions may be further divided into labile and refractory components. DIP is considered bioavailable, whereas organic and PP forms generally must be transformed to inorganic forms before being considered bioavailable.

Routine water quality measurements typically include the determinations of several forms of phosphorus including dissolved/soluble reactive phosphorus (DRP or SRP, respectively), dissolved/soluble organic P (DOP or SOP), and suspended PP. The DRP or SRP, by definition, includes only the DIP in the form of orthophosphate ($H_2PO_4^-$, HPO_4^{2-}, or PO_4^{3-}), depending on the pH (see Figure 9.8). Orthophosphate is the phosphate form most readily available for biological uptake. Specific activity of phosphate ions can be measured by ion-specific electrodes or ion chromatography.

The first step in the determination of DRP is to filter water samples through 0.45 μm membrane filters to remove particulates (APHA, 1999) (Figure 9.8). A solution of ammonium molybdate, antimony potassium tartrate, and sulfuric acid is then added to the filtered water sample; this results in

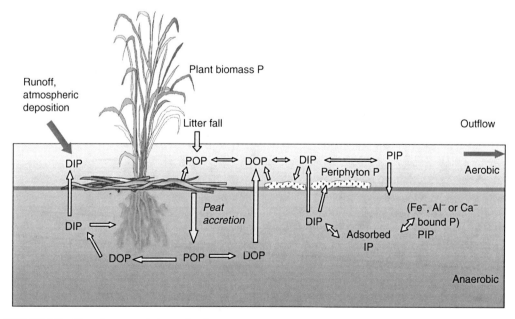

FIGURE 9.7 Phosphorus cycle in wetlands.

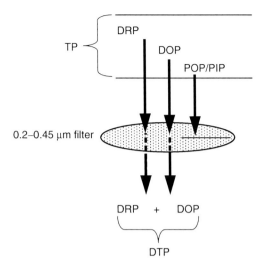

FIGURE 9.8 Schematic showing separation of dissolved reactive phosphorus and particulate phosphorus in water samples.

the formation of an antimony-phospho-molybdate complex. This reaction is specific to orthophosphate. The addition of ascorbic acid to the antimony-phospho-molybdate complex produces an intense blue color, which is proportional to the phosphate concentration. The concentration of orthophosphate is then determined spectrophotometrically or on an automated system using an autoanalyzer.

The results of this analysis are usually reported in an "operationally defined" context because minor quantities of labile dissolved organic P may hydrolyze due to the addition of sulfuric acid and be measured as orthophosphate. For details of this method, see EPA (1990) or APHA (2005).

The DOP fraction is poorly characterized and considered to be diverse with respect to size and structure, ranging from simple organic phosphorus compounds such as sugar phosphates to macromolecules such as phospholipids. The bioavailability and lability of DOP may also differ significantly with respect to bioavailability to aquatic microorganisms and ease of hydrolysis. The first step in the determination of DOP is the filtration of water samples through 0.45 μm membrane filters. The filtered solutions are then digested using potassium persulfate digestion, followed by the determination of the released orthophosphate using the "molybdenum blue" procedure, as described above. Because the phosphorus measured after digestion includes P that was in the form of orthophosphate prior to the digestion procedure, DRP is subtracted from this value. DOP is then calculated as follows:

$$DOP = DTP - DRP \quad (9.6)$$

where DTP is the dissolved total phosphorus of the filtered solutions.

PP is the total phosphorus associated with particulate matter. Total phosphorus in the particulate matter is determined by either wet ashing or perchloric acid digestion methods. PP often includes inorganic forms such as phosphate sorbed onto suspended clay particles, and suspended crystalline and amorphous precipitates of PO_4^{3-} with Ca, Mg, Al, and Fe. Inorganic phosphorus associated with PP is termed as PIP. Particulate matter is extracted with 1 M HCl and filtered solutions are analyzed for orthophosphate as described above.

Organic PP is associated with detrital matter from dead and decomposing bacteria, phytoplankton, zooplankton cells, and periphyton, as well as from vascular plants, and organic phosphorus associated with particulate matter is termed as POP. POP is estimated as follows:

$$POP = TPP - PIP \quad (9.7)$$

where TPP is the total phosphorus in the particulate matter.

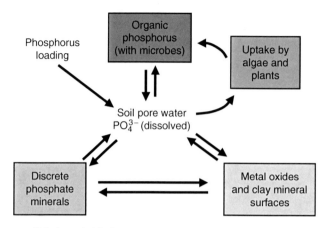

FIGURE 9.9 Influence of biotic and abiotic processes and external loading on soil pore water phosphorus.

9.3.2 SOIL

Phosphorus in wetland soils exists in both organic and inorganic forms. Phosphorus storage in various components of wetlands is shown in Figure 9.9. The relative proportion of each form depends on the nature and origin of these materials. Total phosphorus in soils is determined by either ashing or perchloric acid digestion. Total phosphorus content varies, ranging from 30 to 500 mg P kg^{-1} in wetlands not impacted by anthropogenic phosphorus loading. In impacted wetland areas such as those in watersheds receiving phosphorus from animal operations or other agricultural operations, total phosphorus concentrations of surface soils can exceed 10,000 mg P kg^{-1}.

Inorganic phosphorus compounds are associated with amorphous and crystalline forms of iron, aluminum, calcium, magnesium, and other elements. Soil inorganic phosphorus is determined after extraction of air-dried soil with 1 M HCl at soil to solution ratio of 1:50. After 3 h equilibration under continuous shaking, soil suspensions are centrifuged and supernatant liquid filtered through a 0.45 μm membrane filter. Filtered solutions are analyzed for DRP. For example, Figure 9.10 shows the amount of inorganic phosphorus relative to total phosphorus in several wetland soils of south Florida. Soils in this area are organic soils (histosols). The range in total phosphorus is approximately from 100 to 3,000 mg kg^{-1} of soil, as compared to 10–1,500 mg kg^{-1} of soil for inorganic phosphorus. In these organic soils, inorganic phosphorus accounts for 20–50% of total phosphorus. In mineral wetland soils, inorganic phosphorus content can vary from 50 to 80% of the total phosphorus. Inorganic phosphorus is primarily associated with amorphous and crystalline forms of minerals of iron, aluminum, calcium, and magnesium. Both these forms are acid soluble and thus can be readily extracted with acids (such 1 M HCl) and the filtrates analyzed for phosphorus.

Organic phosphorus forms are generally associated with living organisms and consist of easily decomposable phosphorus compounds (nucleic acids, phospholipids, and sugar phosphates) and slowly decomposable organic phosphorus compounds (inositol phosphates [IHPs], or phytin). Total organic phosphorus (TOP) is estimated as follows:

$$TOP = TP - TIP \tag{9.8}$$

The difference between total phosphorus and inorganic phosphorus is considered organic phosphorus. Organic phosphorus soils typically accounts for 50–80% of total phosphorus. This method of estimating organic phosphorus works well in organic soils. In some mineral soils, an acid extraction such as 1 M HCl may not extract all inorganic phosphorus, thus resulting in overestimation of organic phosphorus. In mineral wetland soils, organic phosphorus content can vary between 20 and 50% of the total phosphorus. See later parts of this chapter for details on characterization of organic phosphorus.

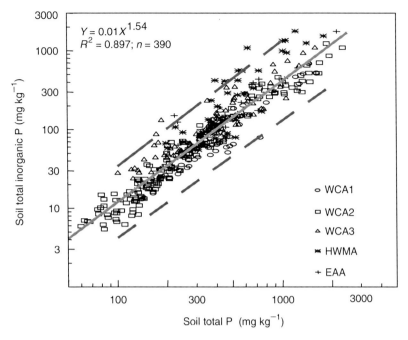

FIGURE 9.10 Relative proportion of inorganic phosphorus in wetlands dominated by organic soils. (From Reddy et al., 1998b.)

9.4 INORGANIC PHOSPHORUS

Inorganic phosphorus in soils is found in combination with aluminum, iron, calcium, and magnesium. Mineral forms of these compounds are often not detected in appreciable quantities in soils as measured by x-ray diffraction techniques. This is probably due to the presence of amorphous or poorly crystalline forms of these compounds in soils that are undergoing dynamic changes in hydrology and redox conditions. Of some 200-plus phosphate minerals identified, only a few are of significance in wetlands, including strengite ($FePO_4$), variscite ($AlPO_4$), vivianite ($Fe_3(PO_4)_2$), and hydroxyapatite ($Ca_5(PO_4)_3OH$). The stability of these minerals is discussed later in this chapter. Complexities in identifying these compounds have led many researchers to use operationally defined chemical fractionation schemes to identify inorganic phosphorus forms. These methods were developed on the basis of differential solubilities in various chemical extractions. The specificity of each chemical extractant was confirmed using pure compounds and minerals.

On the basis of relative bioavailability, inorganic phosphorus forms can be divided into

- Neutral salt (KCl or NaCl or NH_4Cl)–extractable phosphorus or exchangeable phosphorus
- Iron- and aluminum-bound phosphorus
- Calcium- and magnesium-bound phosphorus
- Residual nonreactive inorganic phosphorus

The early fractionation schemes (Chang and Jackson, 1957) developed for terrestrial agricultural soils grouped soil phosphorus into: (i) phosphorus present as orthophosphate ions sorbed onto the surface of P-retaining components (nonoccluded P), (ii) phosphorus present within the matrices of P-retaining components (occluded P), and (iii) phosphorus present in discrete phosphate minerals such as apatite.

The sequential extraction procedure by Chang and Jackson (1957) involved a series of extractants as follows:

- Soluble and exchangeable P extracted with natural salt such as NH_4Cl or KCl or $NaCl$
- Phosphorus bound to Al extracted by a solution of 0.5 M NH_4F (Al-P) at pH 7
- Iron phosphates extracted with 0.1 M NaOH (Fe-P)
- Calcium and magnesium bound extracted with 0.5 M HCl (Ca-Mg-P)
- The residual P extracted with hot nitroperchloric acid

The redox-sensitive P forms, such as occluded and reductant-soluble P, are no longer distinguished from the other forms of inorganic P (Hieltjes and Lijklema, 1980; van Eck, 1982). Several modified schemes have been used to determine the forms of phosphorus present in soils and sediments (Pettersson, 1986; Psenner and Pucsko, 1988; Ruttenberg, 1992; Reddy et al., 1998b; Reynolds and Davis, 2001). The phosphorus removed by the first one or two extracting solutions in a sequential fractionation method is usually considered bioavailable. Depending on the strength of the extracting solution and the fractionation method, bioavailable phosphorus has been estimated from such extracts as 1 M NH_4Cl or 1 M KCl, 0.1 M NaOH, citrate–bicarbonate–dithionite, and nitrilotriacetic acid (NTA). The Ca-bound P (HCl-RP) such as that found in apatite is generally considered to be unavailable, whereas the redox-sensitive Fe-bound P may become available under anaerobic conditions. Slightly modified sequential extraction schemes were used to determine phosphorus pools in marine sediments. This scheme identified five major phosphorus reservoirs of sediments: loosely sorbed or exchangeable P; ferric Fe-bound P; authigenic carbonate fluorapatite plus biogenic hydroxyapatite; and detrital apatite and other forms of inorganic phosphorus (Ruttenberg, 1992).

Modified simple sequential inorganic phosphorus fractionation schemes include (1) extraction of soil with neutral salt such as KCl, NaCl, or NH_4Cl, followed by (2) residual soil extraction with 0.1 M NaOH. The residual soil after alkali extraction is treated with 0.5 M HCl to dissolve Ca- and Mg-bound P (Figure 9.11a) (Reddy et al., 1998b). A few examples of inorganic phosphorus forms determined by chemical fractionation scheme follow.

Sediments obtained from streams in the Okeechobee drainage basin of south Florida were analyzed for select phosphorus forms. DL stream is located closer to a dairy farm, whereas Rucks stream is located in a pasture (Figure 9.11b). The DL stream received heavy phosphorus loading from the adjacent dairy farm, whereas the phosphorus loading to Rucks stream was very low. Phosphorus-loading effects are clearly seen in phosphorus enrichment in sediments. Soils in the drainage basin are spodosols; thus, stream sediments essentially consist of the mineral matter. The pH of DL stream sediments was around 7, whereas that of Rucks stream was around 5. The dominant forms of inorganic phosphorus in DL stream sediments were Ca- and Mg-bound P, followed by Fe- and Al-bound P. Labile phosphorus as indicated by KCl-Pi was in the range of 1–2% of total phosphorus. Dairy effluents discharged into DL stream also contained substantial amounts of calcium and magnesium, thus resulting in precipitation of phosphorus as Ca- and Mg-bound P. In Rucks stream sediments, the dominant forms of inorganic phosphorus were Fe- and Al-bound P. Calcium- and Mg-bound P accounted for 10% of total phosphorus. Low Ca and Mg and the acidic nature of these sediments resulted in very little phosphorus in this pool. These contrasting sediments show that chemical fractionation schemes can be used to differentiate phosphorus forms in soils and sediments.

In intertidal marsh soils, total inorganic phosphorus accounted for 14–40% of total phosphorus in freshwater marsh and 33–85% of total phosphorus in salt marshes. Inorganic phosphorus associated with Fe was most abundant in surface sediments of both freshwater and brackish marshes, whereas Ca-bound P dominated inorganic phosphorus in salt marshes (Paludan and Morris, 1999).

Drainage or fire in peat-based wetlands can result in alteration of soil phosphorus forms. Large areas of wetlands are located in the Everglades of south Florida. These wetlands are dominated

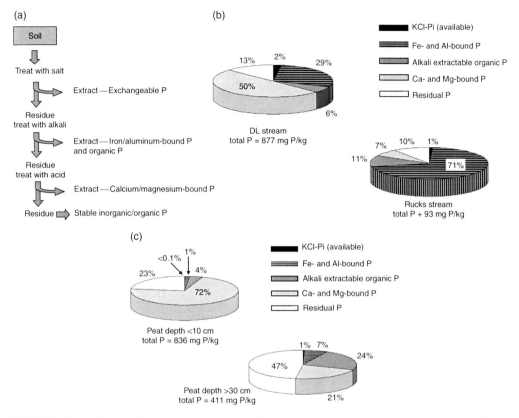

FIGURE 9.11 (a) Schematic showing chemical fractionation scheme used to determine inorganic phosphorus pools in soils. (b) Inorganic phosphorus pools in stream sediments of the Okeechobee drainage basin (from Reddy et al., 1995). (c) Inorganic phosphorus pools in organic soils that are subjected to overdrainage in the northern Everglades (Holeyland Wildlife Management). (From Reddy, K. R., Unpublished Results, University of Florida.)

by organic soils, and historically portions of this ecosystem have been drained. Some areas are also subjected to natural fire. As a result, in some areas, loss of soil organic matter through oxidation has increased the mineral content. Figure 9.11c shows relative proportion of soil phosphorus forms in surface soils (0–10 cm) sampled from two locations: (1) area with severe oxidation of organic matter with peat depths less than 10 cm and (2) area with very little or no oxidation with depths greater than 30 cm. Oxidation of organic matter resulted in conversion of organic forms to inorganic forms. The high calcium content of these soils and near-neutral pH conditions resulted in precipitation of inorganic forms as Ca- and Mg-bound P. In soils with peat depths of less than 10 cm, up to 72% of the total phosphorus was present as Ca- and Mg-bound P. This is in contrast with soils with peat depths greater than 30 cm, where only 21% of the total P was present as Ca- and Mg-bound P. Organic phosphorus content accounted for 27% of total phosphorus in soils with peat depths of less than 10 cm, as compared to 71% of total phosphorus in soils with peat depths greater than 30 cm. Fire can occur in wetlands undergoing long periods of drought. Natural fire often aids in release of organically bound nutrients. This is a major mechanism by which biotic communities in oligotrophic wetlands obtain nutrients. Similar to drainage, fire can influence the conversion of organically bound phosphorus into inorganic phosphorus forms. This is shown in soil samples obtained from Rotenberger Wildlife Management Area of the Everglades, before and after burning

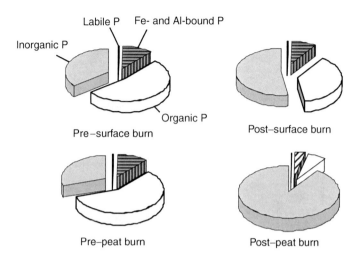

FIGURE 9.12 Proportions of inorganic and organic P fractions in soils (0–2-cm layer) from prefire surface burn, postfire surface burn, prefire peat burn, and postfire peat burn areas. (From Smith et al., 2001.)

(Figure 9.12). Oxidation of organic matter due to fire resulted in conversion of substantial amounts into inorganic phosphorus, specifically to Ca- and Mg-bound P. Inorganic P released during oxidation of organic matter rapidly precipitated and was recovered as Ca- and Mg-bound P.

Most inorganic phosphorus compounds in soils fall into one of two groups: those containing calcium and those containing iron and aluminum. The availability of phosphorus in alkaline soils is determined largely by the solubility of the calcium compounds in which phosphorus is found. In acid soils, solubility of inorganic phosphorus is controlled by iron and aluminum minerals. When soluble phosphates are added to soils, insoluble phosphates are formed with calcium, or iron and aluminum. With time, continuous interactions of soluble phosphorus with Fe and Al minerals (in acid soils) and Ca compounds (in alkaline soils) result in the formation of stable phosphate minerals. This results in a significant decrease in the number of bioavailable forms of phosphorus. The stability of these phosphate minerals is governed by phosphorus loading, pH, and redox potential. Inorganic phosphorus minerals found in acid and alkaline soils are shown in Figure 9.13.

Examples of "anthropogenic" phosphate minerals are shown in Figure 9.14 (Harris, 2002; University of Florida). Figure 9.14a shows patina of black, poorly crystalline apatite on the surface of an oyster shell located near the bottom of an early American mound, which once served as a refuse pile (about 5,000 YBP). Bones of fish and other animals were discarded in the pile and became the source of phosphorus that precipitated on the underlying shell. Note that oyster shells are composed of calcium carbonate, upon which phosphate readily adsorbs and precipitates (scale bar = 5 mm). Figure 9.14b shows vivianite as precipitate on an aggregate from stream sediment near Lake Okeechobee, Florida, collected a short distance downstream from where a dairy barn floor routinely flushed manure. Soils and stream sediments in the area are naturally low in phosphorus (scale bar = 1 mm).

The relative bioavailability of inorganic phosphorus in wetland soils (Figure 9.15) as determined by chemical fractionation scheme can be summarized as follows:

Readily available phosphorus: This form is present in soil pore water and the exchangeable pool. Phosphorus in this pool is continuously replenished from other stable pools at various rates, depending on the solubility of phosphate minerals and the physicochemical properties of soils. Inorganic phosphorus is extracted with neutral salts such as NaCl, KCl, NH_4Cl, and $NaHCO_3$.

Phosphorus

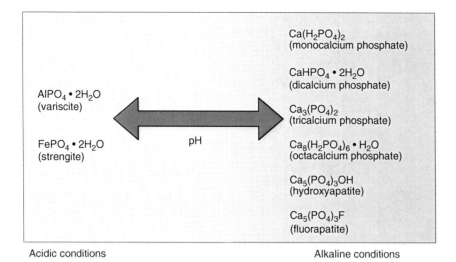

FIGURE 9.13 Stable phosphorus solid phases under acidic and alkaline conditions.

FIGURE 9.14 Patina of black, poorly crystalline apatite on the surface of an oyster shell located near the bottom of an early American mound (a), and vivianite as precipitate on an aggregate from stream sediment near Lake Okeechobee, FL, collected a short distance downstream from where a dairy barn floor routinely flushed manure (b). (From Harris, 2002.)

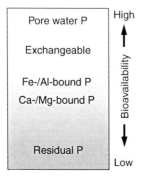

FIGURE 9.15 Schematic showing the bioavailability of various inorganic forms of phosphorus.

Slowly available phosphorus: This form is present as compounds that are recently formed during reactions with Fe, Al, Ca, and Mg compounds. These chemical precipitates are slowly available, and their solubility is regulated by pH and redox potential.

Very slowly available phosphorus: This form is present in discrete mineral forms such as Fe, Al, and Ca phosphates. Recently formed chemical precipitates slowly crystallize and their stability increases. At any time, about 80–90% of the soil phosphorus exists in very slowly available forms. Most of the remainder is in the slowly available forms. Perhaps less than 1% would be readily available.

9.5 PHOSPHORUS SORPTION BY SOILS

Phosphorus sorption refers to the abiotic retention of inorganic phosphorus in soils. In the following text, we will review few concepts and terminology related to inorganic phosphorus retention in soils. For details on the terminology, the reader is referred to basic soil chemistry textbooks (e.g., McBride, 1994; Bohn et al., 1985).

Adsorption: It refers to the movement of soluble inorganic phosphorus from soil pore water to soil mineral surfaces, where it accumulates without penetrating the structure (Figure 9.16). Phosphorus adsorption capacity of soil increases with clay content or minerals. Adsorbed phosphorus maintains equilibrium with phosphorus in soil pore water.

Desorption: It refers to the release of adsorbed inorganic phosphorus from the mineral surfaces into soil pore water. Depletion of phosphorus from soil pore water results in the release of phosphorus from mineral surfaces until the new equilibrium is reached. The balance between phosphorus adsorption and desorption maintains the equilibrium between solid phases and phosphorus in soil pore water. This phenomenon is defined as phosphate buffering analogous to pH buffering.

Anion exchange: Phosphate ions also participate in simple anion exchange (Figure 9.17). Ion exchange results from the electrostatic attraction of phosphate anions to positively charged sites of clay minerals and organic matter. Excess H^+ ions can result in positively charged mineral surfaces $(Al(OH)_2^+)$. The positively charged surface adsorb OH^-, NO_3^-, SO_4^{2-}, $H_2PO_4^-$, and Cl^- ions. Thus, $H_2PO_4^-$ ions can replace other anions adsorbed on the exchange complex.

Reaction with hydrous oxides or ligand exchange: Ligand exchange is a mechanism in which surface OH^- that is coordinated with a metal cation (such as Fe and Al) in a solid phase is replaced by phosphate ion (Figure 9.17). The solid phases involved are Fe- and Al-oxyhydroxides. Organic anions also compete strongly with phosphate for ligand exchange sites.

Fixation by silicate clays: Dissolution of Al^{3+} can take place as a result of kaolinite breakdown, followed by the precipitation of $AlPO_4$ (Figure 9.17).

Precipitation: It refers to the reaction of phosphate ions with metallic cations such as Fe, Al, Ca, and Mg, resulting in the formation of amorphous precipitates (Figure 9.17). These reactions typically occur at high concentration of either phosphate or metallic cations. In acid soils, soluble

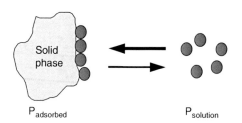

$P_{adsorbed}$ $P_{solution}$

FIGURE 9.16 Schematic showing the partitioning of phosphorus in adsorbed and solution phases.

$$Al^{3+} + H_2PO_4^- + 2H_2O = 2H^+ + Al(OH)_2 H_2PO_4 \text{ (Variscite)}$$

$$\left[\mathord{>} Al\,OH_2^+ \right] OH^- + H_2PO_4^- = \left[\mathord{>} Al\,OH_2^+ \right] H_2PO_4 + OH^-$$

$$Al \genfrac{}{}{0pt}{}{\mathord{-}OH}{\genfrac{}{}{0pt}{}{\mathord{-}OH}{\mathord{-}OH}} + H_2PO_4^- = Al \genfrac{}{}{0pt}{}{\mathord{-}OH}{\genfrac{}{}{0pt}{}{\mathord{-}OH}{\mathord{-}H_2PO_4}} + OH^-$$

$$Al_2SiO_5(OH)_4 + 2H_2PO_4^- = 2Al(OH)_2 H_2PO_4 + Si_2O_5^{2-}$$

FIGURE 9.17 Chemical reactions showing anion exchange, reactions with hydrous oxides or ligand exchange, fixation by silicate clays, and precipitation.

Fe^{3+} and Al^{3+} ions greatly exceed phosphate concentration. This results in the reaction of Al^{3+} with $H_2PO_4^-$ forming insoluble $AlPO_4$. Similar precipitation reactions can take place with soluble Fe and Mn under acidic conditions and Mg and Ca under near-neutral pH conditions.

Following is an example of precipitation reactions under alkaline soil conditions:

$$Ca^{2+} + HPO_4^{2-} \rightarrow CaHPO_4 \text{ [precipitate]} \tag{9.9}$$

Dissolution: This refers to solubilization of the precipitate. This happens when the concentration of any of the reactants decreases below the solubility product of that compound.

$$Ca^{2+} + HPO_4^{2-} \leftarrow CaHPO_4 \text{ [precipitate]} \tag{9.10}$$

Phosphate ion exchange in soils: In soils, phosphate and other ion exchange processes occur on the surfaces of the solid phases (clay minerals, organic matter, and mineral surfaces). Ion exchange reactions are important for maintaining an equilibrium between the solid phase and the solution (soil pore water). Thus, it is important that we understand the origin of surface charge of the solid phase.

Charge can be negative or positive depending on the type of solid phase and its chemical properties.

Permanent charge: Permanent charge on solid surfaces results from the isomorphous substitution in 2:1 clay minerals, for example, in layered silicate minerals, replacement of either the Si^{4+} or the Al^{3+} cations with cations of lower charge can increase negative charges on solid phase. Isomorphic substitution occurs during mineral formation and is largely unaffected by environmental conditions.

Similarly, substitution of higher valence cations (such as Si^{4+} or Al^{3+}) for lower valence cations results in permanent positive charges.

pH-dependent charge on solid phase: pH-dependent charges are primarily the result of dissociation of H^+ from OH^- groups in broken edges of some clays, organic matter, and hydrous oxides of iron and aluminum (Figure 9.18).

Surface OH^- groups on soil particles and organic matter are responsible for the development of pH-dependent charge in wetland soils. Under acidic conditions, availability of excess protons results in development of positive charges on soil particles. Under alkaline conditions, loss of protons results in negative charge. Thus, under acidic conditions, positive charge of soil particles can potentially adsorb phosphate ions. The overall contribution of pH-dependent charge depends strongly on type of minerals, size of soil particles, pH, and ionic composition.

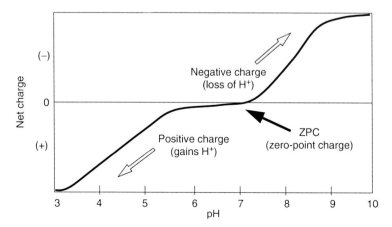

FIGURE 9.18 Influence of pH on net charge on solid phases.

Variable charge on solid phase: Figure 9.19 shows the role of Al–OH and COOH groups in regulating charge on the surfaces of clay minerals and organic matter. In the reaction shown in the upper part of the figure, Al–OH groups obtain negative charge as the pH approaches alkaline. The H^+ ions are replaced with excess OH^- ions, resulting in the formation of a negative charge on the mineral surface.

In Figure 9.19, all three phases of Al–OH groups are shown as a function of pH. Under alkaline conditions, Al-O^- is the stable form with loss of H^+ ions. Under acidic conditions, gain of excess H^+ results in $Al(OH)_2^+$ with positive charge.

Similarly, complex organic molecules associated with soil particles or organic matter can also provide pH-dependent charge. Most of these negative charges are derived from the dissociation of H^+ from carboxylic acid (–COOH) and phenolic (–C_6H_4OH) groups. As pH increases, some of these H^+ ions are combined with OH^- ions, resulting in negative charges on the surfaces of organic matter.

Phosphorus availability in soils is regulated by pH. The relationship between phosphorus solubility and pH is shown in Figure 9.20. In the figure, the y-axis represents the concentration of phosphate

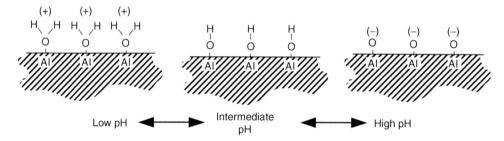

FIGURE 9.19 Schematic showing the influence of pH on positive or negative charges on Al solid phase.

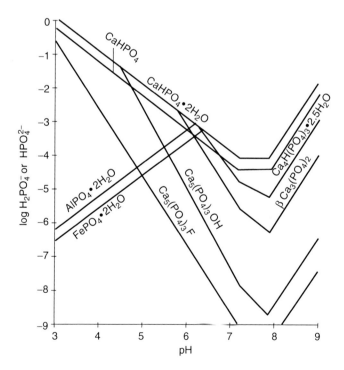

FIGURE 9.20 The solubility of calcium phosphates compared to strengite and variscite when Ca^{2+} is $10^{-2.5}$ M or fixed by calcite and carbon dioxide at 0.0003 atm. (Adapted from Lindsay, 1979.)

ions ($H_2PO_4^-$ or HPO_4^{2-}) in soil solution or pore water, whereas the x-axis represents pH of the soil solution or pore water.

Phosphate reactions in acid soils: Inorganic phosphorus availability in acid soils is regulated by formation of insoluble phosphates of Fe and Al through sorption and precipitation. Common minerals found in acid soils are variscite ($AlPO_4 \cdot 2H_2O$) and strengite ($FePO_4 \cdot 2H_2O$). Each of these minerals supports a certain level of phosphorus concentration in solution, depending on the solubility product of that mineral. Some possible reactions with Al under acid soil conditions are as follows:

$$Al^{3+} + H_2PO_4^- + 2H_2O = Al(OH)_2H_2PO_4 + 2H^+ \tag{9.11}$$

$$Al(OH)_2^+ + H_2PO_4^- = Al(OH)_2H_2PO_4 \tag{9.12}$$

$$Al(OH)_3 + H_2PO_4^- + H^+ = Al(OH)_2H_2PO_4 + H_2O \tag{9.13}$$

Phosphate reactions in alkaline soils: Availability of inorganic phosphorus in alkaline soils is regulated by the solubility of calcium compounds. Chemical reactions of phosphates with $CaCO_3$ are of special significance in calcareous soils and in wetlands receiving Ca inputs. At pH values of >7.5 (typical for Ca-dominated soils), the predominant forms of phosphate ions in soils pore waters are HPO_4^{2-} and $H_2PO_4^-$. In solution, these phosphate ions are adsorbed on $CaCO_3$ surfaces, either by forming $CaHPO_4$ ($Ca^{2+} + HPO_4^{2-} = CaHPO_4$) or by adsorption by displacement of CO_3^{2-} ions by HPO_4^{2-} ions. With increasing concentration of phosphate ions in solution, "multilayer" formation begins on $CaCO_3$ surfaces. This results in the formation of precipitates, which eventually crystallize. Following are some possible reactions with Ca under alkaline soil conditions:

$$Ca^{2+} + 2H_2PO_4^- = Ca(H_2PO_4)_2 \tag{9.14}$$

$$Ca(H_2PO_4)_2 + Ca^{2+} + 2OH^- = 2CaHPO_4 + 2H_2O \tag{9.15}$$

$$2CaHPO_4 + Ca^{2+} + 2OH^- = Ca_3(PO_4)_2 + 2H_2O \tag{9.16}$$

9.5.1 Adsorption—Desorption

Sorption and desorption are chemical reactions by which certain metals (e.g., Fe, Cu, Zn, and Mn) and anions (e.g., phosphate and sulfate) form/break chemical bonds within the coordination shell of atoms comprising the mineral structure. Sorption includes both adsorption and absorption. Physical adsorption refers to the attraction caused by the surface tension of a solid that causes molecules to be held at the surface of the solid. This type can also be reversible. Chemical adsorption (not reversible) involves actual chemical bonding at the solid's surface. Absorption is a process in which the molecules or atoms of one phase penetrate those of another phase.

The sorption process (abiotic retention) is controlled by the concentration of phosphate in soil pore water and the ability of the solid phase to replenish phosphate into pore water. This can be described in terms of two regulating factors: The intensity factor of the sorption process is controlled by the concentration of phosphate in soil pore water, and the capacity factor refers to the ability of the solid phase to replenish phosphate to the soil pore water.

What happens if this system is loaded with additional soluble inorganic phosphorus?

Addition of soluble inorganic phosphorus to soil increases the soil pore water phosphorus concentration. This results in rapid adsorption of phosphorus onto soil surfaces to maintain equilibrium. Soil's capacity to adsorb additional phosphorus dictates the concentration of phosphorus in soil pore water. These adsorption processes occur within a short time period. When soil particles become saturated with phosphorus, there is an increase in phosphorus concentration in soil pore water. Reaction kinetics are on the order of minutes to hours to reach sorption equilibrium. Figure 9.21 illustrates a two-step process in which rapid phosphate exchange takes place between soil pore water and soil particles or mineral surface (adsorption) followed by slow penetration (absorption) of phosphate into solid phase. Similarly, desorption of phosphorus can also

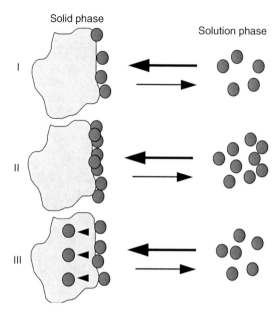

FIGURE 9.21 Schematic showing phosphorus exchange between solid phase and solution phase. (Modified from Froelich, 1988.)

occur in two steps, with rapid release of phosphorus present on surfaces of solid phase followed by a slow release of phosphorus retained within the solid phase. In systems at steady state, the phosphorus adsorbed maintains equilibrium with phosphorus in soil pore water.

With time, some of the adsorbed phosphorus diffuses into solid phase (absorption) where it forms discrete phosphate minerals. This process occurs very slowly over timescales of days to months or years. Decrease in quantity of phosphorus on solid surfaces results in more sites for soil pore water phosphorus adsorption. These conditions can reduce soil pore water phosphorus concentration, resulting in a new equilibrium.

When inorganic phosphorus is added to soil, adsorption process continues until a new state of equilibrium is reached. Adsorption equilibrium is said to be achieved when the concentration of phosphorus in soil pore water does not change between measurements made at different time intervals. If sufficient time is allowed, and biological activity is inhibited, the system eventually reaches equilibrium. At equilibrium, the rates of forward reaction (adsorption) and reverse reaction (desorption) are the same. Adsorption data at equilibrium are presented in the form of an adsorption isotherm as shown in Figure 9.22.

A sorption isotherm describes the equilibrium relationship between the concentrations of adsorbed and dissolved species at a given temperature. Because soil scientists have adapted and modified these functions and used them to describe phosphate adsorption from solution, they have proved to be less than ideal. Phosphorus adsorption increases with increasing soil pore water phosphorus concentration, until all sorption sites are occupied. At that point, adsorption reaches its maximum, as indicated by S_{max}. Similarly, an incremental decrease in soil pore water phosphorus concentration results in desorption of phosphorus from the solid phase. At low pore water phosphorus concentration, the relationship between adsorption and soil pore water phosphorus concentration is linear. The intercept on y-axis (Figure 9.22), as indicated by S_0, suggests that phosphorus is adsorbed at soil pore water phosphorus concentrations approaching near-zero levels. If phosphorus is added to soil at concentrations lower than that of phosphorus in soil pore water, then the soil tends to release phosphorus until new equilibrium is reached. Soils adsorb only when added phosphorus concentrations are higher than the concentration of phosphorus in soil pore water.

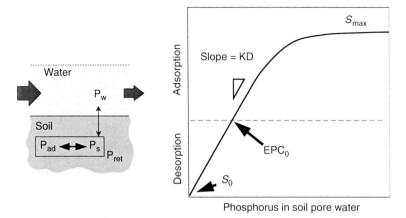

FIGURE 9.22 A typical phosphorus adsorption isotherm showing amount of phosphorus adsorbed or desorbed in relation to phosphorus concentration in soil pore water.

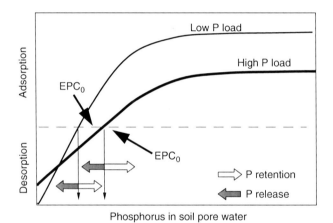

FIGURE 9.23 Phosphorus adsorption isotherms for soils receiving high and low phosphorus loads.

If added solution phosphorus concentration is lower than the concentration of phosphorus in soil pore water (e.g., rainwater), then the soil releases or desorbs phosphorus until new equilibrium is maintained. For any given soil, at some critical concentration, net adsorption equals zero, which means that adsorption equals desorption and the system is at equilibrium as indicated by EPC_0 (equilibrium phosphorus concentration). At this point, soil exhibits maximum capacity for buffering phosphorus in soil pore water. In this region, the system reattains equilibrium conditions, even if soils are loaded with or depleted of phosphorus in soil pore water. If the water entering a wetland has a phosphorus concentration below EPC_0, then that soil releases phosphorus or serves as a source of phosphorus to the water column or soil pore water. If the water entering a wetland has phosphorus concentration higher than EPC_0, then that soil adsorbs or retains phosphorus or serves as a sink for added phosphorus.

An example of phosphorus-loading influence on adsorption characteristics is shown in Figure 9.23. The critical value of EPC_0 increases with phosphorus loading, suggesting that the soil does not have enough buffering capacity to maintain low phosphorus concentration in the soil pore water or the soil solid phase has been saturated with added phosphorus; therefore, phosphorus concentration in solution increases. In general, wetland soils that are heavily loaded function as net source of phosphorus, especially when these soils come in contact with low-phosphorus water (rainwater).

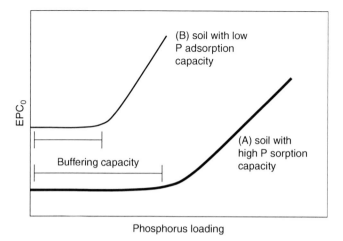

FIGURE 9.24 Influence of phosphorus loading on soil EPC_0 (equilibrium phosphorus concentration at which point adsorption equals desorption) (Clark, 2002).

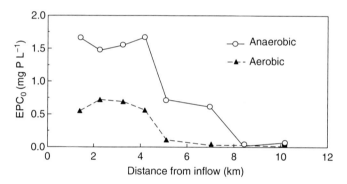

FIGURE 9.25 Influence of soil phosphorus gradients (distance from point source loading) on soil EPC_0 (equilibrium phosphorus concentration at which point adsorption equals desorption) (Clark, 2002).

The EPC_0 can be used as an indicator of wetlands' capacity to function as source or sink for phosphorus. Phosphorus loading can result in the loss of soil's buffering capacity to maintain low EPC_0 values (Figure 9.24). The influence of soil phosphorus enrichment in WCA-2A soils of the Everglades on EPC_0 is shown in Figure 9.25.

9.5.2 Phosphorus Sorption Isotherms

The adsorption isotherms of soils are usually measured by mixing a known amount of soil with a solution containing a known phosphorus concentration. The mixtures are equilibrated for a fixed period (usually 24 h) at a constant temperature under continuous shaking. A wide range of soil-to-water ratios are used in batch isotherm experiments. Phosphorus not recovered in solution is assumed to be adsorbed on solid phase. Phosphate retained by soils is calculated as follows:

$$S^1 = [(C_0 V) - (C_{24} V)]/M \qquad (9.17)$$

where C_0 is the concentration (mg L^{-1}) of phosphorus added to solution, V the volume (L) of electrolyte solution, C_{24} the concentration (mg L^{-1}) of phosphorus in solution after 24 h equilibration period, M the mass (kg) of dry soil, and S^1 the amount (mg kg^{-1}) of added phosphorus adsorbed by soils.

Phosphorus

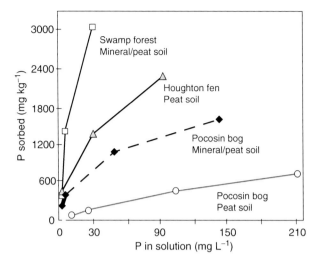

FIGURE 9.26 Phosphorus adsorption characteristics of soils from mineral and peat-based wetlands. (Adapted from Richardson, 1985.)

Equation (9.17) is used primarily to calculate the phosphorus lost (adsorbed) or released (desorbed) in relation to phosphorus added to soil. These calculations do not account for the amount of native soil phosphorus in the adsorbed phase. The total amount (native + added) of phosphorus retained by soil can be calculated as follows:

$$S = S^1 + S_0 \quad (9.18)$$

where S is the total amount (mg kg^{-1}) of phosphorus sorbed, S^1 the amount (mg kg^{-1}) of added phosphorus adsorbed by soils, and S_0 the initial quantity (mg kg^{-1}) of soil phosphorus present in the adsorbed phase.

An adsorption isotherm describes the equilibrium relationship between the concentrations of a chemical constituent such as phosphate ions present in soil pore water and amount adsorbed on the solid phase. The plot showing phosphorus adsorbed (on y-axis) and phosphorus in solution (on x-axis) is called adsorption isotherm or buffer isotherm, as shown in Figures 9.26 and 9.27. The term sorption used in the literature generally refers to both adsorption on the surface of the solid phase (or retaining component) and absorption by the solid phase (diffusion into the retaining component). In many cases, phosphorus sorbed on solid phase is usually not desorbed with a neutral salt such as NaCl or KCl and does not follow adsorption isotherm, an effect called hysteresis. This effect is a function of solid-phase physicochemical characteristics, phosphorus-loading rate, and residence time. This hysteresis effect is due to phosphorus diffusion into solid phase and is observed in soils and sediments having active reaction surfaces. For example, ligand exchange of phosphorus for surface aquo and hydroxyl groups bonded to Al and Fe results in monodentate, bidentate, or binuclear forms of adsorbed phosphorus. The formation of binuclear and bidentate bonds and precipitation are less reversible compared with monodentate bonds. This irreversible pool represents long-term storage in soils. Once the sites are saturated (for example, in soils and sediments with historical phosphorus loading or those low in clay mineral surfaces), the desorption potential increases.

A number of empirical equations have been used to describe these relationships (Berkheiser et al., 1980). The most commonly used phosphorus sorption isotherm equations are as follows.

9.5.2.1 Linear Equation

At low phosphate concentrations in soil solution (soil pore water), the relationship between phosphorus in solution and phosphorus in adsorbed phase can be described by a simple linear equation.

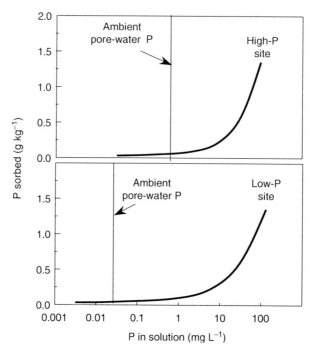

FIGURE 9.27 Phosphorus adsorption characteristics of soils from northern Everglades (Water Conservation Area 2A) (Clark, 2002).

S_0 can be estimated by using a least squares fit of S^1 measured at low solution phosphorus concentrations, using linear sorption model. At these concentrations, the relationship between S^1 and C_{24} is typically linear, and the relationship can be described as follows:

$$S^1 = K_L C_{24} - S_0 \quad (9.19)$$

where S_0 is the y-axis intercept representing the initial quantity (mg kg^{-1}) of soil phosphorus present in the adsorbed phase and K_L the linear adsorption coefficient or partition coefficient (L kg^{-1}) representing binding strength.

High K_L values represent strong phosphorus-adsorbing strength of soils, such as those with high concentrations of iron and aluminum oxides or calcium-based minerals. Soils with high K_L values maintain low phosphorus concentrations in soil solution (or soil pore water). Some examples of K_L values for various wetland soils are presented in Reddy et al. (1998a).

EPC$_0$ is defined as the phosphorus concentration in solution where adsorption equals to desorption ($S^1 = 0$). EPC$_0$ can be estimated as follows:

$$\text{EPC}_0 = \frac{S_0}{K_1} \quad (9.20)$$

9.5.2.2 Freundlich Equation

$$S = K_f C^N - S_0 \quad (9.21)$$

where S_0 is the amount of phosphorus present in sorbed phase at soil pore water concentrations approaching negligible levels (mg kg^{-1}), K_f the Freundlich adsorption coefficient (L kg^{-1}), which is the ratio between log phosphorus sorbed on solid phase and log phosphorus in soil solution, C the phosphorus concentration in soil pore water (mg L^{-1}), N an empirical constant ($N < 1$), and S the quantity of phosphorus sorbed on solid phase (mg kg^{-1}). This equation assumes that the amount of phosphorus adsorbed increases logarithmically with an increase in phosphorus concentration in solution. This equation was originally introduced as empirical expression to describe phosphorus adsorption in a wide range of concentrations in soil solutions. Therefore, it is commonly used by researchers. The linear form of this equation is as follows:

$$\text{Log } S = N \log C + \log K_f \qquad (9.22)$$

The log–log plots are insensitive within the range of experimental data, and the equation contains two empirical constants, K_f and N. Because the Freundlich equation is empirical in nature, it provides very little insight into the mechanism regulating phosphorus adsorption process.

9.5.2.3 Langmuir Equation

This adsorption model was found to be useful in describing adsorption of ions in solution on soil particles, with the assumption that the soil particles have finite capacity to adsorb phosphorus, as described:

$$S = \frac{|S_{max} kC|}{1 + kC} - S_0 \qquad (9.23)$$

where S is the amount of phosphorus in adsorbed phase (mg kg^{-1}), S_{max} the phosphorus adsorption maximum (mg kg^{-1}), k the constant related to bonding energy (L mg P^{-1}), C the concentration of phosphorus in soil pore water (mg L^{-1}), and S_0 the quantity of phosphorus retained under ambient conditions (mg kg^{-1}). Phosphate adsorption maximum (S_{max}) has been widely used to estimate soil's capacity to adsorb phosphorus.

The Langmuir equation was originally developed to describe adsorption of gases on the surfaces of solids. It has been frequently used to describe phosphate adsorption onto solid surfaces. This equation was derived with the following assumptions:

- A constant energy of adsorption that is independent of surface coverage
- Adsorption on specific sites, with no interaction between adsorbate molecules
- Maximum adsorption equal to complete monomolecular layer on all reactive adsorbent surfaces

The Langmuir equation used to describe phosphate sorption by solid phase suggests that the energy of adsorption is constant and is independent of surface coverage. However, Langmuir equation may still describe phosphate sorption over a range of concentrations in the soil pore water typically encountered in natural systems because variation in energy of adsorption is small. As the phosphate concentration in soil pore water increases, phosphorus sorption maximum can occur as a result of complete surface coverage. However, at these high concentrations, phosphate can undergo precipitation reaction, suggesting high sorption capacity. At high phosphorus concentrations, a rapid increase in phosphorus sorption can occur as a result of precipitation reactions at the surfaces of solid phase. This is often referred to as multilayer adsorption. However, at high pore water phosphorus concentrations, it is difficult to discern differences between sorption and precipitation reactions.

9.5.2.4 Single-Point Isotherms

Single-point isotherms were initially used by Bache and Williams (1971) to determine phosphorus sorption capacity of a wide range of soils. The approach involves measuring phosphorus sorption at one fixed concentration after fixed equilibration period (18 h). The phosphorus sorption index (PSI) (Bache–Williams index) is calculated as follows:

$$\text{PSI} = \frac{S}{\log C_t} \qquad (9.24)$$

where PSI is the sorption index and C_t the concentration of phosphorus in solution after an 18 h equilibration period. This index is measured at phosphorus concentrations at the upper end scale of

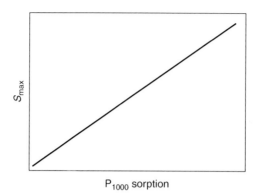

FIGURE 9.28 Relationship between phosphorus adsorption maxima (S_{max}) estimated using Langmuir isotherm equation and saturation maxima estimated from single-point isotherm. (From Reddy et al., 1998a.)

an isotherm. The PSI is calculated as (phosphorus sorbed)/(log P in solution), where phosphorus sorbed is expressed as mg P per 100 g of soil and phosphorus in solution is expressed as mg L^{-1}. This ratio can also be viewed as partition coefficient (K_f), as indicated in Freundlich sorption isotherms. This method was later applied for wetland soils by Richardson (1985) and Walbridge and Struthers (1993). Phosphorus sorption maximum was also estimated using single-point isotherm approach. This procedure involves measuring P sorption at one saturation level concentration, such as 1,000 mg P kg^{-1} added (Reddy et al., 1998a). This approach assumes that single addition of phosphorus at high concentration is adequate to saturate all available sorption sites. Phosphorus sorption measured for various wetland soils at this high phosphorus concentration was highly correlated with S_{max} estimated using Langmuir sorption model. The Langmuir S_{max} values were 1.15 times the values obtained by phosphorus addition of 1,000 mg P kg^{-1}, under both aerobic and anaerobic conditions (Figure 9.28). However, a limitation of single-point isotherms relative to isotherms determined at various phosphorus levels is that they cannot provide parameters such as K, EPC_0, and S_0, which are critical in describing P retention characteristics of soils.

9.5.2.5 Quantity (*Q*)/Intensity (*I*) Relationships

Quantity/intensity relationships are often used to describe soil capacity to buffer phosphorus concentration in soil pore water. The quantity (Q) refers to the amount of phosphorus adsorbed on soil surface, whereas intensity (I) refers to the concentration of P in soil pore water. This ratio can also be viewed as partition coefficient (K_d), as indicated by liner sorption isotherms. The ratio expressed as either Q/I or K_d is influenced by various physicochemical properties of soils, including clay content, high concentration of Fe and Al oxides, $CaCO_3$ content, organic matter content, pH, and redox potential.

9.5.3 Precipitation and Dissolution

Phosphorus adsorption typically occurs at low concentrations and reaches saturation level once all potential sorption sites are occupied. However, if the concentration of soil pore water is increased beyond the capacity of soil to adsorb phosphorus, the precipitation reactions may be involved in retaining phosphorus. When sorption isotherms are measured at high concentrations (Figure 9.26), it is hard to differentiate between adsorption and precipitation reactions. An example of potential precipitation reactions at high solution concentrations is shown in Figure 9.29.

A simple dissolution reaction of a compound (C_aA_b) is shown in Figure 9.30. This compound could be $AlPO_4$, $FePO_4$, or $CaHPO_4$. The letters "C" and "A" represent cation and anion, respectively, in the compound. "K" represents the equilibrium constant and is often referred to as solubility product or K_{sp} at equilibrium. The activity of pure solid phase is assigned a value of 1 by convention.

Phosphorus

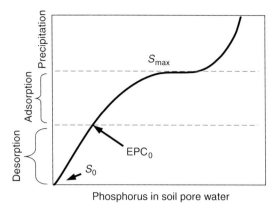

FIGURE 9.29 A typical phosphorus adsorption isotherm showing phosphorus adsorption, desorption, and precipitation in relation to phosphorus concentration in soil pore water.

$$C_aA_b = aC + bA$$

$$K = \frac{[C]^a[A]^b}{[C_aA_b]}$$

$$IAP = [C]^a[A]^b$$

$$SI = \frac{IAP}{K}$$

FIGURE 9.30 A simple chemical reaction showing equilibrium relationships.

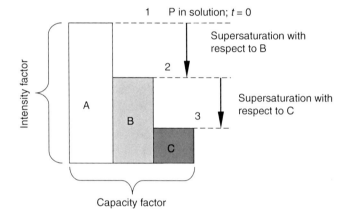

FIGURE 9.31 Relationship between solid phases in equilibrium with phosphorus concentration in solution (From McBride, 1994).

For example, if the concentration of an ion in this reaction is increased, the reaction will be driven to the "left," until a theoretical equilibrium is reached. Precipitation of solid phase will occur only when ion activity product (IAP) exceeds solubility product. What happens when the concentration of an ion in this reaction is decreased? The reaction will be driven to the "right," resulting in slow dissolution of the solid phase until a theoretical equilibrium is reached. Dissolution of solid phase will occur only when IAP is lower than the solubility product.

The status of solid phase can be determined by Saturation Index (SI), which is the ratio of IAP and K_{sp}. If SI > 1, then the solution is supersaturated with respect to the solid phase and if SI < 1, then the solution is undersaturated with respect to the solid phase. At equilibrium, SI = K_{sp}. Any precipitate formed at high concentrations can potentially dissolve, when the concentration of the ion decreases.

Let us assume that there are three types of solid phases of phosphorus in wetland soils (Figure 9.31). Under alkaline conditions, these could be dicalcium phosphate ($CaHPO_4$) (A), octacalcium phosphate ($Ca_8(H_2PO_4)_6$) (B), and hydroxyapatite ($Ca_5(PO_4)_3OH$) (C). The stability of these phosphate solid phases can be explained by intensity and capacity factors. Intensity factor refers to the concentration or activity of ions in solution. Capacity factor refers to the amount and type of solid phase in soil.

Solid phase A is in equilibrium at phosphorus concentration of $(C-1)$, and supersaturated with respect to solid phases B and C. This means at concentrations higher than $(C-1)$, all three solid phases are stable. If the concentration becomes less than $(C-1)$, then the dissolution of solid phase A occurs until a new equilibrium is reached with the formation of solid phase B, which is in equilibrium at concentration $(C-2)$. At this stage, solid phase A is undersaturated and solid phase C is supersaturated with respect to concentration $(C-2)$. If the concentration decreases further to $(C-3)$, then the solid phase B is now undersaturated and continues to dissolve. This results in formation of solid phase C, which is stable at low concentrations such as $(C-3)$. Now, both solid phases A and B are undersaturated with respect to concentration $(C-3)$.

In mineral wetland soils, Al and Fe are major constituents, which regulate phosphorus solubility by forming insoluble phosphate minerals. These minerals are sensitive to changes in soil pH, whereas phosphorus minerals associated with Fe are redox sensitive. Phosphate minerals associated with Al and Fe are stable under acidic soil conditions, and their solubility increases with an increase in pH. Solubility of calcium phosphates decreases with an increase in pH. In soils dominated by $CaCO_3$ (calcareous soils), when Ca^{2+} activity is depressed, phosphorus solubility increases.

Under oxidized conditions in mineral wetland soils, the coating of hydrated ferric oxides on silt or clay particles have occluded in them several forms of phosphate including ferric phosphate, aluminum phosphate, and calcium phosphate (Figure 9.32). As a result of anaerobic conditions, reduction of hydrated ferric oxide to more soluble ferrous hydroxide results in the release of these occluded phosphates. Calcium phosphate released in this manner is available to wetland plants, whereas the occluded ferric phosphate is probably not available to the plants until it has been reduced to more soluble ferrous phosphate.

In calcium-dominated wetlands, high concentrations of Ca^{2+} can result in formation of complex calcium phosphate compounds of varying solubilities such as calcium phosphate, dicalcium phosphate, beta-tricalcium phosphate, octacalcium phosphate, and hydroxyapatite. Under these conditions, the phosphorus concentration in the soil pore water of calcareous soils is a function of Ca^{2+} activity. Solubility of these compounds decreases with an increase in Ca content. Insoluble beta-tricalcium phosphate is more likely to be found at a high pH. Thermodynamically, apatite is the most stable compound in soils. At relatively high phosphate concentrations, dicalcium phosphate or octacalcium phosphate may form and slowly transform to the more stable phase: hydroxyapatite. These precipitation reactions can occur on the surfaces of calcite. The amount of exposed surface will determine the amount of phosphorus precipitated. In a Ca-saturated clay, calcium

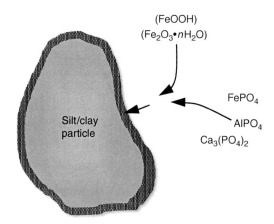

FIGURE 9.32 A schematic showing coprecipitation of iron, aluminum, and calcium phosphate on soil particles.

phosphate will precipitate as a separate phase above pH 6.5. However, this depends on the concentration of phosphate ions in soil pore water.

Photosynthesis and respiration can initiate significant changes in water column pH on a diurnal basis. These processes can increase pH to as high as 10, depending on the buffering capacity of the water column. Under these conditions, a significant portion of water column phosphorus can co-precipitate with $CaCO_3$ where dissolved Ca^{2+} is available. Approximately 75–90% of the phosphorus precipitated was solubilized when pH levels decreased to below 8 as a result of an increase in carbon dioxide levels (Diaz et al., 1994). Retention of inorganic phosphorus by precipitation will be significant in waters with high Ca^{2+} concentration and alkalinity. In high Ca^{2+} concentrations and mildly alkaline waters, House (1990) attributed only 6% of the overall phosphorus removal to coprecipitation, whereas the remainder was due to biological uptake. However, Salinger et al. (1993) found metastable calcium phosphate species to be a predominant form of phosphorus transported in River Jordan. In the water column of wetlands, chemical reactions associated with phosphorus solubility are regulated by dissolved iron and calcium. Others (Hartley et al., 1997) have shown the coprecipitation of phosphorus with calcite in the presence of photosynthesizing green algae, *Chlorococcum* sp. Several studies showed inhibition of calcite growth in the presence of high phosphorus concentrations (Avnimelech, 1980; Dove and Hochella, 1993). However, at low phosphorus concentrations, coprecipitation of phosphorus with calcite was observed in aquatic systems (House and Donaldson, 1986; Giannimaras and Koutsoukos, 1987; Kleiner, 1988).

The precipitation of phosphorus as calcium phosphate has been studied extensively by soil scientists in evaluating applied fertilizer reactions. Initial adsorption of phosphorus onto calcite is followed by precipitation as calcium phosphate. Similar reactions can occur at the soil–floodwater interface of calcareous wetlands. For example, long-term phosphorus accumulation in the Everglades marsh is linearly correlated with calcium accumulation, suggesting the possibility of phosphorus and $CaCO_3$ interactions (Reddy et al., 1993).

9.5.4 Regulators of Phosphorus Retention and Release

Abiotic phosphorus retention by wetland soils is regulated by various physicochemical properties including pH, redox potential, iron, aluminum, and calcium content of soils, organic matter content, phosphorus loading, and ambient phosphorus content of soils.

Iron and minerals such as ferric hydroxide, ferric oxyhydroxide, ferrous hydroxide, and amorphous oxides are known to regulate phosphorus retention in wetland soils dominated by these minerals. Similarly, Al minerals can also regulate phosphorus retention. Figure 9.33 shows the relationship between phosphorus sorption capacity (as estimated by Langmuir equation) or adsorption maximum and oxalate-extractable iron and aluminum. Ammonium oxalate selectively extracts amorphous and poorly crystalline forms of iron and aluminum. The amount of iron and aluminum extracted with ammonium oxalate was highly correlated with phosphorus sorption capacity of wetland mineral soil dominated by iron and aluminum. Similar relationships are shown on a wide range of wetland soils and lake sediments. The relationship shown in Figure 9.33 suggests that approximately 0.17 mmol of phosphorus is retained per mmole of (Fe and Al), suggesting that not all sorption sites are available for phosphorus sorption. Other studies have shown these values to be in the range of 0.4–0.6 for mineral upland soils (Lookman et al., 1995).

Significant correlations were observed between various phosphorus sorption indices (PSI) and physicochemical properties of soils (Figure 9.34). The PSI index exhibited a strong positive correlation with poorly crystalline and amorphous forms of aluminum and iron for mineral wetland soils (Richardson, 1985; Walbridge et al., 1991; Axt and Walbridge, 1999). In wetland and stream bank soils of forested wetlands in Virginia, oxalate-extractable aluminum and organic matter content explained 83% of the variation in PSI (Axt and Walbridge, 1999).

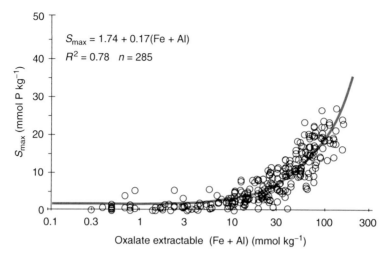

FIGURE 9.33 Relationship between oxalate-extractable iron and aluminum and phosphate sorption maximum of mineral wetland soils. (Adapted from Reddy et al., 1995.)

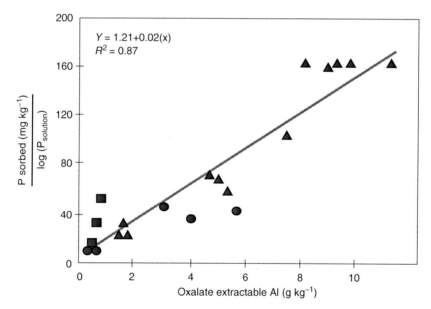

FIGURE 9.34 Relationship between oxalate-extractable aluminum and phosphate sorption in mineral and peat-based wetlands. (Adapted from Richardson, 1985.)

Several researchers reported significant correlations between amorphous and poorly crystalline forms of iron and aluminum, with phosphorus retention maximum of soils (Berkheiser et al., 1980; Khalid et al., 1977; Richardson, 1985; Walbridge and Struthers, 1993; Gale et al., 1994; Reddy et al., 1998a). This correlation suggests that phosphorus sorption in soils/sediments is associated with amorphous and poorly crystalline forms of iron and aluminum. Statistical ($p < 0.01$) correlation of S_{max} (sorption maxima) with total organic carbon (TOC) suggests that organic matter can also play a major role in phosphorus sorption (Reddy et al., 1998a). Iron and aluminum complexed with organic matter may be responsible for phosphorus sorption, suggesting an indirect effect of organic matter that may explain the correlation of S_{max} with TOC (Syers et al., 1973).

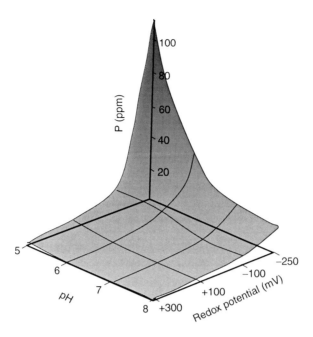

FIGURE 9.35 Influence of Eh and pH on phosphorus solubility in mineral wetland soil. (Adapted from Patrick et al., 1973.)

Solubility of phosphorus is also influenced by soil pH and Eh (Figure 9.35). In a pH range of 5–8, phosphorus solubility is low at 300 mV, resulting in low phosphorus concentration in soil solution (Patrick et al., 1973, Patrick and Khalid, 1974). However, as Eh decreases from 300 to −250 mV, phosphorus solubility increases at all pH levels, reflecting high concentration in soil pore water. Phosphorus solubility is highest under low-pH and low-Eh environments.

$$FePO_4 + H^+ + e^- = Fe^{2+} + HPO_4^{2-} \qquad (9.25)$$

Under alkaline conditions, reduction of $FePO_4$ occurs and P released potentially reacts with Ca^{2+} forming $CaHPO_4$. Similarly, Fe^{2+} reacts with CO_3^{2-} and precipitates as $FeCO_3$.

$$Ca^{2+} + HPO_4^{2-} = CaHPO_4 \qquad (9.26)$$

$$Fe^{2+} + CO_3^{2-} = FeCO_3 \qquad (9.27)$$

These reactions tend to decrease both Fe^{2+} and HPO_4^{2-} concentrations in soil pore water.

In soils dominated by Fe minerals, reduction of the soluble ferrous oxyhydroxide compounds results in amorphous "gel-like" reduced ferrous compounds with larger surface area than the oxidized forms. A reduced soil has many more sorption sites as a result of the reduction of the insoluble ferric oxyhydroxide compounds to more soluble ferrous oxyhydroxide compounds (Patrick and Khalid, 1974). The conversion of insoluble ferric iron compounds to somewhat soluble, surface-active ferrous compounds is one of the major microbial processes affecting phosphorus behavior in wetlands. Even though reduction increases the sorption sites, the larger number of the sites have a lower bonding energy for phosphate than do the smaller number of sites available in the aerobic soil. Thus, a reduced soil will adsorb a large amount of phosphorus with a low bonding energy, whereas an oxidized soil will adsorb less phosphorus but hold what it adsorbs more tightly. Although reduction may create a larger surface area for sorption of phosphorus, the binding energy associated with phosphorus sorption is low, and the desorption potential is high (Patrick and Khalid, 1974). These effects result in higher EPC_0 (of soils or sediments where adsorption equals desorption) values under reduced conditions than under oxidized conditions.

Drying of anaerobic soils and sediments showed contradictory results with respect to phosphorus sorption characteristics. Phosphate buffering capacity of soils and sediments decreases on drying (Twinch, 1987; Qui and McComb, 1994; Baldwin, 1996). However, several other studies showed an increase in the degree of phosphate adsorption upon drying soils (Barrow and Shaw, 1980; Haynes and Swift, 1985, 1986, 1989). In mineral wetland soils, drying potentially decreases the degree of hydration of iron hydroxide gels, hence increasing the surface area, resulting in increased phosphorus adsorption. However, McLaughlin et al. (1981) observed that drying synthetic iron and aluminum oxyhydroxide increased crystallinity and decreased phosphorus sorption capacity. Under flooded-drained conditions, Sah et al. (1989a, 1989b, 1989c) showed an increase in concentration of amorphous iron at the expense of more crystalline forms, suggesting greater surface area and potential for high phosphorus sorption. In floodplain-forested soils, Darke and Walbridge (2000) reported a decrease in aluminum and iron oxide crystallinity during seasonal flooding. These observations are consistent with those made by Patrick and Khalid (1974) for flooded rice soils.

Some of the key concepts on mechanisms regulating phosphorus solubility in wetlands are as follows:

- When ferric phosphorus is reduced to ferrous phosphorus by iron-reducing bacteria, phosphorus is released. We know that phosphate is occluded in hydrated ferric hydroxide coating. This is released by reduction of ferric iron to ferrous iron.

$$FePO_4 = Fe^{2+} + PO_4^{3-}$$

- At high pH, HPO_4^{2-} associated with Fe and Al minerals can be replaced with OH^- ions, resulting in greater availability of phosphate (ligand exchange).
- Organic anions (COO^-) can also displace HPO_4^{2-} from Fe and Al minerals. Microorganisms produce organic acids through fermentation of carbohydrates. These organic acids solubilize phosphorus.
- Phosphate is displaced from ferric and aluminum phosphates due to an increase in pH caused by diurnal fluctuations in photosynthesis. Increase in photosynthesis causes an increase in oxygen concentrations, which causes photosynthetically induced precipitation of ferric and aluminum phosphates.
- Ferrous iron is precipitated with S^{-2}, resulting in greater availability of phosphate.
- In alkaline soils, organic acid release during decomposition of organic matter can decrease pH, resulting in solubilization of calcium phosphate. Similarly, increased CO_2 partial pressure (from decomposition) can also consume some alkalinity, resulting in solubilization of calcium phosphate.
- pH: Phosphorus retention in acid soils is regulated by Fe and Al oxides. Several mechanisms including adsorption, ligand exchange, and precipitation are known to be involved in phosphorus retention. In alkaline soils, calcium-based minerals regulate phosphorus retention.
- Eh: In mineral soils dominated by iron, minerals associated with phosphorus are stable under aerobic or drained soil conditions. As the soil undergoes reduction, a decrease in Eh increases phosphorus solubility through reduction of $FePO_4$. Accumulation of organic acids and dissolved carbon dioxide can also potentially increase the solubility calcium phosphates.
- Clay content: Soils with high clay content retain phosphorus strongly as sorption sites.

- Iron and aluminum oxides: Soils high in clay content typically have high iron and aluminum oxides. These oxides provide surfaces for phosphorus retention.
- Organic matter content: Soils with high organic matter content can provide surfaces for phosphorus sorption.
- Calcium carbonate: In alkaline soils, $CaCO_3$ can provide surfaces for phosphorus sorption and Ca^{2+} ions for coprecipitation as calcium phosphate.
- Reaction time/aging: Phosphate adsorption/absorption increases with time. Precipitation reactions of phosphate with Fe, Al, or Ca ions result in formation of amorphous forms of these compounds. As this material ages, amorphous forms assume crystalline forms and increase stability.
- Temperature: Both adsorption and precipitation rates are accelerated with an increase in temperature. High temperatures decrease the time required to form crystalline forms.

9.6 ORGANIC PHOSPHORUS

Wetland soils are often characterized by high organic matter content; thus, soil properties and factors regulating the breakdown of organic matter determine the long-term storage of phosphorus in the organic pool. Organic phosphorus commonly dominates the total phosphorus in wetlands usually comprising more than half of the soil phosphorus. The proportion (percentage of total phosphorus) and nature of organic phosphorus in wetlands depend on soil type, type of organic loading from external sources, deposition of dead algal cells, and detrital tissue from aquatic vegetation. Plant detrital matter is the major source of organic phosphorus in wetlands. Consequently, peat-dominated wetlands have a higher proportion of organic phosphorus compared to soils with high mineral matter. Although a large proportion of total phosphorus exists in organic forms, only a small portion of this pool may be biologically active. In soils, this labile fraction is rapidly turned over and made available to plants. The relative importance of organic phosphorus to biotic communities (microbes, algae, and macrophytes) increases with phosphorus limitation (oligotrophic wetlands). Thus, to describe phosphorus behavior in wetlands, it is critical to understand the interaction of microbes, fauna, and vegetation and their role in phosphorus cycling.

9.6.1 Forms of Organic Phosphorus

Organic phosphorus in wetlands is derived from various sources including microbes, algae, vegetation, and detritus, and soil organic matter (Figure 9.36). Turnover of organic matter derived from each of these ecosystem components depends on the biological activity and environmental conditions (see Chapter 5) within the wetland. Soil microorganisms are key in regulating the turnover of

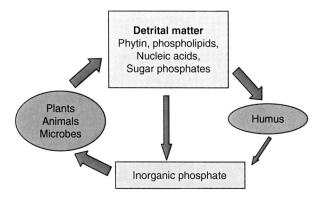

FIGURE 9.36 A schematic showing the sources of organic phosphorus in wetland soils.

organic matter. They can mineralize organic phosphorus, and under phosphorus-limited conditions, any inorganic phosphorus released into the water column or soil solution is rapidly assimilated (immobilized) into microbial biomass. Similarly, vegetation assimilates phosphorus and stores in plant biomass. Synthetic organic phosphorus compounds derived from herbicides, pesticides, and fungicides may also be present in wetlands receiving effluents from agricultural operations. In addition, external loadings of particulate organic matter (POM) can be another source of organic phosphorus to wetlands.

TOP is determined by several operationally defined methods:

1. Difference between total phosphorus (determined by perchloric acid digestion) and inorganic phosphorus (extracted with 1 M HCl or H_2SO_4): This method works well in soils with high organic matter and low mineral matter. Some of mineral forms of phosphorus may not be solubilized in 1 M HCl or H_2SO_4 during the extraction period; thus, organic phosphorus can be overestimated.
2. Ignition method: Organic phosphorus is converted by ashing at 550°C into inorganic phosphorus and extracted with acid. The difference between inorganic phosphorus in the ignited soil and that in unignited soil is considered as organic phosphorus.
3. Acid-alkaline extraction: In this method, inorganic phosphorus is extracted with acid followed by alkaline extraction with strong alkali (0.5 M NaOH). Alkali extracts are digested and analyzed for phosphorus using standard methods (APHA, 2005; U.S. EPA, 1993). This procedure may underestimate organic phosphorus as alkali extracts may not be strong enough to dissolve all forms of organic phosphorus.

In wetlands dominated by organic soils, approximately 50–90% of the total phosphorus is in organic form, as estimated by the difference method (methods described above) (Figure 9.37). In mineral wetland soils, organic phosphorus content may range from 10 to 50% of the total phosphorus.

Phosphorus bonded in some way with organic matter is known as organic phosphorus. This bond is usually covalent. Depending on the nature of the bond, organic phosphorus may be classified into orthophosphate esters, phosphonates, and anhydrides. Orthophosphate esters are esters of phosphoric acid and are classified into several subgroups on the basis of number of ester groups linked to each orthophosphate. For example, orthophosphate esters can be grouped into monoesters, diesters, triesters, and others (Table 9.2). Monoesters include sugar phosphates (e.g., glucose-6-phosphates and phosphophenols, which are intermediates of metabolic pathways), phosphoproteins, mononucleotides, and IHPs. Diesters include nucleic acids, phospholipids, and aromatic compounds.

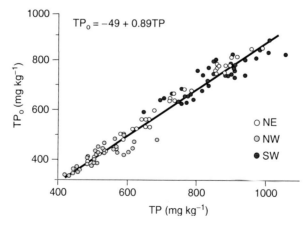

FIGURE 9.37 Relative proportion of organic phosphorus (TP_o) in a wetland dominated by organic soils (Blue Cypress Marsh, Florida).

TABLE 9.2
List of Select Organic Phosphorus Compounds Found in the Environment

Compound	Chemical Formula	Structure	Functional Class
Adenosine bisphosphate: exists in either 2′,5′ or 3′,5′ form.	$C_{10}H_{15}N_5O_{10}P_2$		Phosphate monoester
Adenosine 3′,5′-cyclic monophosphate: guanosine 3′,5′-cyclic monophosphate also exists in the environment	$C_{10}H_{12}N_5O_6P$		Phosphate diester

(*continued*)

TABLE 9.2 (continued)
List of Select Organic Phosphorus Compounds Found in the Environment

Compound	Chemical Formula	Structure	Functional Class
Adenosine 5′ diphosphate: intermediate in energy transfer from AMP to ATP	$C_{10}H_{15}N_5O_{10}P_2$		Organic polyphosphate
Adenosine 5′-monophosphate: exists mainly in 5′ form, although the 2′ and 3′ forms also exist. The ribonucleotides guanosine, cytidine, and uridine 5′-monophosphate also exist and may also be phosphorylated in the 2′ and 3′ positions. In addition, 2-deoxyribonucleotides exists (e.g., 2-deoxyadenosine 5′-monophosphate), which may be phosphorylated in the 3′ or 5′ positions.	$C_{10}H_{14}N_5O_7P$		Phosphate monoester

Structure	Formula	Description	Type
Adenosine 5′-triphosphate structure	$C_{10}H_{16}N_5O_{13}P_3$	Adenosine 5′-triphosphate: involved in energy transfer in biochemical reactions. Uridine, cytidine, guanosine, and thymine triphosphates are also common in biological systems. Nicotinamide adenine dinucleotide phosphate is also an important energy carrier.	Organic polyphosphate
2-Aminoethyl phosphonic acid structure	$C_2H_8NO_3P$	2-Aminoethyl phosphonic acid: most common naturally occurring phosphonate, found in a variety of organisms.	Phosphonate
ELF 97 phosphate structure	$C_{14}H_7Cl_2N_2O_5P$	2-(5′-chloro-2′-phosphoryloxyphenyl)-6-chloro-4-(3H)-quinazolinone: enzyme-labeled fluorescence substrate (ELF 97 phosphate, also known as ELFP) used to detect phosphatase activity by staining the site of cleavage.	Phosphate monoester
		Deoxyribonucleic acid (DNA): polynucleotide containing cellular genetic information.	Phosphate diester
Ethanolamine phosphate structure	$C_2H_8NO_4P$	Ethanolamine phosphate (o-phosphorylethanolamine): originates from the enzymatic cleavage of phosphatidyl ethanolamine by phospholipase C.	Phosphate monoester

(*continued*)

TABLE 9.2 (continued)
List of Select Organic Phosphorus Compounds Found in the Environment

Compound	Chemical Formula	Structure	Functional Class
A-glucose 1-phosphate: sugar phosphate found widely in plants, where it is the precursor of starch.	$C_6H_{13}O_9P$		Phosphate monoester
D-Glucose 6-phosphate: common sugar phosphate. Other sugar phosphates, such as fructose 6-phosphate also exist.	$C_6H_{13}O_9P$		Phosphate monoester
B-glycerophosphate: originates from the breakdown of phosphatidyl choline following loss of the fatty acyl chains and the choline moiety. Thus, it is common in alkaline soil extracts following chemical degradation of phosphatidyl choline.	$C_3H_9O_6P$		Phosphate monoester
myo-Inositol hexakisphosphate (phytic acid): major organic P compound in plant seeds and many soils, where it stabilizes following strong reaction with clays and organic molecules. Regarded as relatively recalcitrant in the environment. The hexakisphosphate and lower esters are detected in soils and aquatic sediments.	$C_6H_{18}O_{24}P_6$		Phosphate monoester
Methylumbelliferyl phosphate: synthetic fluorimetric monoester used as a sensitive substrate in assays of phosphomonoesterase activity.	$C_{10}H_9O_6P$		Phosphate monoester

Compound	Formula	Type	Description
Bis-methylumbelliferyl phosphate: synthetic fluorimetric diester used as a sensitive substrate in assays of phosphodiesterase activity.	$C_{20}H_{15}O_8P$	Phosphate diester	
para-Nitrophenyl phosphate: synthetic colorimetric monoester widely used to determine phosphomonoesterase activity in soils and plant material.	$C_6H_6NO_6P$	Phosphate monoester	
Bis-*para*-nitrophenyl phosphate: synthetic colorimetric monoester widely used to determine phosphodiesterase activity in soils and plant material.	$C_{12}H_9N_2O_8P$	Phosphate diester	
L-α-Phosphatidyl choline (lecithin): phospholipid commonly found in plants, but much less so in microorganisms	$C_{10}H_{19}NO_8P(2R)^a$	Phosphate diester	

(*continued*)

TABLE 9.2 (continued)
List of Select Organic Phosphorus Compounds Found in the Environment

Compound	Chemical Formula	Structure	Functional Class
L-α-Phosphatidyl ethanolamine (cephalin): phospholipid commonly found in microorganisms, but largely absent in plants.	$C_7H_{12}NO_8P(2R)^a$		Phosphate diester
Phosphoenolpyruvate: commonly found in plants; contains a high-energy bond involved in biochemical reactions	$C_3H_5O_6P$		Phosphate monoester
Phosphocreatine: common in vertebrate muscle; contains a high-energy direct N–P bond, as does phosphoarginine.	$C_4H_{10}N_3O_5P$		

Compound	Formula	Structure	Type

N-(Phosphonomethyl) glycine (Glyphosate): common herbicide marketed under the name "Round Up." Numerous other similar compounds exist. — $C_3H_8NO_5P$ — Phosphonate

Polyphosphate (linear): product of microbial activity found in soils and sediments; also exists in cyclic form (metaphosphate) in some bacteria. — $H_{2n}O_{3n+1}P_n$ — Phosphoanhydride (condensed phosphate)

Pyrophosphate: polyphosphate with two phosphate moieties; common in soils and sediments. — $H_4O_7P_2$ — Phosphoanhydride (condensed phosphate)

Ribonucleic acid (RNA): polynucleotide involved in protein synthesis. — Phosphate diester

Note: It also includes complex inorganic phosphates (condensed phosphates) and synthetic organic phosphates commonly used in assays of phosphatase activity. Detailed information on these compounds can be found in Corbridge (2000). For solution phosphorus-31 nuclear magnetic resonance chemical shift values of these compounds, see Turner et al. (2004).

[a] R represents hydrophobic fatty acyl chains, which may not be identical.

Source: Turner et al. (2004).

The range of organic phosphorus forms found in soils includes phospholipids, nucleic acids, IHPs, glucose-6-phosphates, glycerophosphate, phosphoproteins, and polymeric organic phosphorus of high-molecular-weight compounds. Soil organic phosphorus consists of mainly esters of o-phosphoric acid and can be classified into three primary groups:

- Inositol phosphates: The monoesters, comprising mostly IHPs account for approximately 50% of the TOP. In soil organic matter, a major portion of phosphorus is stored in phytin, which is the major source of IHPs. The hexaphosphate ester exists in macrophyte detrital tissues, especially in grains as mixed Ca and Mg salts (Turner et al., 2002).
- Nucleic acids: Diesters of o-phosphate include nucleotides (mainly as RNA and DNA, <2% of organic phosphorus). Phosphorus is a major constituent in nucleic acids (DNA and RNA). Nucleic acids are polymers of five-member ribose sugar linked by phosphate ester group. The two major classes of nucleic acids are RNA and DNA. Nucleic acid accounts for 0.2–2.5% of organic phosphorus. In microbes, more than half of the organic phosphates are nucleic acids. Each nucleic acid contains a ribose sugar, a purine or pyrimidine base, and phosphate group with ester linkage to pentose sugars.
- Phospholipids: Phospholipids (predominantly phosphoglycerides) constitute approximately 1–5% of TOP in soils. Phospholipids are present in cell membranes and are related structurally to fats and oils. Phospholipids are arranged in bilayers in membranes with two nonpolar hydrocarbon tails pointing in and phosphatidyl amine polar head group.
- Metabolic phosphates: They are present in trace quantities. These include ATP, ADP, and NADP. These are primarily monoesters. Phosphorylated sugars (monoesters) are intermediate compounds of carbohydrate metabolism. These include glucose ↔ glucose-6-phosphate ↔ fructose 6-phosphate ↔ fructose 1,6 phosphate ↔ glyceraldehydes 3-phosphate ↔ 1,3-diphosphoglyceric acid ↔ 3-phosphoglyceric acid ↔ 2-phosphoglyceric acid ↔ phosphopyruvic acid ↔ pyruvic acid. The ATP functions as medium of energy storage and transfer, and the energy-rich bond is the result of the pyrophosphate linkage of ATP. The NAD and NADP are coenzymes that are involved in oxidation–reduction reactions within the cell.

Phosphorus is a major constituent of macromolecules, particularly in nucleic acids (DNA and RNA), in phospholipids of membranes, and as monoesters of a variety of compounds, particularly those involved in biochemical pathways. In growing microorganisms, more than half of the organic phosphorus is in nucleic acids, whereas phospholipids and monoesters constitute the remainder in varying proportions (Table 9.3; Magid et al., 1996). Many organisms can store orthophosphate or polyphosphates. In plants, IHP can form a major storage compound for phosphorus, particularly in seeds.

TABLE 9.3
Distribution of Organic Phosphorus (Expressed as % of Total Phosphorus in Soils and Growing Organisms)

Organic Phosphorus	*Escherichia* Cell	Fungi	*Nicotiana*	Soil
Nucleic acids	65	58	52	2
Phospholipids	15	20	23	5
Phosphate monoesters	20	22	25	50

Source: Magid et al. (1996).

Although relatively low in concentration, diesters may contribute to short-term bioavailable pool more than monoesters. Compared to monoesters, diesters are more accessible to microbial attack. Due to reactive o-phosphate groups, the IHP monoester is prone to strong adsorption to humic and fulvic acid components of the stable organic matter. Organic phosphorus associated with humic and fulvic acids represents >40% of total the soil phosphorus. The fulvic acid phosphorus constitutes a large fraction of the organic phosphorus in most soils, and it is likely that this pool is derived from plant litter and recently deposited organic matter (see Section 9.5.2 for details on chemical fractionation of soil organic phosphorus).

9.6.2 Chemical Characterization of Organic Phosphorus

Organic phosphorus forms discussed previously can be generally grouped into (1) easily mineralizable organic phosphorus (nucleic acids, phospholipids, and sugar phosphates) and (2) slowly mineralizable organic phosphorus (IHPs, or phytin). Conventionally, soil organic phosphorus stability is characterized on the basis of its extractability in alkali media. Different alkali media extract different fractions of organic phosphorus on the basis of the mechanism by which the organic phosphorus fraction interacts with other soil components. For example, organic phosphorus turnover studies have shown that $NaHCO_3$-extractable organic phosphorus in upland soils is easily mineralizable and may contribute to plant-available phosphorus during one growing season. More resistant forms of organic phosphorus involved in the long-term transformations of phosphorus in upland soils are extractable with NaOH. Several researchers have used a sequence of alkali extracts to quantify the range of soil organic phosphorus stability. During the last two decades, various combinations of operationally defined chemical extracts have been employed in sequence to obtain more detailed information on organic phosphorus and inorganic phosphorus turnover in soils (Hedely and Stewart 1982; Ivanoff et al., 1998).

A chemical fractionation scheme tested in our earlier studies identifies organic phosphorus in several pools including (1) labile phosphorus (microbial biomass phosphorus [MBP]), (2) acid hydrolyzable phosphorus, (3) fulvic acid–bound phosphorus (FA-P), (4) humic acid–bound phosphorus (HA-P), and (5) residual organic phosphorus. The scheme is operationally defined and was applied successfully for selected wetland soils. The major disadvantages of chemical fractionation schemes relate to the chemical hydrolysis of organic phosphorus compounds during extraction and their inability to provide any direct evidence of the structural components of organic phosphorus compounds. This prevents the schemes from distinguishing clearly and consistently between the major classes of organic phosphorus compounds.

One example of the chemical extraction method traditionally used to characterize TOP, on the basis of their ease of extraction with selected chemicals, is shown in Figure 9.38 (Ivanoff et al., 1998). These fractions include

- Labile organic phosphorus (LOP): Wet soils are extracted with 0.5 M $NaHCO_3$. Extracts are filtered through 0.45 μm membrane filter. Filtered solutions are analyzed for total phosphorus and inorganic phosphorus (labile P_i). The difference between these two fractions is called LOP.
- Total labile organic phosphorus (TLOP): Wet soils are treated with chloroform to lyse microbial cells, followed by extraction with bicarbonate solution for a period of 17 h. Extracted solutions are filtered through 0.45 μm membrane filter. Filtered solutions are analyzed for total phosphorus and inorganic phosphorus. The difference between these two fractions is called TLOP.
- MBP: Organic phosphorus associated with microbial pool. This pool is estimated as follows:

$$MBC = \frac{(TLOP - LOP)}{\text{Efficiency factor}}$$

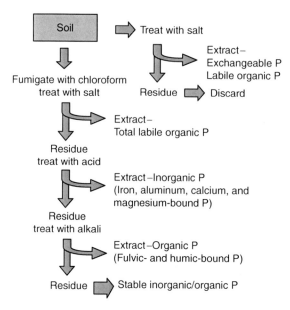

FIGURE 9.38 Schematic showing the chemical fractionation scheme used to determine organic pools in wetland soils. (Ivanoff et al., 1998.)

Depending on the soil type, the efficiency factor (refers to chloroform fumigation extraction efficiency) may vary from 0.3 to 0.4. Caution should be exercised in using these efficiency factors. In recently accreted soils with high labile organic matter content and in soils containing fine roots, use of efficiency factors may overestimate the MBP.

- Moderately labile organic phosphorus (MLOP): The residual soil after chloroform fumigation is extracted with 1 M HCl, followed by centrifugation at 6,000 rpm. Supernatant liquid is filtered through 0.45 μm membrane filter. Filtered solutions are analyzed for total phosphorus and inorganic phosphorus. The difference between these two fractions is called MLOP.
- Moderately resistant organic phosphorus (MROP) or FA-P: Residual soil after acid extraction is subsequently extracted with 0.5 M NaOH for 17 h, followed by centrifugation at 6,000 rpm. Supernatant is filtered through 0.45 μm membrane filter. A portion of the filtered solution is acidified to pH 2.0. The supernatant liquid is digested and analyzed for total phosphorus. This organic phosphorus is associated with fulvic acid. This fraction of phosphorus is dissolved in both alkali and acid.
- Highly resistant organic phosphorus (HROP) or humic acid–bound phosphorus. A second portion of filtered alkaline extracts is digested and analyzed for total phosphorus. The HROP is estimated as follows:

$$HROP = \text{Total P alkali extracts} - MROP$$

This fraction of organic phosphorus is dissolved only in alkali and is insoluble in acid medium.
- Nonreactive organic phosphorus (NROP): The residue after alkaline extraction is either combusted to ash at 550°C or digested with nitric-perchloric acid.

Although these methods are rapid and simple, they only provide indirect evidence of different phosphorus forms on the basis of the ease of extraction with specific chemical reagents. The limitation of sequential fractionation is that it provides no structural information on soil phosphorus. In recent years with availability of different types of instrumentation, a variety of methods including GC, high-performance liquid chromatography (HPLC), thin-layer chromatography, ^{31}P NMR (nuclear magnetic resonance) spectroscopy, mass spectrometry, and reaction with specific enzymes have been used to identify specific forms of organic phosphorus. For details on these methodologies, see Turner et al. (2004). For example, solution ^{31}P NMR spectroscopy was first used to determine the phosphorus composition of soil extracts by Newman and Tate (1980) and has been subsequently used by a number of investigators (Turner et al., 2003). There is little agreement on the most suitable extractant, but those in use include 0.5 M NaOH, the cation exchange resin Chelex in water, 0.5 M NaOH plus Chelex, 0.5 M NaOH plus NaF, 0.25 M NaOH plus 50 mM EDTA, and 0.25 M NaOH plus 50 mM EDTA plus Chelex. These extractants solubilize different concentrations and forms of phosphorus and cations from soils. Chelex and EDTA are both used to release phosphorus from paramagnetic ions, thereby reducing line broadening and improving spectral quality. Similar to traditional chemical fractionation schemes, some of these solvents can alter the chemical forms of organic phosphorus compounds through hydrolysis (Turner et al., 2003a).

The general phosphorus extraction procedure involves shaking the soil with a solution containing 0.25 M NaOH and 50 mM EDTA for 4 h at 20°C. The extracts are then centrifuged and frozen at −80°C, then lyophilized. Freeze-dried extract is redissolved in D_2O and a solution containing 1 M NaOH and 0.1 M EDTA. Solutions are then transferred to a 5 mm NMR tube. Chemical shifts are assigned to individual phosphorus compounds or functional groups. Generally, signals can be identified from phosphonates, inorganic phosphate, phosphate monoesters, phosphate diesters, pyrophosphate, and polyphosphates, and signals are reported as the chemical shift in ppm (Figure 9.39). Although this NMR spectroscopy has been widely used to characterize soil samples in terrestrial ecosystem, few have used it to characterize organic phosphorus in wetland soils. In a book edited by Turner et al. (2004), several papers address the use of NMR techniques and limitations in identifying soil organic phosphorus in a range of ecosystems including wetlands (Turner et al., 2004).

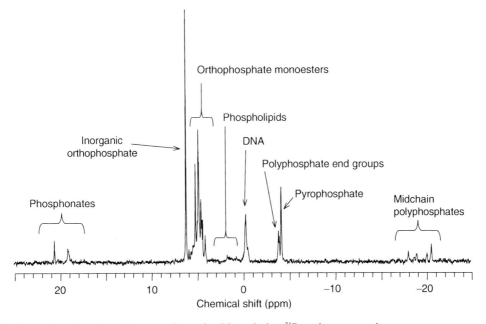

FIGURE 9.39 Phosphorus forms as determined by solution ^{31}P nuclear magnetic resonance spectroscopy (Turner et al., 2004).

9.7 PHOSPHORUS UPTAKE AND STORAGE IN BIOTIC COMMUNITIES

9.7.1 Microorganisms

Phosphorus content of microbial tissue is in the range of 1.5–2.5% for bacteria and 5% for fungi. Fungi are typically restricted to the aerobic environment of wetlands, including detrital matter in the water column and aerobic microzones at the soil–floodwater interface and rhizosphere. MBC accounts for approximately 2–5% of soil organic carbon, as compared to 5–30% of total phosphorus recovered as MBP. Soil microbial biomass is relatively labile. For example, in phosphorus-enriched areas (eutrophic wetlands) approximately 5–10% of total phosphorus may be present as MBP, as compared to 10–20% of total phosphorus in soils from phosphorus-limited areas (oligotrophic wetlands) (Figure 9.40). As described previously, MBP is estimated using chloroform fumigation techniques. Estimated MBP is adjusted for extraction efficiency. In soils with high organic matter content, correcting for extraction efficiency may overestimate the amount of phosphorus present in microbial biomass pool. In the example shown in Figure 9.40, if the values are corrected for extraction efficiency, up to 60% of the total phosphorus is shown to be in microbial pool of organic soils. It is unlikely that such a high proportion of phosphorus is present in microbial pool of organic soils. However, in phosphorus-limited areas, a larger proportion of phosphorus is assimilated in the microbial biomass, indicating greater efficiency of phosphorus assimilation of remineralized phosphorus. Microbes in these systems are very efficient in scavenging any available phosphorus.

Soil redox potential can influence MBP with fungi and aerobic bacteria dominating soils with higher redox conditions. In the example shown in Figure 9.41, organic soils obtained from a wetland were incubated under various redox conditions, and after equilibration of several days at respective redox potential, soils were analyzed for MBP (McLatchey and Reddy, 1998). Under aerobic conditions, MBP accounted for up to 20% of total phosphorus, as compared to <10% under anaerobic conditions. In addition to high phosphorus demand by aerobic microorganisms (bacteria and fungi), labile inorganic phosphorus can also be precipitated with iron oxides.

Microorganisms also can play a major role in retaining phosphorus in wetlands with organic matter inputs or those producing large quantities of detrital matter internally as a result of high primary productivity. Microorganisms incorporate dissolved phosphorus into cellular constituents,

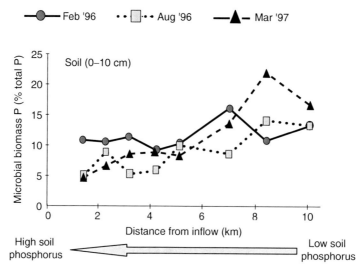

FIGURE 9.40 Influence of phosphorus loading on microbial biomass phosphorus in soils of northern Everglades. (From Chua, 2000.)

Phosphorus

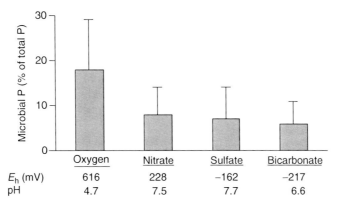

FIGURE 9.41 Influence redox potential and electron acceptors on microbial biomass in organic soil. (Adapted from McLatchey and Reddy, 1998.)

which then become integral parts of the particulate matter. Through the formation of polyphosphate compounds, microorganisms are able to survive in alternating oxidation–reduction environments. This luxury consumption of phosphorus (polyphosphate formation) can significantly influence the concentration of phosphorus in the water column by removing phosphorus at high water column concentrations, and contributing phosphorus when water column concentrations are low.

9.7.2 Periphyton

Periphyton generally refers to the microbial mat communities of attached microorganisms, both floral and faunal, that grow on submerged surfaces. These communities are formed by complex assemblages of cyanobacteria, eubacteria, diatoms and eukaryotic algae. Periphyton exists in wetlands and littoral habitats as (1) contiguous mats, loose material, or motile cells associated with the benthos (epipelon, BP), (2) thin films or more extensive growths ("sweaters") growing on submerged and emergent macrophytes (epiphyton, EP), and (3) floating or subsurface mats that may become entangled with floating and submerged vegetation (metaphyton, MP).

Benthic periphyton utilizes phosphorus from both soil and the water column, whereas floating periphyton receives phosphorus largely from the water column. Epiphytic periphyton receives phosphorus largely from the water column and phosphorus released from senescent macrophyte tissue. A small amount of phosphorus is also derived by epiphytic periphyton from living tissue of macrophytes. Chemical composition of periphyton varies depending on water column phosphorus, water flow, and type of macrophytes. Phosphorus content of periphyton is in the range of 0.1–4.5 mg g^{-1} (dry weight), but the phosphorus content of 1.1–2.2 mg g^{-1} is necessary for normal growth. High phosphorus concentrations reported are the result of luxury uptake, which is commonly observed in enriched water. Periphyton may reach phosphorus saturation at relatively low water column phosphorus concentrations (<0.025 mg L^{-1}).

The epiphytic periphytic community, which grows on submerged portions of living macrophytes and organic detritus, acquires phosphorus both from the water within and passing through the wetlands and from the supporting "host" macrophyte tissues. Although relatively little of the total phosphorus pool within actively growing macrophytes is released, this released phosphorus can be important to certain epiphytic species that grow attached to the macrophyte tissue (Moeller et al., 1988). Even when phosphorus concentrations in the overlying water are high, some nutrients are obtained from the macrophyte simply because diffusion within the complex epiphytic community is too slow to meet phosphorus demands (Wetzel, 1993a). The periphyton, rather than the macrophytes, functions as the primary scavenger for limiting nutrients such as phosphorus from the water column.

During dormancy and macrophyte senescence of macrophyte tissues, the release of organic phosphorus is readily utilized by the periphytic community, which develops profusely on plant tissues and detritus. Much of the released phosphorus is rapidly retained and recycled by epiphytic bacteria and algae (Moeller et al., 1988; Wetzel, 1990, 1993a). The scavenging and retention capacity of the attached microflora can exceed if the loading of phosphorus is very high or the rate of water flushing through this biological sieve is too rapid to allow for uptake and retention. Rapid water movements through wetlands can occur by natural storm events or from artificial human-induced runoff. Once phosphorus enters the macrophyte-detritus-periphyton community, however, it has a high probability of being retained and recycled.

9.7.3 Vegetation

Three types of macrophytes are present in wetlands, including emergent macrophytes, submerged macrophytes, and floating macrophytes. Emergent macrophytes in marshes and trees in bottomland hardwood-forested wetlands and riparian wetlands assimilate most of their phosphorus by roots and rhizomes from the soil pore water, where bioavailable phosphorus concentrations are several orders of magnitude greater than in the overlying water. Submerged macrophytes can assimilate phosphorus from both soil pore waters and the water column. Floating macrophytes, epiphytic algae, bacteria, and other organisms attached to the foliage assimilate nearly all of their phosphorus directly from the water column.

Phosphorus concentrations in vegetation exhibit wide inter- and intraspecific variability (Table 9.4 and Figure 9.42). Tissue phosphorus concentrations vary significantly among wetland species even when these plants are cultured under identical nutrient conditions. In addition, tissue phosphorus concentrations are also influenced by soil nutrient availability and growth stage of the plant. Low tissue phosphorus levels are found in older and matured plants or plants grown under low-nutrient conditions, whereas high tissue phosphorus concentrations reflect plants cultured in

TABLE 9.4
Range of Phosphorus Concentrations in Wetland Vegetation

Plant	Phosphorus (g kg^{-1})	
Herbaceous vegetation		
Typha (cattail)	0.5–4	
Juncus (rush)	2	
Scirpus (bulrush)	1–3	
Phragmites (reed)	2–3	
Eleocharis (spike rush)	1–3	
Saururus	1–5	
Forested vegetation	Leaves	Wood
Taxodium	0.5–2.6	0.03–0.06
Nyssa	1.01	0.22
Acer	1.1–2.6	0.22
Magnolia	0.9–2.0	0.30

Source: Reddy and DeBusk (1987).

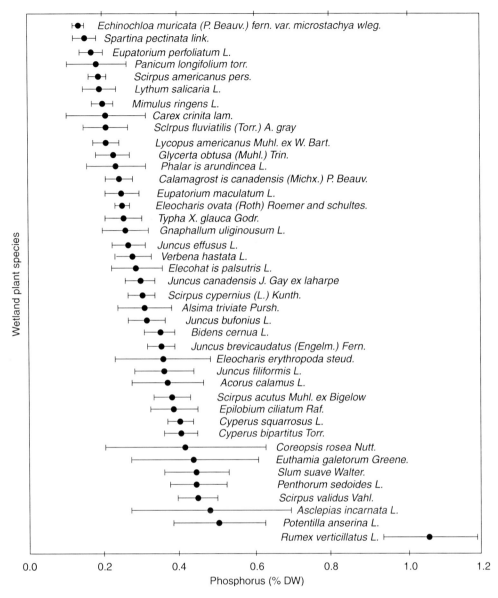

FIGURE 9.42 Mean plant tissue phosphorus concentration (percentage dry weight) of emergent wetland plants grown under conditions of sustained fertilization. (From McJannet et al., 1995.)

nutrient-rich environments or those in early stages of growth. Tissue phosphorus content of wetland plants can range from <1 (dry-weight basis) to 7 g kg^{-1}.

The range of phosphorus concentration in live biomass and litter varies with wetland types (Figure 9.43). Phosphorus concentration ranged from 0.2 to 2.8 g kg^{-1} for live biomass and from 0.1 to 1.7 g kg^{-1} for litter. For live tissue, phosphorus concentration was in the order of bogs < rich fens < moderate-rich fens < swamps, whereas for litter, phosphorus concentrations were in the order of poor fens < rich fens < moderate-rich fens < marshes and swamps (Bedford et al., 1999).

Phosphorus uptake by macrophytes is maximum during the peak growing season, followed by a decrease or even cessation in the fall/winter. Typically, uptake rates of phosphorus by many

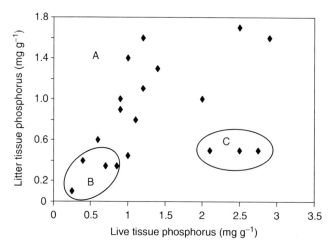

FIGURE 9.43 Relationship between phosphorus concentration of live tissue and litter (A, wetland plants of temperate northern wetlands [from Bedford et al., 1999]; B, *Typha* and *Cladium* from northern Everglades; C, bogs and fens [from Kellog and Bridgham, 2003]).

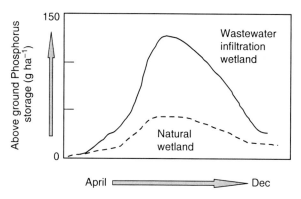

FIGURE 9.44 Seasonal changes in phosphorus storage in aboveground biomass of *Phragmites australis*. (From Meuleman et al., 2002.)

aquatic macrophytes are highest during early-spring growth, before maximum growth rate is attained (Figures 9.44 and 9.45). This early uptake and storage of nutrients provides a competitive edge for growth-limiting nutrients during the periods of maximum demand. Phosphorus concentrations are higher in young plants, but as the biomass increases, tissue phosphorus concentration decreases (Figure 9.44), and during the same period, total phosphorus storage in the plant tissues increases.

Another important response to seasons is the translocation of nutrients within the plant. Prior to fall senescence, the majority of important ions are translocated from shoot portions to the roots and rhizomes. These stored nutrients are used during early-spring growth. Phosphorus storage in vegetation can range from short to long term, depending on type of vegetation, litter decomposition rates, leaching of phosphorus from detrital tissue, and translocation of phosphorus from aboveground to belowground biomass (Figure 9.46).

Plants contribute to soluble phosphorus not only after their death but also while they are still alive, as older tissues leach. Uptake of phosphorus by vegetation maintains low soluble phosphorus concentration in the soil profile. A large portion of phosphorus stored in belowground biomass is usually not accounted for in mass balance studies. Most of the emphasis is placed on aboveground

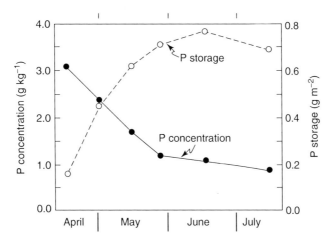

FIGURE 9.45 Seasonal changes in phosphorus concentration and storage in the aboveground biomass of *Typha latifolia*. (From Boyd, 1970.)

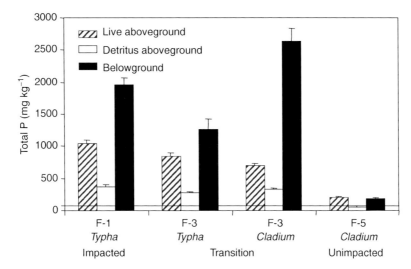

FIGURE 9.46 Phosphorus concentration of *Typha* and *Cladium* grown along eutrophic gradient in the northern Everglades. (From Reddy, K. R., Unpublished Results, University of Florida.)

biomass and its fate. As plants produce detrital matter, phosphorus is rapidly released as a result of leaching and decomposition.

Phosphorus assimilation and storage in plants depends on vegetative type and growth characteristics. Floating and submerged vegetation has limited potential for long-term phosphorus storage. Because of rapid turnover, phosphorus storage in biomass is short term, and much of the phosphorus is released back into water column upon vegetative decomposition. Emergent macrophytes have an extensive network of roots and rhizomes and have great potential for phosphorus storage. They have more supportive tissue than floating macrophytes and have a high ratio of belowground biomass (roots and rhizomes) to aboveground biomass (stem and leaves), providing ideal anatomical structures for phosphorus storage. Phosphorus storage in the aboveground biomass of emergent macrophytes is usually short term, with large amount of phosphorus being released during decomposition of detrital tissue. The aboveground parts of most macrophytes grow and die on a cycle ranging from the annual growing season in northern climates to faster cycles in southern climates. Shoot turnover times in subtropical zones are as short as 2–3 months (Davis, 1994). Root turnover times are largely unstudied but are estimated at 1.5–3 years. In addition, many macrophytes possess the ability to translocate nutrients from aboveground parts to their roots and to use these stored materials to foster growth during the early part of the growing season. Approximately 50% of the phosphorus lost from living emergent macrophyte shoots is translocated to roots and rhizomes. Belowground storage of phosphorus is often underestimated, and its fate and stability are not clearly understood. Partitioning of phosphorus in live, detritus, and belowground tissue for *Typha* and *Cladium* in phosphorus-enriched and phosphorus-limited sites of the Everglades is shown in Figure 9.46. High concentration of phosphorus in the belowground portions of the plant suggests mobilization of phosphorus from aboveground tissue.

Unlike emergent vegetation, trees in forested wetlands provide long-term storage. Reported phosphorus uptake rates in forested wetlands are 1–15 kg ha^{-1} year^{-1} (Reddy and DeBusk, 1987). Net assimilation of phosphorus in woody tissue is small, as compared to leaves; thus, a substantial portion of phosphorus is recycled in litter fall. Aboveground vegetation produces detrital material as a result of senescence and frost damage. Most of this detrital material is either attached to the primary plant or detached and deposited on the soil surface. In forested wetlands, detrital production occurs primarily through litter fall.

Aboveground plant parts return leached phosphorus to the water column upon death and decomposition and deposit refractory residuals on the soil surface. However, dead roots decompose underground and therefore add refractory compounds to subsurface soils, and leachate to the pore water in the root zone. Thus, the aboveground portion of the macrophyte cycle returns phosphorus to the water, whereas the belowground biomass returns phosphorus to the soil. Nutrient loading to wetlands produces a larger standing crop and stimulates the entire biocycle. Incremental storage of phosphorus in biomass occurs due to the stimulation of growth, producing larger quantities of leaves, roots, and litter. Relative rates of biological processes regulating phosphorus cycle are speeded up in a system receiving sustained addition of nutrients.

9.8 MINERALIZATION OF ORGANIC PHOSPHORUS

The net accumulation of organic phosphorus from vegetation depends on type of vegetation, root:shoot ratio, turnover rates of detrital tissues, C:P ratio of the detrital tissue, type of metabolic pathways, and physicochemical properties of the water column. Phosphorus is cycled with in the soil–water–plant system as follows: herbaceous vegetation rooted in soil obtains most of its phosphorus needs from the soil pore water and translocates it to aboveground vegetation to support active vegetative growth. On maturity and senescence, a substantial portion of phosphorus present in the aboveground vegetation is translocated into belowground biomass (roots and rhizomes). In nutrient-rich systems, up to 80% of the phosphorus stored in some aquatic macrophyte detrital tissue is released into the water column either by initial leaching or as a result of decomposition (see Section 9.5.6). The residual detrital material is deposited on the soil surface and becomes an integral part of the soil, thus providing long-term storage. But over the short term, rapid turnover rates and cycling can contribute bioavailable phosphorus to the water column and influence water quality.

Detrital organic phosphorus is decomposed and transformed during burial, when there is generally a shift from aerobic to anaerobic conditions in the profile (Figure 9.47). Mineralization rates are drastically reduced under anaerobic conditions, which leads to accumulation of moderately decomposable compounds, along with lignin and other recalcitrant fractions. Thus, the accumulation of soil organic phosphorus in wetlands is typically characterized by a stratified buildup of partially decomposed plant remains, with relatively little humification. However, the biodegradability of organic phosphorus decreases with depth because material in deeper layers is older and more organic phosphorus is humified compared to recently accumulated material near the surface.

Phosphorus

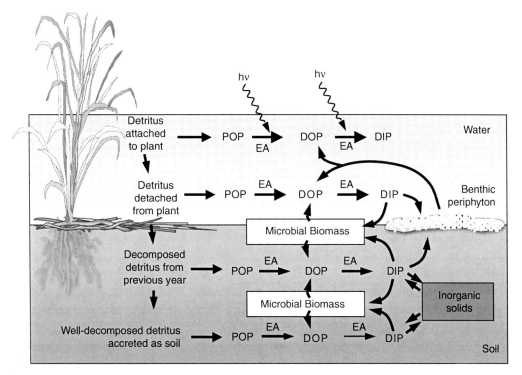

FIGURE 9.47 A schematic showing organic phosphorus cycling in aerobic and anaerobic layers of wetland soils.

9.8.1 Abiotic Degradation and Stabilization of Organic Phosphorus

DOM and associated DOP produced from POM (detrital matter and soil organic matter) are the major components transported from wetlands to adjacent aquatic systems. Within wetlands, DOM and associated DOP can have significant influence on heterotrophic production and nutrient cycling. DOP can be produced and mineralized by the following abiotic processes:

- Physical leaching of soluble organic phosphorus from senescing plant tissue
- Noncatalyzed hydrolysis of phosphate esters
- Photolysis

9.8.1.1 Leaching of Soluble Organic Phosphorus

Abiotic degradation of DOP and POP in the water column is mediated by physical leaching of soluble organic phosphorus compounds from partial or completely senesced leaves. During growing season, with constantly maturing and gradual, partial senescence of leaves of wetland plants, particularly among submersed macrophytes, phosphorus stored in plant tissue can be leached into the water column. Leaching of phosphorus compounds from senescent macrophyte tissue is often rapid and can release from 20 to 50% of total phosphorus content in a few hours and 65–85% over longer periods. Rates of leaching are often higher from roots than from leaves.

9.8.1.2 Noncatalyzed Hydrolysis of Phosphate Esters

The oxygen–phosphorus bond (ester linkage) in organic phosphorus can be cleaved by the addition of water. Both abiotic and biotic hydrolytic reactions are involved in cleaving this bond. It is often difficult to differentiate relative roles of abiotic and biotic pathways. The significance of abiotic hydrolysis of these phosphate esters is not documented in wetlands and aquatic ecosystems.

9.8.1.3 Photolysis

Photochemical reactions are another important mechanism involved in the breakdown of DOP in wetlands and aquatic systems. Several organic phosphorus compounds can be degraded through photolytic reactions, including humic-phosphorus complexes, nucleic acids, and other simple organic phosphorus molecules. Photolysis, often referred to as photooxidation, photomineralization, and photochemical degradation, is a process where high-energy solar radiation (UV light) breaks down DOM into simple compounds (see Chapter 5 for details). The role of photolysis in breaking down organic phosphorus compounds has not been documented in wetlands. Photochemical reactions are active only in the photic zones of the water column. Shading resulting from vegetation can significantly influence the photolysis of DOP. The significance of photodegradation in cycling of DOP in aquatic environments (such as lakes) has been documented. However, when the same water samples were incubated with alkaline phosphatase and no UV light, phosphorus was not released, providing additional evidence for abiotic degradation on phosphorus release. An increase in phosphorus release in these waters during photodegradation also resulted in accumulation of Fe(II), suggesting the abiotic reduction of Fe(III) (see Chapter 10 for additional details). Photolytic reactions can have an indirect effect in releasing enzymes complexed with humic materials (see Section 8.5.5 for details on enzymatic hydrolysis of organic phosphorus).

9.8.1.4 Stabilization of Organic Phosphorus

Several abiotic processes can regulate the bioavailability of organic phosphorus in soils and sediments. These processes include:

- Sorption of phosphate monoesters on iron and aluminum oxide surfaces and clays
- Ability to form insoluble complexes with polyvalent cations
- Incorporation into soil organic matter

Organic phosphorus is readily sorbed on clays and soil organic matter. IHPs are sorbed to clays to a greater extent than simple sugar phosphates, nucleic acids, and phospholipids. The extent and the rate of sorption depend on soil physicochemical properties and molecular size of organic phosphorus. In acid soils, IHP is regulated by the amount of iron and aluminum oxides, whereas in neutral and alkaline soils, it is controlled by the amount of organic matter, clays, and calcium minerals. Montmorillonite clays have greater affinity to sorb organic phosphorus than illite and kaolinite. Nucleic acid sorption in soils is regulated by montmorillonite and the molecular weight of DNA.

Organic phosphorus compounds can form complexes with metallic cations. However, the ability to form complexes depends on the number of phosphate groups in the organic phosphorus compound. For example, IHPs have greater ability to form complexes than nucleic acids and phospholipids. Like humic and fulvic acid complexes with metals (see Chapter 5), organic phosphorus compounds can form stable complexes in the order of: $Cu^{2+} > Zn^{2+} > Ni^{2+} > Mn^{2+} > Fe^{2+} > Ca^{2+}$. These complexes are stable at neutral pH. Although stability of these complexes in wetland soils is not known, soil anaerobic conditions typically increase DOM and associated organic phosphorus, thus increasing complexation with metal cations. In minerals soils dominant with iron redox couple, reduction of ferric iron to ferrous iron can decrease potential complexation of organic phosphorus. However, oxidation of ferrous iron to amorphous ferric oxyhydroxide can increase organic phosphorus complexation.

9.8.2 Enzymatic Hydrolysis of Organic Phosphorus

Microbial biomass regulates storage and transformation of phosphorus, and flow of phosphorus through the soil microbial pool can be substantial. In phosphorus-limited wetlands containing significant quantities of organic phosphorus, bioavailability of phosphorus may be regulated by

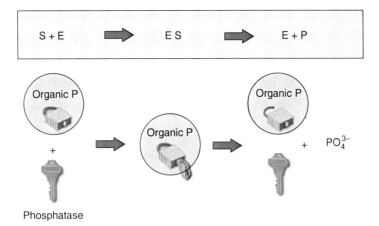

FIGURE 9.48 A schematic showing enzyme-mediated chemical reactions involved in breakdown of organic phosphorus. S = substrate; E = enzyme. (Newman, S., South Florida Water Management District, West Palm Beach, Florida.)

the mineralization of organic phosphorus. Microbial communities are central to the mineralization of organic phosphorus. Soil organic matter and detrital plant tissue contain complex structural compounds (high molecular weight), which are first hydrolyzed through the activity of extracellular enzymes into simple organic molecules. These low-molecular-weight compounds can be directly transferred into the cell where they are oxidized and used as an energy source. Nutrient enrichment can significantly influence the activity of extracellular enzymes, which are critical in the first step in degradation of soil organic matter and detrital plant tissue. Measurement of enzyme activity can be linked to litter characteristics and decomposition rates. Several studies reported in the literature suggest that enzyme activities might serve as indices for the first step in breakdown of complex polymers (see Chapter 5). Thus, changes in enzyme activity may prove to be a sensitive indicator of the bioavailability of organic phosphorus.

In enzyme-mediated reactions, the first step involves linking substrate (S) and the enzyme (E) with weak bonds (Figure 9.48). This is called the enzyme–substrate complex (ES). During the second step, the substrate is activated and a product is separated as a result of breaking bonds. The product and enzymes are now released. In this example, phosphate is linked to organic compound through ester linkage. This linkage is broken by hydrolytic enzymes such as phosphatases, resulting in the release of DIP.

Enzymes play a key role in regulating phosphorus release from organic phosphorus, especially in P-limited wetland systems. Some of the enzymes involved are:

- Phosphatases or monoesterases (acid and alkaline phosphatases)
- Phosphodiesterases

9.8.2.1 Phosphatases or Monoesterases

Organic phosphorus availability from monoesters requires enzymatic cleavage of the ester linkage. This occurs primarily through several extracellular enzymes such as phosphatases through hydrolysis of phosphate esters. Common monoesters in soils include glycerophosphate, glucose-6-phosphate, fructose-6-phosphate, and IHPs. This hydrolysis is achieved by phosphomonoesterases bound to or within the cell membrane, or present in the soil pore water or in the water column.

Soil microorganisms mineralize organic phosphorus through the secretion of various phosphatases. Similarly, plants can utilize organic phosphorus by means of phosphatase activity. Phosphatase activity in the rhizosphere originates from plant roots, fungi, ectomycorrhizae or from bacteria. However, phosphatases have been detected in soils without the influence of plant roots in addition to root surfaces and rhizosphere soil. Regardless of the source of phosphatases, their activities are modulated by the biochemical changes. The limiting factor of the plant utilization of organic phosphorus is the availability of hydrolyzable organic phosphorus sources. In most studies, a positive correlation has been found between phosphatase activity and soil organic phosphorus and organic matter contents. Plants that are grown in P-deficient soils possess higher root phosphatase activity. Several field investigations have supported the hypothesis that alkaline phosphatases are, to a large extent, produced during P-deficiency periods.

The activity of extracellular enzymes, particularly phosphatases or monoesterases, may be useful in predicting the impacts of phosphorus on wetland or other aquatic ecosystems. Limitation of DIP induces the phosphatase production by microbes. In the next two examples, usefulness of alkaline phosphatase activity (APA) as an indicator of phosphorus impacts is clearly demonstrated. The measurements in periphyton mats collected along the nutrient gradient in the northern Everglades (Figure 9.49) consistently showed increased APA levels with distance from the nutrient inflow, with high values measured in periphyton samples obtained in the interior phosphorus-limited marsh. Typically, APA is inversely related to the DIP concentration of surface waters. Similarly, the APA levels of detrital plant tissue and of soil samples collected in the same system were at background levels for samples collected up to 5 km from the inflow and then increased dramatically between 5 and 10 km, indicating phosphorus limitation in the interior marsh (Figure 9.50). These results

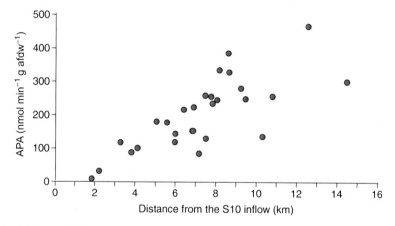

FIGURE 9.49 Influence of phosphorus loading on alkaline phosphatase activity (APA) in periphyton mats in the northern Everglades. (Newman, S., South Florida Water Management District, West Palm Beach, Florida.)

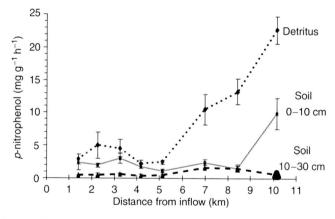

FIGURE 9.50 Influence of phosphorus loading on alkaline phosphatase activity (APA) in floc and surface soils from northern Everglades. (From Wright and Reddy, 2001.)

suggest that APA can be used as an indicator of phosphorus enrichment or phosphorus limitations in an ecosystem.

IHPs, another group of monoesters, constitute approximately 10–50% of the total phosphorus in terrestrial soils. At present, no information is available for wetland soils. It is likely that a larger proportion of total phosphorus may consist of IHPs in wetlands soils, resulting from a high rate of organic matter accumulation. IHPs are relatively recalcitrant and are strongly sorbed to clays and organic matter and well protected from enzymatic cleavage by monoesterases.

9.8.2.2 Phosphodiesterases

Phosphodiesters include nucleic acids, phospholipids, and others. These compounds are labile and actively involved in biological cycling of phosphorus and are known to be important sources of DIP in aquatic environment (Chrost and Siuda, 2002). Turnover time of these compounds is short because of their rapid enzymatic hydrolysis by phosphodiesterases. The absolute quantity of DIP release during the hydrolysis of these compounds has not been determined in wetland soils. In aquatic environment, phosphodiesters were found to be one of the major sources of DIP in the water column. For example, enzymatically hydrolysable extracellular DNA contributes approximately 10–60% of phosphorus to the total DOP (Chrost and Siuda, 2002).

Surface floc samples in wetlands typically contain benthic periphyton and detrital material colonized by microbes. Data in Table 9.5 indicate that phosphomonoesterase and phosphodiesterase activities in floc samples from the interior marsh (10.1 km from the inflow, phosphorus-limited system) of the northern Everglades were high as compared to soils collected from the impacted site (phosphorus-enriched system), indicating phosphorus limitation in the interior marsh. The ratio of monoesterase to diesterase was close to 1 in the interior marsh as compared to 2–3 in the impacted site. High diesterase activity in the unimpacted sites suggests that a significant source of organic phosphorus in the floc is of microbial origin.

9.8.3 MICROBIAL ACTIVITIES AND PHOSPHORUS RELEASE

The processes and the rate at which organic phosphorus compounds are hydrolyzed into bioavailable pool (labile pool) are controlled by various physical, chemical, and biological factors coupled with environmental influences in the soil-water column of a wetland. Large quantities of organic phosphorus are immobilized in soils, especially phosphorus associated with IHPs, which play minimal role in the biological cycling of phosphorus. However, simple sugar phosphates, nucleic acids, and phospholipids are bioavailable and actively participate in the cycling of phosphorus.

TABLE 9.5
Phosphatase Activity of Floc and Soils in the Water Conservation Area (WCA-2A) of the Northern Everglades

Site	Depth	Phosphomonoesterase Activity (mg p-nitrophenol kg^{-1} h^{-1})	Phosphodiesterase Activity (mg p-nitrophenol kg^{-1} h^{-1})
2.3 km	Floc	305±20	156±29
5.1 km	Floc	1,579±246	457±188
10.1 km	Floc	2,109±805	2,052±1,256
2.3 km	0–10 cm	89±30	23±2
5.1 km	0–10 cm	608±197	164±55
10.1 km	0–10 cm	693±355	134±69

Source: Chua (2000).

The role of microorganisms in the transformation of organic phosphorus to inorganic phosphorus in soils has long been recognized. Microorganisms can influence DIP release in soils as follows:

- Decomposition of organic phosphorus compounds through enzymatic hydrolysis and catabolic activities (mineralization)
- Assimilation of DIP into microbial biomass (immobilization)
- Altering physicochemical environment in soil, which may result in solubilization of insoluble phosphate minerals

Active cycling of organic phosphorus is largely mediated by microbial metabolism. The catabolic activities regulate the mineralization of organic phosphorus. During their growth, microorganisms assimilate and transiently store phosphorus in their biomass. Net release of phosphorus from decomposing organic matter depends on C:P ratio of the substrate. This is similar to C:N ratio concept for organic nitrogen mineralization discussed in Chapter 8. Organic carbon content of most organic materials is relatively constant, but the phosphorus content can be variable. Thus, C:P ratios of the plant tissue can vary substantially, depending on the trophic conditions in wetlands.

As the detrital matter or soil organic matter undergoes decomposition, any DIP released will be first assimilated by microbes involved in the decomposition process. Once their needs are met, excess phosphorus is released. Microbes are very efficient in scavenging the DIP as compared to algae and higher plants. Mineralization of organic phosphorus is regulated by phosphorus content of the decomposing tissue. For terrestrial systems, the relationship between C:P ratios of detrital residues and DIP is as follows (Stevenson, 1986):

C:P ratio <200 = net mineralization − net gain of DIP
C:P ratio 200–300 = neither gain nor loss of DIP
C:P ratio >300 = net immobilization − net loss of DIP

For wetlands, the above-mentioned relationship may be applicable only to aerobic portions of the soil and water column, where aerobic bacteria and fungi are actively involved in decomposing the detrital matter and soil organic matter. Generally, phosphorus requirements of aerobic microorganisms are higher than those of anaerobic bacteria. Thus, for the detrital matter decomposing under anaerobic conditions, the critical C:P ratios will be higher.

Bacterial biomass C:P ratio of <20 indicates that phosphorus is not limiting in the system. The biomass C:P ratios were found to be influenced by the redox condition and the presence of selected electron acceptors. For example, MBC:P ratios of 56, 43, 9, and 6 were obtained in a wetland soil maintained under aerobic, nitrate-reducing, sulfate-reducing, and methanogenic conditions, respectively (McLatchey and Reddy, 1998). High MBC:P ratios under aerobic and nitrate-reducing conditions suggest phosphorus limitation, resulting from high rate of microbial growth and possible precipitation of bioavailable P with iron oxides.

Organic phosphorus cycling is governed by microbial metabolic activities functioning in the soil profile, which include aerobic, facultative anaerobic, and obligate anaerobic. The rate of microbial breakdown of organic phosphorus depends not only on substrate characteristics but also on availability of electron acceptors such as oxygen (aerobic); nitrate and manganese and iron oxides (facultative anaerobic); and sulfate and carbon dioxide (obligate anaerobic). Studies have shown the effect of alternate electron acceptors and the associated redox conditions on organic phosphorus mineralization, with approximately threefold higher rates under aerobic conditions, as compared to nitrate-, sulfate-, and bicarbonate-reducing conditions. Similar response was also observed for microbial biomass (McLatchey and Reddy, 1998). Phosphatase activity is directly influenced by redox potential, with high activities found under aerobic conditions than anaerobic conditions. Under anaerobic conditions, low microbial activity and greater availability of soluble inorganic phosphorus result in

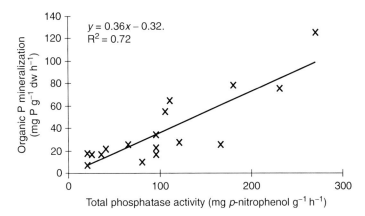

FIGURE 9.51 Relationship between phosphatase activity and organic phosphorus mineralization in a wetland soil. (Adapted from McLatchey and Reddy, 1998.)

lower phosphatase activity. Organic phosphorus mineralization is directly related to the phosphatase activity in soils (Figure 9.51).

Heterotrophic degradation of organic phosphorus by microbes depends not only on the availability of phosphorus but also on labile organic carbon, which serves as the energy source for microbes. Most of this carbon is respired and lost as carbon dioxide, whereas DIP is assimilated into microbial biomass and accumulates in the system. Increased microbial activity resulting from the availability of labile carbon can create demand for available DIP. This may lead to synthesis of extracellular enzymes, followed by cleavage of bond between oxygen and phosphorus in organic compounds. In phosphorus-limited wetlands, both carbon and phosphorus cycles are tightly coupled and DIP is efficiently conserved and cycled within the system. However, majority of wetlands in the world are not phosphorus-limited. Under these conditions, heterotrophic microbes assimilate phosphorus beyond their needs and store in cells as polyphosphates.

Decaying plant detritus can act as an internal source during early stages of decomposition and upon burial of litter can serve as long-term storage (Figure 9.48). Leaching of soluble cellular phosphorus is the first step by which phosphorus is released from the detrital matter attached to the macrophyte or detached and deposited on the soil surface. The total phosphorus content of detrital matter attached to the plant can be significantly lower than that of the photosynthetic tissue. As macrophytes senesces, the detrital matter attached to the plant can be subjected to leaching of soluble phosphorus, as a result of water movement and rainfall. In addition, this material can also undergo decomposition by microbial communities colonizing on this substrate. Once the detrital matter is detached from the plant, it is deposited on the soil surface. At this stage, both aerobic and anaerobic microorganisms (depending on oxygen availability) regulate the decomposition of detrital material deposited on the soil surface (see Chapter 5 for the decay continuum of detrital matter in wetlands). With time, the undecomposed organic matter is accreted as stable organic matter. The turnover time of the detrital material depends on various factors including C:P ratios of the substrate, microbial activities, and physicochemical environment in the soil and water column.

Varieties of methods are used to determine the mineralization of organic phosphorus. These include

- Litterbag field methods
- Basal mineralization of organic phosphorus or potentially mineralizable organic phosphorus (PMP)
- Mineralization of added organic substrates
- Substrate-induced organic phosphorus mineralization (SIPM)

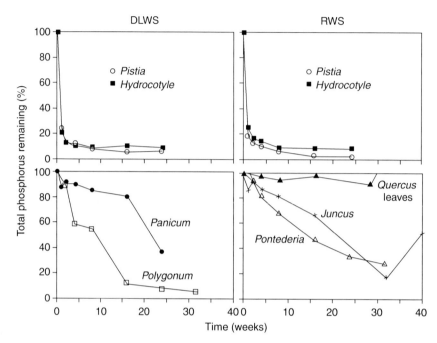

FIGURE 9.52 Phosphorus release during decomposition of plant detrital matter as determined by litterbag techniques (Reddy et al., 1995).

9.8.3.1 Litterbag Method

Detrital matter collected from the field is cut into 2–5 cm long pieces and a known amount (20–30 g, fresh-weight basis) is placed in nylon bags with a mesh spacing of 0.5–1 mm. The bags are returned to the field and either allowed to float on the soil surface or placed at the soil–floodwater interface. Bags are retrieved at select time intervals and analyzed for total phosphorus. On the basis of the dry weight loss and tissue phosphorus content, phosphorus released from detrital tissue is estimated. An example of phosphorus release from different detrital tissues is shown in Figure 9.52. In plant tissues with low C:P ratios (e.g., *Pistia* and *Hydrocotyle*), approximately 74–82% of the phosphorus was released during the first week, followed by 91–97% released by 24-week decomposition. A highly significant correlation was observed between dry weight loss as a result of decomposition and phosphorus release. Phosphorus release from aquatic vegetation was in the order of *Pistia* and *Hydrocotyle* > *Polygonum* > *Pontederia* > *Panicum* > *Juncus* > oak leaves.

9.8.3.2 Basal Mineralization of Organic Phosphorus

Few studies have been conducted to determine the cumulative organic phosphorus mineralization in wetland soils. Soils are incubated under either aerobic or anaerobic conditions at 30°C for a predetermined period. At the end of each incubation period, soils are extracted for inorganic phosphorus. An example of a study on mineralization of organic phosphorus in northern wetlands is shown in Figure 9.53 (Bridgham et al., 1998). As shown in figure, organic phosphorus mineralization was higher under anaerobic conditions than under aerobic conditions. The ratio of aerobic to anaerobic ranged from 0.28 in tamarack swamps to 1.12 in the acidic fens. Organic phosphorus turnover rates were highest in bogs and acidic fens and lowest in fens and meadows (Bridgham et al., 1998).

9.8.3.3 Potentially Mineralizable Phosphorus

To determine the PMP, field moist soils are incubated under anaerobic conditions at 30°C for a period of 15 days. This method is similar to the one described for potentially mineralizable nitrogen

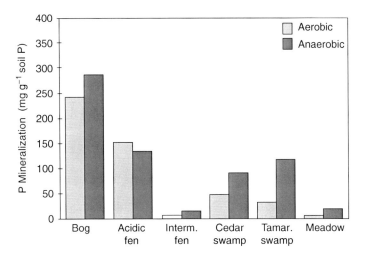

FIGURE 9.53 Organic phosphorus mineralization potentials in select northern temperate wetlands. (From Bridgham et al., 1998.)

(PMN), with some modifications (Chua, 2000). At the end of incubation, soils are extracted with 1 M HCl (3 h), followed by filtration through 0.45 μm membrane filter. Filtered solutions are analyzed for DRP using standard methods. The PMP is calculated as follows:

$$\text{PMP} = \frac{|\text{HCl-P}_i|_{t=15} - |\text{HCl-P}_i|_{t=0}}{t}$$

where $[\text{HCl-P}_i]_{t=15}$ is the inorganic phosphorus extracted after 15-day incubation (mg kg^{-1}), $[\text{HCl-P}_i]_{t=0}$ the inorganic phosphorus extracted at the beginning of the incubation (mg kg^{-1}), and t the incubation period in days.

PMP is significantly correlated with LOP, MBP, and TOP. Turnover rate constant for each of these organic pools can be calculated as follows:

$$k = \frac{1}{\text{PMP (mg kg}^{-1}\text{)/Organic pool (mg kg}^{-1}\text{)}}$$

where k is the rate constant (day^{-1}). Estimated turnover times for labile organic P, MBP, and TOP were 100, 62, and 370 days, respectively, for soils in the northern Everglades (Chua, 2000).

9.8.3.4 Mineralization of Added Organic Phosphorus

Plant residues or other synthetic organic substrates are added to soils, and mineralization of these materials is measured over a period of time under laboratory conditions. Typically, soils are incubated in the dark under either aerobic or anaerobic conditions at 30°C. At the end of incubation, soils are extracted as described above, and the DIP release is determined.

DIP released from added substrate = [DIP released in amended soil] − [DIP released in unamended soil]

9.8.3.5 Substrate-Induced Organic Phosphorus Mineralization

Phosphatase activities in soils and surface waters have been linked to organic phosphorus mineralization, which is regulated by enzymatic hydrolysis by microbes. To determine the enzymatic activities of ambient microbial populations, field moist samples are spiked with nonlimiting concentration of LOP and incubated for a period of 2 h, followed by extraction of soils with 0.5 M NaHCO$_3$. Soil suspensions are filtered through a 0.45 μm membrane filter and analyzed for DRP

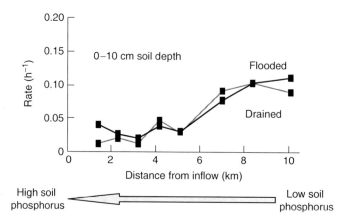

FIGURE 9.54 Substrate induced organic phosphorus mineralization in northern Everglades wetland soil. (From Chua, 2000.)

using standard methods. Short-term incubation assumes that there is no shift or increase in microbial populations as a result of organic phosphorus additions.

$$\text{SIPM} = \frac{|\text{NaHCO}_3\text{-P}_i|_{t=4} - |\text{NaHCO}_3\text{-P}_i|_{t=0}}{t}$$

where $[\text{NaHCO}_3\text{-P}_i]_{t=4}$ is the inorganic phosphorus extracted after 4 h incubation (mg kg^{-1}), $[\text{NaHCO}_3\text{-P}_i]_{t=0}$ the inorganic phosphorus extracted at the beginning of the incubation (mg kg^{-1}), and t the incubation period in hours.

Using this approach, enzymatic hydrolysis of phosphorus was measured by adding a substrate (glucose-6-phosphate) to soils collected along the nutrient gradient in the northern Everglades wetland soils (Chua, 2000). Glucose-6-phosphate includes monoester phosphates that are found in living cells, where they are intermediates in carbohydrate metabolism (Figure 9.54). As discussed earlier, phosphatase activity is highest under aerobic and phosphorus-limiting conditions. Demand for phosphorus is high under aerobic conditions, because of higher activities of bacterial and fungi, as compared to anaerobic conditions. Mineralization rates of glucose-6-phosphate increase as the system becomes more phosphorus limiting (oligotrophic). Phosphorus released as result of hydrolysis of ester bond can be potentially assimilated by microbial communities as a nutrient source. Areas of interior marsh (>6 km from inflow) are phosphorus limiting, whereas areas closer to nutrient inflow (<6 km from inflow) are phosphorus enriched and eutrophic.

In phosphorus-enriched systems (eutrophic), microbes do not have a need to produce the enzyme phosphatase to hydrolyze ester linkages. In phosphorus-limited systems, organisms are dependent on organic phosphorus, resulting in induction of phosphatase activity. High rates of glucose-6-phosphate mineralization suggest high monoesterase (phosphatase) activity in phosphorus-limited areas of the Everglades. This substrate-induced mineralization provides an indication of the activity of ambient microbial population.

In addition to mineralization of organic phosphorus, microbial decomposition of detrital matter and soil organic matter can have indirect effect on phosphorus solubility, as described below:

- Anaerobic decomposition of organic matter and detritus can result in production of organic acids, which can result in solubilization of phosphorus bound to iron, aluminum, calcium, and magnesium.
- Organic anions produced during decomposition can compete with phosphate ions for adsorbing surfaces, thus decreasing phosphorus retention.

- DOM can form protective coating on minerals such as Fe and Al oxides and calcite, thus resulting decreased phosphorus retention.
- Wetland soils are known to produce DOM as a result of partial decomposition of organic matter under anaerobic conditions. This can result in greater complexation of DOM with phosphorus-sorbing metals.
- High rates of carbon dioxide production and formation of carbonic acid can result in solubilization of calcium- and magnesium-bound phosphorus.
- Production of DOM can increase the complexation of phosphate with humic and fulvic acids.

9.8.4 Regulators of Organic Phosphorus Mineralization

Organic P mineralization is regulated by a number of biotic and abiotic factors shown below:

- Phosphorus content of organic substrate or C:P ratios
- Microbial activities and synthesis of hydrolytic enzymes
- Supply of electron acceptors and redox conditions
- Humic and fulvic acid content of soil organic matter
- Presence of metallic cations
- Type of clay minerals
- Soil and water column pH
- Temperature

As discussed earlier, phosphodiesters are easily biodegradable as compared to monoesters such as IHP. The relative proportion of these compounds in decomposing substrate (detrital matter and soil organic matter) regulates the amount of DIP release. Heterotrophic microbial activities are dependent on the availability of not only DIP but also labile organic carbon, which serves as energy source. Mineralization of organic phosphorus and release of DIP are inversely related to the C:P ratio of the decomposing substrate.

Redox potential and the availability of electron acceptors can have an indirect effect on organic phosphorus breakdown because enzymatic hydrolysis of ester linkages is not directly linked to catabolic activities of microbes. As discussed earlier, after phosphate groups are removed from organic phosphorus, the remaining compound can be used by microbes as an energy source. Some simple organic phosphorus compounds can be directly assimilated by microbes, where intracellular phosphatases may be involved. Thus the influence of Eh and electron acceptors is primarily on catabolic activities of microbes (see Chapter 5), with an indirect effect on organic phosphorus mineralization. In mineral wetland soils and sediments dominated by Fe redox couple, phosphatase activity is high under aerobic conditions and decreases with a decrease in Eh. Under aerobic conditions, ferricoxyhydroxide removes DIP from pore waters and precipitates as ferric phosphate. Low DIP concentration and high phosphorus requirements of aerobes create phosphorus limitation and increase phosphatase activity. Under anaerobic conditions, reduction of ferric phosphate to Fe(II) releases DIP. Low phosphorus requirements of anaerobic bacteria and high DIP concentration inhibit the phosphatase synthesis.

Dissolved organic compounds, particularly phenolic compounds (humic and fulvic acids), resulting from breakdown of higher plants complex chemically with many enzymes, particularly phosphatases, and their activity is inhibited. The humic enzyme complexes can be stored in an inactivated state for some time, relocated within the wetland (Wetzel, 1993b). Soils with high levels of humic substances can result in lower rates of organic phosphorus mineralization. In soils with high concentrations of metallic cations, humic metal complexes can aid in reducing this inhibitory effect, resulting in increased rates of organic phosphorus mineralization. Some organic compounds, such as IHP, can be adsorbed to oxides of iron and aluminum and to clay minerals, resulting in decreased availability of these compounds for mineralization. Reduction of iron oxides can result in mobilization of sorbed organic phosphorus.

Depending on pH, phosphomonoesterases have been grouped into acid phosphatase and alkaline phosphatase. Optimum pH range of 4–6 for acid phosphatase and 8.3–9.5 for alkaline phosphatase

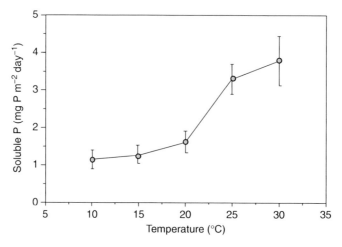

FIGURE 9.55 Influence of temperature on soluble phosphorus release in organic soil–dominated wetland (Kadlec and Reddy, 2001).

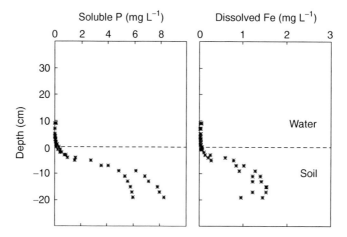

FIGURE 9.56 Dissolved iron and phosphorus profiles in a wetland soil showing the relationship between iron reduction and phosphorus release. (From D'Angelo and Reddy, 1994.)

has been reported (Hoppe, 2003). The release of soluble phosphorus from flooded organic soil was shown to increase with temperature, with rates of release doubled for every 10°C rise in a temperature range of 10–30°C (Figure 9.55). These results suggest that the decomposition processes and associated phosphorus release in wetland soils are influenced by temperature.

9.9 BIOTIC AND ABIOTIC INTERACTIONS ON PHOSPHORUS MOBILIZATION

9.9.1 Phosphorus–Iron–Sulfur Interactions

The reduction and dissolution of iron and its reprecipitation to form ferrous minerals are thought to be the dominant processes controlling phosphorus solubility in anaerobic systems. Soil pore water profiles shown in Figure 9.56 show that the concentrations of soluble phosphorus and dissolved iron are low in the water column and increase with depth. Similar patterns in these profiles suggest that the solubilities of phosphorus and iron are strongly coupled. In the water and in the aerobic soil layer, iron is in the immobile ferric form, which reacts with phosphate and precipitates. Thus, under aerobic conditions, oxidation and precipitation of iron control phosphorus solubility and limit

Phosphorus

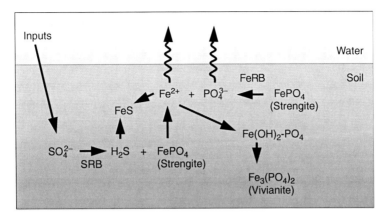

FIGURE 9.57 Interactions of sulfate on the reduction of Fe(III) and release of Fe(II) and phosphate. SRB, sulfate-reducing bacteria; FeRB, iron-reducing bacteria. (Modified from Roden and Edmonds, 1997.)

phosphorus flux into the overlying water. Below aerobic layer, anaerobic conditions reduce ferric phosphate, thus increasing soluble ferrous iron. In anaerobic zones, phosphorus solubility is controlled by more soluble ferrous iron or calcium phosphate precipitation.

Phosphorus solubility can also be affected by microbial reduction of $FePO_4$ and sulfate, although reduction of these electron acceptors occurs in different redox zones (Figure 9.57). We know that the reduction of ferric iron in $FePO_4$ to ferrous iron results in the release of phosphate. Similarly, at lower Eh values, sulfate is reduced to sulfide. Both sulfide and ferrous iron can participate in various reactions, which potentially result in greater phosphate availability. The ferrous iron can react with sulfide, eventually forming FeS_2 (see Chapter 10). The ferrous iron can also react with phosphate to form $Fe_3(PO_4)_2$ (vivianite). Some of the lithotrophic bacteria can use sulfide as an electron donor and reduce ferric iron in $FePO_4$, resulting in release of ferrous iron and phosphate. Removal of ferrous iron through reaction with sulfide results in greater availability of phosphate. Biotic and abiotic interactions shown in Figure 9.57 can occur in wetland soils with Fe minerals and sulfate inputs. These characteristics are typical of coastal wetlands with mineral soils.

In sulfate-dominated wetlands, production of sulfide (through biological reduction of sulfate) and formation of ferrous sulfides may preclude phosphorus retention by ferrous iron in regulating phosphorus bioavailability (Caraco et al., 1991). In iron- and calcium-dominated systems, Moore and Reddy (1994) observed that iron oxides likely control the behavior of inorganic phosphorus under aerobic conditions, whereas calcium phosphate mineral precipitation governs the solubility under anaerobic conditions. This difference is in part due to a decrease in pH under aerobic conditions as a result of oxidation of ferrous iron compounds, whereas an increase in pH occurs under anaerobic conditions as a result of reduction of ferric iron compounds. The juxtaposition of aerobic and anaerobic interfaces promotes oxidation–reduction of iron and its regulation of phosphorus solubility.

Sequential reduction of electron acceptors can have a significant effect on soluble phosphorus release. After a soil is flooded, it is expected that the amount of soluble P will increase. This is attributed to the anaerobic conditions occurring in the flooded soil and the various mechanisms of releasing phosphorus under those conditions. As shown in Figure 9.58, the amount of soluble phosphorus starts increasing after the third day of inundation, when almost the entire nitrate pool has been reduced, and consequently the reduction of manganese and iron contained in oxide minerals is already in process. On reduction of ferric oxide minerals, water-soluble and exchangeable concentrations of ferrous iron increase markedly. Thus, the dissolution of iron minerals is accompanied by increases in concentrations of both adsorbed and water-soluble phosphorus. Some of the ferrous ions react with the released phosphorus and precipitate to form new ferrous phosphate minerals. As the soil continues to be under anaerobic conditions, ferric ions are soon depleted and the reduction

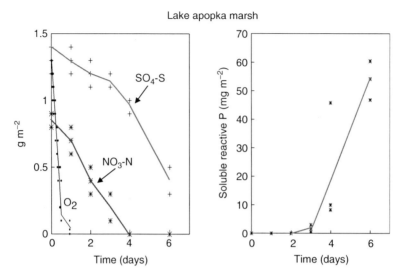

FIGURE 9.58 Relationship between sequential reduction select electron acceptors and soluble phosphorus release in a wetland soil. (Modified from D'Angelo and Reddy, 1994.)

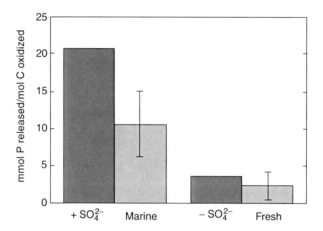

FIGURE 9.59 Maximum amount of dissolved phosphate released per unit of carbon mineralized during anaerobic decomposition. (Adapted from Roden and Edmonds, 1997.)

of sulfate to sulfide starts. Sulfide reacts with ferrous iron to form FeS. Because the solubility of FeS is lower than that of ferrous oxide minerals, precipitation of FeS drives the dissolution of these latter minerals, resulting in the release of sorbed phosphorus once again to pore waters at very low redox potentials. These are some of the most important mechanisms governing the increase in the amount of soluble phosphorus after inundation. Other reactions that may be important in releasing phosphorus upon flooding are the hydrolysis of ferric and aluminum phosphates and the release of phosphorus sorbed to clays and hydrous oxides by the exchange of anions.

Release of phosphorus during decomposition of fresh organic matter (freeze-dried cyanobacteria) was sevenfold higher in sulfate-amended soils than sulfate-free soils (Roden and Edmonds, 1997). This also shows that Fe-rich anaerobic soils can immobilize substantial amounts of phosphorus under ferric iron–reducing conditions, through formation of ferrous phosphate complex. However, extensive phosphorus release can occur when ferrous iron is precipitated with sulfide. Thus, coupling of ferric iron reduction, ferrous iron oxidation, and sulfate reduction in wetland soils can have significant effect on phosphorus mobilization from soil to the overlying water column (Figure 9.59).

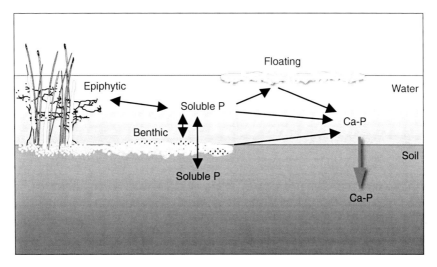

FIGURE 9.60 A schematic showing phosphorus cycling in periphyton mats of a wetland.

9.9.2 Periphyton–Phosphate Interactions

The influence of biotic and abiotic processes on phosphorus in wetlands is depicted in Figure 9.60. The biotic communities include vegetation, periphyton, and microbial communities. Periphyton plays an important role in removing DIP from the water column in wetlands and littoral habitats. Periphyton mats have high affinity for phosphorus and respond more rapidly to phosphorus inputs relative to other wetland components. Periphyton can be highly productive and strongly influence nutrient cycling, particularly in nutrient-deprived environments. The conspicuous nature of highly productive periphyton communities in phosphorus-deprived environments has been linked to increased efficiency of uptake and rapid and efficient release and remineralization due to the close association of autotrophic and heterotrophic microbial components in these compressed communities. This high productivity and intense nutrient cycling can therefore influence many of the biological and physicochemical aspects of these environments.

The photosynthetic activity of the periphyton creates environmental conditions (high pH, low carbon dioxide partial pressure) that are favorable to the precipitation of $CaCO_3$. DIP can be coprecipitated with this $CaCO_3$ during periods of high photosynthetic activity or under conditions that mimic this activity. This abiotic adsorption of phosphorus has been suggested as an effective mechanism for the removal of phosphorus from the water column, but to our knowledge, quantification of this process in periphyton has not been studied. Calcite formed in the mats is eventually incorporated into the surface soil and can be important to long-term phosphorus storage. This is an important characteristic of these mats in the Everglades and is known to regulate soluble P concentration of the water column. The Figure 9.61 shows a closeup of benthic periphyton tissue with $CaCO_3$ and without $CaCO_3$ (Scinto and Reddy, 2003). Visual inspection of periphyton tissue under scanning electron microscope (SEM) showed filaments encrusted with $CaCO_3$ spicules. These spicules were absent in periphyton tissue treated with 0.01 M HCl. The $CaCO_3$ encrustation on tissues plays a critical role in inorganic P adsorption. Additional support for this discussion is shown in Figure 9.62. In a study by Scinto and Reddy (2003), radioactive P (^{32}P) was used to trace phosphorus uptake by periphyton. Phosphorus present in $CaCO_3$ layer was removed by acid treatment followed by the analysis of biotic tissue for phosphorus. At ambient concentrations, >20% of $^{32}P_i$ activity was in the abiotic compartment ($CaCO_3$) after 1 h. In 12 h, ambient abiotic phosphorus pool ($CaCO_3$) had <10% $^{32}P_i$ activity. Phosphorus uptake in biotic compartment increased from 74% in 1 h to 88% in 12 h (Figure 9.62). Ratios of biotically incorporated ^{32}P activity were similar between light and dark incubations.

FIGURE 9.61 Scanning electron micrograph of typical filament from epipelon showing $CaCO_3$ encrustation (left) and a typical filament after 0.01 M HCl extraction (right). Note lack of encrustation after 2 h treatment with dilute (pH = 2.5–3.0) acid. (Adapted from Scinto and Reddy, 2003.)

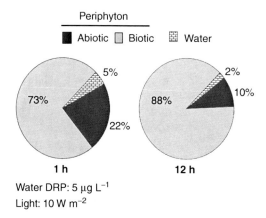

FIGURE 9.62 Biotic and abiotic uptake of added ^{32}P by periphyton mats obtained from northern Everglades. (From Scinto and Reddy, 2003.)

In Figures 9.63 and 9.64, mechanisms of inorganic phosphorus uptake by periphyton are illustrated. Phosphorus can move directly to cell surfaces where it is taken up biotically, or in the case of DOP it is first hydrolyzed and then incorporated. Inorganic phosphorus is rapidly but weakly sorbed to the surface of $CaCO_3$ crystals. A portion of this adsorbed phosphorus is incorporated into the biotic portion with time. Some of the adsorbed phosphorus may become incorporated into more stable Ca-P if the adsorption occurs during active $CaCO_3$ crystal growth. The phosphorus uptake by periphyton may be influenced by pH regulation during photosynthesis and respiration. During daytime, removal of carbon dioxide during photosynthesis can result in elevation of pH, creating conditions for possible precipitation of soluble phosphorus on the surface of periphyton with calcium. During nighttime, absence of photosynthesis can result in accumulation of carbon dioxide (due to respiration), which can result in a decrease in pH and possible solubilization of precipitates. Biotic incorporation of phosphorus is by far the leading phosphorus removal mechanism in this phosphorus-limited system. Although the study by Scinto and Reddy (2003) did not determine the long-term stability of this phosphorus fraction, it is possible that as the biotic portion of the periphyton degrades, this $CaCO_3$-associated phosphorus is deposited in the surficial soil. Significant correlations between historical accumulations of phosphorus and calcium in the northern Everglades wetland soils suggest the significance of calcium in long-term accretion of phosphorus in calcareous wetlands (Figure 9.65).

COLOR FIGURE 3.15 Soil profiles showing (a) oxidized and (b) reduced forms of soils.

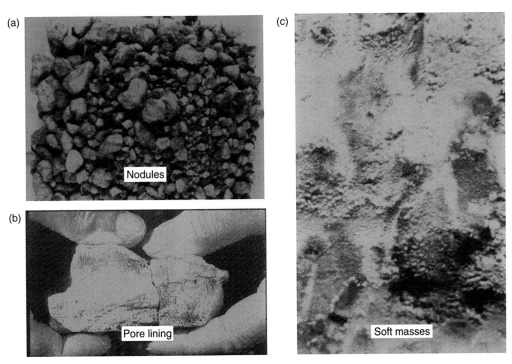

COLOR FIGURE 3.17 Redox concentrations shown as (a) nodules and concretions, (b) pore linings, and (c) soft bodies within the soil matrix. (USDA–NRCS, 2006.)

COLOR FIGURE 3.20 This soil meets the requirements of indicator A1. Muck (sapric soil) material is about 0.5 m thick. The shovel is approximately 1 m. (USDA–NRCS, 2006.)

COLOR FIGURE 3.21 Indicator A5 (Stratified Layers) in sandy soil material. (USDA–NRCS, 2006.)

COLOR FIGURE 3.22 Indicator A6 (Organic Bodies). Scale is in inches. (USDA–NRCS, 2006.)

COLOR FIGURE 3.23 Indicator A14 (Alaska Redox) in a gleyed matrix with reddish redox concentrations around pores and root channels. (USDA–NRCS, 2006.)

COLOR FIGURE 3.24 Indicator S5 (Sandy Redox) with redox concentrations occurring almost at the surface. The soil slice is about 40 cm long. (USDA–NRCS, 2006.)

COLOR FIGURE 3.25 Indicator S6 (Stripped Matrix) occurs at a depth of about 12 cm. The knife blade is 15 cm long. (USDA–NRCS, 2006.)

COLOR FIGURE 3.26 Indicator F2 (Loamy Gleyed Matrix) occurs at the soil surface in this example of the indicator. (USDA–NRCS, 2006.)

COLOR FIGURE 3.27 Indicator F8 (Redox Depression) occurs at the soil surface in this example of the indicator. (USDA–NRCS, 2006.)

COLOR FIGURE 7.27 Photographs showing oxidized roots of rice plants.

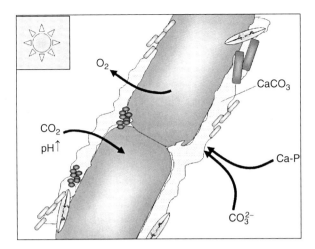

FIGURE 9.63 A schematic showing phosphorus cycling in periphyton mats during daytime.

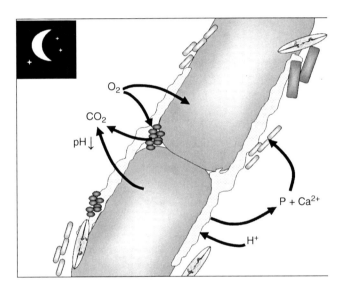

FIGURE 9.64 A schematic showing phosphorus cycling in periphyton mats during nighttime.

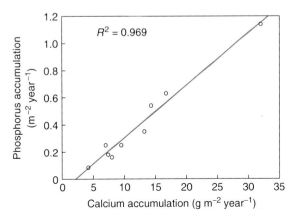

FIGURE 9.65 Relationship between long-term phosphorus and calcium accumulation in northern Everglades. (From Reddy et al., 1993.)

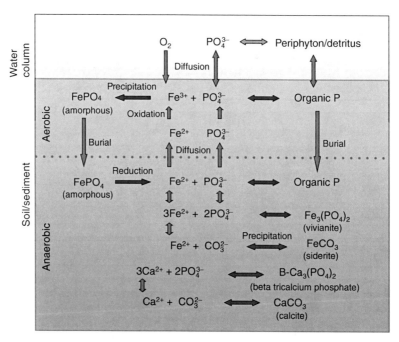

FIGURE 9.66 Chemical reactions regulating phosphorus solubility in aerobic and anaerobic layers of wetland soils. (Adapted from Moore and Reddy, 1994.)

9.9.3 BIOTIC AND ABIOTIC INTERACTIONS OF FE AND CA WITH PHOSPHORUS

Biotic and abiotic interactions between organic and inorganic phosphorus play a major role in the wetland phosphorus cycle. A summary of select processes controlling phosphorus in aerobic and anaerobic layers of a wetland soil profile or a lake sediment profile is depicted in Figure 9.66. Various types of minerals can form in wetland soils depending on the physicochemical conditions. In aerobic soil layers, iron is typically present as Fe-oxides, which can readily precipitate any DIP as amorphous ferric phosphate. This precipitate is stable under aerobic conditions. However, under anaerobic conditions, ferric iron can be reduced to ferrous iron, thus liberating phosphorus, especially under near-neutral and acidic conditions. Under anaerobic conditions, ferrous iron can react with phosphorus and form $Fe_3(PO_4)_2 \cdot 8H_2O$ (vivianite). This typically can happen in soils with high iron and phosphorus content. In soils with near-neutral and alkaline conditions, ferrous iron can also react with carbonates to form $FeCO_3$ (siderite). Similarly, calcium present in soil solution can react in two ways: calcium can react with carbonates to form $CaCO_3$ and with phosphorus to form various forms of calcium phosphate depending on IAPs and solubility equilibria. In Figure 9.66, the phosphorus present in solution was shown to form beta-tricalcium phosphate, which is stable under alkaline conditions. As shown in Figure 9.66, several competing reactions regulate phosphorus solubility and availability in wetland soils or lake sediments. In addition to these chemical reactions, flux of dissolved ferrous iron and phosphorus also regulates phosphorus availability and solubility. For example, ferrous iron and soluble phosphorus can diffuse from anaerobic soil layer to aerobic soil layer and into overlying water column. Ferrous iron can be readily oxidized to ferric oxides in the presence of oxygen. Dissolved phosphorus can be precipitated with ferric oxides or assimilated by macrophytes and periphyton. As discussed earlier, under alkaline pH conditions in the water column, dissolved phosphorus can also be precipitated as calcium phosphates. Deposition of detrital matter at soil–floodwater interface can increase the demand for electron acceptors, thus creating anaerobic conditions and resulting in reduction of iron oxides. In anaerobic soil layer, organic matter decomposition process mediated by microbes can result in accumulation of organic acids, which can potentially dissolve some of the calcium phosphates. Reactions regulating phosphorus solubility are complex and are regulated by solubility products of each compound, biotic processes, and physicochemical conditions in soil or sediment.

9.9.4 Gaseous Loss of Phosphorus

The existence of reduced volatile phosphorus compounds in aquatic systems has been in question for several decades (Morton and Edwards, 2005). Similar to nitrate reduction to ammonia (dissimilatory nitrate reduction, see Chapter 8) and sulfate reduction to sulfide, thermodynamic reduction of phosphate to phosphine is possible. Under highly anaerobic conditions, phosphate (oxidation number of +5) can be reduced by obligate anaerobes to phosphine (oxidation number of −3).

$$H_2PO_4^- + 9H^+ + 8e^- = PH_3 + 4H_2O, \qquad E° = -0.78 \text{ V (pH = 7.0)}$$

Phosphine has been detected only at trace levels in various systems including sewage treatment facilities, floodplain wetlands, rice fields, and freshwaters. Using GC and MS techniques, Dévai et al. (1988) were the first to detect phosphine in sewage treatment facilities. They estimated that approximately 25–50% of the total phosphorus removal might be accounted for by phosphine gas emission. The first quantifiable emissions of phosphine from wetlands to the atmosphere were reported in Louisiana brackish (0.42–3.03 ng PH_3 m^{-2} h^{-1}) and salt marsh (0.91–6.52 ng PH_3 m^{-2} h^{-1}) (Dévai and DeLaune, 1995). Similarly, laboratory experiments have demonstrated that Everglades marsh soil can also produce phosphine (16 ng PH_3 kg^{-1} of wet sediment per day), and phosphine production was enhanced (380 ng PH_3 kg^{-1} of wet sediment per day) by the addition of phosphate and an energy source. Under oxic conditions, phosphine is highly unstable and can be oxidized back to phosphate in the aerobic zones of wetlands, thus preventing emissions to the atmosphere.

9.10 PHOSPHORUS EXCHANGE BETWEEN SOIL AND OVERLYING WATER COLUMN

Transport processes between soil and the overlying water column affect the availability of phosphorus for assimilation by biota and retention by soils. Phosphorus is transported through wetlands in soluble and particulate forms, including DIP, PIP, DOP, and POP (Figure 9.67). Chemical and biological interactions between dissolved and particulate forms are discussed in earlier sections of this chapter. Although these fractions do not describe the functional, chemical, or biological behavior, they are useful in quantifying the physical behavior during transport. Soluble organic and inorganic phosphorus fractions are not only transported with the water, but also mobilized in response to the concentration gradients between soil and overlying water column. The chemical characteristics of

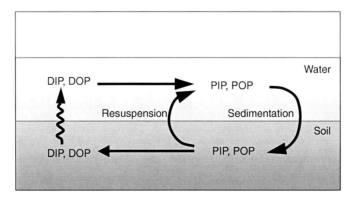

FIGURE 9.67 A schematic showing exchange processes between water column and soil of a wetland.

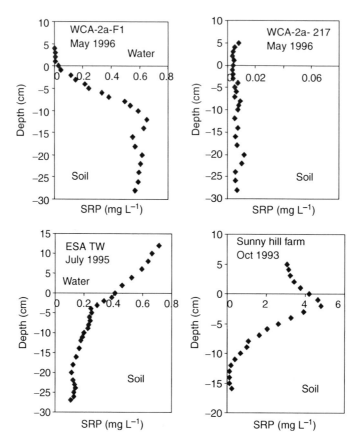

FIGURE 9.68 Soil pore water phosphorus profiles in select wetlands in Florida. WCA-2a = water conservation area-2 of the Everglades; ESA = constructed wetland, Orlando, Florida; Sunny Hill Farm = agricultural land converted to wetland. (Reddy, K. R., Unpublished Results, University of Florida.)

these fractions are altered as they are transported with water through different reaches of wetlands and associated aquatic systems. The transport processes involved in mobilization of phosphorus between sediment or soil and overlying water column are advection, dispersion, diffusion, seepage, resuspension, sedimentation, and bioturbation (see Chapter 14).

Dissolved inorganic and organic phosphorus concentrations in soils are typically much higher than the overlying water column; thus the flux of these dissolved components is always from soil to overlying water column. Typical soluble phosphorus profiles of select wetlands and aquatic systems are shown in Figure 9.68. In all these systems, sharp gradients are noted between the soil and overlying water column, suggesting that flux of soluble phosphorus is from soil to the overlying water column. PP generated in the water column either by detrital tissue or by precipitation reactions results in settling on soil surface. The flux of particulate matter is always from water column to soil. Settling of PP provides long-term retention by wetlands, whereas the flux of dissolved components into water column provides bioavailable phosphorus to biotic communities. Thus, overall net P flux is always from the water column to the soils or sediments. This process results in phosphorus accretion in soils/sediments. So, if the net flux is to soils/sediments, then what is the problem? Continuous accretion of phosphorus in soils increases the dissolved phosphorus concentrations of soil pore waters, which results in flux from sediments to the water column. This dissolved phosphorus is bioavailable and maintains eutrophic conditions in the water column. Although the flux of dissolved phosphorus is small, it is very critical in regulating water quality.

Phosphorus flux across the soil/sediment–water interface occurs in two different modes, depending on meteorological and hydrodynamic conditions. These processes can be more prevalent in open shallow water bodies such as lakes and rivers than wetlands. During calm days when vertical turbulent mixing and bottom shear stress are insufficient for resuspension of surface sediment, dissolved phosphorus moves via passive diffusion and advection. Although sediments act as a net sink for phosphorus, release of DIP into the overlying water can occur if the concentration of interstitial phosphorus exceeds that of the overlying water. This is especially true if surface sediments are anaerobic, reducing the ability of iron to bind to inorganic phosphorus, thus increasing the amount of interstitial phosphorus that can be released to the water column.

The processes affecting phosphorus exchange at the soil/sediment–water interface include (i) diffusion and advection due to wind-driven currents, (ii) diffusion and advection due to flow and bioturbation, (iii) processes within the water column (mineralization, sorption by particulate matter, and biotic uptake and release), (iv) diagenetic processes (mineralization, sorption, and precipitation dissolution) in bottom sediments, (v) redox conditions (O_2 content) at the soil/sediment–water interface, and (vi) phosphorus flux from water column to soil mediated by evapotranspiration by vegetation. In shallow lakes during windy periods, resuspension and deposition of sediments may be an important mode of phosphorus transfer to the water column. Because sediment resuspension events are transitory, phosphorus flux by this process may occur at shorter timescales but at more rapid rates compared to diffusive flux. In wetlands, this process may not be a significant factor. For details on phosphorus transport processes in wetlands see Reddy et al. (1999, 2005) and Chapter 14.

9.11 PHOSPHORUS MEMORY BY SOILS AND SEDIMENTS

Wetlands and aquatic systems are often the final recipients of nutrients discharged from adjacent terrestrial ecosystems. Because many freshwater systems are phosphorus limited, loading of this nutrient is of particular concern to environmental managers. Nonpoint sources of phosphorus dominate eutrophication processes of many wetlands and aquatic ecosystems. Historically, wetlands, ditches, and other depressions in the watershed tend to accumulate phosphorus discharged from adjacent uplands. The majority of this phosphorus accumulates in soils. Alternative improved land use management practices in uplands can potentially reduce the overall load to receiving water bodies. Once the external phosphorus loading from uplands is curtailed through the implementation of best management practices (BMPs) and other improved practices, the question is how well wetlands and aquatic systems would respond to external load reduction and reverse eutrophication.

Following are the key questions often asked: (i) Will wetlands and aquatic systems respond to phosphorus load reduction? (ii) If so, how long will it take for these systems to recover and reach their background condition? (iii) Are there any economically feasible management options to hasten the recovery process? The "phosphorus memory" (phosphorus retained in soils and sediments and potential release from these sources) can extend the time required for a wetland or an aquatic system to reach an alternate stable condition to meet environmental regulation such as TMDLs (total maximum daily loads). Phosphorus memory is also referred to as "legacy phosphorus." This time lag for recovery should be considered in developing management strategies for reducing phosphorus loads to wetlands and aquatic systems (Figure 9.69). Decisions regarding management and restoration of wetlands and aquatic systems are often difficult and controversial as they involve regulating phosphorus loads from both internal and external sources.

State and federal agencies are now acquiring lands adjacent to sensitive aquatic systems. These acquired agricultural lands are now converted into wetlands. However, upon flooding, resulting anaerobic soil conditions increase phosphorus solubility in a number of ways. The processes involved in regulating phosphorus solubility are discussed in earlier sections of this chapter. In the example shown in Figure 9.70, phosphorus flux from flooded agricultural land (organic soils) to overlying water column was highest during first few months after flooding, followed by a steady decrease within next 30 months. Prior to conversion to wetland, this particular land was intensively fertilized and

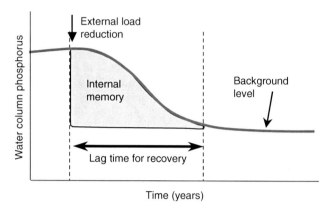

FIGURE 9.69 A schematic showing the influence of phosphorus memory in wetlands on lag time for recovery or for the system to reach alternate stable condition.

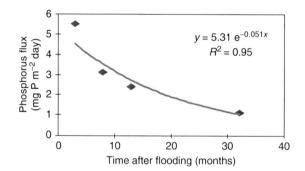

FIGURE 9.70 Phosphorus flux from a flooded agricultural land in central Florida.

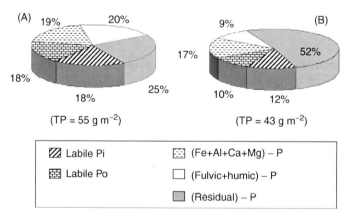

FIGURE 9.71 Influence of flooding of an agricultural land on changes in organic and inorganic pools. Flooded agricultural land Lake Griffin Flow-Way, Florida (A = before flooding; B = 6 months after flooding).

used for vegetable production. Flooding altered the proportion of organic and inorganic phosphorus forms (Figure 9.71). Labile inorganic and organic phosphorus decreased substantially, probably due to flux into overlying water column. However, a significant amount of phosphorus was incorporated into stable residual fraction, which could not be extracted by either acid or alkali. The mechanisms involved in transforming labile pools phosphorus into stable pools are not known at this time.

This phosphorus flux from internal memory is a major concern, when intensively used agricultural lands are converted to wetlands. The phosphorus retention capacity of soils can be improved

Phosphorus

by application of chemical amendments such as $CaCO_3$ (calcite), $Ca(OH)_2$ (lime), $CaMg(CO_3)_2$ (dolomite), alum, and $FeCl_3$ or water treatment residuals (WTRs), which may include many of these metals. These chemical amendments are effective in decreasing soluble phosphorus concentration of soil pore water. Effective amounts of each chemical amendment to minimize phosphorus release to overlying floodwater are soil specific. Lower rates are typically needed in low-organic-matter soils (mineral soils), where complexation of added metal with DOM is minimal. However, higher rates of chemical amendments may be needed in high-organic-matter soils (such as histosols). This is due to potential complexation of added metals with organic matter. One example of addition of chemical amendments on soluble phosphorus flux from flooded organic soils is shown in Figures 9.72 and 9.73 (Ann et al., 2000a). Effective amounts of each chemical amendment to minimize phosphorus release to overlying floodwater are approximately 6–9 t ha^{-1} for calcite and lime. Dolomite was not effective in immobilization of phosphorus. An effective reduction in phosphorus flux is observed in soils

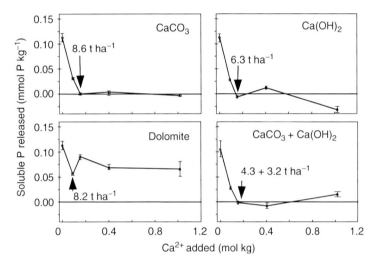

FIGURE 9.72 Effects of Ca materials ($CaCO_3$, $Ca(OH)_2$, $CaMg(CO_3)_2$, and mixture of $CaCO_3$ and $Ca(OH)_2$) on immobilizing soluble P in the floodwater. (From Ann et al., 2000a.)

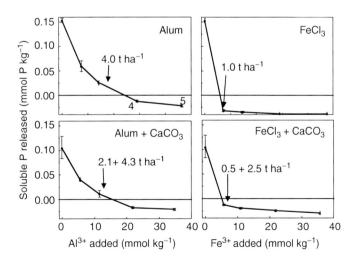

FIGURE 9.73 Effects of alum, $FeCl_3$, and the mixtures of $CaCO_3$ with alum or $FeCl_3$ on immobilizing water-soluble P in the floodwater. (From Ann et al., 2000a.)

amended with alum at a rate of approximately 4 t ha^{-1} and with FeCl$_3$ at a rate of approximately 1 t ha^{-1}. Combinations of alum and calcite or FeCl$_3$ and calcite were found to be effective for *in situ* immobilization of soluble phosphorus.

The two preceding examples demonstrate the role of inorganic amendments in regulating phosphorus solubility in wetland soils. Possible sequential reactions after application of chemical amendments are follows:

Calcium carbonate (CaCO$_3$) and calcium hydroxide (Ca(OH)$_2$)

$$CaCO_3 = Ca^{2+} + CO_3^{2-}$$
$$CO_3^{2-} + H_2O = HCO_3^- + OH^-$$
$$Ca^{2+} + HPO_4^{2-} + 2H_2O = CaHPO_4 \cdot 2H_2O$$
$$Ca^{2+} + 3HPO_4^{2-} + 2.5H_2O = Ca_4H(PO_4)_3 \cdot 2.5H_2O + 2H^+$$
$$Ca(OH)_2 = Ca^{2+} + 2OH^-$$
$$Ca^{2+} + CO_3^{2-} = CaCO_3 \text{ [calcite]}$$

Liming soils can increase the pH resulting in the formation of calcium carbonate, which can co-precipitate phosphate. Under anaerobic conditions CaCO$_3$ amendment can significantly decrease Fe(II) concentration in soil pore water as shown below:

$$Fe^{2+} + CO_3^{2-} = FeCO_3 \text{ [siderite]}$$

However, increased alkalinity due to liming of soils potentially can result in dissolution of Fe and Al-P compounds, and the phosphorus release can precipitate with Ca.

Alum (aluminum sulfate) and FeCl$_3$ (ferric chloride) are common amendments used to precipitate phosphorus.

Aluminum Sulfate

$$Al_2(SO_4)_3 \cdot 8H_2O = 2Al^{3+} + 3SO_4^{2-} + 8H_2O$$
$$Al^{3+} + 2H_2O = Al(OH)_2^+ + 2H^+$$
$$Al(OH)_2^+ + H_2PO_4^- = Al\,PO_4 \cdot 2H_2O$$

Ferric Chloride

$$FeCl_3 = Fe^{3+} + 3Cl^-$$
$$Fe^{3+} + 3H_2O = Fe(OH)_3 + 3H^+$$
$$Fe(OH)_3 + 3H^+ + H_2PO_4^- = FePO_4 \cdot 2H_2O + H_2O$$

Inorganic chemical amendments are effective in reducing excess dissolved phosphorus from water column or soil pore water. The rates of application depend on soil type (especially, amount of organic matter content) and amount of soluble phosphorus. Critical chemical amendment to phosphate ratios needs to be established for the site under remediation. Caution should be exercised in using inorganic chemical amendments in wetlands because excessive application may have an adverse effect on biotic communities. Chemical amendments can be integrated into design of constructed wetlands used to increase their longevity for phosphorus removal.

Phosphorus retention in wetlands is an important function in watershed nutrient cycling particularly in drainage basins with significant nonpoint nutrient contributions from agriculture and urban sources. Phosphorus retention in wetlands involves complex intercoupled physical, chemical, and biological processes that ultimately retain phosphorus in organic and inorganic forms. These biogeochemical processes interact with the phosphorus loaded to the wetland ecosystem within various internal storage compartments (soils, vegetation, detritus, microorganisms, and fauna). When evaluating phosphorus retention by wetland ecosystems, both short-term storage (assimilation into vegetation, translocation within aboveground and belowground plant tissues, microorganisms, periphyton, and detritus) and long-term storage components (retention by inorganic and organic soil particles and net accretion of organic matter) should be taken into account.

Numerous published articles evaluated the potential of wetlands to retain phosphorus. Most of this information is related to inflow and outflow characteristics of the water, with very limited information on internal processes and mass balances of phosphorus in wetlands. However, a wealth of quantitative information on phosphorus biogeochemistry of upland and of other aquatic ecosystems provides extensive process-level information. These data on chemical and biological reactions regulating phosphorus availability and mobility in soils and sediments form a basis for much of the phosphorus biogeochemistry research related to phosphorus retention in wetlands. For example, early research on flooded soils showed the regulation of phosphorus solubility and retention by oxidation and reduction of iron. Similarly, iron regulation of phosphorus solubility was also shown in lake sediments. More recent research in wetlands also confirmed that these principles are applicable to wetland ecosystems.

Phosphorus retention by wetland soils includes surface adsorption on minerals and organic particles, precipitation, microbial immobilization, and plant uptake and sequestering. Phosphorus sorption and precipitation in soils involves removal of inorganic phosphorus from the soil solution into a more stable solid phase. When plants and microbes senesce and die, phosphorus contained in cellular materials either may recycle within the wetland or may be interred with relatively refractory organic compounds. Accretion of phosphorus associated with organic and inorganic matter is putatively a major sink for phosphorus in wetlands. The genesis of the accreted soil is a relatively slow process and can affect the phosphorus retention characteristics of the wetland and associated water quality. Hence, it is important to understand the composition and stability of the newly accreted materials to determine the long-term efficacy of phosphorus retention in wetlands. Therefore, although the biogeochemistry of wetlands can function effectively in the collective removal of influent inorganic and organic nitrogen compounds, the processes are less effective for phosphorus. The retentive and release processes are highly dynamic and variable on daily and seasonal time scales. Over long periods (years) in relatively undisturbed wetlands, limited phosphorus loading can be sequestered and stored, largely in recalcitrant organic matter.

9.12 SUMMARY

Wetlands and aquatic systems are recipients of phosphorus loads from upland systems. Increased loading of phosphorus to a system can cause nitrogen limitations. The phosphorus enrichment of P-limited systems leads to eutrophication and ecosystem stress. The phosphorus cycle does not have a significant gaseous loss mechanism. Thus, most of the added P accumulates in the systems.

- Phosphorus exists in two basic forms, dissolved and particulate, and is available from two sources, organic and inorganic. Common inorganic phosphorus pools include loosely bound fractions, fractions associated with Al, Fe, and Mn oxides and hydroxides, Ca- and Mg-bound fractions, and minerals.
- Phosphate is not commonly used as an oxidant but is affected by redox dynamics. Oxidized forms of iron can react with phosphorus and form insoluble compounds. Chemical extraction methods are used to determine inorganic phosphorus pools in soils.
- Adsorption and precipitation are the retention mechanisms for inorganic phosphorus. Adsorption initially happens very quickly, and then the rates decrease. Inorganic phosphorus sorption by soils is described by several models: Langmuir, Freundlich, and linear

equations. Phosphorus adsorption occurs by both ion exchange and ligand exchange. The slow phase of phosphorus sorption is attributed to diffusion and precipitation.
- Reduction of ferric iron in $FePO_4$ to the soluble ferrous iron is one of the dominant process controlling phosphorus solubility in anaerobic systems. The charge of the solid phase is pH dependent. Phosphate adsorbs more strongly under acidic conditions because the positive charge from the hydrogen ion attracts the negatively charged phosphate anion.
- Soil's capacity to adsorb phosphorus is regulated by the EPC_0, at which point adsorption equals desorption. Each soil buffers a threshold concentration. If the water entering the soil has a concentration of $P > EPC_0$, that soil will have a tendency to adsorb phosphorus. If the water entering the soil has a concentration of phosphorus $<EPC_0$, then that soil will have a tendency to release phosphorus.
- The phosphorus retention abilities of mineral soils are directly related to amorphous and poorly crystalline forms of Fe and Al. Phosphate bound to Fe and Al can be displaced by organic anions.
- When sulfate reduction occurs, sulfide can react with iron to form pyrite. This decreases the amount of phosphorus that is bound with iron and results in more available phosphorus.
- Inorganic phosphorus retention is regulated by pH, Eh, phosphate concentration (there is a limited amount of substrate for adsorption), concentrations of Fe, Al, and calcium carbonate, and temperature.
- Much of the organic phosphorus is present as monoesters and diesters. Monoesters include sugar phosphates (e.g., glucose-6-phosphates and phosphenols, which are intermediates of metabolic pathways), phosphoproteins, mononucleotides, and inositol phosphates. Diesters include nucleic acids, phospholipids, and aromatic compounds.
- Many organisms can store orthophosphate or polyphosphates. In plants, inositol hexaphosphate (IHP) can form a major storage compound for phosphorus, particularly in seeds. Although relatively low in concentration, diesters may contribute to short-term bioavailable pool more than monoesters. Compared to monoesters, diesters are more accessible to microbial attack. This fraction was shown to be rapidly made available to plants.
- Organic phosphorus associated with humic and fulvic acids represents >40% of the total soil phosphorus. The fulvic acid phosphorus constitutes a large fraction of the organic phosphorus in most soils and it is likely that this pool is derived from plant litter and recently deposited organic matter.
- Microorganisms incorporate dissolved phosphorus into cellular constituents, which then become integral parts of the particulate matter. Through the formation of polyphosphate compounds, microorganisms are able to survive in alternating oxidation–reduction environments.
- Vegetation stores a significant amount of phosphorus in above ground and below ground biomass and is a major source of organic phosphorus in wetlands.
- Several abiotic processes including leaching, fragmentation, and photolysis are involved in the mineralization of organic phosphorus.
- Biotic process is the dominant pathway by which organic phosphorus is mineralized in wetlands. Organic phosphorus availability from monoesters requires enzymatic cleavage of the ester linkage. This occurs primarily through several extracellular enzymes such as phosphatases through hydrolysis of phosphate esters.
- Common monoesters in soils include glycerophosphate, glucose-6-phosphate, fructose-6-phosphate, and inositol phosphates. This hydrolysis is achieved by phosphomonoesterases bound to or within the cell membrane or present in the soil pore water or in the water column.
- Several abiotic and biotic processes are involved in mobilizing phosphorus between soil and overlying water column.

- Phosphorus is transported through wetlands in soluble and particulate forms, including dissolved inorganic phosphorus (DIP), particulate inorganic phosphorus (PIP), dissolved organic phosphorus (DOP), and particulate organic phosphorus (POP).
- The "phosphorus memory" (phosphorus retained in soils and sediments and potential release from these sources) can extend the time required for a wetland or an aquatic system to reach an alternate stable condition to meet environmental regulation such as TMDLs (total daily maximum daily loads).

STUDY QUESTIONS

1. Identify inorganic pools of phosphorus in wetlands and rank them in the order of potential bioavailability.
2. Draw a phosphorus cycle for a constructed wetland by showing all biogeochemical processes in soil and water column involved in regulating phosphorus retention.
3. Assume that phosphate adsorption by soils is the dominant biogeochemical process regulating DIP retention by a wetland and then estimate the time required to saturate phosphate sorption sites of 30 cm soil profile under the following conditions:

    Mineral wetland soils high in Fe and Al oxides
    DIP loading to wetland = 10 mg P/m^2 day
    Soil bulk density = 1 g/cm^3
    Soil depth = 30 cm
    Phosphorus sorption maximum as estimated by sorption experiments and Langmuir sorption isotherms = 1000 mg P/kg
    Note: P added to this wetland is all in DIP. Adsorption process is the dominant process and other processes are not significant.
    DIP is added continuously and is uniformly distributed in the soil profile.

4. Briefly discuss the factors regulating DIP retention by wetlands. How can you increase the longevity of this wetland to retain DIP?
5. Identify organic pools of phosphorus in wetlands and rank them in the order of bioavailability.
6. What is the role of enzymes in organic phosphorus mineralization? Identify types of enzymes involved. What are some of the key factors that regulate the activity of enzymes?
7. Why does added phosphorus accumulate in wetlands?
8. What is phosphorus memory (also referred to as legacy phosphorus) in wetlands? What role does it play in restoration of wetlands?

FURTHER READINGS

Bohn, H. L., B. L. McNeal, and G. A. O'Connor. 1985. Soil Chemistry. Wiley, New York, NY. 341 pp.
Burns, R. G. and R. P. Dick. 2002. *Enzymes in the Environment*. Marcel Dekker, New York, NY. 614 pp.
Essington, M. E. 2004. *Soil and Water Chemistry*. CRC Press, Boca Raton, FL. 534 pp.
Harris, W. G. 2002. Phosphate minerals. In *Soil Mineralogy with Environmental Applications*. SSSA Book Series No. 7. Soil Science Society of America, Madison, WI. pp. 637–665.
Heath, R. T. 2004. Microbial turnover of organic phosphorus in aquatic environments. In B. L. Turner, E. Frossard, and D. S. Baldwin (eds.) *Organic Phosphorus in the Environment*. CAB Publishing, Cambridge, MA. pp. 185–204.
Maher, W. and L. Woo. 1998. Review: procedures for the storage and digestion of natural waters for the determination of filterable reactive phosphorus, total filterable phosphorus, and total phosphorus. *Anal. Chim. Acta* 375:5–47.

Reddy, K. R., R. G. Wetzel, and R. Kadlec. 2005. Biogeochemistry of phosphorus in wetlands. In J. T. Sims and A. N. Sharpley (eds.) *Phosphorus: Agriculture and the Environment.* Soil Science Society of America, Madison, WI. pp. 263–316.

Reynolds, C. S. and P. S. Davis. 2001. Sources and bioavailability of phosphorus fractions in freshwaters: a British perspective. *Biol. Rev.* 76:27–64.

Rhue, R. D. and W. C. Harris. 1999. Phosphorus sorption/desorption reactions in soils and sediments. In K. R. Reddy, G. A. O'Connor, and C. L. Scheske (eds.) *Phosphorous Biogeochemistry in Subtropical Systems.* Lewis Publishers, Boca Raton, FL. pp. 187–206.

Sims, J. T. and A. N. Sharpley. 2005. *Phosphorus in Agriculture and the Environment.* American Society of Agronomy, No. 46, Madison, WI. 1021 pp.

Stevenson, F. J. 1986. *Cycles of Soil: Carbon, Nitrogen, Phosphorus, Sulfur, Micronutrients.* Wiley, New York, NY. 380 pp.

Turner, B. L., E. Frossard, and D. D. Baldwin. 2004. *Organic Phosphorus in the Environment.* CAB Publishing, Cambridge, MA. 389 pp.

10 Iron and Manganese

10.1 INTRODUCTION

Iron and manganese represent the 4th and 12th most abundant elements, respectively, in the earth's crust. Iron and manganese are two of the key elements involved in oxidation–reduction reactions in soils of the biosphere and are essential trace elements for living organisms. The biogeochemistry of iron and manganese is complex and is regulated by various biotic and abiotic processes. For example, microbial reduction of iron (III) and manganese (IV) coupled with organic matter decomposition greatly influences biogeochemical cycles of many metals and nutrients in the environment. The biogeochemistry of iron and manganese in wetland soils and aquatic sediments follows several similar pathways. This is why we have chosen to present a discussion on the biogeochemistry of these two elements in the same chapter.

Both iron and manganese change their oxidation status depending on redox conditions in soils and sediments. The ecological and environmental significance of iron and manganese oxidation and reduction reactions can be summarized as follows:

- Are important in decomposition of organic matter and nutrient regeneration
- Have adverse effects on plant growth; excessive concentrations of dissolved manganese can be toxic to plants
- Decrease the availability of certain plant nutrients through precipitation reactions
- Reduce sulfide levels in soils through precipitation reactions
- Suppress other microbial processes that regulate organic matter decomposition (e.g., sulfate reduction and methanogenesis)
- Help in the mobilization/immobilization of phosphate and trace metals
- Alter pH/alkalinity
- Help in the formation of minerals (e.g., siderite, vivianite, pyrite, and others)
- Cause mottling and gleying, which serve as indicators of hydric soils
- Oxidizes toxic organic contaminants

10.2 STORAGE AND DISTRIBUTION

Iron and manganese are widely distributed in soils and sediments. Owing to a lack of gaseous and minimal soluble phases, both iron and manganese cycles are largely uncoupled from hydrologic and atmospheric cycles. However, within earth's surface layers, iron and manganese play a significant role in various biogeochemical cycles. The aerobic and anaerobic environments in wetlands promote oxidation and reduction reactions of iron and manganese. Anaerobic zones can promote mobilization of iron and manganese into aerobic zones.

Iron is one of the dominant elements in mineral soils and is typically present as oxides in aerobic environments. Iron is an essential element in all organisms and is involved in various essential functions. Total iron in soils and sediments ranges from 0.1% to as high as 10%. Iron occurs predominantly in two oxidation states, trivalent (Fe^{3+}) and divalent (Fe^{2+}), and depending on environmental conditions iron can form stable compounds (Table 10.1).

In nature, manganese can exist as common minerals, such as carbonates, oxides, hydroxides, silicates, and to a limited extent as sulfides (Table 10.1). Next to iron, manganese is the most abundant element in the earth's crust and is found in variable concentrations in living organisms, water,

TABLE 10.1
Some Important Iron and Manganese Solid and Dissolved Phases in Soils and Sediments

Mineral	Formula	Oxidation States of Iron and Manganese
Solid phases		
Rhodochrosite	$MnCO_3$	+2
Pyrolusite	MnO_2	+4
Goethite	$\alpha\text{-FeOOH}$	+3
Siderite	$FeCO_3$	+2
Hematite	Fe_2O_3	+3
Magnetite	Fe_3O_4	+3
Troilite	FeS	+2
Pyrite	FeS_2	+2
Strengite	$FePO_4$	+3
Vivianite	$Fe_3(PO_4)_2$	+2
Dissolved phases		
Manganic manganese	Mn^{4+}	+4
Manganous manganese	Mn^{2+}	+2
Ferric iron	Fe^{3+}	+3
Ferrous iron	Fe^{2+}	+2

and soils. Mineral deposits have a limited geographical range and are sedimentary in origin. These deposits are the result of the large-scale conversion of soluble manganese into a relatively insoluble precipitate, meditated by various biotic and abiotic processes. Manganese is essential in several biochemical reactions, including activation of a number of enzymes. Manganese concentrations in soils and sediments vary with values in the range between 20 and 6,000 mg kg^{-1}. Plants assimilate considerable amounts of manganese with concentrations in the range of 20–500 mg kg^{-1} of dry matter. Levels in excess of 500 mg kg^{-1} of dry matter can be toxic to plants. Manganese also occurs in two oxidation states, tetravalent (Mn^{4+}) and divalent (Mn^{2+}), which are typically present as oxides (Table 10.1).

Dissolved manganese and iron are released as a result of dissolution or weathering of minerals and microbial reduction (under anaerobic conditions). These can be either redeposited as oxides or hydroxides or remain in a solution (depending on redox and pH of soils). In addition, both dissolved manganese and iron have a tendency to form stable complexes with organic materials, and remain in solution for long periods of time (days). Iron and manganese can cycle rapidly at aerobic–anaerobic interfaces of wetlands, bottoms of eutrophic and stratified lakes, ponds and reservoirs, streams and river bottoms, poorly drained mineral soils, and any other flooded environment containing minerals of these two elements.

The biogeochemical cycling of iron and manganese in wetland soils is shown in Figures 10.1 and 10.2, respectively. Cycling of iron and manganese involves various oxidation and reduction reactions mediated by a range of microbial groups. In addition, abiotic reactions are also involved in the cycling of iron and manganese. Both iron and manganese occur in the dissolved phase and solid phase. The major difference between these two metals is their relative tendencies to form insoluble precipitates with sulfides. In sulfate-dominated wetlands, reduced iron can be continuously removed from the environment through precipitation by sulfides, whereas precipitation of manganese with sulfide is insignificant. Thermodynamically, oxidized forms of iron and manganese serve as excellent electron acceptors for facultative and anaerobic bacteria, with reduction potentials below nitrate reduction and well above sulfate reduction. Oxidized forms of iron and manganese

Iron and Manganese

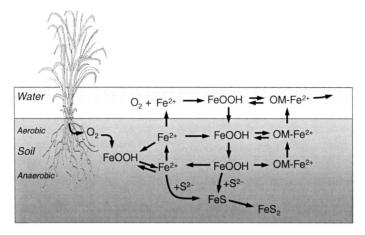

FIGURE 10.1 Schematic showing iron cycling in wetlands and aquatic systems.

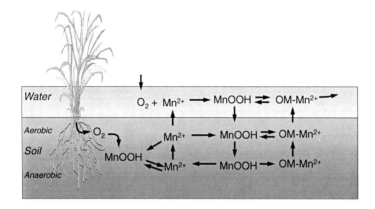

FIGURE 10.2 Schematic showing manganese cycling in wetlands and aquatic systems.

are present in solid phases as precipitates and mineral forms, which makes them immobile in the environment as compared to oxygen, nitrate, and sulfate. The organisms that use these compounds as electron acceptors must have unique physical and biochemical characteristics to grow on these substrates (see detailed discussion in later sections of this chapter).

Iron and manganese concentrations can vary with soil type. Mineral soils with high clay content typically have higher concentrations of iron and manganese. Organic soils typically have lower iron and manganese content than mineral soils. Soil anaerobic conditions result in increase in soluble iron and manganese. Sodium acetate or ammonium acetate is typically used to extract both soluble and exchangeable iron and manganese (see Section 10.4.3 for additional details). Extractable iron and manganese in 67 flooded Mississippi River floodplain soils of Louisiana were reported to be in the range of 225–3,365 mg kg^{-1} (average = 1,472) and 11–767 mg kg^{-1} (average = 225), respectively (Redman and Patrick, 1965; Reddy et al., 1980).

10.3 Eh–pH RELATIONSHIPS

Oxidation–reduction reactions involving iron and manganese include changes in oxidation states, relative stabilities of iron and manganese compounds, and the energetics and kinetics of oxidized and reduced compounds. Using thermodynamic data (from free energies of formation, see Chapter 4), it is possible to predict the stability of various inorganic phases of iron and manganese. Iron has two oxidation states, Fe(III) and Fe(II), whereas manganese has three oxidation states, Mn(IV), Mn(III), and Mn(II). The stability of manganese and iron solid phases in soils is regulated by both redox potential and pH.

Stability diagrams developed using thermodynamic data for iron and manganese are shown in Figures 10.3 and 10.4. These diagrams indicate the dominant stable species of iron and manganese

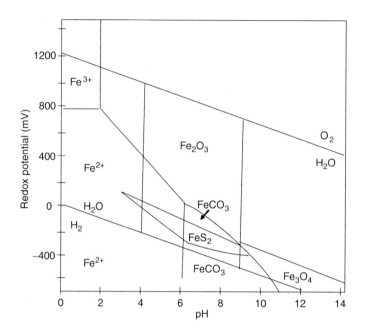

FIGURE 10.3 A Eh–pH diagram for Fe(III)–Fe(II) redox couple.

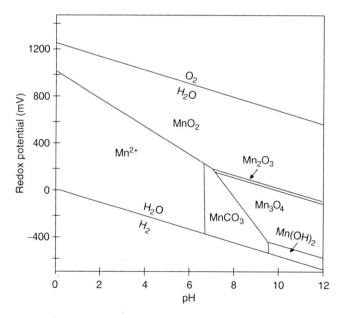

FIGURE 10.4 A Eh–pH diagram for Mn(IV)–Mn(II) redox couple.

Iron and Manganese

at any specified redox potential and pH. The predicted pH level needed to precipitate iron or manganese is dependent on the Eh of the system. Similarly, the Eh level required to oxidize or to reduce iron and manganese from one oxidation state to another also depends on the pH of the system. The pH level for the stability of Fe(II) is much lower than Mn(II) at any Eh value, suggesting that manganese solubility exceeds that of iron at any given Eh and acidic pH. It should be noted that these stability diagrams assume equilibrium conditions between dissolved and solid phases, with no consideration given to the kinetics of these reactions. Thus, caution must be exercised in the interpretation of these diagrams, as they may not represent the complexities associated with biotic and abiotic interactions.

10.3.1 Iron

The principal chemical reactions for select redox couples involving reduction of Fe(III) oxides are as follows. Relationships between Eh and pH were calculated using the Nernst equation (see Chapter 4). The principal iron (Fe) forms in soils are likely to be $Fe(OH)_3$, α-FeOOH (goethite), γ-FeOOH (lepidocrocite), Fe_2O_3 (hematite), $Fe_3(OH)_8$ (ferrosoferric hydroxide), $FeCO_3$ (siderite), and FeS_2 (pyrite) in equilibrium with Fe^{2+}. Of the various Fe(III) oxides identified in soils, the most stable form is goethite, from which hematite can be formed upon dehydration (Gotoh and Patrick, 1974). The pFe and pH slope together with $E°$ value from the Nernst equation can be used to evaluate a particular iron system that might be operating in soil.

$Fe(OH)_3$ *(ferric hydroxide)*–Fe^{2+} *(ferrous iron) redox couple:*

(amorphous form)

$$Fe(OH)_3 + 3H^+ + e^- = Fe^{2+} + 3H_2O \quad E° = 1.057 \text{ V} \tag{10.1}$$

For which

$$Eh = 1.057 + 0.059 pFe^{2+} - 0.177 pH \tag{10.2}$$

$FeOOH$ *(ferric oxyhydroxide)*–Fe^{2+} *(ferrous iron) redox couple:*

(crystalline form)

$$FeOOH + 3H^+ + e^- = Fe^{2+} + 3H_2O \quad E° = 0.693 \text{ V} \tag{10.3}$$

For which

$$Eh = 0.693 + 0.059 pFe^{2+} - 0.177 pH \tag{10.4}$$

$Fe_3(OH)_8$ *(ferrosoferric hydroxide)*–Fe^{2+} *(ferrous iron) redox couple:*

$$Fe_3(OH)_8 + 8H^+ + 2e^- = 3Fe^{2+} + 8H_2O \quad E° = 1.373 \text{ V} \tag{10.5}$$

For which

$$Eh = 1.373 + 0.0885 pFe^{2+} - 0.236 pH \tag{10.6}$$

Fe_2O_3 *(hematite)*–Fe^{2+} *(ferrous iron) redox couple:*

(amorphous form)

$$Fe_2O_3 + 6H^+ + 2e^- = 2Fe^{2+} + 3H_2O \quad E° = 0.728 \text{ V} \tag{10.7}$$

For which

$$Eh = 0.728 + 0.059 pFe^{2+} - 0.177 pH \quad (10.8)$$

Fe(OH)₃ (ferric hydroxide)–Fe₃(OH)₈ (ferrosoferric hydroxide) redox couple:

$$Fe(OH)_3 + H^+ + e^- = Fe_3(OH)_8 + H_2O \quad E° = 0.429 \text{ V} \quad (10.9)$$

For which

$$Eh = 0.429 - 0.059 pH \quad (10.10)$$

Controlled laboratory microcosm experiments by Gotoh and Patrick (1974) were used to confirm the observed relationship among redox potential, pH, and the activity of Fe^{2+} to the theoretical relationships derived using equations of redox equilibria to determine the specific redox couple functioning in the soil. The $Eh - 0.059 pFe^{2+}$ was plotted as a function of pH for select solid phases (Equations 10.2, 10.4, and 10.8) and for experimental values (Figure 10.5). Similarly, the $Eh - 0.0885 pFe^{2+}$ was plotted as a function of pH for solid phase (Equation 10.6) and for experimental values (Figure 10.5). The experimental values tended to cluster within the area delineated by theoretical lines of amorphous and crystalline $Fe(OH)_3$–Fe^{2+}, whereas the experimental values fell below the theoretical lines of $Fe_3(OH)_8$ (ferrosoferric hydroxide)–Fe^{2+} (ferrous iron) redox couple, indicating undersaturation with respect to the Fe^{2+} activity. This is true even at high pH conditions that are favorable for $Fe_3(OH)_8$ formation and stable under anaerobic conditions (Gotoh and Patrick, 1974). The lack of agreement between theoretical and experimental values suggests that Fe^{2+} activity may be governed by mixed redox couples. However, the results presented in Figure 10.5 suggest that the Fe^{2+} activity in the flooded Crowley silt loam of Louisiana is governed by the solubility of ferric oxyhydroxide.

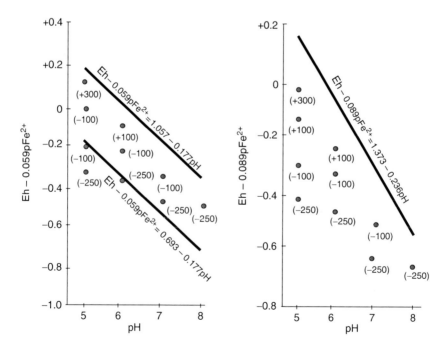

FIGURE 10.5 The $Eh - 0.059 pFe^{2+}$ and $Eh - 0.0885 pFe^{2+}$ plotted as a function of pH for select solid phases (Equations 10.2, 10.4 and 10.8, and Equation 10.6, respectively) and for experimental values. (Redrawn from Gotoh and Patrick, 1972.)

Iron and Manganese

10.3.2 MANGANESE

The principal chemical reactions for select redox couples involving reduction of Mn(IV) oxides are as follows. Relationships between Eh and pH were calculated using the Nernst equation (see Chapter 4). The principal manganese forms in soils are likely to be MnO_2, Mn_2O_3, Mn_3O_4, $MnCO_3$, and $Mn(OH)_2$ in equilibrium with Mn^{2+}. The pMn and pH slope together with $E°$ value from the Nernst equation can be used to evaluate a particular manganese system that might be operating in soil.

MnO_2–Mn^{2+} redox couple:

$$MnO_2 + 4H^+ + 2e^- = Mn^{2+} + 2H_2O \quad E° = 1.225 \text{ V} \tag{10.11}$$

For which

$$Eh = 1.225 + 0.0295 \text{pMn} - 0.118 \text{pH} \tag{10.12}$$

Mn_2O_3–Mn^{2+} redox couple:

$$Mn_2O_3 + 6H^+ + 2e^- = 2Mn^{2+} + 3H_2O \quad E° = 1.497 \text{ V} \tag{10.13}$$

For which

$$Eh = 1.497 + 0.059 \text{pMn} - 0.177 \text{pH} \tag{10.14}$$

Mn_3O_4–Mn^{2+} redox couple:

$$Mn_3O_4 + 8H^+ + 2e^- = 3Mn^{2+} + 4H_2O \quad E° = 1.824 \text{ V} \tag{10.15}$$

For which

$$Eh = 1.824 + 0.0885 \text{pMn} - 0.236 \text{pH} \tag{10.16}$$

Redox potential and pH have been used extensively by biogeochemists to evaluate equilibrium relationships between Mn(II) activity in solution and various manganese compounds (Equations 10.11–10.16). General conclusions drawn are that between pH 6 and 8 the conversion of Mn(IV) to soluble Mn(II) was dependent on both Eh and pH, whereas at pH 5 the effect of acidity was so marked that changes in Eh had little effect on manganese solubility (Gotoh and Patrick, 1972). Agreement between theoretical relationships (Equations 10.11–10.16) and experimentally derived Mn^{2+} activity was poor, suggesting that not any one redox couple, but rather mixed redox couples are controlling the Mn^{2+} activity. In addition, manganese reactions are also affected by chelation with organic compounds produced in wetland soils. Mn(IV) oxides involved in oxidation–reduction reactions in wetland soils are complex and respond to both abiotic and biotic reduction.

10.4 REDUCTION OF IRON AND MANGANESE

Microorganisms involved in oxidation–reduction reactions of iron and manganese can be grouped into the following:

- Bacteria that are able to use Fe(III) and Mn(IV) as electron acceptors and derive energy for their growth from organic and inorganic substrates. These organisms must have unique characteristics, which might include the ability to (1) solubilize substrates, (2) attach to substrate and directly transfer electrons to it, or (3) transport the substrate into the cell as a solid (Nealson and Myers, 1992).

- Bacteria that are able to use Fe(II) and Mn(II) compounds as electron donors and derive energy for their growth from oxidation.
- Bacteria producing distinct precipitates of iron and manganese associated with cells.

Below the soil depth of oxygen penetration, Fe(III) and Mn(IV) oxides are reduced to more soluble Fe(II) and Mn(II) forms. All oxidized forms of iron and manganese can function as electron acceptors for several facultative and obligate anaerobic bacteria during their catabolic activities. Reduction of Fe(III) and Mn(IV) can occur enzymatically (linked to microbial activity) and nonenzymatically in wetlands and aquatic systems. However, enzymatic reduction of Fe(III) and Mn(IV) is the dominant process. Studies have shown that oxidation of organic matter is coupled to the reduction of Fe(III) and Mn(IV). As this process is linked to bacterial catabolic activities, we can define this process as the *catabolic reduction of iron and manganese*, more commonly known as dissimilatory iron and manganese reduction. During this process Fe(III) and Mn(IV) are reduced to Fe(II) and Mn(II), respectively. A small fraction of iron and manganese is also assimilated during cell synthesis, a process known as assimilatory iron and manganese reduction.

Thermodynamically, manganese reduction occurs at higher reduction potentials than iron reduction. For example, on a theoretical basis, reduction of MnO_2 using glucose as an electron donor can result in an energy yield of $-2,027$ kJ mol^{-1} of glucose (pH = 7), as compared to a reduction of $Fe(OH)_3$ that can result in -441 kJ mol^{-1} of glucose (pH = 6).

Oxidation: $C_6H_{12}O_6 + 6H_2O = 6CO_2 + 24H^+ + 24e^-$
Reduction: $12MnO_2 + 48H^+ + 24e^- = 12Mn^{2+} + 24H_2O$
Oxidation–reduction: $C_6H_{12}O_6 + 12MnO_2 + 24H^+ = 6CO_2 + 12Mn^{2+} + 18H_2O$

Oxidation: $C_6H_{12}O_6 + 6H_2O = 6CO_2 + 24H^+ + 24e^-$
Reduction: $24Fe(OH)_3 + 72H^+ + 24e^- = 24Fe^{2+} + 72H_2O$
Oxidation–reduction: $C_6H_{12}O_6 + 24Fe(OH)_3 + 48H^+ = 6CO_2 + 24Fe^{2+} + 66H_2O$

To oxidize 1 mol of glucose, 12 mol MnO_2 and 24 mol of $Fe(OH)_3$ are needed, as compared to 6 mol of oxygen, 4.8 mol of nitrate, and 3 mol of sulfate. The amount of energy yield generated by these processes accounts for approximately 70% of aerobic energy yield (using MnO_2) and only 20% of aerobic energy yield (using $Fe(OH)_3$) (Figure 10.6). Based on thermodynamics, it is likely that Mn(IV) oxides are the preferred electron acceptor over Fe(III) oxides.

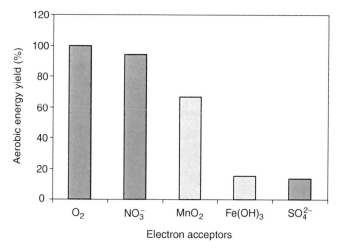

FIGURE 10.6 Relative energy yield during reduction of select electron acceptors with glucose as the electron donor. Values expressed as percent energy yield under aerobic (oxygen as electron acceptor) conditions.

TABLE 10.2
Iron and Manganese Respiring Bacteria

Microorganisms	Electron Acceptor	Electron Donor
Bacillus, Clostridium, Desulfovibrio, Escherichia, Ferribacterium, Geobacter, Geothrix, Lactobacillus, Pelobacter, Pseudomonas	Fe(III) and Mn(IV)	Fermentative, organic substrates such as sugars and amino acids
Sulfolobus, Thiobacillus	Fe(III) and Mn(IV)	Elemental sulfur
Pseudomonas, Shewanella	Fe(III) and Mn(IV)	Hydrogen
Bacillus, Clostridium, Desulfovibrio, Escherichia, Ferribacterium, Geobacter, Lactobacillus, Pseudomonas, Shewanella	Fe(III) and Mn(IV)	Acetate, butyrate, ethanol, lactate, propionate, pyruvate, and valerate
Geobacter	Fe(III)	Aromatic compounds—benzoate, phenol, and toluene
Alternaria, Fusarium	Fe(III)	

Source: Ehrlich (2002), Lovley (1991, 2004), and Nealson and Myers (1992).

10.4.1 MICROBIAL COMMUNITIES

Several heterotrophs and autotrophs are capable of using Fe(III) and Mn(IV) as electron acceptors (Table 10.2). The microbial communities included in this group are those that are capable of respiring using Fe(III) and Mn(IV) as electron acceptors, while oxidizing a range of organic and inorganic substrates (electron donors). For additional information, the reader is referred to reviews by Lovley (1991) and Ehrlich (2002).

The oxidation of all of these electron donors can potentially yield enough energy for ATP generation via electron transport linked to Fe(III) and Mn(IV) reduction. This is based on investigations using pure cultures. Depending on the electron donor utilization, iron and manganese reducers are divided into following groups (Lovley, 1991):

- Fermentative Fe(III) and Mn(IV) reducers
- Sulfur-oxidizing Fe(III) and Mn(IV) reducers
- Hydrogen-oxidizing Fe(III) and Mn(IV) reducers
- Organic acid-oxidizing Fe(III) and Mn(IV) reducers
- Aromatic compound-oxidizing Fe(III) reducers

Microorganisms with primary fermentative metabolism are also capable of using Fe(III) and Mn(IV) as electron acceptors. Many studies have focused on Fe(III) and Mn(IV) reduction with fermentable substrates such as glucose. Thermodynamically, complete oxidation of these substrates to carbon dioxide with the transfer of electrons to Fe(III) and Mn(IV) is possible. However, electron flow through this pathway is limited as compared to organic electron acceptors used during fermentation. These organisms derive very little or no energy through the reduction of Fe(III) and Mn(IV). It is likely some of the reduction of Fe(III) and Mn(IV) may be due to the nonenzymatic pathway (for details see later part of this chapter). This suggests that thermodynamically it may not be efficient for fermenters to use Fe(III) and Mn(IV) as electron acceptors and to oxidize fermentable sugars and amino acids all the way to carbon dioxide. Sugars are first fermented into simple organic compounds, which are subsequently used as electron donors by iron and manganese reducers (see Chapter 5).

Some autotrophs such as *Thiobacillus thioxidans, T. ferrooxidans, Sulfolobus* spp., and others listed in Table 10.2 are capable of using Fe(III) as an electron acceptor with elemental sulfur serving as an electron donor.

$$H_2S + 2Fe(III) + 6H^+ = S^0 + 2Fe(II) + 4H_2O$$

$$S^0 + 6Fe(III) + 4H_2O = HSO_4^- + 6Fe(II) + 7H^+$$

Several bacteria can reduce Fe(III) and Mn(IV) using hydrogen as their electron donor. Both *Pseudomonas* and *Shewanella* are known to use hydrogen as their electron donor with Fe(III) as an electron acceptor.

$$H_2 + 2Fe(III) = 2H^+ + 2Fe(II)$$

$$H_2 + 2Mn(IV) = 2H^+ + 2Mn(II)$$

For hydrogen to be available in wetlands, soils need to have a highly reduced environment and favorable conditions for fermentation. The presence of oxygen or other electron acceptors with a higher reduction potential can inhibit fermentation, thus preventing hydrogen production. In wetland soil profiles, hydrogen production may occur in redox environments favorable to sulfate reduction and methanogenesis. Any hydrogen produced under these conditions can be used by sulfate reducers and methanogens. Although iron and manganese reducers may be present in this redox environment, most of the Fe(III) and Mn(IV) may already be reduced and under these conditions, it is likely that microbial activity is limited by the availability of reducible iron and manganese.

Several heterotrophic iron- and manganese-reducing bacteria are known to effectively utilize organic acids as electron donors, while reducing Fe(III) and Mn(IV). The following examples are for complete oxidation of acetate and formate:

(Note: reactions are not balanced)

$$Acetate + 8Fe(III) + 4H_2O = 2HCO_3^- + 8Fe(II)$$

$$Acetate + Mn(IV) + H_2O = HCO_3^- + Mn(II)$$

$$Formate + Fe(III) = HCO_3^- + Fe(II)$$

$$Formate + Mn(IV) = HCO_3^- + Mn(II)$$

The following examples are for incomplete oxidation of lactate and pyruvate:

$$Lactate + 4Fe(III) + 4H_2O = Acetate + 2HCO_3^- + 8Fe(II)$$

$$Lactate + Mn(IV) + 4H_2O = Acetate + 2HCO_3^- + Mn(II)$$

$$Pyruvate + 2Fe(III) + 4H_2O = Acetate + 2HCO_3^- + 8Fe(II)$$

$$Pyruvate + Mn(IV) + 4H_2O = Acetate + 2HCO_3^- + Mn(II)$$

Iron-reducing bacteria, *Geobacter*, was shown to obtain energy from the oxidation of aromatic organic compounds, including toluene, phenol, *p*-cresol, benzoate, and others (Lovley, 1991).

Iron and Manganese

The following reactions involving *Geobacter* have been documented for complete oxidation of aromatic organic compounds to carbon dioxide, while using Fe(III) as sole electron acceptor (Lovley, 1991):

$$\text{Benzoate} + 30\text{Fe(III)} + 19\text{H}_2\text{O} = 7\text{HCO}_3^- + 30\text{Fe(II)}$$

$$\text{Toluene} + 36\text{Fe(III)} + 21\text{H}_2\text{O} = 7\text{HCO}_3^- + 306\text{Fe(II)}$$

$$\text{Phenol} + 28\text{Fe(III)} + 17\text{H}_2\text{O} = 6\text{HCO}_3^- + 28\text{Fe(II)}$$

$$p\text{-Cresol} + 34\text{Fe(III)} + 20\text{H}_2\text{O} = 7\text{HCO}_3^- + 34\text{Fe(II)}$$

Fungi such as *Alternaria* and *Fusarium* were shown to respire using Fe(III) as an electron acceptor. However, conclusions drawn from these earlier studies are inconclusive (Ehrlich, 2002). Currently, there is no experimental evidence on the use of Mn(IV) as an electron acceptor by fungi.

10.4.2 Biotic and Abiotic Reduction

The reduction of Fe(III) and Mn(IV) oxides in soils has long been recognized, especially in paddy soils, waterlogged soils, poorly drained soils, and in various wetland soils. Biogeochemists have recognized the importance of iron and manganese cycling in soils and sediments and its linkages to organic matter decomposition.

10.4.2.1 Biotic Reduction

In wetland soils, Fe(III) and Mn(IV) may be reduced by microorganisms to Fe(II) and Mn(II). Biotic degradation involves two enzymes: iron reductase and manganese reductase. Some of these organisms are also capable of using nitrate as an electron acceptor and reducing it to nitrogen gas. In some cases, Fe(III) and Mn(IV) reducers and nitrate reducers use the same enzyme system nitrate reductase. Using pure cultures, several studies have documented microbial growth and respiratory pathways (terminal electron-accepting process) to the reduction of Fe(III) and Mn(IV). The resulting reduced compounds may be present in soluble form or in insoluble forms (Table 10.1).

Oxidized forms of iron and manganese are present as insoluble solid phases either as amorphous or as crystalline forms (Figure 10.7). However, at low pH both oxidized and reduced forms of iron and manganese are present in the soluble phase (Figure 10.7). Some of the common insoluble

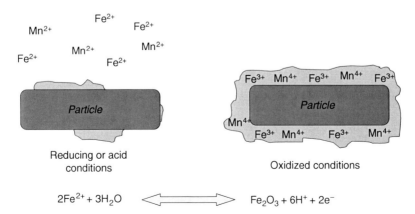

FIGURE 10.7 Schematic showing the portioning of iron and manganese under reducing and oxidizing conditions.

oxidized forms include strengite ($FePO_4$), goethite ($FeOOH \cdot H_2O$), hematite (Fe_2O_3), pyrolusite (MnO_2), and manganite (MnOOH).

Microbial Fe(III) and Mn(IV) reduction in soils presents a unique problem relative to other terminal electron-accepting reactions (e.g., oxygen, nitrate, and sulfate reduction) because Fe(III) and Mn(IV) oxides are highly insoluble at circumneutral pH. As a result, the reduction process involves physical contact and the interaction of bacterial cell surfaces with particulate oxide phases that are not transported into the cell (Lovley, 1987; Ghiorse, 1988). The need for this interaction depends on the organism, the electron donor, and the form of Fe(III).

Bacterial reduction of Fe(III) and Mn(IV) depends on

- The ability of organisms to solubilize solid phases and increase the mobility of the oxidized form
- The ability of organisms to create conditions to form organic Fe or Mn complexes and increase the mobility of iron and manganese
- The ability of organisms to attach directly to solid phases and directly transfer electrons
- The ability of organisms to transport solid phases directly into their cells as solids and solubilize these within the cell
- The amount of bioavailable reductant (organic and inorganic substances) available for microbial utilization
- The amount of bioavailable oxidant (manganese or iron) available for microbial utilization

Microbial communities involved in iron and manganese reduction can use a wide range of organic substrates as electron donors. Some of these bacterial groups are facultative, are adapted to the assimilation of monomers, and use glycolysis and TCA pathways to completely oxidize organic substrates to carbon dioxide, while using iron and manganese as electron acceptors. However, several bacteria involved in Fe(III) and Mn(IV) reduction require a fermentative step, where fermenters oxidize monomers into a range of fatty acids. During organic matter degradation, often these initial steps are limiting factors.

Oxidation of organic matter using Fe(III) and Mn(IV) as electron acceptors follows the steps described in Figure 10.8. Complex polymers of organic matter are enzymatically hydrolyzed into

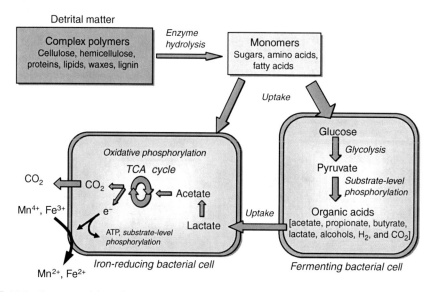

FIGURE 10.8 Decomposition of organic matter and associated metabolic pathways involved in the use of Fe(III) and Mn(IV) oxides as electron acceptors.

simple monomers (see Chapter 5 for details). These monomers can be directly assimilated by Fe(III) and Mn(IV) reducers and subsequently oxidized to carbon dioxide. However, some of these monomers are fermented into simple organic compounds before Fe(III) and Mn(IV) reducers can use them. As discussed previously, several studies have linked the reduction of Fe(III) and Mn(IV) directly to the complete oxidation of low-molecular-weight organic substrates into carbon dioxide. Other bacteria must ferment higher-molecular-weight organic matter into simpler fermentation products that can be used by Fe(III)- and Mn(IV)-reducing bacteria. Thus, in anaerobic soils there is an assembly of fermentors and Fe(III) and Mn(IV) reducers, which decompose organic carbon to carbon dioxide. Some of the reduced fermentation products such as fatty acids (butyrate, formate, propionate, and long-chain fatty acids) and lactate are oxidized by Fe(III) and Mn(IV) reducers to acetate and CO_2. Electron donors used in driving Fe(III) and Mn(IV) reduction include lactate, pyruvate, ethanol, fumarate, malate, propionate, acetate, butyrate, and other fatty acids.

10.4.2.2 Abiotic Reduction

Abiotic reduction of Fe(III) can be mediated by many reducing agents, some of which are typical excretion products of anaerobic bacteria, as well as lowering the Eh and pH of soils and creating a stable environment for reduced over-oxidized species. Many of the more reactive microbial metabolic end products such as citrate, formate, oxalate, malate, and pyruvate and anthropogenic aromatic compounds are capable of nonenzymatic reduction of Fe(III) under low pH conditions (Lovley, 1991). However, nonenzymatic reduction may not occur in many wetlands as pH of these soils approaches near neutral under anaerobic conditions. Oxidation of organic compounds through this process is incomplete.

Following are some examples of the nonenzymatic reduction of Fe(III) by organic compounds.

$$2Cysteine + 2Fe(III) = Cysteine + 2Fe(II) + 2H^+$$

Several organic compounds (such as oxalate, pyruvate, and reducing sugars such as glucose and xylose) are capable of the nonenzymatic reduction of Mn(IV) compounds at circumneutral pH (Lovley, 1991). During this process, many of these organic compounds are incompletely oxidized.

$$Pyruvate + MnO_2 = Acetate + MnCO_3$$

In this reaction, pyruvate can be oxidized to acetate with the reduction of Mn(IV). Acetate is not capable of nonenzymatic reduction of Mn(IV); therefore, it is subsequently utilized by microbes as an electron donor, which enzymatically reduce Mn(IV).

In soils and sediments rich in sulfides, abiotic reduction of Fe(III) and Mn(IV) is possible. For example, sulfides produced during sulfate reduction can reduce Fe(III) to Fe(II) and Mn(IV) to Mn(II). Manganese oxides are known to be more reactive with reduced sulfur compounds than with Fe(III) oxides.

$$2FeOOH + 3H_2S = FeS + FeS_2 + 4H_2O$$

$$MnO_2 + H_2S + 2H^+ = Mn^{2+} + S^0 + 2H_2O$$

$$2FeOOH + H_2S + 4H^+ = 2Fe^{2+} + S^0 + 4H_2O$$

Other potential abiotic reactions may involve the reduction of Mn(IV) with Fe(II), hydrogen peroxide, and nitrite.

$$2MnO_2 + 2Fe^{2+} + 4H_2O = 2Mn^{2+} + 2Fe(OH)_3 + 2H^+$$

$$MnO_2 + NO_2^- + 2H^+ = Mn^{2+} + NO_3^- + H_2O$$

$$2MnO_2 + NO_2^- + 2H^+ = Mn_2O_3 + NO_3^-$$

$$MnO_2 + H_2O_2 + 2H^+ = Mn^{2+} + O_2 + 2H_2O$$

In recent years, the reduction of Mn(IV) oxides has been linked to the benthic nitrogen cycle in marine sediments. Now there is some evidence that ammonia can be directly oxidized to nitrogen gas or to nitrate by Mn(IV) oxides (Anschutz et al., 2005).

$$3MnO_2 + 2NH_4^+ + 4H^+ = 3Mn^{2+} + N_2 + 6H_2O$$

$$4MnO_2 + NH_4^+ + 6H^+ = 4Mn^{2+} + NO_3^- + 5H_2O$$

$$6MnOOH + 2NH_4^+ + 10H^+ = 6Mn^{2+} + N_2 + 12H_2O$$

$$8MnOOH + NH_4^+ + 14H^+ = 8Mn^{2+} + NO_3^- + 13H_2O$$

Recent studies have shown that iron-reducing bacteria can use humic acids as electron acceptors during the oxidation of simple electron donors such as lactate, acetate, and hydrogen. These organisms have the ability to transfer electrons onto quinone moieties in humic substances. The hydroquinone moieties that are produced during this process can function as an electron shuttle between cells and extracellular Fe(III) oxides, reducing them to Fe(II) and converting humic substances to oxidized forms (Figure 10.9). In the presence of humic substances, Fe(III)-reducing microorganisms no longer have to establish direct physical contact with Fe(III) oxides to reduce them, thus greatly enhancing the reduction of Fe(III) oxides (Lovley et al., 1996; Kappler et al., 2004).

Based on critical reviews, Lovley (1991, 2004) concluded that there are potential mechanisms for the abiotic reduction of Fe(III) and Mn(IV), but the significance of this process is minimal as compared to biotic reduction catalyzed by microbial activities. Typically, the end products of Fe(II) and Mn(II) are measured as indicators of the biotic and abiotic reduction of Fe(III) and Mn(IV) in anaerobic environments. The reduction of Fe(III) and Mn(IV) as a function of Eh is shown in Figures 10.10 and 10.11. Sodium acetate extractable iron and manganese in anaerobic soils represents Fe(II) and Mn(II), end products of reduction. As expected, extractable Mn(II) and Fe(II) concentrations are low under oxidized conditions and increase with a decrease in the Eh of soil. The accumulation of Mn(II) occurs at higher Eh values than the accumulation of Fe(II), suggesting Mn(IV) reduction precedes Fe(III) reduction. Because the reduction of Fe(III) and Mn(IV) occurs

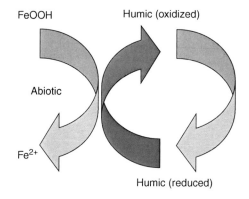

FIGURE 10.9 Schematic showing the role of humic acid during abiotic reduction of Fe(III) oxide.

Iron and Manganese

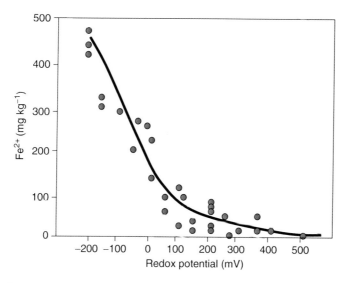

FIGURE 10.10 Influence of redox potential on reduction of Fe(III) oxide as indicated by the accumulation of Fe(II) in a flooded Crowley silt loam soil. (Redrawn from Patrick and Henderson, 1981.)

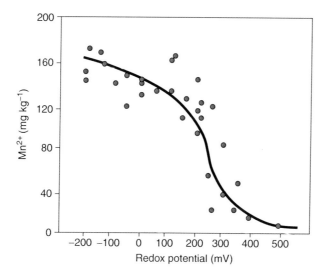

FIGURE 10.11 Influence of redox potential on reduction of Mn(IV) oxide as indicated by the accumulation of Mn(II) in a flooded Crowley silt loam soil. (Redrawn from Patrick and Henderson, 1981.)

as a function of time, accumulation of reduced end products can be influenced by pH and secondary reactions involving precipitation with carbonates and sulfides. Once the soil is flooded, as discussed in previous chapters, oxygen is reduced first, followed by nitrate, and then by Mn(IV) and Fe(III). Reduction of these compounds is inhibited by the presence of oxygen and to some extent nitrogen oxides such as nitrate. The abundance of reduced Fe(II) and Mn(II) in soil solution is favored by low soil pH and anaerobic soil conditions. Thus, the measuring end products of reduction in the soluble and exchangeable phase may underestimate overall reduction rates.

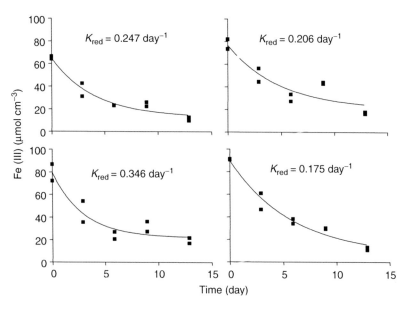

FIGURE 10.12 Kinetics of Fe(III) oxide reduction in wetland soils. (Redrawn from Roden and Wetzel, 2002.)

Although a vast amount of literature exists on the reduction of Fe(III) and Mn(IV) in soils and sediments, little information is available on the overall kinetics of reduction. However, the kinetics of reduction of other electron acceptors such as oxygen, nitrate, and sulfate have been widely studied and reported (see Chapters 6, 7, and 11). The kinetics of Fe(III) and Mn(IV) reduction are influenced by interaction of bacterial cells with solid phases of these compounds (Roden and Wetzel, 2002). Microbial reduction of Fe(III) oxides was reported to follow first-order kinetics in wetland soils and rates of reduction were directly linked to organic matter mineralization (Roden and Wetzel, 2002). The first-order rate constants were in the range of 0.175–0.346 day^{-1}, suggesting a turnover rate of 2.9–5.7 days (Figure 10.12).

Amorphous forms of Fe(III) and Mn(IV) are reduced preferentially, leaving more resistant crystalline forms. Thus, Fe(III) and Mn(IV) reduction rates will decrease on removal of respective bioavailable iron and manganese. In redox zones where Mn(IV) reduction occurs, lack of bioavailable Mn(IV) can promote the reduction of Fe(III) oxides in the presence of crystalline forms of Mn(IV) oxides. Similarly, in redox zones where Fe(III) reduction occurs, depletion of bioavailable Fe(III) oxides can promote the reduction of electron acceptors of lower reduction potentials.

In laboratory studies conducted using wetland soils, the reduction of Mn(IV) oxides occurs at redox potentials of 200–300 mV (pH = 7) (Figure 10.13). This is the same range that is observed for nitrate reduction. At this time, it is not known whether organisms would preferentially utilize nitrate over manganese in the same redox range if both electron acceptors are available at the same time. However, nitrate (if present) is more readily available than Mn(IV) oxides, because nitrate is an anion and is highly mobile in soils. In contrast, manganese becomes mobile and more available only under acid conditions.

Reduction of Fe(III) oxides (at pH = 7) occurs in an Eh range of 0–100 mV (Figure 10.13). Typically, iron oxides are not reduced in the presence of nitrate, whereas they can be reduced in the presence of manganese oxides. However, if manganese oxides are present in bioavailable form, it is likely that facultative bacteria would prefer Mn(IV) over Fe(III). The reduction of Fe(III) and Mn(IV) can also occur in different redox zones of the soil profile. In permanently waterlogged soils, most of the bioavailable Fe(III) and Mn(IV) are already present in reduced forms. Under these

Iron and Manganese

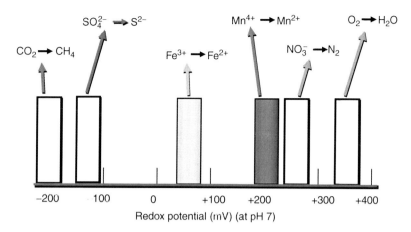

FIGURE 10.13 Critical redox potentials for reduction of Fe(III) and Mn(IV) oxides.

conditions, reduction of Fe(III) and Mn(IV) plays a minimal role in supporting overall microbial activities. However, as wetlands undergo wet and dry cycles as a result of water-table fluctuations, some of this reduced manganese and iron can be oxidized to oxides and hydroxides. These oxidized forms can be subsequently reduced by microbes when soils are flooded.

In summary, the reduction of Fe(III) and Mn(IV) has an indirect impact on wetland soil chemistry as follows:

- Water-soluble Fe(II) and Mn(II) increases
- Soil pH tends to approach near neutrality (see Chapter 4) as a result of the consumption of protons during reduction
- Other soil cations (such as calcium, magnesium, and ammonium) are displaced from the exchange complex to soil solution (see Chapter 4)
- Solubility of phosphorous increases (see Chapter 8)
- Sulfide levels decrease as a result of precipitation with Fe(II) (see Chapter 11)
- New minerals form

10.4.3 Forms of Iron and Manganese

Iron and manganese in wetland soils are present in at least four forms:

- Water soluble (dissolved, pore water)
- Exchangeable (sorbed on cation exchange complex)
- Reducible (insoluble ferric and ferrous compounds)
- Residual (crystalline stable pool)

Operationally defined chemical extraction schemes are used to determine these forms. Water-soluble or pore water forms are extracted under oxygen-free conditions (to avoid oxidation of Fe(II) and Mn(II)) by centrifuging soil slurries and supernatant filtered through 0.45 μm filter. The residual soil is extracted under oxygen-free conditions with either 1 M sodium acetate or ammonium acetate adjusted to pH of soil suspensions. Sodium or ammonium ions displace adsorbed Fe(II) and Mn(II) from the exchange complex into solutions. Extracted solutions are filtered through 0.45 μm filter and analyzed for iron and manganese. The water-soluble pool represents iron and manganese ions present in soil pore water, whereas the exchangeable pool represents iron and manganese ions adsorbed on the soil cation exchange complex.

10.4.3.1 Iron

Both Eh and pH have a significant effect on the distribution patterns of iron and manganese (Gotoh and Patrick, 1974). The increase in water-soluble and exchangeable pools is favored by a decrease in pH and Eh. Mississippi River sediments incubated at different pH and Eh levels showed the highest water-soluble iron at pH 5 and Eh of −150 mV (Figure 10.14). The exchangeable pool of iron followed similar trends (Figure 10.15). Under reducing conditions, low water-soluble iron concentrations at near-neutral and alkaline pH conditions are probably due to the precipitation of Fe(II) with carbonates and sulfides. The flooding of drained paddy soils increases water-soluble iron, rapidly reaching peak concentrations, followed by a steady decrease resulting from secondary reactions such as precipitation with carbonates and sulfides (Figure 10.16). In another example, water-soluble iron concentrations in flooded organic soil also increased steadily with decreases in Eh values between 0 and −200 mV (Figure 10.17). Subsequent increases in Eh decreased the water-soluble iron as a result of oxidation. In flooded Crowley silt loam soil maintained at an Eh value of −250 mV,

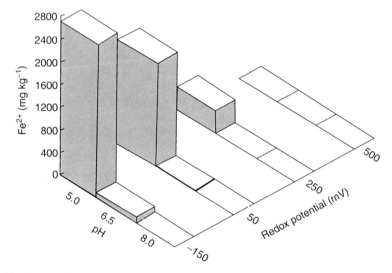

FIGURE 10.14 Water-soluble iron accumulation in Mississippi River sediments incubated at a range of pH and Eh levels. (From Gambrell et al., 1975.)

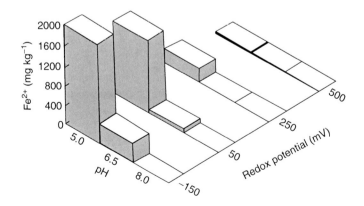

FIGURE 10.15 Exchangeable iron accumulation in Mississippi River sediments incubated at a range of pH and Eh levels. (From Gambrell et al., 1975.)

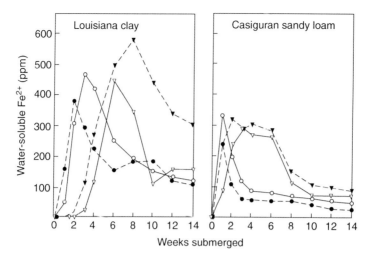

FIGURE 10.16 Kinetics of ferrous iron accumulation in paddy soils. (Redrawn from Ponnamperuma, 1972.)

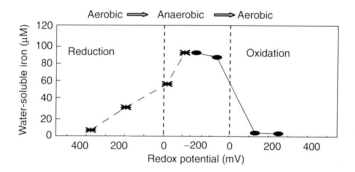

FIGURE 10.17 Influence of redox potential on soluble iron accumulation in flooded organic soil. (Redrawn from Ann et al., 2000.)

water-soluble and exchangeable Fe accounted for 28% of total Fe at pH 5, followed by 21% at pH 6, and 6 and 1% at pH 7 and 8, respectively (Figure 10.18). At the same Eh, reducible Fe accounted for 40, 47, 63, and 66% of total Fe at pH 5, 6, 7, and 8, respectively. Approximately 32% of the total Fe was not affected by pH and Eh. The adsorption coefficient (ratio of exchangeable fraction to water-soluble fraction) increased with increase in pH, with high values observed at pH 7 and 8. Low adsorption coefficients for iron at low pH may be due to H^+ and Al^{3+} ions displacing Fe^{2+} from the exchange complex into soil solution.

10.4.3.2 Manganese

Soluble and exchangeable manganese pools increase with a decrease in both pH and Eh. Between pH 6 and 8, the conversion of insoluble soil manganese to water-soluble and exchangeable forms is dependent on both pH and Eh, whereas at pH 5 the effect of acidity was so marked that changes in Eh had little effect on manganese solubility (Gotoh and Patrick, 1972). At an Eh of −150 mV and pH 5, 6, 7, and 8, water-soluble plus exchangeable forms accounted for approximately 80, 70, 50, and 30% of reducible manganese, respectively. Large proportions of manganese in water-soluble plus exchangeable pools were not only due to reduction of Mn(IV) but also due to solubilization under acidic conditions. Thus, at pH 5, it is likely that the water-soluble fraction may contain both Mn^{2+} and Mn^{4+} ions. This is expected under acidic soil conditions where H^+ and Al^{3+} dominate

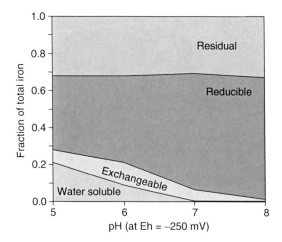

FIGURE 10.18 Distribution of various pools of iron accumulation in Crowley silt loam soil incubated under anaerobic conditions (−250 mV) and a range of pH values. (Redrawn from Gotoh and Patrick, 1974.)

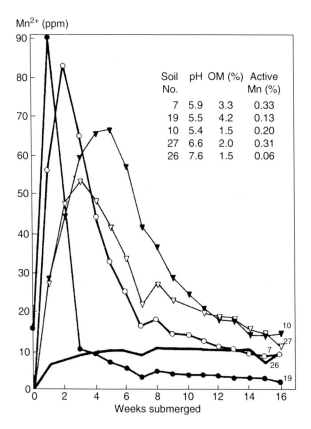

FIGURE 10.19 Kinetics of Mn(IV) reduction and accumulation of Mn(II) in paddy soils. (Redrawn from Ponnamperuma, 1972.)

the cation exchange complex, resulting in the displacement of manganese ions into the soil solution (Gotoh and Patrick, 1972). As observed for iron, as a result of flooding of drained paddy soils, water-soluble manganese increases rapidly to peak concentrations, followed by a steady decrease resulting from secondary reactions such as precipitation with carbonates (Figure 10.19).

10.4.3.3 Complexation of Iron and Manganese with Dissolved Organic Matter

Wetland soils are known to accumulate high concentrations of DOM, which may include several fermentative organic acids and low-molecular-weight fulvic and humic acids. This provides an opportunity for soluble iron and manganese to form complexes with DOM, thus maintaining stability of these dissolved species in a wide range of redox conditions. The stability of the metal–organic complex cannot be predicted by a typical Eh–pH stability diagram. For example, the complexation of Fe(II) with DOM can maintain Fe(II) in soluble form for long periods (several days) even under aerobic conditions. The extent of metal–organic complexation is higher under anaerobic conditions. This is due to higher levels of DOM accumulation under anaerobic conditions. Iron is known to form a more stable organic–metal bond than manganese.

10.4.3.4 Mobile and Immobile Pools of Iron and Manganese

Mobile pools of iron and manganese are present in water-soluble or dissolved forms in soil pore water. Immobile forms include solid phases such as insoluble precipitates and mineral phases (amorphous and crystalline forms) present both in aerobic and anaerobic soil layers. The flux of dissolved iron and manganese is typically from anaerobic soil layers to aerobic soil layers, where it is oxidized to insoluble precipitates. This results in the establishment of concentration gradients across the aerobic–anaerobic soil interface. Mobilization is also regulated by pH and CEC. Manganese is more soluble in moderately acidic conditions (between pH 5 and 6) than iron.

The reduction of the oxidized forms of iron and manganese is regulated by

- Quality of organic substrates (electron donor)
- Bioavailability and crystallinity of iron and manganese minerals
- Presence of electron acceptors with higher reduction potentials
- pH
- Temperature

The accumulation of Fe(II) in wetland soils is coupled to Fe(III) reduction and organic matter decomposition, as shown for several Mississippi River floodplain soils (Figure 10.20). The addition

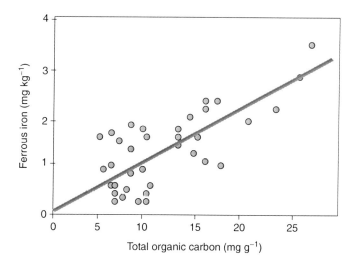

FIGURE 10.20 Relationship between soil organic matter content and accumulation of Fe(II) in several soils from the Mississippi River floodplain. (Redrawn from Reddy et al., 1980.)

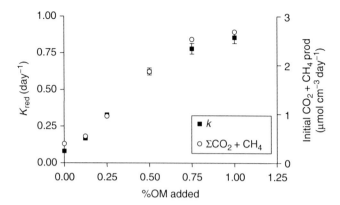

FIGURE 10.21 First-order Fe(III) reduction rate constants and initial cumulative carbon dioxide and methane production in a wetland soil amended with different amounts of labile organic matter. (Redrawn from Roden and Wetzel, 2002.)

of organic substrates was also shown to increase the rate of Fe(III) oxide reduction in wetland soils (Figure 10.21), with reduction rates directly linked to carbon dioxide production, suggesting that the reduction of Fe(III) oxides is mediated by heterotrophic bacteria. At present similar relationships are not established for Mn(IV) reduction in wetland soils and aquatic sediments.

The forms and amounts of Fe(III) and Mn(IV) oxides are important factors in controlling the extent of organic matter decomposition with these metals as electron acceptors. The most available forms of these electron acceptors for bacterial reduction are dissolved Fe(III) and Mn(IV) forms, which include Fe(III) and Mn(IV) in solution under acidic pH conditions; Fe(III) and Mn(IV) complexes; and Fe(III) and Mn(IV) complexes (chelates) with DOM. Insoluble forms include amorphous and a range of crystalline forms of Fe(III) and Mn(IV) oxides. In addition, particle size and available surface area may also influence the bioavailability of Fe(III) and Mn(IV) oxides. These oxides also occur as a complex mixture of each other and as coatings on clay, silt, and sand particles. Iron oxides can also be present as occluded coprecipitates on soil particles. Thus, the bioavailability of Fe(III) and Mn(IV) oxides is in the order of dissolved Fe and Mn >> metals complexed with DOM >> amorphous forms > crystalline forms.

The physical location of Fe(III) and Mn(IV) oxides in the soil profile can also influence their utility as electron acceptors by microorganisms. As discussed before, oxidized forms of iron and manganese are insoluble/immobile and they are usually present in oxidized/aerobic portions of the soil. Thus, the decomposition of organic matter with Fe(III) and Mn(IV) as electron acceptors will be restricted to this zone. If this layer is disrupted as a result of organic matter loading, oxygen depletion may promote the reduction of Fe(III) and Mn(IV) oxide precipitates by facultative bacteria. Microbes growing directly on Fe(III) and Mn(IV) oxide mineral surfaces may solubilize and utilize them as electron acceptors.

Several studies have shown that the more crystalline the Fe(III) and Mn(IV) oxides, the slower the rate of their reduction (reviewed by Lovley, 1991, 2004). In laboratory experiments, Ottow and Klopotek (1969, cited by Lovley, 1987) reported that the reduction capacity of select iron minerals was found to be in the order of $FePO_4$ > $Fe(OH)_3$ > $FeOOH$ > Fe_2O_3. Amorphous and poorly crystalline forms of Fe(III) and Mn(IV) oxides dominate wetlands that undergo frequent wet and dry cycles. These systems are dynamic, and repeated oxidation–reduction reactions involving iron and manganese will not allow time for stable crystalline forms of Fe(III) and Mn(IV) oxides to form in these systems. Strong relationships are observed between Fe(III) reduction rates and poorly crystallined forms of Fe(III) oxides (as determined by hydroxylamine extraction) in fresh and brackish water sediments (Lovley and Philips, 1987).

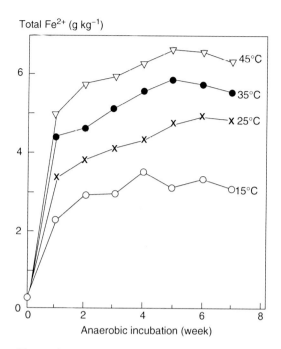

FIGURE 10.22 Kinetics of ferrous iron accumulation in paddy soils as influence by temperature. (Redrawn from Ponnamperuma, 1984.)

The presence of electron acceptors with higher reduction potentials (such as oxygen and nitrate) can maintain iron and manganese compounds in an oxidized stable state. In the presence of oxygen and nitrate, Fe(III) oxide is stable and will not be reduced until the concentration of oxygen and nitrate reaches negligible levels. However, Mn(IV) oxide can be reduced in the absence of oxygen but in the presence of nitrate. Some of the Fe(III) and Mn(IV) reducers are facultative and are capable of preferentially using oxygen as an electron acceptor during their respiration. Oxygen can be toxic to some of the obligate anaerobes involved in the reduction of Fe(III) and Mn(IV). Under acidic pH conditions, iron and manganese are present in soluble form. This increases the mobility of Fe(III) and Mn(IV) and make them readily available for use by microbes as electron acceptors. Near-neutral and alkaline pH conditions, Fe(III) and Mn(IV) may be present as insoluble precipitates, thus decreasing overall bioavailability.

As the reduction of Fe(III) and Mn(IV) is associated with microbial activity, an increase in temperatures will increase the reduction rates. Water-soluble iron in flooded soils increased with an increase in temperature (Figure 10.22). Similar to other biological reactions, reaction rates may double for every 10°C rise in temperature.

10.5 OXIDATION OF IRON AND MANGANESE

Reduced iron and manganese compounds can be oxidized in both anaerobic and aerobic environments of wetlands and aquatic systems. Reduced Fe(II) is stable in soil zones where sulfate reduction and methanogenesis dominate, and is rapidly oxidized in oxygen and nitrate reduction zones. These compounds are oxidized enzymatically (by bacteria in the domains Bacteria and Archaea) or nonenzymatically (spontaneous reactions with oxidants such as oxygen and nitrate). Few bacteria are capable of using Fe(II) and Mn(II) compounds as an energy source. However, only a small amount of energy is released during the oxidation of these compounds, so to grow on these compounds bacteria must oxidize a large amount of Fe(II) or Mn(II) compounds. Much of the discussion presented in this section is based on iron-oxidizing bacteria and very little is known about the biotic oxidation of Mn(II) compounds. A detailed historical perspective of biotic and abiotic oxidation of Fe(II) and Mn(II) is presented by Emerson (2000). It has been assumed that Mn(II) and Mn(IV) were the only important oxidation states in sediment. But it has been recently shown that during the oxidation of Mn(II) by bacteria, Mn(III) can be formed as an intermediate (Johnson, 2006). Mn(II), which is both a strong oxidant and reductant, was previously thought to rapidly convert to either Mn(II) or MnO_2. Mn(III) can, however, persist because it can be stabilized by dissolved ligands (Trouwborst et al., 2006).

10.5.1 Microbial Communities

A limited group of iron and manganese-oxidizing bacteria are able to derive energy from oxidation of Fe(II) and Mn(II) compounds, while using oxygen and nitrogen oxides as electron acceptors. Some of these bacteria are able to produce distinct precipitates of iron and manganese generally associated with cells. Iron-oxidizing bacteria are commonly referred to as iron bacteria. A partial list of microorganisms involved in the oxidation of reduced iron and manganese compounds is presented in Table 10.3. For additional information on bacterial physiology and strains, the reader is referred to Emerson (2000) and Ehrlich (2002).

Based on the physicochemical environment, iron bacteria are classified as acidophilic and neutrophilic. Acidophilic bacteria may include autotrophs, mixotrophs, and heterotrophs and are adapted to grow under acidic environments. Neutrophilic bacteria may include a range of bacterial groups that are adapted to grow under circumneutral pH conditions. The Eh–pH diagram (Figure 10.23) shows the natural domain for main groups of bacteria involved in the oxidation of iron compounds. Most commonly reported bacteria in this group belong to genus *Thiobacillus*. These bacteria are capable of using various inorganic compounds such as Fe(II), sulfides and elemental sulfur, and hydrogen. These bacteria are autotrophic and use carbon dioxide as their carbon source and ammonia as their

TABLE 10.3
A Partial List of Microorganisms Involved in the Oxidation of Iron and Manganese Compounds

Microorganisms	Electron Donor	Electron Acceptor
Iron-oxidizing bacteria		
Neutrophilic		
Leptothrix, Gallionella, Ferroglobus,	$FeCO_3$, FeS_2	Oxygen
Chromatium, Rhodobacter	Fe^{2+}	Carbon dioxide
Acidophilic		
Thiobacillus, Ferrobacillus, Sulfobacillus, Ferroglobus,	Fe(II): $FeSO_4$, FeS_2	Oxygen, nitrate
Ferroplasma, Sulfolobus		
Heterotrophs		
Naumanniella, Ochrobium, Siderocapsa, Siderococcus,	Organic complexes of Fe	Oxygen
Sphaerotilus		
Manganese-oxidizing bacteria		
Arthrobacter, Bacillus, Leptothrix, Metallogenium,	Mn(II): $MnCO_3$	Oxygen
Flavobacterium, Pseudomonas		

Source: Emerson (2000).

Iron and Manganese

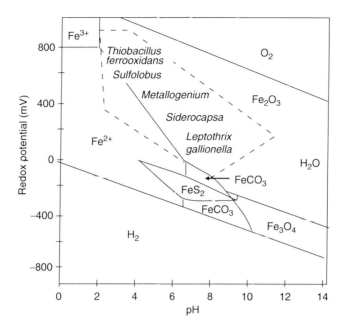

FIGURE 10.23 Eh–pH iron stability diagram showing the natural domains of main groups of the iron bacteria. (Redrawn from Lundgren and Dean, 1979.)

nitrogen source. All strains of this genus grow in environments with a pH of <2. Under low pH conditions Fe(II) is present in ionic form and may be readily available for bacterial groups involved in oxidation of iron. Some of these environments may include acid peat bogs, drained coastal wetlands, and constructed wetlands used to treat acid mine drainage.

10.5.2 BIOTIC AND ABIOTIC OXIDATION

Wetland soils and aquatic sediments are uniquely characterized by aerobic and anaerobic interfaces at the soil–floodwater interface or in the root zone of wetland plants (see Chapter 4 for details). Aerobic oxidation of Fe(II) and Mn(II) is restricted to the thin aerobic layer at the soil–floodwater interface or in the root zone. Thus, the extent of aerobic oxidation of Fe(II) and Mn(II) is dependent on the flux of dissolved species from anaerobic soil layers to aerobic zones. At circumneutral pH, concentrations of dissolved Fe(II) and Mn(II) are very low, thus restricting flux into aerobic portions of the soil. At this pH level, the majority of Fe(II) and Mn(II) compounds are present as immobile solid phases such as $FeCO_3$, $MnCO_3$, FeS_2, $Fe(OH)_2$, and $Mn(OH)_2$. These compounds can be oxidized only when the water table is lowered, thus exposing top portion of the soil profile to aerobic conditions.

The oxidation of Fe(II) and Mn(II) is regulated by a number of factors:

- Soil or sediment pH and redox potential
- Oxygen flux and the thickness of aerobic layer
- Presence of oxidants with higher reduction potentials
- Flux of soluble Fe(II) and Mn(II)
- Soil CEC
- Metal–DOM complexation

10.5.2.1 Iron

The oxidation of Fe(II) is well studied by geochemists and microbiologists. Basic conclusions drawn from these studies include (1) iron is oxidized spontaneously with oxygen under neutral pH conditions, (2) acidophilic and neutrophilic bacteria use Fe(II) compounds as electron donors during aerobic respiration, (3) some bacterial groups can use nitrate or sulfate or Mn(IV) as electron acceptors and oxidize Fe(II) compounds, and (4) Fe(II) compounds can be chemically oxidized in the absence of oxygen. Although limited information is available for Mn(II) oxidation, similar conclusions can be drawn for the biotic and abiotic oxidation of manganese.

The electron flow in *Thiobacillus* with Fe(II) as an electron donor is shown in Figure 10.24 (Madigan and Martinko, 2006). The difference in the reduction potential of the Fe^{3+}–Fe^{2+} couple ($E° = +0.77$ V) and the O_2–H_2O redox couple ($E° = +0.88$ V) is small, suggesting that the route for electron transport between these two couples will be short. The electron transport in this system is not the same as the typical aerobic terminal electron transport process in aerobic bacteria. The electrons donated from the Fe(III) reduction cannot reduce NAD^+, FAD^+, or many of the other electron transport chains (Madigan and Martinko, 2006). *Thiobacillus* uses preexisting proton gradients of their environment for energy generating, while maintaining cytoplasm pH near neutrality. The pH (or proton gradient) across the cytoplasmic membrane plays a key role in driving ATP synthesis via proton-linked ATPase (Nealson, 1983). Protons entering the cell are consumed during oxidation–reduction reactions coupling Fe(III) oxidation and oxygen reduction.

$$4Fe^{2+} + O_2 + 4H^+ = 4Fe^{3+} + 2H_2O$$

The Fe(II) oxidation occurs near the outer face of the cytoplasmic membrane, whereas oxygen reduction occurs on the inner face of the membrane. The electrons donated from Fe(II) oxidation are accepted by rusticyanin in the periplasm, followed by a transfer of electrons to cytochrome *c*

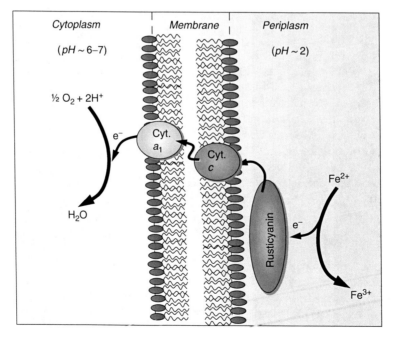

FIGURE 10.24 Electron flow in *Thiobacillus* grown in medium with ferrous iron. (Modified from Madigan and Martinko, 2006.)

and a_1, and then finally to oxygen, thus completing the terminal electron-accepting process (Madigan and Martinko, 2005). *Thiobacillus* is a chemoautotrophic bacteria that uses the Calvin cycle to fix carbon dioxide as a sole carbon source for growth. The redox potentials of NAD$^+$/NADH and NADP$^+$/NADPH are low ($E° = -320$ and 325 mV, respectively), whereas that of Fe^{2+}/Fe^{3+} is high, and under normal redox conditions iron cannot reduce these dinucleotides. To reduce carbon dioxide, electrons from the oxidation of Fe(II) are shunted to NAD$^+$ and NADP$^+$ via the cytochrome system, against the electropotential gradient, by using ATP. This process is called *reverse electron transport*, which involves oxidation of Fe(II) to satisfy the energy requirement (Aleem et al., 1963).

At near-neutral pH, Fe(II) is not stable and subject to rapid chemical oxidation by dissolved oxygen to form insoluble precipitates such as ferric hydroxide. The relative importance of biotic oxidation of Fe(II) over abiotic oxidation is not clear. Few organisms use Fe(II) compounds as electron donors and derive energy through oxidation while using oxygen as their terminal electron acceptor. For example, FeCO$_3$ and FeS$_2$ are common Fe(II) compounds at near-neutral pH and may serve as energy sources for *Gallionella* and *Leptothrix*. These bacteria are found in wetland soils and in lake and marine sediments and have been implied to oxidize Fe(II) compounds. *Gallionella* and *Leptothrix* grow best in moderately reduced soils with redox potential in the range of 200–300 mV (Figure 10.23), pH of 6–7.5, and low oxygen concentrations (<1 mg L^{-1}). In wetlands, aerobic Fe(II) oxidation may be restricted to the soil–floodwater interface and in plant rhizosphere, where oxygen diffusion is limited and dissolved oxygen concentrations are low. Iron oxidizers can grow both autotrophically and mixotrophically. Under these conditions, abiotic oxidation may be slower because of low oxygen concentrations in relation to Fe(II) concentrations, resulting in possible biotic oxidation by these organisms. *Leptothrix* are filamentous and their sheaths are often heavily encrusted with hydrated Fe(III) and Mn(IV) oxides, formed as a result of abiotic oxidation.

In the presence of light, anaerobic oxidation of Fe(II) is possible by photosynthetic bacteria within anoxic environments (Widdel et al., 1993). These bacteria were isolated from freshwater and marine sediments and were affiliated with different genera of purple or green phototrophic bacteria (Straub et al., 2001). No iron oxidation was found in the dark with bacteria or in light without bacteria. These bacteria use the reducing power generated by Fe^{2+}–Fe^{3+} redox couple to support carbon dioxide fixation (Widdel et al., 1993). Other studies have shown that nitrate-reducing bacteria are capable of using Fe(II) as an electron donor (Straub et al., 1996; Benz et al., 1998). The coupling of Fe(II) oxidation (Fe(III) oxide accumulation) to nitrate reduction (nitrate disappearance from culture) under lithotrophic conditions in freshwater enrichment cultures is shown in Figure 10.25.

$$10FeCO_3 + 2NO_3^- + 24H_2O = 10Fe(OH)_3 + N_2 + 10HCO_3^- + 8H^+$$

Nonenzymatic oxidation of Fe(II) compounds is the most common pathway at circumneutral pH. The overall abiotic oxidation of Fe(II) can be described as follows:

$$4Fe^{2+} + O_2 + 4H^+ = 4Fe^{3+} + 2H_2O$$

$$4Fe^{3+} + 12OH^- = 4Fe(OH)_3 \text{ (precipitate)}$$

$$4Fe^{2+} + O_2 + 4OH^- + 2H_2O = 4Fe(OH)_3 \text{ (precipitate)}$$

In these three reactions, during the first step, Fe(II) oxidation with oxygen consumes protons with the formation of ferric iron, which leads to a rise in pH. In the second step, hydrolysis of ferric iron results in the formation of ferric hydroxide. This reaction consumes hydroxyl ions, thus decreasing the pH. The overall process creates the acidification process.

Wetlands and aquatic systems are known to contain high concentrations of DOM in pore waters and the water column. Some of the Fe(II) ions can complex with DOM, thus slowing down the

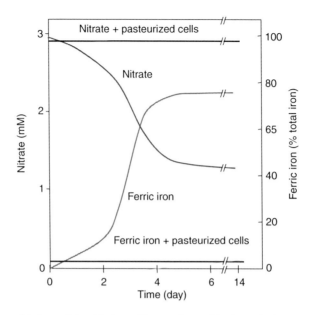

FIGURE 10.25 Anaerobic bacterial oxidation of ferrous iron with nitrate in freshwater enrichment culture under lithoautotrophic conditions. (Redrawn from Straub et al., 1996.)

overall abiotic oxidation. In fact the Fe(II)–DOM complex can maintain ferrous iron in solution in oxygenated waters. Oxidation of DOM by heterotrophs releases ferrous iron into solution, which is subsequently oxidized to Fe(III) oxides. At circumneutral pH, some of the Fe(III) can complex with DOM, preventing some of the precipitation reactions. Uncomplexed Fe(III) can hydrolyze to form Fe(OH)$_3$ and settle down as a precipitate in the aerobic zones of soil or sediment.

10.5.2.2 Manganese

The oxidation of Mn(II) to Mn(IV) involves a two-electron transfer that occurs in the soil profile where oxygen and nitrate reduction occurs. Manganese (II) is soluble in moderately acidic conditions and neutral pH conditions, where Fe(II) may be present as insoluble precipitates. Greater solubility of Mn(II) than Fe(II) makes it more bioavailable for biotic oxidation. Aerobic oxidation of Mn(II) involves the energy yield of –71 kJ mol^{-1}.

$$2Mn^{2+} + 2O_2 + 2H_2O = 2MnO_2 + 4H^+$$

10.6 MOBILITY OF IRON AND MANGANESE

The mobility of oxidized forms of iron and manganese is increased greatly by reduction to soluble Fe(II) and Mn(II). As the intensity of anaerobiosis increases, the effectiveness of the soil exchange complex in adsorbing cations will be reduced due to a greater total concentration of cations. Because iron concentrations are higher than manganese, the majority of the soil cation exchange complex in anaerobic soils is occupied by Fe(II) ions, thus leaving Mn(II) ions in solution along with other cations. In general, Fe(II) does not appear to be as mobile as Mn(II), probably as a result of the easier reduction of Mn(IV) oxides. This condition results in the appearance of Mn(II) ions in the solution of a wetland soil before Fe(II) ions appear in it. It also means that at a given redox potential, a greater fraction of manganese is going to be in reduced soluble form and subject to mobility. Several studies show maximal concentrations of Fe(III) and Mn(IV) oxides in aerobic soil layers and the concentrations decrease with depth, whereas concentrations of Fe(II) increase with depth (anaerobic soils layers) (Howler and Bouldin, 1971; Roden and Wetzel, 2002).

Wetlands exhibit distinct redox gradients between the soil and overlying water column and in the root zone (Chapter 4), resulting in aerobic interfaces. For example, the aerobic layer at the soil–floodwater interface is created by a slow diffusion of oxygen and the rapid consumption at the interface. The thin aerobic layer at the soil–floodwater interface and around roots functions as an effective zone for aerobic oxidation of Fe(II) and Mn(II). Below this aerobic layer there exists the zone of anaerobic oxidation of Fe(II) and Mn(II) and reduction of Fe(III) and Mn(IV). The juxtaposition of aerobic and anaerobic zones creates conditions of intense cycling of iron and manganese mediated by both biotic and abiotic reactions.

In the anaerobic soil layer, a significant amount of iron and manganese is present in soluble form in soil pore water. This form of iron and manganese is mobile in response to concentration gradients and leaching. Dissolved forms can flux from anaerobic soil layer to aerobic layer in response to the concentration gradient created across these two layers. Typically, steep gradients exist in the anaerobic soil layer adjacent to the aerobic soil layer. Depth distribution of dissolved Fe(II) and oxygen in a freshwater wetland located in Alabama is shown in Figure 10.26. Oxygen and Fe(II) show opposing gradients with minimum overlap. Steep gradients of dissolved Fe(II) are seen with short distances (<10 cm). These steep gradients are the result of dissolved Fe(II) flux from anaerobic soil layer to aerobic soil layer, where it is oxidized either abiotically or biotically to Fe(III) oxides. For some wetland soils, opposing profiles are seen for solid phase Fe(II) and for amorphous Fe(OH)$_3$ as determined by 0.5 M HCl extraction (Roden and Wetzel, 2002) (Figure 10.26). A steady flux of dissolved Fe(II) into aerobic soil layers can result in an accumulation of Fe(III) oxide precipitates. The depth distribution of soluble and insoluble Fe showed opposing gradients in short distances

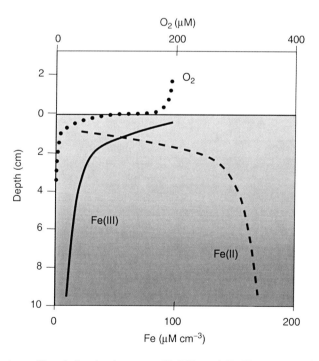

FIGURE 10.26 Depth profile of dissolved oxygen, Fe(III), and Fe(II) concentrations in wetland soil. (Redrawn from Roden and Wetzel, 2002.)

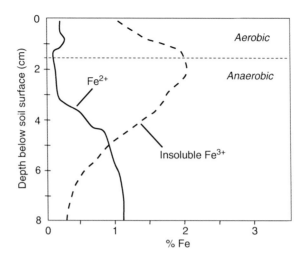

FIGURE 10.27 Mobile (Fe^{2+}) and immobile (Fe-oxides) forms of iron in a flooded soil profile. (Redrawn from Howler and Bouldin, 1971.)

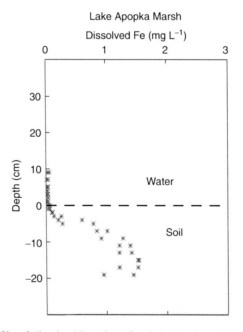

FIGURE 10.28 Depth profile of dissolved iron in a flooded organic soil. (Redrawn from D'Angelo and Reddy, 1994a.)

(<4 cm) in flooded soils, as shown in Figure 10.27 (Howler and Bouldin, 1971). Similarly, dissolved Fe(II) profiles in a flooded peatland also showed steep gradients in the top 10 cm of soil profile (Figure 10.28). Similar profiles are reported for pore water manganese and extractable manganese (Figure 10.29). In the Black Sea, steep gradients of dissolved Fe(II) and Mn(II) occur at depths of 50–100 m, with a clear separation between Fe and Mn (Figure 10.30).

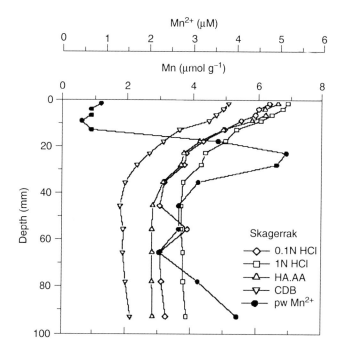

FIGURE 10.29 Vertical profile of pore water Mn(II) and manganese extracted with CDB at the Skagerrak site. (Redrawn from van der Zee and van Raaphorst, 2004.)

Wetland plants provide an oxygenated environment in the root zone, which can function as an effective sink by oxidizing dissolved Fe(II) and Mn(II). Several studies have shown root plaque formation on surface roots are due to the precipitation of Fe(III) and Mn(IV) oxides (see Section 10.7 for additional details).

In summary, the mobility of Fe(II) and Mn(II) is controlled by (1) diffusion and mass flow to aerobic soil layers, where they are oxidized and precipitated; (2) diffusion and mass flow toward plant roots, where they are precipitated as Fe(III) and Mn(IV) oxides and deposited as encrustations around roots; and (3) mass flow with downward-percolating water to the oxygenated subsoil layer, where they are oxidized and precipitated.

10.7 ECOLOGICAL SIGNIFICANCE

10.7.1 Nutrient Regeneration/Immobilization

Oxidation–reduction reactions of iron and manganese on nutrient release/retention can occur as follows:

- Organic matter breakdown and ammonium release
- Phosphorous release and retention
- Coprecipitation of trace elements with iron and manganese oxides

10.7.1.1 Organic Matter Decomposition and Nutrient Release

Iron and manganese reduction in wetland soils and aquatic sediments is linked to organic matter decomposition. In the absence of oxygen, microbial communities use a wide range of electron acceptors, including nitrate, Mn(IV), Fe(III), sulfate, carbon dioxide, and several simple organic compounds.

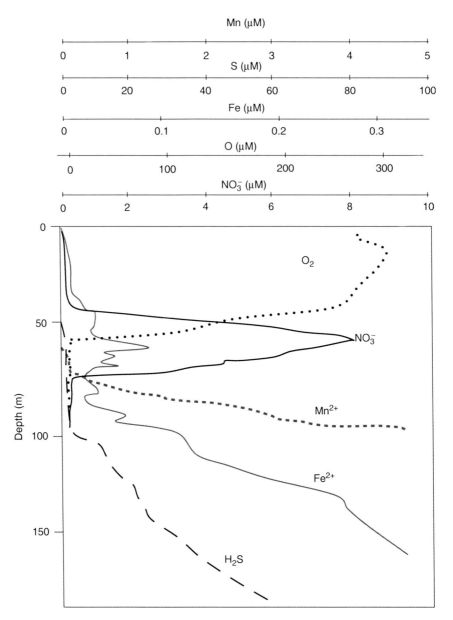

FIGURE 10.30 Distribution of select oxidants and reductants in the water column of Black Sea. (Redrawn from Nealson and Myers, 1992.)

In addition, several trace metals can also function as electron acceptors. Under all these conditions, the consumption of electron acceptors is directly linked to organic matter decomposition. In most wetlands, nitrate and Mn(IV) concentrations are low compared to other inorganic electron acceptors, so they may not play a significant role in organic matter decomposition. However, where nitrate inputs are higher (such as constructed wetlands used for nitrate removal or riparian wetlands receiving nitrate inputs from adjacent watershed), nitrate reduction can play a significant role in organic matter turnover. In wetlands dominated by mineral soils, Fe(III) concentrations are typically several-fold higher than Mn(IV) concentrations and the role of manganese reduction in organic matter decomposition may be minimal. The relative importance of manganese versus iron reduction in organic matter decomposition is not established. The processes involving inorganic electron acceptors in wetlands and aquatic

Iron and Manganese

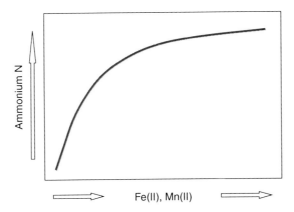

FIGURE 10.31 Relationship between accumulation of Fe(II) and Mn(II) and ammonium production, showing the relationship between bacterial reduction of electron acceptors and organic nitrogen mineralization.

systems are generally considered mutually exclusive (see Chapter 5 for details). Thermodynamically, in the absence of oxygen, nitrate and Mn(IV) are the most preferred electron acceptors, followed by Fe(III). All these electron acceptors are utilized by facultative and obligate anaerobes.

The relative importance of iron and manganese in organic matter turnover is not well documented, as compared to other electron acceptors such as oxygen, sulfate, and carbon dioxide. Facultative and some of the obligate heterotrophic bacteria can use oxidized forms of iron and manganese as their electron acceptors, while using organic matter as their electron donor. Studies have shown that up to 35–65% of the carbon dioxide production was estimated to be due to decomposition of organic matter through Fe(III) reduction (Lovley, 1987). The extent of organic nitrogen mineralization and ammonium release using Fe(III) and Mn(IV) reduction has not been documented. In the following equation, hypothetical organic matter (using Redfield ratios) is mineralized using Fe(III) and Mn(IV) oxides as electron acceptors.

$$(CH_2O)_{106}(NH_3)_{16}(H_3PO_4) + 212MnO_2 + 424H^+ = 106CO_2$$
$$+ 212Mn^{2+} + 16NH_3 + H_3PO_4 + 318H_2O$$

$$(CH_2O)_{106}(NH_3)_{16}(H_3PO_4) + 424FeOOH + 848H^+ = 106CO_2$$
$$+ 424Fe^{2+} + 16NH_3 + H_3PO_4 + 742H_2O$$

These two equations show that almost twice the amount of Fe(III) is needed to decompose 1 mol of organic carbon compound, indicating the inefficiency of iron as an electron acceptor. However, mineral soils have much higher concentrations of iron than manganese, which may compensate for low efficiency. In wetland soils, the reduction of iron and manganese oxides and the accumulation of Fe(II) and Mn(II) can be directly related to the accumulation of ammonium resulting from the mineralization of organic nitrogen mediated by iron and manganese reduction (Figure 10.31).

A highly significant, positive relationship has been shown between iron oxide reduction and organic nitrogen mineralization in tropical wetland soils (Sahrawat, 2004). The following empirical relationships were reported:

$$N_{min} = 16.4 + 1.32TOC + 0.037EDTA\text{-}Fe \quad R^2 = 0.85$$

$$N_{min} = 11.1 + 1.81TOC + 0.0047Fe_{ox} \quad R^2 = 0.81$$

where N_{min} is the ammonium nitrogen produced during the anaerobic mineralization of organic N (mg kg^{-1}), TOC the total organic carbon (g kg^{-1}), ETDA-Fe the reducible iron extracted by EDTA (mg kg^{-1}), and Fe_{ox} the reducible iron extracted with ammonium oxalate (mg kg^{-1}).

10.7.1.2 Phosphorous Release or Retention

It is well established that iron reduction is coupled to soluble phosphorous release in soils dominated by iron redox couples (Figure 10.32) (see Chapter 9 for details). Although phosphorous itself is not normally involved in redox reactions, it does undergo reactions that have a pronounced effect on its reactivity. Most of this change in the reactivity of phosphorous in wetland soils and aquatic sediments is associated with the oxidation–reduction of iron and manganese. The reduction of ferric phosphate compounds results in the release of phosphorous, a major solubility mechanism in wetlands and aquatic systems.

$$FePO_4 \cdot 2H_2O + 2H^+ + e^- = Fe^{2+} + H_2PO_4^- + 2H_2O$$

$$Eh = E° - 0.059 \log[Fe^{2+}] - 0.059 \log[H_2PO_4^-] - 0.118pH$$

Under aerobic conditions, the oxidation of ferrous iron results in an accumulation of Fe(III) oxides, which can subsequently precipitate soluble phosphorous.

$$Fe^{3+} + PO_4^{3-} = FePO_4$$

Phosphorous precipitation in wetlands can occur in the following zones in the soil: (1) the aerobic soil layer at the soil–floodwater interface, (2) the aerobic layers in benthic periphyton mats, and (3) the aerobic interface in the root zone. The precipitation of phosphorous in this boundary layer at the soil–floodwater interface can decrease soluble phosphorous flux into the water column. Similarly, Fe(III) and Mn(IV) oxide precipitates on oxidized root surfaces can immobilize phosphorous, thus decreasing its availability to plants. The phosphorous and iron precipitation reactions can be very dynamic especially at soil–floodwater interfaces in benthic periphyton mats, as precipitation may occur during the day (oxygen production during photosynthesis) and dissolution during the night (reduction of ferric iron in the absence of oxygen). Similar precipitation reactions can occur in the root zone, potentially creating phosphorous limitation, as observed in rice plants.

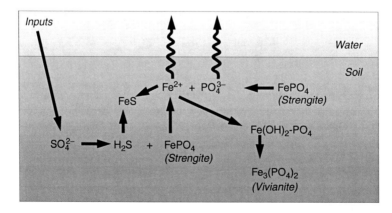

FIGURE 10.32 Schematic showing the interaction between Fe(III) reduction and phosphate solubility in wetland soils.

10.7.1.3 Coprecipitation of Trace Elements with Iron and Manganese Oxides

Trace elements are often immobilized by Fe(III) and Mn(IV) oxides present in aerobic portions of the soil profile and in aerobic root zones. For example, Fe(III) and Mn(IV) oxides may contain trace elements such as nickel, cadmium, zinc, copper cobalt, chromium, and others. Some of the Fe(III) oxides can retain manganese. Trace elements precipitate as coatings on the surface of minerals. Amorphous and crystalline forms of Fe(III) and Mn(IV) oxides can either sorb or coprecipitate trace elements. Although, Mn(IV) oxides are less abundant in soils than Fe(III) oxides, they are proven to have a higher trace element retention capacity. The zero point of charge (ZPC) for Mn(IV) oxides is in the range of 2–5, as compared to 6.8–8 for Fe(III) oxides, suggesting that at pH levels normally encountered in soils, Mn(IV) oxides have a negative surface charge, which favors the retention of positively charged trace elements. The affinity of Mn(IV) oxides to trace metals is in the order of lead > copper > cobalt > zinc > nickel (McKenzie, 1989).

The solubility and reactivity of Fe(III) and Mn(IV) oxides are regulated by pH and Eh. They are soluble under reducing or acidic conditions, thus resulting in the release of trace elements. As discussed before, Mn(IV) oxides are present in a dissolved form at a higher Eh and pH than Fe(III) oxides. Thus, any trace elements released as a result of the dissolution of Mn(IV) oxides can be retained by Fe(III) oxides.

10.7.2 Ferromanganese Nodules

Concretions, nodules, or mottles rich in iron and manganese are collectively referred to as ferromanganese nodules. Nodules were first collected on the seafloor during the Challenger Expedition in the 1870s (Lundgren and Dean, 1979). However, ferromanganese nodules have been found in a wide range of environments including wetland soils and aquatic sediments (Tables 10.4 and 10.5). These nodules are hard, generally spherical-shaped bodies, and range in size from less than 1 mm to sometimes up to 5 cm. The color of nodules may range from brown to reddish brown, with hues of 5YR to 10YR. Nodules are formed in most soils that undergo wet and dry cycles. During wet periods, Fe(III) and Mn(IV) oxides are reduced to more soluble Fe(II) and Mn(II) forms; whereas during dry periods reduced species are oxidized and precipitated by cementing soil particles. Differences in oxidation rates of Mn(II) and Fe(II) result in variable concentrations and the formation of Fe-rich

TABLE 10.4
Elemental Composition of Ferromanganese Nodules Collected from Freshwater and Marine Sediments

Sediments	Manganese (%)	Iron (%)
Oneida Lake	24.9	15.1
English Lake	1.4–13	5.9–39
Lake Michigan	1.3–14	8–32
Canadian Lakes	15.7–35.9	11.7–40.2
Lake Ontario	17	20.6
Pacific Ocean	8.2–50.1	2.4–26.6
Baltic Sea	23.8	14.7

Source: Lundgren and Dean (1979) and Hawatsch et al. (2002).

TABLE 10.5
Elemental Composition of Ferromanganese Nodules Collected from Soils

Soils	Manganese (%)	Iron (%)
Mississippi River Basin	2.6	5.8
Alfisols, China	6.8	12.2
France	0.3–1.3	8.4–18
Wetland soils, Oregon	0.4–5.7 (CDB)	8.6–20.5 (CDB)
Ulitsol, Taiwan	0.03–0.05	3.8–7.1

Sources: Cornu et al. (2005), D'Amore et al. (2004), Liu et al. (2002), and Manceau et al. (2003).

and Mn-rich layers in nodules (White and Dixon, 1996). For example, in nodules from Sicilian soils, calculated enrichment ratios (ratio of metal concentration in nodule to metal concentration in whole soil) for iron and manganese were 296 and 2.6, respectively (Palumbo et al., 2001). Repeated wet and dry cycles can eventually result in the formation of nodules that range from a few millimeters to less than 1 mm. Wetlands with seasonal water-table fluctuations provide an ideal environment for nodule formation.

A wide range of iron and manganese concentrations is reported for nodules collected from soils and sediments (Table 10.5). The accumulation of iron and manganese in nodules depends on the sources of ambient concentrations in soils. The flux of soluble Fe(II) and Mn(II) into aerobic environments and redox fluctuations in soils are mediated by hydroperiod.

Unique characteristics of ferromanganese nodules and associated oxidation–reduction reactions have been used by soil scientists as morphological indicators to help identify hydric soils (see Chapter 3). These characteristics are termed by soil scientists as redoximorphic features; however, various terms such as redox concentrations, redox depletions, and reduced matrix are synonymously used for the oxidation–reduction of iron and manganese and their respective concentrations. We prefer not to define these characteristics as redoximorphic features because oxidation–reduction reactions not only involve iron and manganese but also a range of elements that support biotic communities in the biosphere.

10.7.3 ROOT PLAQUE FORMATION

Soluble Fe(II) and Mn(II) are oxidized and precipitated on root surfaces, resulting in plaque formation. Although the actual site for Fe(II) and Mn(II) oxidation in the root zone varies with plant species, the plaque formation typically follows oxygen release from roots. Details of plaque formation in the root zone of wetland plants are discussed in Chapter 7. In this section, consequences of root plaque formation will be discussed. Factors controlling iron plaque formation on roots have been reviewed in detail by Mendelssohn et al. (1995). Plaque on root surfaces can have both positive and negative effects on plants. These include

- Protection from reduced phytotoxins
- Interference with nutrient uptake

The oxidation and precipitation of reduced Fe(II) and Mn(II) in the root zone can decrease activity, thus reducing the potential toxicity of these elements. In addition, some of toxic trace metals can be coprecipitated in the plaque, thus reducing their availability to the plant. Several studies have shown lower tissue concentrations of iron and manganese in plants with defined plaque on root surfaces. Ecologically, the increased availability of Fe(III) oxides due to the reoxidation of Fe(II) is influenced by roots, resulting in a shift in the electron flow from methanogenesis to Fe(III) reduction. In situations of oxygen stress, the Fe(III) and Mn(IV) oxides in plaque can be reduced to Fe(II) and reoxidized in the presence of oxygen. Root plaque can potentially create a barrier for nutrients to diffuse through this layer, thus creating nutrient limitation. Amorphous forms of Fe(III) and Mn(IV) oxides on the root surfaces can be very reactive and potentially sorb soluble phosphate, thus decreasing its bioavailability (see Chapter 6).

10.7.4 Ferrolysis

Ferrolysis is a soil-forming process and is defined as the clay destruction process involving the disintegration of clay lattice resulting from alternate reduction and oxidation of iron in acid and seasonally wet soils (Brinkman, 1970; Van Ranst and De Coninck, 2002). The process of ferrolysis can be summarized as follows:

- Flooding or saturated soil conditions results in the reduction of Fe(III) to Fe(II).
- Accumulation of Fe(II) ions displaces base cations from soil CEC into soil pore waters.
- Drainage of soils removes base cations from soil pore waters through leaching.
- Oxidation of Fe(II) to Fe(III) oxides results in the release of H^+ ions, thus decreasing soil pH to acid conditions (pH < 3).
- Protons produced during oxidation reaction occupy a majority of the soil CEC and saturate clay with protons, displacing Fe(II), Al(III), and other cations into soil pore waters.
- High proton activity and low pH disintegrate aluminum-silicate minerals. This increases Al(III) concentrations in soil pore water, creating toxicity to plants.
- Repeated cycles of flooding and draining (redox and leaching) decrease base cations and increase acidity in soils.

Partial neutralization of exchangeable Al(III) along with the reduction of Fe(III) oxides during flooded soil conditions produces relatively stable ferrous-aluminum hydroxide interlayers in 2:1 clays, leading to a further decrease in CEC. Ferrolysis typically occurs in surface soil horizons, where there is an abundant supply of bioavailable organic matter to support the microbial reduction of Fe(III) oxides.

10.7.5 Methane Emissions

Wetlands and rice paddies are major sources of methane and emit approximately up to 50% of annual methane to the atmosphere. Microbial pathways involved in production of methane from wetlands and aquatic systems are discussed in detail in Chapter 5. Until all electron acceptors (oxygen, nitrate, iron and manganese oxides, and sulfate) with higher reduction potentials are exhausted, no methane will be produced. Potentially all these electron acceptors can be present in the same soil profile with electron acceptors with higher reduction potentials utilized in surface layers and the electron acceptors with lower reduction potentials utilized in lower depths (Figure 10.33).

In most freshwater wetlands and aquatic systems, sulfate concentrations are present at low concentrations to make them ineffective as electron acceptors to support organic matter decomposition. The suppression of methane production during Fe(III) oxide reduction was demonstrated in laboratory experiments by Roden and Wetzel (1996) (Figure 10.34). In their study, methane was not produced until a major portion of the Fe(III) oxide was reduced, as evidenced by Fe(II) accumulation. Iron oxides are present in high concentrations in mineral wetland soils such as those

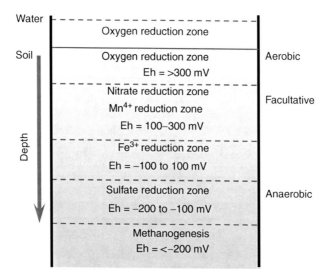

FIGURE 10.33 Vertical distribution of reduction of range of electron acceptors in wetland soil profile.

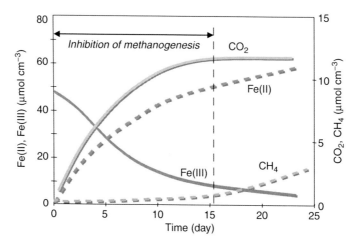

FIGURE 10.34 Fe(III) reduction, carbon dioxide production, methane production, and accumulation of Fe(II) in wetland soils. (Redrawn from Roden and Wetzel, 1996.)

in Mississippi River floodplain soils, where Fe(III) oxides can play a significant role in organic matter decomposition. Studies have shown that Fe(III) oxide reduction could markedly limit rates of methane production in the wetland soils of the southeastern United States, where iron concentrations are substantial (2–6% dry weight basis) (Roden and Wetzel, 1996).

Studies have shown an inhibition of methane production in sediments with a high availability of Fe(III) oxides. In these systems, Fe(III)-reducing bacteria and methanogens compete for common substrates such as acetate and hydrogen. Iron reducers are capable of utilizing these substrates at levels far below (low K_m values) those of methanogens. Low hydrogen and acetate concentrations are maintained in sediments where Fe(III) reduction was the predominant terminal electron acceptor, as compared to sediments where either sulfate reduction or methanogenesis was dominant (Lovley and Philips, 1987; Van Bodegom et al., 2004). In paddy soils, the presence of Fe(III) oxides resulting from the leakage of oxygen from roots can result in a shift in electron flow from methanogenesis to Fe(III) reduction, thus reducing overall methane emissions (Frenzel et al., 1999). Iron reducers can outcompete methanogens for substrates, thus reducing overall methane production. Similarly, manganese reducers can also outcompete methanogens for substrates.

In soils, methane production rates are inhibited at soil depths with intense cycling of iron (Ratering and Schnell, 2000). However, the utilization of Fe(III) oxides depends on the appropriate form that is readily available or on microbial communities. In soils with a limited availability of Fe(III) oxides and other electron acceptors, methanogenesis is the dominant pathway to regulating organic matter decomposition and ultimately a major methane source to the atmosphere.

10.8 SUMMARY

- Iron and manganese are widely distributed in wetland soil and sediments. Iron and manganese transformation involves various oxidation–reduction reactions mediated primarily by a range of microorganisms.
- Oxidized forms of iron and manganese serve as electron acceptors for facultative and anaerobic bacteria at reduction potential below nitrate reduction and above sulfate reduction.
- Iron has two oxidation states, Fe(III) and Fe(II). Manganese has three oxidation states, Mn(IV), Mn(III), and Mn(II). Stability of manganese and iron phases in wetland soil is regulated by both redox potential and pH.
- Biotic reduction involves microbial-mediated enzymatic reduction. Some abiotic reduction of Fe(III) and Mn(IV) can be mediated by reducing agents such as microbial metabolic end products, including cysteine and pyruvate.
- Iron and manganese in wetland soils are found in several phases: water-soluble, exchangeable, reducible, and residual (crystalline). Both Eh and pH have a significant effect on distribution patterns among the various phases or pools. An increase in water solubility or exchangeable pool is favored by a decrease in Eh and pH.
- Soluble iron and manganese can form complexes with DOM. Typical Eh–pH diagrams cannot easily predict the stability of the organic complexes of iron and manganese.
- Regulators of iron and manganese reduction include the quality of organic substances (electron donors), the bioavailability of iron and manganese minerals, and the quantity of electron acceptors with higher reduction potential (e.g., nitrate) and sediment pH and temperature.
- Reduced iron and manganese can be oxidized in both aerobic and anaerobic environments of wetland and aquatic systems. Fe(II) and Mn(II) can be oxidized enzymatically (microbial mediated) or nonenzymatically (spontaneous reaction with oxidants such as oxygen).
- Only a limited group of iron- and manganese-oxidizing bacteria can derive energy from oxidation of the reduced iron and manganese compounds. In wetland soils, the aerobic oxidation of Fe(II) and Mn(II) is primarily restricted to the aerobic surface layer at the oxidized soil–water interface and in the root rhizosphere. It has also been shown that in the presence of light, anaerobic oxidation of Fe(II) is possible by photosynthetic bacteria in anoxic or reducing environments.
- Oxidation–reduction reactions of iron and manganese are also involved in nutrient release in flooded soil and sediments. Fe(III) and Mn(IV) serve as electron acceptors for organic matter decomposition or turnover. Organic nitrogen mineralization results in the release of nutrients such as ammonium nitrogen. Iron reduction is also coupled to phosphorous release in soils dominated by iron redox couples.
- Iron and manganese can influence the chemistry of heavy metals in wetland soils and sediment. Amorphous Fe(III) and Mn(IV) oxides have the ability to absorb or coprecipitate heavy metals. Under reducing or acidic conditions, the resultant increase in solubility of iron and manganese results in the release of heavy metal precipitate or absorption by iron and manganese oxide.

- The oxidation and precipitation of reduced Fe(II) and Mn(II) in the root zone (a result of oxygen transport by wetland plants) results in iron and manganese plaque formation on the root surfaces. Iron plaque on root surfaces can protect plants from reduced phytotoxins such as sulfide, but it can also potentially create a barrier limiting nutrient diffusion into the root.
- Iron and manganese, which serve as important electron acceptors, especially in mineral soils, can regulate methane production (an important greenhouse gas). Until electron acceptors with higher reduction potential (e.g., Fe(III), Mn(IV)) are exhausted, no methane will be produced. In wetland soils, methane production is inhibited in zones where intense cycling of iron occurs.
- Methane production rates are inhibited in soils with intense cycling of iron and manganese. In soils with limited availability of Fe(III) and Mn(IV) oxides and other electron acceptors, methanogenesis is the dominant pathway in regulating organic matter decomposition, and ultimately a major methane source to the atmosphere.

STUDY QUESTIONS

1. List the oxidation states of iron and manganese. Which oxidation states are present under reducing conditions? Describe the sequential reduction of iron and manganese as related to the presence of other electron acceptors.
2. What factors regulate iron and manganese solubility in wetland soils and sediment?
3. Explain the difference between biotic and abiotic iron and manganese reduction. Which is the dominant process?
4. List the various pools or fractions of iron and manganese in wetland soil. Explain how sediment Eh–pH affects the distribution in the various pools.
5. List examples of enzymatic and nonenzymatic oxidation of iron and manganese.
6. In what zone of the flooded soil does oxidation of iron and manganese occur?
7. Under what conditions can anaerobic oxidation of Fe(II) by photosynthetic bacterial occur?
8. How can oxidation–reduction reactions of iron and manganese influence nutrient release in flooded soil and sediment? Under what conditions is phosphorous available?
9. Discuss the relationship of Fe and Mn transformation to heavy metal chemistry.
10. How is iron plaque formed on the surface of roots of wetland plants? List the processes the iron plaque can impact in the root rhizosphere.
11. If a chemical analysis of a coastal wetland soil indicates that soluble iron is present as Fe(II), is it likely an appropriate analysis would also find: (1) some soluble manganese in the pore water, (2) nitrate in the pore water, (3) sulfate in the pore water?
12. Free reactive iron tends to be in the Fe(III) state and thus insoluble in oxidized environments. Floodwater in a swamp will normally contain some dissolved oxygen, but may contain soluble iron as well. Why?
13. Manganese as Mn(IV) compound and nitrate can both be used as terminal electron acceptors at about the same redox potential. If both are present in bottomland hardwood forest wetland soil that is undergoing a decrease in redox potential as a consequence of seasonal flooding, why might nitrate be more readily available to microbes than the Mn(IV) compounds as a terminal electron acceptor for respiration?
14. Why do upland soils tend to be more limited in plant-available phosphorous than wetland soils?
15. Under what pH and Eh conditions can iron compounds play a role in minimizing the mobility and biological availability of some trace and toxic metals such as copper and lead? Briefly describe the process.
16. In wetlands, what factor causes nonenzymatic reduction of Fe(III) to be comparatively low relative to enzymatic reduction?

FURTHER READINGS

Burdige, D. J. 1993. The biogeochemistry of manganese and iron reduction in marine sediments. *Earth Sci. Rev.* 35:249–284.

Ehrlich, H. L. 2002. *Geomicrobiology.* Chapters 15 and 16. Marcel Dekker, New York.

Emerson, D. 2000. Microbial oxidation of Fe(II) and Mn(II) at circumneutral pH. In D. R. Lovley (ed.) *Environmental Metal-Microbe Interactions.* ASM Press, Washington DC. pp. 31–52.

Lovley, D. R. 2004. Dissimilatory Fe(III) and Mn(IV) reduction. *Adv. Microbiol. Physiol.* 49:219–286.

Madigan, M. and J. Martinko. 2006. *Brock Biology of Microorganisms.* 11th Edition. Benjamin Cummings.

Nealson, K. and D. Saffarini. 1994. Iron and manganese in anaerobic respiration. *Annu. Rev. Microbiol.* 48:311–343.

Thamdrup, B. 2000. Bacterial manganese and iron reduction in aquatic sediments. In B. Schink (ed.) *Advances In Microbial Ecology.* Vol. 16. Kluwer Academic/Plenum Publishers, New York. pp. 41–84.

Vepraskas, M. J. 2001. Morphological features of seasonally reduced soils. In J. L. Richardson and M. J. Vepraskas (eds.) *Wetland Soils* Lewis Publishers, Boca Raton, FL. pp. 163–182.

11 Sulfur

11.1 INTRODUCTION

Sulfur cycling has been extensively studied in coastal wetlands and marine ecosystems. Cycling of various forms of sulfur governs many microbial communities that regulate oxidation and reduction reactions. The large inputs of organic matter into wetland soils, along with aerobic and anaerobic zones where sulfur transformation occurs, allow sulfur to play a critical role in the biogeochemistry of wetlands (Figure 11.1). Sulfur is closely linked to the carbon and nitrogen cycle. An important distinction between sulfur and nitrogen cycling is that there is a plenty of sulfate for organisms to utilize. By contrast, the major reservoirs of nitrogen atoms (N_2) and carbon atoms (CO_2) are gases that must be pulled from the atmosphere.

Sulfur is a ubiquitous element present in many sources in the environment (Brown, 1982). Sulfur biogeochemistry is an important regulator of redox chemistry and has strong linkages with nutrient cycling, trace-metal redox behavior, and microbiological energetics, plant growth, and mineralogical composition of sediments. Sulfur transformations are complex due to the number of redox states for sulfur and the rapid transformations that can occur.

Sulfur is a constituent of amino acids such as cysteine and methionine, which are essential for the structure and function of living cells. Inorganic compounds of sulfur are the essential source of sulfur to microorganisms and plants. Animals that require preformed sulfur-containing amino acids obtain only a small amount of inorganic sulfur compound from the environment. Microorganisms and plants utilize inorganic sulfur-containing compounds from soil and water to form sulfur-containing amino acids required for their survival. Only ruminants can utilize the inorganic compounds of sulfur through symbiotic association with microorganisms in their digestion tract.

11.2 MAJOR STORAGE COMPARTMENTS

Sulfur is a widely distributed and abundant element on the earth's surface. Sulfur is among the ten most abundant elements in biological materials and is essential to all known organisms. Inorganic sulfur forming on the earth's surface occurs largely as mineral sulfide, sulfate, hydrogen sulfide, sulfur dioxide, and elemental sulfur. Approximately 90% of the sulfur occurs in deep oceans and sedimentary rocks (lithosphere), whereas a majority of the remainder is accounted for as sulfates in the ocean (hydrosphere). Table 11.1 shows major sulfur reservoirs. In a global context only a small amount of sulfur is present in soils, living matter, and the atmosphere. Sulfur undergoes continuous cycling within and between these reservoirs through well-established oxidation–reduction reactions.

Conceptually, the major features of the sulfur cycle are similar in all wetlands. Sulfur is involved in a number of important biogeochemical processes such as sulfate reduction, pyrite formation, metal cycling, energy transport, and atmospheric sulfur emissions. Figure 11.1 depicts a general sulfur cycle in wetlands. In wetlands, sulfur is present in both inorganic (pyrite, iron and hydrogen sulfide, monosulfides, sulfate, and elemental sulfur) and organic forms (animal and microbial plant tissue). In addition, there are sedimentation sinks or burial that remove sulfur from active surface processes. There is also sulfur exchanges between wetlands and the atmosphere, including deposition and emission, and tidal exchange transfers of both dissolved and particulate forms of sulfur in tidally influenced systems.

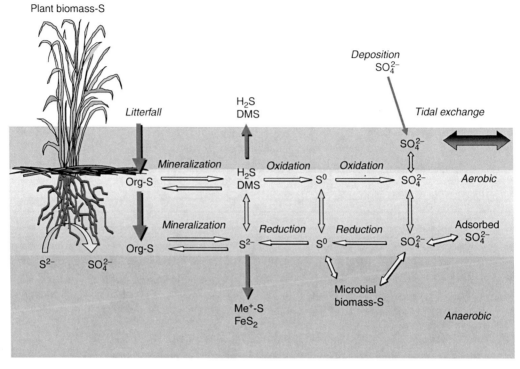

FIGURE 11.1 Sulfur cycling in wetlands is closely associated with carbon and nitrogen cycling. (From Krairapanond et al., 1992.)

TABLE 11.1
Major Reservoirs of Sulfur

Atmosphere	4.8×10^9 kg
Lithosphere	24.3×10^{18} kg
Hydrosphere	
Sea	1.3×10^{18} kg
Freshwater	3.0×10^{12} kg
Pedosphere	
Soil	2.6×10^{14} kg
Soil organic matter	0.1×10^{14} kg
Biosphere	8.0×10^{12} kg

Source: Freney et al. (1983), Trudinger (1975), and Stevenson (1986). The lithosphere refers to the crust of the earth.

11.3 FORMS OF SULFUR

Sulfur occurs in both inorganic and organic forms as shown in Table 11.2. The general forms of inorganic sulfur in the environment fall into three major categories:

- Oxidized inorganic sulfur (sulfate, sulfite, and thiosulfate)
- Reduced inorganic sulfur (elemental sulfur and sulfide)
- Gaseous sulfur compounds (SO_2, H_2S, DMSO, and DMS)

Minerals containing sulfur include galena (PbS_2), gypsum ($CaSO_4$), jarosite (Fe_2S), barite ($BaSO_4$), and pyrite (FeS_2).

Sulfur

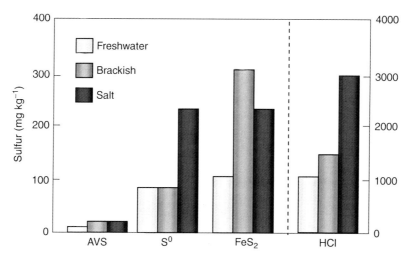

FIGURE 11.2 Organic sulfur constituents in Louisiana marsh soils, present as acid-volatile sulfides (AVS); elemental sulfur (S^0); pyrite (FeS_2), and HCl-soluble forms. (Modified from Krairapanond et al., 1992.)

TABLE 11.2
General Forms of Sulfur in the Environment

Minerals	Galena (PbS_2)
	Gypsum ($CaSO_4$)
	Jarosite (Fe_2S)
	Barite ($BaSO_4$)
Fossil fuels	Petroleum (0.1–10%)
	Coal (1–20%)
Organic S	In plant, animal, and microbial tissue (as essential components of amino acids and proteins)
	In soil and sediments as humic material (naturally occurring soil and sediment organic matter)
Gaseous S compounds	Sulfite (SO_2)
	Sulfide (H_2S)
	DMSO
	DMS
Oxidized inorganic S	Sulfate (SO_4) (seawater contains about 2,700 mg L^{-1} of sulfate)
Reduced inorganic S	Elemental sulfur (S^0)
	Sulfide (S^{2-})

Various operationally defined chemical fractionation schemes can be used to determine various sulfur species in soils. These methods group sulfur into the following fractions:

- AVSs (dissolved sulfide, iron monosulfides, and FeS)
- HCl-soluble sulfur (pore water sulfates, thiosulfates, polythionates, polysulfides, soluble organic sulfur, and HCl-hydrolyzable organic sulfur such as sulfate polysaccharides and amino acids)
- Pyrite sulfur (FeS_2)
- Elemental sulfur (S^0)
- Ester sulfates (see example below)
- Carbon-bonded sulfur (see example below)

Sulfur mainly exists as organic sulfur in wetland soils. Organic sulfur is present as both ester sulfates and carbon-bonded sulfur as shown in Figure 11.3. Living organisms contain an average of 1% sulfur in their tissues on dry weight basis. Dissolved sulfate concentrations in wetlands can span several orders of magnitude. Sulfate concentrations in marine systems can be as high as 28 mM (900 mg L$^-$). By contrast, concentrations in freshwater systems can be below 1–2 mM (32–64 mg L^{-1}). In salt marshes and brackish swamps, the majority of the total sulfur is often present as reduced inorganic sulfur minerals, although organic sulfur is the dominant species in some brackish water marshes and swamps such as estuarine systems with large freshwater inputs. In salt marshes the majority of reduced inorganic sulfur is present as pyrite with elemental sulfur and FeS typically making up a smaller percent of the reduced inorganic sulfur. In contrast to marine and brackish wetland soils, most of the sulfur in freshwater wetland soils is typically present as organic rather than inorganic sulfur, with carbon-bonded sulfur much more abundant than ester sulfate (Table 11.3). Of the reduced inorganic sulfur in freshwater wetlands, pyrite is often the dominant form, although concentrations are much lower than that in salt marshes (Figure 11.2).

Organic sulfur in soils is poorly characterized. In soil organic matter, organic sulfur constitutes 93% of the total sulfur and is present in two major forms (Table 11.3):

- Carbon-bonded sulfur compounds. These are primarily sulfur-containing peptides, proteins, and amino acids such as cysteine, cystine, and methionine. Approximately 41% of the total sulfur in soil organic matter is present in this form.

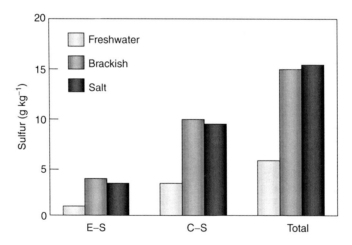

FIGURE 11.3 Organic sulfur in Louisiana fresh brackish and salt marsh soils present as ester sulfur (E-S), carbon bonded sulfur (C-S), and total sulfur (Total). (Modified from Krairapanond et al., 1992.)

TABLE 11.3
Distribution of S in Soil Organic Matter

Organic Sulfur [93%]	Inorganic Sulfur [7%]
Carbon-bonded sulfur (cysteine and methionine) 41%	Adsorbed + soluble sulfates 6%
Non-carbon-bonded sulfur (ester sulfates) 52%	Inorganic compounds less oxidized than sulfates and reduced sulfur compounds (e.g., sulfides) 1%

Source: Data from Krairapanond et al. (1992).

- Non-carbon-bonded sulfur compounds. These are primarily ester sulfates (–C–O–S, sulfur bonded to oxygen) such as phenolic sulfates, sulfamic acid (–C–N–S), sulfated polysaccharides, and adenosine-5′-phosphosulfate (APS). Approximately half of the total sulfur in soil organic matter is present in this form.

In Louisiana coastal marsh soils, inorganic sulfur constitutes 13–30% of the total sulfur pool, with HCl-soluble sulfur representing 78–86% of the inorganic sulfur fraction in freshwater, brackish, and salt marsh (Pezeshki et al., 1991; Krairapanond et al., 1992). AVS accounted for <1% of the total sulfur pool. Pyrite sulfur and elemental sulfur together accounts for 8–33% of the inorganic sulfur pool (Krairapanond et al., 1992).

Organic sulfur, in the form of ester sulfate and carbon-bonded sulfur, is the major constituent (76–87%) of the total sulfur in all Louisiana marsh soils. Organic sulfur is partly derived from sulfate assimilated into proteins and other organic forms by plants and microorganisms, some of which are subsequently incorporated into humic materials during decay. However, incorporation of sulfate into organic materials independently of assimilatory pathways is probably the more important process under anaerobic conditions. Nonassimilatory metabolism of sulfate entering peat by diffusion from overlying surface water could account for sulfur accumulation in the peat profile. Sulfur gases and reduced sulfur compounds are only a small portion of the total sulfur in wetland soils.

Sulfur exists as various stable isotopes including ^{32}S (95.02%), ^{33}S (0.75%), ^{34}S (4.21%), and ^{36}S (0.02%). By convention, stable isotope ratios are expressed as heavy to light isotopes. The ratio of $^{34}S/^{32}S$ is often used to investigate sulfur biogeochemical processes in soils. In most biogeochemical reactions, a lighter sulfur isotope is preferentially utilized by microbes over a heavier isotope. The most important process for variation in isotopic composition of sulfur is the reduction of sulfate ions by anaerobic bacteria such as *Desulfovibrio desulfuricans*. These bacteria in sediment split oxygen from sulfate ions and form H_2S, which is enriched in ^{32}S relative to sulfate. Sulfur isotopic composition has been used by the U.S. Geological Survey to trace the source of sulfate contaminates entering Florida Everglades. Variation in ^{34}S abundance can be used to evaluate soil–plant–atmosphere interactions. In addition, ^{35}S, a man-made radioactive isotope, is commonly used to determine sulfur transformations in wetland soils and sediments.

11.4 OXIDATION–REDUCTION OF SULFUR

The stable sulfur-containing species in a solution at an ordinary temperature and pressure are depicted in an Eh–pH diagram (Figure 11.4). The only important sulfur-containing ion in an oxidizing solution is SO_4. Reducing solutions generally contain H_2S at pH less than 7 and HS^- at pH greater than 7 (Figure 11.5). Sulfide (S^{2-}) is never a major constituent of any solution found in the environment. It should be pointed out that the Eh–pH boundary condition shown in Figure 11.4 may take considerable time for sulfur species to reach equilibrium unless bacteria are present. The diagram would be more complicated if heavy metals were present which form complex stable sulfide with various reduced sulfur species.

The form of sulfide present in the soil and sediment depends on pH, as shown below:

$$H_2S \text{ [acidic pH]} = HS^- \text{ [neutral pH]} = S^{2-} \text{ [alkaline pH]}$$

Under acidic conditions H_2S is stable, and it will change to HS^- and to S^{2-} as pH is increased:

$$H_2S = HS^- + H^+, \quad K_{H_2S} = 10^{-7}$$
$$HS^- = S^{2-} + H^+, \quad K_{HS} = 10^{-14}$$

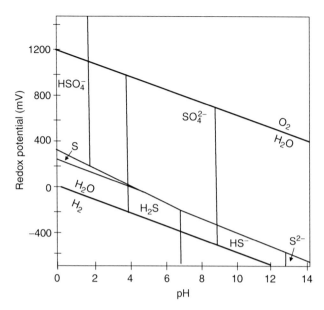

FIGURE 11.4 Eh–pH diagram of distribution of sulfur species solution at 1 atm pressure and 25°C. (Modified from Krauskopf, 1979.)

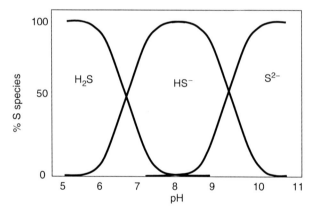

FIGURE 11.5 Effect of pH on sulfide speciation.

The activities of $[H_2S] = [HS^-]$ when $K_{H_2S} = [H^+]$ and the activities of $[HS^-] = [S^{2-}]$ when $K_{HS} = [H^+]$. The equilibrium constants for K_{H_2S} and K_{HS} are 10^{-7} and 10^{-14}, respectively. The Nernst equation (see Chapter 4) can be used to determine the relationship between Eh and pH:

$$H_2S + 4H_2O = SO_4^{2-} + 10H^+ + 8e^-$$

$$Eh = 0.303 - 0.074\,pH + 0.0074 \log \frac{[SO_4^{2-}]}{[H_2S]}$$

$$HS^- + 4H_2O = SO_4^{2-} + 9H^+ + 8e^-$$

$$Eh = 0.252 - 0.066\,pH + 0.0074 \log \frac{[SO_4^{2-}]}{[HS^-]}$$

$$S^{2-} + 4H_2O = SO_4^{2-} + 8H^+ + 8e^-$$

$$Eh = 0.303 - 0.074\, pH + 0.0074 \log \frac{|SO_4^{2-}|}{|S^{2-}|}$$

Sulfur exists in wetland soils in a variety of oxidation states between +6 and −2 and may be present in gaseous, soluble, and solid forms. Its most common oxidation states in wetlands are −2 (sulfhydrl, R-SH) and sulfide (HS⁻), 0 (elemental sulfur), and +6 (sulfate) (Table 11.4).

Several other oxidation states are possible since sulfur can substitute for oxygen in anions and form divalent anions consisting of groups of two or more sulfur atoms, such as thiosulfate (S_2O_3) in which sulfur has a valence of +2. Polysulfide, which occurs in pyrite (FeS_2) and marcasite, is assigned a valence of −1. Sulfur plays two key roles in wetlands: first, it is an essential element required by all plants and microorganisms for cell synthesis. In organisms, most of the sulfur occurs in reduced state in proteins, mainly as part of cysteine and methionine, and oxidized sulfur can be present as sulfate esters. Second, sulfur acts as a significant electron carrier in wetland environments. The ability of sulfur to undergo changes in oxidation states over an eight-electron shift is similar to the oxidation–reduction reactions in carbon and nitrogen cycling. Sulfur cycling and associated oxidation–reduction reactions bear resemblance to reactions involved in nitrogen cycling.

The overall important reactions involved in the sulfur cycle are shown in Table 11.5. Mineralization of organic sulfur is a process mediated by many heterotrophic microorganisms under both aerobic and anaerobic conditions. The process is linked to overall organic matter decomposition. These reactions include dissimilative sulfate reductions where obligate anaerobes reduce sulfate (SO_4) to sulfide (S_2) using sulfate as an electron acceptor with sulfate metabolism linked to energy production. Assimilative sulfate reduction is where sulfate is reduced to organic sulfhydryl groups (R-SH) by plant fungi and various prokaryotes that assimilate sulfate and sulfide as a nutrient source for the synthesis of cellular constituent. Desulfuration is a process in which organic molecules containing sulfur are desulfurated, producing hydrogen sulfide gas (H_2S). Sulfide oxidation is a process where H_2S is oxidized to elemental sulfur (S^0). This process is conducted by photosynthetic green and purple bacteria and some chemolithotrophs. Sulfur oxidation is where sulfur oxidizers produce sulfate. Dissimilatory sulfur reduction is where elemental sulfur is reduced to hydrogen sulfide. Microbial groups involved in sulfur cycles are shown in Table 11.6.

Sulfur cycling processes are analogous to nitrogen cycling processes. Sulfur and nitrogen are an integral part of organic matter and coupled in many ways. Sulfur is present as −SH in most reduced state, whereas nitrogen is present as −NH₂ in most reduced state. Upon decomposition, organic S and N are mineralized to sulfides and ammonium, respectively. Both sulfides and ammonium are

TABLE 11.4
Oxidation States of Sulfur

Oxidation State	Example	Formula
+6	Sulfate	SO_4^{2-}
+5	Dithionate	$S_2O_6^{2-}$
+4	Sulfite	SO_3^{2-}
+4	Disulfite	$S_2O_5^{2-}$
+3	Dithionite	$S_2O_4^{2-}$
+2	Thiosulfate	$S_2O_3^{2-}$
0	Sulfur (elemental)	S
−2	Sulfide (H_2S)	S^{2-}
−2	Organic S (R-SH)	S^{2-}

TABLE 11.5
Sulfur Cycling Processes

Mineralization of organic sulfur
Dissimilatory sulfate reduction
Assimilatory sulfate reduction
Desulfurylation
Sulfide oxidation
Sulfur oxidation
Dissimilatory S^0 reduction

TABLE 11.6
Microbial Groups Involved in Sulfur Cycles

- Assimilatory sulfate reduction
  Bacteria, fungi, algae, and plants
- Heterotrophs
  Desulfovibrio, Desulfotomaculum, Desulfobacter, Desulfuromonas
- Chemolithotrophs
  Thiobacillus, Beggiatoa
- Phototrophs
  Chlorobium, Chromatium

used as electron donors under aerobic conditions and subsequently oxidized to sulfate and nitrate, respectively. Under anaerobic conditions, sulfate and nitrate are used as electron acceptors and are linked to microbial catabolic processes.

11.5 ASSIMILATORY SULFATE AND ELEMENTAL SULFUR REDUCTION

Bacteria, cyanobacteria, fungi, eukaryotic algae, and vascular plants assimilate sulfate and elemental sulfur as nutrient from the environment. These compounds are reduced inside the cell and transformed into sulfur-containing amino acids and other organic sulfur compounds. Sulfur is assimilated primarily as a nutrient source and used for cell biosynthesis. Sulfate is a stable ion. Within the cells, sulfate is activated by ATP. The enzyme ATP sulfurylase catalyzes the attachment of sulfate ion to phosphate ion of ATP to form adenosine-5-phosphosulfate (APS) and subsequently phosphorylated (by the addition of P) to form 3-phosphodenosine-5-phosphosulfate (PAPS). The sulfate then undergoes subsequent reduction to sulfite (SO_3^-) and then to sulfide, which is finally transferred to amino acids as sulfhydryl groups (R-SH). The majority of the reduced sulfur is fixed intracellularly and only a small amount is released as volatile-reduced sulfur gases. This process occurs in both aerobic and anaerobic environments. However, once an organism dies and decomposes, the reduced sulfide can be reoxidized to elemental sulfur and sulfate under aerobic and anaerobic conditions, respectively.

As an example, for the green alga *Chlorella*, the assimilation of sulfate and subsequent reduction to cysteine (the first organic metabolite formed) is a multistage process. Cysteine serves as the beginning compound for biosynthesis of all other sulfur metabolites, including amino acids such as methionine. Cysteine and methionine represent a large fraction of the sulfur content of biota. The quantity of MBS is estimated using chloroform fumigation techniques, similar to the technique described for MBC and MBN with one exception. After fumigation, soils are extracted with either 0.1 M $NaHCO_3$ or 0.01 M $CaCl_2$. The MBS is calculated as follows:

$$MBS = \frac{[\text{S released from CHCL}_3 - \text{fumigate soil} - \text{S released from nonfumigate soil}]}{K_s}$$

where K_s represents extraction efficiency, which ranges from 0.35 to 0.41. The amount of sulfur present in MBS represents an active pool and is directly linked to bioavailable sulfur. The microbial biomass accounts for approximately 1–3% of the organic sulfur. This is similar to the amount of nitrogen present in microbial biomass.

11.6 MINERALIZATION OF ORGANIC SULFUR

The sources of organic sulfur in wetlands include detrital matter from vegetation and periphyton and soil organic matter. Organic sulfur constitutes 93% of the total sulfur in the organic matter.

Sulfur

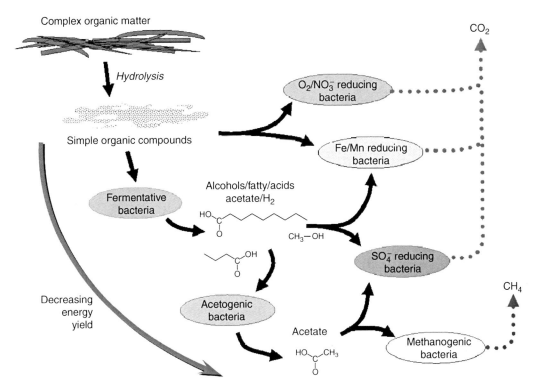

FIGURE 11.6 Metabolic pathways regulating organic matter decomposition and nutrient release.

Thus, the availability of sulfur depends on the mineralization of organic matter. The mineralization of organic sulfur is linked to microbial catabolic activity, where microbial communities use organic matter as their energy source (Figure 11.6). This process is strongly coupled to organic carbon decomposition, since a major portion of sulfur in soil organic matter and detrital matter is bonded to carbon. As organic matter is decomposed, organic sulfur is mineralized to sulfides and under aerobic conditions subsequently oxidized to sulfates. Depending on the microbial sulfur requirements, mineralized sulfide is rapidly assimilated and unutilized sulfide is released. The carbon-to-sulfur ratio of bacteria is approximately 85 and for fungi the ratio ranges from 180 to 230 (Dick, 1992). The sulfur content of most microorganisms is in the range of 0.1–1% on dry weight basis.

The microbial biomass regulates the storage and transformation of sulfur. The flow of sulfur through the soil microbial fraction can be substantial. In sulfur-limited wetlands containing significant quantities of organic sulfur, bioavailability of sulfur may be regulated by the mineralization of organic sulfur. Soil organic matter and detrital plant tissue contain complex structural compounds (high molecular weight) that are first hydrolyzed through the activity of extracellular enzymes into simple organic molecules. These low-molecular-weight compounds can be directly transferred into the cell where they are oxidized and used as an energy source. Nutrient enrichment can significantly increase the activity of extracellular enzymes, which are critical in the first step in degradation of soil organic matter and detrital plant tissue. Most models for plant litter decomposition link degradation rates to variations in climate or composition of litter, rather than microbial activity. The measurement of enzyme activity can be linked to litter characteristics and the decomposition rates. Several studies suggest that enzyme activities might serve as indices for the first step in breakdown of complex polymers. Thus, changes in extracellular enzyme activity may prove to be a sensitive indicator of bioavailability of organic sulfur.

Approximately half of the total sulfur present in soils is in non-carbon-bonded form, where sulfate is linked to an organic compound (R) through ester linkage. This linkage is broken by hydrolytic enzymes, resulting in the release of dissolved sulfate. The enzymes that play a key role in regulating sulfate release from organic sulfur are arylsulfatases. Arylsulfatases are involved in cleaving ester linkages and release of sulfate from non-carbon bonded organic sulfur compounds.

Arylsulfatases (EC 3.1.6.1) catalyze the irreversible reaction:

$$R \cdot OSO_3^- + H_2O \Rightarrow R \cdot OH + H^+ + SO_4^{2-}$$

Sulfur present as carbon bonded in amino acids such as cysteine is mineralized as follows:

$$HSCH_2NH_2CH-COOH \Rightarrow H_2S + NH_3 + CH_3COOH$$

Microbial degradation of the organic matter and associated detrital matter can result in the production of several reduced sulfur gases (Table 11.7).

The mineralization of organic sulfur follows a similar pattern as organic nitrogen. This does not mean that the relative rates of organic sulfur and nitrogen mineralization are similar and the end products are released in the same ratio as they are found in the organic matter. This depends on the occurrence of sulfur and nitrogen in same organic compound and in such a case both sulfur and nitrogen are released in proportion to their ratios. The net release of sulfur (as sulfates under aerobic conditions) and sulfide (under anaerobic conditions) depends on the balance between mineralization (catabolic) and immobilization (assimilatory). Under aerobic conditions, the C:S ratios of plant residues have been used as indicators of mineralization or immobilization processes (Table 11.8).

Assuming 40% carbon in the plant tissue, the C:S ratio of 200 to 400 approximately corresponds to 0.1–0.2% of sulfur in the plant tissue. MBS turnover rates are fairly rapid. The soil MBS fraction of the total organic sulfur in soil is relatively labile. The microbial biomass in soils range from 3 to 300 µgS dry soil^{-1} (Banerjee and Chapman, 1996).

TABLE 11.7
Origin of Reduced Sulfur Gases Produced in Soils by the Microbial Degradation of Organic Matter under Aerobic and Anaerobic Conditions

Volatile Sulfur Compounds	Biochemical Precursors
H_2S (hydrogen sulfide)	Proteins, polypeptides, cystine, cysteine, glutathione
CH_3SH (methyl mercaptan)	Methionine, methionine sulfoxide, methionine sulfone, S-methyl cysteine
CH_3SCH_3 (dimethyl sulfide)	Methionine, methionine sulfoxide, methionine sulfone, S-methyl cysteine, homocysteine
CH_3SSCH_3 (dimethyl disulfide)	Methionine, methionine sulfoxide, methionine sulfone, S-methyl cysteine
CS_2 (carbon disulfide)	Cysteine, cystine, homocysteine, lanthionine, djenkolic acid
COS (carbonyl sulfide)	Lanthionine, djenkolic acid

Source: Warneck (1988).

TABLE 11.8
Carbon/Sulfur Ratios of Plant Residues as Indicators of Mineralization and Immobilization Processes

C:S Ratio = <200	C:S Ratio = 200–400	C:S Ratio = >400
Mineralization >> Immobilization	Mineralization = Immobilization	Immobilization > Mineralization
Net gain of inorganic sulfur	Neither a gain nor a loss of inorganic sulfur	Net loss of inorganic sulfur

Sulfur

11.7 ELECTRON ACCEPTOR—REDUCTION OF INORGANIC SULFUR

Obligate anaerobes use inorganic sulfur compounds such as sulfate and elemental sulfur as electron acceptors. Since this process is a linked bacterial catabolic activity, we can define this process as *catabolic sulfur reduction*, which is most commonly known as dissimilatory sulfate reduction (Figure 11.7). In anaerobic environments, organisms incorporate a small fraction of sulfate for cell biosynthesis, but utilization of majority of sulfate is linked to bacterial catabolic activities. Sulfate is the most oxidized form of sulfur and is the most abundant in the environment, especially in coastal and marine ecosystems. During this process, sulfate is reduced to hydrogen sulfide. Sulfide participates in various biogeochemical reactions in wetlands, including as an energy source for microbes and in precipitation of metals.

11.7.1 Dissimilatory Sulfate Reduction

Similar to assimilatory sulfate reduction, dissimilatory sulfate reduction to sulfide involves eight-electron transfer from reduced compounds (organic carbon sources) to sulfate. Dissimilatory sulfate reduction plays a major role in the organic matter oxidation and nutrient mineralization in wetland environment. The key requirements for inorganic sulfur reduction in a wetland ecosystem are

- Anaerobic conditions (low redox potentials)
- Electron donors (simple organic compounds)
- Microbial groups capable of utilizing inorganic sulfur compounds as electron acceptors
- Inorganic sulfur compounds (as electron acceptors)

When soils are flooded, the oxygen supply is greatly restricted. Depending on the hydroperiod and organic matter inputs, wetlands exhibit a wide range of redox conditions. During drought, water levels can fall below the soil, which can result in the introduction of oxygen into the soil profile. Oxygen and its intermediates super oxide (O_2^-) and hydrogen peroxide (H_2O_2) can be toxic to sulfate reducers because they lack enzymes that can aid in detoxifying intermediate compounds. In the absence of oxygen, some microorganisms are able to use alternate electron acceptors (such as nitrate, oxides of manganese and iron, sulfate, and carbon dioxide) for the oxidation of organic matter. Reduction of each of these electron acceptors occurs in a specific redox range (see Chapter 3), as sulfate reduction occurs under highly reducing conditions with redox potentials lower than

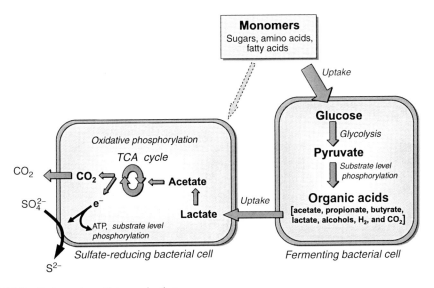

FIGURE 11.7 Pathway in sulfate respiration.

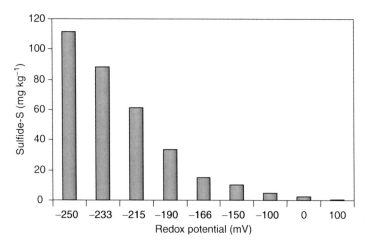

FIGURE 11.8 Effect of soil redox potential on sulfate reduction or sulfide production. (From Connell and Patrick, 1968.)

−100 mV at pH 7. Typically, sulfate reduction is dominant in soils with redox potentials less than −100 mV (Figure 11.8). Addition of electron acceptors with higher reduction potentials (such as nitrate or amorphous ferric oxyhydroxide) to soils can inhibit sulfate reduction. Sulfate is not reduced to sulfide until most of the iron has been reduced to the ferrous form. Inhibition of sulfate reduction is explained by (1) redox buffered above levels where sulfate reduction occurs and (2) competition for electron donor between microbial groups.

Complex polymers of organic matter are enzymatically hydrolyzed into simple monomers (see Chapter 5 for details). These monomers must be fermented into simple organic compounds before sulfate reducers can utilize the carbon source. Sulfate-reducing bacteria metabolize only low-molecular-weight organic substrates. Other bacteria must ferment higher-molecular-weight organic matter into these simpler fermentation products that can be used by sulfate-reducing bacteria. Thus, in anaerobic environments there is an assembly of bacteria involved in the fermentation and reduction of sulfate, which decompose organic carbon. Some of the reduced fermentation products such as fatty acids (butyrate, formate, propionate, lactate, and long-chain fatty acids) are oxidized to acetate, H_2, and CO_2. This pathway to some extent is regulated by H_2 partial pressure in fermenting bacterial cell. An increased H_2 pressure may result in more reduced products such as propionate. The low H_2 partial pressure maintained in wetland environment is due to close coupling via interspecies hydrogen transfer between fermentors and sulfate-reducing bacteria or methanogens. The electron donors used in driving sulfate reduction include lactate, pyruvate, ethanol, fumarate, malate, propionate, acetate, butyrate, and other fatty acids. On the basis of substrate utilization, sulfate-reducing bacteria are grouped into nonacetate users and acetate users (Table 11.9).

The pathways involved in dissimilatory sulfate reduction are shown in Figure 11.7. The sulfate reducers have a plenty of electron and hydrogen carriers. The electron transport is carried out by cytochrome c_3 and other coenzymes, the electrons are transferred from the energy source to sulfate ion in APS. The c_3 cytochrome is unique to sulfate-reducing bacteria and is not found in the bacteria using other electron acceptors. The end products of fermentation such as propionate, lactate, alcohols, and higher fatty acids are degraded to acetate by incomplete oxidizers (*Desulfovibrio* sp.). These sulfate reducers belong to Group I (Table 11.9). For Group I sulfate reducers, lactate oxidation to acetate proceeds via pyruvate and acetyl-CoA, with the production of H_2. Electrons from H_2 and other organic electron donors are transferred to the enzyme *hydrogenase* present in the membrane along with the cytochrome. Protons resulting from the oxidation of H_2 remain outside the membrane and the electrons are transferred across the membrane. This creates proton gradient across the membrane and proton motive force, and synthesis of ATP. The electrons are transferred to APS.

TABLE 11.9
Inorganic Sulfur-Reducing Bacteria and Some of Their Properties

Bacterial Groups—Genus	Substrates Utilized
Group I—Dissimilatory sulfate reducers: nonacetate oxidizers	
Desulfovibrio	Lactate, ethanol, malate, and H_2
Desulfomonas	Lactate
Desulfotomaculum	Lactate, ethanol, and H_2
Desulfobulbus	Propionate
Group II—Dissimilatory sulfate reducers: acetate oxidizers	
Desulfotomaculum	Acetate and H_2
Desulfobacterium	Acetate
Desulfococcus	Acetate, propionate, and H_2
Desulfonema	Acetate and formate
Desulfosarcina	Acetate, propionate, and H_2
Dissimilatory sulfate reducers—cannot reduce sulfates	
Capable of reducing elemental sulfur, sulfite, and thiosulfate	Acetate, succinate, propionate, and ethanol
Desulfuromonas Campylobacter	Variety of low-molecular-weight organic compounds including acetate

Source: Gottschalk (1996).

The final acetate production is a result of ATP synthesis. Acetate is excreted from these organisms and used by Group II sulfate reducers. However, certain groups of organisms can oxidize some of these fermentative products to carbon dioxide, provided acetate is excreted from these organisms. This group is known as complete oxidizers. The acetate oxidation to carbon dioxide involves the TCA cycle, with a series of oxidation–reduction reactions involving electrons.

As discussed above, sulfate is activated by ATP to form APS. Electrons from donors are transferred to APS. The sulfate ion of APS is reduced to sulfite and then to sulfide, which is excreted from cells. The first step, sulfate to sulfite reduction, involves two-electron transfer followed by six-electron transfer during the sulfite to sulfide reduction. The bacteria are capable of performing catabolic sulfate reduction.

Sulfate reduction products can be divided into acid-volatile and non-acid-volatile fractions. The former fraction consists mainly of H_2S, HS^-, S^{2-}, and FeS. The latter fraction contains FeS_2, elemental sulfur, and organic sulfur. Pyrite is a major short-term end product of sulfate reduction. Under some conditions AVSs (H_2S and FeS) and elemental sulfur are the major end products of sulfate reduction, and less than 15% is defined as pyrite (FeS_2).

A basic reduction reaction is as follows:

$$SO_4^{2-} + 8e^- + 10H^+ \Rightarrow H_2S + 4H_2O$$

Thermodynamically, sulfate and other inorganic sulfur compounds are the least preferred electron acceptors, in relation to other inorganic electron acceptors. The reactions in Table 11.10 show the oxidation of glucose with various inorganic electron acceptors. Since many sulfate reducers cannot use glucose as their electron donors, it must be hydrolyzed or fermented into simple fatty acids. To oxidize one mole of glucose, organisms would need three moles of sulfate, resulting in a net energy yield of 381 kJ mol^{-1}. During this process the energy is retained in the three moles of sulfide formed, suggesting high-energy requirements for the reduction of sulfate to sulfide. Sulfate reduction accounts for less than 15% of energy yield under aerobic conditions, making it a very inefficient electron acceptor.

TABLE 11.10
Glucose Oxidation by Various Inorganic Electron Acceptor Found in Wetland Soils

Oxidation–Reduction Reaction	kJ mol^{-1} Glucose
$C_6H_{12}O_6 + 6O_2 = 6CO_2 + 6H_2O$	2,877
$5C_6H_{12}O_6 + 24NO_3^- + 24H^+ = 30CO_2 + 12N_2 + 42H_2O$	2,721
$C_6H_{12}O_6 + 12MnO_2 + 24H^+ = 6CO_2 + 12Mn^{2+} + 18H_2O$	2,027
$C_6H_{12}O_6 + 24Fe(OH)_3 + 48H^+ = 6O_2 + 24Fe_2^+ + 66H_2O$	441
$C_6H_{12}O_6 + 3SO_4^{2-} = 6CO_2 + 3S_2^- + 6H_2O$	382

The following biochemical reactions show the utilization of select electron donors during inorganic sulfur reduction by obligate anaerobic bacteria. Listed below are few examples of reactions common to all wetland soils where sulfate is present as an electron acceptor:

$$4H_2 + SO_4^{2-} + H^+ \Rightarrow HS^- + 4H_2O, -152 \text{ kJ per reaction}$$

$$4H_2 + S_2O_3^{2-} \Rightarrow \updownarrow HS^- + 3H_2O, -174 \text{ kJ per reaction}$$

$$H_2 + S^0 \Rightarrow HS^- + H^+, -28 \text{ kJ per reaction}$$

$$2CH_3CHOHCOO^- \text{ (lactate)} + SO_4^{2-} \Rightarrow 2CH_3COO^- \text{ (acetate)} + HS + H^+ + 2HCO_3^-,$$
$$-160 \text{ kJ per reaction}$$

$$2CH_3CHOHCOO^- \text{ (lactate)} + SO_4^{2-} \Rightarrow 3HS + H^+ + 6HCO_3^-, -255 \text{ kJ per reaction}$$

$$2CH_3CH_2COO^- \text{ (propionate)} + 3SO_4^{2-} \Rightarrow 4CH_3COO^- \text{ (acetate)} + 3HS + H^+ + 4HCO_3^-,$$
$$-152 \text{ kJ per reaction}$$

$$4 \text{ Pyruvate} + SO_4^{2-} \Rightarrow 4 \text{ Acetate} + H_2S + 4HCO_3^-, -356 \text{ kJ per reaction}$$

$$CH_3COO^- + SO_4^{2-} + 3H^+ \Rightarrow HS^- + 2HCO_3^-, -48 \text{ kJ per reaction}$$

$$CH_3COO^- + 4S^0 \Rightarrow HS^- + 2HCO_3^-, -39 \text{ kJ per reaction}$$

$$4CH_3OH + 3SO_4^{2-} \Rightarrow 3HS^- + 4HCO_3^- + 4H_2O + H^+, -364 \text{ kJ per reaction}$$

In the absence of sulfate, some sulfate reducers such as *Desulfovibrio* species can grow on ethanol or lactate in the presence of H$_2$ scavenging methanogens. This is likely the case in freshwater wetlands with low sulfate levels. Using these syntrophic relationships, sulfate-reducing bacteria can function as hydrogen-producing bacteria. In coastal wetlands, sulfate is not limiting, and any of these electron donors formed during fermentation are rapidly utilized by sulfate reducers. Sulfate reduction by molecular hydrogen and other reducing agents is thermodynamically favorable at 25°C. However, the kinetics of these reactions are too slow to produce measurable amounts of hydrogen sulfide. No measurable abiotic reduction of sulfate is reported at temperatures less than 200°C and at high partial pressure.

The reaction rates of dissimilatory sulfate reduction depend primarily on the interplay between several factors: sulfate concentration, availability of organic matter and other nutrients, and the size and density of the bacterial population. Temperature can also affect sulfate reduction rates. Sulfate concentrations can have a greater effect on the sulfate reduction rate than on the available organic substrate concentration. In sulfate-deficient systems, the growth of the sulfate-reducing organism *D. desulfuricans* is linearly related to sulfate concentrations.

Aerobic and facultative bacteria outcompete sulfate reducers for electron donors. These organisms can use simple monomers (such as glucose) as their electron donor. They do not need fermentation to use simple monomers. As discussed before, sulfate reducers need fermentation to utilize

Sulfur

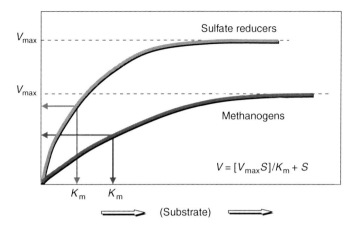

FIGURE 11.9 Comparison of substrate utilization kinetics by sulfate-reducing bacteria and methanogens.

organic substrates as electron donors. Competition for electron donors (acetate and H_2) can be explained by consumption kinetics using Michaelis–Menten equations (Figure 11.9). In most cases, sulfate reducers have higher V_{max} and lower K_m values than methanogens. Thus, sulfate reducers maintain low levels of acetate or H_2 and these levels are too low for methanogens to effectively use them for their growth. Reported values of K_m for H_2 consumption by sulfate reducers are in the range of 0.7–1.9 µmol L^{-1} as compared to 2.5–6.6 for methanogens (Wieder and Lang, 1988) (Table 11.7). Similarly, values for acetate consumption by sulfate reducers are in the range of 64–330 as compared to 33–3,000 µmol L^{-1} (Widdel, 1988).

11.7.2 Role of Sulfur in Energy Flow

In aerobic environments, the energy trapped in organic matter is released during decay and follows the flow of carbon. In contrast, during anaerobic decomposition, all the energy present in the organic carbon is not released; the remaining energy is conserved in high-energy reduced compounds such as hydrogen sulfide. During dissimilatory sulfate reduction approximately 25% of the energy present in the organic matter is available to microorganisms and 75% of the energy remains trapped in sulfides. During anaerobic decomposition, carbon and energy flow are decoupled. Sulfur recycling is important in the recycling of the energy flow of salt marsh. Sulfate reduction in salt marshes is reported to account for a significant portion of the total respiration in the salt marsh sediment (Howarth, 1984). Large amounts of energy from the respired organic matter are stored in inorganic-reduced sulfur compounds such as soluble sulfides, iron monosulfides, pyrite, elemental sulfur, thiosulfate, and polythionite. When these reduced inorganic compounds are subsequently reoxidized, energy is released. This energy can be trapped by a variety of organisms, which thereby fix CO_2 as organic biomass. Energy flow involving reduced sulfur compounds can be significantly and should be included in the energy budget of a wetland ecosystem especially in salt marsh.

Sulfate reduction is tied closely to methane production and flux from wetland soils and sediments (Lovley and Phillips, 1978). Methanogenesis is the dominant process in organic matter degradation in anaerobic soils with low levels of sulfate, especially in freshwater wetlands. However, there is competition for electron donors between methanogens and sulfate reducers. Sulfate reduction results in the liberation of more energy than methanogenesis; therefore, no appreciable methane is formed until all sulfate is reduced (Figure 11.10). Methane flux is inversely correlated with sulfate concentration. In general at sulfate concentration greater than approximately 30 µm, sulfate reducers maintain high acetate levels for methanogenesis to occur. There is little methane production/flux in salt marsh environments because they contain high levels of sulfate. In coastal regions, methane flux is inversely correlated with soil salinity. The addition of sulfate-containing fertilizer to rice soil can also reduce methane emission.

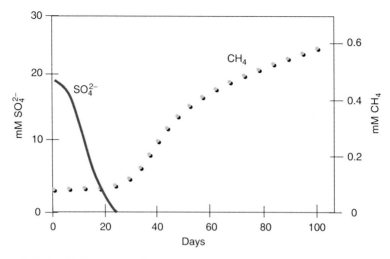

FIGURE 11.10 Relationship between sulfate reduction and methane production. (From Jorgensen, 1983.)

11.7.3 MEASUREMENT OF SULFATE REDUCTION IN WETLAND SOILS

Sulfate reduction rates in wetlands have been estimated using numerous techniques. These include the measurement of sulfate disappearance in soil cores and the measurement of sulfide concentration. The direct and most sensitive method used to measure sulfate reduction rates is the use of radioactively labeled sulfate (^{35}S) (Howarth and Merkel, 1994). This method involves injecting $^{35}SO_4^{2-}$ into cores and measuring the amount of ^{35}S found. The procedure in which $^{35}SO_4^{2-}$ is injected into the soil cores has been used by numerous researchers. Measuring sulfate reduction using this procedure is based on the assumption that the $H_2^{35}S$ produced from the added $^{35}SO_4^{2-}$ is retained as free sulfide, and FeS can be liberated and assayed when the soil is acidified. Measurement of the incorporation of ^{35}S into acid-volatile samples may underestimate sulfate reduction rates because some of the ^{35}S enters sulfur pools that are not acid volatile. It has been shown that an appreciable portion of the reduced ^{35}S is also incorporated into non-acid sulfur compounds such as pyrite and elemental sulfur.

Following stripping of AVS, aqua regia has been used to extract pyrite and elemental forms from sediment. However, it has been determined that the use of aqua regia to analyze for ^{35}S in pyrite and elemental sulfur can overestimate rates of sulfate reduction if unreduced $^{35}SO_4^{2-}$ is not adequately removed. The chromium reduction technique that uses Cr^{2+} in acid solution to reduce pyrite and elemental sulfur to H_2S eliminates the contamination problem associated with the use of aqua regia. This procedure is currently the most used method for quantifying ^{35}S end product of sulfate reduction. Cr^{2+} does not reduce sulfate to sulfide and the presence of any unreduced sulfate in the sample is not of any concern. The H_2S produced by the Cr^{2+} reduction is purged and quantified by analyzing for radioactivity. Using the two-step procedure the sequential distillation determines the radioactivity of AVS and chromium reduction sulfide (CRS). The fraction of $^{35}SO_4^{2-}$ reduced during incubation is calculated for the sum of ^{35}S in AVS and CRS. The procedure has been refined into a single-step distillation procedure (Fossing and Jørgensen, 1989), which compares favorably with the two-step procedure for determining sulfate reduction in sediment.

Some examples of sulfate reduction rates as a function of depth in marine sediment cores are shown in Figure 11.11. Connell and Patrick (1968) demonstrated that addition of 200 mg kg^{-1} sulfate to an anaerobic soil slurry doubled the reduction rate observed when 100 mg kg^{-1} sulfate was added.

The sulfate reduction rate varies greatly among various wetland ecosystems. Reported sulfate reduction rates in Louisiana coastal marshes range from 10^{-6} to 10^{-8} mol SO_4^{2-} cm^{-3} day^{-1}

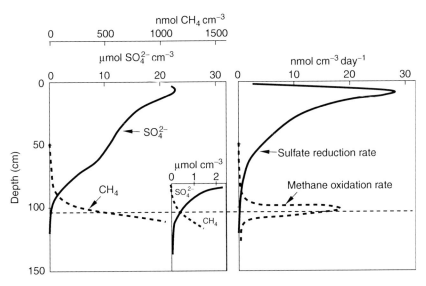

FIGURE 11.11 Sulfate reduction rates in marine sediment profile. (From Iversen and Jorgensen, 1985.)

TABLE 11.11
Sulfate Reduction in Select Wetlands and Aquatic Ecosystems

Wetland Ecosystem	Sulfate Reduction Rate (nmol g^{-1} day^{-1})
Low carbon wetland	20 (15°C)
Peaty wetland (neutral pH)	130 (10°C)
Acid peat lands	700–1,700 (8°C)
Oligotrophic lake sediments	700 (10°C)
Eutrophic lake sediments	1,200 (12–16°C)
Marine and salt marsh sediments	700–24,000 (25–32°C)

(DeLaune et al., 2002a). Table 11.11 shows the sulfate reduction rates in selected wetland and aquatic systems.

There is a link between sulfate levels and methyl mercury production in wetland ecosystems (Figure 11.12). High sulfate levels (or salinity levels) inhibit methylation whereas methylation increases when the sulfate levels are low. The increased sulfide formation at high sulfate levels actually inhibits the MeHg production. The sulfide produced reacts with mercury to form H_2S and makes mercury less available for the sulfate-reducing bacteria, which methylate the mercury present in wetland soils.

In the Everglades, sulfate contamination and input of the new mercury appears to be the major factors influencing MeHg production (see Chapter 17). Heavy sulfate-contaminated areas have low MeHg production (due to inhibition by sulfide). And the pristine areas have low MeHg production because the low sulfate concentration limits sulfate reduction.

The ability of sulfate-reducing bacteria to methylate inorganic mercury depends on sulfide-to-sulfate ratios; sulfate concentrations less than 1 mg L^{-1} starves sulfate-reducing bacteria and thus inhibits methylation. Sulfate concentration greater than 20 mg L^{-1} will result in excess sulfide production, which will also inhibit mercury methylation.

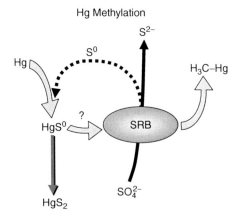

FIGURE 11.12 Relationship between sulfate-reducing bacteria (SRB) to Hg methylation in sediment.

TABLE 11.12
Regulators of Sulfate Reduction

Presence of electron acceptor with higher reduction potentials
Oxygen is toxic to sulfate reducers
Sulfate concentration
 Freshwater (<0.1 mM)
 Marine (20–30 mM)
Substrate/electron donor
Temperature
Microbial populations

11.7.4 Regulators of Sulfate Reductions

The primary regulators of sulfate reduction in wetland soils and sediments are supply of inorganic sulfur (sulfate), presence of oxidants with higher reduction potential, organic matter content of sediment, temperature, and seasonal fluctuation associated with temperature (Table 11.12).

There is a distinct seasonal and temperature influence on sulfate reduction, with high rates during summer and low rates during winter (Howarth and Giblin, 1983) (Figure 11.13). During winter, surface sediments were oxidized due to greater penetration of oxygen into the sediment profile, thus influencing sulfate reduction rates. During summer, much of the oxygen was consumed in surface layers, thus promoting greater activity of sulfate reducers. The sulfate reduction rate is highly correlated with temperature, maximum rates generally occurred during summer and minimum rates during winter (DeLaune et al., 2002a).

Sulfur reduction is the major form of microbial respiration in salt marsh sediments. However, it is difficult to estimate with precision what percentage of respiration is mediated by sulfate reduction, since total respiration and aerobic respiration are both poorly known in salt marsh sediments. The estimates on the percentage of total microbial respiration that is mediated by sulfate reduction are based on the comparison of measured rates of sulfate reduction with estimates of inputs and the decomposition of organic matter in these marsh sediments. Unfortunately, the inputs of organic carbon to marsh sediments are not easily measured and not well known.

Sulfate reduction rates are also controlled by the availability of sulfate and carbon as an energy source. Salt and brackish marsh generally have more sulfate present as compared to the fresh marsh or swamp forest soils due to saline water containing sulfate. Significant sulfate reduction can occur when adequate organic or label carbon is available to support intense reduction or low sediment

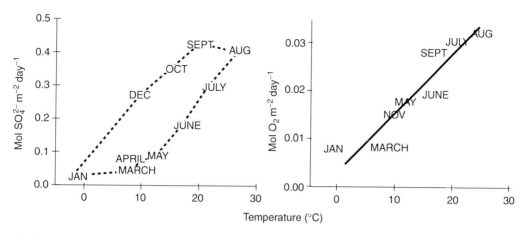

FIGURE 11.13 Seasonal and temperature effects on sulfate reduction. (From Howarth and Teal, 1979.)

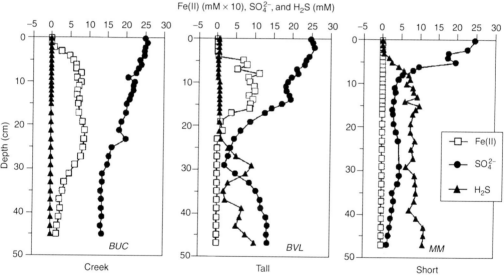

FIGURE 11.14 Relationship between the iron content of sediment and the H_2S levels. (From Kostka et al., 2002.)

redox condition. Sulfate reduction is generally less at depth in the soil profile where there is less labile organic carbon.

The presence of soil oxidants such as oxygen nitrate and oxidized form of iron and manganese, which become reduced at levels above where sulfate reduction occurs, tends to retard sulfate reduction. The presence of oxidized iron can inhibit sulfate reduction in the zone of ferric iron reduction in sediments. Fe(III) reducing bacteria can outcompete sulfate-reducing and methanogenic food chains for organic matter in sediments. The addition of amorphic iron (III) oxyhydroxide to the sediment where sulfate reduction was the predominant terminal electron-accepting process has been shown to inhibit sulfate reduction by 86–100% (Lovley and Phillips, 1987). In a recent study, Kostka et al. (2002) showed a strong relationship between iron content of sediment and sulfide production (Figure 11.14).

11.8 ELECTRON DONOR—OXIDATION OF SULFUR COMPOUNDS

Reduced sulfur compounds can be oxidized by both biotic and abiotic processes. The bacterial communities involved can be grouped into chemolithotrophs (inorganic chemicals as energy source) and phototrophs (light as energy source) (Table 11.13). Several reduced sulfur compounds are used as electron donors by a wide spectrum of bacteria known as "colorless sulfur bacteria." The term colorless is used primarily to distinguish this group of bacteria from other groups containing color. The other group of bacteria that also use reduced sulfur compounds as electron donors are called "phototrophic sulfur bacteria," commonly known as green and purple bacteria due to the pigments that they contain. Some of these bacteria are autotrophic and use light energy converted to chemical energy in the form of ATP, whereas the electrons for reduction of carbon dioxide form various reduced sulfur compounds including sulfide. The reduced sulfur compounds used by these bacteria as electron donors include sulfide, thiosulfate, thiocyanate, elemental sulfur, and tetrathionate. Oxidation of these compounds is carried out by both the chemolithotrophic sulfur oxidizers, which may use reduced sulfur as an energy source and fix CO_2, and the phototrophic green and purple bacteria, which use reduced sulfur as an electron donor in the fixation of CO_2. The following are some examples of oxidation reactions and respective energy yields per mole of reduced compound oxidized:

$$H_2S + 2O_2 = SO_4^{2-} + 2H^+, \quad -171 \text{ kJ per mole of sulfide}$$

$$S^0 + 1.5O_2 + H_2O = SO_4^{2-} + 2H^+, \quad -589 \text{ kJ per mole of elemental sulfur}$$

$$S_2O_3^{2-} + 2O_2 + H_2O = SO_4^{2-} + 2H^+, \quad -409 \text{ kJ per mole of thiosulfate}$$

In the presence of oxygen, abiotic oxidation of hydrogen sulfide can occur. However, this reaction can be catalyzed by sulfur-oxidizing bacteria (Figures 11.15 and 11.16). Oxidation of reduced sulfur occurs at interfaces: the oxic–anoxic interface for abiotic and chemolithotrophic oxidation and, in the anoxic zone, the light–dark interface for phototrophic oxidation.

TABLE 11.13
Inorganic Sulfur-Oxidizing Bacteria and Some of Their Properties

Bacterial Groups/Genus	Characteristics and Substrates Utilized
Autotrophic sulfur bacteria	Present in environments with hydrogen sulfide
Chlorobium	Green sulfur bacteria. Strict anaerobes. Sulfur metabolism is linked by energy fixation. Elemental sulfur is deposited intracellularly during sulfide oxidation
Chloromatium	Purple sulfur bacteria. Elemental sulfur is deposited intracellularly during sulfide oxidation
Oscillatoria	Blue-green bacteria
Heterotrophs	
Arthrobacter, Bacillus, Micrococcus, Pseudomonas, Mycobacterium	Sulfur-oxidizing bacteria
Colorless sulfur bacteria	
Thiobacillus	Electron donors: hydrogen sulfide, elemental sulfur, and thiosulfate. Freshwater and marine environments. Aerobic but several species can use nitrate and ferric iron as electron acceptor under oxygen-free conditions
Thiomicrospira	Electron donors: hydrogen sulfide and thiosulfate. Aerobic
Thiothrix	Electron donors: hydrogen sulfide and thiosulfate. Filamentous, aerobic
Beggiatoa	Electron donors: hydrogen sulfide and thiosulfate. Filamentous, aerobic. Oxidize hydrogen sulfide to elemental sulfur. Found in habitats rich in hydrogen sulfide. Present in the root zone of wetland plants

Sulfur

Sulfide oxidation

$$2H_2S + O_2 = 2S^0 + 2H_2O$$
−204 kJ per reaction

$$2S^0 + 3O_2 + 2H_2O = 2SO_4^{2-} + 4H^+$$
−583 kJ per reaction

$$H_2S + 2O_2 = SO_4^{2-} + 2H^+$$
−786 kJ per reaction

$$H_2S \Leftrightarrow S^0 \Leftrightarrow SO_3^{2-} \Leftrightarrow SO_4^{2-}$$

FIGURE 11.15 Examples of sulfide oxidation by oxygen.

Sulfide oxidation

- $2H_2S + 8NO_3^- = SO_4^{2-} + 4N_2 + 4H_2O + 2H^+$
  −736 kJ per reaction
  Chemolithotrophic bacteria (e.g., *Thiobacillus*)

- $3H_2S + 6CO_2 + 6H_2O = C_6H_{12}O_6 + 3SO_4^{2-} + 6H^+$
  +506 kJ per reaction
  Lithoautotrophic (phototrophic) bacteria (e.g., *Chromatium*)

FIGURE 11.16 Examples of sulfide oxidation by bacteria.

Bacteria and other microorganisms contribute to the oxidation of sulfur. *Beggiatoa* and *Thiothrix* are species of bacteria that are capable of oxidizing hydrogen sulfide using oxygen as electron acceptor:

$$2H_2S + O_2 = 2S + 2H_2O$$

Beggiatoa is a filamentous bacteria and is capable of oxidizing hydrogen sulfide to sulfate, accumulating sulfur in their cells. Such bacteria are widely distributed in the wetland and marine environments, which contain hydrogen sulfide. They are also found in the oxidized rhizosphere of wetland plants. *Beggiatoa* and the other genera of bacteria that oxidize sulfide to sulfate with the formation of sulfur globular are aerobic organisms:

$$2H_2S + CO_2 + light \rightarrow (CH_2O) + H_2O + 2S$$

where (CH_2O) represents a component of bacteria in all organic matter. These bacteria are active in the absence of free oxygen and occur at the sediment–water interface of shallow water bodies where hydrogen sulfide emerges.

Oxidation of elemental sulfur to sulfate results in the production of hydrogen ions that can lower pH. *Thiobaccillus* species such as *T. thiooxidans* grow well under acidic soil conditions, oxidizing sulfur to sulfate under aerobic conditions:

$$2S + 3O_2 + 2H_2O \rightarrow 2H_2SO_4$$

The bacteria is found in natural environments where soil or sediment pH is below 6.0.

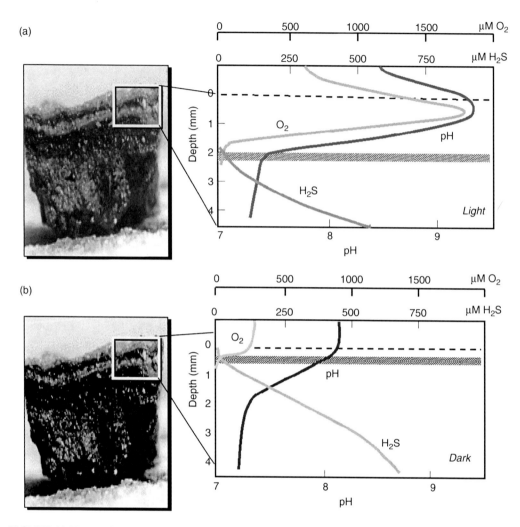

FIGURE 11.17 Profiles of hydrogen sulfide, oxygen, and pH in periphyton mats and sediments. (Redrawn from Jorgensen, 1983.)

T. ferrooxidans is another acid-tolerant bacteria that is capable of oxidizing both thiosulfate and ferrous iron:

$$N_2S_2O_3 + 2O_2 + H_2O \rightarrow Na_2SO_4 + H_2SO_4$$

$$12FeSO_4 + 3O_2 + 6H_2O \rightarrow 4Fe_2(SO_4)_3 + 4Fe(OH)_3$$

T. thiooxidans bacteria are active in drained marine environments high in elemental sulfur. *Thiobacillus denitrificans* can oxidize elemental sulfur in the presence of nitrate under anaerobic conditions as depicted below:

$$5S + 6NO_3 + 2H_2O \rightarrow 3SO_4 + 2H_2SO_4 + 3N_2$$

A variety of heterotrophic bacteria are capable of oxidizing elemental sulfur to sulfate.

A number of chemolithotrophic bacteria can oxidize reduced sulfur compounds using oxygen or nitrate as electron acceptors. This group of bacteria are found at the soil–floodwater interface of wetlands and aquatic systems where reduced sulfur and oxygen exist in the aerobic/anaerobic surface layer. Sulfur can also be oxidized by a variety of phototrophic bacteria in the absence of oxygen and nitrate if light is present. Purple and green sulfur bacteria and some cyanobacteria use sulfide as an electron donor in anoxygenic photosynthesis. These bacteria can occur in large numbers in oxygen-depleted water columns where there is sufficient sunlight to support their photosynthesis, especially in wetlands and lakes with little reactive iron for sulfide precipitation. Photosynthetic sulfur bacteria are also found in bacterial mats located on the surface of sediment that receives sunlight (Figures 11.17a and 11.17b).

In salt marsh sediment profiles, much of the sulfide formed by sulfate reduction is very quickly precipitated as pyrite (Howarth, 1979; Howarth and Teal, 1980). Sulfide not directly precipitated is mobile and can be exported as H_2S or oxidized at the soil surface or in the root rhizosphere. Some volatile sulfides are released to the atmosphere, whereas other reduced sulfur may be exported to tidal creeks or be carried away by the tidal water. Pyrite is the major form of inorganic sulfur in marsh sediments. However, pyrite does not just accumulate; rather, it is a very dynamic constituent (Luther and Church, 1988). Most of the pyrite in marsh sediments is present as very small microcrystals with diameters of 0.2–2 μm (Figure 11.18). The crystals have a high surface-to-volume ratio, and thus, the pyrite is probably reactive and more easily oxidized than larger crystals. Pyrite forms throughout the year, but during summer, when the marsh grasses are most active and capable of oxidizing the sediments, there is a net loss of pyrite. During fall, winter, and spring, when the grasses are less able to oxidize the sediments, there is a net accumulation of pyrite, although the net rate of accumulation may still be less than the gross rate of formation. The mechanism of oxidation is not yet clear, but may involve O_2 transport through the internal gas spaces of the marsh grasses or advection through the sediments as they drain. Sulfide diffusing from a reducing soil environment to oxygen-enriched surface waters and sediment can be rapidly chemically oxidized to elemental sulfur. Elemental sulfur is a predominant material in the surface

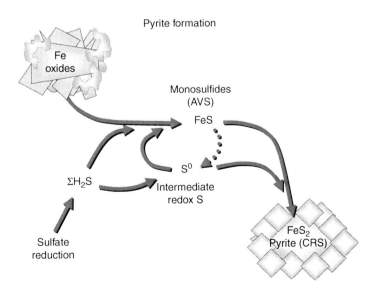

FIGURE 11.18 Schematic of pyrite formation in wetland soils.

film commonly observed on water surfaces in Louisiana Gulf Coast brackish and salt marshes (Whitcomb et al., 1989).

11.9 BIOGENIC EMISSION OF REDUCED SULFUR GASES

As a result of bacterial sulfate reduction and decomposition of organic material, large amounts of reduced sulfur gases are emitted into the atmosphere annually from wetland systems, especially tidally influenced marine sediments and salt marshes (Adams et al., 1979, 1981).

Coastal marshes have the highest aerial gaseous sulfur emissions with rates 10–100 times greater than that from oceans and inland soils. The range of reported emission rates is large, and extreme variability at a given sampling site is common.

In addition to hydrogen sulfide, volatile organic sulfur gases such as dimethyl sulfide (DMS), methanethiol, dimethyl disulfide (DMDS) emitted from natural aquatic ecosystem are important sources of sulfur to the atmosphere and play a significant role in the global sulfur cycle (Steudler and Peterson, 1984; DeLaune et al., 2002b). Biogenic estimates of reduced sulfur gas atmosphere emissions are reported to be equivalent to anthropogenic emissions. The production of volatile sulfur compounds by soil microorganisms and aquatic plants in wetlands is important in the global sulfur cycle. The coastal environments such as salt marshes, mangrove swamps, and intertidal zones are a major source of reduced sulfur gases to the atmosphere. The reported atmospheric emission rates for reduced sulfur gases span several orders of magnitude, creating a large uncertainty in the global budget.

Sulfur emission to the atmosphere contributes to the acidity of precipitation. The reduced sulfur compounds can undergo chemical and photochemical oxidation to yield compounds such as methanesulfanic and sulfuric acids. Through the formation of aerosol particles that can increase reflectivity of clouds, reduced sulfur gases entering the atmosphere can affect the earth's radiative balance suggesting a link to climate change. Sulfur compounds are very efficient relative to mass emissions of other greenhouse gases in regulating global temperatures.

Gaseous sulfur compounds produced in wetlands are either intermediate metabolites or end products of biological processes. Hydrogen sulfide (H_2S) produced by dissimilatory sulfate reduction in anaerobic environment was originally thought to be the primary gaseous sulfur source emitted to the atmosphere (Rodhe and Isaken, 1980).

Wetlands are significant sources of not only H_2S but also DMS, methyl mercaptane (methanethiol-MeSH), carbonyl sulfide (COS), DMDS, and carbon disulfide (CS_2).

Reported rates of emission of biogenic sulfur compounds from *Spartina alterniflora* salt marsh vary among coastal regions (DeLaune et al., 2002b). Both H_2S and DMS are the dominant emissions from salt marsh ecosystems. Other reduced organic sulfur gases are emitted from salt marsh but in lower concentrations. Frequencies of emission measurements and absolute detection limits varied greatly among the previous salt marsh studies.

High rate of DMS emissions from *S. alterniflora* is attributed to the presence of high concentrations of the DMS precursor dimethylsulfoniopropionate (DMSP in the plant tissue). Enzymatic cleavage of this compound produces DMS plus acrylic acid. *S. alterniflora* is one of only three plant species containing DMSP, the others being *S. anglica* and *S. foliosa*. *Spartina patens* does not contain DMSP. In general, emissions to the atmosphere are lower than that reported for salt marshes.

Measured reduced sulfur gas emissions from wetlands along a salinity gradient in Louisiana Gulf Coast salt, brackish, and freshwater marshes show that reduced sulfur gas emission was strongly associated with habitat and salinity gradient (Figure 11.19). The dominant emission component was DMS in the Louisiana salt marsh. Hydrogen sulfide was dominant in the brackish marsh and COS in the freshwater marsh. The emission of total reduced sulfur gaseous species decreased with decrease in both salinity and distance inland from the coast. Emission of total reduced sulfur

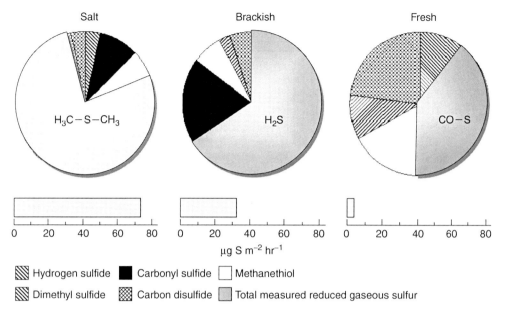

FIGURE 11.19 Summary data showing species and quantity evolved from freshwater, brackish, and salt marsh sites. (From DeLaune et al., 2002b.)

gases over the study averaged 73.3 μg S m^{-2} h^{-1} for the salt marsh and 32.1 μg S m^{-2} h^{-1} for the freshwater marsh (DeLaune et al., 2002b).

11.10 SULFUR–METAL INTERACTIONS

Dissolved sulfides (H$_2$S and HS$^-$), formed from sulfate reduction, can react with Fe(II) and precipitate as iron sulfides (Berner, 1970). Under conditions of early diagenesis of amorphous iron sulfides (FeS), mackinawite (FeS$_{0.9}$) is formed. One of the first minerals formed is hydrotrolite (FeS · nH$_2$O), which is labile under acidic conditions. During diagenesis the hydrotrolite converts to pyrite, which is a more stable mineral. The reaction can be simply represented as

$$SO_4^{2-} \Leftrightarrow S^{2-} + Fe^{2+} \Leftrightarrow FeS \cdot nH_2O \Leftrightarrow FeS_2$$

In the presence of elemental sulfur (S^0), FeS is transformed to gregite (Fe$_3$S$_4$) and pyrite (FeS$_2$):

$$3FeS + S^0 \rightarrow Fe_3S_4 \tag{11.1}$$

$$Fe_3S_4 + 2S^0 \rightarrow 3FeS_2 \tag{11.2}$$

In most wetland or coastal systems, iron monosulfide is thermodynamically less stable than pyrite, but its formation is kinetically favored over pyrite.

The formation of pyrite can occur following two mechanisms: (i) single pyrite crystals are formed rapidly through direct precipitation of Fe(II) and polysulfides (S$_n^{2-}$) and (ii) framboidal pyrite is produced by a slower reaction of FeS with S^0 to produce a gregite intermediate (Equations 11.1 and 11.2). The direct precipitation of pyrite requires the prior oxidation of H$_2$S to either S^0 or S$_n^{2-}$ for reactions like the following (Luther, 1991; Rickard and Luther, 1997):

$$Fe^{2+} + S^0 + H_2S \rightarrow FeS_2 + 2H^+ \tag{11.3}$$

$$Fe^{2+} + S_n^{2-} + HS^- \rightarrow FeS_2 + S_{n-1}^{2-} + H^+ \tag{11.4}$$

or

$$Fe^{2+} + S_5S^{2-} + HS^- \rightarrow FeS_2 + S_4S^{2-} + H^+ \qquad (11.5)$$

The two mechanisms for pyrite formation can proceed at the same time in the coastal marsh profiles. The rapid pyrite formation (Equation 11.3, 11.4, or 11.5) occurs in the upper 20 cm of marsh profile (more oxidized), displaying a larger variation in pyrite content, whereas the slow pyritization (Equations 11.1 and 11.2) tends to exist below 20 cm depth of marsh profile (more reduced) where maximum concentration of pyrite is normally found.

In coastal marsh areas, where marsh plants can affect the dynamic transformation of pyrite, a large percentage of sedimentary pyrite is converted into an oxidized Fe mineral and sulfate by the oxidation around plant roots during the marsh plant growing season. Oxygen diffusing into the surface oxidized layer can induce pyrite oxidation in marsh soil sediments, releasing excess dissolved Fe(II) and SO_4^{2-} to the pore waters, and precipitating an authigenic solid Fe phase.

Sulfur is not only involved in the cycling of Fe or pyrite in marsh soils and estuarine sediments, but can also play an important role in retaining heavy metal and trace elements in these systems (Boulegue et al., 1982).

The production of reduced S compounds during sulfate reduction has a controlling influence on the chemistry of trace metals in anoxic sediments. In the presence of sulfide, the extremely low solubility products of most metal sulfides predict that metal will precipitate as stable sulfide minerals in the reducing soil environment.

The concentrations of Cu, Zn, Hg, Cd, and other metals in sediment pore water are controlled by the solubility of metal sulfides in the reduced zone where sulfate reduction and sulfide formation is dominant. Change in redox condition upon burial results in a system where the growth of diagenetic copper, zinc, and arsenic sulfides control the distribution and partitioning of metals and arsenic in the sediment. In polluted sediments, sulfate reduction plays a key role in the formation and retention of sedimentary S as metal sulfides. The majority of the iron and manganese in coastal lake sediments is associated with sulfidic forms, especially in saline area high in available sulfate that is reduced to sulfide.

Metals such as Hg, Cu, Zn, Cd, Pb, and Ni form insoluble sulfides in reducing sediment (Eh less than −150 mV) where sulfide is produced. For example, the stability constant for HgS is approximately 10^{-52} which shows that Hg is strongly immobilized (Table 11.14). ZnS and CuS are also very insoluble. In contrast, NaS and KS are weakly stable.

Although other thermodynamically stable metal sulfides, such as those of Zn, Pb, Ni, and Hg, should form in wetland soils and sediment, they seldom form simply because these metals are not present in sufficient concentrations in most sediments to allow the formation of precipitated metal sulfides. However, at sites that are contaminated with heavy metals, metal sulfide formation is an important process in reducing heavy metal availability.

TABLE 11.14
Stability Constant of Selected Metal Sulfides

Metal Sulfide	K_{sp}
Na_2S	1.0×10^{-1}
MnS	1.4×10^{-15}
FeS	3.7×10^{-19}
ZnS	1.2×10^{-22}
NiS	1.0×10^{-26}
CdS	1.0×10^{-28}
CuS	3.5×10^{-36}
HgS	1.0×10^{-52}

Sulfur

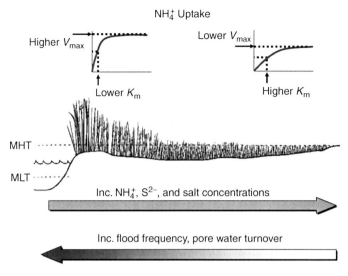

FIGURE 11.20 Relationship of soil sulfide levels to tall and short *Spartina alterniflora*.

11.11 SULFIDE TOXICITY

The soluble sulfide species in wetlands, including H_2S, HS^-, and S^{2-}, can be toxic to wetland plants. High levels of sulfide can affect wetland plant distribution in salt marshes. The so-called short form of *Spartina* is affiliated with high sulfide levels (Figure 11.20). Sulfide may affect productivity through toxicity or by inhibition of nutrient uptake. Linthurst (1979) reported that sulfide adversely affected growth in *S. alterniflora*. Howes et al. (1981) reported that sulfides at relatively low concentrations of 0.2 mM caused death of hydroponically grown *S. alterniflora*. Pezeshki et al. (1988) studied sediment concentrations of sulfide and reported that the net photosynthetic rate of *S. alterniflora* declined significantly when H_2S concentrations exceeded 34 mg L^{-1}. The net carbon assimilation responses of *Panicum hemitomon* and *Spartina patens* to the combined sediment anaerobiosis and hydrogen sulfide concentrations were also measured (Pezeshki et al., 1991). DeLaune et al. (1983) demonstrated that sulfide in the sediment limited root development; the root distribution was observed to be inversely related to the sulfide concentrations in the inland site of Louisiana coastal marshes.

11.12 EXCHANGE BETWEEN SOIL AND WATER COLUMN

The surface of flooded soil and sediment normally contains a thin oxidized layer overlying the reduced sediment column. Oxygen diffusion from the oxidized water column into the surface layer of sediment maintains the oxidized layer. The rate of oxygen diffusion through the interstitial water in the sediment pores is approximately 1/10,000 of the rate of gaseous diffusion. Due to slow diffusion of oxygen into the sediment, the oxidized layer at the surface is very thin (several millimeters). Beneath the oxidized layer, the sediment is usually highly reduced. Soluble sulfate from tidal water can diffuse down into the reduced layer and be reduced to sulfide. Sulfide diffusion up into the surface oxidized layer can be oxidized to elemental sulfur or sulfate (Figure 11.21).

Wetland plants growing in flooded soil possess well-developed aerenchyma (airspace tissue) systems that act as conduits for the diffusion of oxygen from the atmosphere through the plant leaves and stems to the roots. The root rhizosphere contains an oxidized area where reduced sulfide can be oxidized to sulfate or elemental sulfur (Figure 11.20). Sulfide can be oxidized either chemically or microbiologically to elemental sulfur. *Beggiatoa* organisms that oxidize H_2S to elemental sulfur are abundant in the root zone.

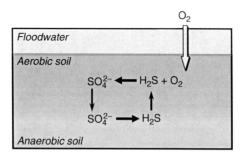

FIGURE 11.21 Schematic showing sulfur oxidation–reduction at the sediment–water interface.

TABLE 11.15
Rates of S Accumulation in the Louisiana Barataria Basin Marsh Soils (Mean ± SE, N = 15)

	Freshwater	Brackish	Salt
Mm year^{-1}			
Vertical accretion rate	8.5	9.5	10.5
gS m^{-2} year^{-1}			
AVS S	0.111	0.456	0.683
HCL-soluble S	1.077	2.309	9.030
Elemental S	0.086	0.126	0.726
Pyrite S	0.104	0.479	0.713
Ester sulfate S	1.27	5.37	9.57
C-bonded S	3.5	15.3	25.9
Total S	6.1	23.7	46.0

11.13 SULFUR SINKS

Wetlands are known as sinks or reservoirs for sediments, nutrients, and metals. In coastal areas wetland marsh soil vertically accrete to keep pace with subsidence or sea level rise. Coastal marsh must accumulate sufficient organic matter (peat) or mineral sediment to keep pace with sea level rise, thereby maintaining the marsh surface water level relationship. Wetlands are known to sequester large amounts of organic carbon. Organic S that is related to organic carbon is a dominant S species in wetland soil horizons. Sulfide (including pyrite) and elemental sulfur are also found in significant quantities in coastal or wetland soil profiles. The accumulation of sulfur is related to the rate of marsh accretion.

The accumulation of all S forms follows a seaward increase as a function of vertical accretion. In Louisiana Barataria Basin (Krairapanond et al., 1992), the rate of sedimentation of inorganic and organic sediments is a rapidly continuing process and increases with increased hydraulic energy toward the Gulf Coast. The accumulation rate of inorganic S associated with Fe (AVS-S and FeS$_2$), for example, was greatest in the salt marsh, where the sedimentation rate of mineral input is also greatest. (Table 11.15) Because of the high accretion in Louisiana Coast marshes, the accumulation of S was dependent largely on the sedimentation rate of both inorganic and organic sediments, rather than the chemical reduction process such as sulfate reduction. The actual level of sediment S accumulation may depend on a combination of biotic, abiotic, and other environmental factors.

A significant fraction of marsh energy flow is entrapped in these accreted reduced S forms. Pyrite S and elemental S sequestered in the soil profile also retain appreciable energy. In rapidly accreted coastal marsh, a significant portion of marsh energy flow enters this sedimentation pool.

11.14 SUMMARY

- Sulfur is involved in a number of biogeochemical processes in wetland, including sulfate reduction, pyrite formation, metal cycling, energy transport, and gaseous emissions to the atmosphere.
- Sulfur in wetlands exists in both organic and inorganic forms.
- The reduction of sulfate to sulfide is a dominant process in wetland soils. Sulfide produced can be toxic to wetland plants.
- There is a competition for electron donors between methanogens and the sulfate reducers that govern methane flux.
- Sulfate reduction liberates more energy than methanogenesis and as a result no methane is formed until all sulfate is reduced. In salt marshes containing high sulfate levels, little methane is produced.
- High sulfate levels also limit high methylation of mercury in wetland environments.
- Energy flow involving reduced sulfur species formed through sulfate reduction also occurs in salt marshes.
- Sulfate reduction accounts for a portion of the total respiration in salt marshes and, when the reduced, inorganic sulfur in oxidized energy is released and exported from the system.
- Wetlands also emits reduced sulfur gases to the atmosphere. DMS is the dominant form emitted from salt marshes.
- Hydrogen sulfide and various reduced organic compounds are the primary species evolved from freshwater system.

STUDY QUESTIONS

1. Explain the linkages of sulfur to carbon and nitrogen cycles.
2. List the major sulfur reservoirs and the forms found in wetland and aquatic environment.
3. What is the distribution of organic and inorganic sulfur in wetland soils?
4. List the oxidation states of sulfur.
5. List the major sulfur cycling processes.
6. What are the main microbial groups involved in sulfur cycling?
7. At what soil redox level does sulfate reduction occur?
8. Explain the relationship of sulfate reduction to competition with methanogenesis.
9. List the electron donors used during sulfate reduction.
10. What are the major regulators of sulfate reduction in wetland soils?
11. Which are the reduced sulfur gases that are emitted from wetland to the atmosphere? Which is the major source emitted from salt marshes?
12. How does Eh–pH of soils influence sulfide formation and sulfide speciation?
13. Discuss pyrite formation and iron sulfide reactions.
14. What is the relationship of sulfate reducers to mercury methylation in wetland soils and sediment?
15. What is the relation of sulfur cycling to energy flow in salt marshes?
16. What is the difference between dissimilatory and assimilatory sulfate reduction?
17. Can sulfate-reducing bacteria utilize complex carbon compounds as electron donors during sulfate reduction?

FURTHER READINGS

Canfield, D. E., and R. Raiswell. 1999. The evolution of sulfur cycle. *Amer. J. Sci.* 299:697–723.

Howarth, R. W., J. W. B. Stewart, and M. V. Ivanov. (eds.) 1992. *Sulfur Cycling on the Continents: Wetlands, Terrestrial Ecosystems and Associated Water Bodies*, SCOPE Report 48. Wiley, Chichester, England. 350 pp.

Ingvorsen, K. and B. B. Jorgensen. 1982. Seasonal variation in H_2S emission to the atmosphere from intertidal sediments in Denmark. *Atmos. Environ.* 16:855–865.

Jorgensen, B. B. 1982. Mineralization of organic matter in the seabed: the role of sulfate reduction. *Nature* 296:643–645.

Jorgensen, B. B. 1990. The sulfur cycle of freshwater sediments: Role of thiosulfate. *Limnol. Oceanogr.* 35:1329–1342.

King, G. M. 1988. Patterns of sulfate reduction and sulfur in a South Carolina salt marsh. *Limnol. Oceanogr.* 33:376–390.

King, G. M., B. L. Howes, and W. H. Dacey. 1985. Short-term end products of sulfate reduction in a salt marsh: formation of acid volatile sulfides, elemental sulfur, and pyrite. *Geochim. Cosmochim. Acta* 49:1561–1566.

12 Metals/Metalloids

12.1 INTRODUCTION

Heavy metals in wetlands have both natural and anthropogenic sources and can be delivered by either eolian, fluvial, or tidal sources. Once within a wetland or aquatic environment, the ability of a metal to be transported depends on its chemical properties. The chemical properties of metal pollutants also influence toxicity. The species of metal or metal speciation determine the behavior in aquatic and wetland environments. Valence, the formation of oxyanions, sorption to the particulate or sediments, complexation with organic matter, precipitation, and interaction with microorganisms are processes governing the availability or toxicity of heavy metals in wetlands.

The sources of heavy metals to wetlands, while in some part from natural sources, are dominated by human activity. Natural weathering of rocks introduces some metals into wetlands. However, the majority of elevated heavy metal inputs are from industrial sources. Humans have allowed the surface water to be the prime repository of waste materials including industrial sources. These waste materials include heavy metals that are toxic to aquatic plant and animal life. As a result, there are now concerns with secondary impacts: the bioaccumulation and bioconcentration of metals (e.g., Hg) through the food chain that result in toxicity to the nonaquatic species. The atmosphere can also contribute large amounts of heavy metals through emissions from industrial sources which are deposited by both dry aerosol fallout and wet scavenging as precipitation into watersheds.

Some heavy metals if present in the environment at excessive concentrations can be toxic to aquatic organisms and inhibit the activity or role of microbes in key biogeochemical processes in wetlands. Metal can exhibit toxicity through interference with the natural functions of enzymes. The toxicity is related to the form in which the metal is found in wetlands. Toxic metals may precipitate or chelate essential metabolites, act as antimetabolites, or displace essential metals in metalloenzymes. An elevated heavy metal concentration in soils can influence key microbial processes such as microbial respiration, organic nitrogen mineralization, denitrification, and methanogenesis. Heavy metals can also influence the degradation of toxic organics, including petroleum hydrocarbon entering the wetland environment. Some of the metal/metalloids (e.g., arsenic and chromium) can also function as alternate electron acceptors during microbial respiration.

A simplified diagram of the fate of heavy metals in a wetland environment is presented in Figure 12.1. Partitioning of metals in wetlands is subject to seasonal flooding, drainage cycles, and hydroperiod. Drained wetland soils contain a surface-oxidized horizon and a low pH. Periods of flooding result in anaerobic soil conditions with a shift in pH to a near-neutral state.

12.2 FACTORS GOVERNING METAL AVAILABILITY AND TRANSFORMATION

Metals can be present in solid or aqueous (including pore water) phases. In the aqueous phase, metals can exist as free cations, which is generally the most toxic form, or as complexes with either inorganic elements or organic compounds.

Toxic metals associated with wetland soils are present in various forms: dissolved, adsorbed, bound to carbonates, to Fe and Mn oxides, to sulfides, and insoluble organic matter forms, and within the crystalline structure of primary minerals (Shannon and White, 1991). The amount of organic matter and clay minerals, the soil acidity (pH), and the sediment oxidation–reduction status (Eh) of soils are very important physicochemical properties influencing the mobility of toxic metals.

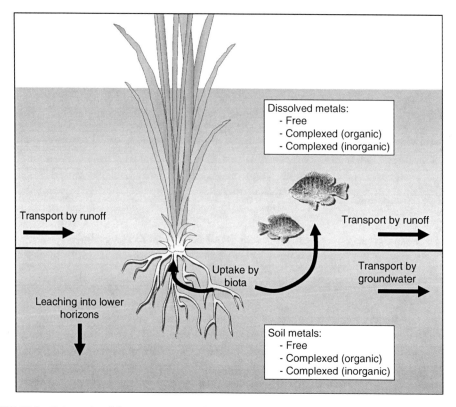

FIGURE 12.1 Schematic of the process governing heavy metal transport in wetland environments.

Humus is an amorphous, hydrophillic, acidic, partly aromatic, generally dark colored, and structurally complex material, resulting from the microbial degradation of plant detritus. Humus can be further classified as follows: (a) humic acids: a fraction that is soluble in alkali but precipitates on acidification of the solution, (b) fulvic acids: a fraction that remains in solution after the extraction is acidified, and (c) humin: a fraction that cannot be extracted by either alkali or acid (see Chapter 5 for details).

The reactions of metals with organics are particularly important in wetland soils containing soluble metals. Organics in soils can form stable complexes with the reduced soluble form of some metals such that the metal is maintained in a water-soluble form for several hours or days in an oxidizing environment (see Chapter 10 for additional details).

Humic materials have a higher capacity than clay to hold water and most metals, because they have a high density of active functional groups such as $-COOH$, phenolic-, alcoholic-, enolic-OH, and $C=O$ structures, as well as amino and imino groups. They are colloidal in nature and exhibit a very large surface area and negative charge. This charge arises from exposed $-COOH$ and $-OH$ groups, which have H^+ ion available for exchange with metals. Because of their chemical functional groups, they can form stable water-soluble and -insoluble complexes with many metal ions.

Insoluble high-molecular-weight humic acids are very effective in immobilizing most trace and toxic metals. Humic acids form fairly insoluble complexes with metal ions, precipitating them under conditions that otherwise would promote migration. Humic acids can also reduce certain oxidized metal species in such a way that they make it easier for the metal to be fixed to the humic matter and make it unavailable for further mobilization or plant uptake (Stevenson, 1982).

Clays also play an important role in the mobility of the metals. Clays are basically alternating layers of Si and O tetrahedra, and Al and O tetrahedra. Since oxygen charges are not entirely balanced, the particle has a net negative charge associated with it. Broken edges of the laminar structure and isomorphous substitution within the crystal lattice enhances this effect. Isomorphous substitution leaves one negative charge unsatisfied. Because this charge originates from within the lattice, it is permanent in nature, and depending on its magnitude, it will strongly influence the metal adsorptive capacity of the clay particle.

Clays exhibit a very large surface area because they are extremely small colloidal particles and the structures of some clays provide them with inner surfaces. With this negatively charged surface area, the clay particle is surrounded by an ionic double layer. The cations in the double layer are subject to interchange with other cations in pore water, giving rise to what is known as cation exchange capacity (CEC). The cation exchange will occur only if the ion in pore water can be held more strongly than the ion already at the surface, or if a large excess of a cation enters the zone.

Soils usually contain a low concentration of toxic metals from natural sources. These background levels vary widely depending on the parent minerals, sedimentation processes, and other soil-forming processes. Usually, it is the anthropogenic sources that increase metal content in wetland soils, resulting in potential ecological risks. However, elevated total concentrations of metals do not necessarily result in problem releases to water or excessive bioavailability. In addition to the particular metals present and the amount of these metals, the chemical forms of metals present and the processes affecting transformations between these forms are important in assessing risk (Gambrell, 1994).

There are a number of general chemical forms of metals in soils and these differ in their mobility and bioavailability. A listing of some of the common chemical forms of metals ranging from most available to least available is as follows (Gambrell and Patrick, 1991; Shannon and White, 1991): (a) readily available: dissolved and exchangeable forms; (b) potentially available: metal carbonates, metal oxides and metal hydroxides, metals adsorbed on or occluded with iron and manganese oxides, metals strongly adsorbed or chelated with insoluble high-molecular-weight humic materials and metals precipitated as sulfide; and (c) unavailable: metals within the crystalline lattice structure of clay and other residual minerals. Table 12.1 shows the general chemical forms or fractions in which metals can be found in wetland soils.

Metals dissolved in pore water are the most mobile and bioavailable. Adsorbed (exchangeable) metals are also bioavailable due to equilibrium between exchangeable and dissolved metals. Both dissolved and exchangeable metals are readily mobilized and bioavailable. On the opposite extreme are metals bound with the crystalline lattice structures of clay and other residual minerals. Metals in this form are essentially permanently immobilized and thus unavailable. Only under long period of mineral weathering, residual metals would become mobile and bioavailable. Between these two extremes are potentially available metals. In metal-contaminated soils, excess metals become primarily associated with these potentially available forms rather than the readily available soluble and exchangeable forms (Feijtel et al., 1988). By contrast, in uncontaminated soils or sediments, only background levels of metals exist in these forms.

TABLE 12.1
General Chemical Forms of Metals in Wetland Soils and Sediments

Water-soluble metals
    Soluble free ions, e.g., Fe^{2+}
    Soluble as inorganic complexes
    Soluble as organic complexes
Exchangeable metals
Metals precipitated as inorganic compounds
Metals complexed with high-molecular-weight humic materials
Metals adsorbed or occluded to precipitated hydrous oxides
Metals precipitated as insoluble sulfides

TABLE 12.2
Typical Fate of Potentially Available Metals in a Changing Chemical Environment in Soils

Metal Bound to	Initial Soil Condition	Environmental Change in Soil	Resultant Change in Metal Solubility
Carbonates, oxides, and hydroxides	Metal salts in the sediments	Reduction of pH	Release of the metals as the salts dissolve
Bound by iron and manganese oxides	Metals absorbed in sediment	Sediment becomes reducing or acidic	Manganese and iron oxides become unstable and release metals
Chelated to insoluble high-molecular-weight humic	Metal in chelated form	Strongly immobilized metal in both oxidizing and reducing conditions (however, there is some indication that the process is less effective if a reduced sediment becomes oxidized)	
Sulfides	Insoluble metal precipitates	Sediment becomes oxidized	Sulfides become unstable, oxidize to sulfates, and release metals

Metals bound to oxides, hydroxides, and carbonates are effectively immobilized at near neutral to somewhat alkaline pH conditions. If, however, pH becomes moderately to strongly acidic, as can sometimes occur when reduced soils (containing sulfides) become oxidized, these metals may be transformed into readily available forms (Table 12.2).

Metals complexed with insoluble high-molecular-weight humic compounds are effectively immobilized. There is some evidence that these metals are less effectively immobilized if reduced soils are oxidized. The long-term oxidation of wetland soils will result in the significant release of Cd and Zn (Gambrell and Patrick, 1988; Gambrell et al., 1991).

It is well established that oxides of Fe and perhaps Mn and Al effectively adsorb or occlude most toxic metals. These oxides exist in mineral soils in large quantities. When soils become reducing, the metals bound to Fe and Mn oxides are transformed into readily available forms due to dissolution of Fe and Mn oxides. During flooding and drainage cycles of wetlands, the formation of iron oxyhydroxides is important in retaining metals in surface soils (Gambrell, 1994).

Under strongly reducing conditions (Eh <-100 mV), the sulfates in pore water are reduced to sulfides, which quickly react with toxic divalent metal ions to form highly insoluble metal sulfides. The metals will remain in this form as long as wetland soils remain under reducing conditions. However, when reduced soils are drained or oxidized, all sulfides are oxidized into sulfates with the release of metals. Metals are removed from pore water by sulfide precipitation during anaerobic cycles and are released during aerobic cycles. The release is attributed to the lowering of soil pH during oxidation of sulfides.

12.2.1 Soil/Sediment Redox–pH Conditions

Soil redox conditions or Eh status governs the oxidation and reduction of some trace metals found in wetlands. Trace metals are present in various oxidation states, for example, chromium can exist in several oxidation states from Cr(0), the metallic form, to Cr(VI). The most stable oxidation states of chromium in the environment are Cr(III) and Cr(VI). Besides the elemental metallic form, which is extensively used in alloys, chromium has three important valence forms: Cr(II), Cr(III), and Cr(VI). The trivalent Cr(III) and the hexavalent Cr(VI) are the most important forms in the environment.

Metals/Metalloids

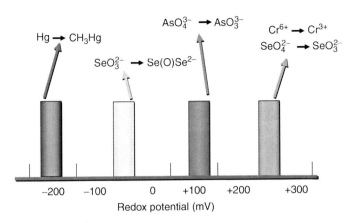

FIGURE 12.2 Soil redox (mV) condition influencing metal and metalloid transformations.

The presence of Cr(VI) is of particular importance because in this oxidation state Cr is water soluble and extremely toxic. They are also the only forms that undergo valence changes in the Eh–pH ranges encountered in natural systems. The solubility and potential toxicity of the chromium that enters into wetlands and aquatic systems is governed, to a large extent, by oxidation–reduction reactions. Cr(VI) reduces to Cr(III) at approximately +300 mV (Figure 12.2).

Oxidation–reduction processes also play a role in arsenic (As) and selenium (Se) chemistry (Figure 12.2). For the pH and redox conditions in most soils and sediments, arsenic exists as an oxyanion in the arsenate (As(V):$H_2AsO_4^-$ and $HAsO_4^{2-}$) or arsenite (As(III): H_3AsO_3) form. Monomethylarsenic acid (MMAA) and dimethylarsenic acid (DMAA) are important organo-arsenic chemical species. Selenium occurs as selenate (Se(VI): SeO_4^{2-}), selenite (Se(IV): $HSeO_3^-$ and SeO_3^{2-}), elemental selenium (Se(0)), and selenide (DMSe), which is the most important form. At higher soil redox levels (+200 to +500 mV), As(V) is the predominant As species present. The reduction of As(V) to As(III) occurs at redox levels corresponding within the nitrate-reducing zone of soils characterized by a soil redox level of approximately +300 mV. When Fe reduction starts and soil redox levels drop below +50 mV, selenite is reduced to elemental Se or metal selenides. Reoxidation of reduced As and Se species occurs at similar soil redox levels as those identified for reduction reactions. Both microbial-mediated and chemical oxidation processes are involved in the oxidation of As and Se species. Methylation of mercury and arsenic occurs in low soil redox conditions (Figure 12.2). The methylation occurs at or below the redox level where sulfate reduction occurs.

The Eh and pH of sediment also play an important role in regulating the solubility and chemical transformations of trace or heavy metals. Low pH and redox potential in sediment–water systems tend to favor the formation of soluble species of many metals, whereas in oxidized, nonacid systems, slightly soluble or insoluble forms tend to predominate. However, pH, and particularly Eh, may regulate other processes such as sulfide formation, which indirectly influence the solubility of metals. In wetlands with high sulfate inputs and reduced environments, the formation and accumulation of sulfides occur. The solubility of divalent metal sulfides in these systems is extremely low. Where large amounts of sulfide are present, sulfide precipitation is thought to be a very effective process for immobilizing trace metals in reduced sediments. Thus, a reducing environment, which causes a metal to be present in a soluble ionic form, may also contribute to its being effectively immobilized by sulfide precipitation. Sparingly soluble metal sulfides, which are stable in reduced environments, can oxidize to relatively soluble metal sulfates in aerobic environments, or when soil becomes oxidized.

Iron and manganese transformations are another redox potential–pH regulated process in sediment–water systems, which can affect heavy metal availability (see Chapter 10). The reduced forms of iron and manganese, when oxidized, form amorphous hydrous oxides with large surface area, which have a large sorptive capacity for trace or toxic heavy metals. These hydrous oxides are known to be effective scavengers of many trace metals. Freshly formed hydrous oxides of iron and manganese (typically in amorphous forms) are more effective scavengers than aged oxides (typically on crystallized forms).

In simple aqueous systems consisting of a limited number of elemental components, the regulation of metal forms by pH and redox potential is characterized by distinct redox potential–pH boundary conditions. However, in natural sediment–water systems typically consisting of a heterogeneous mixture of both inorganic and organic compounds, the factors influencing the immobilization or release of metal are much more complex than theoretical. Thermodynamic calculations based on simple systems cannot be extrapolated to natural systems for use in predicting metal availability. Additional factors, which complicate the understanding of metal chemistry, include cation exchange and other surface absorption reactions as well as complexation with organic matter.

Owing to the diversity of organic materials capable of binding metals in sediment–water systems, the fixation or release of metals in soil–water systems does not exhibit the rather precise, predictable Eh–pH boundary conditions shown by simple aqueous systems. However, Eh and pH have been shown to influence metal–organic complex formation and stability. The effects of soil/sediment Eh and pH on trace metal transformations in natural systems must therefore be studied empirically.

Using Eh–pH diagram for particular metal and measured soil pH and redox (Eh) condition, we can obtain an estimate of the metal species present under a specific set of soil or sediment conditions. Some of the most comprehensive data detailing Eh–pH diagram for elements can be obtained in the studies by Pourbaix (1966), Garrels and Christ (1965), Krauskopf (1979), Berner (1971), and Stumm and Morgan (1981).

12.3 MERCURY—METHYL MERCURY

Mercury is a naturally occurring metallic element and is one of the least abundant elements found in the earth's crust. It is present in air, water, soils/sediments, and biota, and is unique with its ability to exist in gas, liquid, and solid forms. In elemental state, it exists as a liquid at the surface of earth and as such will vaporize into the atmosphere and condense as determined by its vapor pressure and the barometric pressure and temperature of its environment. Mercury can exist in natural systems in three different oxidation states: elemental mercury (Hg^0), mercurous mercury (Hg^+), and mercuric mercury (Hg^{2+}), and as a variety of organic compounds, including the most significant form, monomethyl mercury (Ullirich et al., 2001). Mercury forms depend on pH and oxidation–reduction potential (Figure 12.3). Under oxidized conditions, mercuric mercury species are the dominant mercury forms. Mercury forms strong associations with organic material in wetland soils. Both inorganic and organic anions are reported to form mercuric complexes. Under anaerobic conditions, insoluble mercuric sulfide may form. Using thermodynamic data, Stumn and Morgan (1981) illustrated the importance of hydroxyl and chloride ions under oxidized conditions and sulfide under reducing conditions in the speciation of mercury. Divalent mercury has a very high affinity for a large number of organic substances, especially for those containing SH– groups (Nagase et al., 1984). Dissolved organic matter (DOM) increased mercury solubility significantly. Mercury complexed with high-molecular-weight organics controlled Hg chemistry under both oxidized and reduced conditions (Gambrell et al., 1980; Patrick and Verloo, 1998).

Under moderately reduced conditions, elemental mercury (Hg^0) is thermodynamically stable over a wide pH range. Because of its low solubility, Hg^0 is rapidly lost from aqueous solution and released into the atmosphere. Strong reducing conditions or anaerobic conditions (such as lake bottoms and stream beds) would form insoluble HgS in sediments.

In addition, mercury is found in several organic compounds. Some organo-mercurial substances are manufactured because of their commercial importance, and some are known to be produced in natural ecosystems. Organo-mercurial compounds are environmentally important because of their

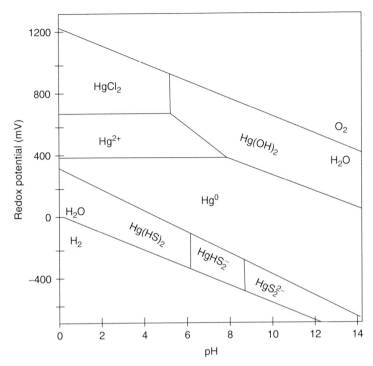

FIGURE 12.3 The stability of various mercury forms under varying pH and redox conditions.

mobility and well-known toxicological properties. The stability of volatile mercury species allows widespread and long-term dispersion of mercury in the environment. In the atmosphere, particular phases of mercury from sources such as power plants can be transported similarly to other metals. Precipitation washes mercury out of the atmosphere into soil and surface waters. Mercury washed into soil from the atmosphere is bound to the mineral and humic material in the upper few centimeters of soil. Mercury is naturally distributed in groundwater and surface water as a result of contact with water in the atmosphere and during surface and subsurface drainage. Presently, atmospheric mercury inputs represent approximately one-half of the mercury entering ecosystems.

Because of the tendency of mercury to be sorbed readily by a variety of earth materials, bottom sediments and suspended particulate matter in water are more likely to contain higher concentrations of mercury than the associated overlying water. The suspended material in surface waters may contain from 5 to 25 times the mercury found dissolved in the surrounding water in areas of industrial pollution.

The accumulation of unacceptable concentrations of mercury in fish occurs largely as a result of bioaccumulation of methyl mercury up the food chain to the top predator fish, which are the most desired species. Methyl mercury forms largely in the anaerobic sediments of water bodies and then moves up through several trophic levels.

Research in the Everglades and elsewhere has shown that uptake of mercury by fish can occur at low levels of mercury if proper biogeochemical conditions exist in soil and sediment. The total mercury content of soil and sediment is less important than the soil biogeochemical conditions controlling methyl mercury formation.

Wetlands provide a unique interface between soil substrate, water, and biota, which supports various mercury transformations. Methylation of mercury occurs through chemical (abiotic) and biochemical (biotic) processes. Abiotic reactions involve transmethylation and photochemical processes (Ullirich et al., 2001). Biotic processes involve enzymatic and nonenzymatic metabolic methylations by microorganisms (Choi and Bartha, 1993). The relative importance of abiotic versus biotic methylation has not been established but in wetland mercury methylation is primarily a microbial-mediated process (Ullirich et al., 2001).

The conversion of inorganic mercury to organic form in wetlands is governed by microbial methylation. Formation of soluble methyl mercury in anaerobic sediments by obligate anaerobic bacteria suggests that wetlands may serve as a source of bioavailable mercury. Sulfate-reducing bacteria are key participants in the methylation of mercury. The rate of mercury methylation is coupled with the rate of sulfate reduction (King et al., 2001). Methylation is important because it enhances the mobility of mercury and greatly increases its toxicity (Ullirich et al., 2001). The organo-mercurials of environmental interest are monomethyl mercury and dimethyl mercury. Both methylated forms are lipophilic, which causes increased mobility and accumulation in organisms as compared to inorganic mercury.

Because of its low solubility in water, dimethyl mercury will readily volatilize into the atmosphere. However, when monomethyl mercury is formed, it will be released from soils to the overlying water column and is accumulated in living organisms. Monomethyl mercury present in the aquatic environment is taken up by lower organisms and is accumulated readily in the food chain (Ullirich et al., 2001). Flooding of reservoirs that contain ample organic matter has been shown to stimulate bacterial methylation that converts inorganic mercury to methyl mercury and increases the methyl mercury content of zooplankton up to 10-fold (Paterson et al., 1998). This increase is passed up the food chain to the top predator fish used for human consumption (Figure 12.4).

The microbial-mediated processes of methylation and demethylation are influenced by factors such as redox potential, pH, sulfate concentration, and microbial activity. In a study conducted by DeLaune et al. (2004), methylation of added Hg in sediment was greater under reduced conditions

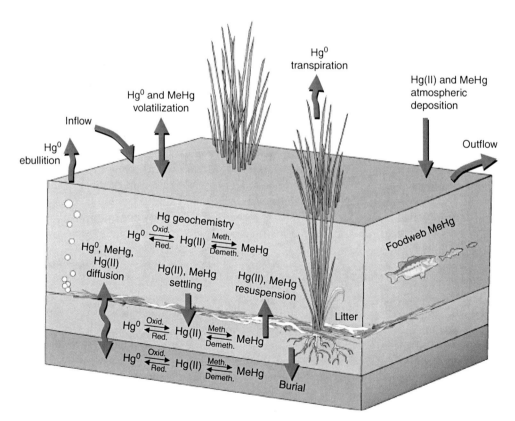

FIGURE 12.4 Conceptual model of mercury cycles in the wetlands. (From Krabbenhoft, D. P., Hurley, J. P., Aiken, G., Gilmour, C., Marvin-DiPasquale, M., Orem, W. H., and Harris, R., *Verh. Int. Verein. Limnol.*, 27, 1657, 2000.)

as compared to oxidized conditions. Further research indicated that 95% of mercury methylation in an estuarine sediment was attributable to anaerobic sulfate-reducing bacteria such as *Desulfovibrio* (Compeau and Bartha, 1985). These researchers also found that high salinity (sulfate) levels inhibited methylation, and methylation increased when sulfate levels were low and concentrations of organic fermentation products were high (Compeau and Bartha, 1987). The presence of sulfide anions allows the formation of HgS. The extremely low solubility of this compound helps prevent methylation under these conditions. The formation of cinnabar (HgS) has been reported in sulfur-rich sediments and could be interpreted as a detoxification process.

Demethylation is another mercury speciation process. There are several pathways for methyl mercury to be demethylated. Oxidative demethylation is a dominant process in anaerobic sediment. Demethylation of methyl mercury (to elemental mercury and methane) was found to occur at higher redox potentials (+110 mV) in estuarine sediments (Compeau and Bartha, 1984). Therefore, demethylation may be the important transformation in the aerobic or moderately aerobic zone in soils and sediments. The dynamics of methylation and demethylation of Hg in sediments may be a key factor in the flux of methyl mercury from bottom sediments to the overlying water column. When methyl mercury is released from anaerobic conditions through a zone of higher redox or aerobic conditions, the methyl mercury may be demethylated to elemental Hg, lowering the flux of methyl mercury to the water column. Because the size of the aerobic layer is dependent on the sediment oxygen demand (organic matter content), the lake sediments with high organic matter content have very small aerobic zones, which may limit demethylation. Little quantitative information is known about the dynamics of these processes, in wetland ecosystem.

12.4 ARSENIC

12.4.1 Sources of Arsenic

Arsenic (As) in the environment includes both natural and anthropogenic sources. Natural sources of arsenic include the weathering of rocks and soils. The geologic history of soils determines arsenic content. Volcanic activity introduces arsenic into the atmosphere as gases, which return to the earth as dust or in precipitation. The yearly contribution of arsenic to soil is small, but has added significant amount to the sedimentary column over geological times. The abundance of arsenic in the continental crust of the earth is in the range of 1.5–2 mg kg^{-1} (Smith et al., 1998).

Anthropogenic sources of arsenic in the environment include contamination from industrial sources and pesticides, medicine, or feed additive. Another important source of arsenic contamination originates from the burning of fossil fuels, coal and petroleum by-products, or as a by-product of the smelting of ores. It is estimated that the emission of arsenic into the atmosphere from anthropogenic sources is 28,060 T year^{-1} (Chilvers and Peterson, 1987). Aquatic organisms exhibit a range in sensitivity to different chemical forms of arsenic. In general, inorganic forms of arsenic are more toxic than organic forms. Arsenite is more toxic than arsenate. Arsenic found in treated lumber used in the construction of piers and bulkheads has also been a source of arsenic to environment. Drinking water from unpolluted source normally contains a small amount of arsenic. However, well water can be contaminated in areas (e.g., Bangladesh) where groundwater is in contact with natural arsenic from minerals.

12.4.2 Dissolution of Primary Minerals

Arsenic present in parent material is usually in the form of chemically reduced minerals such as real agar, orpiment, or arsenopyrite (Oremland and Stolz, 2003). Weathering can oxidize arsenic to arsenite, which is further oxidized to arsenate minerals. As a result, arsenic found in secondary minerals is mainly composed of arsenate. Because of the relative high solubility of metal–arsenite compounds (e.g., Ca, Fe, Mn, and Al), some As(V) or As(III) can be released into aqueous phase through mineral dissolution. The kinetics of the oxidation and reduction of arsenic-bearing sulfide minerals are influenced by several environmental conditions (dissolved oxygen, pH, temperature, carbonate, etc.). In general, the dissolution of arsenic from sulfide minerals is a slow process that can last for thousand of years.

Attention has been focused on arsenic in wetlands and aquatic environment for some time because of the intensive past use of arsenical compounds as pesticides that enter wetland as a result of agricultural runoff.

12.4.3 Biotransformation

Microorganisms in soils and natural waters have significant implications for the speciation and behavior of arsenic. Many bacteria, fungi, and algae organisms are capable of reducing or oxidizing arsenic. A comprehensive review on biotransformation of arsenic can be found in the book *Environmental Chemistry of Arsenic* by Frankenburger (2001). Macur et al. (2004) showed that bacteria capable of either oxidizing As(III) or reducing As(V) coexist and are ubiquitously present in the soil environment.

12.4.3.1 Thermodynamics

Eh and pH diagrams are commonly used to describe the distribution of arsenic species under various soil conditions (Figure 12.5). Bohn (1976) provided an Eh–pH diagram of solid and gaseous arsenic state in the presence of oxygen, water, and sulfur. At equilibrium As_2O_5, As_4O_5, and As_2S_3 are stable solids, whereas H_3AsO_4, $HAsO_2$, and $As_2S_3^{3-}$ are stable solution species over the range of possible soil redox conditions. Reducing soil conditions (Eh < 0 mV) greatly enhances the solubility of arsenic, and the majority of soluble arsenic is presented as As(III).

12.4.4 Oxidation–Reduction

Soil and sediment oxidation–reduction reactions play an important role in determining arsenic solubility, mobility, bioavailability, and toxicity. Under natural environmental conditions, arsenate

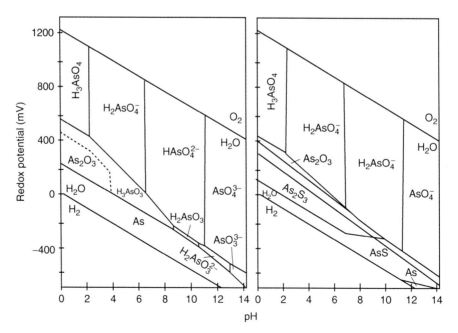

FIGURE 12.5 Eh–pH diagram for part of the As–S–O–H system.

(As(V)) and arsenite(As(III)) are the most abundant forms of arsenic (Smith et al., 1998). In soils and water systems, As(V) is dominant under aerobic condition and As(III) under anoxic and anaerobic conditions. But, because the redox reactions between As(V) and As(III) are relatively slow, both oxidation forms are also found in soils regardless of the pH and Eh (Masscheleyn et al., 1992). Reducing soil conditions (Eh < 0 mV) greatly enhances the solubility of arsenic, and the majority of soluble arsenic is present as As(III).

12.4.5 Reductive Dissolution of Metal Oxides

Under acidic and strongly reducing conditions, iron oxides may be dissolved either biotically or abiotically. Arsenic associated (adsorbed or precipitated) with oxides can be released to the aqueous solution during the iron oxide dissolution (Smedley and Kinniburgh, 2002). Pedersen et al. (2006) observed the release of arsenic during the reduction of ferrihydride, lepidocrocite, and goethite. For ferrihydride and goethite, the release of arsenic with the desorption process was attributed to the reduced surface area of iron oxides during dissolution.

12.4.6 Kinetics of Arsenic Oxidation–Reduction in Soils

Because of the slow kinetics of redox reaction of arsenic, both As(V) and As(III) are often found in the soil environment regardless of the redox conditions. Massacheleyn et al. (1991a) reported the persistence of arsenate under reducing conditions and arsenite under oxidizing conditions, which was attributed to slow reaction kinetics. McGeehan and Naylor (1994) demonstrated that the reduction of As(VI) to As(III) was highly dependent on the sorption processes of arsenic. Manning and Suarez (2000) showed that the heterogeneous oxidation of As(III) to As(V) was controlled by soil properties including pH and content of Al, Fe, and Mn oxides. Takahashi et al. (2004) determined that arsenic was quickly released from flooded soils as a result of reductive dissolution of Fe hydro(oxide) accompanied with the reduction of As(V) to As(III).

12.4.7 Importance of As Speciation

Elemental As is not considered poisonous, but many arsenic compounds are extremely toxic. Arsenic differs in toxicity in different oxidation states. Trivalent arsenic (As(III)) is considerably more toxic than pentavalent arsenic (As(V)). The toxicity of arsenic is not only dependent on its chemical form but also on its solubility and mobility in soils. Understanding the speciation and species transformation of arsenic is essential to determining its toxic effect on plants and animals. The chemical form determines the availability to animals and plants.

In natural waters, soils, and sediments, the As species of interest are the arsenate oxyanions, As(V); the arsenite oxyanions, As(III); monomethylarsonic acid, As(III); and dimethylarsinic acid, As(I). Arsenic chemistry is governed by many factors. The solubility of their salts, the complexing ability of solid and soluble ligands, the biological reactions, the pH and redox potential, and the presence of other ions are all reported to control As concentration and speciation.

Upon flooding, the solubility of As in soil/sediment increases. Under oxidized soil conditions, arsenic solubility is low and most of the As in solution is present as As(V). Upon reduction to As(III) arsenic, solubility increases considerably (Figure 12.6). Information about the behavior of organic arsenicals under reduced conditions is limited. Methanearsonates can be broken down by soil microorganisms with the residual As retained in the soil in its inorganic form under aerobic conditions. Methanearsonates are reduced to alkylarsine form under anaerobic conditions. The reverse of this process (the methylation of As from arsenate and arsenite) can also occur in flooded soils. Under highly reducing conditions (low Eh values), organic arsenical compounds are stable (Smith et al., 1998).

A wide range of interactions including ion exchange, surface complexation, and precipitation contribute to the removal of arsenic from aquatic solution by soil and sediments. The majority of arsenic present in soils is sorbed onto the surface of the solid matrix. Adsorption processes,

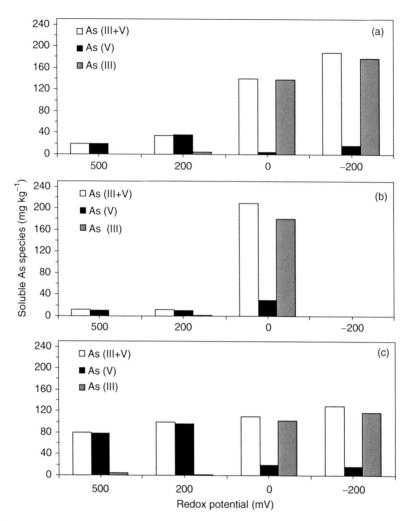

FIGURE 12.6 Distribution of soluble As species under controlled redox conditions. (a) Equilibrations at pH 5.0. (b) Equilibrations at natural pH (5.2 for 500 mV, 6.7 for 200 mV, 7.0 for 0 mV, and 7.2 for −200 mV). (c) Equilibrations at pH 7.5. (Redrawn from Masscheleyn, P. H., DeLaune, R. D., and Patrick, W. H., Jr., *Environ. Sci. Technol.*, 25(8), 1414, 1991.)

especially sorption onto metal oxide surfaces, control the arsenic distribution in contaminated soils (Smith et al., 1998). Soil minerals such as metal oxides, clay minerals, and calcite have the capacity of adsorbing arsenic anions. Iron and aluminum oxides and hydroxides have much higher sorption capacity and stronger bond strength for arsenic than other soil constitutes. Generally, arsenate and arsenite anions are strongly adsorbed as inner-sphere surface complexes through a ligand exchange mechanism. Inner-sphere surface complexes are formed through strong chemical bond between surface functional group and As(V) or As(III) anions without a water molecule between them. Arsenic adsorption on clay surfaces was highly dependent on pH but independent of ionic strength (Goldberg and Glaubig, 1988).

12.4.7.1 Competition with Other Anions

Many anions found in soils can compete with arsenic for adsorption sites by ligand exchange mechanism. The most competitive anion is phosphate, because arsenate and phosphate have similar mechanical properties. The presence of phosphate can substantially suppress the sorption of arsenate on minerals and soils (Smith et al., 1998).

Arsenate and phosphate are specifically adsorbed on a similar set of surface sites, although evidence showed some sites are only available for either As(V) or P (Hingston et al., 1967). Both As(V) and As(III) can be adsorbed onto the surface of Fe and Al oxides by forming inner-sphere complex. These two arsenic oxyanions compete with each other for adsorption sites.

12.4.7.2 Coprecipitation with Metal Oxides and Sulfide

Generally, precipitation contributes only a small portion of the arsenic retention except in highly contaminated soils (e.g., soils around acid mines). If present at very high concentrations, direct precipitation or coprecipitation of arsenic with solid-phase Al, Fe, Mn, Mg, and Ca can occur. For example, Masscheleyn et al. (1991a) indicated that under reduced conditions, the formation of $Mn_3(AsO_4)_2$ may control dissolved As(V). Under highly reduced conditions and in the presence of sulfides, arsenic sulfide precipitation may occur.

12.4.8 CHEMICAL OXIDATION AND REDUCTION OF ARSENIC

12.4.8.1 Oxidation by Metal Oxides

There is considerable evidence that mineral surfaces can play an important role in the transformation between As(V) and As(III). Manganese (IV) oxides can effectively oxidize As(III) to As(V).

12.4.8.2 Reduction by Sulfides

Under highly reducing conditions, sulfide minerals can reduce As(V) to As(III). Rochette et al. (2000) showed that arsenate reduction by dissolved hydrogen sulfide is a rapid reaction that follows second-order kinetics with rate constant, $k = 8.9 \times 10^{-2}$ s^{-1}, which is more than 300 times greater at pH 4 than at pH 7.

12.5 COPPER

Copper is an essential element, being active in many enzymes and hemocyanin. Copper is an essential nutrient element to animals and plants. However, high Cu accumulation in animals and plants can be toxic. Copper is found in three oxidation states including cupric (+2), cuprous (+1), and elemental Cu (0). Cu^+ and Cu^{2+} are the most important forms and are involved in oxidation–reduction reactions in soils and sediments (Figure 12.7). Cu^+ and Cu^{2+} can exist in aqueous systems, although the latter is much more dominant. Copper is widely distributed in nature in its elemental state and in the form of sulfide, arsenite, chloride, and carbonates. The earth's crust on an average contains approximately 50 ppm copper. Soil and sediment contain approximately 20–50 ppm of copper. Normal concentration of copper in plants is between 20 and 50 ppm. The copper toxicity is proportional to the concentration of Cu^{2+} rather than total Cu in aquatic ecosystems. Cu ranges as low as 40 μg L^{-1} in water have been found to be toxic to many fishes. The common Cu concentration in soil solution ranges from 25 to 140 μg L^{-1}, whereas the Cu found in groundwater could be 10–2,800 μg L^{-1}. The 3 μg L^{-1} of Cu concentration has been considered as the level of reference for freshwater, whereas Cu found in "typical" seawater is about 0.9 μg L^{-1}. However, the Cu concentration in estuaries can be high due to contributions from industrial sites and municipal waste treatment facilities. The Cu in surface water is often strongly associated with organic colloids so that almost no free Cu ion concentration would exist when dissolved or particulate organic carbon (POC) is present. The Cu associated with organic colloids is also significantly impacted by the salinity level of a water body such as a lake due to relatively easy flocculation of Cu-associated colloidal phases.

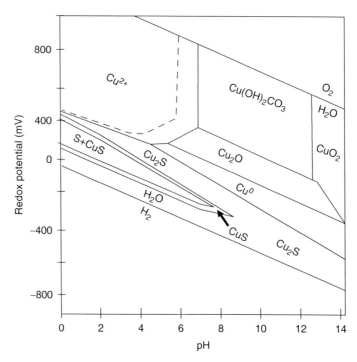

FIGURE 12.7 Eh–pH diagram for the $Cu-H_2O-O_2-S-CO_2$ system at 25°C and 1 atm pressure ($Cu^{2+} = 10^{-6}$ M, $P_{CO_2} = 10^{-3.5}$ M, total dissolved sulfur $= 10^{-1}$ M). (Adapted from Garrels, R. M. and Christ, C. L., *Minerals, Solutions and Equilibria*, Harper and Rowley, New York, 1965.)

The stability fields of various forms of copper can be predicted from thermodynamic considerations and an Eh–pH diagram (Figure 12.7). The system includes copper, water, oxygen, sulfur, and CO_2 or carbonate. In acid-oxidizing area of the diagram, we note the stable field of cupric ion. Generally, these conditions can be found in acid mine drainage, where there is copper and the pyrite has been oxidized to sulfuric acid. It is rare to find copper in its cuprous form above about 10^{-7} to 10^{-8} M, as it is not stable. Various other stability fields, including oxides and hydroxides, dominate the copper Eh–pH diagram. It should be emphasized, however, that the diagram is strictly an inorganic system. If we include the chelation of copper by organic materials, we put another complete dimension of complexity into the system. Even if the critical redox potential for the reduction of cupric to cuprous is identified, secondary reactions that remove copper from solution ensure that very little copper is present in ionic forms. These secondary reactions include sorption–complexation reactions with organic matter and to some extent adsorption by the negatively charged clay particles. The natural changes in redox and pH, however, can alter the make-up of these secondary reactions and affect the bioavailability of copper.

Secondary reactions of metals result in various forms in which toxic heavy metals (including copper) occur in soils. The first phase, and probably the smallest as far as copper is concerned, is dissolved free cations (e.g., cuprous or cupric). This is a small pool for copper because it is one of the most prone cation to be complexed with other systems. Copper cations include the copper–EDTA complex or more importantly, copper complexed with the soluble DOM, a fraction of soil, which is very reactive to copper. In this case, the free ions exist as a positive charge in the soil solution of the soil–water system. When they are complexed, they are surrounded usually by ions of charged or uncharged materials. The positive charge on the copper is neutralized and sometimes results in a negative charge. These complexes can be differentiated in several ways. The simplest way is to pass the solution through a cation exchange resin, which is negatively charged, designed to adsorb cations. Negatively charged materials such as Chelex 100 are common resins used for this purpose. Theoretically, the uncharged or negatively charged complexed material will pass through the resin and the positively charged copper will be absorbed. This can be used to differentiate between complexed and ionic dissolved cations. The dissolved complexed metal (e.g., copper) can be thought of as occurring in a position close to several of the reactive groups in organic material such as the carboxyl group. In a reducing environment, these complexed metal–organic matter systems are difficult to oxidize. Bacteria cannot easily attack this humic acid fraction in anaerobic sediments. Therefore, the complexes formed here are much more stable than in oxidized environments where organic material can be broken down into smaller units that lose their complexing ability. Anaerobic systems that have high concentrations of complexing DOM in solution are very important in regulating the bioavailability of copper.

Other potential forms of copper include sulfide, hydroxide, oxide, and carbonate forms. Under reducing conditions, copper sulfide is formed. Sulfides are stable under anaerobic conditions. Sulfides can be oxidized with electron acceptors of higher reduction potentials by sulfide-oxidizing bacteria to form sulfate, releasing the metal ion back into the solution. Under alkaline conditions, the hydroxide, oxide, or carbonate can be formed so that dissolved copper ions can be removed from the solution and precipitated in this form.

Coprecipitation of copper can occur within the larger mass of ferric oxyhydroxide. Most mineral soils contain significant amounts of active iron, up to 1%. The iron can be reduced to its soluble ferrous form when the system becomes anaerobic or it can be oxidized to the insoluble ferric form when the system is oxidized. As the ferrous ion is precipitated into more insoluble ferric form, it forms coatings on clay and silt particles often as a ferric oxyhydroxide layer, which can potentially coprecipitate trace metals. This is an important sink apparently for copper as well. The copper trapped in this material is released only under reducing conditions.

Copper is another primary metal found in clay particles or primary minerals and can only be released by slow weathering process involving dissolution of the clay particles or primary mineral. This process functions over a timescale of centuries and is unimportant in the bioavailability of copper over the short term.

Studies have documented the effect of different redox conditions on the stability of copper complexes (Reddy and Patrick, 1977). Copper complexed as EDTA was placed in soil suspensions that had been maintained under either oxidized conditions (+500 mV) or under a range of reducing conditions (−100 to −200 mV). Under oxidizing conditions, the copper chelates were stable and remained in solution. As the system became more reducing, however, the copper disappeared from solution (Figure 12.8). Several possible mechanisms explain this observation. One is that as the soil became more reducing, it released large amounts of iron (II) and manganese (II). The ferrous ion, in particular, has a strong affinity for EDTA and the mass action would tend to displace the copper

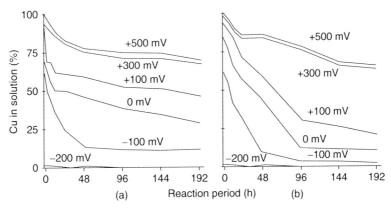

FIGURE 12.8 Effect of redox potential on the stability of (a) copper–EDTA complexes and (b) copper–DTPA complexes. (From Reddy, C. N. and Patrick, W. H., Jr., *Soil Sci. Soc. Am. Proc.*, 38, 66, 1977.)

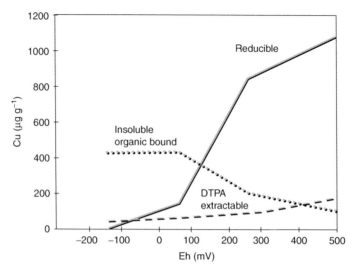

FIGURE 12.9 The effect of redox potential on the distribution of copper among selected chemical forms in Calcasieu River sediment suspensions at pH 8.0 (total copper = 21.4 mg kg^{-1}).

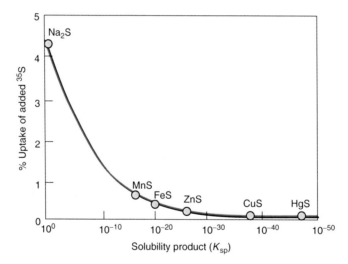

FIGURE 12.10 Percentage uptake of the rice plant of ^{35}S from sulfide salts in relation to the solubility product constant. (Adapted from Engler, R. M. and Patrick, W. H., Jr., *Soil Sci.*, 119, 217, 1975.)

from this system. Another possible mechanism is due to the formation of insoluble copper sulfides at these low redox potentials (−100 to −200 mV). Also, some of the humic acid fractions of soil organic matter may have higher affinity for the copper than the EDTA.

Figure 12.9 shows the effect of oxidizing and reducing conditions (redox potential) on the distribution of copper among several selected forms in river sediment. Under oxidized conditions, the copper was primarily tied up with iron because the extractant used to remove the reducible fraction primarily extracts the iron found in the insoluble ferric form that is microbially reduced to a more soluble ferrous form, which would release bound copper. The insoluble organic copper decreases as the amount of copper coprecipitated with the increase in iron oxidation. The diethylene triamine pentaacetic acid (DTPA) extractable Cu showed little change with redox condition.

Another aspect involved in the release and availability of copper is the formation of sulfide minerals (Engler and Patrick, 1975). Figure 12.10 shows uptake of sulfur by rice plant from several metal sulfides, that is, from sodium sulfide, which is very soluble in water, to manganous sulfide whose K_{sp} is 10^{-14}, to ferrous sulfide, a reduced form of iron ($K_{sp} = 10^{-18}$). Also shown are metal uptakes from some very insoluble metal sulfides including zinc sulfide ($K_{sp} = 10^{-23}$), copper sulfide ($K_{sp} = 10^{-37}$), and mercury sulfide ($K_{sp} = 10^{-50}$), one of the most insoluble inorganic compounds in nature.

In aquatic systems such as lakes, water column oxygen can regulate the availability of copper (Balistrieri et al., 1992). During periods of the year when the water column was aerobic, copper was dominated by $CuCO_3$ and Cu^{2+} species in the water column. When all the oxygen disappeared, the copper was dominated exclusively by a cuprous sulfide form. From this study it is clear that the change in oxygen status from aerobic to anaerobic has some marked effects on the copper chemistry.

The chemistry of copper in water and soils is very complex. Potential chemical forms include insoluble organic complexes, sulfide minerals, and solid copper phases. It is clear that changes in soil redox condition (from oxidized to reduced) change the relative distribution of copper in various soil phases.

12.6 ZINC

12.6.1 Distribution in Soils and Sediments

The average amount of zinc in the earth's crust is a little over 100 mg kg^{-1}. Average zinc concentration in soils and sediment is approximately 30 mg kg^{-1}. In sediments, the reported concentrations are variable and have been found to range from 10 to over 200 mg kg^{-1}. Anthropogenic activities seem to be an important factor in the increase of zinc in the environment. High concentrations of zinc in aquatic environment are detrimental to fish and aquatic life.

Zinc is one of the most mobile of the heavy metals. The zinc compounds formed with the common anions found in surface waters are soluble in neutral and acidic conditions. In reducing environments, zinc sulfide (ZnS) is a relatively insoluble and stable compound, which may oxidize in the presence of dissolved oxygen. Zinc carbonate ($ZnCO_3$) is assumed to be less stable than zinc sulfide, though still relatively insoluble. Zinc ions are dominant up to pH values of about 9 in simple aqueous systems. In basic solutions, zinc hydroxide ($Zn(OH)_2$) precipitates if the concentration of zinc is 10.4 M. Zinc hydroxide shows minimal solubility at pH 9.5 and dissolves at higher pH values as the zincate anion, $Zn(OH)_4^{2-}$.

The redox potential–pH stability diagram (Figure 12.11) indicates that between pH 7 and 8, zinc carbonate ($ZnCO_3$) is formed when the concentration of dissolved carbon dioxide (CO_2) is 10^{-3} mol L^{-1}. At low redox values, zinc sulfide is the most stable combination. Zinc precipitation by the hydrous metal oxides of manganese and iron is the principal control mechanism for zinc in wetland soils and freshwater sediments. The occurrence of these oxides as coatings on clay and silt enhances their chemical activity in excess of their total concentration. The uptake and release of the metals is governed by the concentration of other heavy metals, pH, organic and inorganic compounds, clays, and carbonates.

The cycling of sediment-bound Zn among various geochemical forms is strongly influenced by changes in pH and oxidation–reduction (redox) potential in sediment–water systems (Khalid et al., 1978). A study was conducted on Mississippi River sediment material under conditions of controlled pH (5.0, 6.5, and 8.0) and redox potential (−150, 50, 250, and 500 mV) to determine the effect of these parameters on chemical forms and distribution of added zinc. The results of this study indicate that adsorption by or coprecipitation with oxides and hydroxides of iron and manganese was the important regulatory process governing the availability of zinc in this sediment–water system.

Retention of added zinc by sediment solids was 56–60% at pH 5.0, >97% at pH 6.5, and essentially 100% at pH 8.0. Most of the zinc was present in the exchangeable and reducible fractions. Only a small proportion of added zinc was associated with insoluble organic material, as evidenced by the low recovery in DTPA and residual organic fractions (Table 12.3). The reducible phase is believed to consist of Zn strongly adsorbed to or coprecipitated with oxides and hydroxides of iron

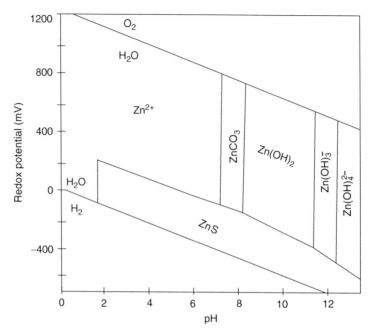

FIGURE 12.11 Eh–pH stability diagram for Zn.

TABLE 12.3
The Effect of pH and Redox Potential on the Chemical Form and Distribution of Added Zinc in Mississippi River Sediment Suspensions

	% Zn Recovered at pH 5.0				% Zn Recovered at pH 6.5				% Zn Recovered at pH 8.0			
	Redox Potential (mV)											
Fraction	−150	50	250	500	−150	50	250	500	−150	50	250	500
Total water soluble	41.6	41.1	43.6	39.9	0.3	2.9	2.1	1.8	0.1	0.2	0.1	0.1
Soluble complexed	0.5	0.7	0.3	0.3	0.3	2.9	2.1	1.8	ND	ND	ND	ND
Exchangeable	45.4	52.0	40.5	46.2	20.2	23.3	24.9	25.1	1.2	3.8	1.6	0.9
Reducible	13.9	19.6	16.8	22.6	81.3	85.3	73.1	81.4	97.0	88.0	91.6	108.2
DTPA extractable	0.0	0.0	0.0	0.1	1.7	1.5	1.7	1.9	2.3	2.1	2.2	2.7
Residual organic bound	4.4	0.4	0.6	0.3	4.5	1.9	4.1	1.8	0.1	0.0	0.0	0.0

ND, not determined.
Source: Modified from Khalid, R. A., Gambrell, R. P., and Patrick, W. H., Jr., in *Environmental Chemistry and Cycling Processes*, U.S. Department of Energy, 1978.

and manganese and is considered potentially bioavailable. Hydrous oxides of iron and manganese are present as partial coatings on the silicate minerals in soils and freshwater sediments and that their large surface area results in a high potential for Zn adsorption.

12.7 SELENIUM

Selenium is a rare element in the earth's crust, but is of considerable interest due to its ready incorporation in the food chain. Selenium is most frequently found in base metal ores of lead, copper, and nickel. Selenium abundance in the earth's crust is estimated to be 0.05 mg kg^{-1}, and selenium

Metals/Metalloids

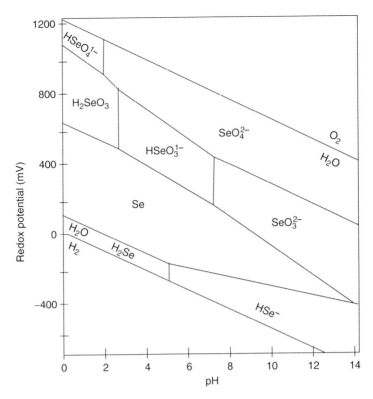

FIGURE 12.12 Eh–pH stability diagram for Se–O–H system. The activity of dissolved Se = 10^{-6}. (Modified from Brookins, D. G., *Eh–pH Diagrams for Geochemistry*, Spring-Verlag, Heidelberg, 1988.)

content in igneous rocks ranges from 0.004 to 1.5 mg kg^{-1}. The concentration of selenium in seawater is reported to be 3.0 µg L^{-1}. The Eh–pH diagram for selenium has large stability field for HSeO^{3-}, SeO$_2^{2-}$, and SeO$_4^{2-}$ (Figure 12.12).

Environmental pollution by selenium results from burning of fossil fuel and trash, mining, and also as a contaminant in the air with sulfur dioxide. Selenium concentration in river water in the United States is normally less than 0.5 µg L^{-1}. However, certain alkaline streams draining seleniferous lands in the western United States contain elevated levels of selenium.

The forms and concentration of selenium in the soil solution are governed by various physical–chemical factors expressed in terms of chemical ligands, pH, dissociation constants, solubility products, and oxidation–reduction states. In acid soils (pH 5–6.5), selenium is usually bound as a basic ferric selenite of extremely low solubility and is essentially unavailable to plants. In alkaline soils (pH 7.5–8.5), selenium may be present as selenate ions and become water soluble.

Selenium can exist in four different oxidation states: selenide (Se(II)), elemental Se (Se(0)), selenite (Se(IV)), and selenate (Se(VI)). The chemistry of selenium is comparable to that of another Group VIA element, sulfur. Again, major factors affecting the fate and transport of selenium in soils and sediments are its oxidation–reduction reactions and the subsequent effects on retention processes (e.g., sorption). Studies completed by Masscheleyn et al. (1991b) on contaminated Kesterson reservoir sediment document some important features of Se geochemistry, including the effect of Eh (Figure 12.13). The hydride generation/trapping/detection apparatus utilized for selenium analyses distinguishes between selenate (Se(VI)), selenide (Se(IV)), dimethyl selenide, oxidized methylated Se compounds, and reduced selenium (Se(0–II)) (Masscheleyn et al., 1991b).

The solubility of Se under reduced conditions is lower (controlled by an iron selenide phase) than under oxidizing conditions. When soil Eh or redox potential is increased, oxidation of Se(II, 0)

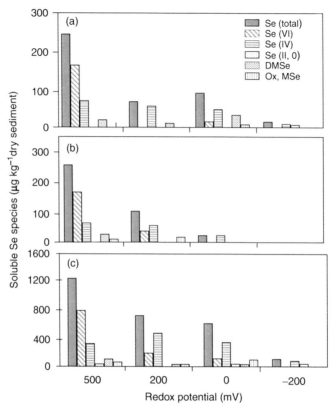

FIGURE 12.13 Distribution of soluble Se species under controlled redox conditions. (a) Equilibrations at natural pH (4.0 for 500 mV, 5.3 for 200 mV, 6.1 for 0 mV, and 6.9 for −200 mV). (b) Equilibrations at pH 5.0. (c) Equilibrations at pH 7.5. Note changes in scale. (Redrawn from Masscheleyn, P. H., DeLaune, R. D., and Patrick, W. H., Jr., *Environ. Qual.*, 20, 1991.)

to selenite (Se(IV)) is rapid (timescale of several days). Subsequent oxidation to the more oxidized selenate (Se(VI)) species is much slower (i.e., several weeks). The critical Eh for the selenium transformations were +50 mV for the oxidation of Se(II, 0) to selenide (Se(IV)) and +200 mV for the oxidation of selenide to selenate (Se(VI)). The microbial biomethylation of Se species is important under oxidized and moderately reduced conditions (500, 200, and 0 mV).

Selenium redox chemistry demonstrates several features regarding changes in metal speciation in response to changes in Eh or redox condition. Differences in kinetics of transformation between different species affected the predictability of speciation. Release of high percentages (e.g., >50%) of total selenium into soluble forms was observed when sediment was oxidized. Conversion of selenium to organic forms was also observed. Dramatic differences in the solubility of oxidized species such as the selenium oxyanions and the reduced species (metal selenides) emphasize the importance of considering speciation changes in the biogeochemistry of this metalloid.

12.8 CHROMIUM

Chromium content varies with the textural composition of soils and sediments. Sandy soils and sediments contain lower chromium content than fine textured soils and sediments. Mean chromium content of the soil in the United States and Western Europe is reported to be 53 and 56 mg kg^{-1}, respectively. When its concentration reaches 0.1 g kg^{-1} body weight, chromium can ultimately become lethal.

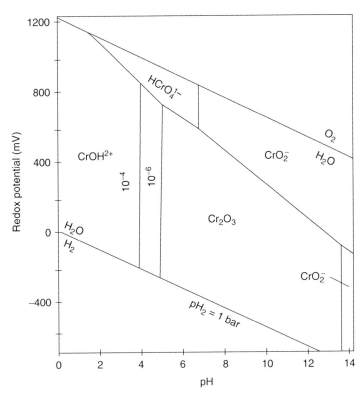

FIGURE 12.14 Eh–pH stability diagram for chromium.

The Eh–pH diagram for chromium is shown in Figure 12.14. Much of the diagram is occupied by Cr_2O_3, which dissolves to form $Cr(OH)^{2+}$ at pH values below 5.

Chromium can exist in several oxidation states from Cr(0), the metallic form, to Cr(VI). The most stable oxidation states of chromium in the environment are Cr(III) and Cr(VI). Besides the elemental metallic form, which is extensively used in alloys, chromium has three important valence forms. The trivalent chromic (Cr(III)) and the tetravalent dichromate (Cr(VI)) are the most important forms in the environmental chemistry of soils and waters. The presence of chromium (Cr(VI)) is of particular importance because in this oxidation state Cr is water soluble and extremely toxic. The solubility and potential toxicity of chromium that enters wetlands and aquatic systems are governed to a large extent by the oxidation–reduction reactions. In addition to the oxidation status of the chromium ions, a variety of soil/sediment biogeochemical processes such as redox reactions, precipitation, sorption, and complexation to organic ligands can determine the fate of chromium entering a wetland environment.

The speciation of Cr(VI) and Cr(III) is regulated by the oxidation–reduction status of soils and sediments. Masscheleyn et al. (1990) determined the rate of Cr(VI) reduction as affected by the soil redox status. Soil suspensions amended with Cr(VI) were equilibrated under controlled redox conditions (−200, 0, +200, and +500 mV). Figure 12.15 shows the critical soil redox values for Cr(VI) reduction. Cr(VI) reduction occurred at approximately the same redox levels as nitrate. Water-soluble Cr(VI) decreased with a decrease in Eh values from 500 to 300 mV.

Dissolved oxygen can oxidize Cr(III) to Cr(VI); however, the oxidation is slow at normal temperature in aqueous environments. Due to the slow oxidation, Cr(III) is involved in faster sorption and precipitation reactions. Similar results were obtained by Masscheleyn et al. (1992) investigating Cr(III) oxidation as affected by the soil redox potential. In this reported soil study, Cr(III) was

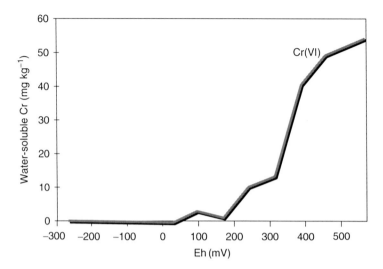

FIGURE 12.15 Critical redox level for chromate reduction.

added at a rate of 50 mg of Cr kg^{-1} of dry soil to a reduced (−200 mV) soil suspension. After 14 days of equilibration, a soil suspension was oxidized stepwise in 100 mV increments, to +500 mV, and dissolved Cr species were determined at preselected soil redox level. The stepwise oxidation of the soil suspension containing Cr(III) did not result in the production of Cr(VI) until +200 mV. Cr(VI) reduction dominated over the Cr oxidation reaction. The lack of Cr(III) oxidation in the soil suspension was attributed to the precipitation (as Cr(OH)$_3$ or Cr$_2$O$_3$) or sorption in the soil. Cr(III) sorption–precipitation reactions were too rapid to measure, being complete in less than 1 min.

Trivalent chromium is readily oxidized to the hexavalent form under select conditions prevalent in wetland soils. The major reaction by which Cr(III) is oxidized to the Cr(VI) form is chemical oxidation by manganese dioxide (MnO$_2$) (Kim et al., 2002). As oxygen (O$_2$) and nitrate (NO$_3^-$) are higher on the redox scale than MnO$_2$, it is unusual that MnO$_2$ is more effective than either of these better oxidants in oxidizing Cr(III). The reason is that the capacity of MnO$_2$ to chemically oxidize Cr(III) is rapid, whereas NO$_3^-$ and O$_2$ are largely involved in slower biological oxidation reactions.

The oxidation of Cr(III) to Cr(VI) has been reported in the water column overlying reduced sediment by Masscheleyn et al. (1992). Adsorption/precipitation reaction was slow enough to deter increase in Cr(VI) in the water column spiked with 1 and 10 mg L^{-1} Cr(III). In this study, the majority of the Cr(III) initially added to the water column was precipitated, but some remained in solution as Cr(III)–organic ligand complex. The observed oxidation was as rapid with 12 and 57% of the 1 and 10 mg L^{-1} added Cr(III) being oxidized within 24 h. After 1 day, the Cr oxidation rate was exceeded by Cr(VI) removal (adsorptive difference). Table 12.4 depicts the quantities of Cr(VI) measured with time.

The solubility of chromium is strongly dependent on its oxidation state, hexavalent chromium being more soluble than Cr(III). Cr(III), like other cationic metals, is rapidly adsorbed by soil Fe and Mn oxides and clay minerals. Generally, added chromium is sorbed within 24 h by iron oxides and clay minerals in soil. Cr(III) adsorption increases with pH.

The soil redox status did not affect the capacity of the soil to retain Cr(III). An increase in the Cr(III) content of soil from 0 to 1,500 mg Cr kg^{-1} did not affect the amount of water-soluble Cr in soil suspension maintained at +500, +200, 0, and −200 mV (DeLaune et al., 1998).

Cr(III) is more strongly adsorbed to Mn, Al, and Fe oxides and clay and organic colloids than Cr(VI). Adsorption of hexavalent chromium is reported to be a surface reaction between aqueous chromates and hydroxyl-specific surface sites. The reaction is pH dependent with Cr(VI) adsorption being favored on adsorbents that are positively charged at low to neutral pH.

TABLE 12.4
Removal of Chromium from Floodwater of Soil Water Core Spiked with Cr(III)

	Soluble Cr (mg L^{-1})			
	1 mg Cr(III) L^{-1} Added		10 mg Cr(III) L^{-1} Added	
Days	Total Cr	Cr(VI)	Total Cr	Cr(VI)
0.25	0.54	0.12	5.5	4.0
3.00	0.25	0.07	6.0	3.6
6.00	0.19	0.05	4.0	2.2

Source: Selected data from Masscheleyn, P. H., Pardue, J. H., DeLaune, R. D., and Patrick, W. H., Jr., *Environ. Sci. Technol.*, 26, 1217, 1992.

Competing anions have a drastic effect on Cr(VI) adsorption. Chromate sorption in the natural environment is highly influenced by electrostatic conditions imposed by common anions (Cl$^-$, NO$_3^-$, SO$_4^{2-}$, HCO$_3^-$, etc.), which are bound to the soil or sediment surface. Accordingly, adsorption of Cr(VI) is minimal in soil.

In contrast to Cr(III), the soil redox condition strongly influences sorption of Cr(VI). Under oxidized and moderately reduced (+500 to +100 mV) soil conditions, chromium behavior is dominated by Cr(VI) sorption and reduction of Cr(VI) to Cr(III) (DeLaune et al., 1998). Under more reduced soil redox levels (<+100 mV), chromium chemistry and solubility is controlled by the reduction of Cr(VI) by soluble ferrous iron.

Low pH favors rapid reduction of Cr(VI) to Cr(III). For example, James and Bartlett (1983) reported greater reduction of soluble Cr(VI) in unlimed (pH 5.3) soils than limed (pH 6.5) soils. Soil pH can also affect the form of Cr(VI) adsorbed on soil colloids. Below pH 6.4, HCrO$_4^-$ is the dominant chromium form in aqueous solution. Above pH 6.4, HCrO$_4^-$ dissociates to CrO$_4^{2-}$. For their adsorption studies, James and Bartlett (1983) reported that increasing soil pH from 5.4 to 7.0 by liming decreased the amount of chromium removed as phosphate exchangeable form from 7.2 to 2.6 kg ha^{-1}. Cr(III) is generally insoluble except under very acidic condition, below approximately pH 4.0, which occurs only in soils and waters where pyrite oxidation has occurred. For this reason, soluble Cr(III) is usually restricted to acid mine drainage and acid sulfate soils.

Speciation and solubility of chromium in wetlands and aquatic systems is governed by the competition among chromium oxidation states, adsorption/desorption mechanism, and soil/sediment redox–pH conditions. Chromium (VI) is reduced to chromium (III) at approximately +350 mV in soils and sediment. Reduced Cr(III) can be rapidly oxidized to the tetravalent chromate and dichromate forms by manganese compounds. Cr(III) is much less soluble in natural system than the hexavalent form and has a much lower toxicity. Chromium is less likely to be a problem in wetlands than in nonwetlands because the reducing conditions cause its reduction or conversion to the more insoluble Cr(III) form. This is depicted in Figure 12.15, which shows changes in water-soluble chromium as affected by the soil redox potential.

12.9 CADMIUM

Cadmium is a relatively rare metal and its abundance in the earth's crust is estimated at less than 1 mg kg^{-1}. Cadmium is closely related to zinc in its chemical properties and in nature is always found associated with zinc. It is a divalent metal, readily forming halides of which chloride is the most reactive. It is soluble in most inorganic and some organic acids, but insoluble in alkalies. In the presence of sulfide, cadmium precipitates out of solution as cadmium sulfide (Figure 12.16). Cadmium is toxic to living organisms in virtually all of its chemical forms. The environmental presence of cadmium is

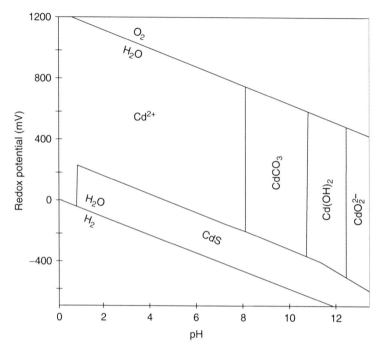

FIGURE 12.16 Eh–pH stability diagram of the Cd–C–S–O–H system. The assumed activities for dissolved species are Cd = 10^{-8}, C = 10^{-3}, and S = 10^{-3}.

TABLE 12.5
Total Cadmium Content of the Sediment Materials from Selected Sites in Louisiana and Alabama

Sediment Material	Total Cadmium (mg Cd kg^{-1} solids)
Barataria Bay	0.41
Mobile Bay	0.34
Mississippi River	0.47
Calcasieu River	0.32

normally linked to that of zinc because of their geochemical kinship and incomplete technical separation. Cadmium enters the environment from smelting of metals other than zinc; attrition of automobile tires; and the combustion of petroleum products, coal, wood, paper, and urban organic trash. The cadmium concentration of seawater is reported to vary from 0.075 to 0.32 µg L^{-1}. The average cadmium concentration in the main streams and lakes draining 16 major U.S. watersheds, measured between 1962 and 1967, was 9.5 µg L^{-1}. The cadmium content in Barataria Bay, Mobile Bay, Mississippi River, and Calcasieu River sediment is shown in Table 12.5.

The contamination of wetlands with Cd has been a cause of great concern because biological accumulation may cycle Cd into the food chain. Waterways are important in the accumulation, transport, and geochemical cycling of Cd, because this element, when discharged from industrial sources, becomes associated with sediment solids at the bottom of the waterways. Cd sinks in soils and sediments include high-molecular-weight organics, sulfides, carbonates, clay minerals, and oxides and hydroxides of Fe and Mn. Biological availability of Cd depends on the mobilization

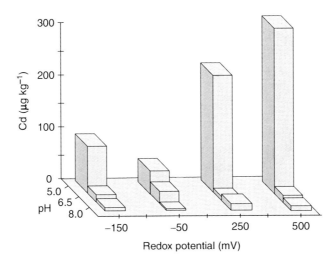

FIGURE 12.17 Effects of pH and redox potential on total water-soluble cadmium in Mississippi River sediment suspensions measured by flameless atomic absorption. (From Khalid, R. A., Gambrell, R. P., and Patrick, W. H., Jr., *J. Environ. Qual.* 10, 523, 1981.)

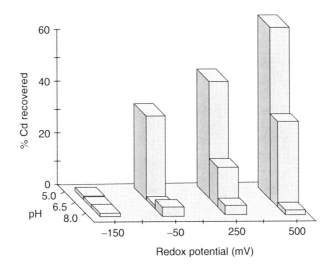

FIGURE 12.18 Effects of pH and redox potential on exchangeable ^{109}Cd in Mississippi River sediment suspensions. (From Khalid, R. A., Gambrell, R. P., and Patrick, W. H., Jr., *J. Environ. Qual.* 10, 523, 1981.)

of Cd from various sinks and is affected by several environmental factors. Sediment oxidation–reduction potential (Eh) and pH are probably the most important physicochemical parameters influencing cadmium transformation and availability to biota (Figures 12.17 and 12.18). The changes in the physicochemical conditions occurring in soils and sediments can increase chemical mobility and hence bioavailability of Cd. Cadmium is more soluble under oxidized and acidic conditions.

12.10 LEAD

The average concentration of lead in the earth's crust is about 15 ppm by weight. Lead is the most abundant of the heavy elements having atomic numbers greater than 60. The relative abundance of lead is because its three predominant isotopes are the end products of naturally occurring neutron

TABLE 12.6
Total Lead Content of the Sediment Materials from Selected Sites in Louisiana and Alabama

Sediment Material	Total Lead (μg Pb g^{-1} solids)
Barataria Bay	37
Mobile Bay	36
Mississippi River	44
Calcasieu River	53

capture processes, which result in the formation of lead from the radioactive decay of other elements. The slow radioactive decay of uranium and thorium is reported to account for about a third of the lead currently found in the earth's crust.

Prior to the use of unleaded gasoline, the combustion of lead-containing fuel was reported to be the primary source of lead in the atmosphere. Much of the lead emitted by automotive exhaust fell on or near the roadway and through rainfall was transported to adjacent stream and water bodies. Much of the lead is in fine particulate form and is dispersed as an aerosol over a wide area by wind. The lead is eventually removed from the atmosphere by precipitation or aggregation and falls on the earth's surface. Thus, soils serve as an intermediate recipient of airborne lead deposits. Soil lead may cycle through wetland plants and enter aquatic food webs, or it may enter waterways through surface and subsurface drainage. This is in addition to lead entering waterways via industrial and municipal waste discharges and direct fallout into water from lead-containing aerosols. As for many other potentially toxic materials, soils and sediments are the final recipient of much of the lead discharged into the environment. Unless there is a nearby source of contamination, surface sediment generally contains little lead (Table 12.6).

The chemistry of Pb in natural systems is immensely complex. The solubility product constant (K_{sp}) of lead carbonate, hydroxide, phosphate, sulfide, and sulfate is 10^{-14}, 10^{-15}, 10^{-42}, 10^{-28}, and 10^{-8}, respectively. In the presence of these ligands, lead is only sparingly soluble. Thus, in natural systems, it is usually not possible to explain trace metal levels by a straightforward application of the solubility data. In the lead–water system, lead was shown to exist in the divalent soluble state (presumably, this includes the charged hydroxide of lead, $PbOH^+$) up to a pH of about 8. Above pH 8, the insoluble lead oxide (PbO) would form in the system. At near neutral, lead can also form lead carbonate ($PbCO_3$) (Figures 12.19).

Redox potential–pH diagrams can be expanded to cover more complex systems when the concentration of all components are known. For instance, chloride, sulfate, phosphate, and other ions may complex with lead under specified redox potential–pH conditions. The forms of lead in complex water systems can be determined where the concentrations and chemistry of all components are known. However, in natural sediment–water systems, the factors affecting lead chemistry may be in a dynamic state, and the chemistry of all the components is not known. Such is the case with interactions between organic matter and metals.

The bioavailability of lead is determined to a large extent by the solubility of the metal species. As shown previously, pH and redox potential of sediments can affect the solid phase of metals and metalloids. Organo-metal complexes are also important in determining heavy metal solubility. Organo-metal complexes are generally less toxic than ionic forms of the element. In a study of complexed Pb in large size organic fraction, the majority of added labeled lead was immobilized in the >0.45 μm fraction (Patrick and Verloo, 1998) (Figure 12.20).

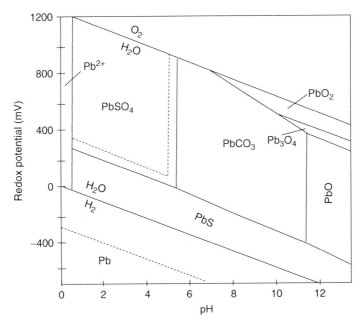

FIGURE 12.19 Eh–pH diagram for the Pb–S–C–O–H system. The activities of the dissolved species are Pb = 10^{-6}, S = 10^{-3}, and C = 10^{-3}.

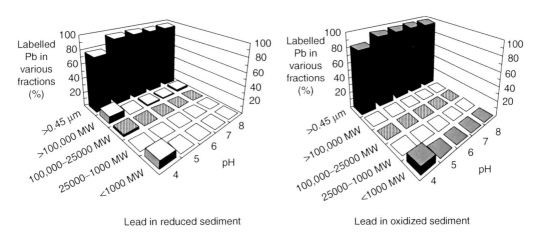

FIGURE 12.20 Size distributions of soluble organo-complex Pb in sediment under different pH and redox conditions. (Redrawn from Patrick, W. H., Jr. and Verloo, M., *Water Sci. Technol.* 37, 165, 1998.)

12.11 NICKEL

Nickel, the 24th most abundant element in the earth's crust, is found in combination with other elements in soils and sediments. Nickel is present in detectable amounts in most types of rocks, and it may be mobilized in the surface environment during the chemical and mechanical weathering of rock to form soil.

In addition to natural source, nickel can be released in the environment by industries that use nickel or nickel alloys and by oil and coal burning power plants. Nickel found in soils and sediments is strongly attached to soil/sediment particles or embedded in mineral component. Sediments generally contain between 5 and 80 ppm nickel. Although Ni has −1, 0, +1, +2, and +3 oxidation states, the most common valence state in the environment is Ni^{2+}. In soils and sediments, it exists primarily as oxides and sulfides.

Most of the nickel in sediments and suspended solids is distributed among organic materials, precipitated and coprecipitated particle coatings, and crystalline particles. In soil and sediment, Ni is preferentially adsorbed on iron and manganese oxides and can substitute for magnesium in the lattice of soil clay minerals. Water solubility and thus bioavailability to plants are affected by soil pH, with decreases in pH below 6.5 generally increasing mobilizing of nickel and other metals.

In aquatic environment, Ni is transported in both particulates and dissolved forms. The pH, oxidation–reduction potential, ionic strength, types, concentration of organic and inorganic ligands (in particular humic and fulvic acids), and the presence of solid surfaces of adsorption (in particular hydrous iron and manganese oxides) can all affect the transport and biological availability. Under reducing conditions and in the presence of sulfur, relatively insoluble nickel sulfide is formed. Under aerobic condition and pH < 9, the compounds that nickel forms with hydroxide, carbonate and sulfate, and naturally occurring organic ligands are sufficiently soluble to maintain aqueous Ni^{2+} concentrations above 60 µg L^{-1} (Callahan et al., 1979). Figure 12.21 shows the Eh–pH diagram.

The microbial activity or changes in some of the physical and chemical parameters described above (e.g., decreasing pH or increasing concentrations of organic ligands) may result in the desorption of nickel from suspended particulate material or sediment into water column (Di Toro et al., 1986).

Nickel does not appear to concentrate to toxic levels in fish and plants at levels normally found in the sediment. There is no significant concentration in wetland unless there are industrial or other source entering the environment.

Microorganisms in sediment can influence the chemistry of nickel entering the aquatic food chain; for example, the reducing condition in sediment and the conversion of sulfate to sulfide can result in the formation of insoluble nickel sulfide.

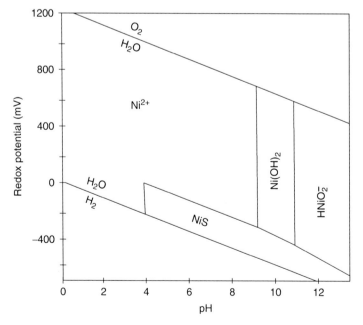

FIGURE 12.21 Eh–pH diagram for the Ni–O–H–S system. Assumed activities for dissolved species are Ni = 10^{-4}, 10^{-6} and S = 10^{-3}.

Benthic organisms accumulate higher concentrations of nickel than pelagic organisms. However, reported bioconcentrations factors for sediment-associated biota are generally low. Acute or chronic toxicity to freshwater organisms is in the range of 24–10,000 µg L^{-1} (Keller and Zam, 1991).

12.12 SUMMARY

- Heavy metals that are toxic to aquatic plants and animals can enter wetlands from several sources. Toxic metals in wetland soils exist in various forms and may undergo numerous transport and transformation processes when they enter wetlands.
- Dissolved metals may be taken up by biota or sorbed to particle surfaces. Metals may dissolve precipitate, desorb, or be involved in redox reactions. The chemical properties, concentrations, and availability determine their fate and toxicity.
- In wetland soils, the amount of organic matter, clay mineral, soil acidity (pH), and sediment oxidation–reduction status govern the solubility and mobility of toxic metals.
- The Eh and pH of wetland soils play an important role in regulating metal transformation and bioavailability.
- The speciation of metal forms found in wetland soil includes
  1. Water-soluble metals including soluble free ions, (e.g., Cr^{6+}) and soluble organic and inorganic complexes
  2. Exchangeable metals
  3. Metals precipitated as inorganic compounds
  4. Metals complexed with high-molecular-weight humic materials
  5. Metals precipitated as insoluble sulfides or carbonates
  6. Metals bound in crystalline lattice of clay mineral and parent sediment

STUDY QUESTIONS

1. List the sources of heavy metals entering wetland soils.
2. What are the key biogeochemical processes that elevated levels of heavy metals can impact?
3. What are the general chemical fractions or forms in which metals are found in wetland soils and sediment? Which forms are more bioavailable? Which forms are effectively immobilized?
4. What are the primary factors governing heavy metal availability and transformation in wetlands?
5. How does clay and organic matter content influence metal solubility and transport?
6. Explain how sediment Eh–pH condition can be used in predicting stability of heavy metals. Provide examples.
7. Which metals/metalloids are subjected to microbial methylation? Explain the difference between biotic and abiotic methylation. What is demethylation?
8. Why is methyl mercury readily accumulated in the aquatic food chain?
9. List the factors governing forms of arsenic found under reducing conditions in sediment.
10. How does sulfate reduction and sulfide production influence heavy metal availability?
11. How does ferric oxyhydroxide influence metal availability? What is coprecipitation? How does changes in soil reduction condition or iron transformation influence metal adsorption or availability?
12. Which oxidation states can selenium exist in soils and sediment? What are the major transformations? List the oxidation states which selenium can exist in sediment.
13. How does sediment reduction status regulate the speciation of chromium? What is the critical redox value for Cr(VI) reduction?

14. Under what sediment condition is cadmium soluble?
15. What is the average concentration of Pb in the earth's crust? List the solubility product constant for lead carbonate, hydroxide, and sulfide. Which form is the least soluble?
16. What is the most common valence state of Ni? What is the primary form of nickel in soils and sediment? Can under normal conditions nickel accumulate to toxic levels in fish and plants in wetlands?

FURTHER READINGS

Adriano, D. C. (ed.). 1992. *Biogeochemistry of Trace Metals*. CRC Press, Boca Raton FL.

Giblin, A. E., G. W. Luther, III, and I. Valiela. 1986. Trace metal solubility in salt marsh sediments contaminated with sewage sludge. *Estuar. Coast. Shelf Sci.* 23:477–498.

Griffin, R. A., A. K. Au, and R. R. Frost. 1977. Effect of pH on adsorption of chromium from landfill-leachate by clay minerals. *J. Environ. Sci. Health* A12:431–449.

Hebert, E. A., A. W. Garrison, and G. W. Luther, III (eds.). 1998. *Metals in Surface Waters*. Ann Arbor Press, Chelsea, MI.

Masscheleyn, P. H. and W. H. Patrick, Jr. 1993. Biogeochemical processes affecting selenium cycling in wetlands. *Environ. Toxicol. Chem.* 12:2235–2243.

13 Toxic Organic Compounds

13.1 INTRODUCTION

Approximately 70,000 different organic compounds are now used in the day-to-day life of human beings. These compounds include pesticides, fungicides, herbicides, industrial compounds, explosives, dyes, phenols, organic compounds in landfill leachate, and petroleum hydrocarbons. These compounds are often referred to as xenobiotics (xeno means foreign; bios means life) as they are foreign to life. Xenobiotic compounds are chemically synthesized and do not exist naturally in biological systems. The most common and widely distributed xenobiotics are pesticides. These compounds are often referred to as toxic organic compounds. These toxic organic compounds when found at levels higher than background can be toxic to biotic communities. They are also referred to as organic pollutants (Alexander, 1994). Recent advances in analytical techniques and the use of advanced instrumentation have aided in detecting low levels of toxic organics in the environment. Toxic organics can occur as single compounds or as complex mixtures. Widespread pollution of soils and surface and groundwaters occurs as a result of improper use and disposal, and accidental release of organic chemicals into the environment. In recent years, numerous strategies and technologies have been developed for remediation of contaminated areas.

Toxic organic compounds enter wetlands and aquatic systems in a number of ways. Concentrations of many organic compounds in wetlands and aquatic systems are low (approximately in the order of 1 µg L^{-1}), but even at these concentrations some of these compounds can be toxic to biotic communities. To evaluate organic chemicals in wetland environments, one must understand factors that influence the translocation, transformation, and persistence of the chemical. Many processes affect transformation, transport, and accumulation of organic chemicals in wetlands. These include both biotic and abiotic transformations. Abiotic transformations in wetlands may include photochemical reactions in water, or soil surface, volatilization, sorption onto mineral and organic materials, and partitioning onto colloidal organic matter in the water column and the soil surface. Exchange between water column and the soil, plant uptake, and burial of the contaminant in the soil profile are also important. Biotic transformations are probably most effective in degrading these compounds to inorganic forms. These transformations involve enzymes that act as catalysts and increase the rate of biochemical reactions. The capacity of microbes to transform organic compounds for remediation of contaminated sites is often referred to as biodegradation. Both abiotic and biotic transformations of organic compounds occur simultaneously in wetland environments. However, it is difficult to estimate the relative importance of abiotic transformations under conditions where both biotic and abiotic transformations occur. Schematic of processes governing retention of toxic organics in wetlands is shown in Figure 13.1.

Major sources of organic pollutants entering wetlands and aquatic systems are agricultural and urban runoff, pesticide and herbicide usage in agricultural ecosystems, herbicides used to control aquatic weeds and algae, landfill leachate and industrial chemical spillage, residual pesticides in agricultural lands converted into wetlands, and atmospheric deposition. Agricultural runoff is a major source of nonpoint source pollutants to wetlands. Wetlands are often the receiving bodies of nonpoint source agricultural pollution, which includes pesticides associated with runoff from adjacent farmland. The extent of contamination in wetlands depends on proximity relative to input source. It is estimated that pesticides impact approximately 5,000 wetlands and aquatic systems in the United States. Usage of conventional pesticides on U.S. farms was about 350 million kg

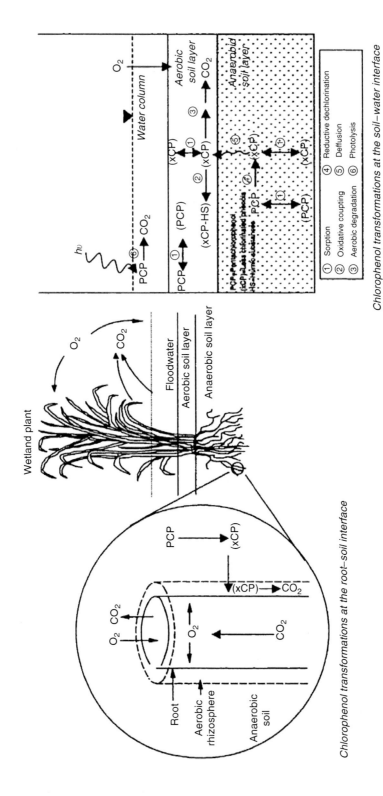

FIGURE 13.1 Processes governing retention of toxic organics in wetlands (D'Angelo, 1998).

(active ingredient basis) in 1995. The most widely used herbicide is atrazine (32 million kg) followed by metolachlor (28 million kg) and alachlor (about 10 million kg) (USEPA, 1997). The U.S. Environmental Protection Agency has set maximum acceptable levels for most pesticides in drinking water but no set standards for pesticides on aquatic life are currently available. Pesticides can affect aquatic life in wetland and estuarine ecosystems (USEPA, 1996). Over 40 pesticides have been detected in the Mississippi River and most of the pesticides came from agricultural states (Antweiler et al., 1995). Atrazine and metolachlor were detected in more than 95% of the river samples collected. Alachlor was detected in 75% of the samples analyzed.

Mississippi River water contains levels of s-triazine that are of environmental concern (USEPA, 1994). The source of the herbicide is nonpoint runoff from the vast agricultural area the Mississippi River drains. Atrazine loads up to 2,000 kg day^{-1} in the Mississippi River have been reported in 1989 and 1992 near the river mouth (Pereira et al., 1990; USGS, 1994). During a 1-year period, discharge of atrazine and alachlor to the Gulf of Mexico was estimated at 365,700 and 33,700 kg, respectively (Goolsby et al., 1993). Very little is known about concentrations and total amounts of pesticides entering wetland areas from numerous watersheds, which receive agricultural runoff. Atrazine, metolachlor, and alachlor, and many other pesticides, are used extensively as preemergence and postemergence herbicides to control weeds. Atrazine and alachlor represent more than 45% of total herbicide application in coastal U.S. counties (Aelion and Cresi, 1999). Pesticides in the Mississippi River water pose a potential threat to the Gulf Coast ecosystems receiving water from this river. Atrazine can be phytotoxic to submerged vascular plants at 5 ppb (Kemp et al., 1985). Alachlor and metolachlor can also be toxic at ppb levels (Fairchild et al., 1998). McMillin and Means (1996) reported that atrazine was ubiquitous over the entire northwestern Gulf of Mexico coastal shelf in spring, summer, and fall. Dissolved atrazine as high as 1,000 pg mL^{-1} was measured in water off the Louisiana coast.

The recovery, processing, and transport of petroleum hydrocarbons have also resulted in spillage or pollution in both freshwater and coastal wetlands. Although much attention to oil pollution in wetlands has been associated with major spills, chronic continuing input from municipal/industrial sources and storm water runoff from urban areas are also sources of petroleum hydrocarbons and other toxic organics to drainage and watersheds that ultimately empty into wetlands.

Once in wetland environments, organic compounds are subjected to biological and nonbiological transformation processes. Among biological processes, microbial metabolism is the primary force in transformation and degradation of organic compounds. In many cases, microbes are more important in the degradation of organic compounds than are physical or chemical mechanisms. Table 13.1 lists common organic compounds considered toxic to biotic communities that are found in wetland environment. Basic chemical structures of selected organic compounds are shown in Figure 13.2.

Several abiotic and biotic processes that degrade naturally occurring dissolved and solid organic matter are also capable of degrading toxic organic compounds ranging from photochemical reactions such as photolysis to biological reactions such as mineralization (conversion of the contaminant to carbon dioxide). These reactions are important because they generally serve to destroy contaminants, converting them to more innocuous forms. A list of some important reactions is presented in Table 13.2.

Wetlands have the potential for degradation of toxic organics. Wetlands have the potential for a variety of reactions that may not occur in other systems. They provide aerobic–anaerobic interfaces that can impact the degradation of certain compounds. Reduced soil conditions are known to enhance the degradation of organic compounds such as pentachlorophenols (PCPs) and polychlorinated biphenyls (PCBS). However, biodegradation of organic compounds such as benzene, octane, toluene, xylene, and polycyclic aromatic hydrocarbons (PAHs) derived from petroleum products is enhanced under aerobic conditions. Alternating aerobic and anaerobic conditions often provide an ideal environment for enhanced biodegradation of organic compounds. Wetland substrates are characterized by a wide range of redox zones (aerobic to anaerobic conditions). Certain degradation reactions are favored under anaerobic conditions (e.g., reductive dechlorination) and others under

TABLE 13.1
Organic Compounds Toxic to Wetland Biotic Communities

Aromatic compounds	Halogenated compounds	Halogenated aromatic compounds
Benzene	Carbon tetrachloride	Polychlorinated biphenyls
Toluene	Chloroform	
Napthalene	Vinyl chloride	Organochlorine insecticides
Naphthol	1,2-Dichloroethane	DDT, toxaphene
Phenol	Trichloroethylene	Chlorinated herbicides
Biphenyl	Tetrachloroethylene	2,4-D, 2,4,5-T
	Benzoates	Chlorinated phenols
		Pentachlorophenol
		2,4-Dichlorophenol
		2,4,5-Trichlorophenol
		2,3, and 4-Nitrophenol

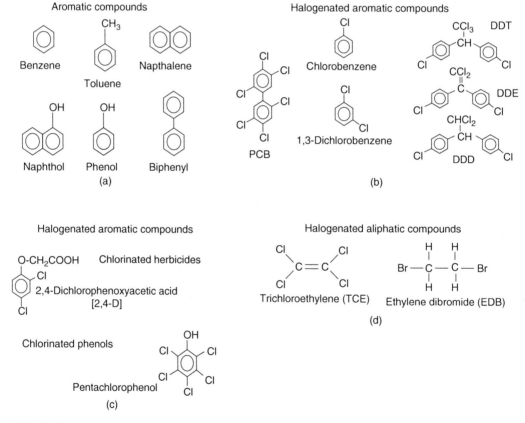

FIGURE 13.2 Basic chemical structures of selected organic compounds.

aerobic conditions (e.g., oxidation). The high productivity of wetland systems also supports large populations of detrital microorganisms, which can biodegrade certain toxic organics. Other features of wetlands may inhibit certain reactions. Contaminants can sequester in redox zones where organisms lack the enzymes to degrade them.

Rates of degradation in wetland systems depend on the interplay of a number of environmental factors including chemical characteristics of organic compounds (water solubility, oxidation

TABLE 13.2
Examples of Abiotic and Biotic Processes Regulating the Fate of Organic Contaminants in Wetlands and Aquatic Ecosystems

Hydrolysis	Microbial reduction
Mineralization	Methylation
Volatilization	Dehydrohalogenation
Photolysis	N-dealkylation
Sorption	Chemical oxidation
Reduction dehalogenation	Microbial oxidation
Cometabolism	Epoxidation
Chemical reduction	Oxidative deamination

state, types of functional groups), redox potential, pH, nutrient and carbon availability, contaminant bioavailability and concentration, electron acceptors, temperature, salinity, and microbial consortia and biomass (D'Angelo, 2002). Reaction rates can vary over several orders of magnitude depending on these environmental factors. Studies have documented the effects of several of these factors on rates of mineralization of contaminants in wetland substrates. Redox potential, a measure of the electron availability and an indirect measure of the oxygen status, has been used to show certain compounds degrade favorably under aerobic conditions (e.g., naphthalene), others under anaerobic conditions (e.g., DDT), and still others under moderately anaerobic conditions (e.g., polychlorobiphenyls [PCBs]).

13.1.1 PHARMACEUTICALS

Introduction of pharmaceuticals into environment is rapidly becoming an emerging issue in aquatic and wetland ecosystems. Pharmaceuticals due to their amount and application for human and veterinary cares are entering streams and water bodies. Pharmaceutically active compounds are complex molecules with various physicochemical and biological properties. The molecular weights of these compounds range from 300 to 1,000. Some pharmaceuticals are largely metabolized whereas others are only moderately or poorly metabolized and others are excreted intact.

Antibiotics, antineoplastics, hormones (compounds with endocrine effects), and various other pharmaceuticals and metabolites are found in sewage plant effluent. The occurrence of pharmaceuticals in streams and other water bodies is directly related to municipal wastewater sources. Pharmaceuticals and metabolites not totally biodegraded during sewage treatment may enter the aquatic environment. There is only limited amount of data on the occurrence and amount of veterinary pharmaceuticals entering aquatic ecosystem. Antibiotics are used in medicine, veterinary medicine, disease prevention in aquaculture, and as an antimicrobial substance to improve nutrient uptake in the gastrointestinal tract of feedlot animals. Antibacterial substances can be introduced into the environment when manure wastes (liquids and solids) are spread on fields. If not degraded, these antibiotics enter the aquatic environment through runoff. In a U.S. waterway survey, the majority of 139 streams examined contained pharmaceuticals (Kolpin et al., 2002). Being produced and applied with the purpose of causing a biological effect, their occurrence in wetland and aquatic environment is of interest (Table 13.3). Pharmaceuticals can impact reproduction capacity of aquatic organisms (Hugget et al., 2002; Brooks et al., 2003).

Advances in sample extraction and analytical instrumentation now allow the identification of trace levels of pharmaceutical compounds and their metabolites in wastewater effluent, surface and groundwater. Currently there is a need for additional or expanded toxicological studies to determine the effect of a wider range of pharmaceuticals on aquatic and wetland organisms.

TABLE 13.3
Selected Pharmaceuticals Compounds Found in Surface and Wastewater

Compound	Use
Bezafibrate	Lipid regulator
Caffeine	Stimulant
Carbamazepine	Antiepileptic, psychiatric drug
Clofibric acid	Lipid regulator (active metabolite)
Cotinine	Metabolite of nicotine
Cyclophosphamide	Antineoplastic
Diclofenac	Analgesic/antiinflammatory
Fenoprofen	Analgesic/antiinflammatory
Fluoxetine	Psychiatric drug (Prozac)
Gemfibrozil	Lipid regulator
Ibuprofen	Analgesic/antiinflammatory
Indomethacin	Analgesic/antiinflammatory
Ketoprofen	Analgesic/antiinflammatory
Naproxen	Analgesic/antiinflammatory
Norfluoxetine	Metabolite of fluoxetine
Pentoxifylline	Vasodilator
Trimethoprim	Antibiotic
Amoxicillin	Antibiotic
Sulfamethazine	Antibiotic
Erythromycin	Antibiotic
Chlortetracycline	Antibiotic
Tetracyclin	Antibiotic
Bisoprolol	Beta blocker
Metoprolol	Beta blocker

The ability of wetlands to process pharmaceuticals is not clear. Research is needed for determining the sorption, degradation, and transport of pharmaceuticals in wetlands. Wetlands contain unique properties that determine transformation and rate of removal (White et al., 2006). These include

- High soil organic matter content, which provides sorption medium for removal from surface water
- High microbial biomass, which promotes degradation
- Existence of aerobic–anaerobic interface at the soil–water interface and in plant rhizosphere, which provides condition for removal of a wide range of pharmaceuticals, some of these degrade best under aerobic condition and others under anaerobic condition
- Microbial biofilm found on plant and detritus surfaces, which is effective in removing pharmaceuticals
- Shallow surface water, which is conducive to photodegradation of pharmaceuticals
- Hydraulic retention time in wetland and high soil surface area to water depth ratio, which increases opportunity for pharmaceutical removal

(Refer to the book *Pharmaceuticals in Environment: Soils Fate Effects and Risks* for additional reading.)

Toxic Organic Compounds

13.2 BIOTIC PATHWAYS

Microbes in wetlands degrade organic compounds for the following reasons (see Chapter 5):

- Source of energy
  - Electron acceptor
  - Electron donor
- Source of carbon
- Substitution for a similar "natural" compound
- Cometabolism. Organisms mediating the mineralization of certain compounds obtain no apparent benefit from the process

In general, microbial transformations of organic pollutants can be classified as (1) biodegradation, (2) cometabolism, (3) accumulation, (4) polymerization or conjugation, and (5) secondary effects of microbial activity. Although these transformations are considered to be mediated by microorganisms, abiotic transformations are also involved, especially in transformations related to factors (4) and (5).

The degradation of organic compounds through microbial processes is the primary mechanism in wetlands. In wetlands, the microbial degradation of the toxic organics is governed by the relative amount in the bioavailable pool as depicted in Figure 13.3. Biodegradation of organic chemicals is associated primarily with heterotrophic bacteria and certain autotrophic bacteria, fungi including basidiomycetes and yeasts, and specific protozoa. Biodegradation can occur under both aerobic and anaerobic conditions.

The microbial transformations of organic pollutants can involve more than one type of mechanism, and under different conditions, several products can be derived from the same initial compound depending on the environmental parameters. Transformation can be mediated by one organism or through combined effects of several organisms (Alexander, 1994). The following terminologies are often used to describe the degradation of organic compounds:

1. *Acclimation.* Time required for microorganisms to adapt to a new compound or lag period
2. *Biodegradation.* It can serve as a substrate for growth
3. *Cometabolism.* It is transformed by metabolic reactions but does not serve as an energy source for the microorganism

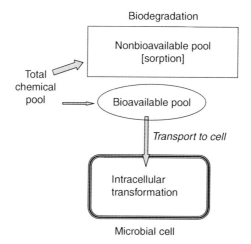

FIGURE 13.3 Microbial degradation of the toxic organics governed by the amount in the bioavailable pool.

4. *Accumulation.* Toxic organics are incorporated into the microorganism
5. *Polymerization or conjugation.* Molecules are linked together with other pesticides, or with naturally occurring compounds
6. *Secondary effects of microbial activity.* Xenobiotics are transformed because of changes in the pH, redox conditions, reactive products, etc., in terrestrial or aquatic environments brought about by microorganisms

13.2.1 Acclimation

Acclimation is defined as the length of time required for microorganisms to adapt to organic chemical addition to the soil. During this period, concentration of organic chemicals does not change. The length of acclimation period varies from less than one hour to several months. The time required for acclimation depends on various factors including concentration of organic chemicals, pH and redox status of soils or sediments, temperature, microbial activity, and other unidentified factors. Once the indigenous soil microbial community has become acclimated to the degradation of a given organic compound, the community may or may not maintain its activity after depletion of organic compounds. This depends on the type of organic compounds and diversity of the microbial community.

13.2.2 Biodegradation

An important aspect of microbial transformation of toxic organics is the complete degradation of the compounds. If the toxic organics are used by one or more interacting microorganisms, they will be metabolized to carbon dioxide and other inorganic compounds, and the microorganisms can obtain their requirements for growth and energy from the toxic organic molecules. From an environmental point of view, the complete metabolism of toxic organics is desired if one is interested in avoiding the generation of potentially hazardous intermediates.

13.2.3 Cometabolism

The degradation of synthetic compounds in the environment largely involves cometabolism; it is the prevalent form of microbial metabolism. In cometabolism, microorganisms, while growing at the expense of a growth substrate, are able to transform a compound without deriving any nutrient or energy for growth from the process. Thus, cometabolism is a fortuitous process, in which enzymes involved in catalyzing the initial reaction often lack substrate specificity. Cometabolic transformation may lead to the accumulation of intermediate products with decreased or increased toxicity, and thus cause some adverse environmental impacts and, in some cases, inhibit the microbial growth as well as their metabolism.

13.2.4 Microbial Accumulation

Cellular accumulation of organic compounds by target and nontarget microorganisms represents another type of microbial interference with organic pollutants. Primary microbial uptake of pesticides is attributed to a passive physical process of absorption rather than active metabolism. The rate of accumulation differs in various organisms and depends on the type and concentration of organic compounds in the surrounding medium.

13.2.5 Polymerization and Conjugation

Polymerization or conjugation is a microbially mediated process. Polymerization is an oxidative coupling reaction by which toxic organic (or its intermediate product) combines with itself or with other naturally occurring organics found in wetland soils to form larger molecular organic polymers. Microbial polymerization plays an important role in the incorporation of toxic compounds into soil–sediment organic matter. Figures 13.4a and 13.4b show biotic pathways for polymerization of 2,4-dichlorophenol and TNT.

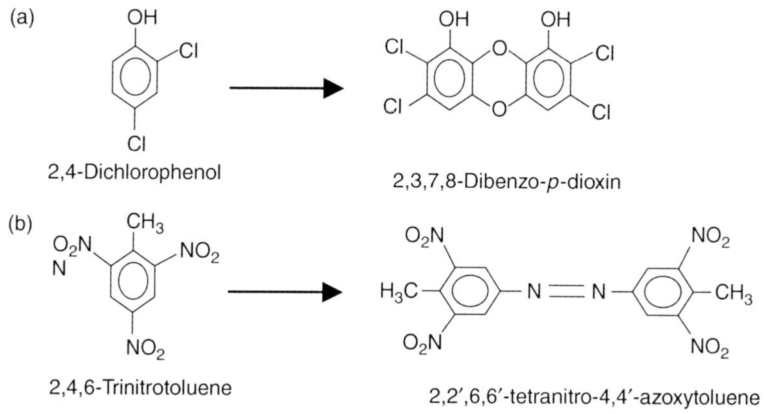

FIGURE 13.4 Biotic pathways for polymerization of (a) 2,4-dichlorophenol and (b) 2,4,6-trinitrotoluene.

13.3 METABOLISM OF ORGANIC COMPOUNDS

Microorganisms can convert toxic organics into nontoxic and inactive compounds. This process is referred to as detoxication or detoxification. The inactive compounds can undergo the following: (1) they are excreted as waste products; (2) after one or two enzymatic steps, the resultant compounds may become available for microorganisms to metabolize; and (3) they could be used by microorganisms as electron donors and oxidized to carbon dioxide (Alexander, 1994). Several processes (as listed below) may be involved in converting toxic organic compounds into inactive organic compounds or mineralized to inorganic compounds such as carbon dioxide:

- Hydrolysis [$+H_2O$]
- Oxidation [$+O_2$]
- Reduction [$+e^-$]
- Synthesis [+functional groups]

13.3.1 Hydrolysis

Hydrolysis is a major step in microbial degradation of organic compounds in wetlands. The process involves the addition of H_2O to a molecule accompanied by cleavage of the molecule into two compounds. The microbial enzymes that bring about hydrolysis are hydrolase enzymes. Hydrolysis involves the reaction between an organic compound and water that results in the breaking of one bond while forming a new C–O bond. Hydrolysis is an important reaction that determines the fate of many toxic organics in aquatic environments. However, many organic chemicals do not have functional groups susceptible to hydrolysis (Bollag and Liu, 1990).

Ether hydrolysis
- R–C–O–C–R + H_2O → R–C–OH + HO–C–R

Ester hydrolysis (chlorpropham)
- R–C–O–C=O + H_2O → R–C–OH + HO–C=O

Phosphate ester hydrolysis (parathion)
- R–C–O–P=O + H_2O → R–C–OH + HO–P=O

Amide hydrolysis (propanil)
- R–N–C=O + H_2O → O=C–OH + H–N–R

Hydrolytic dehalogenation (PCP)
- R–C–Cl + H_2O → –C–OH + HCl

13.3.2 Oxidation

Oxidation of organic compounds that occur in microorganisms is one of the most important and basic metabolic reactions. Enzymes involved in toxic organic oxidation reactions belong to various groups of known oxidative enzymes such as peroxidases, laccases, and mixed-function oxidases. The major oxidation reactions are presented in Table 13.4. Key to detoxification of toxic organics and subsequent mineralization through oxidation include:

- Presence of molecular oxygen
- Presence of selected facultative aerobic microbial groups (fungi or bacteria)
- Presence of aromatic rings without functional groups (e.g., benzene, toluene, and naphthalene) (Figure 13.5)

13.3.2.1 Hydroxylation

Microbial hydroxylation is often the first step in degradation of organic compounds. The addition of a hydroxyl group into the toxic organic molecule makes the compound more polar and hence more soluble in water. This in turn makes the compound more biologically reactive. Enzymes catalyzing this reaction are hydroxylases, monooxygenases, or mixed-function oxidases.

TABLE 13.4
Oxidation Reactions Involved in Microbial Metabolism of Toxic Organics

Hydroxylation
$$RCH \rightarrow RCOH$$
$$ArCH \rightarrow ArOH$$

N-dealkylation
$$RNCH_2CH_3 \rightarrow RNH + CH_3CHO$$
$$ArNRR' \rightarrow ArNH_2$$

β-Oxidation
$$ArO(CH_2)_nCH_2CH_2COOH \rightarrow ArO(CH_2)_nCOOH$$

Decarboxylation
$$RCOOH \rightarrow RH + CO_2$$
$$ArCOOH \rightarrow ArH + CO_2$$
$$Ar_2CH_2COOH \rightarrow Ar_2CH_2 + CO_2$$

Ether cleavage
$$ROCH_2R' \rightarrow ROH + R'CHO$$
$$ArOCH_2R' \rightarrow ArOH + R'CHO$$

Expoxidation
$$RCH=CHR' \rightarrow \overset{O}{\underset{RCH - CHR'}{\triangle}}$$

Oxidative coupling
$$ArOH \rightarrow (Ar)_2(OH)_2$$

Sulfoxidation
$$RSR' \rightarrow RS(O)R' \text{ or } RS(O_2)R'$$

Note: R = organic moiety, Ar = aromatic moiety.
Source: Bollag and Liu (1990).

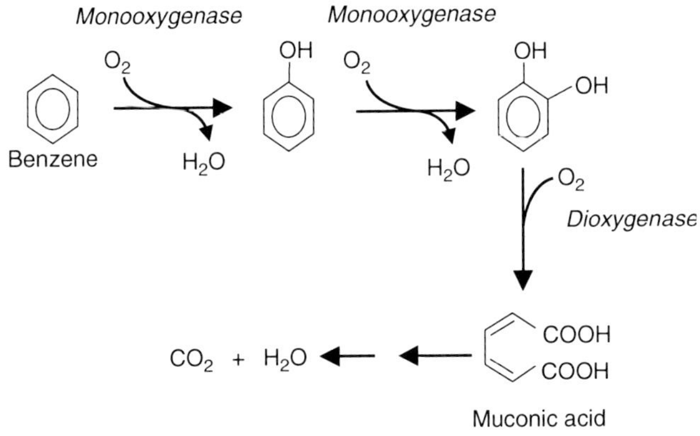

Toxic Organic Compounds

FIGURE 13.5 Oxidation of benzene to carbon dioxide and water.

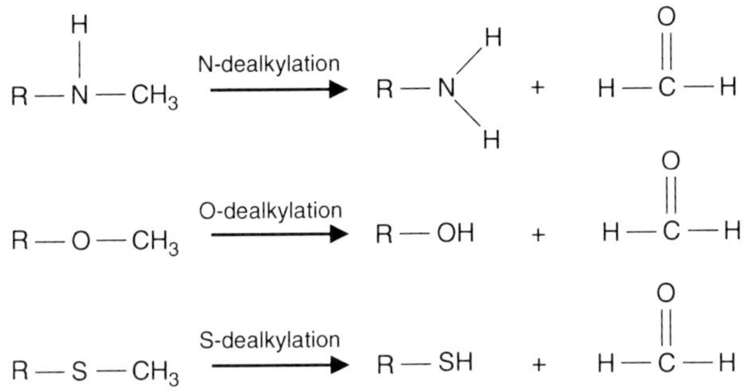

FIGURE 13.6 Metabolic dealkylation reaction for the removal of CH_3 from N, O, and S atoms in organic compounds.

Addition of OH to aromatic rings or aliphatic groups makes these compounds less toxic. For example, replacement of H by OH can inactivate organic compounds such as 2,4-D:

$$RH \rightarrow ROH$$

13.3.2.2 Dealkylation

Many toxic organics contain alkyl groups, such as the methyl ($-CH_3$) group, attached to atoms of O, N, and S. Microbial metabolism of many toxic organics involves dealkylation, or replacement of alkyl groups by H. Such reactions include O-dealkylation of methoxychlor, N-dealkylation of carbaryl, and S-dealkylation of dimethyl mercaptan. Alkyl groups microbially removed by dealkylation usually are attached to oxygen, sulfur, or nitrogen atoms. Metabolic dealkylation reactions are shown for the removal of CH_3 from N, O, and S atoms in organic compounds. An example is in Figure 13.6.

13.3.2.3 β-Oxidation

Many aromatic organics, such as phenoxyalkanoate herbicides, contain fatty acid side chains that can be metabolized by β-oxidation. β-Oxidation proceeds by the stepwise cleavage of two-carbon fragments from a fatty acid. The short-chain fatty acid is further decomposed until the chain length is four or two carbons.

13.3.2.4 Decarboxylation

Replacement of a carboxyl group by H is an enzymatic reaction achieved by microorganisms. For aliphatic carboxylic acids, the carboxyl group may be readily degraded depending on the influence of the configuration and substituents of the molecule. Decarboxylation is a widespread, microbially catalyzed reaction for naturally occurring compounds, and has been utilized in pesticide metabolism. Decarboxylation of benzoic acid herbicides by soil microorganisms is an important degradation pathway.

13.3.2.5 Cleavage of Ether Linkage

Cleavage of ether linkages in pesticides may significantly decrease their toxicity to organisms. Such cleavage, the separation of a hydrocarbon from an oxygen atom that functions to link it with the other moiety of a molecule, is important in the microbial transformation of pesticides. The specific mechanism of this reaction is not clear. It is suggested that the cleavage is catalyzed by mixed-function oxidases in the presence of reduced pyridine nucleotides and molecular oxygen. O-dealkylation also represents an ether cleavage reaction in which only an alkyl group is removed as opposed to larger molecular groups.

Pesticides, such as benzoic acid herbicides, organophosphates, carbamates, methoxy-*s*-triazines, phenylureas, and phenoxyalkanoates, have an ether linkage or an alkoxy group.

13.3.2.6 Epoxidation

Epoxidation or the insertion of an oxygen atom into a carbon–carbon double bond can frequently result in the formation of products with greater environmental toxicity. Various microorganisms can catalyze the reaction of the chlorinated cyclodiene insecticides aldrin, isodrin, and heptachlor to their more toxic epoxide derivatives.

13.3.2.7 Oxidative Coupling

Oxidative coupling or condensation reactions involving toxic organics are catalyzed by phenol oxidases such as laccases or peroxidases. In the oxidative coupling of phenol, for example, aryloxy or phenolate radicals are formed by removal of an electron and a proton from the hydroxyl group. The resulting phenolate radicals then couple with phenolic or other compounds to yield dimerized or polymerized products (Brown, 1967).

13.3.2.8 Aromatic Ring Cleavage

Many toxic organics have one or more aromatic rings, and consequently, their catabolism is possible if ring cleavage can take place. Numerous microorganisms, especially bacteria, are capable of benzene ring cleavage. However, many aromatic compounds with multiple substitutions are resistant to microbial degradation, depending on the molecular configuration.

13.3.2.9 Heterocyclic Ring Cleavage

Compounds with aromatic carboxylic molecules with heterocyclic rings are subject to metabolism by microorganisms. In pesticides with such heterocyclic rings, the path followed by the degradation process is complicated by the heteroatoms, usually N, O, and S, which contribute individual characteristics to the decomposition reactions. These compounds may contain one or more rings (mostly aromatic), and the rings generally have five to six members.

13.3.2.10 Sulfoxidation

The sulfoxidation reaction involves the enzymatic conversion of a divalent compound to a sulfoxide or to a sulfone:

$$R(S) \rightarrow R(SO) \rightarrow R(SO_2)$$

The oxidation of organic sulfides (thioethers) and sulfites to sulfoxides and sulfates can be catalyzed by minerals (Crosby, 1976), which makes it difficult to distinguish between a biological and

TABLE 13.5
Reduction Reactions in Microbial Metabolism of Toxic Organics

Reduction of nitro group

$$RNO_2 \rightarrow ROH$$
$$RNO_2 \rightarrow RNH_2$$

Reduction of double bond or triple bond

$$Ar_2C-CH_2 \rightarrow Ar_2CHCH_3$$
$$RC\equiv OH \rightarrow RCH-CH_2$$

Sulfoxide reduction

$$RS(O)R' \rightarrow RSR'$$

Reductive dehalogenation

$$Ar_2CHCCl_3 \rightarrow Ar_2CHCHCl_2$$

Note: R = organic moiety, Ar = aromatic moiety.
Source: Bollag and Liu (1990).

---

a chemical reaction. The measurement of a sulfoxidation reaction in soil cannot be attributed to microbial activity unless it is shown that the catalyst is of biological origin.

13.3.3 Reduction

Microbial reduction of toxic organics is carried out by reduction enzymes. Major reduction reactions of selected toxic organics are shown in Table 13.5. The reduction of the nitro group to amine involves the formation of a nitro and a hydroxyamino group. This type of reduction reaction occurs during the microbial metabolism of various pesticides. Organophosphorous pesticides such as parathion, paraoxon, or fenitrothion are often reduced to nontoxic amino compounds (Miyamoto et al., 1966; Matsumura and Benezet, 1978).

13.3.3.1 Reductive Dehalogenation

Reductive dehalogenation is a mechanism for the anaerobic biotransformation of chlorinated hydrocarbons such as hexachlorobenzene (HCB). In reductive dehalogenation, the halogenated compound serves as the electron acceptor rather than the donor that requires a separate carbon source. In a microbially catalyzed reaction, a halide ion is replaced by a hydrogen ion (Figure 13.7). The removal of halide ions results in compounds that are generally easier to degrade, and, in some instances, are completely mineralized.

Dehalogenation is a microbially mediated reaction in which a halogen atom is replaced with –OH. It is an important step in the metabolism of the many toxic organics that contain covalently bound halogens (F, Cl, Br, I). Dehalogenation is a major pathway for the biodegradation of organohalide compounds. Microorganisms do not utilize the organohalide compound as a sole carbon source for its degradation. Bacterial degradation of small amounts of organohalide compounds occurs while the microorganisms involved metabolize larger quantities of another substance, a process known as cometabolism. Reductive dehalogenation results in sequential replacement of Cl ions with H ions and in accumulation of toxic intermediates, and is promoted under highly reduced conditions (low redox potentials) and high microbial activity.

Bioconversion of DDT to DDD involves a process in which Cl is replaced after hydrolysis. DDD is more toxic than DDT and is even manufactured as a pesticide (Figure 13.8). The same situation applies to microbially mediated conversion of aldrin to dieldrin.

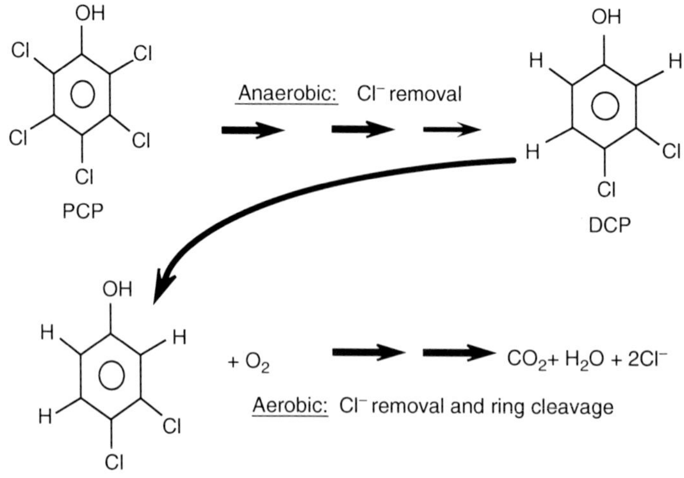

FIGURE 13.7 Example of reduction dechlorination.

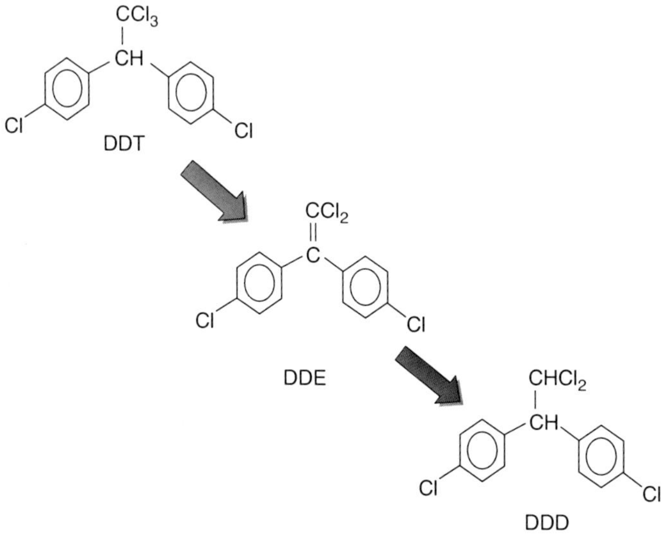

FIGURE 13.8 Dechlorination of DDT.

13.3.4 SYNTHESIS

Toxic organics or their intermediates can be linked to themselves or with other organic compounds. Binding takes place at functional groups that may be present either in the original molecule, in its intermediates, or at a substituent added during metabolic or chemical transformation. The results of such synthesis reactions are the formation of larger organic compounds.

Soil microorganisms play an important role in the binding of toxic residues to soil organic matter. In wetland soils, phenolic and quinonoid compounds originating from either lignin or other organic residues, as well as those synthesized by microorganisms, are oxidatively coupled either alone or with other substances possessing a specific functional group to form polymers that constitute the base of humus material (Sjoblad and Bollag, 1981). Many toxic organics or pesticides, when degraded, yield phenolor aniline-like chemicals analogous to naturally occurring compounds. As a result many toxic organics are cross-coupled into soil humus.

13.4 PLANT AND MICROBIAL UPTAKE

Plant uptake is potentially an important assimilation process for removing organics from wetlands. Few studies have examined plant uptake of toxic organic contaminants. Accumulation of toxic chemicals in biota is usually quantified by a bioconcentration factor (BCF):

$$\text{BCF} = \frac{\text{concentration in fresh plant tissue}}{\text{concentration in dry soil}}$$

This factor reflects the tendency of a compound to accumulate in plant tissue and is a basis for comparison between different plant species and organic contaminants.

Most organics can be readily taken up by plant roots and foliage. The chemicals can be transported in living plant tissue (symplast) and nonliving tissue (apoplast). Enzymes in the symplasts may metabolize the chemicals into less toxic compounds. Toxic organics are translocated in cell walls and xylem (apoplast) that form on interconnected continuum within plants.

Microorganisms in wetlands also accumulate toxic organics. Cellular accumulation of toxic organics by microorganisms actually involved in degradation and other microorganisms not involved in degradation is the mechanism through which pesticides/toxic organics are removed from soils and water column. The primary process involved in microbial uptake is primarily attributed to physical absorption rather than actual metabolism. The rate of accumulation of organic compounds depends on their type and concentration in wetlands. Accumulation also differs among organisms. DDT is rapidly accumulated in the bacterial tissues. Also, mycelia of fungi and actinomycetes can accumulate DDT, deldrine, and pentachloronitrobenzene. Some forms of bacteria have a large capacity for absorption of toxic organics.

Microorganisms as a result of their ability to accumulate xenobiotics can translocate pesticides and other toxic organics into the food chain. Microorganisms are a food source for many filter feeding organisms. The presence of toxic organics containing microorganisms in aquatic environments can be translocated into the food chain of fish and higher invertebrates. These higher organisms can, in turn, transport these organic compounds into other environments.

There may also be a delayed impact on the environment associated with the accumulation of toxic organics by microorganisms and plants. Dead plant tissue and microbial cells containing organic pollutants are degraded over time by biological and nonbiological processes and thus can release toxic organics back into the environment.

13.5 ABIOTIC PATHWAYS

In wetland systems, abiotic reactions can take place in solution and at the sediment–water interface. Hydrolysis, photolysis, sorption, and redox reactions are the most common abiotic reactions.

13.5.1 Redox–Potential–pH

Oxidation reactions that take place in aquatic environments can be mediated by direct or indirect photolysis reactions, which depend on the organic chemicals and substrates present. Nonphotolytic oxidation of organic chemicals can occur directly by reactions involving ozone, or via catalytic pathways with certain metals. Abiotic reduction reactions that influence organic chemical transformation in wetlands include Fe and Mn species and sulfides.

Soil redox potential (Eh) and the pH parameters are closely related. Production of carbon dioxide, an end product of the reduction of oxygen, has considerable influence on the soil's pH. When a reducing wetland soil system becomes oxidized, its pH may decrease drastically due to the oxidation of iron to Fe(III) and the subsequent hydrolysis of the iron or the oxidation of sulfite to sulfate, which is accompanied by the release of protons. Lowering of the Eh of the soil due to flooding will result in a rise of pH, because many reduction reactions (such as the reduction of sulfate to sulfide, Fe^{3+} to Fe^{2+}, and Mn^{4+} to Mn^{2+}) involve the uptake of protons or the release of hydroxyls.

The reaction of a redox couple that controls the system will directly cause a change in Eh. pH can affect the redox reactions by determining the concentrations of members of the redox couple in the soil solution. Decrease in soil pH will increase the solubility of trivalent iron and of other oxidized transition metal species, but will have a smaller effect on the solubility of the reduced species of these metals. The redox reactivity of a xenobiotic in soil is dependent on the pH. Altering the pH of the soil can affect redox reactions of toxic organics, just as it affects other pH-dependent reactions such as hydrolysis.

The reduced forms of cations such as Fe and Mn are much more soluble than other oxidized forms. As a result, in wetland soils, appreciable concentrations of reduced cationic species, such as Fe^{2+}, can be present in the soil solution. This reduction and consequent dissolution can have a strong influence on the abiotic transformations of toxic organics in the liquid phase through the capacity of metals to catalyze abiotic transformation. Under stronger reduction conditions, sulfate is reduced, producing such S species as sulfides that can also be involved in the degradation of certain toxic organics.

13.5.2 Hydrolysis

Hydrolysis reactions are important steps in the degradation of many pesticide compounds. For some toxic organics, hydrolysis reactions are nonbiological and are enhanced in soil; that is, hydrolysis reactions in some cases occur more rapidly in soil than in comparable soil-free aqueous systems due to catalysis of the reaction by sorption.

Evidence exists that chemical hydrolysis plays an important role in the degradation of the chloro-s-triazines in soil and other compounds. Such evidence is based on studies using sterilized and nonsterilized soil systems.

13.5.3 Sorption to Suspended Solids and the Substrate Bed

Sorption of organic contaminants to solid surfaces (e.g., suspended material in the water column or surface of soil sediments) is an important assimilation process because it lowers soluble concentrations of toxic organics into a phase that is less available. Sorption of nonionic organic compounds onto solids and sediments can be best described as a partitioning process, where the toxic organic "partitions" onto the sediment organic matter. Sorption of cationic toxic organics is dominated by ion-exchange reactions but the large majority of priority pollutants are nonionic. Figure 13.9 shows schematic of relationship of adsorption/desorption to toxic organics in solution.

Adsorption isotherms are used for determining the sorption of organic chemicals by wetland soil. Sorption data are generally described by using either the Freundlich or the Langmuir equation.

The Freundlich adsorption equation is

$$\frac{x}{m} = K_f C^{1/n} \tag{13.1}$$

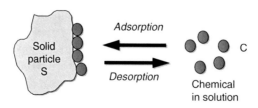

FIGURE 13.9 Adsorption/desorption to toxic organics in solution.

… Toxic Organic Compounds

where x/m is mass of organic chemical bound to the sediment (µg kg^{-1}), K_f and n are empirical constants, and C is the equilibrium concentration of the organic chemical. The value of K_f is a measure of the extent of sorption.

The linear form of the Freundlich equation is represented by logarithmic transformation:

$$\log\left(\frac{x}{m}\right) = \log K_f + \frac{1}{n}\log C \tag{13.2}$$

A plot of $\log(x/m)$ versus $\log C$ would produce a straight line, with $1/n$ equal to the slope and $\log K_f$ to the intercept.

The Langmuir adsorption equation is

$$\frac{x}{m} = \frac{K_l b C}{1 + K_l C} \tag{13.3}$$

where x/m and C are as described above, K_l is an adsorption constant related to binding strength, and b is the maximum amount of xenobiotics that can be adsorbed by the soil.

A linear form of the Langmuir equation is

$$\frac{C}{x/m} = \frac{1}{K_l b} + \frac{C}{b} \tag{13.4}$$

If a plot of $C/(x/m)$ versus C is a straight line, then the adsorption data conform to the Langmuir equation, and b can be calculated from the slope and K_l from the intercept.

The distribution constant (K_d) for nonpolar toxic organics can be calculated using a modified Freundlich equation that considers $1/n$ to be approximately equal to 1:

$$K_d = \frac{(x/m)}{C} \tag{13.5}$$

where x/m is the toxic organics adsorbed on sediment (mmol kg^{-1}) and C the equilibrium concentration of toxic organics (mmol L^{-1}).

Using the K_d and sediment organic C (OC), another constant can be determined that is independent of soil type and is specific for the toxic organics investigated:

$$K_{OC} = \frac{K_d}{f_{OC}} \tag{13.6}$$

where f_{OC} is the fraction of soil OC, for example, %OC/100. K_{OC} is a coefficient that describes the distribution of the organic between the water and sediment organic matter phases.

Sorption of toxic organic contaminants in wetlands is high because of high organic matter (high f_{OC}). The f_{OC} of highly organic marsh soils can exceed 0.5 as compared to 0.05 for an average mineral-dominated upland soil. This significantly increases potential for sorption of toxic organic in highly organic wetland substrates. Bottomland hardwood forests or riverine wetlands containing sandy substrates dominated by mineral rather than organic matter have a lower sorption potential.

- For organic chemicals not adsorbed by soils, K_d distribution constant is equal to zero.
- For a given organic chemical, sorption (K_d) is greater in soils with larger organic matter content. Chemicals with high K_d values move slowly in soils.
- For a given soil, organic chemicals with smaller K_d values are sorbed to lesser extent and are highly mobile.
- Bioavailability of toxic organics to degradation is strongly influenced by sorption.
- Chemicals with low sorption coefficients are generally more soluble and are more readily degraded.
- Sorption of chemicals increases with the amount of soil organic matter.
- Sorption may protect microbes from toxic levels of chemicals.
- High levels of DOC may increase the mobility of chemicals.
- Chemicals with high sorption coefficients are not generally less mobile.

13.5.3.1 Effect of Colloidal Organic Matter in Surface Water on Sorption in Wetlands

Surface waters and wetlands contain high concentrations of dissolved organic carbon. These high concentrations of DOC are due to high plant productivity and the anaerobic condition, which prevents complete oxidation of organic matter.

Colloidal organic matter has a significant effect on the solubility and sorption potential of toxic organics, particularly hydrophobic contaminants such as PCBs. The mechanism involved is a "partitioning" process of the toxic organic contaminant on to the colloidal organic matter. The process results in a decrease in the observed partition coefficient between the aqueous and particulate phases, K_p. Since K_p no longer represents a true partitioning between particulate and aqueous phases, the new coefficient is defined as

$$K_{\text{p-obs}} = \frac{K_p}{1 + K_{\text{doc}} \text{DOC}} \quad (13.7)$$

where $K_{\text{p-obs}}$ is the observed partition coefficient (L kg^{-1}), K_{doc} the partition coefficient between dissolved organic carbon and water (L kg^{-1}), and DOC the concentration of DOC (kg L^{-1}).

The partition coefficient between DOC and water (K_{doc}) can be measured independently by a number of techniques (Hassett and Milicic, 1985; McCarthy and Jimenez, 1985) and can be used in Equation 13.7 to estimate the contribution of DOC to the value of the observed partition coefficient.

Solubility of toxic organics in water is influenced by temperature, ionic strength, pH, and presence of other organic chemicals.

The most frequently used parameters to estimate solubility are (1) octanol–water partition coefficient and (2) chemical structure of the chemical. Figure 13.10 shows the relationship between octanol–water partition coefficient (K_{ow}) and solubility of a range of organics. The octanal–water coefficient is determined by the following equation:

$$K_{\text{ow}} = \frac{\text{amount of organic in octanol (mg L}^{-1})}{\text{amount of organic in water (mg L}^{-1})}$$

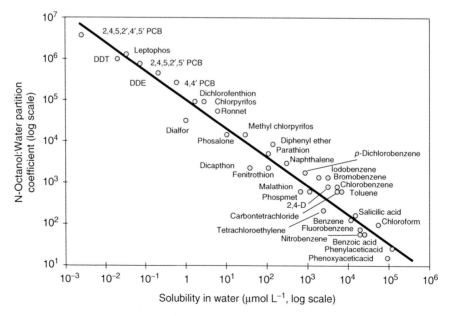

FIGURE 13.10 Relationship between octanol–water coefficient (K_{ow}) and solubility of several organic chemicals showing the range in solubilities of the organic chemicals. (From Chiou et al., 1977.)

13.5.4 EXCHANGE BETWEEN SOIL AND WATER COLUMN

Diffusion between the overlying water and the soil is an important assimilation process because it involves the transfer of toxic organics from a region of low density (the water column) to a region of high density (the sediment), where sorption can occur. Diffusion is particularly important in wetlands where long residence and low flow velocities are normal.

Diffusive flux is driven by the concentration gradient of the toxic organic at the sediment–water interface. This process is usually described as a simple Fickian process (Equation 13.8) although the flux involves multiple steps that are best approximated using a two-resistance model (Thibodeaux, 1979). The flux equation (Fick's first law) is given as

$$J = -D\left(\frac{dc}{dx}\right) \tag{13.8}$$

where J is the flux of chemical (g m^{-2} day^{-1}), D the diffusion coefficient (m^2 day^{-1}), and dc/dx the concentration gradient (see Chapter 14 for details).

Diffusive flux is a significant transfer process of toxic organics from water column to the sediment. However, under certain cases, the reverse flux, that is, diffusion from the sediment bed to the water column, can be important. This process occurs when a concentration gradient in the reverse direction occurs due to changes in loading or partitioning of the compound. A typical example is the influx of relatively "clean" water into a system with contaminants present in the sediment bed. In this case, the sediment can serve as an internal source of toxic organics to the surface water.

In wetlands, diffusive flux of toxic organics associated with colloidal organic matter may also be important. The sediment serves as a reservoir of dissolved organic carbon. Diffusion of toxic organics absorbed to organic carbon compounds into and out of the sediment bed may be an important transport process in highly organic soils.

13.5.5 SETTLING AND BURIAL OF PARTICULATE CONTAMINANTS

Organics bound to particulates can be removed from the water column by settling and sedimentation processes. Vegetation in wetlands reduces water velocity, causing the sediment to be deposited on the surface of the wetland. Sedimentation from the water column to the surface sediment layer is called the settling velocity. The quantification of the settling of particles in aquatic systems has a strong theoretical basis (e.g., Stoke's law). Estimating settling can be difficult due to the heterogeneity of particle size and composition. Nevertheless, sedimentation is an important removal process for contaminants in wetlands. Natural wetlands can remove a significant portion of the sediment load of incoming waters.

Particles dropping out of the water column is a distinct process from burial of contaminants in the bed. Burial describes the removal rate of particle-bound toxic organics to deeper sediment layers from where they can no longer diffuse back into the water column. Burial is an important assimilation process for removing contaminants from biologically active zones.

Sedimentation operates over a scale of months to years as opposed to other transport and assimilation processes.

13.5.6 PHOTOLYSIS

Photolysis is the process in which ultraviolet or visible light results in the transformation of chemical compounds. Sunlight is a factor affecting the loss of toxic organics from wetland environments. It is an environmental factor that can chemically transform toxic organics to less toxic compounds, initiating the process of mineralization. Photolysis of toxic organics in water occurs by either direct or indirect photolysis.

Direct photolysis requires absorption of sunlight by the toxic organics. Light wavelengths below 290 nm are absorbed by ozone in the earth's atmosphere. Thus, toxic organics that can absorb

TABLE 13.6
Half-Life Values of Selected Toxic Organics That Undergo Photolysis

Class	Organic Chemical	Half-life ($t_{1/2}$)
PAHs	Pyrene	0.1 h
	Benz(a)anthracene	3.3 h
	Phenanthrene	8.4 h
	Fluoranthene	21 h
	Naphthalene	70 h
Pesticides	Trifluralin	~1 h
	Malathion	15 h
	Carbaryl	50 h
	Sevin	11 days
	Methoxychlor	29 days
	2,4-D; methyl ester	62 days
	Mirex	1 year

radiation above 290 nm may undergo direct photolysis. Determining the direct photolysis of a toxic organic requires measuring the extinction coefficient of the compound occurring in solar radiation reaching the surface of wetlands.

Indirect photolysis may occur when a compound other than the toxic organics absorbs the sunlight and initiates the reaction that results in the transformation of the toxic organics. Organic and inorganic species in wetlands and aquatic environments, humic substances, clay, and transition metals are reported to be involved in indirect photolysis. Oxidation is the predominant photolytic process in toxic organics breakdown. These oxidants are activated by energy transfer to O_2-producing singlet oxygen or radical production, which results in formation of radical and nonradical active oxidants. Singlet oxygen ($^1O_2$) is formed when an excited sensitizer (^3S) transfers energy to ground-state triplet oxygen ($^3O_2$) as shown below:

$$S \rightarrow {}^1S^* \rightarrow {}^3S^*$$

$$^3S^* + O_2 \rightarrow S + {}^1O_2$$

The aromatic or chromophoric DOM in wetland or aquatic systems is generally related to dissolved humic substances.

Photolysis of toxic organics in wetland soil is limited. Determining the rate of photolysis on wetland soil surfaces requires an estimate of depth of light penetration and distribution of xenobiotic concentration in the surface soil. Light is more than 90% attenuated within 0.2 mm soil surface. As a result, the rate of direct photolysis of organic compounds in wetland soil is substantially lower in relation to that of photolysis in surface water.

Typical phototransformation products of organic compounds are often similar to products derived from metabolic and other abiotic processes. The significance of photolysis should be determined only in relation to other degradation processes in wetlands. Table 13.6 shows the half-life of selected organics that undergo photolysis (Miller and Orgel, 1974).

13.5.7 Volatilization

Volatilization of pesticides and other toxic organics from the water column to the atmosphere is a process that can remove some organics from aquatic ecosystems. Volatilization can be quantified using a resistance model, for example, two-film model, which describes the limiting factor in the mass transfer as the resistance present at the air–water interface. The two-film model states that volatilization is dependent on the mass transfer through both the liquid and gas film at the air–water interface, the temperature, and the tendency of a compound to partition to the air phase (usually represented by a Henry's law constant). The mass transfer coefficient from water to air, v_v (m year^{-1}), is equal to

$$v_v = \frac{K_l H_e}{[H_e + RT_a (K_l/K_g)]} \quad (13.9)$$

where K_l and K_g are the liquid-phase and gas-phase mass transfer coefficients (m year^{-1}), R the universal gas constant, T_a the absolute temperature (K), and H_e the Henry's constant (atm m^3 mol^{-1}). As expected, soluble compounds ($H_e = 10^{-6} - 10^{-7}$) volatize at a negligible rate compared to insoluble compounds ($H_e = 10^{-2} - 10^{-3}$).

Depending on the relative size of K_l and K_g, volatilization can be controlled by transfer in the gas phase, the liquid phase, or both phases. This can be demonstrated graphically comparing the fraction of the resistance in the liquid phase at various values of K_l/K_g. The magnitudes of K_l and K_g are proportional to the wind speed and inversely proportional to the molecular weight of the compound. In aquatic systems sheltered from the wind (e.g., wetlands), volatilization tends to be controlled by the gas phase (K_g).

Most volatilization occurs shortly after toxic organics (e.g., aquatic herbicide) enter a wetland. The importance of volatilization in removal of strongly adsorbed toxic organic existing in a sediment water column is less.

13.5.8 Runoff and Leaching

The hydrologic cycle governs the rate of water transfer and movement in wetlands. Water transfer also governs the removal of toxic organics from wetlands. Water enters wetlands primarily from rainfall and surface runoff from the adjacent watersheds or upland areas. Tidal waters are a major water source in coastal wetlands. Generally, over time the quantity of water leaving a wetland is compensated by water entering the system, especially in lakes, streams, and estuaries.

Toxic organics found in water in wetlands, where the mean residence time (MRT) of water is low, would be rapidly removed or transported to another ecosystem. If the MRT of water in a wetland is high (e.g., backwater swamp), water-soluble toxic organics would remain in the system for extended periods of time. The solubility of toxic organics in water and the MRT of water in wetlands govern the rate of transport and removal from the system.

The MRT for wetlands can be calculated if it is assumed that the input is equal to the output and if the mass of the water in the wetland and the rates at which water is entering and exiting the wetland are known:

$$\text{MRT} = \frac{\text{mass}}{\text{flux}}$$

Transport of toxic organics from wetlands to surface water is dependent on MRT and solubility of the organic compound. Solubility of an organic can vary a billion fold. Solubility of xenobiotics in water is influenced by temperature, ionic strength, pH, and presence of other organic chemicals.

Toxic organics entering wetlands, which are strongly adsorbed to suspended sediment, would remain in wetlands for a longer period of time. Organics absorbed to clay particles would settle out and enter the sediment pool. The rate of diffusion back into the surface water would be slow. A greater number of flushing events would be required to remove significant quantities of organics during surface runoff from wetlands. Figure 13.10 shows the relationship between octanol–water coefficients and solubility of several organic chemicals.

Groundwater movements in wetlands are a function of gravitational forces and porosity of sediment substrate. Groundwater movements are faster in coarse-textured sediment (e.g., sand) compared to heavy clay substrates.

Groundwater transport of xenobiotics into subsurface environments is distinguished by two transport mechanisms:

1. Transport without flow of soil solution: molecular diffusion only
2. Transport with flow of soil solution: convection–dispersion

Convection, a gravity flow, is the major mechanism through which xenobiotics in water are leaching into subsurface environments. Organic compounds soluble in water can be rapidly transported into the subsurface environments. Molecular diffusion accounts for movement over only a short distance. Molecular diffusion is also governed by adsorption of the organic material and clay particles.

13.6 REGULATORS

13.6.1 Effect of Electron Acceptors on Toxic Organic Degradation

Many bacteria in wetland environments are capable of anaerobic respiration to conserve energy and reoxidize their reduced electron carriers in the absence of oxygen. Depending on the redox potential of the alternative electron acceptor used, more or less energy can be conserved by anaerobic respiration. This causes a strong hierarchical and spatial gradient of organisms competing for electron acceptor compounds with the highest redox potentials available. Anaerobic respiration occurs in habitats that have a depletion of oxygen. The hierarchy of anaerobic electron acceptors used in reducing environment starts with nitrate (+433 mV) and extends via oxidized metal ions, such as ferric iron or Mn^{4+} (at ca. +200 mV), to sulfate (ca. −200 mV); see Chapter 4. The successive utilization of different electron acceptors for anaerobic respiration becomes visible in the gradients of metabolic activity formed in anaerobic sediments. The different types of anaerobic respiration occur in defined layers in the order of the redox potentials of the acceptors.

Energy yield of acetate oxidation with different electron acceptors is shown in Table 13.7. Also shown is energetics of aerobic and anaerobic degradation of benzoate with various electron acceptors found in wetland soils and sediments. Theses electron acceptors include O_2, NO_3^-, Fe^{3+}, and SO_4^{2-}.

13.6.2 Denitrifying Bacteria

Denitrifying bacteria are frequently encountered as anaerobic degraders of diverse xenobiotics. They are facultatively anaerobic and frequently also degrade xenobiotics aerobically. However, the pathways as well as the specific patterns are different under aerobic and denitrifying conditions. Nitrate is reduced to N_2 via nitrate, NO, and N_2O. Iron (III) or manganese (IV) reduction is a geochemically important alternative pathway of anaerobic respiration and has been shown repeatedly to be linked

TABLE 13.7
Energetics of Benzoate Oxidation Using Different Electron Acceptors

Reactions	ΔG (kJ)
Benzoate + $7.5O_2 \rightarrow 2CO_2$	−3,175
Benzoate + $6NO_3^- \rightarrow 3N_2 + 7CO_2$	−2,977
Benzoate + $8NO_3^- \rightarrow 14NH_4^+ + 7CO_2$	−1,864
Benzoate + $30Fe^{3+} \rightarrow 30Fe^{2+} + 7CO_2$	−303
Benzoate + $SO_4^{2-} \rightarrow 7CO_2 + 3.75HS^-$	−185
Benzoate + $S^0 \rightarrow 7CO_2 + 15HS^-$	−36

Source: Thauer et al. (1977).

to anaerobic degradation of xenobiotics. Sulfate-reducing bacteria are organisms that can degrade a broad range of organic compounds. However, the degradation seems to be more restricted within denitrifying or iron-reducing bacteria.

The source and availability of electron acceptors present in the wetland soil and sediment at various redox levels govern the rate of degradation of toxic organics. In saturated soil or flooded soil, oxygen may be limited or absent and degradation will be dependent on other electron acceptors. In the absence of oxygen, microbial degradation of toxic organisms proceeds by the ability of organisms to use alternate electron acceptors such as nitrate, sulfate, and iron under methanogenic conditions. A wide variety of biodegradation reactions have been shown to occur under conditions of little or no oxygen. For instance, certain hydrocarbons such as phthalic acids and naphthalene can be decomposed under denitrifying conditions by mixed populations using nitrate (Aftring, 1981). In addition, microbial mineralization of toluene, phenol, and p-cresol has been shown to be coupled to Fe(III) reduction (Lovley and Lonergan, 1990). A very important transformation, reductive dechlorination, has been shown to occur under methanogenic (e.g., Freedman and Gossett, 1989) and possibly sulfate-reducing conditions (Bagley and Gossett, 1990). Reductive dechlorination is important in reducing toxicity of chlorinated aromatic and aliphatic compounds in addition to opening spots on the chemical structures for further attack by other microorganisms.

13.6.3 Effect of Sediment Redox–pH Conditions on Degradation

The effectiveness of microbial degradation of chemicals in the environment is dictated primarily by environmental physicochemical conditions. Like other organisms, toxic organic-degrading microbes have certain requirements for growth. These optimal growth conditions can vary considerably from species to species. The conditions in the media, therefore, largely determine the composition of the mixed population of microorganisms.

Soil–water systems provide a wide range of microhabitats for microbial activity. Sediments range from saline to fresh, well oxidized to strongly reduced, and nutrient poor to nutrient rich. Oxygen availability, pH, temperature, salinity, and nutrient status are a few parameters that can vary considerably within a coastal sediment system. Temperature has been shown to be a limiting environmental factor in biodegradation in aquatic systems and soil systems. Temperature is significant since microbial activity increases exponentially over the 0–20°C range. Salinity has also been shown to be a limiting factor for biodegradation in estuarine environments. One of the most significant environmental factors affecting microbial degradation is the oxygen status of the sediment, often indirectly measured as the oxidation–reduction potential.

Various electron acceptors in wetland soil also buffer or control the redox levels in the soil. Some organics degrade faster under aerobic conditions (>+350 mV) and others degrade more quickly under reducing conditions. To enhance *in situ* degradation of organics found in wastes, it would be useful to determine the influence of adding alternate electron acceptors and the manipulation of soil or subsoil redox conditions in increasing rates. Bioremediation of contaminated sediments can therefore be limited when electron acceptors are not available. Figure 13.11 shows the effect of nitrate and sulfate on anaerobic degradation of TNT in sediments.

Considerable evidence exists that microbial activities contribute to the formation of reactive products and also to the alteration of environmental parameters such as pH, redox potential, or other factors that are conducive to the secondary or nonenzymatic transformation of pesticidal molecules. Incorporation of pesticide molecules or their intermediates into soil humus often takes place by interaction between enzymatic and nonenzymatic processes.

Through microbial activities, a reducing environment can be created, particularly in a flooded soil. Many pesticides are degraded by reductive reactions that proceed nonenzymatically under anaerobic conditions. DDT, methoxychlor, and heptachlor readily break down in an anaerobic flooded ecosystem (Sethunathan, 1973).

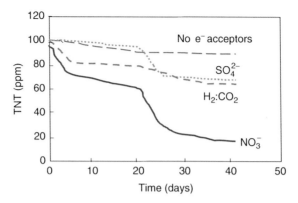

FIGURE 13.11 Effect of nitrate and sulfate on the anaerobic degradation of TNT in sediments.

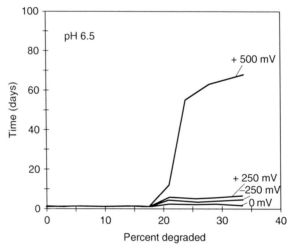

FIGURE 13.12 Degradation rates of ^{14}C-labeled pentachlorophenol at pH 6.5 under controlled redox potentials. (From DeLaune et al., 1983.)

A study on degradation of PCP (a wood-preserving compound) showed that degradation was strongly influenced by sediment pH and oxidation–reduction potential conditions (DeLaune et al., 1983). Mineralization rates decreased with decreasing sediment oxidation–reduction potential (Figure 13.12).

Maximum degradation occurred at pH 8.0. Sediment pH and redox potential also determined the rate of PCB mineralization (Pardue et al., 1988). In contrast to PCP, which degrades faster under strong aerobic conditions, mineralization rates of PCB were higher under moderate aerobic conditions (+250 mV) than under strong aerobic conditions (+500 mV), or anaerobic conditions (−200 mV) being 30–40-fold higher than anaerobic sediments (Figure 13.13). Results demonstrated that the PCB-degrading microbial population was not comprised entirely of aerobic microorganisms but was based on a coupled anaerobic–aerobic degradation.

Redox potential and pH have been shown to control the degradation of many classes of organics in Louisiana bottom sediment. Most of the hydrocarbons investigated degrade more rapidly under high redox (aerobic) conditions, although there are exceptions (e.g., DDT and PCBs). Some of these compounds, due to their slow degradation in anaerobic sediment, may persist in the system for decades. Of particular interest is whether these bottom sediments are truly a "sink" for these toxic compounds or whether they will act as a large source of contamination in the future.

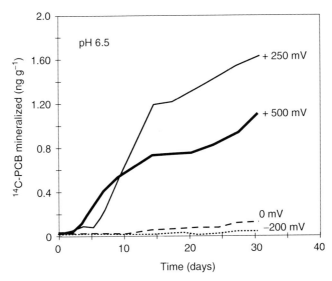

FIGURE 13.13 Microbial mineralization rates of ^{14}C-labeled polychlorinated biphenyls at pH 6.5 under controlled redox potential showing degradation was greater under intermediate redox potential +250 mV.

One of the major fates of toxic organics that are released in or transported to wetland environments is their incorporation into bottom sediment. Bottom sediments are viewed as the ultimate "sink" for many toxic organics. The oxygen status of coastal region sediments is variable and supports a wide range of aerobic, facultatively aerobic, and anaerobic microbial communities. Bottom sediment redox potentials range from +700 mV in highly aerobic sediment to −300 mV in highly anaerobic sediment (DeLaune et al., 1980). Microbial degradation appears to be the major process through which petroleum hydrocarbons and other toxic organics are removed from bottom sediments.

Sediments are usually characterized by a distinct surface-oxidized layer. Petroleum hydrocarbons and other toxic organics moving into the sediments column will enter through this oxidized layer. Subsequent burial of this oxidized layer through sedimentation will move the sediment-bound hydrocarbon into the reduced zone.

Studies have shown that sediment oxidation–reduction potential and pH were important factors governing the activity of petroleum hydrocarbon-degrading microorganisms and subsequent mineralization rates in sediment from Louisiana's Barataria basin (Hambrick et al., 1980). Highest mineralization rates occurred at pH 8.0 and the lowest occurred at pH 5.0. At all pH levels, mineralization decreased with decreasing oxidation–reduction potential (increasing sediment anaerobiosis) (Figure 13.14). Generally, mineralization rates for octadecane were greater than those for naphthalene.

Aerobic microorganisms in the oxidized sediment were more capable of degrading hydrocarbons than anaerobic microorganisms in reduced sediment of the same pH. Likewise, degradation of benzo(a)pyrene increased with increasing redox potential (DeLaune et al., 1981). Up to a 100-fold difference in benzo(a)pyrene degradation rates could be attributed to changes in sediment pH and redox potential. This is shown in Figure 13.15, which records the recovery of $^{14}CO_2$ from the mineralization of benzo(a)pyrene during a 37 day incubation period.

The herbicides 2,4-dichlorophenoxyacetic acid (2,4-D) and trifluralin are mineralized or degraded at slower rates under anaerobic conditions. Pronounced differences in degradation rates of added, labeled 2,4-D as affected by sediment redox potential were observed (DeLaune and Salinas, 1985). Degradation of 2,4-D was approximately six times faster under aerobic conditions (+500mV) as compared to anaerobic conditions (−200 mV) in sediments. The effect of redox

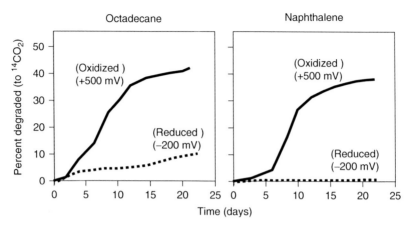

FIGURE 13.14 Degradation rates of ^{14}C-labeled octadecane and naphthalene in oxidized and reduced sediments from the Leesville, Louisiana oil field area. (From DeLaune et al., 1980.)

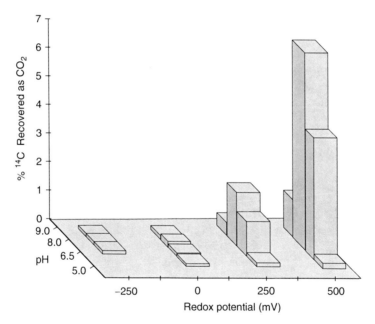

FIGURE 13.15 Effect of pH and redox potential on degradation of ^{14}C-labeled benzo(a)pyrene in Airplane Lake, Louisiana sediment after 37 days incubation period. (From DeLaune et al., 1981.)

potential on the degradation of a number of insecticides in sediment–water systems has been studied in the laboratory (Gambrell et al., 1984). Some degrade more quickly under reducing conditions (e.g., 1,1,1-trichlor-2,2-bis(4-chlorophenyl) [DDT]), others degrade more quickly under oxidizing conditions (e.g., Guthion and Permethrin), whereas the degradation of some others may be little affected by soil or sediment oxidation–reduction conditions (e.g., Kepone, Table 13.8) (DeLaune et al., 1990).

Degradation of PCP in a spectrum of wetland soils including manipulations of electron donors and acceptors and PCP concentrations showed maximum transformation rates at PCP

TABLE 13.8
Optimal Redox Conditions for Degradation of Various Classes of Organic Compounds

Compound	Reducing	Moderately Oxidizing	Highly Oxidizing	Location	Reference
Octadecane	—	—	X	Leesville, Louisiana	Hambrick et al. (1980)
Naphthalene	—	—	X	Leesville, Louisiana	Hambrick et al. (1980)
Benzo(a)pyrene	—	—	X	Airplane Lake, Louisiana	DeLaune et al. (1981)
2,4-D	—	—	X	Lake Palourde, Louisiana	DeLaune and Salinas (1985)
DDT	X	—	—	Mobile Bay, Alabama	Gambrell et al. (1984)
Permethrin	—	—	X	Oliver Soil, Louisiana	Gambrell et al. (1984)
Kepone	—	—	X	James River, Virginia	Gambrell et al. (1984)
PCP	—	—	X	Lake Borgne, Louisiana	DeLaune et al. (1983)
PCB	—	X	—	Capitol Lake, Louisiana	Pardue et al. (1988)
Dioxin	—	—	X	Baton Rouge, Louisiana	Pardue et al. (1986)

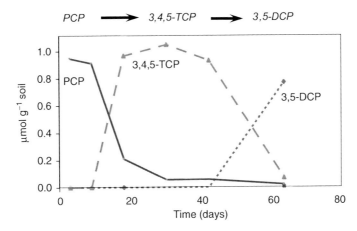

FIGURE 13.16 Degradation of pentachlorophenol. (Redrawn from D'Angelo and Reddy, 2000.)

concentrations <10μm (methanogenic conditions) and <6 μm to >23 μm (aerobic conditions) (D'Angelo and Reddy 2000). PCP degradation occurred within a 30 day period (Figure 13.16).

13.6.4 BURIAL

In wetlands, any xenobiotics that are not rapidly degraded in the oxidized surface sediment layer will be buried by incoming sediment. Sedimentation rates determined by ^{137}Cs dating have shown that certain areas of Louisiana coastal wetlands are accreting at rates as great as 1 cm year^{-1}. Any buried xenobiotics will be in an anaerobic zone where there is much slower degradation. However, burial may enhance degradation of some chlorinated organics (e.g., PCB and DDT) where anaerobic reductive dechlorination is the important transformation.

13.7 SUMMARY

- Toxic organics, which can be toxic to the biotic community, enter wetland and aquatic environment through spills and runoff from agricultural, industrial, and urban sources. These organics include pesticides, industrial organics, and petroleum hydrocarbons.

- Once in a wetland environment, toxic organics are subject to both biological and nonbiological processes.
- Many processes affect the transformation, transport, and accumulation of toxic organics. The processes included biotic and abiotic transformations. Both these processes can occur simultaneously in wetlands.
- Biotic transformation or microbial metabolism is generally the most effective in degrading toxic compounds in wetlands.
- Wetlands can efficiently degrade or process toxic organics. The rate of degradation and turnover is governed by the nature of the organic chemical and soil–sediment biogeochemical properties.
- Rates of degradation in the wetland systems depend on a number of environmental factors including redox, pH, nutrient availability, temperature, contaminant bioavailability, and microbial biomass density.
- Wetlands provide aerobic and anaerobic interfaces that govern the rate of degradation of specific compounds.
- The degradation of toxic organics through microbial processes is the primary removal process. Highly reduced condition is favorable for the degradation of some chloronated hydrocarbons.
- Biodegradation of benzene, octane, toluene, PAH derived from petroleum, and most other manufactured organic compounds is enhanced under aerobic condition, and PCB degrades faster under moderately reducing conditions.

STUDY QUESTIONS

1. List the major sources of toxic organics entering wetlands and aquatic ecosystems.
2. What is the difference between biotic and abiotic transformations. List an example of each process that regulates the fate of organic contaminants in wetlands.
3. What are the major processes involved in/governing the microbial transformation of toxic organics in wetlands?
4. What are the reasons microbes degrade or consume organic compounds?
5. What is cometabolism?
6. What is mineralization?
7. List the major metabolic processes involved in the conversion of toxic organics into inactive organic compounds.
8. What is hydrolysis? Provide examples.
9. What are oxidation reactions? List different types of oxidation reactions involved in the metabolism/degradation of toxic organisms.
10. Which reduction metabolic reactions are used in degradation of toxic organics? Provide an example of reducing dehalogenation.
11. Can toxic organics be taken up or accumulate in plants and microorganisms? Explain under what conditions the toxic organics can be released back into the enrichment.
12. List important abiotic pathway or processes involved in transformation of toxic organics in wetlands.
13. What is photolysis?
14. How can sorption influence availability or amount of toxic organic in solution? What sediment properties influence adsorption?
15. Explain what is meant by diffusion flux or exchange of organics between sediment and water column. What drives the diffusion flux?
16. How does solubility of a particular compound govern transport and degradation in wetlands?
17. How do sediment redox–pH conditions govern toxic organics degradation? Under what redox conditions do PCB degrade faster?

18. Explain how electron acceptors found in reducing soil environments influence degradation of toxic organics? List the electron acceptors.
19. Can sedimentation serve as a sink for toxic organics? Explain.

FURTHER READINGS

Alexander, M. 1994. *Biodegradation and Bioremediation*. Academic Press, New York. 302 pp.

Cheng, N. N. (ed.). 1990. *Pesticides in the Soil Environment: Processes Impacts and Modeling*. SSSA Book Series No. 2. Soil Science Society of America, Madison, WI. 530 pp.

D'Angelo, E. M. 2002. Wetlands: biodegradation of organic pollutants. In G. Bitton (ed.) *Encyclopedia of Environmental Microbiology*. Wiley, New York.

Field, J. A., A. J. M. Stams, M. Kato, and G. Schraa. 1995. Enhanced biodegradation of aromatic pollutants in cocultures of anaerobic and aerobic bacterial consortia. *Antonie van Leeuwenhoek*. 67:47–77.

Klaus, K. (ed.) 2004. *Pharmaceuticals in the Environment: Sources, Fate, Effects and Risk*. 2nd Edition. Springer. 528 pp.

Kummerer, K. (ed.) 2004. *Pharmaceuticals in the Environment: Sources, Fate, Effects and Risk*. 2nd Edition. Springer, Berlin. 528 pp.

Manahan, S. E. 2000. *Environmental Chemistry*. 7th Edition. Lewis Publishers, Boca Raton, FL. 898 pp.

14 Soil and Floodwater Exchange Processes

14.1 INTRODUCTION

This chapter focuses on biogeochemical processes that regulate the exchange of solutes and other chemical constituents between soil, water, and air in wetlands. Geologists working in aquatic ecosystems refer to some of these exchange processes as diagenesis, which is defined as the sum total of processes that bring changes in a sediment subsequent to particulate matter deposition and exchange processes between sediment and water. Soil scientists working in terrestrial ecosystems refer to pedogenesis as the sum total of processes and factors that influence soil formation. As wetlands possess the characteristics of both terrestrial and aquatic systems, both diagenetic and pedogenic processes influence the forming of wetland soils and associated exchange processes between soil and water. Many of the transport processes discussed in this chapter are applicable to both wetlands and aquatic systems, such as shallow lakes and streams. For processes with limited information on wetland ecosystems, examples from select aquatic systems will be used to demonstrate basic concepts related to exchange processes. Much of the basic principles presented in this chapter follow from two textbooks (Lerman, 1979; Berner, 1980); the reader should refer to these for an excellent overview on geochemical processes and sediment–water interactions in aquatic systems.

Wetlands are known to accumulate particulate matter in soils and benthic sediments and retain/release dissolved substances. Particulate matter associated with wetland inflows may settle on soil surface via sedimentation or accretion processes (Figure 14.1). Owing to low relief in relation to the landscape plus dense vegetation stands within a wetland, marshes, bogs, and forested wetlands accumulate sediment and associated particulate matter (Figure 14.2). As sediment-laden water flows through wetlands, sediments and other particulate matter are deposited onto the wetland surfaces. The particulate matter deposited on the soil surface can be in organic or inorganic form and may contain both nutrients and contaminants.

Erosion is not a major transport mechanism in typical wetland environments, and wetlands can in fact act as barriers to erosion. The root system of wetland vegetation stabilizes soil and enhances the particulate matter accumulation. The aboveground wetland vegetation can reduce erosion by dampening wave action in the water column and slowing water current speed. Sedimentation in general is a relatively irreversible mechanism in low-lying wetlands. Sediments and particulate organic matter (POM) accumulation on the soil surface create a distinct horizon, which we define as *recently accreted soil* (RAS), also referred to as "floc" in the literature.

The soil–water interface separates solid materials (mineral and organic matter of soils and sediments) and pore waters from the overlying water column. Biogeochemical processes in wetland soils can produce large gradients of various dissolved substances across the soil–water interface. Such gradients can also develop as a result of anthropogenic inputs of nutrients and contaminants to the water column. These solutes are then subject to transport across the soil–water interface into the water column.

The rate of transfer of solutes between soil and overlying water column and from one physical or chemical state to another is defined as flux. The dimensions of flux are $M L^{-2} T^{-1}$, where M is the mass of material transferred by flux, L is the distance or length, and T is the time. The processes associated with flux are advection, diffusion, and dispersion. Diffusive and advective flux between soil and overlying water and elemental uptake by rooted wetland vegetation are the major transport

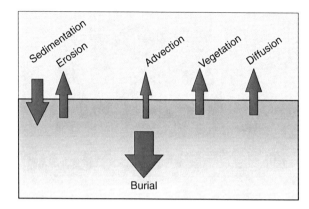

FIGURE 14.1 Schematic showing major exchange processes between soil and water column.

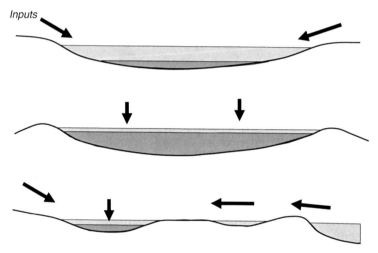

FIGURE 14.2 Schematic showing the low relief in wetlands, which results in the accumulation of sediment and organic matter.

mechanisms by which nutrients, metals, and toxic organic compounds are removed from soil and the water column.

Two characteristic features distinguish the zone of soil–water interface from the water column above and the soil below. These are residence time of reactive solutes and biological activities. Wetland ecosystems include a biologically active layer in the water column, such as benthic periphyton mats (as seen in the Everglades and other wetland ecosystems), and vegetation in the water column, such as a bed of submerged, floating, and emergent aquatic vegetation. The thickness of this active layer in the water column to some extent is controlled by the rate of sedimentation and organic matter deposition (affected by external loading and internal primary productivity). Thus, the exchange of solutes across the soil–water interface is affected not only by the physicochemical processes at the interface but also by the activities of biotic communities such as periphyton and aquatic vegetation. The primary driving forces of solute distribution in wetlands and aquatic systems are the water movement either by surface or subsurface flow, and associated hydrodynamic processes.

The major components of the hydrological cycle and budgets influence the overall transport of solutes across the soil–water interface in wetlands. These include precipitation (all wetlands), surface

Soil and Floodwater Exchange Processes

inflows and outflows (all wetlands except ombrotrophic bogs and riparian wetlands), groundwater (potentially all wetlands except ombrotrophic bogs and other perched wetlands), evapotranspiration (all wetlands), and tides (in tidal freshwater and salt marshes, and mangrove swamps) (Mitsch and Gosselink, 2000).

In this chapter we discuss the one-dimensional transport processes affecting the distribution of particulate matter including sediments and dissolved substances between the soil and the water column. The major processes responsible for transport of solutes across soil or sediment–water interface can be grouped into the following (Lerman, 1979):

- Sedimentation flux (mineral and POM)
- Flux of dissolved material and water into soil (downward flux)
- Upward flow of dissolved materials and pore water flow upward, caused by hydrostatic pressure gradients of groundwaters
- Molecular diffusional fluxes in pore water
- Mixing of soil or sediment and water at the interface (bioturbation and water turbulence)
- Evapotranspiration by vegetation and downward flux of solutes from water to soil

If the rates of material input into a wetland reservoir and removal are equal, then the material in the wetland does not change with time and a steady state is maintained. The mean residence time (τ) of material flowing through a wetland is defined as

$$\tau = \frac{M}{\text{Sum of } F} \tag{14.1}$$

where M is the amount of material in the reservoir and F the sum of all input or removal rates. The flow of materials through wetland reservoirs is controlled by inputs, biogeochemical processes within the reservoir, water flow velocity, vegetation density, and outflow.

14.2 ADVECTIVE FLUX

14.2.1 Advective Flux Processes

Advective flux refers to the bulk flow of solids or pore water relative to an adopted frame of reference such as the soil–water interface of wetlands (Berner, 1980). Advection is associated with the flow of material with either the velocity of its own or the velocity of medium (water) through which the material is transported (Lerman, 1979). During advective transport, solutes are typically transported at the same velocity as water or air. Advective fluxes can include sediments and solutes carried in surface water flows, or solutes in groundwater and pore water flow. The flow of material of a given density $\rho(\text{M L}^{-3})$ with velocity $U(\text{L T}^{-1})$ results in the advective sediment flux (J_m) as described in the following equation:

$$J_m = \rho U (\text{M L}^{-2} \text{T}^{-1}) \tag{14.2}$$

Similarly, the flux of solute (J_s) with a concentration $C(\text{M L}^{-3})$ transported in water at a velocity U is given as

$$J_s = CU (\text{M L}^{-2} \text{T}^{-1}) \tag{14.3}$$

The flow velocity U is considered as the driving force for the flux of material.

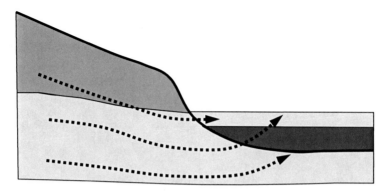

FIGURE 14.3 Advective flux processes in wetlands.

For example:
If nitrate concentration in inflow water of a constructed wetland is 10 mg L^{-1} and the average water velocity is 10 cm s^{-1}, what is the average flux of the nitrate in the outflow (assume no biological transformations or downward flux into soil)?

$$C = 10 \text{ mg nitrate N L}^{-1}$$

$$U = 10 \text{ cm s}^{-1}$$

Convert concentration units to stay consistent with velocity

$$[10 \text{ mg L}^{-1}][1 \text{ L}/1{,}000 \text{ cm}^3] = 0.01 \text{ mg cm}^{-3}$$

$$J = 0.01 \text{ mg cm}^{-3} \times 10 \text{ cm s}^{-1}$$

$$\text{Flux } (J) = 0.1 \text{ mg cm}^{-2} \text{ s}^{-1}$$

Advective flux of solutes in wetlands can result from pressure gradients that force pore water from soil pores to overlying water, carrying solutes and fine particulate matter with it across the soil–water interfaces. The flux is influenced by hydraulic gradients, associated water, and adjacent upland area (Figure 14.3). Pore water movement and its transport could be significant in sandy, permeable soils and sediments. In low-permeability soils such as those with high silt and clay contents, molecular diffusion (see Section 14.3) and bioturbation (see Section 14.4) can be the major transport processes.

14.2.2 Measurement of Advective Flux

Advective flow of solutes can be significant in sandy sediments, which have greater permeability than cohesive muddy sediments. The flow of water and solutes through soil pore spaces is an important advective transport mechanism that affects the biogeochemical processes occurring within the soil profile. The movement of water and solutes through porous media such as soils and sediments is regulated by hydrostatic pressure differential as described by Darcy (1856). Darcy's law states that water flow rate per unit area of porous bed is proportional to hydrostatic pressure difference across the bed. The pressure gradients may be created by bottom currents and water density changes. Darcy's law is defined as

$$q = \frac{Q}{A} = -K\frac{dh}{dx} \tag{14.4}$$

where q is the discharge of water per unit area (A) of porous bed in unit time, Q is the water flow rate ($L^3\ T^{-1}$), K is the hydraulic conductivity ($L\ T^{-1}$), dh/dx is the hydraulic gradient ($L\ L^{-1}$), x is the length of porous bed (L), and h is the hydraulic head (L). The specific discharge of q has dimensions of $L^3\ L^{-2}\ T^{-1}$, which reduces to $L\ T^{-1}$. The preceding equation can be modified to yield the flow rate Q or discharge.

$$Q = qA \quad (14.5)$$

where A is the cross-sectional area of the porous bed (L^2).

The hydraulic conductivity (K) is influenced by permeability, viscosity of fluid, gravitational force, and density of the medium.

$$K = \frac{k\rho g}{\mu} \quad (14.6)$$

where k is the coefficient of hydraulic permeability (L^2), ρ is the density of water at 25°C ($M\ L^{-3}$), g is the acceleration due to gravity ($g = 980\ \text{cm s}^{-1}$), and μ is the viscosity of the fluid (water = 0.01 g cm^{-1} s^{-1}). For water at standard temperature and pressure, the above can be reduced to

$$K = 10^{-5}k \quad (14.7)$$

The values of k are determined experimentally and vary with soil type. The values of k for sandy sediments are typically higher than 10^{-12} cm^2. The values of K are in the range of 1 cm s^{-1} for coarse gravels, 10^{-3} cm s^{-1} for fine sands, and 10^{-8} cm s^{-1} for fine clays.

Advective fluxes can be measured using indirect and direct methods:

1. Direct measurements include
   - Seepage meters
   - Piezometers
2. Indirect measurements include
   - Tracers
     – Salinity/conductivity
     – Radium/radon isotope
     – Dyes

14.2.2.1 Seepage Meters

A variety of methods exist for quantifying advective flux in wetlands. A direct approach involves use of seepage meters, which are the enclosures placed on the sediment surface to measure the flow over a small area.

Seepage meters can be used in making direct measurement of water flux across the sediment–water interface. The half-barrel seepage meter (Lee, 1977) has been used for measuring seepage flux in lakes, wetland, estuaries, and river bottoms. A plastic bag and tubing are attached to each meter through a rubber stopper inserted into the top of the barrel. The change in water volume is converted to units of liters per square meter per day. The Lee half-barrel seepage meter consists of a cylinder that covers the interface and connects to a variable volume chamber, usually a plastic bag. The rate of seepage is calculated by measuring the change in volume of water in the bag over time. There are several limitations in using this system (see Rosenberry, 2005) (Figure 14.4). These include very low seepage rates requiring substantial time for measurement and high labor costs and time in setting the system. When flow is heterogeneous, it may be difficult to extrapolate seepage meter results to a larger area.

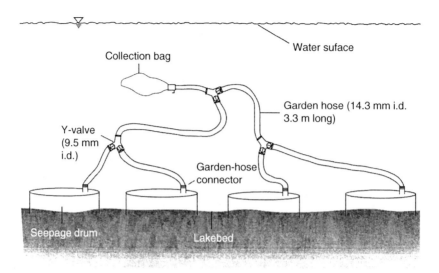

FIGURE 14.4 Schematic showing seepage cylinders placed together with one collection bag. (From Rosenberry, 2005.)

14.2.2.2 Piezometer

A piezometer (which is commercially available) can also be used for measuring advective flux. It can be placed at various depths in the soil for measuring groundwater elevations and pore water pressure. The piezometer is a stainless steel capsule that incorporates porous filter stone at one end which allows fluid to pass through but prevent soil particles from entering. The piezometer contains a pressure-sensitive diaphragm. The fluid pressure acting on the outer face of the diaphragm causes deflections of the diaphragm and changes in tension can be measured. Some piezometer modules can be thread into a rod and pushed directly into soft soil sediment with signal cable located inside the rod. Other piezometers are suitable for installation inside wells or bore holes.

14.2.2.3 Salinity/Conductivity

The freshwater updwelling or submarine groundwater discharge in nearshore or coastal region can be determined by measuring the change in salinity or conductivity of surface water or groundwater. In bay and estuarine environments, the input of freshwater can be large. In the nearshore marine environments, the source of subsurface groundwater discharge or discharge from land to ocean is induced by gradient and regulated by conductivity in terrestrial and aquatic environments. The salinity differences between groundwater and surface water measurements can be used in estimating advective flux.

14.2.2.4 Radium/Radon Isotopes

Groundwater fluxes include geochemical approaches that integrate advection flux over much larger areas. Two primary geochemical tracers, radon and radium isotopes, are strongly enriched in groundwater compared to surface water. Radon is an excellent tracer of groundwater discharge to the coastal ocean because it is enriched in groundwater by 1–3 orders of magnitude over activities in surface waters; behaves conservatively upon mixing of freshwater and seawater; is produced continuously from naturally occurring, long-lived U-series parent isotopes in sediments, soils, and rocks; decays with a 4-day half-life, ensuring that excess radon is only found near a site of groundwater discharge; and reflects discharge over a wide area. The flux of groundwater to coastal waters can be determined from an estimate of the radon flux to surface waters and the mean activity of radon in groundwater, using corrections for radon loss due to gas exchange and mixing (Burnett and Dulaiova, 2003). A similar approach can be used with radium isotopes (Moore, 1996, 2003). The interpretation of such fluxes based on radium isotopes is complicated by the strong dependence of radium activity on salinity in groundwater. These groundwater fluxes can be translated into fluxes of nutrients or other contaminants simply by multiplying the water flux by nutrient (and contaminant) concentration. Furthermore, the groundwater flux estimates can be compared to predictions based on groundwater flow models. See papers by Moore (2003) and Burnett et al. (2003) for methods used in calculating flux of groundwater discharges as determined by radium isotope and radon.

14.2.2.5 Dyes

Fluorescent and Rhodamine dyes can be used as tracers for the measurement of advection and longitudinal dispersion. Dye is injected into the soil at a selected depth and its distribution is surveyed. Net advection and longitudinal dispersion are then measured by drawing the water sample from the wells in the soil profile where the dye was injected. On the basis of dye distribution, quantitative estimates of transport processes (net advection and longitudinal dispersion) are then calculated. See paper by Ho et al. (2006) for the method of calculating diffusion.

14.3 DIFFUSIVE FLUX

14.3.1 Diffusive Flux Processes

Diffusive flux refers to material transport in response to a chemical potential gradient (concentration gradient) in a medium (gas or water or solids) that is stationary. If the medium is turbulent, then the transport of material is referred to as turbulent or eddy diffusion (Lerman, 1979). The flux (J_D) of solutes diffusing through stationary gas or liquid medium is proportional to the concentration gradient (dC/dx):

$$J_D = -D_0 \frac{dC}{dx} \tag{14.8}$$

where C is the concentration of chemical species (M L^{-3}), x is the length or distance (L), D_0 is the diffusion coefficient (L^2 T^{-1}), and J is the flux (M L^{-2} T^{-1}) (Figure 14.5). The preceding equation is known as Fick's first law and was developed in the 1850s by the German chemist Adolf Fick. Diffusion coefficients for selected ions of interest to wetlands and aquatic systems are presented in Table 14.1. Molecular diffusion is much faster in gases than in liquids. For example, the oxygen diffusion in still air is approximately 10,000 times faster than that in water.

The diffusion of solutes in soils is slower than that in an equivalent volume of water. Solutes diffusing through soils/sediments encounter irregular paths as they move through pore water. For example, the diffusion of the solute from point A to point B is not in a straight line as solutes have to travel around soil/sediment particles, thus increasing the overall diffusive path length (Figure 14.5). Therefore, both porosity and tortuosity of diffusive path regulate molecular diffusion. Porosity (ϕ) is defined as the ratio of volume of pore water to total volume of the soil and has values between 0 and 1.

The path length for diffusion increases due to tortuosity (θ), which is defined as the square of the ratio of the mean path length L_p of the path through the soil pore space between two points (A and B as shown in Figure 14.6) to the straight line distance (L) between the same points (Lerman, 1979).

$$\text{Tortuosity} = \theta = \left(\frac{L_p}{L}\right)^2 > 1 \tag{14.9}$$

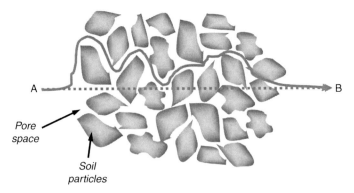

FIGURE 14.5 Schematic showing irregular path for solutes diffusing through soils.

TABLE 14.1
Diffusion Coefficients in Water ($D_0 = 10^{-6}$ cm^2 s^{-1}) for Ions of Interest to Wetlands and Aquatic Systems

Ion	0°C	18°C	25°C
H^+	56.1	81.7	93.1
Na^+	6.27	11.3	13.3
K^+	9.86	16.7	19.6
NH_4^+	9.8	16.8	19.8
Mg^{2+}	3.56	5.94	7.05
Ca^{2+}	3.73	6.73	7.93
Mn^{2+}	3.05	5.75	6.88
Fe^{2+}	3.41	5.82	7.19
Fe^{3+}		5.28	6.07
Al^{3+}	2.36	3.46	5.59
OH^-	25.6	44.9	52.7
Cl^-	10.1	17.1	20.3
Br^-	10.5	17.6	20.1
HS^-	9.75	14.8	17.3
SO_4^{2-}	5.00	8.90	10.7
NO_3^-	9.78	16.1	19.0
HCO_3^-			11.8
$H_2PO_4^-$		7.15	8.46
HPO_4^{2-}			7.34
PO_4^{3-}			6.12

To express D_0 as cm^2 day^{-1}, multiply the values given in the table by 8.64×10^4.
Source: Li and Gregory (1974).

Porosity dictates the extent of tortuous diffusive path. Soils of low porosity exhibit longer diffusive paths than those with high porosity (Sweerts et al., 1991). Boudreau (1997) suggested the following empirical relationship between tortuosity and porosity to estimate tortuosity factor.

$$\theta^2 = 1 - \ln(\varphi^2) \tag{14.10}$$

The diffusion coefficient of sediment or soil, in terms of tortuosity is expressed as:

$$D_s = \varphi D_0 / \theta^2 \tag{14.11}$$

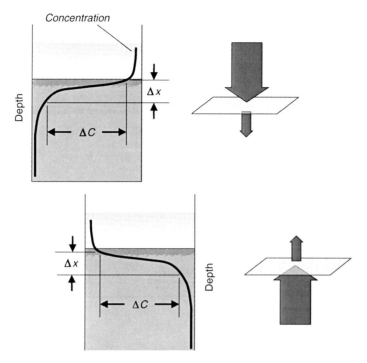

FIGURE 14.6 Schematic showing solute concentration gradients between soil and water column.

Where D_s is the whole sediment or soil diffusion coefficient in terms of area of sediment or soil per unit time, and D_0 is free ion diffusion coefficient measure in bulk solution (Berner, 1980).

The net flux of solutes due to diffusion across the soil or sediment–water interface can then be described as follows:

$$J_D = -\phi D_s \frac{dC}{dx} \tag{14.12}$$

If the concentration gradient in soil pore water is positive, $(dC/dx) > 0$, then the concentration increases with depth, as observed for many solutes including pore water concentration profiles for ammonium, phosphate, Fe(II), Mn(II), sulfide, methane, carbon dioxide, and other dissolved species. This indicates that the diffusive flux is upward from soil to the overlying water column. The value for "x" represents the depth where concentration of a solute shows linear gradient across the sediment–water interface (Figure 14.6). This depth can vary depending on soil type and the relative rates of biogeochemical processes. In wetlands and lake sediments, the depth of this linear gradient can range from 1 to 20 cm. If the concentration gradient is negative, $(dC/dx) < 0$, then the concentration decreases with depth, as observed for many solutes including pore water concentration profiles for nitrate and sulfates. This indicates that the diffusive flux is downward from water column to underlying soil/sediment.

The sample calculations for ammonium, phosphate, and sulfate fluxes are shown as follows.

14.3.1.1 Ammonium Flux

Soil porosity $(\phi) = 0.8$

Soil depth: $x_w = 0$ cm; $x_s = 10$ cm

ΔSoil depth with linear gradient in ammonium concentration $= x_s - x_w = (dx) = 10$ cm

Ammonium concentration: $C_w = 0$ μg cm^{-3}; $C_{pw} = 30$ μg cm^{-3}, where C_w is the ammonium concentration in the water and C_{pw} is the ammonium concentration in the soil pore water.

ΔAmmonium concentration $(C_{pw} - C_w) = (dC) = 30$ µg cm^{-3}
Free ion diffusion coefficient (at 25°C) = 19.8×10^{-6} cm^2 s^{-1} or 1.71 cm^2 day^{-1}
Calculated tortuosity factor = 1.45

$$J = 0.8/1.45 \times 1.71 \text{ cm}^2 \text{ day}^{-1} \times 30 \text{ µg cm}^{-3} \times 1/10 \text{ cm}$$
$$J = 2.83 \text{ µg cm}^{-2} \text{ day}^{-1}$$
$$J = 28.3 \text{ mg m}^{-2} \text{ day}^{-1}$$

Ammonium flux is upward from soil to water column as the flux value is positive.

14.3.1.2 Phosphate Flux

Soil porosity $(\phi) = 0.8$
Soil depth: $x_w = 0$ cm; $x_s = 10$ cm
ΔSoil depth with linear gradient in phosphate concentration = $x_s - x_w = (dx) = 10$ cm

Dissolved reactive P concentration: $C_w = 0$ µg cm^{-3}; $C_{pw} = 3$ µg cm^{-3}, where C_w is the phosphate concentration in the water and C_{pw} is the phosphate concentration in the soil pore water.

ΔDissolved reactive P concentration = $(C_{pw} - C_w) = (dC) = 3$ µg cm^{-3}
Free ion diffusion coefficient for $H_2PO_4^-$ (at 25°C) = 8.46×10^{-6} cm^2 s^{-1} or 0.731 cm^2 day^{-1}
Calculated tortuosity factor = 1.45

$$J = 0.8/1.45 \times 0.731 \text{ cm}^2 \text{ day}^{-1} \times 3 \text{ µg cm}^{-3} \times 1/10 \text{ cm}$$
$$J = 0.121 \text{ µg cm}^{-2} \text{ day}^{-1}$$
$$J = 1.21 \text{ mg m}^{-2} \text{ day}^{-1}$$

14.3.1.3 Sulfate Flux

Soil porosity $(\phi) = 0.8$
Soil depth: $x_w = 0$ cm; $x_s = 10$ cm
ΔSoil depth with linear gradient in sulfate concentration = $x_s - x_w = (dx) = 10$ cm

Sulfate concentration: $C_w = 30$ µg cm^{-3}; $C_{pw} = 0$ µg cm^{-3}, where C_w is the sulfate concentration in the water and C_{pw} is the sulfate concentration in the soil pore water.

ΔSulfate concentration $(C_{pw} - C_w) = (dC) = -30$ µg cm^{-3}

TABLE 14.2
Estimated Diffusive Flux of Ammonium, Phosphate, and Sulfate as Influenced by Soil or Sediment Porosity

Porosity (ϕ)	Tortuosity Factor (θ^2)	Ammonium Flux (mg m^{-2} day^{-1})	Phosphate Flux (mg m^{-2} day^{-1})	Sulfate Flux (mg m^{-2} day^{-1})
0.8	1.45	28.3	1.21	−15.3
0.6	2.02	15.2	0.65	−8.2
0.4	2.83	7.3	0.31	−3.9
0.2	4.21	2.4	0.11	−1.3

The dC/dx values are 3, 0.3, and −3 µg cm^{-4} for ammonium, phosphate, and sulfate, respectively. Positive flux indicates upward diffusion from soil or sediment to overlying water column, whereas negative flux is from water column into soil or sediment.

Free ion diffusion coefficient (at 25°C) = 10.7×10^{-6} cm^2 s^{-1} = 0.924 cm^2 day^{-1}
Calculated tortuosity factor = 1.45

$$J = -0.8/1.45 \times 0.924 \text{ cm}^2 \text{ day}^{-1} \times 30 \text{ μg cm}^{-3} \times 1/10 \text{ cm}$$

$$J = -1.53 \text{ μg cm}^{-2} \text{ day}^{-1}$$

$$J = -15.3 \text{ mg m}^{-2} \text{ day}^{-1}$$

The estimated diffusive flux values for soils with the range of porosities are shown in Table 14.2. The molecular diffusion of ammonium, phosphate, and sulfate is affected by soil porosity, with greater flux observed in soils of high porosity and low tortuosity.

14.4 BIOTURBATION

Bioturbation is a process of disruption of soils at the soil–water interface, resulting from macrobenthos burrowing/irrigation, feeding, and reworking of surficial sediments (Figures 14.7 and 14.8). In certain wetlands, the movement of animals such as cattle (in agricultural wetlands) and other wildlife can create physical disturbance in the soil and affect flux across soil–water interface. This effect on exchange processes is not documented. Macrobenthos are known to modify sediment properties due to their size as compared to sediment grains, their population density, and their ability to move freely across sediment–water interface. Very limited information is available for wetland ecosystems. The concepts presented in this section are primarily derived from marine and estuarine ecosystems. The bulk sediment mixing caused by random movement of benthos and solute flux is often lumped to describe bulk exchange of solutes between soil and overlying water column (Berner, 1980).

The intensity of bioturbation is expressed as biodiffusivity (D_B) and the effect of bioturbation versus the molecular diffusion is described as the ratio between D_B and D_S. A vast amount of literature is available on the role of macrobenthos on the exchange of solutes across sediment–water interface in freshwater and marine sediments. It should be noted that in wetlands, rooting of vegetation in soils creates additional complexity on exchange of solutes across soil–floodwater interface. The macrobenthos can influence the vertical distribution of sediments and POM, and the

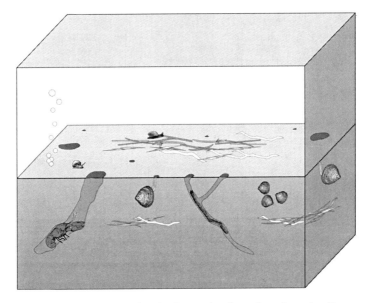

FIGURE 14.7 Schematic representation of major burrowing fauna in soils and sediments. (Redrawn from Aller, 1982.)

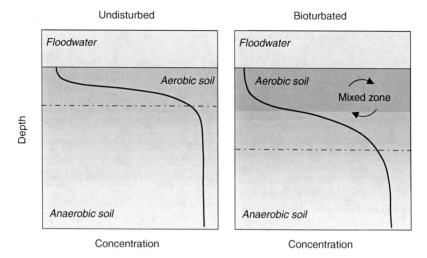

FIGURE 14.8 Schematic showing solute concentration gradients influenced by bioturbation.

stratification of electron acceptor distribution in sediments and its maintenance in four major ways (Aller, 1982; Kristensen, 2000):

- Material is transported continuously between soil/sediment and water column during feeding, burrowing, and tube construction.
- Burrow and fecal pellet formation alters reaction and solute diffusion geometries, creating a mosaic of biogeochemical microenvironments rather than a vertically stratified distribution (see Chapter 4).
- New reactive organic substrates in the form of mucus secretions may be introduced into the sediment independent of sedimentation processes, which may increase microbial activities.
- Tubes and burrows are irrigated with oxic surface water by ventilation.
- Feeding and mechanical disturbance may influence redox gradients and microbial populations that mediate biogeochemical reactions.

14.4.1 Macrobenthos Communities

Benthic invertebrates are important to the biodiversity of wetlands and aquatic systems, food webs, and nutrient cycling (Batzar et al., 1999). Their distribution, density, biomass, and species diversity is influenced by vegetation density, plant species composition, water depth, and hydroperiod of wetlands. Benthic invertebrates are diverse in their reproductive behaviors, growth, feeding, and respiratory activities, and they vary spatially. (McCall and Tevesz, 1982). This section is summarized from an excellent review presented by McCall and Tevesz (1982). For additional references on description of benthos, the reader is directed to their work. Freshwater and marine sediments are dominated by chironomid insect larvae, oligochaete worms, bivalve mollusks, and amphipods. Chironomid larvae feed on detrital particles from water and they inhabit U-shaped burrows lined with a transparent, fibrous salivary secretion in sediments. These larvae are likely to be 10–20 mm in length and weigh 1–2 mg. They live in the upper 8–10 cm depth and move freely between sediments and overlying water column creating channels. Chironomid larvae are known to be the most mobile of all benthic invertebrates and these larvae are able to leave sediments and swim up freely in the water column of 1-m depth. Oligocheates do not burrow into soils extensively and feed on soil particles at the soil–water interface. Bivalve mollusks move freely in sediments and feed on filter particles in the water column. Amphipods feed on bacteria and algae attached to detrital matter and inhabit in surface sediments. Adult amphipods may weigh 1–2 mg and are 5–10 mm long. Their density may range from a few hundred to 14,000 m^{-2}. Like chironomids, amphipods can also move freely between the sediment and the overlying water column. Amphipods live in burrows in the top 2 cm of the sediment.

14.4.2 Benthic Invertebrates and Sediment–Water Interactions

External loading of mineral sediments and POM and internally produced organic matter accumulate vertically on the soil surface of wetlands. In the absence of bioturbation, spatial variation in soil/sediment physicochemical properties is predominantly in the vertical direction. The simplest influence of benthic invertebrate activity is nonselective homogeneous mixing of surface sediments during burrowing, feeding, and construction of burrows. The extent of bioturbation and sediment mixing is influenced by species type, population density, size of specimen, feeding and respiratory behavior, and burrowing depth (Kristensen, 2000). The population density is spatially variable and density is influenced by sediment physicochemical properties (texture and organic matter content). Some benthos are capable of selecting particles on the basis of their size, shape, texture, and position. For example, some benthos can ingest fine particles from greater depths and excrete at the soil–floodwater interface, whereas some other benthos can rework the organic matter deposited on the soil surface and incorporate them into deeper layers. Tube-dwelling animals can irrigate sediments and enhance the transport of solutes by pumping pore water from sediments into the water column (Aller, 1983; Pelegri and Blackburn, 1994; Van Rees et al., 1996).

Natural populations of meiofauna (~100 cm^{-3}; nematodes, juvenile bivalves, and polychaetes) have been shown to increase solute transport by approximately twice the molecular diffusive flux, as shown for chloride and bromide (Figure 14.9) (Aller and Aller, 1992). These values are lower than

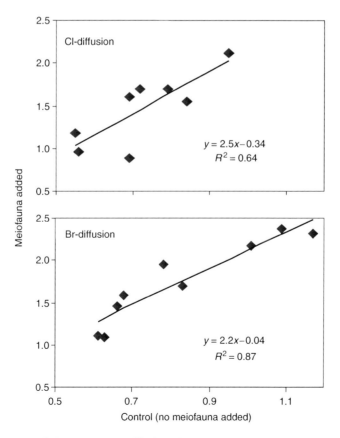

FIGURE 14.9 Influence of bioturbation on diffusion of chloride and bromide in the sediments. (Data from Aller and Aller, 1992.)

those observed for other systems with macrofaunal populations that increased solute transport by upto 10 times the molecular diffusion (Aller, 1982). Aller and Aller (1992) also found that the solute transport was significantly influenced by temperature, with Q_{10} values of 1.8. In shallow eutrophic lakes of the Danube Delta, bioturbation by oligochaetes accounted for nitrogen release rates of 378 and 960 mg N m^{-2} year^{-1} for lakes dominated by macrophytes and phytoplankton, respectively (Geta et al., 2004). In the same study, nitrogen excretion rates by these benthos were 60 and 154 mg N m^{-2} year^{-1} for lakes dominated by macrophytes and phytoplankton, respectively. Increased solute transport as a result of benthic invertebrate activity may influence nutrient cycling in wetlands and aquatic systems as follows:

- Nutrient sequestration in biomass and transfer of nutrients through food webs to other trophic levels
- Excretion of nutrients
- Steady flux of nutrients from sediments or soil to the overlying water column through channels created during bioturbation
- Transport of oxygen from the water column into anaerobic sediment or soil layers and subsequent oxidation of reductants (e.g., nitrification–denitrification, sulfide oxidation, and sulfate reduction)

14.5 WIND MIXING AND RESUSPENSION

The wind-induced mixing of surface waters with sediments can increase the flux of solutes across the sediment–water interface. This process can be significant in aquatic systems such as shallow lakes and open areas of wetlands. The significance of wind-driven resuspension on solute flux in lakes has been documented by several studies (Simon, 1988a, 1988b; Reddy et al., 1996), but no information is available on the importance of this process in wetland ecosystems. Vegetation and benthic mats of periphyton and detrital matter on soil surface can decrease the wind-driven resuspension of sediment and POM particles. A schematic of the processes involved in solute exchange across the sediment–water interface is shown in Figure 14.10. The depth of the surface sediment layer involved in mixing with water column may be highly variable depending on the shear stress and sediment texture.

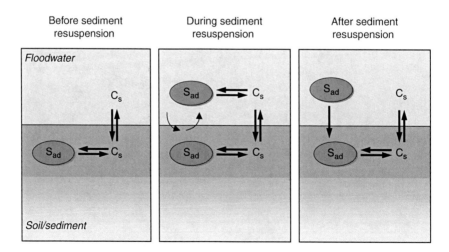

FIGURE 14.10 Schematic showing adsorption–desorption regulating solute concentration in the water column, as a result of resuspension and diffusive flux from sediment. S_{ad} is solute adsorbed on sediment particles and C_s is solute in solution.

Soil and Floodwater Exchange Processes

For shallow aquatic systems, the depth of surface sediment involved in resuspension is estimated to be about 10 cm. For shallow lakes, several studies have shown that wind-induced sediment resuspension can significantly increase soluble phosphorus and ammonium flux from sediments to the water column (Reddy et al., 1996). Sediment resuspension was the major mechanism for the cycling of ammonium between the sediments and the overlying water column in the transition zone of the Potomac River Estuary (Simon, 1988a, 1988b). During the periods of sediment resuspension, desorption of ammonium and phosphorus from the sediments into the solution can be a major pathway for enriching the water column. For example, in the Potomac River Estuary, the magnitude of ammonium flux during sediment resuspension resulted from the desorption of the suspended sediment solids rather than the pore water ammonium, which comprised only a small fraction of the total ammonium in the sediment (Simon, 1988b). Ammonium desorbed to the water column from resuspended sediment could be a greater source to the water column than the diffusive flux (see Section 14.6). However, this flux would occur in pulses (depending on the wind events) rather than a constant rate as it would occur for diffusive flux. Simon (1988b) estimated that 10–1,000 days are needed for the diffusive flux to add similar amounts of ammonium to the water column relative to one resuspension event lasting a few minutes.

In Lake Okeechobee, Florida, Reddy and Fisher (1991) estimated that under normal wind events, the resuspension flux of soluble phosphorus can account for approximately 6–18 times the diffusive flux. In comparison, resuspension of loosely bound, organically rich sediments into the water column was found not to be a major factor for delaying the recovery of shallow Danish lakes (Jeppesen et al., 2003). However, the resuspension of nutrient-enriched benthic sediments (accumulated during eutrophication process) along with the diffusive flux can result in the release of nutrients supporting phytoplankton productivity and delay the overall recovery process.

14.6 EXCHANGE OF DISSOLVED SOLUTES BETWEEN SOIL/SEDIMENT AND THE WATER COLUMN

The exchange of dissolved solutes across the soil–floodwater interface is calculated using Fick's first law (see Section 14.3.1) (Berner, 1980) and Equation [14.12]

$$J_D = -D_s \left(\frac{dC}{dx} \right) \tag{14.12}$$

In general, the total solute flux from sediments to overlying water column is the result of the sum of several mixing processes, which include molecular diffusion (D_S), biodiffusion (D_B), diffusion due to irrigation (D_I), and mixing due to wave and current mixing (D_{WC}). The sum of all these effects constitutes effective diffusion (D_E) from soil to the overlying water column.

$$D_E = D_S + D_B + D_I + D_{WC} \tag{14.13}$$

In soil pore water, solute transport as a result of D_B, D_I, and D_{WC} may be similar in effect to D_S but not in mechanisms (i.e., flux is not proportional to concentration gradient) (Van Rees et al., 1991). In deep-sea sediments solute flux due to D_B, D_I, and D_{WC} is smaller than D_S, whereas in shallow, organically rich sediments, D_I and D_{WC} values can be much greater than D_B and D_S (Berner, 1980). In addition to the direct flux, Li et al. (1997) reported that the solute flux also occurs as a result of the indirect processes: (1) the net release of solutes of particle-bound constituent due to bottom sediment entrainment and associated redeposition of eroded material and (2) the net dissolved solute flux due to the bottom pore water release associated with the exchanges at the sediment–water interface. However, below the zone of physical or biological disturbance, D_S is the dominant solute process. Therefore, molecular diffusion is an important process in the exchange of solutes and pollutants between soil and the overlying water column (Van Rees et al., 1991) (see Section 14.3.1 for details).

There are many methods available to quantify the extent of solute flux across the soil–water interface. Simple one-dimensional (depth) diagenetic models such as Fick's first law of diffusion are often used to estimate the vertical flux based on pore water concentration gradients. Solute gradients can be measured using the pore water equilibration devices or by the soil core sectioning and determining solute concentration changes with soil depth. The other methods for determining the solute flux include measuring the changes in water column concentrations of intact sediment cores through time or placing benthic flux chambers over the soil and monitoring changes in nutrient concentration in the water trapped above the soil. Examples of gradient-based and intact core methods are listed as follows.

- Gradient-based measurements:
  - Coring
  - Pore water equilibrators
  - Multisamplers
- Overlying water incubations with intact cores:
  - Benthic flux chambers
  - Intact soil core incubations

14.6.1 Gradient-Based Measurements

The *in situ* pore water sampler also called "pore water equilibrator" or dialyzer is based on the principle that given enough time, a contained quantity of water in the sampler and adjacent pore water will diffuse and equilibrate through a dialysis membrane, or other materials such as porous Teflon, with the surrounding water and its dissolved solutes. The *in situ* equilibrator can either be removed from the soil to collect pore water or the pore water can be collected through an attached tubing while the device remains in the sediment.

The pore water equilibrators are described by Hesslein (1976) and are now widely used in wetlands and aquatic systems to determine solute profiles in pore waters (Fisher and Reddy, 2001). The equilibrators are 2 cm thick, 10 cm wide, 50 cm long blocks of clear acrylic that consist of discrete 8 mL cells that are spaced at 1 cm vertical intervals (Figure 14.11). The cells are filled with deionized water, overlain with a 0.2 µm pore size polyethersulfone membrane, and a slotted, acrylic faceplate is fastened to the device. The membrane and faceplate are held tightly to the equilibrator with stainless steel screws to ensure that there is no mixing between the vertical cells. They are then sealed in acrylic cases and purged with nitrogen gas to ensure anaerobic conditions. In the field, the equilibrators are removed from the case, pushed into soil, and left for a period of 10–15 days. This allows time for dissolved constituents in pore water to equilibrate with the water inside the cells. At the end of equilibration period, the device is withdrawn from soils/sediments and individual cells are immediately sampled by withdrawing their contents with a syringe. The samples are stored at 4°C until analyzed for dissolved solutes such as ammonium, phosphate, sulfate, calcium, iron, magnesium, manganese, and dissolved gases such as carbon dioxide and methane. Examples of pore water profiles are shown in several chapters of this book. Other pore water samples have been proposed by Watson and Frickers (1999) and Steinman et al. (2001).

14.6.2 Overlying Water Incubations

14.6.2.1 Benthic Chambers

The exchange of solutes such as nutrients, dissolved organic substances, and metals and their associated processes at soil–floodwater interfaces are measured using laboratory-incubated cores or *in situ* pore water equilibrators as has been described. But these methods often underestimate fluxes because they do not account for processes such as bioturbation and bioirrigation at the soil–floodwater interface. To overcome these limitations, autonomous benthic chambers installed on top

Soil and Floodwater Exchange Processes

FIGURE 14.11 Pore water equilibrators used to determine *in situ* solute concentration gradients.

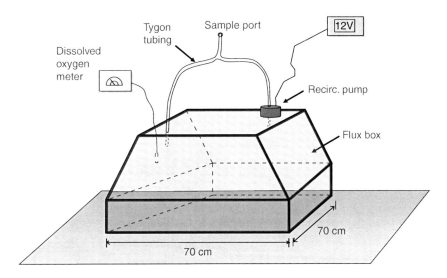

FIGURE 14.12 Schematic showing benthic chamber used to measure *in situ* solute flux.

of the soil surface are used. This method consists of isolating a certain amount of water and soil surface and measuring the change in concentration of a target solute over a predetermined time.

Benthic chambers are constructed in various sizes. For example, benthic chambers used by Fisher and Reddy (2001) were constructed of 6.35 mm (0.25 in.) thick acrylic and enclosed a soil surface area of 0.5 m^2 (Figure 14.12). Each of the chambers was equipped with a recirculation pump and ports for installing electrodes and other related sensors.

14.6.2.2 Intact Cores

Core tubes are made of clear polycarbonate, measuring about 7–10 cm internal diameter or higher and approximately 30–50 cm in length. Intact soil cores are obtained by using a specialized wetland soil core sampler or simply by mallet-driving core tubes into soil. A typical incubation setup is shown in Figures 14.13a and 14.13b. Soil cores are placed into a temperature-controlled water bath that is completely covered with a foil to exclude light, thus diminishing the chances of photosynthesis and the algal uptake of nutrients. Typically, flux measurements are taken over 7–14 days.

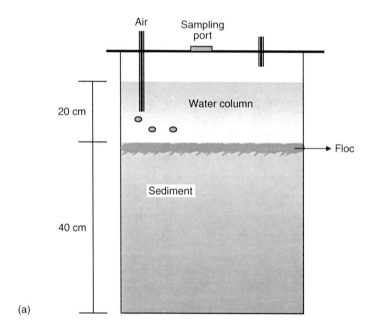

FIGURE 14.13 (a) Schematic showing intact soil core incubations used to determine the solute exchange between soil and water column. (b) A photograph showing intact soil core incubations used to determine the solute exchange between soil and water column.

Soil and Floodwater Exchange Processes

The temperature of the water bath is controlled as needed, but under most conditions *in situ* water column temperatures are maintained. Prior to starting the core flux measurements, the overlying water is removed from each core, filtered through a 0.45 μm pore size polyethersulfone filter and gently replaced into the respective cores to a final water column depth of approximately 15 cm. The water column is slowly bubbled with ambient air throughout the incubation period, both to ensure an aerobic water column and to completely mix the water column. Small aliquots (~20 mL) of water samples are collected and prepared for analysis of desired solutes.

Solute fluxes are calculated from the change in concentration with time by using the following equation:

$$J_i = \frac{dC_i}{dt} \frac{V}{A} \tag{14.14}$$

where J_i is the flux of component i (M L^{-2} T^{-1}), C_i is the concentration of i in water column (M L^{-3}), V is the water column volume (L), A is the soil surface area (L^2), and t is the time (T). Only the initial linear portion of the release curve is used to calculate solute release or retention rates. Two exchange processes are shown in Figures 14.14 and 14.15. The flux of oxygen from the water column to the sediment is measured using the intact core method (Figure 14.14). By this method, the soil or sediment oxygen can be measured. Data in Figure 14.15 show the soluble phosphorus flux from the sediment to the overlying water column. The change in concentration of soluble phosphorus concentration in the water column is used to calculate the net flux from the sediment to water column.

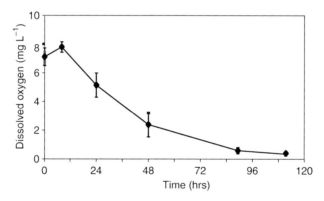

FIGURE 14.14 Flux of dissolved oxygen from the water column to the sediment, measured using the intact sediment cores and laboratory incubations. (Reddy, K. R., Unpublished Results, University of Florida.)

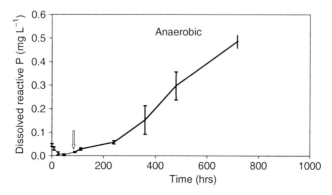

FIGURE 14.15 Flux of dissolved phosphorus oxygen from the sediment to water column, measured using the intact sediment cores and laboratory incubations. (Reddy, K. R., Unpublished Results, University of Florida.) Arrow at 100 hours indicates oxygen depletion in the water column (see Figure 14.14) and anaerobic conditions.

14.7 SEDIMENT TRANSPORT PROCESSES

Sediment deposition can vary in wetlands and is influenced by surface inflows such as overland flow and channelized stream flow. The capacity of flowing water to transport sediment into wetlands depends on velocity and the size of sediment particles being transported. The general size of sand, silt, and clay particles is 0.06–6 mm, 0.002–0.06 mm, and <0.002 mm, respectively. The small-size clay particles can be transported to a greater distance with less velocity than the large-size silt and sand particles. The channelized stream flowing into and out of wetlands can be described as the product of average stream velocity and cross-sectional area of the stream, using the following equation:

$$Q = AU \tag{14.15}$$

where Q is surface channelized flow ($L^3 T^{-1}$), A is cross-sectional area (L^2), and U is average velocity ($L T^{-1}$). The stream velocity can be calculated by means of the Manning equation:

$$U = \frac{1}{n} R^{2/3} s^{1/2} \tag{14.16}$$

where R is the hydraulic radius, m is the cross-sectional area divided by wetted perimeter (L), s is the stream gradient or channel slope (dimensionless), and n is the roughness coefficient or Manning coefficient (s m$^{-1/3}$) (T L$^{-1/3}$). The velocity decreases with the increase in hydraulic radius, a decrease in stream gradient, or an increase in the roughness coefficient. When a mass of solute is introduced at a point near the inflow of a wetland, the solute moves downstream at the average velocity of water. The average amount of time it takes for a solute to move from point A to point B in the stream channel (length or reach in the stream) is called travel time, which can be estimated as follows:

$$\tau = \frac{L}{U} \tag{14.17}$$

where τ is the travel time (T), L the length (L), and U the average velocity (L T^{-1}). The mass of solute transported by a stream or channel past a given point per unit time is the product of discharge Q(L^3 T^{-1}) and the average solute concentration C(M L^{-3}), with the units of M T^{-1}.

When a stream floods a wetland, the resultant decrease in water velocity allows stream sediment to be deposited on the wetland surface. Depositional fluxes (1) result from the settling of particulate matter from the water column and (2) are influenced by sediment properties, chemical properties, hydraulic gradients, water velocity, and hydraulic retention time (HRT). As velocity decreases, sand, silt, clay, and other particles begin to settle out of the water and form layers on wetland surfaces. This process is called deposition. Not all sediments settle out at the same rate. The resultant force (gravity versus friction) on sediment grain determines the rate of settling. The smaller the grain, the bigger the surface area per mass. In the case of clay which has high surface area per mass values, there is more friction, and therefore a slower settling rate than for either silt- or sand-size particles. The movement of sediment and POM through water under the influence of gravity is described by the Stokes equation, which relates the terminal settling velocity of a smooth, rigid sphere in a viscous fluid of known density and viscosity to a diameter of the sphere when subjected to a known force field. The rate of settling of particles from the water column is calculated as follows:

$$U_f = \frac{2}{9} \frac{(gr^2)(\rho_p - \rho_f)}{\mu} \tag{14.18}$$

where U_f is the velocity of fall (L T^{-1}), g is the acceleration of gravity (L T^{-2}), r is the "equivalent" radius of particle (L), ρ_p is the density of particle (M L^{-3}), ρ_f is the density of fluid medium (M L^{-3}), and μ is the viscosity of medium (g cm^{-1} s^{-1}). U_f is the particles' settling velocity (vertically downward if $\rho_p > \rho_f$ and upward if $\rho_p < \rho_f$). The clay-size particles are slow to settle out as compared to the silt- and sand-size particles. As a result, the clay particles can be transported a greater distance into the wetlands than the sand sediment particles. The settling of sediment/organic particle has important implications with the dynamics of nutrients, metals, and toxic organic compounds in wetlands.

14.7.1 SEDIMENT/ORGANIC MATTER ACCRETION IN WETLANDS

Studies of sediment accretion in wetlands focused on different interests and viewpoints such as sediment stratigraphy, heavy metal accumulation, historic shoreline change, and plant ecology. Geologists often address wetland stability from a long-term sedimentary viewpoint, studying sediment stratigraphy and erosion rates. Hydrologists focus on the transport of material in and out of the wetland. Ecologists are interested in sedimentation from the perspective of habitat change, plant stress, nutrient cycling, and accretion balances. Policy analysts also study the effects of sea level rise on wetlands, determining rates of loss and potential societal and economic impacts of wetland stability. Sediment accretion in wetlands is regulated by a range of physical, chemical, and biological processes. Typically, the water movement in a wetland is slower, which aids in the physical settling of the sediment particles. Physical processes such as the collision of sediment particles with stems of vegetation and the trapping of sediment particles in detrital matter aid the settling of larger particles near wetland inflows and smaller particles are transported farther from inflow point (Figure 14.16). In addition to anthropogenic inputs of sediment and particulate matter, wetlands also generate sediments by detrital matter decomposition processes.

Much of the coastal sedimentation accretion work has been conducted in the Mississippi River deltaic plain, where vertical accretion rates are large, and where there are also very high rates of subsidence and coastal land loss. In other coastal areas of the United States and Europe, a wide range of results have been found, with sedimentation rates varying from 0 to 1.5 cm year^{-1}. Many coastal marshes are not accreting at a rate sufficient to compensate for the present rates of sea level rise. Many factors affect accretion rates in wetlands. These include plant community, density of vegetation, tidal elevation, sediment input from riverine, estuarine and marine sources, proximity to sediment sources, total organic matter input from primary productivity of wetland, and relative sea level rise.

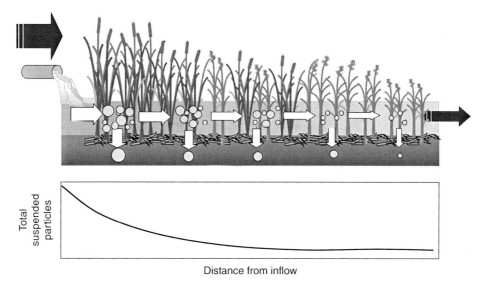

FIGURE 14.16 Schematic showing settling of suspended solids as a function of distance from the inflow of a wetland.

In an environment of increasing water level, wetlands maintain elevation by vertically accreting mineral sediment or organic matter. The relative contribution of organic and inorganic material is related to geographic, geological, and hydrological factors. In tidal wetlands, the tidal currents resuspend the near-shore sediments and transport them onto the salt marsh surfaces (Pethick, 1992). The result is a highly inorganic marsh soil. In microtidal areas, storm events, which resuspend the sediment in inland bays and other bodies, replace tides as the mechanism of mineral sediment input into wetlands. Organic soil production is related to climate and geology, which also affects large river systems supplying freshwater and sediments to coastal wetland areas.

Along the Louisiana Gulf coast, due to leveeing of the Mississippi River deltaic plain marsh has been hydrologically isolated from riverine sediment sources. Sediment supply is primarily from the reworked sediment or tidal source. Extending inland along a salinity gradient from the coast mineral sediments thus constitutes a progressively decreasing fraction of soil solids along a transect extending inland from the coast. This seaward gradient is a direct consequence of the hydrological regime. The high-energy marine processes provide an abundance of reworked mineral sediments to the salt marshes near the coast; this energy is progressively attenuated and dissipated inland from the coast, with a consequent decrease in the suspended load of floodwaters. The influx of fluvial sediments is limited in the upper reaches of the basin due to leveeing of the Mississippi River. As a result, plant remains constitute an increasing fraction of soil solids as the marine influence diminishes inland from the coast, and as such are of greatest structural significance in low-density freshwater and intermediate environments.

Louisiana coastal marsh soil bulk density ranged from more than 0.3 g cm^{-3} in salt marshes to as little as 0.05 g cm^{-3} in freshwater and brackish marsh environments receiving little sediment input (Hatton et al., 1983) (Figure 14.17). Organic carbon (dry weight %) varies significantly along transects from saline to freshwater marsh environments, with progressively higher concentrations in the less saline environments (Figure 14.18). The values ranged from a maximum of almost 40% carbon (or 86% organic matter) in fresh marsh to as little as 5% carbon in salt marsh in lower Barataria Basin. The bulk density is related to organic carbon content of the sediment. The bulk density decreases with an increase in organic carbon content of sediment (Figure 14.19).

The relative distribution of organic and mineral sediment accretion is related to pulses of water and sediments from tidal floods, river floods, or storm events (Hensel et al., 1999). The river discharge into a coastal region provides nutrients and reduces salinity stress, thereby increasing

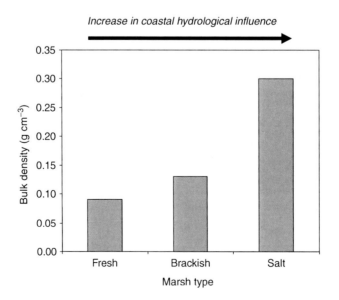

FIGURE 14.17 Variation of soil bulk density with marsh type. (Modified from Hatton et al., 1983.)

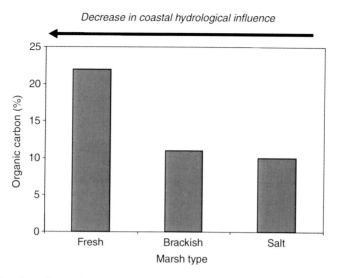

FIGURE 14.18 Variation of organic matter content with marsh type. (Modified from Hatton et al., 1983.)

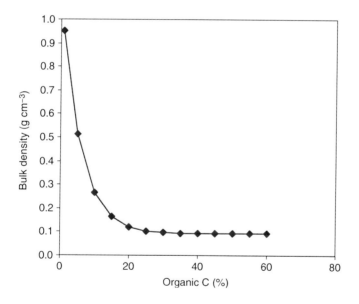

FIGURE 14.19 Relationship between bulk density and soil organic carbon content in Louisiana coastal marsh soil. (Data from Hatton et al., 1983.)

plant productivity and subsequent peat formation. The river discharge and over bank flooding supports sediments, which add nutrients to soil. The relative contributions of primary production and sediment pulses determine the importance of organic and inorganic matter in soil formation. The wetland vegetative growth supported by nutrient inputs enhances the organic matter in the soil structure. Vegetation traps sediments by baffling currents and enhancing deposition.

Even though geological, hydrological, and ecological factors strongly govern sedimentation, human impacts also influence sediment input into wetland. Impoundments restrict water and material fluxes in wetlands restricting mineral sediment input (Boumans and Day, 1993). Canals and associated spoil banks also reduce sediment input into wetlands. Dams reduce sediment discharge in many rivers reducing the supply to river deltas where major wetlands exist.

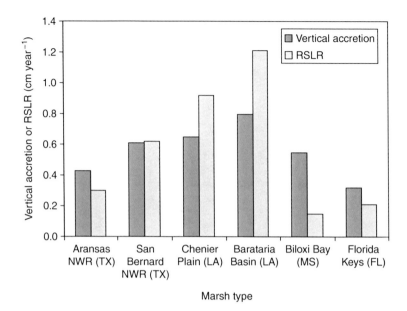

FIGURE 14.20 Vertical marsh accretion–relative sea level rise (RSLR) at selected marsh locations along the northern shore of the Gulf of Mexico. (Modified from Callaway et al., 1997 which includes data from DeLaune et al., 1983 and Hatton et al., 1983.)

In the Mississippi River deltaic plain, the suspended load of the Mississippi River presently reaching the Gulf is approximately half of the sediment reaching the Gulf prior to the 1950s when dams were placed on many tributaries of the river and extensive channel control works were implemented (Meads and Parker, 1985). Artificial levees, which line the Mississippi River, prevent sediment from being dispersed into the adjacent floodplain and wetland. Mississippi River sediments are now funneled to the mouth of the river and discharged off the continental shelf. The isolation of the Louisiana coastal wetland from riverine sediments has impaired their ability to keep pace with increase in relative sea level rise. Accretion by organic soil formation is very important in microtidal wetlands if storm frequency is low or wetlands are far from active river mouths or sediment sources. In sediment-deprived wetlands, organic matter accumulation is important in maintaining marsh elevation in wetlands distant from active riverine sediment sources (Hatton et al., 1983; Nyman et al., 1990).

Extensive data on accretion rates in the Louisiana Mississippi deltaic plain clearly document that marsh accretion is not keeping pace with submergence. Accretion rates were determined from Cs-137 dating range from 0.55 to 1.4 cm year^{-1}. Accretion rates were greater in Louisiana marshes as compared to other marsh sites along the Gulf of Mexico (Figure 14.20). However, the marsh locations along the Gulf of Mexico outside of Louisiana show that accretion is keeping pace with relative sea level rise. Even though there are regional differences in accretion rates in other U.S. Gulf Coast marshes, in contrast to Louisiana's Gulf Coast marsh, vertical accretion has been in general sufficient to compensate for the sea level rise.

Along the east coast of the United States, eustatic sea rise is generally the primary factor governing marsh accretion and stability. By comparison, along the Louisiana Gulf Coast, subsidence is of far greater significance than global sea level rise in maintaining coastal marsh stability. Table 14.3. shows accretion rates for other coastal areas of the United States and Europe.

14.7.2 Measurement of Sedimentation or Accretion Rates

Many methods have been used to measure accretion and sedimentation in wetlands. Each method gives a result relative to a different time period. Most measurements use some variation of a marker horizon or radiometric dating. The time period over which the rates are measured varies from a few

TABLE 14.3
Sediment Accretion Rates from Coastal Wetlands beyond the Gulf Coast

Location	Method Used	Sediment Accretion Rates (cm year^{-1})	Reference
East coast			
Multiple sites, Maine	Marker horizon	0–1.3	Wood et al. (1989)
Nauset Marsh, Massachusetts	^{210}Pb, ^{137}Cs, marker	0–2.4	Roman et al. (1997)
Waquoit Bay, Massachusetts	^{210}Pb	0.3–0.5	Orson and Howes (1992)
Narragansett Bay, Rhode Island	^{210}Pb	0.2–0.6	Bricker-Urso et al. (1989)
Barn Island, Connecticut	^{210}Pb, ^{137}Cs, marker	0.1–0.4	Orson et al. (1998)
Farm River, Connecticut	^{210}Pb	0.5	McCaffrey and Thomson (1980)
Long Island Sound, Connecticut	^{210}Pb, ^{137}Cs	0.1–0.7	Anisfeld et al. (1999)
Flax Pond, New York	^{210}Pb	0.5–0.6	Armentano and Woodwell (1975)
Great Marsh, Delaware	^{210}Pb	0.5	Church et al. (1981)
Delmarva Peninsula, Virginia	^{210}Pb	0.1–0.2	Kastler and Wiberg (1996)
Chesapeake Bay, Maryland	^{210}Pb	0.2–0.4	Stevenson et al. (1985)
Chesapeake Bay, Maryland	^{210}Pb	0.5–0.7	Kearney and Ward (1986)
Chesapeake Bay, Maryland	^{210}Pb, ^{137}Cs	0.3–0.8	Kearney and Stevenson (1991)
Chesapeake Bay, Maryland	^{210}Pb, ^{137}Cs	0.3–0.8	Kearney et al. (1994)
Chesapeake Bay, Maryland	^{210}Pb	0.4–0.8	Griffin and Rabenhorst (1989)
Pamlico Sound, North Carolina	^{137}Cs	0–0.5	Craft et al. (1993)
Sapelo Island, Georgia	Marker horizon	0.2–0.7	Letzsch (1983)
Everglades, Florida	^{210}Pb, ^{137}Cs	0.1–0.8	Craft and Richardson (1998)
Everglades, Florida	^{137}Cs	0.1–1.2	Reddy et al. (1993)
Upper St. Johns River Basin, Florida	^{137}Cs	0.3–0.5	Brenner et al. (2001)
Pacific coast			
Tijuana Estuary	Marker horizon	0.1–8.5	Cahoon et al. (1996)
San Francisco Bay, California	^{137}Cs	0.4–4.2	Patrick and DeLaune (1990)
Multiple sites, Oregon and Washington	^{137}Cs	0.2–0.7	Thom (1992)
Europe			
Severn Estuary, England	^{210}Pb	0.4	French et al. (1994)
Scolt Head Island, England	Marker horizon	0.1–1.4	Stoddart et al. (1989)
Scolt Head Island, England	Marker horizon	0.1–0.8	French and Spencer (1993)
Eastern Scheldt, Netherlands	^{137}Cs, marker horizon	0.4–1.6	Oenema and DeLaune (1988)
Island of Sylt, Germany	^{210}Pb, ^{137}Cs	0.6–1.5	Kirchner and Ehlers (1998)
Multiple sites, England, Netherlands, Poland	^{137}Cs	0.3–1.9	Callaway et al. (1996a, 1996b)
Floodplain wetland, Las Tablas de Daimiel, Spain		1.6–3.8	Sanchez-Carillo et al. (2001)

Accretion rates were measured using ^{210}Pb (100-year timescale), ^{137}Cs (20–35-year timescale), and marker horizons (1–10-year timescale).
Source: Modified from Callaway (2001) and DeLaune et al. (2001).

months or years (feldspar markers) to 30–40 years (^{137}Cs), 100 years (^{210}Pb), or thousands of years (^{14}C). Other techniques evaluate changes in relative elevation with time. These techniques include surveying and sedimentation–erosion tables (SETs), which measure the net changes in sediment surface elevation relative to some sort of a benchmark. The components of subsidence may be a part of these measurements depending on the type of benchmark used.

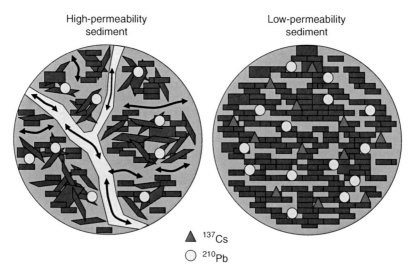

FIGURE 14.21 Radionuclide dating can be successfully used in low-permeability soils and sediments (high in clay or organic matter) when there is little migration.

The tools used to quantify depositional rates are

- Coring followed by ^{7}Be, ^{137}Cs, ^{210}Pb, and ^{14}C dating
- Artificial marker horizon, sediment traps, and SETs

Additional techniques, such as the elevation of geomorphic features and pollen analysis have also been used to cover much longer periods.

The most common physical methods for measuring vertical marsh accretion is placement of artificial marker horizons on the marsh surface such as the use of filter pad traps and SETs. Radioisotopes such as beryllium-7, carbon-14, cesium-137, and lead-210 are used to date the recently deposited sediment. Dating requires no transport in the sediment profile. Low permeability in sediments high in clay and organic matter retain radioisotope with little transport in sediment profiles and therefore are best for use in dating. High permeability sediments (sands) in general are not suitable for radioisotope dating using Cs-137 and Pb-210 (Figure 14.21). Several techniques for quantifying the rates of vertical accretion and sedimentation in wetlands are described in the following paragraphs.

14.7.2.1 Filter Pad Traps

The short-term sediment accumulation can be measured by collecting sediment on preweighed glass fiber filters (GF/F) with an underlying Petri dish base (Reed, 1989). Sediment traps are pinned to the sediment at the level of marsh surface with a wire mesh cage anchored to the marsh around the filter pads for protection from interference by fauna and large detritus. Filter pads can be collected at weekly intervals for determining short-term sediment deposits during the intervals. The pads are dried in an oven at 60°C to obtain the dry weight of sediments.

14.7.2.2 Artificial Marker Horizons

Many materials are used for establishing a marker horizon. These include brick dust, sand, kaolin, glitter, and feldspar clay. White feldspar clay is generally the best marker material because it is easily distinguishable from the surrounding sediment (Figure 14.22). Feldspar clay works well in most wetland environments. The feldspar marker may not work well in high-energy areas (marker

FIGURE 14.22 Wetland soil core showing white feldspar clay marker horizon.

gets washed away) or in highly organic porous marsh soils. Bioturbation, in some cases, may be a problem. Marker horizon material is generally spread over the marsh surface in replicated 0.5 m² plots. The plots are sampled with time to determine the depth of burial of the feldspar marker. When used simultaneously with the SET techniques, the information on the belowground processes that influence elevation can also be measured. Differences between rates of vertical accretion and elevation change can be attributed to processes occurring below a marker horizon and above the bottom of the SET pipe representing a zone of shallow subsidence.

14.7.2.3 Sedimentation–Erosion Table

SETs offer a method of measuring wetland surface elevation with precision, for example within 1.5 mm (Boumans and Day, 1993; Cahoon et al., 1995). Measurements of elevation changes with SET, when simultaneously combined with short-term accretion marker horizons, allow for the determination of shallow compaction (Cahoon et al., 1995). If no compaction is occurring at the site, short-term accretion rates will equal the changes in elevation. When compaction occurs, it will serve to reduce elevation, and compaction is calculated as sediment accretion minus the change in elevation. The use of both methods together can yield short-term estimates of sediment dynamics plus compaction rates.

The SET has a supporting aluminum base pipe placed permanently at each site that is designed to receive the upper portable part of the SET (Figure 14.23). This core pipe is driven into the soil to refusal using either a vibracorer or a hand-held pile driver as close to vertical as possible. The core pipe is then cut off at a few centimeters above the sediment surface and filled to within a meter of the surface with quick-setting cement. The elevation of the top of the pipe after cutting may vary depending on water depth at the site and the tidal range. Next, an aluminum base support pipe (diameter 7 cm, length 60 cm, wall thickness 3 mm) is cemented into the top of the core pipe with mortar mix. No cement is placed inside the top 30 cm of the base support pipe to allow the portable part of the SET to be inserted. The support pipe extends about 5 cm above the core pipe and is leveled to vertical. The portable part of the SET has four components: a vertical arm, a horizontal arm, a flat plate or table, and pins.

The horizontal arm of the SET can be leveled in two planes. A bubble level determines when the horizontal arm is leveled in both planes. When leveled, the table on the end of the horizontal arm provides a constant reference plane in space. The distance to the sediment surface is measured with pins passing through holes in the table on to the marsh surface. The length of each pin above the table is measured with a ruler to the nearest millimeter. This procedure is repeated for each of the four directions to yield an elevation measurement.

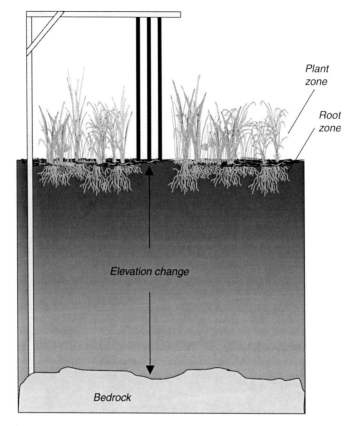

FIGURE 14.23 Diagram of sedimentation–erosion table.

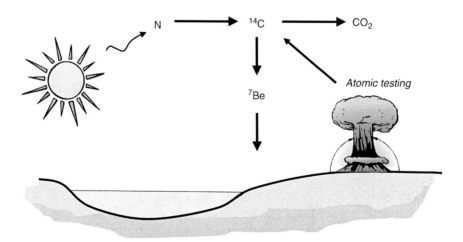

FIGURE 14.24 Atmospheric sources of beryllium-7 and carbon-14.

14.7.2.4 Beryllium-7 Dating

Beryllium-7 is a naturally occurring radioisotope that is produced by cosmic ray bombardment of atmospheric nitrogen (N) and oxygen (O) (Figure 14.24). Precipitated in the atmosphere of the earth, beryllium is a highly reactive element that is tightly bound to the sedimentary substrate. It has a half-life of 53 days, which makes its effective range of applicability for dating sediment about

Soil and Floodwater Exchange Processes

1 year. Thus, detection of its presence in surface sediment is a reliable indicator that the sediment was in contact with the atmosphere within the past year. The presence of beryllium in a surface profile is a good indicator of recently deposited sediment. As a result, beryllium is used to determine regional short-term sediment distribution patterns.

14.7.2.5 Lead-210 Dating

Lead-210 with half-life of 22.3 years can be used to date sediment for the last 150 years. A member of ^{238}U family, ^{210}Pb forms by the decay of its intermediate gaseous parent, radon-222. ^{222}Rn, formed by the decay of radium, escapes into the atmosphere by diffusion, and rapidly decays to form ^{210}Pb. This isotope has about 10 days of residence time in the atmosphere before it is removed by precipitation (Figure 14.25). Its concentration in the rainwater remains constant over a very long period, and its rate of accumulation in the uppermost layer of soil is constant for any given wetland system. The highly reactive lead is then rapidly adsorbed to and incorporated into the depositing sediment. This flux produces a concentration of "unsupported" ^{210}Pb (lead whose activity in the sediment is higher than that of its radium grandparent, ^{226}Ra). ^{210}Pb decays by β-particle emission with a half-life of 22 years (Figure 14.26). The decay rate and the accuracy make the method suitable for sediment ages up to 150 years. The dates of sediment deposition are calculated by determining the decrease in ^{210}Pb activity at each selected sediment interval as a function of time (Figure 14.27).

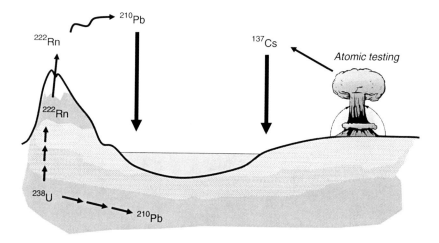

FIGURE 14.25 Atmospheric sources of lead-210 and cesium-137.

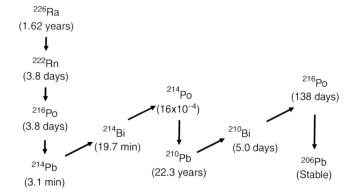

FIGURE 14.26 Decay pathways and products of radium-226 and lead-210.

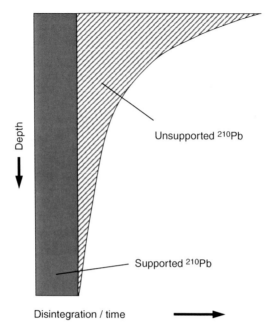

FIGURE 14.27 Schematic showing the vertical distribution of lead-210 activity (supported and unsupported Pb) in sediments.

Supported ^{210}Pb is based on the average measurement of ^{210}Pb activity determined by measuring the activity of a ^{210}Pb as a decay product such as Pb in the lowest section of the sediment profile where ^{210}Pb activity is constant. Sedimentation or accretion rates are estimated from the excess (unsupported) ^{210}Pb profiles in the sediment profile using the constant initial concentration method (Goldberg et al., 1977). Linear regression analysis is used to solve for (λ/s) in the log-transformed equation for radioactive decay:

$$\ln A_x = \ln A_0 - \frac{\lambda}{s} x \tag{14.19}$$

where x is the depth of section (cm), A_x is the excess activity of ^{210}Pb at a depth of x (dpm g^{-1}), A_0 is the excess activity of ^{210}Pb at surface (intercept in the regression), λ is the decay constant for ^{210}Pb = 0.03114 year^{-1}, and s is the vertical accretion rate (cm year^{-1}).

The slope of the regression in the ^{210}Pb profiles is equal to ($-\lambda/s$). More negative slope indicates lower accretion rates.

14.7.2.6 Cesium-137 Dating

Cesium-137 is a fallout radioisotope from nuclear weapons' tests and does not occur naturally. The first significant appearance of Cs-137 in the atmosphere was in the early 1950s, and it was present in peak quantities in 1963–1964. Like Pb-210, Cs-137 is carried down by rainwater and accumulates in the sediment, where it decays radioactively with a half-life of 30 years. Thus, a profile of Cs-137 concentration with depth (Figure 14.28) shows a maximum activity at a depth corresponding to 1963 and a "tail", where cesium-137 is first detectable associated with deposition in the 1950s. The early 1980s peak is associated with the fallout from the 1986 accident at Chernobyl. This peak is also used to determine sediment accumulation rates in selected Northern European wetlands. In these wetlands, the Chernobyl peaks can be used to measure sedimentation ratios from 1963 to 1986 and 1986 to present.

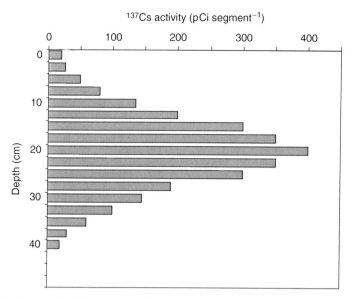

FIGURE 14.28 Typical vertical distribution of cesium-137 used to determine sediment and organic matter accretion in wetlands.

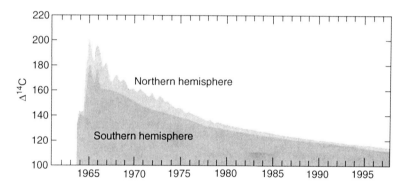

FIGURE 14.29 Tropospheric ^{14}C observations. (From Clark and Fritz, 1997.)

14.7.2.7 Carbon-14 Dating

Carbon-14 is produced in the earth's atmosphere by the interaction of cosmic ray particles with nitrogen (N), oxygen (O), and carbon (C). Nitrogen is the most important in terms of the amount of ^{14}C produced. ^{14}C was also produced by thermonuclear activity (bomb testing), which contributed significantly to the atmospheric ^{14}C levels, reaching peaks in 1963 (Northern Hemisphere) and 1964 (Southern Hemisphere) (Figure 14.29). The ^{14}C produced is oxidized to carbon dioxide and assimilated into the global carbon cycle. When carbon dioxide is incorporated in organic material a balance is established between the intake, respiration, and decay. ^{14}C has a half-life of 5,730 years and has an effective range of 500–50,000 years for dating organic material in a soil/sediment profile. The amount of bomb-produced carbon is determined by comparing the present radiocarbon activity to that in the year 1950, the date established by convention as the baseline for all radiocarbon dating. The values of ^{14}C in the sediment profile are reported as a percentage of modern (i.e., 1950) carbon, and denoted as $\Delta^{14}C$.

14.7.2.8 Application of Sediment Dating

Once the date/depth reference points or curve have been obtained from the dating studies as outlined, it is possible to relate the past history to the sediment soil profile. By chemical, biological, or microscopical examination of cores to the timescale of their formation, the history and the changes that have occurred over a given period can be documented in some detail. The comparison of recent changes to past changes in environmental conditions can be assessed to determine if there have been changes in anthropogenic inputs of pollutants, in the rate of carbon accumulation, and in the sediment input in wetland marsh soil profile.

Sediment accumulation determined by various profile analyses is calculated using sedimentation and bulk density of the wetland soil profile. The flux rate is calculated from the vertical accretion rate (R) and the bulk concentration of the material (C_v) using the expression

$$A = C_v R = (C^D D) R \tag{14.20}$$

where C^D is the steady-state dry weight concentration, A is the sediment accretion rate, and D is the bulk density. Accumulation of nutrients, organic matter, metals, etc., can be determined by the analysis of contents in the sediment profile and using appropriate calculations based on the bulk density and sediment accretion rate.

14.8 VEGETATIVE FLUX/DETRITAL EXPORT

Vegetation influences the diagenetic processes in a number of ways:

- Detrital and elemental accumulation in surface soils
- Organic matter and elemental accumulation in subsurface soils
- Mining of subsurface nutrients by vegetation and deposition in surface soils (through detrital accumulation)
- Transpiration by vegetation increases downward flux of solutes from water column to soils
- Particulate and dissolved organic matter export from wetlands to downstream aquatic ecosystems

Owing to the large standing crops and high plant productivity in wetlands, nutrient and metal uptake can be significant. High net productivity (2–4 kg dry matter m^{-2} $year^{-1}$) and low rate of herbivory in *Spartina* marshes can result in accumulation of large pools of labile organic matter in soils (Craft et al., 1989). Some of the element removal is sequestered in woody vegetation. Nutrients absorbed into the herbaceous material, generally, are recycled as the vegetation senesces and detrital matter deposited on the soil surface. A significant portion of nutrients assimilated by plants is flushed from the wetland as detritus or dissolved nutrient that is released from decaying vegetation (see Chapters 5, 8, and 9). For example, in estuarine environments, tidal export of detrital and dissolved organic matter can be significant in regulating secondary productivity of these ecosystems. The net accumulation of organic matter in surface soil layers varies by wetland type (emergent marshes versus forested wetlands), primary productivity, climate, hydroperiod, and decomposition processes (see Chapter 5).

The nutrient uptake by vegetation contributes to nutrient reduction in the soil profile with time. In low-nutrient systems, plants can sequester nutrients from the subsurface soil layers and deposit them on soil surfaces through detrital accumulation and increasing the connectivity of nutrients with water. Vegetative water uptake and transpiration can increase the solute flux from water column into the soil (Figure 14.30). For example, Martin et al. (2003) showed a greater reduction of surface water nitrate concentration in experimental *Typha* mesocosms with greater rates of evapotranspiration,

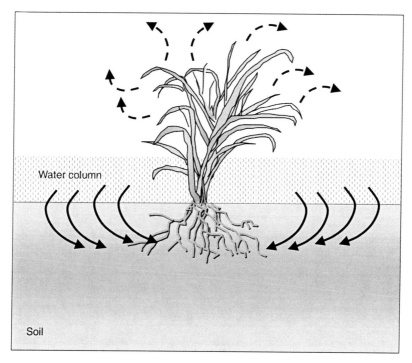

FIGURE 14.30 Flux of solutes as influenced by evapotranspiration in wetlands. Solute flux from water column to soil is influenced by hydraulic mass flux caused by water uptake and transpiration by vegetation. (Modified from Martin et al., 2003.)

supporting the hypothesis that vegetation increases the nitrate removal from wetland surface waters by increasing the soluble nitrate flux with vegetative water uptake (Martin and Reddy, 1997).

14.9 AIR–WATER EXCHANGE

Similar to the exchange of dissolved substances between the soil and overlying water column, exchange of gases between water and air can play a significant role in biogeochemical cycling of carbon, nitrogen, sulfur, and certain volatile toxic organic compounds. Several gases of interest in wetlands include oxygen, carbon dioxide, methane, ammonia, nitrous oxide, hydrogen sulfide, and several volatile organic compounds. For biogeochemical reactions involving dissolved gases, the activity of gas is expressed either in terms of its partial pressure in gas phase in the units of atmospheres, or as concentration in the units of moles per liter. The concentration of dissolved gas in water (C_{aq}) in equilibrium with atmosphere is determined by the concentration of gas in air (C_g), and the Henry's law constant (K_H):

$$C_{aq} = \frac{C_g}{K_H} \qquad (14.21)$$

The units for C_{aq} and C_g are expressed as moles per liter and Henry's law constant is dimensionless. The preceding equation is also expressed in terms of partial pressure of gas (atmospheres at a given temperature).

$$C_{aq} = \frac{P_g}{K'_H} \qquad (14.22)$$

where P_g is the partial pressure of gas (atm at a given temperature), C_{aq} is the equilibrium concentration of gas in water (mol L^{-1}), and K'_H is atm mol^{-1} L^{-1} of water. Using the ideal gas law, C_g can be related to partial pressure of a gas, gas constant, and temperature.

$$PV = nRT \tag{14.23}$$

or

$$\frac{P}{RT} = \frac{n}{V} = C_g \tag{14.24}$$

The Henry's constants expressed in Equations 14.22 and 14.23 can be related to each other as follows:

$$K_H = \frac{K'_H}{RT} \tag{14.25}$$

Both Henry's constants will provide similar results, as long as the reader pays careful attention to how they are defined and how the appropriate equations are used. The reader should refer to standard chemistry textbooks for additional details. Excellent discussion on the use of Henry's constants in a range of environmental conditions is provided by Schwarzenbach et al. (1993).

If the concentration of gas in water (C_{aq}) is higher than C_g/K_H, then the movement of gas from water to atmosphere occurs. The flux of gas can be described using a first-order mass transfer relationship as follows:

$$J_g = -k \left(\frac{C_{aq} - C_g}{K_H} \right) \tag{14.26}$$

where J_g is the flux [M L^{-2} T^{-1}], k is the gas transfer coefficient [L T^{-1}], C_{aq} is the concentration of the gas in the water [M L^{-3}], and C_g is the equilibrium concentration of the gas in the atmosphere [M L^{-3}] (Hemond and Fechner, 1994). The magnitude of the gas transfer coefficient (k) depends on a number of factors including water movement, plant density, air movement, and temperature. The flux J is positive if the flux of a gas is into the water column or $C_g > C_{aq}$.

A schematic depicting the exchange of dissolved gases across the air–water interface is shown in Figure 14.31 (Schwarzenbach et al., 1993). This simple model describes four layers: (1) turbulent air, (2) a quiescent thin air layer, (3) a quiescent thin water layer, and (4) turbulent well-mixed water layer. The bulk movement in air in region (1) or in the water layer (4) is significant enough to create eddy turbulent movement of gases. This situation creates a rapid movement of gases except in a very thin layer at the water surface. Owing to high viscosity of fluid on smaller length scales (0.1 mm in water and 1 mm in air), the energy required to drive eddies at these short distances decreases. These conditions create a "boundary layer" (zones 2 and 3) that are different from zones 1 and 4 (Figure 14.31). This model assumes that dissolved gases are present essentially well mixed in zones 2 and 3, except in thin layer where gases are transported via molecular diffusion in response to concentration gradient. The significance of gaseous exchange across air–water interface and emission of gases and other chemicals into the atmosphere is discussed in various chapters (see Chapters 5, 8, 11, and 13).

14.10 BIOGEOCHEMICAL REGULATION OF EXCHANGE PROCESSES

Physical, chemical, and biological processes regulate the production of dissolved solutes in soils and flux into the water column. The decomposition processes associated with the reduction of oxygen, nitrate, oxides of manganese and iron, sulfate, and carbon dioxide have been shown to be important with organic matter input from internal and external sources (see Chapters 4–6, 8, 10, and 11). The biogeochemical processes regulating the distribution and exchange of dissolved substances between soil and water column, and water column and air are discussed in various chapters of this book.

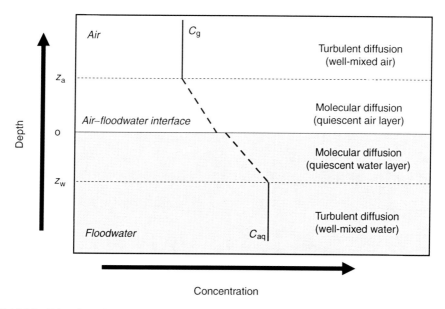

FIGURE 14.31 Direction of physical processes regulating exchange of gases and volatile compounds across air–water interface. (From Schwarzenbach et al., 1993.)

In this section, we provide a brief summary of selected physical, chemical, and biological processes related to the cycling of carbon, nitrogen, sulfur, iron, manganese, trace metals, and selected toxic organic compounds. These processes may include microbial metabolic reactions regulating oxidation and reduction reactions, sorption and equilibrium reactions, precipitation and dissolution reactions, volatilization of gaseous end products of reactions, and others involved in coupled biogeochemical cycles. These processes are the driving forces for most diagenetic changes in soils and sediments.

Anaerobic soil condition in wetland soils influences the pore water chemistry and various other chemical and microbial processes. The thin surface aerobic layer at the soil–floodwater interface forms a boundary between overlying water and soil phases. Processes operating at the soil–water interface affect the net release of dissolved substances into the water column. Addition of organic matter from detrital input from vegetation and algae and external POM to the water column can increase the oxygen demand and result in the loss of surface-oxidized layer normally found at the soil–floodwater interface. The Fe(III) and Mn(IV) oxides found in oxidized layers are reduced to more soluble Fe(II) and Mn (II) forms, which can readily diffuses into water column (see Chapter 10). The loss of oxidized surface layer due to increased oxygen demand can also increase soluble phosphorous and ammonium flux from soil to water column (see Chapters 8 and 9). Because of the reduction of Fe(III) and Mn(IV) oxides, and sulfate, concentrations of dissolved Fe(II), Mn(II), sulfides, and nutrients in pore water often exceed the concentrations in water column leading to upward diffusion from soil to water.

Temporal variations in wetlands can have direct and indirect effects on nutrient fluxes from soil to the overlying water. Nutrient concentrations and distribution in the soil profile exhibit distinct seasonal patterns (Morris, 2000). Studies have shown there is typically a midsummer peak in dissolved reactive phosphorous and ammonium concentrations in surface waters, attributed to temperature-regulated reaction processes, as well as changes in sediment redox conditions resulting in benthic regeneration (Kemp, 1989). The depletion of dissolved oxygen in the water column can result in remobilization of metals and nutrients into the underlying water; when dissolved oxygen concentration reaches levels of less than 2 mg L^{-1} the release of ammonium, soluble reactive

TABLE 14.4
Select Examples of Dissolved Reactive Phosphorus Fluxes from Sediments to Overlying Water Column Maintained in Anoxic or Oxic Conditions

Study Area	Redox	P Flux (mg P m^{-2} day^{-1}) Minimum	Maximum	Reference
Lake Okeechobee, USA	Anoxic	0.16	2.22	Moore et al. (1998)
	Oxic	0.33	1.89	
Lake Pepin, USA	Anoxic	8.6	24	James et al. (1995)
	Oxic	1.9	9.3	
St. Johns River, USA	Anoxic	2.35	11.7	Malecki et al. (2004)
	Oxic	−0.13	0.6	
Swan-Canning Estuary, Australia	Anoxic	2	53	Lavery et al. (2001)
	Oxic	0.5	5.4	
Lake Teganuma, Japan	Anoxic	90	109	Isikawa and Nishimura (1989)
	Oxic	4.0	8.7	
Everglades—WCA-2A	Anoxic			Fisher and Reddy (2001)
	Oxic			

phosphorus, Fe(II), and Mn(II) increases. Some examples of phosphorus fluxes under oxic and anoxic water column conditions are presented in Table 14.4.

14.11 SUMMARY

- Wetlands accumulate particulate matter in soils and sediments and release dissolved substances into water column. The particulate matter settles on soil surface via sedimentation or by accretion process.
- Biogeochemical processes can produce large gradients in the concentration of various dissolved substances across the soil–water interface. The rate of transfer of solute between the soils and water column and from one physical-chemical state to another is defined as flux.
- The major processes governing flux of solutes in wetlands are sedimentation, advection, diffusion, bioturbation, water flow, and evapotranspiration.
- Several methods are now available to measure accretion, advection, flux, and diffusion flux in wetland.
- Sedimentations or accretion are measured by radiochemical procedures (Cs-137, Be-7, Pb-210, and C-14 dating), feldspar marker horizons, sediment traps, and SETs.
- Advective flux measurements can be conducted using seepage meter, piezometer, dye tracers, and radium or radon isotopes.
- Diffusion flux can be made by measuring solute gradients using pore water equilibrator, benthic flux chamber, and core incubation.
- Vegetative flux also influences digenetic and nutrient cycles in wetlands. Due to high plant productivity, nutrient and metal uptake by plants can be significant.
- Nutrients incorporated into herbaceous material are deposited on soil surface or exported from the wetland as detritus or dissolved nutrients released by decaying vegetation.
- Air–water exchange also plays an important role in biogeochemical cycling of carbon, nitrogen, and sulfur. Wetlands emit methane, carbon dioxide, nitrous oxide, and reduced sulfur gases to the atmosphere.

STUDY QUESTIONS

1. Why do wetlands tend to accumulate sediment and serve as sink for nutrients?
2. Define flux.
3. What is the difference between advection flux and diffusion flux?
4. Which two characteristics in wetland distinguish the zone of soil–water interface from the water?
5. List the major components of the hydrological cycle that influence the transport of solutes in wetland and aquatic systems.
6. What are the major processes responsible for the transport of solute across the soil–water interface?
7. Define Darcy's law.
8. List the methods used in the measurement of advection flux.
9. What is Fick's law? Explain how it is used in measuring diffusion flux.
10. What is bioturbation? How does bioturbation influence vertical distribution of nutrients in sediment?
11. What roles do benthic invertebrates play in wetland food web, natural cycling, and sediment–water interface?
12. Under what condition is wind-induced mixing important?
13. Which methods are available in quantifying the extent of solute flux across the soil–water interface?
14. Briefly describe the methods used in measuring accretion or sedimentation rates in wetland. What are the methods used in estimating short-term accretion rates (1–2 years)?
15. List the various ways through which vegetation influences diagenitic processes.
16. List important gaseous flux from wetland to atmosphere.
17. List the important biogeochemical process regulating the exchange of dissolved substances between soil and water columns, and water column and atmosphere.

FURTHER READINGS

Aller, R. C. 1982. The effects of macrobenthos on chemical properties of marine sediment and overlying water. In P. L. McCall and M. J. S. Tevesz (eds.) *Animal-Sediment Relations*. Plenum Press. pp. 53–102.

Berner, R. A. 1980. *Early Diagenesis. A Theoretical Approach*. Princeton University Press. Princeton, NJ. 481 pp.

Boudreau, B. P. 1997. *Diagenic Models and their Implementation*. Springer-Verlag, Berlin. 414 pp.

Boudreau, B. P. and B. B. Jorgensen. 2001. *The Benthic Boundary Layer*. Oxford University Press, New York. 403 pp.

Hemond, H. F. and E. J. Fechner. 1994. *Chemical Fate and Transport in the Environment*. Academic Press, New York. 338 pp.

Lerman, A. 1979. *Geochemical Processes; Water and Sediment Environments*. Wiley, New York. 481 pp.

Winter, T. C. 1999. Relation of streams, lakes, and wetlands to groundwater flow systems. *Hydrogeol. J.* 7:28–45.

15 Biogeochemical Indicators

15.1 INTRODUCTION

Wetlands and aquatic ecosystems are sensitive to disturbances, from both natural and man-made activities. When the environment is adversely affected, the physical, chemical, and biological properties and processes may become stressed. To determine whether a biological community within wetland and aquatic systems has been negatively impacted, scientists are developing means to compare characteristics of impacted communities within wetlands to those of reference communities (those believed to be naturally or relatively undisturbed by anthropogenic sources). Many species depend on wetlands for successful completion of their life cycle and most require, or benefit from, nearby aquatic habitats. Changes in the structure and function of a wetland can impact the biota of the wetland, the surrounding uplands, and the nearby aquatic habitat. Monitoring of wetland ecosystems can provide information on environmental change, including changes in community structure and function, in both the wetland and upland watersheds.

The development of sound concepts and methodologies for wetland ecosystem monitoring requires an understanding of the ecosystem structure and function. However, given the high cost of environmental monitoring in terms of time, human resources, and funding, methods developed should be simple and efficient, yet scientifically rigorous and ecologically meaningful. One of the most attractive approaches to developing scientifically rigorous methods is based on the concept of using physical, chemical, or biological properties or processes as indicators of wetland condition, change, or response to anthropogenic impacts.

Wetlands host complex microbial communities, including bacteria, fungi, protozoa, and viruses. The size and diversity of microbial communities are directly related to the quality and quantity of resources available in the system. Many of the water and soil parameters that influence the ecosystem are the end products of biogeochemical processes that are microbiologically mediated. Microbial processes and populations often have more rapid turnover times than higher trophic stages and, due to their size, are often more responsive to lower thresholds of environmental change. These characteristics make microbial processes good candidates as efficient indicators of wetland condition because they are potentially very sensitive to perturbations such as external nutrient loading, hydrologic alterations, and fire.

Biogeochemical processes provide direct inference on ecological changes at a fundamental level that affects all species utilizing the ecosystem. Changes at higher trophic levels, such as a decline or shift in plant communities, may be due to factors that affect only a small portion of the biota, whereas changes in biogeochemical processes signify comprehensive alteration of the biota. Thus, it is critical to evaluate the water and soil ecosystem components in an integrated framework by linking processes and associated biogeochemical indicators to describe the structure and function of a wetland ecosystem. Biogeochemical processes are sensitive and reliable indicators of wetland condition, but measurements can be time-consuming and expensive. However, measurement of concentrations of certain chemical substrates, intermediates, and end products of ecologically potent biogeochemical processes can provide rapid and inexpensive indicators of the rates of those processes. Hence, simple measurements of biogeochemical processes within a wetland could serve as efficient indicators to infer on its environmental condition and surrounding landscape. Furthermore, relationships between indicators and processes may provide more reliable estimates of ecosystem health for assessment at landscape levels.

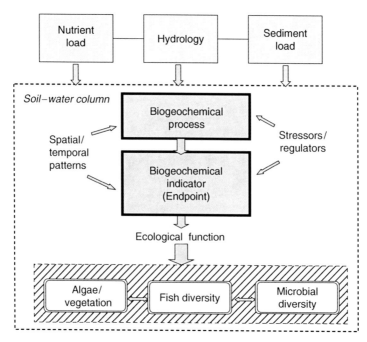

FIGURE 15.1 Schematic showing the effect of anthropogenic impacts on biogeochemical processes in soil and water column.

Prior to the development of biogeochemical indicators, a fundamental understanding of the biogeochemical processes regulating the functions of an ecosystem is imperative for evaluating nutrient impacts and the recovery of a wetland ecosystem from nutrient impacts. The certainty associated with an assessment decreases if the factors that affect the biogeochemical processes regulating the fate and transport of nutrients in wetlands are not well understood; that is, the risk assessment is only as good as the information or knowledge available at the time. Therefore, it is imperative to develop sound linkages between biogeochemical indicators and assessment of structural and functional impacts.

Ecosystem impairments such as eutrophication of wetlands can be attributed to (i) increased external inputs of nutrients from point and nonpoint sources or (ii) accelerated nutrient cycling within the ecosystem associated with change in environmental conditions of the soil and water column. Typically, eutrophication is linked to increased external inputs of nutrients. However, internal nutrient sources are equally important, especially in highly impacted wetlands or wetlands that have large reserves of organic and inorganic bound nutrients. Anthropogenic nutrient loading from point or nonpoint sources to a nutrient-limited wetland system can alter physical, chemical, and biological properties and processes in soil and water, which in turn can influence ecological function and productivity (Figures 15.1 and 15.2). Wetlands, as low-lying areas in the landscape, receive inputs from hydrologically connected uplands. Many wetlands are open systems receiving inputs of carbon and nutrients from upstream portions of the watershed, including agricultural and urban areas. Prolonged nutrient loading from surrounding lands to wetlands can result in distinct nutrient gradients in surface waters and soils. Mass loading and hydraulic retention time are important factors that determine the degree of nutrient enrichment. Continual point source nutrient loading to an oligotrophic wetland results in a zone of high nutrient availability or nonlimiting nutrient conditions near point source inputs, and low nutrient availability or nutrient-limiting conditions furthest from input points. Between these two extremes, there exists a gradient in quality and quantity of organic matter, nutrient accumulation, microbial and macrobiotic communities, composition, and biogeochemical cycles. This enrichment effect can be seen in many freshwater wetlands, most notably in the subtropical Everglades (see Chapter 17 for details).

Biogeochemical Indicators

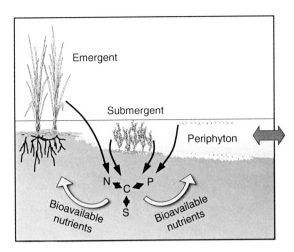

FIGURE 15.2 Schematic showing basic nutrient cycles in the soil–water column of a wetland.

One of the most attractive approaches to measuring and quantifying nutrient availability is based on using physical, chemical, or biological properties or processes in the soil and water column as indicators of change or response to anthropogenic impacts. In this chapter we describe simple-to-measure indicators with reasonable scientific rigor and reproducibility that will assess nutrient conditions in wetlands.

15.2 CONCEPT OF INDICATORS

Worldwide concern about eutrophication has led to increased efforts to monitor and assess status and trends in environmental condition of wetlands and aquatic systems. Monitoring of wetland ecosystems is initially focused on obvious, discrete sources of stress including pollutant concentrations, such as nutrients, trace metals, and toxic organics, or measurements of physicochemical changes in the water column, such as dissolved oxygen and pH. This is followed by examining ecological receptors, such as biotic communities. To characterize the condition of these ecosystems and associated biotic communities, simple indicators that respond to change are needed.

An indicator is a sign or signal that relays a complex message, potentially from numerous sources, in a simplified and useful manner. An ecological indicator is defined as a measure, an index, or a model that characterizes an ecosystem or one of its critical components (Jackson et al., 2000). An indicator may reflect biological, chemical, or physical attributes of ecological conditions. The primary uses of indicators are to characterize current ecosystem status and to track and predict future significant changes. Indicators are designed to easily identify the positive or negative changes in wetlands and aquatic systems. They communicate information about conditions, and, over time, about changes and trends. As it is neither possible nor economically feasible to measure all parameters, it is important to identify critical physical, chemical, and biological components of wetlands and aquatic systems that respond to change and use these properties as indicators. Developing indicators, and monitoring them over time, can help to determine whether ecosystem changes are occurring, whether any action is required, and what action might yield the best results, and also to assess the success of restoration efforts. Indicators can be used for assessing the impacts of both, external (nutrient and sediment loads, fire, and other natural disturbances) and internal stresses, including chemical and physical condition, biological (microbes, algae, plants, and animals) communities, and ecosystem productivity.

Wetlands and aquatic systems are the recipients of contaminant loads from uplands and should serve as indicators for the success (or lack) of management in upland systems. There is an immediate need for sensitive, reproducible, accurate, rapid, and inexpensive indicators of ecological integrity in order to facilitate management and restoration of large-scale ecosystems. The development of sensitive and accurate tools in this area allow for the quantification of ecosystem restoration and succession over time.

This chapter focuses on indicators of physical, chemical, and biological processes (as discussed in various chapters in this book) and their usefulness as indicators to monitor wetland condition. The following key questions are addressed:

- What biogeochemical processes are affected by environmental perturbation in wetlands?
- What biogeochemical indicators are suitable to describe impacts on wetlands?
- Is there a sufficient range of observation values so that selected biogeochemical parameters may serve as sensitive indicators of impact or recovery?
- Does the distribution and central tendency of biogeochemical indicators discriminate between natural spatial variability and anthropogenic impact in wetlands?

15.3 GUIDELINES FOR INDICATOR DEVELOPMENT

The U.S. Environmental Protection Agency provides the following guidelines to develop ecosystem indicators (Jackson et al., 2000). These guidelines should be applicable for employing biogeochemical indicators for wetlands and aquatic systems. The guidelines as reported by Jackson et al. (2000) are functionally related and allow users, such as regulators and environmental managers, to focus on four fundamental questions:

Conceptual relevance. Is the indicator relevant to the assessment question (management concern) and to the ecological resource or function at risk?
Feasibility of implementation. Are the methods for sampling and measuring the environmental variables technically feasible, appropriate, and efficient for use in a monitoring program?
Response variability. Are human errors of measurements and natural variability over time and space sufficiently understood and documented?
Interpretation and utility. Will the indicator convey information on ecological conditions that is meaningful to environmental decision-making?

15.3.1 CONCEPTUAL RELEVANCE

The indicator(s) selected must provide relevant information to determine the ecological condition of a wetland or an aquatic system in question. The indicator must have conceptual relevance to a given biogeochemical process, particularly when it is a surrogate for the measurement of a given biogeochemical process or multiple interacting processes. A selected indicator should be well understood and based on well-established scientific principles. First, it must be demonstrated in concept that the proposed indicator(s) is responsive to address individual or multiple assessment questions and will provide information useful to a management decision. It must be demonstrated that the proposed indicator is conceptually linked to the biogeochemical function of a wetland in question. Most often the indicator selected may be an end product of a given biogeochemical process. A conceptual relationship between indicators and associated biogeochemical processes should be established on the basis of current understanding of elemental cycles in the system (discussed in several chapters of this book).

15.3.2 FEASIBILITY OF IMPLEMENTATION

Any indicator selected for a routine monitoring program must be feasible to measure and practical. Sampling, processing, and analytical methods and quality assurance must be evaluated and documented for all measurements related to the indicator. The logistics and costs associated with training, travel, equipment, and field and laboratory work should be evaluated before the selection of an indicator. Methods to be used should be compatible with the monitoring design of the program for which the indicator is intended. Sampling strategies including location, number of observations and

Biogeochemical Indicators

replicates, design and measurements should be appropriate for the spatial scale of analysis. Needs for specialized equipment and expertise should be identified.

15.3.3 RESPONSE VARIABILITY

It is important to understand the variability in indicator measurements to distinguish extraneous factors from a true environmental effect. Variability is contributed by natural variation and by measurement error introduced by field and laboratory activities. Natural variability includes temporal (within field season and long-term variation over multiple years) and fine, medium, and coarse spatial variations. The selected indicators must be robust enough to provide significant responses at distinct points along an environmental condition gradient. If an indicator is composed of multiple measurements, variability and accuracy should be evaluated for each measurement as well as for the resulting indicator.

15.3.4 INTERPRETATION AND UTILITY

The selected indicator must produce results that are clearly understood and accepted by scientists, environmental managers, and policy makers. A range of values should be established that define wetland condition as acceptable/unacceptable or marginal in relation to indicator results. Finally, the indicator results should be used to develop specific management practices to minimize impacts and determine the recovery after restoration.

15.4 LEVELS OF INDICATORS

Indicator levels are determined based on the ease of measurement and its ability to respond to change. We describe three levels of indicators: level I indicators are easily measurable, whereas level II and III indicators provide more scientific rigor and are used to support the validity of easily measurable indicators. For routine monitoring of a wetland, only selected level I indicators are used to assess the level of impacts. As defined, assessment endpoints are explicit expressions of an environmental value to be protected, whereas measurement endpoints are measurable responses of an assessment endpoint to a stressor (USEPA, 1992; Suter, 1990). Level I indicators are of low cost, are easily measurable but less sensitive to stress/impact, and show a weak spatial variability and have long response time. Level II indicators are moderately complex and sensitive, show moderate spatial variability, and have medium response time. Level III indicators are highly complex and sensitive, show high spatial variability, and have short response time (Figure 15.3).

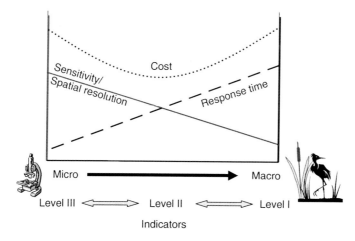

FIGURE 15.3 Relationship between biogeochemical indicator measurement and cost, sensitivity, spatial resolution, and response time.

TABLE 15.1
Causal and Response Variables for Indicator Evaluation in Wetlands

Causal Variable	Response Variable
Level I indicators: external loads—nutrients and sediments	Level I indicators: physicochemical properties of water column, detritus, and soils
Level I indicators: physicochemical properties of water column	Level I indicators: physicochemical properties of detritus and soils
Level I indicators: physicochemical properties of water column, detritus, and soils	Level II indicators: water column, detritus, and soils
Level I indicators: physicochemical properties of water column, detritus, and soils	Level III indicators: water column, detritus, and soils

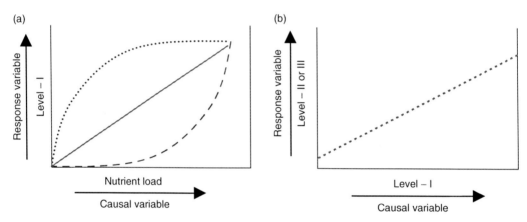

FIGURE 15.4 Linear relationship between causal and response variables. Nonlinear relationships do exist but are not shown in this figure.

Several biogeochemical indicators can be used as causal and response variables to evaluate impacts in wetlands (Table 15.1). Causal variables are parameters that can affect biological, chemical, and physical characteristics of a wetland. For example, nutrient and sediment loads and hydraulic load to a wetland can affect several physical, chemical, and biological properties of wetlands (Figure 15.4a). Anthropogenic-induced nutrient (nitrogen and phosphorus) loads can increase the concentration of these nutrients in water column, detritus, and soil. Water column responds rapidly to anthropogenic nutrient load as compared to detrital matter and soils. Nutrient enrichment of water, detritus, and soils can in turn affect microbial, periphyton, and vegetation communities. The associated biogeochemical processes in a wetland serve as response variables (Figure 15.4b).

Level I soil and water quality indicators are relatively easy to measure, and many are now routinely used either in monitoring of aquatic systems such as streams, rivers, lakes, and estuaries or, in terrestrial ecosystems. Level II indicators are relatively complex to measure, although many of these are now also measured in wetland systems. Level II indicators provide a better understanding of the influence of nutrient enrichment on soil processes and its ultimate effect on ecological function. However, both processes and indicators are regulated by various controls, including nutrient loads, hydrology, fire, and spatial and temporal variability. Many of the level I water and soil indicators can be used as independent response variables that may influence the biogeochemical processes functioning in soil and water of a wetland. The dependent variables linked to specific biogeochemical processes can be used as key response variables affected by nutrient loading. A list of level I, II, and III indicators are presented in Table 15.2.

TABLE 15.2
Potential Water and Soil Quality Indicators for Assessing Nutrient Impacts in Wetlands

Level I Indicators	Level II Indicators	Level III Indicators
Water Column		
Color[a]	Total and dissolved metals (site-specific situation)	Microbial and algal diversity
Temperature[a]		Cellular fatty acids
Water depth[a]	Total and dissolved organic carbon (TOC and DOC)	rRNA sequence analysis
Salinity/conductivity[a]		
Turbidity[a]	Total and dissolved organic nitrogen (TON and DON)	
Total suspended solids[a]		
Dissolved oxygen	Elemental ratios (Si:C:N:P)	
pH and alkalinity[a]	Enzyme assays	
Nitrogen (TKN)[a]	Heterotrophic respiration	
Phosphorus (TP)[a]	UV absorbance	
Nitrogen (NH_4-N, NO_3-N + NO_2-N)	Biological N_2 fixation	
Phosphorus (TDP and DRP)	Periphyton community composition	
Biochemical oxygen demand	Primary productivity	
	Diel pH and dissolved oxygen	
Detritus/Soil		
Soil bulk density[a]	Cation exchange capacity	Soil–water nutrient exchange rates
Soil pH and Eh[a]	Soil oxygen demand	Substrate-induced respiration
Total nitrogen and phosphorus[a]	Acid-volatile sulfides	Arginine mineralization
Organic matter content[a]	Microbial biomass carbon, nitrogen, and phosphorus	Microbial diversity
Total carbon and labile carbon		Cellular fatty acids
Particle size distribution	Detrital decomposition—litter bag	rRNA sequence analysis
C:N:P ratios[a]	Enzyme activities	Organic matter accretion rates: Cs-137 and Pb-210 profiles
Extractable nutrients[a]	Microbial respiration	
	Potentially mineralizable nitrogen and phosphorus	Phosphate sorption index
	Nitrification potential	Equilibrium phosphorous concentration (EPCo)
	Denitrification potential	Phosphorous sorption coefficients
	Manganese and iron reduction	Phosphorous sorption maximum
	Sulfate reduction	Stable isotopes
	Methanogenesis	Soil mineralogical composition
	Soil pore water nutrients	
	Single-point phosphate sorption isotherm	
	Degree of phosphorous saturation	
	Oxalate extractable metals	

Note: Standard methods are available to determine level I indicators and some of the level II indicators. Research methods are available in the literature to determine level II and level III indicators. References to many of these methods are presented in various chapters of this book. Some of the key methods are described in the following books: APHA (2002); Wetzel and Likens (1990); *Methods of Soil Analysis*, Book Series 5, Parts 1–4.

[a] Denotes minimum data required for each site.

15.5 WETLAND ECOSYSTEM REFERENCE CONDITIONS

To determine the impact of anthropogenic nutrients or contaminants and natural disturbances (fire, hurricanes, and droughts) on wetlands, the background conditions must be established using a wetland that is not impacted by nutrients, contaminants, and natural disturbances. In many regions, it

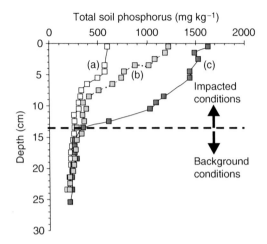

FIGURE 15.5 Total soil phosphorus of soils as a function of depth. A = soil core from reference condition; B = soil core from transition zone; C = soil core from nutrient-impacted zone. (From Reddy et al., 1993.)

may be difficult to locate a wetland that is not impacted by anthropogenic inputs and that is similar to the wetland in question. Under these conditions, an area that is least impacted may be suitable as a reference site. All these impacts can also influence the nutrient profiles in wetlands. It is critical to determine the background levels of biogeochemical indicators and processes that can be used to determine the change in wetland conditions. Wetland sites monitored for background levels will function as "reference" sites, which can be used to establish natural (historic) wetland condition. Reference sites for sampling and monitoring can be identified on the basis of the historical site information available for an area. At the minimum, three to four reference sites for a given wetland type within the same watershed may be adequate; however, several in a watershed would be better. Once reference sites are established, spatial and temporal variability of selected indicators should be monitored to determine the ranges in parameter values. This initial parameter database is essential to set up numeric nutrient or contaminant criteria for a wetland.

In certain watersheds, all wetlands may be impacted and reference wetland sites may not be available. For these sites, the underlying native soil physicochemical characteristics can be used as a background condition. This can be accomplished by taking intact soil cores and determining the nutrient profiles from overlying impacted site to underlying unimpacted site. Typically, impacted wetlands will have high nutrient levels in surface layers, which decrease with depth and reach steady levels at lower soil depths. Therefore, nutrient levels at lower soil depths can be used as an indication of background levels for that site. The depth of nutrient impact can also be estimated by determining the age of the material, using Cs-137 dating techniques (Reddy et al., 1993; Ritchie and McHenry, 1990) (see Figure 15.5).

15.6 SAMPLING PROTOCOL AND DESIGN

Before an effective method for evaluating wetland biogeochemical characteristics can be established, one must identify a portion of a wetland that responds rapidly, represents accurately the impact of external loading, and provides early warning signals of decline in ecosystem health. Changes in vegetative communities are often slow to respond to nutrient inputs relative to other ecosystem components (algae and microbes). The ecosystem may be severely degraded by the time vegetative communities' shifts are observed. Water column nutrient concentrations are often used as indicators in aquatic systems and are useful indicators for determining downstream effects of impacted wetlands. However, one prominent feature of wetland ecosystems is that water levels often fluctuate and some wetlands have little or no period of inundation. Further, due to variable

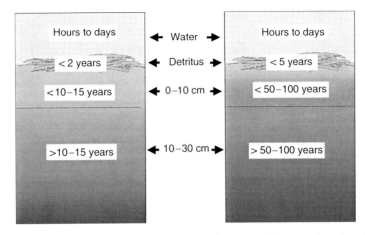

FIGURE 15.6 Approximate response and turnover times of the material present in soil and water column.

hydrologic inputs, nutrient concentrations in the water column of wetlands can change rapidly and can be highly variable. This makes sampling of wetland water column somewhat unpredictable and requires high-frequency sampling to quantify indicator values. Therefore, water column indices are useful and are often used to evaluate inflow and outflow conditions from a wetland using a mass balance approach, but do not indicate overall nutrient conditions within a particular wetland. As water column nutrients are in direct contact with the microbial communities associated with periphyton, plant detritus, and surficial soils, they change the composition and activities of these ecosystem communities, thus indicating recent impact from added nutrients (Figure 15.6). Plant detritus and soil components function as the major storages of essential nutrients and serve as integrators of nutrient impacts. This integration is mediated in two ways: (1) direct assimilation of nutrients by microbes and algae colonizing on detrital plant tissues in the water column, (2) assimilation of nutrients by plant communities and enriched detritus on the soil surface. It should be recognized that microbes are dependent on organic substrates provided by macrophytes, whereas macrophytes are dependent on microbes to transform organically bound nutrients into more bioavailable forms. This mutual dependency between microbes and macrophytes is a key regulator of biogeochemical processes in wetlands. Therefore, the biogeochemical processes and associated microbial communities that respond rapidly to nutrient loading and related physicochemical properties of soil, detritus, and water column have the ability to integrate the high variability of nutrient loading in space and time, which is often associated with pulsed runoff events providing well-suited indicators for nutrient impacts. Water column indicators provide short-term responses, detrital matter serves as indicators over time scales of <5 years, and soils serve as long-term indicators of impacts (Figure 15.6).

Nutrient inputs to a wetland can occur at various geographic locations from point and nonpoint sources. The effects of nutrient loading are usually patterned with impacted zones adjacent to inflow points and unimpacted zones furthest away from them. Thus, monitoring stations should be located in both impacted and unimpacted zones to accurately quantify differential nutrient loads. Biogeochemical indicator selection and evaluation requires systematic steps, before the selected indicators can be incorporated into routine monitoring programs.

Wetlands identified as "at risk of being degraded" should be evaluated through a sampling program to characterize the degree of degradation. Once characterized, wetlands should be placed in one of the following categories (USEPA, 2006):

1. Degraded wetlands. Wetlands in which the level of anthropogenic perturbance interferes with designated uses.
2. High-risk wetlands. Wetlands where anthropogenic impacts are high but do not significantly impair designated uses. In high-risk systems impairment is prevented by one or a

few factors that could be changed by human activities or natural disturbances such as fire, flooding, and drought.
3. Low-risk wetlands. Wetlands where many factors prevent impairment, and stressors are maintained below problem levels.
4. Reference wetlands. Wetlands where ecological characteristics closely represent pristine or minimally impaired conditions.

Numerous sampling designs can be implemented to (i) monitor indicator variables across a wetland or (ii) compare environmental conditions among numerous wetlands to reference sites (controls). Monitoring is defined as collecting information on ecosystem components (e.g., water, soil, periphyton, vegetation, and detritus samples) through repeated or continuous observation to determine possible changes at specific sites (de Gruijter et al., 2006). We can distinguish between (i) sampling in time, that is, repeated observations of ecosystem components at specific landscape positions (e.g., transect sampling along a nutrient or vegetation gradient, sampling of select sites in multiple contrasting wetlands) and (ii) sampling in space, that is, spatially distributed observations across a given wetland. Both types of sampling require different sampling strategies to capture the temporal and spatial variability of indicator variables (level I, II, or III). The most complex approach is space–time monitoring that addresses the variability of indicator variables across a given wetland area, along different profiles (water column, periphyton, detritus, and soil layers) and through time. Commonly, preferential sampling in space (i.e., one or few base level I indicator variables are measured densely at multiple geographic locations across a wetland) or time (i.e., sparse sampling of level II and III indicators at few critical geographic locations) is performed due to cost and labor limitations that are imposed on exhaustive space–time monitoring programs (Figure 15.7).

Dixon and Chiswell (1996) identified differences among status (ambient), trend (effect), and regulatory (compliance) monitoring on the basis of their aims. *Status monitoring* aims at characterizing the status or condition of ecosystem component(s), and following this over time, whereas *trend monitoring* aims at studying possible effects of a natural event or an anthropogenic-induced impact on the ecosystem components. In *compliance monitoring*, the aim is to decide whether the

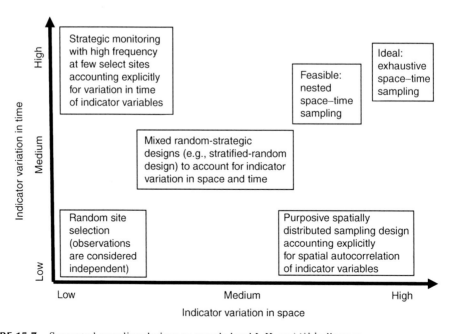

FIGURE 15.7 Suggested sampling designs to sample level I, II, and III indicators.

ecosystem components comply with a given regulatory standard (e.g., reference or control wetland). Some of the possible types of sampling design that can be used in sampling of wetlands are

1. Convenience sampling
2. Purposive (targeted, strategic) sampling
3. Probability (or random) sampling
4. Before-after-control-impact (BACI) design

Convenience sampling may be one option if accessibility to sites is limited due to field conditions in wetlands (e.g., alligator holes, dense stands of vegetation with access). Although this design saves time and cost, the estimates are highly biased. In contrast, *purposive sampling* attempts to select observation sites such that a given purpose (e.g., contrast between impacted and unimpacted sites) is served. This sampling method assumes some knowledge of the wetland sampled, such as evidence of degradation. Independent variables measured in degraded sites are compared to variables measured in a predetermined reference site in the same types of wetland. In this sampling design sites are selected strategically in a subjective manner, using experience, readily available information (e.g., data from previous sampling events, soil or vegetation maps or topographical maps), and visible landscape features (e.g., extracted from a remote sensing image), or systematically to optimize sampling. Other criteria of purposive sampling may include distributing sampling points throughout the survey area with different spacing between sampling sites to account for short-, medium-, and long-range spatial correlation of indicator variables (Grunwald, 2006). These criteria are important if the purpose of sampling is to characterize spatial variability and distribution of biogeochemical indicator variables throughout a wetland (Grunwald et al., 2007b).

Probability sampling, unlike the other modes, involves selection of sampling locations at random (de Gruijter et al., 2006). Collectively, this approach to sampling is referred to as the design-based approach followed in classical survey sampling. This is opposed to the model-based approach, where the sampling sites are fixed instead of random, and statistical inference is based on a model of the variation within a wetland. The latter one is adopted in geostatistics and time series analysis (Brus and de Gruijter, 1997). As level I, II, and III indicator variables have different spatial autocorrelation lengths and variability in time, nested sampling designs have been suggested (van Meirvenne and Cleemput, 2006). Such nested designs aim to sample those indicator variables that show high spatial variability with dense spacing in geographic space, whereas indicator variables that show high temporal variability are measured with high frequency. If functional relationships among nested level I, II, and III indicator variables can be established, those measured at only few locations may be predicted across larger wetlands on the basis of predictive models that link spatially exhaustive measurement (e.g., level I indicators) to sparsely measured ones (e.g., level II and III indicators) (Figure 15.6).

BACI design involves collection of data prior to wetland impact and compares the data after the impact (Figure 15.8). However, in many instances, there may be little or no data available prior to impact. An ideal BACI design consists of the following features (Green, 1979; USEPA, 2006):

1. Type of impact, time of impact, and place of occurrence should be known in advance. This feature allows for effective sampling to account for changes in the ecosystem.
2. Impact should not have occurred. This feature allows the collection of baseline data prior to impact.
3. Unimpacted experimental control areas should be available. This feature enables the differentiation between impacted and unimpacted areas. Changes unrelated to impacts may include natural variations and temporal effects.

The basic approach used in BACI design is to collect data over a brief period (such as 1–2 years with monthly sampling before and after) and treat time series data as independent samples and compare the samples between impacted and unimpacted sites using appropriate statistical methods (Smith, 2002). Any difference found between these two groups of samples is attributed to the impact on

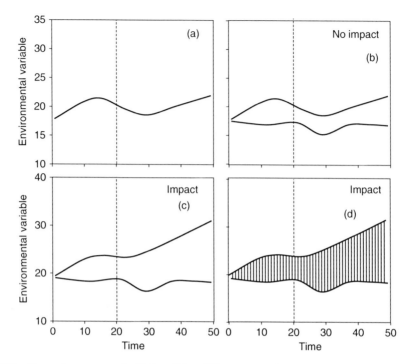

FIGURE 15.8 Plots of patterns at site(s) describing different scenarios to evaluate impact. (a) Example data for before–after analysis. Dashed line represents time of impact. (b) Example data for BACI analysis, assuming no impact. (c) Example data for BACI analysis, assuming impact. (d) Example data for BACI, assuming impact. Vertical lines indicate pairs of samples. (Adapted from Smith, 2002.)

wetland under monitoring. The trends observed before and after may not be due to anthropogenic impacts on wetland, but rather due to impacts of extreme natural events such as drought, fire, and hurricanes. To overcome this limitation, the BACI design should include a control or reference site (Green, 1979). Thus, there are two sites designated as the (1) impacted site and (2) control site where there is no anthropogenic (or natural) impact on the wetland. The data are collected before and after at both locations over space and time. Green (1979) suggested the use of a two-factor analysis of variance (ANOVA) for the analysis of the data collected at both sites. For examples of BACI applications in natural systems the reader is referred to Green (1979), Osenberg et al. (1994), Underwood (1994), Benedetti-Cecchi (2001), Smith (2002), and de Gruijter et al. (2006).

The approaches described above are designed to allow one to obtain a significant amount of data for statistical analyses with relatively minimal sampling effort. Sampling should be designed to collect information that will help answer management questions in a way that will allow robust statistical analysis. In addition, site selection, characterization of reference sites or systems, and identification of appropriate index periods are of particular concern when selecting an appropriate sampling design.

Case study of a sample design: One example of a purposive sampling design used to survey riparian and non-riparian wetlands throughout the southeastern United States implemented a two-zone composite sampling scheme (Clark, M., Unpublished, University of Florida). Each wetland was divided into two general zones, referred to as a Center zone "A" and an Edge zone "B" (Figure 15.9). In riverine systems, the Center zone is adjacent to the stream, but landward of any natural levees that may have formed. The Edge zone of riverine wetlands was located parallel to uplands, approximately one-third the distance between upland and stream. In nonriverine wetlands, a similar zoning criteria was applied where the center third of the wetland was designated as Center and the driest third as the Edge zone.

A typical soil core retrieved from a wetland is shown in Figure 15.10.

Biogeochemical Indicators

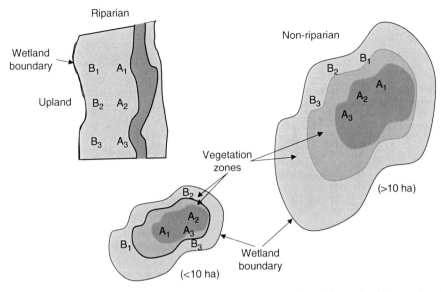

FIGURE 15.9 Example of sampling scheme used to partially quantify within-wetland biogeochemical gradients while minimizing within-zone variability and samples for analysis.

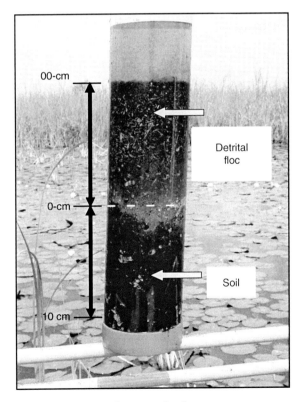

FIGURE 15.10 A typical soil core retrieved from a wetland.

Subsampling within Center and Edge zones varied slightly depending on the wetland size but in general consisted of a triplicate composite sample of soil, litter, and water at each sampling station. In riverine wetlands and large (>10 ha) nonriverine wetlands, samples were collected along one side of the wetland by collecting three subsamples approximately 50 m apart along a transect parallel to the wetland topographic boundary. In small (<10 ha) nonriverine wetlands, samples were collected within Edge and Center zones at three equidistant points around the wetland. Although information about within-zone variability was not retained due to compositing, greater confidence in the true central tendency of values was provided without having to run three times as many samples in the laboratory. When conducting a broad baseline survey for reference or monitoring purposes this method may provide a suitable compromise between data gained and available resources.

15.6.1 Water Quality Indicators

Water quality monitoring protocols to assess trophic conditions of natural wetlands have not been established. However, wetlands used for treating wastewaters are heavily monitored, and the protocols required to monitor these systems are well established. The objective of monitoring treatment wetlands is often to evaluate nutrient removal efficiency. These wetlands are typically monitored (flow and concentrations) at inflow and outflow points for various water quality parameters. Similarly, water-sampling stations in natural wetlands could be established to capture nutrient inputs, vegetation, and periphyton communities. Because of high temporal variability of water quality parameters, continuous monitoring at high frequencies (seconds–minutes–hours) at sampling stations is very important. Rainfall and hydraulic loading affect the water depth and concentration of various water column constituents, and also influence biotic communities. Thus, it is critical to measure water depth and stage in conjunction with selected water quality parameters at water sampling stations to calculate loads. Water samples should be collected using the protocols approved by respective state and federal agencies. Some of these protocols are described by APHA (1992) and USEPA (1993).

In addition to nutrients, anthropogenic loads of organic matter, suspended solids, trace metals, and pesticides can also impair water quality and associated biotic communities within a wetland. A number of physical, chemical, and biological water quality parameters are now used as indicators of water body impairment. Water quality variables should be evaluated critically to obtain the most cost-effective information required to assess wetland impairment. Specifically, water quality monitoring should determine the range of values that would significantly impair the ecological health of a wetland. Once the data are obtained, relationships between water quality variable and response variables should be established. For example, relationships between nutrient concentrations and response variables such as algae, vegetation, and benthic invertebrates are useful in evaluating impacts on wetlands. However, sampling of the water column is restricted to periods when the wetland is flooded. Although the collection of water quality data is restricted to certain times of the year, it provides useful information on the short-term temporal effects on wetland biota resulting from nutrient loading. Detailed methodologies for the determination of water quality parameters can be obtained from the following sources: APHA (1992), Clesceri et al. (1998), and Wetzel and Likens (1990).

15.6.2 Soil Quality Indicators

Wetlands exhibit a high degree of spatial/temporal heterogeneity in composition of detrital and soil layers. For some parameters, areas impacted by nutrients may exhibit a higher degree of variability as compared to unimpacted sites of the same wetland. Thus, soil-sampling protocols should aim to capture spatial/temporal variability. Sampling design and strategies are closely linked to quantitative analyses of biogeochemical datasets; therefore, a work plan that outlines step-by-step procedures and expected outcomes is essential to implement successful monitoring program (EPA, 2006).

Soil samples are usually obtained using either a grab sampling approach or collection of intact soil cores. Grab samples are not suitable for characterizing bulk density of soils, as this parameter is essential for the determination of nutrient storage in a defined soil depth. Soil samples obtained from a constant depth (e.g., 0–5 or 0–10 cm) are more useful in characterizing soils for nutrient enrichment. To make comparisons between and within wetlands it is important that both nutrient concentrations and bulk density are determined. For example, nutrient concentrations (expressed per unit dry weight of soil) in wetlands with light soils (such as organic soils) will be high in nutrients as compared to wetland dominated by mineral soils, even though both wetlands may have similar impacts. This problem can be corrected by expressing soil nutrient concentrations per unit volume of soil, which requires the measurement of soil bulk density. Nutrient concentrations measured at a predetermined depth along with bulk density data can be used to calculate the total nutrient storage on an areal basis.

To take intact cores, various types of coring tubes are used. These include the use of PVC, acrylic, and aluminum tubing. Core diameter is critical to avoid compaction. Coring tubes with an internal diameter of <10 cm typically cause soil to compact as compared to cores with an internal diameter of 12–15 cm or higher. Standard coring probes used in upland soils are not suitable for wetlands, because of saturated soil conditions and low bulk densities of wetland soils. Typically, organic-rich wetland soils have bulk densities in the range of 0.1–0.3 g (dry weight) cm^{-3}.

Intact soil cores with little or no detectable soil compaction can be obtained by a PVC, acrylic, or aluminum cylinder (15 cm diameter) with sharpened lower edge that can be twisted through fibrous marsh soils to a depth of 60 cm. The top of the cylinder is sealed with a PVC cap or a stopper to provide suction, and the bottom of the cylinder after soil is extracted from soil is sealed with a rubber stopper. Soil cores can then be sectioned into desired depth increments, either in the field or in the laboratory. Surface detritus (distinguishable plant litter) is removed from the soil and saved for chemical analysis. Typically, soil cores are sectioned into 0–10, 10–30, and 30–60 cm for routine characterization. Selecting soil depth increments should be based on site-specific conditions and soil profile characteristics. For routine monitoring of soil properties, typical root zone depth (0–10 and 10–30 cm) may be adequate to characterize the system.

Soil can include both native soil and detrital/litter components. Over time, the detrital/litter matter becomes an integral part of soil organic matter. A number of physical, chemical, and microbial parameters measured on soils and detrital matter can be used as potential indicators of nutrient enrichment and recovery of an ecosystem. Thus, biogeochemical processes representing soil indicators are directly linked to turnover and storage of nutrients and other elements in biotic and abiotic compartments of the ecosystem, and thus related to biological productivity in the soil as well as in floral and faunal components of an ecosystem. Studies in terrestrial and wetland ecosystems demonstrated the potential for using microbial communities and process measurements as sensitive indicators of environmental perturbations in soils (Torstensson et al., 1998), including heavy metals (Frostegard et al., 1993), physical disturbance (Findlay et al., 1990), and nutrient impacts (DeBusk and Reddy, 1998; Reddy et al., 1999).

Decomposition of organic matter is the primary ecological role of heterotrophic microorganisms in soils, as this process releases potentially growth-limiting nutrients and forms recalcitrant organic compounds (e.g., humus) that contribute to chemical stability of soils (Swift, 1982; Jorgensen et al., 1999; Middelboe et al., 1998). Soil microbes also exert a significant influence on ecosystem energy flow, since mineralization of organically bound nutrients is a regulator of nutrient availability for both primary production and decomposition. Therefore, most of the net ecosystem production passes through microbial compartment at least once and typically several times; however, microbial biomass comprises only a small fraction of soil organic matter (Jorgensen et al., 1999; Ruttenberg and Goni, 1997; Seitzinger and Sanders, 1997). Microbial processes also induce or facilitate ecosystem production by increasing nutrient availability. Many of the process-level parameters are tedious and difficult to measure and require specialized training and equipment; thus, they may not be suitable as indicators for routine monitoring in wetlands and may be replaced by proxies.

Monitoring soil properties as indicators may provide long-term integrated effects of nutrient impacts on wetlands, but may not be suitable to determine short-term temporal changes in the system. Detailed methodologies for determining soil indicators can be obtained from the following sources: Klute (1986), Weaver et al. (1994), Sparks et al. (1996), and Dane and Topp (2002).

15.6.3 Minimum Monitoring Requirements

Nutrient-related data on water and soil quality indicators in wetlands is limited as compared to terrestrial and aquatic ecosystems. To date, much of the data collection is at the experimental scale for site-specific conditions. Systematic data collection at a large spatial scale, using comparable techniques, is required to assess wetland eutrophication. Webster and Oliver (2001) and de Gruijter et al. (2006) provide recommendations for spatial surveys that highlight the importance of large datasets of >100 observations per wetland site to quantify spatial variability and distribution of ecosystem properties. It should be recognized that under most conditions adequate resources might not be available to obtain detailed data even for level I indicators but are highly desirable. We present a simple, systematic approach to minimum data collection of soil and water quality indicators in order to determine the change in ecological function resulting from anthropogenic impacts or recovery after restoration of degraded systems. Level I indicators required for minimum data needs are listed in Table 15.1.

Location of suitable field sites should be coordinated through local academic or governmental agencies to determine appropriate reference sites for sampling (USEPA, 2006). Criteria for reference site selection should be based on areas of least cultural impact within a particular region as determined by local knowledge sources. When the sites are identified, they are characterized using Cowardin et al.'s (1979) classification scheme or suitable methods and community characteristics. Latitude and longitude of each sampling site should also be collected using global positioning systems for use in relocation and cross-referencing the site with other GIS (geographic information system) layers.

Sampling at selected sites should consist of a minimum of three composite samples collected from the water column, detritus, and soil, respectively. Water samples (when available) should be collected at a mid-water depth, filtered, homogenized (with other replicates), and then stored on ice at 4°C or preserved until analysis. Intact soil should be collected to a depth of 10 cm below the litter/soil interface. Depending on the objective of monitoring, soil-sampling depths may vary. Litter from these cores, as defined by easily distinguishable plant fragments lying on the surface of the core, should be collected, air-dried, and then combined with other detritus samples from the site. The remaining upper 10 cm of soil from each core should be air-dried and then combined with site replicates.

Laboratory analysis of water samples should include total nitrogen and total phosphorus. Soil and detritus composite samples should be air dried at 25–30°C, ground, and homogenized, and subsamples should be analyzed for organic matter content, pH (to be determined on ambient wet sample), total nitrogen, total phosphorus, extractable ammonium nitrogen (to be determined on ambient wet sample), and extractable phosphorus, iron, aluminum, calcium, and magnesium. Moisture content of air-dried detrital matter and soils should be determined after oven drying samples at 70°C for 2–3 days or until constant weights are recorded. All nutrient concentrations determined on air-dried samples should be normalized using an oven-dried basis.

15.7 DATA ANALYSIS

The selection of indicator variables, sampling design, data collection and statistical/geostatistical methods to analyze datasets are inherently linked. A schematic plan to develop indicators for a given site is presented in Figure 15.11. Components of a monitoring program should include (1) objectives/hypotheses (expected outcomes), (2) data collection/sampling, (3) (geo)statistical methods/models, (4) results: assessment of magnitude/accuracy/precision/distribution/variability/ relationships of indicator variables, and (5) interpretation of indicator variables in the context of impacts/stresses, biogeochemical cycling, and relationships to environmental landscape variables. Ideally, this sequence of steps should not be confused or replaced by *ad hoc* field sampling without thorough planning. In the following section we provide an overview of scenarios for indicator studies and provide suggestions for data analysis and examples.

Biogeochemical Indicators

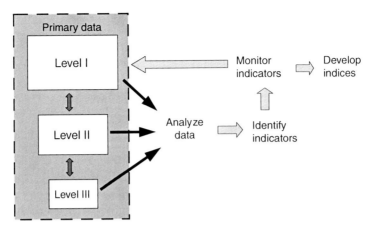

FIGURE 15.11 Schematic showing a plan to develop indicators for a given site.

Scenario I—Univariate method: In the univariate case, one indicator variable of interest $z(x_i)$ is measured at sites x with $i = 1, 2, \ldots, n$, with n representing the number of observations, measured at time t. It is assumed that sites i are independent of each other. Classical statistical methods such as t tests and ANOVA can be used to test for significant differences among sites and a reference site (control) (Ott and Longnecker, 2001). For example, multiple wetlands in a watershed are sampled using within-wetland composite sampling with the objective to compare observations (e.g., total soil phosphorus) among wetlands. This scenario requires that observations are independent. In contrast, dense sampling of an indicator variable within a wetland along a nutrient gradient (e.g., transect) would not represent spatial independence, as all samples are collected within the wetland. The assumptions of independence are met only if the spacing between sampling sites exceeds the spatial autocorrelation range of an indicator variable. A simple exploratory data analysis to derive standard statistical metrics (minimum, maximum, range, mean, standard error of the mean, mode, median, standard deviation, skewness, and kurtosis) of $z(x_i)$ should be included in scenario I as well as scenarios II–VI. Demonstrations of scenario I type analysis were given by Burdt et al. (2005) and Prenger and Reddy (2004).

Scenario II—Multivariate method: This approach involves measurement of multiple indicator variables of interest $z1(x_i)$, $z2(x_i)$, $z3(x_i)$, ..., $zm(x_i)$ at sites x with $i = 1, 2, \ldots, n$ and at time t. Assuming that sites i are independent, functional relationships between indicator variables can be derived using scatterplots, correlations, simple or multivariate regression analysis (Ott and Longnecker, 2001), or regression variants such as Classification and Regression Trees (CART) (Breiman et al., 1984), Discriminant Analysis, Generalized Linear Models, or Mixed Models with fixed and random effects (Schabenberger and Pierce, 2002). For example, envision that multiple level I and II variables (e.g., total nitrogen, total phosphorus, potentially mineralizable nitrogen and phosphorus) are measured at multiple independent wetland sites and empirical relationships are identified among variables.

To reduce the dimensionality of multivariate datasets, PCA or similar ordination methods are commonly used to reduce the number of variables in a dataset with minimal information loss (Wackernagel, 2003). Canonical correlation analysis (CCA) (Goovaerts, 1994; Wackernagel, 2003) is another method suited for multivariate indicator analysis with the aim to analyze relationships between sets of variables.

Scenario II assumes that indicator variables are *collocated*, meaning they are sampled at the same sites x_i, that is, sample sites are shared (isotopic data) (Wackernagel, 2003). A special case arises if an indicator variable of interest (e.g., level III indicator that is difficult to measure, labor-intensive or costly to measure—phosphate sorption index) is known at a few sites and an auxiliary variable known at many sites (e.g., level I indicator that is easy and cheap to measure—total phosphorus).

Such heterotrophic data are suited to derive predictive models using simple covariance function models (Wackernagel, 2003). Assuming that successful and robust functional relationships are derived, models can be used to predict a target variable (e.g., level III indicator variable) at unsampled locations across a wetland(s). The prediction range should match the model range to avoid extrapolations with high uncertainties.

Demonstration studies of a multivariate approach were presented by Grunwald et al. (2007a), Mitsch et al. (2005), and Kennedy et al. (2006).

Scenario III—Univariate spatial method: In the univariate spatial case, one indicator variable of interest $z(x_i)$ is measured at sites x with $i = 1, 2, ..., n$ at time t where $z(x_i)$ is spatially correlated with $z(x_i + h)$ with h representing a distance vector. The underlying assumption here is that observed values at points close together in space are more likely to be similar than points further apart. Spatial autocorrelation is defined as the degree of relatedness of a set of spatially located data, the extent to which adjoining or neighboring spatial units influence particular variables recorded on those units. When data are spatially autocorrelated, the assumption that they are independently random is invalid excluding the use of many classical statistical methods (compare assumptions under I and II). Local, deterministic methods (e.g., nearest neighbor interpolation, inverse distance weighting, and splines) focus on interpolations within a restricted neighborhood around the point being predicted. Geostatistics, which is the statistics of spatially correlated data (Burrough and McDonnell, 1998), incorporates spatial autocorrelation explicitly into the modeling process.

The conceptual idea of geostatistics is that spatial variation of any variable Z can be expressed as the sum of three major components (Equation 15.1): (i) a structural component, having a constant mean or trend that is spatially dependent, (ii) a random, but spatially correlated component, and (iii) spatially uncorrelated random noise or residual term (Webster and Oliver, 2001):

$$Z(x_i) = m(x_i) + \varepsilon'(x_i) + \varepsilon'' \tag{15.1}$$

where $Z(x_i)$ is the value of a random variable at x_0; $m(x_i)$ the deterministic function describing the "structural" component of Z at x_i; $\varepsilon'(x_i)$ the stochastic, locally varying but spatially dependent residual from $m(x_i)$—the regionalized variable; ε'' a residual, spatially independent noise term having zero mean and variance; and x_i the geographic position.

The spatially dependent component $\varepsilon'(x_i)$ is described by the semivariance γ. If γ is plotted as a function of distance h between sampling locations, the semivariogram $\gamma(h)$ is obtained:

$$\gamma(h) = \frac{1}{2N(h)} \sum_{i=1}^{N(h)} [Z(x_i) - Z(x_i + h)]^2 \tag{15.2}$$

where γ is semivariance, h distance (lag), and N total number of data pairs.

A semivariogram provides input for kriging (e.g., ordinary kriging—OK), which is a weighted interpolation technique to create raster prediction maps. Limitations of the univariate geostatistical technique of kriging are assumptions of stationarity and the large amount of data required to generate robust variograms (>50 observations according to Chilès and Delfiner; >100 observations at minimum, but ideally >150 according to Webster and Oliver, 2001). For a comprehensive overview of geostatistical methods see Chilès and Delfiner (1999), Goovaerts (1999), McBratney et al. (2000), Wackernagel (2003), and Grunwald (2006). Examples for the univariate, spatial approach to model various biogeochemical properties for select hydrologic units of the Everglades, Florida, are given by DeBusk et al. (2001), Bruland et al. (2006), Corstanje et al. (2006), and Rivero et al. (2007b).

Scenario IV—Multivariate spatial: In the multivariate spatial case multiple indicator variables $z1(x_i), z2(x_i), z3(x_i), ..., zm(x_i)$ are measured at sites x with $i = 1, 2, ..., n$ at time t, where variables are spatially cross-correlated. This is an extension of scenario III where multiple variables are used to predict an indicator variable (the same assumptions as outlined under III apply). Cokriging (CK) is the multivariate extension of kriging that combines a sparsely measured primary variable (or target variable) with a denser set of ancillary data as secondary variable (e.g., remote sensing data) that are spatially cross-correlated (Grunwald, 2006). Collocated CK addresses a heterotrophic situation when the variable of interest is known at a few points and the auxiliary variable is known everywhere across the wetland (Wackernagel, 2003). Regression kriging (RK) combines advantages of traditional regression methods to model the trend or drift in conjunction with geostatistical methods that model residuals from regression analysis (Grunwald, 2006). In RK, the deterministic component is modeled using multiple linear regression or other regression variants. The stochastic component, which represents the spatially varying but dependent variables, is modeled using OK or simple kriging of the regression residuals (Odeh et al., 1995; Goovaerts, 1997). Other methods that account for secondary variables to predict a target indicator variable are simple kriging with varying local means, kriging with an external drift (Goovaerts, 1997), or stochastic simulations (Chilès and Delfiner, 1999), which have been rarely applied to wetlands, although they have much potential to improve predictive modeling of biogeochemical properties. Multivariate spatial methods are powerful because they can implicitly describe relationships between environmental factors (e.g., topography, water table depth, remote sensing–derived spectral indices—Normalized Difference Vegetation Index) and indicator variables. To develop spatially explicit models that quantify relationships between site-specific biogeochemical properties and environmental variables, these methods offer many opportunities for future applications (Grunwald and Lamsal, 2006).

An overview of multivariate geostatistical methods as applied to mapping of biogeochemical properties in wetlands is provided by Grunwald et al. (2006, 2007b). Rivero et al. (2007a) demonstrated the usefulness of scenario IV models to predict soil phosphorus within a subtropical wetland. McBratney et al. (2000) compared 24 different univariate and multivariate statistical, geostatistical/hybrid geospatial soil prediction models at field, subcatchment, and regional scale, and found that geostatistical/hybrid methods outperformed all other methods at all scales.

Scenario V—Univariate and multivariate (temporal): In the univariate, temporal case one indicator variable of interest $z_t(x_i)$ is measured at sites x with $i = 1, 2, …, n$, where n represents the number of observations and time t is measured repeatedly at times $t = 1, 2, …, v$. Observations are considered independent in space. Potential methods to analyze such datasets are provided by time series analysis and variants that are based on an empirical statistical approach. In the simplest case the measured time series is treated as a stationary process $Z(t)$:

$$Z(t) = m_g + \varepsilon(t) \tag{15.3}$$

where $Z(t)$ is the realization of a stationary random process; m_g global mean; $\varepsilon(t)$ temporally autocorrelated random residual with a mean of zero and variance characterized by its autocovariance function $C(s) = \text{Cov}[\varepsilon(t), \varepsilon(t + s)]$, where s denotes the lag; and t time.
The multivariate, temporal case is just an extension of scenario V into the multivariate realm.

Scenario VI—Univariate and multivariate spatiotemporal: This is the most complex case where multiple dimensions are considered to model change of one or multiple indicator variables in space and time. Several methods have been proposed to integrate both spatial and temporal variability into one model. Spatial-temporal semivariogram modeling has been described by Hoosbeek et al. (2000). But studies that use statistical/geostatistical methods for space–time modeling are rare. Instead mechanistic (process-based) simulation models offer to model the evolution of biogeochemical indicator variables in space and time (deterministic approach). Some of the biogeochemical processes are well understood and can be modeled mechanistically, while others are complex and are based on empirical knowledge, thus departing from the mechanistic paradigm. The advantage is that simulation models are holistic, integrating various landscape components (hydro-, topo-, pedo-, litho-, and biospheres). A limiting factor is that simulation models require detailed data on several input parameters. Some examples demonstrating this method are given by Tilley and Brown (2006) and Bruland et al. (2007).

For all data analysis (scenarios I–VI) it is important to include replicate sampling to assess the precision of measurements. In addition, it is essential to identify prediction errors to assess the uncertainty associated with a specific method or model.

15.7.1 IMPACT/RECOVERY INDICES

For comparison purposes, a simple impact index can be calculated for each process or parameter measured:

$$\text{Impact index} = \log\left[\frac{\text{IS}}{\text{RS}}\right] \quad (15.4)$$

where [IS] is the rate or concentration of a parameter measured at an impacted site and [RS] the rate or concentration of a parameter measured at a reference site. The log [IS/RS] provides an index value of 0, which indicates no change; a negative value indicates a decrease and a positive value indicates an increase. For example, a value of "1" represents a tenfold change in the concentration of a parameter or rate of a process, relative to the reference site (Figure 15.12). This approach allows ranking of indicator values most affected by nutrient loading/disturbance. Impact indices should be viewed in the context of spatial variability within impacted and unimpacted sites. Field replicate variability in relatively small areas ($<m^2$) is process or parameter dependent, and can be highly variable. For example, variability of enzyme activities in wetland detrital layers can range from 27 to 100%, whereas variability in microbial respiration rates can be <20% (Wright and Reddy, 2001a, 2001b).

Impact index values (Table 15.3) aid in normalizing rates of various biogeochemical processes and concentrations of various parameters based on their sensitivity to P loading. Because of variability of uncertainty, the index values were grouped into four broad groups of positive and negative impacts. We assume that impact index values in the range of –0.1 to 0.1 are within the experimental variability of many of the processes and parameters measured. For example, the variability of the impact index for total P was <0.2 for the detrital layer and <0.1 for the surface soil of the reference site. However, for some parameters such as microbial parameters (usually reported on log-scale), variability could be much higher. Biogeochemical processes and parameters measured on the detrital layer are more sensitive to nutrient loading than those measured on surface soil (Table 15.3). The detrital component probably represents most recent (nonsteady state) impacts, whereas the surface soil may represent long-term impacts (Figures 15.13 and 15.14). Microbial communities associated with the detrital layer are in direct contact with water column nutrients, and should respond rapidly to changes in water chemistry. However, impact index values obtained on soils may be more reliable because they represent long-term steady-state conditions.

The index approach presented provides a simple strategy for integrated evaluation of impacts on wetlands, using soil biogeochemical processes and parameters as potential indicators. Although

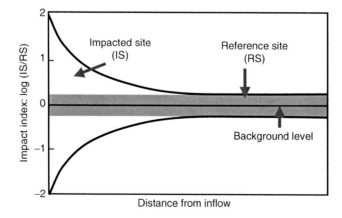

FIGURE 15.12 Schematic showing the concept of impact index as a function of distance from inflow (or impact). (Reddy et al., 1999.)

TABLE 15.3
Impact Index Values for Select Biogeochemical Indicators Measured in Detritus and Soil Layers at the Impacted Site and Interior Reference Site

Indicator	Level	Detritus/Floc	0–10 cm Soil
Total phosphorus	I	1.35	1.17
Total inorganic phosphorus	I	0.60	0.88
Labile inorganic phosphorus	I	1.07	0.52
Loss on ignition (LOI)	I	0.10	−0.07
Total carbon	I	0.08	0.03
Total nitrogen	I	0.21	−0.03
C/P ratio	I	−0.51	−0.51
Extractable ammonium nitrogen	I	0.38	0.87
Pore water ammonium nitrogen	I	nd	1.06
Pore water soluble phosphorus	I	nd	3.98
Pore water sulfate	I	nd	1.00
Microbial biomass carbon (MBC)	II	1.73	1.08
Microbial biomass nitrogen (MBN)	II	1.31	0.55
Microbial biomass phosphorus (MBP)	II	1.05	0.42
MBC/TC	II	1.65	1.06
MBN/TN	II	1.09	0.59
MBP/TP	II	−0.29	−0.93
Beta-D-glucosidase	II	0.56	0.40
Protease	II	0.10	0.07
Alkaline phosphatase	II	−0.70	−0.37
Aryl sulfatase	II	−0.10	−0.12
Phenol oxidase	II	0.19	0.16
Aerobic respiration	II	0.21	0.23
Anaerobic respiration	II	0.41	0.28
Potentially mineralizable nitrogen (PMN)	II	0.40	0.25
Substrate-induced nitrogen mineralization (SINM)—Alanine	II	0.62	0.45
Nitrification	II	0.17	0.20
Denitrification	II	0.31	0.31
Biological nitrogen fixation	II	1.0	
Potentially mineralizable phosphorus (PMP)	II	0.35	0.03
Substrate-induced phosphorous mineralization (SIPM)—G-6-P	III	0.05	−0.41
Substrate-specific microbial communities	III	nd	
Lactate	III	nd	4.62
Acetate	III	nd	4.15
Propionate	III	nd	3.84
Butyrate	III	nd	5.0
Formate	III	nd	4.34
Stable isotope—$\delta^{15}N$	III	0.69	0.69

Note: nd = not determined.
Source: Chua (2000), Inglett et al. (2004), Inglett and Reddy (2006), White and Reddy (1999, 2000, 2001, 2003), and Wright and Reddy (2001a, 2001b).

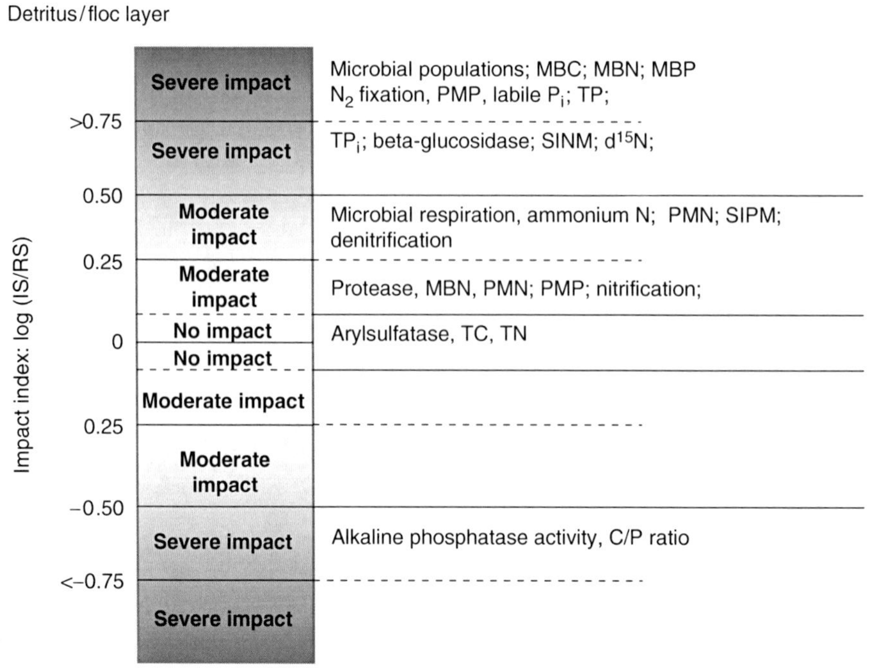

Detritus/floc layer

Impact index: log (IS/RS)	Impact Level	Indicators
>0.75	Severe impact	Microbial populations; MBC; MBN; MBP N_2 fixation, PMP, labile P_i; TP;
	Severe impact	TP_i; beta-glucosidase; SINM; $d^{15}N$;
0.50 / 0.25	Moderate impact	Microbial respiration, ammonium N; PMN; SIPM; denitrification
	Moderate impact	Protease, MBN, PMN; PMP; nitrification;
0	No impact / No impact	Arylsulfatase, TC, TN
0.25	Moderate impact	
	Moderate impact	
−0.50	Severe impact	Alkaline phosphatase activity, C/P ratio
<−0.75	Severe impact	

FIGURE 15.13 Impact index values for a range of biogeochemical indicators and processes measured in detrital/floc layer of the Everglades.

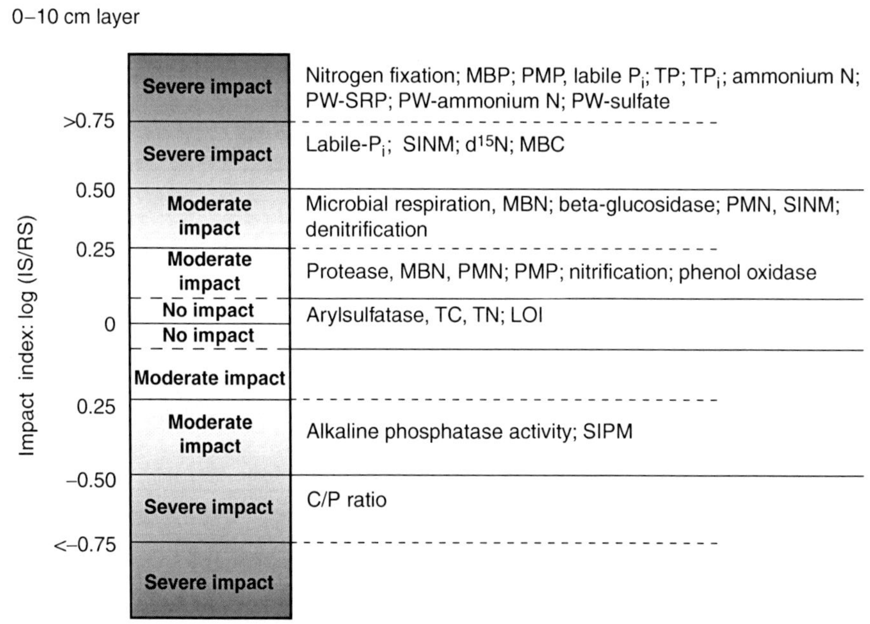

0–10 cm layer

Impact index: log (IS/RS)	Impact Level	Indicators
>0.75	Severe impact	Nitrogen fixation; MBP; PMP, labile P_i; TP; TP_i; ammonium N; PW-SRP; PW-ammonium N; PW-sulfate
	Severe impact	Labile-P_i; SINM; $d^{15}N$; MBC
0.50 / 0.25	Moderate impact	Microbial respiration, MBN; beta-glucosidase; PMN, SINM; denitrification
	Moderate impact	Protease, MBN, PMN; PMP; nitrification; phenol oxidase
0	No impact / No impact	Arylsulfatase, TC, TN; LOI
0.25	Moderate impact	
	Moderate impact	Alkaline phosphatase activity; SIPM
−0.50	Severe impact	C/P ratio
<−0.75	Severe impact	

FIGURE 15.14 Impact index values for a range of biogeochemical indicators and processes measured in surface soil (0–10 cm) of the Everglades.

biogeochemical processes may be sensitive and reliable indicators of wetland integrity, their measurement can be time-consuming and expensive. However, concentrations of certain chemical substrates, intermediates, and end products can function as surrogates for biogeochemical processes (Reddy and D'Angelo, 1996; Reddy et al., 1998a, 1998b; D'Angelo and Reddy, 1999; DeBusk and Reddy, 1998). Relationships between indicators and processes may provide reliable estimates of ecosystem condition. Simple strategies presented in this chapter provide tools to normalize process-level information into a common format for possible integration into models. To evaluate impacts, a reliable database is needed for several reference sites (background level of impacts) in various geographical regions as classified by wetland vegetation, soils, and hydrology. The strategies presented could be used in different types of wetlands to assess restoration and remediation efforts, and potentially be used as a screening tool to choose how best to utilize limited restoration resources.

15.8 SUMMARY

- Wetlands are sensitive to stress associated with natural and man-made activities. To characterize the condition and impact to these ecosystems and associated biotic communities, indicators that respond to changes are needed.
- Such ecological indicators can be defined as measures or indices that characterize the ecosystem or a critical component.
- Indicators may reflect biological, chemical, or physical attributes that can be used to characterize current status or to predict impact or change.
- Such indicators of wetland ecosystem integrity should be sensitive, reliable, accurate, rapid, and inexpensive.
- Indicators should be clearly understood and accepted by scientists, environmental managers, and policy makers. Indicator levels in accessing impact to wetland ecosystem may be based on the ease of measurement and the ability to respond to change.
- Level I indicators are easily measurable whereas level II and III indicators provide more scientific rigor and are used to support easily measurable indicators.
- Level I indicator for water column and soil may include salinity, conductivity, oxidation-reduction potential bulk density, turbidity, BOD, and suspended solid and nutrient contents.
- Level II and III indicators may include more comprehensive measurements such as enzyme activity, denitrification, microbial and algal diversity, microbial biomass, primary production, and acid volatile sulfide and gaseous flux measurements to name a few.
- Wetland can exhibit a high degree of spatial heterogeneity; thus, the sampling protocol should aim to capture the underlying spatial variability of wetland indicators.
- The selection of indicator variable, sampling design, data collection, and statistical/geostatistical methods to analyze dataset is linked.
- Before using any selected biogeochemical indicator, a sampling design must be established that represents unimpacted and impacted portions of the wetland to assess early warning signals of decline in ecosystem health.
- Numerous established sampling designs are available for use in monitoring selected indicator variables in wetlands. Each sampling protocol, if possible, should compare environmental conditions to a reference or control site.

STUDY QUESTIONS

1. Define ecological indicator. What should the indicator reflect?
2. List the qualities desired in selecting a biogeochemical indicator.
3. Should the indicators have conceptual relevance to a more detailed biogeochemical process? Explain.

4. What are the differences among level I, II, and III indicators?
5. List the selected level I indicators used for water, water column, and sediment assessment.
6. List the selected level II and III indicators used for assessing water column and sediment.
7. Explain the importance of sampling design. What is the purpose of a reference (control) site?
8. What are the differences between convenience, purposive, and probability sampling? Explain.
9. What are the categories of wetlands at risk?
10. What feature should an ideal before-after-control-impact (BACI) survey include?
11. Should soil nutrients be expressed on unit volume or concentration basis? What measurement is required?
12. How is compaction minimized when collecting soil cores?
13. Describe what factors should be considered when designing a water quality monitoring protocol. Discuss the impact of temporal variability.
14. What is spatial autocorrelation and how can it be incorporated into wetland indicator studies?
15. Provide a brief description of various scenarios that can be used for data analysis of the indicators studied.
16. What is an impact index?

FURTHER READINGS

Childress, W. M., C. L. Coldren, and T. McLendon. 2002. Applying a complex, general ecosystem model (EDYS) in large-scale land management. *Ecol. Model.* 153:97–108.

D'Angelo, E. M. and K. R. Reddy. 1999. Regulators of heterotrophic microbial potentials in wetland soils. *Soil Biol. Biochem.* 31:815–830.

Dick, R. P. 1994. Soil enzyme activities as indicators of soil quality. In J. W. Doran, D. C. Coleman, D. F. Bezdick, and B. A. Stewart (eds.) *Defining Soil Quality for a Sustainable Environment.* SSSA Book Series No. 35. Soil Science Society of America, Madison, WI. pp. 107–124.

Grunwald, S. (ed.) 2006. *Environmental Soil-Landscape Modeling: Geographic Information Technologies and Pedometrics.* CRC Press, New York. 488 pp.

Jackson, L. E., J. C. Kurtz, and W. S. Fisher (eds.) 2000. *Evaluation Guidelines for Ecological Indicators.* EPA/620/R-99/005. U.S. Environmental Protection Agency, Office of Research and Development, Research Triangle Park, NC. 107 pp.

National Research Council (NRC). 2000. Committee to Evaluate Indicators for Monitoring Aquatic and Terrestrial Environments. *Ecological Indicators for the Nation.* National Academy Press, Washington DC. http://www.nap.edu/catalog/9720html.

Reddy, K. R. and E. M. D'Angelo. 1994. Soil processes regulating water quality in wetlands. In W. Mitsch (ed.) *Global Wetlands: Old World and New.* Elsevier, New York. pp. 309–324.

Reddy, K. R., J. R. White, A. Wright, and T. Chua. 1999. Influence of phosphorus loading on microbial processes in soil and water column of wetlands. Phosphorus in Florida's Ecosystems: Analysis of Current Issues. In K. R. Reddy, G. A. O'Connor, and C. L. Schelske (eds.) *Phosphorus Biogeochemistry in Subtropical Ecosystems: Florida as a Case Example.* CRC Press/Lewis Publishers, Boca Raton, FL. pp. 249–273.

The Heinz Center Report, 2002. The State of the Nation's Ecosystems: Measuring the lands, waters, and living resources of the United States. The H. John Heinz III Center for Science, Economics, and the Environment, Washington, DC. Cambridge University Press, New York. 279 pp. http://www.heinzctr.org/ecosystem.

USEPA (U.S. Environmental Protection Agency). 2006. *Draft Nutrient Criteria Technical Guidance Manual: Wetlands.* EPA-823-B-05-003. Washington, DC. 179 pp.

White, J. R. and K. R. Reddy. 2001. Influence of selected inorganic electron acceptors on organic nitrogen mineralization in Everglades soils. *Soil Sci. Soc. Am. J.* 65:941–948.

White, J. R. and K. R. Reddy. 2003. Nitrification and denitrification rates of everglades wetland soils along a phosphorus-impacted gradient. *J. Environ. Qual.* 32:2436–2443.

16 Wetlands and Global Climate Change

16.1 INTRODUCTION

Increased industrial emissions, fossil fuel combustion, deforestation and related burning of biomass, and changes in land use and management practices have changed the gaseous composition of the earth's atmosphere. These activities have resulted in increased emissions of naturally occurring radiatively active trace gases (e.g., carbon dioxide, methane, and nitrous oxide), known as "greenhouse gases."

The "greenhouse effect" or "global warming" is also impacted by biogenic trace gases evolving from wetlands. Greenhouse gases are transparent to the incoming short-wave solar radiation, but absorb and reflect long-wave infrared radiation emitted from the warm surface of the earth. As a result, a part of the energy associated with the long-wave radiation, reflected by the earth's surface and lower atmosphere, is trapped within the atmosphere. This maintains the earth's surface temperature at a higher level than when trace gas concentrations were low.

Methane (CH_4) and nitrous oxide (N_2O) are two important atmospheric components that contribute to global warming, besides carbon dioxide which is the major contributor. Both methane and nitrous oxide absorb infrared radiation, while nitrous oxide is photochemically oxidized to nitric oxide (NO) resulting in ozone depletion in the stratosphere. The nitric oxide formed also governs the concentration of other greenhouse gases. The amount of nitrous oxide and methane in the atmosphere has been continuously increasing since industrialization, with current increase rates of 0.25 and 1.2% per year, respectively.

Ozone destruction

$$\left.\begin{array}{l} O_3 + h\nu \rightarrow O + O_2 \\ O + NO_2 \rightarrow 2NO \\ O + N_2O \rightarrow 2NO \\ NO + O_3 \rightarrow NO_2 + O_2 \end{array}\right\} 2O_3 \rightarrow 3O_2$$

Ozone formation

$$\left.\begin{array}{l} R + O_2 \rightarrow RO_2 \\ RO_2 + NO \rightarrow RO + NO_2 \\ NO_2 + h\nu \rightarrow NO + O \\ O + O_2 \rightarrow O_3 + M \end{array}\right\} RO_2 + O_2 \rightarrow RO + O_3$$

R (CH_4, CH_3CO, H)

TABLE 16.1
Major Greenhouse Gases

	CO_2	CH_4	N_2O
Preindustrial concentration (ppm)	~280	~0.70	~0.27
Concentration in 1998 (ppm)	365	1.7	0.3
Range of concentration change (year^{-1})	1.5 ppm	7.0 ppb	0.8 ppb
Increase (ppb year^{-1})	1,750	19	0.75
%Biotic	30	70	90
Residual time (year)	100	8–12	100–200
Atmospheric lifetime (years)	5–200	12	114
Global warming potential[a]			
20 years	1	62[a]	275
100 years	1	23[a]	296
500 years	1	7[a]	156

[a] The methane GWPs include an indirect contribution from stratospheric H_2 and O_3 production.

Because methane concentration increases more rapidly than nitrous oxide concentration, its relative importance in modifying the climate also changes. Global increases of greenhouse gases in the atmosphere over the last 200 years are about 20% for carbon dioxide, 8% for nitrous oxide, and over 200% for methane (Table 16.1). The heat absorption potential of these gases is dependent on their relative concentration in the atmosphere, infrared absorption profile, and atmospheric lifetime. Global warming potentials (GWPs) are a measure of the relative effect of a given substance compared to carbon dioxide, integrated over a chosen time horizon (IPCC, 2001). Methane and nitrous oxide are more effective atmospheric trace gases than carbon dioxide in contributing to global radiative forcing. Increased atmospheric concentrations of these trace gases, and their atmospheric derivatives, will affect climate, which in turn will influence ecosystems since the atmospheric lifetimes of carbon dioxide, methane, and nitrous oxide are relatively long.

Wetland soils are able to exhibit both production and consumption of greenhouse gases and play an important role in regulation of climate change. Natural wetlands account for approximately 20–25% of global methane emissions. After carbon dioxide, methane is the second most important greenhouse gas. Its atmospheric concentration is only about 1/200th that of carbon dioxide, but its thermal absorption is more effective. Methane strongly influences the photochemistry of the atmosphere, accounting for approximately 15% of the current increase in greenhouse gas contribution to global warming. Methane, which strongly absorbs infrared radiation, has a relatively short atmospheric lifetime (8–12 years).

Nitrous oxide is much more reactive than methane and carbon dioxide. Since biogenic nitrous oxide is produced during denitrification and nitrification, wetland soils are an important source of atmospheric nitrous oxide. Nitrous oxide contributes to the greenhouse effect by absorbing radiation in the infrared band leading to the depletion of O_3. On a molecular basis, the relative potential for thermal absorption of nitrous oxide is 150 times that of carbon dioxide (IPCC, 2001). Nitrous oxide has an atmospheric lifetime of ≈100–200 years, which is much greater than that of other nitrogen gases. It is estimated that nitrous oxide accounts for 5–10% of the greenhouse effect. Nitrous oxide is chemically inert in the troposphere but plays an important role in the depletion of stratospheric ozone.

Wetlands, including tropical and subtropical irrigated rice, have soil conditions suitable for both methane and nitrous oxide formation and, as a result, are major anthropogenic sources of atmospheric methane and nitrous oxide. Aerobic and anaerobic environments existing in wetland soil–plant systems provide conditions for both production and consumption of methane and nitrous oxide. There is an aerobic surface layer in wetland soils maintained by oxygen in the water column and an underlying anaerobic layer into which the oxygen does not penetrate (see Chapter 6). The close interface between these layers favors both chemical and especially biological redox reactions. The plant rhizosphere forms the other aerobic–anaerobic interface in the soil profile, which receives oxygen internally from the atmosphere. The oxidation–reduction (redox) reactions including nitrification–denitrification, redox cycling of iron and manganese compounds, sulfate reduction and sulfide oxidation, and methane formation and oxidation are influenced and often controlled by these aerobic–anaerobic interfaces. Although most of methane from natural sources is thought to originate from biological processes in anoxic environments, recent studies have suggested that there may be significant methane emissions from terrestrial plants under aerobic conditions (Keppler et al., 2006). The significance of land plants being a source of methane is less investigated at the moment.

Many physical, chemical, and biological factors of soil influence the production and emission of nitrous oxide and methane. Wetland hydrology and hydroperiod determine whether soil aerobic or anaerobic conditions exist. Redox status is a quantifiable measurement of the reduction process occurring in wetlands. It is well known that nitrous oxide is mainly produced through denitrification and nitrification at moderately reducing conditions, but methanogenesis occurs only under strictly anaerobic conditions.

Global warming will also impact coastal regions of the world. Glacial melting and thermal expansion of oceans caused by global warming will increase the sea level, thereby flooding the coastal wetlands and shifting the wetland area in landward direction. Increased rainfall in global regions is another greenhouse effect that could result in higher levels of water table and development of more extensive wetland systems.

16.2 POTENTIAL IMPACT OF GLOBAL CHANGE TO WETLANDS

Natural wetland areas, a major source of greenhouse gases, are widely distributed throughout the world; however, they are most concentrated in the far north and within the tropics. Globally, wetlands cover approximately 7% of the total land surface. In high latitudes above 45°N, the landscape is dominated by frozen tundra most of the year. The permafrost prevents soil drainage and thus deprives the soil environment of oxygen. In the mid-latitudes between 30°S and 20°N, precipitation levels are quite high. Therefore, in this region known as the tropics, the soil is often completely saturated with water. Globally, the total wetland area has been estimated as 5.3×10^{12} m^2 (Matthews and Fung, 1987) to 5.7×10^{12} m^2 (Aselman and Crutzen, 1989; Bartlett and Harriss, 1993). The largest wetland areas are located between 50°N and 70°N and are classified as bogs and fens. Wetlands located between 10°N and 20°S are classified as swamps and floodplains. While these two general regions represent a major portion of the global wetland areas, there are also a wide variety of wetland regimes that reside in the temperate latitudes.

Climate change associated with increased concentrations of carbon dioxide and other greenhouse gases will significantly alter many of the world's wetland ecosystems. Impacts of climate change will vary depending on the types, magnitudes, and rate of changes in temperature, precipitation, runoff, atmospheric carbon dioxide concentration, and other factors. Warmer climate accompanied by changes in precipitation patterns will affect wetland ecological functions through changes in hydrology, biogeochemistry, and biomass accumulation. Temperature and precipitation are strong determinants of wetland ecosystem structure and function (Mulholland et al., 1997).

Sea-level rise is one of the more certain consequences of increased global temperature. During the past 100 years sea level has risen at a global average rate of about 1–2 mm year^{-1} (Gornitz, 1995). A projected two- to fivefold acceleration of global average sea-level rise during the next 100 years (IPCC, 1996) would inundate low-lying coastal wetland habitats that cannot transgress inland or vertically accrete at a rate that equals or exceeds sea-level rise.

Increased mean global temperature of 1–3.5°C over the next century (IPCC, 1998) in combination with either stable, or reduced, or even slightly increased total precipitation would seriously impact freshwater wetlands. Relatively small changes in precipitation, evaporation, or transpiration that alter surface and groundwater level by only a few centimeters will be enough to reduce or expand many wetlands in size, convert some wetlands to dry land, or shift one wetland type to another (Burkett and Kusler, 2000).

Coastal and estuarine wetland habitats will be severely impacted if sea-level rise exceeds the rate of vertical sediment accretion and inland migration is not possible. Loss of wetlands in Louisiana, Maryland, and other parts of the low-lying Gulf of Mexico and Mid-Atlantic coastal margin has been attributed, in part, to sea-level rise (Penland and Ramsey, 1997). Increasing submergence will initially favor greater methane emission and carbon accumulation in coastal regions. However, over time, this may not be the case if marshes cannot keep accretion with water level and inland migration is restricted through the construction of levees for protection of coastal resources.

Vast expansions of tundra, marshes, and wet meadows underlain by permafrost (ground material below freezing) may be altered by changes in temperature and hydrology. A warming of 4–5°C in Alaska could melt a significant portion of the subarctic permafrost (Gorham, 1995). Bogs, fens, and other largely organic wetlands in the arctic, subarctic, and temperate regions are highly vulnerable to changes in groundwater, which plays a crucial role in the accumulation and decay of organic matter. IPCC (1996) climate scenarios for 2020 and 2050 project a temperature increase of 1–2°C and a decrease in soil moisture for areas of boreal and subarctic peatlands. Areas of peat accumulation will move northward if temperature rises as predicted. On the contrary, increases in summer drought, despite overall increasing precipitation, would cause a degradation of southern peatlands (Gorham, 1995). Peatlands are important carbon reservoirs as they store more carbon than other terrestrial ecosystems (Clair et al., 1995). Estimates of carbon stored in the world's boreal peatlands range from 20% (IPCC, 1996) to 35% of global terrestrial carbon (Patterson, 1999).

Reductions in wetland area and distribution can be expected with increases in temperature or reduced precipitation in prairie pothole regions, which provide habitat for about half of the nation's waterfowl (Poiani et al., 1995). Virtually all wetlands provide storage of carbon in trees, shrubs, grasses, organic debris, and soils. Due to anaerobic conditions, carbon is stored in wetlands for much longer periods than in most upland systems (see Chapter 5).

Hydrology generally determines the rates of aerobic and anaerobic decomposition in wetland soils (see Chapter 5). Lowering of the water level of the highly organic wetland soils will increase decomposition rates and elevate fluxes of carbon dioxide to the atmosphere (IPCC, 1996). A decrease in water availability to wetlands can lead to a decrease in methane formation in wetlands since methane formation in the soil is dependent on anaerobic conditions. Increased temperatures in the peat profile will lead to increased methane production in soils that remain flooded (Clair et al., 1995).

16.3 METHANE

16.3.1 Wetlands as a Source of Methane

Wetlands are transitional between terrestrial and aquatic systems and include swamps, marshes, bogs, and similar areas (Burkett and Kusler, 2000). Wetlands are formed where soils are naturally or artificially inundated or saturated by water due to high groundwater or surface water for part or all of the year. Wetlands are common in river deltas, estuaries, floodplains, and tidal areas. Global wetlands, which cover about 7% of the total land surface, contribute about 10% of the total global net primary productivity (NPP). Many systems have a high turnover rate, indicating that loss and

TABLE 16.2
Estimated Sources of Methane

Methane Sources	Tg year^{-1}
Natural	
Wetlands	120
Lakes, rivers	20
Oceans	10
Termites	10
Total	160
Anthropogenic	
Mining, processing, and use of coal, oil, and natural gas	100
Enteric fermentation (mainly cattle)	80
Flooded rice fields	50
Landfills	30
Animal waste	30
Domestic sewage	20
Total	340
Total source	500

Source: Neue et al. (1993).

export rates are high. Wetlands including paddy soils and natural wetlands produce over 20% of the global methane emissions. The degree to which wetlands produce methane is related to hydrology, vegetation, and management practices. Systems such as rice paddies, which are saturated much of the time, have greater methane emission rates than natural wetlands. Estimates of methane sources are shown in Table 16.2. Anthropogenic sources (340 Tg year^{-1}) dominate over natural sources (160 Tg year^{-1}), and 80% of the total methane emission is of modern biogenic origin. Only 20% is due to fossil carbon sources (Wahlen et al., 1989).

In both natural wetlands and rice paddies, flooding cuts off the oxygen supply from the atmosphere to the soil, resulting in anaerobic decomposition of soil organic matter. Methane is a major end product of anaerobic decomposition (see Chapter 5 for details). Methane is released from submerged soils and sediment to the atmosphere by diffusion and ebullition through roots and stems of plants. For example, global estimates of emission rates from wetland rice fields range from 20 to 100 Tg year^{-1} (IPCC, 1990), which corresponds to 6–29% of the total annual anthropogenic methane emission.

16.3.2 METHANE PRODUCTION IN WETLANDS

Microbial processes regulating methane production in wetlands are discussed in detail in Chapter 5. In this section, a summary is presented to reemphasize the processes responsible for methane production. Methane is primarily biogenic in origin, released through the degradation of organic material in oxygen-depleted environments. The end products of organic matter decomposition in wetlands are carbon dioxide and methane. Both natural and agricultural wetlands are important sources of atmospheric methane. Estimates of the total annual global emission of methane are 120 Tg year^{-1} from natural wetlands (range 100–200) and 60 Tg year^{-1} from rice fields (range 20–100) (Houghton et al., 1992). Methane emissions from all sources are estimated to be about 500 Tg year^{-1}.

Under aerobic conditions the remineralization of organic carbon by microbial action primarily results in the evolution of carbon dioxide. However, when soils flood for prolonged periods of time, oxygen and other inorganic electron acceptors are consumed. The microorganisms begin to sequentially utilize other available substrates as terminal electron acceptors for respiration. Methane production, the terminal step in the anaerobic breakdown of organic matter in wetland soils, is exclusively carried out by methanogenic bacteria that can metabolize only in the absence of free oxygen and at redox potentials of less than −150 mV (Wang et al., 1993). Most methanogens are neutrophilic with an optimum pH of 6–8 and mesophilic with temperature optima of 30–40°C. Methanogens rely on other microorganisms to provide them with few substrates such as hydrogen, carbon dioxide, formate, acetate, methanol, methylamines, and methylsulfides (Conrad, 1989; Garcia, 1990). In wetland soils, methane is largely produced by transmethylation of acetic acid and, to a small extent, by the reduction of carbon dioxide (Takai, 1970). The formation of methane is preceded by the production of volatile fatty acids.

Methane-forming bacteria comprise a family of microorganisms known as Methanobacteriacae. Although these bacteria are biochemically related, they are divided into different genera on the basis of dissimilar cellular morphology: *Methanobacterium* (rods), *Methanococcus* (cocci), and *Methanosarcina* (clusters of cocci). These anaerobes need other fermentative bacteria to metabolize celluloses, sugars, and proteins in detritus into short-chained fatty acids and simple alcohols that can be used by the methanogens.

Methane may be produced via several metabolic pathways; however, in freshwater environments, the main mode of production comes from the splitting of acetate:

$$CH_3COOH \rightarrow CO_2 + CH_4$$

Most of this acetate has its origin in the fermentation of organic matter. Methane can also be produced by simple reduction of carbon dioxide:

$$CO_2 + 4H_2 \rightarrow CH_4 + 2H_2O$$

This reaction explains why less hydrogen is emitted from wetland soils. This hydrogen is readily available due to the fermentative action of bacteria on cellulose substrate (Schutz et al., 1989). Methanogens are very influential in cycling carbon in freshwater wetland areas as they remove a significant amount of organic carbon from these systems. Through interspecific hydrogen transfer by methanogenic associations, the efficiency of anaerobic fermentation and remineralization is increased.

16.3.3 Methane Emission

Emission of CH_4 from soils to the atmosphere is a balance between methane oxidation, production, and transport within the soil systems (Chan and Parkin, 2000; Bradford et al., 2001). Methane is released from anaerobic wetland soils to the atmosphere through diffusion of dissolved methane, through ebullition of gas bubbles, and through wetland plants that develop aerenchyma tissue (Figure 16.1). Large portions of methane formed in an anaerobic soil remain trapped in the flooded soil. Entrapped methane can be oxidized to carbon dioxide when the floodwater is drained or when the soil dries. Entrapped methane can escape to the atmosphere immediately after the floodwater is removed or recedes.

The low solubility of methane in water limits its diffusive transport in the flooded soil, and most methane is oxidized to carbon dioxide. The aerenchyma of plants mediates the transport of air (oxygen) to the roots and methane from the anaerobic soil to the atmosphere. The flux of gases in the aerenchyma depends on concentration and total pressure gradients and internal structure, including openings of the aerenchyma (see Chapter 7 for details).

There are three major processes that regulate methane emission into the atmosphere. The first of these processes is the ebullition (bubbling) of gases produced and trapped in flooded sediments. This process may account for anywhere from 30 to 85% of the total release (Byrnes et al., 1995).

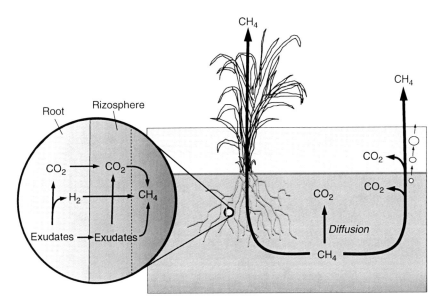

FIGURE 16.1 Fate of methane in wetlands.

This process of ebullition can be affected by changes in atmospheric pressure and hydrostatic pressure, due to changes in water level.

Ebullition is facilitated through the creation of "vesicular structures" within wetland soils. This process arises when there is an increase in pressure due to the continued production of gases in soils in conjunction with the impairment of gas exchange due to the inundation by water. The trapped gases are subsequently erratically released from soils in the form of bubbles.

The flooded surface soil layer maintained by the oxygen in the flooding water and the plant rhizosphere maintained by the oxygen diffusing through the plant have been identified as being important in regulating methane flux. A significant amount of methane produced in wetland soils is oxidized in these two zones before it escapes from the soil into the atmosphere. The movement of methane from anaerobic soils through the leaves, stems, and flowers of aquatic plants and into the atmosphere was found to provide a significant pathway for the emission of CH_4 (Sebacher et al., 1985).

A number of wetland plants have developed adaptive systems to survive and flourish in these extreme environments (see Chapter 7). The most basic of these is the aerenchyma that allows aquatic macrophytes to exchange gases between the atmosphere and sediment (Figure 16.2). Wetland plants not only stimulate methane emissions to the atmosphere by providing a gas conduit and releasing organic compounds through root exudation to increase methane production, but also reduce methane emissions by delivering oxygen to the rhizosphere that oxidizes methane (Ding et al., 2005). Communities of methane-oxidizing bacteria exist in sediment where methane produced meets oxygen transported to the rhizosphere through plant aerenchyma.

Approximately 90% of the methane transport from soils to the atmosphere in rice paddies and freshwater marshes is through aerenchyma portion of roots and stems of the plants. Gases are transported according to their concentration gradient, not only for CH_4 but also for N_2O (Yu et al., 1997). The importance of these gas conduits in aquatic macrophytes of natural wetland areas has been well documented. Methane flux to atmosphere by vascular transport is related to soil redox conditions in which the plant grows (Figure 16.3).

The third major release mechanism for the regulation of methane emission to the atmosphere is diffusion. However, this mode of transport does not appear to be as important as ebullition or vascular transport to net emissions. There appears to be two major reasons for this according to the literature reviewed. First, diffusion of gases into water is slower than into air by a factor of 10^4.

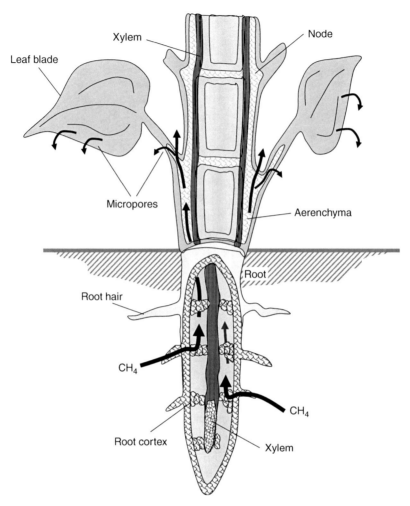

FIGURE 16.2 Pathway of methane transport from the rhizosphere to the atmosphere through wetland plants.

The second reason is that as soil gases are transported to the surface they must pass through the thin aerobic soil–floodwater interface where methane oxidation occurs.

Past studies of global methane emissions have estimated emission rates from freshwater wetlands ranging from 11 to 300 Tg year^{-1} (Aselman and Crutzen, 1989; Batjes and Bridges, 1992; Bartlett and Harriss, 1993). The annual global methane emission rates from wetlands are estimated to be 115 Tg year^{-1} (Cicerone and Oremland, 1988). It is estimated that approximately 20% of all methane emitted annually to the atmosphere comes from natural wetlands. In an assessment of methane emissions conducted by Bartlett and Harriss (1993), flux from both temperate and subtropical wetlands ranged over several orders of magnitude from −7.9 to 3,563 mg CH_4 m^{-2} day^{-1}. The average fluxes from forested freshwater swamps ranged from 39.8 to 155 mg CH_4 m^{-2} day^{-1}. Estimated methane emissions from the coastal wetlands of northern Gulf of Mexico are 1.5×10^{12} g CH_4-C year^{-1} (DeLaune et al., 1983).

Methane emissions from coastal wetlands are strongly governed by salinity. Decreasing CH_4 emission rates have been demonstrated along a gradient of increasing salinity (fresh, brackish, and saline vegetation types; DeLaune et al., 1983) (Table 16.3). The reduction in methane emission is attributed to the effect of SO_4^{2-} in seawater, which serves as a more readily available alternate e$^-$ acceptor, and the formation of SO_4^{2-} reduction products, which inhibit CH_4 formation.

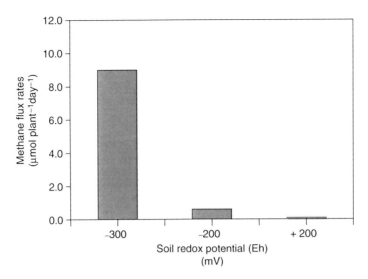

FIGURE 16.3 Rates of CH_4 emission from rice plants as a function of soil redox potential after 50 days of soil incubation. (Data from Kludze et al., 1993.)

TABLE 16.3
Methane Emissions from Louisiana Gulf Coast Fresh, Brackish, and Salt Marshes

Marsh Type	pH	Soil Organic C (%)	Chlorinity (%)	Sulfate ($\mu g\ cm^{-3}$)	Methane Emission ($g\ cm^{-2}\ year^{-1}$)
Spartina alterniflora (Salt)	7.2	8.6 ± 1.8	10	20	4.3
Spartina patens (Brackish)	7.1	18.6 ± 3.4	1	<1	73
Panicum hemitomon (Fresh)	6.3	23.1 ± 4.2	0.2	<1	160

Source: DeLaune et al. (1983).

16.3.4 REGULATORS OF METHANE EMISSION

The production and emission of methane is greatly affected by physical, chemical, and biological properties of soils. Some of these properties include organic matter content, nutrient composition, redox, pH, and salinity. Factors such as regional hydrology, temperature, substrate availability and quality, and salinity conditions affect not only the production of this trace gas but also its consumption by methanotrophs or its release from the soil to the atmosphere.

Hydrology plays a key role in these phenomena: wetland soils that are inundated and anoxic favor the production of methane. Conversely, wetland soils that are aerated to some degree may serve as a sink with aerobic methanotrophs consuming methane (Bartlett et al., 1985; Schutz et al., 1990). Methane flux is related to floodwater height. Moderate to high water levels result in relatively constant fluxes of greenhouse gases. Low water levels, which are more subject to fluctuations, often result in reduced emission rates. Fluctuating water levels dramatically affect the release of methane and nitrous oxide through ebullition or bubbling. Moore et al. (1993) demonstrated the relationship between methane flux and water table in northern peatlands (Figure 16.4).

Vegetation also plays a key role in the dynamics of methane production and emission. Generally the presence of macrophytes enhances emissions of methane. Emergent macrophytes supply important

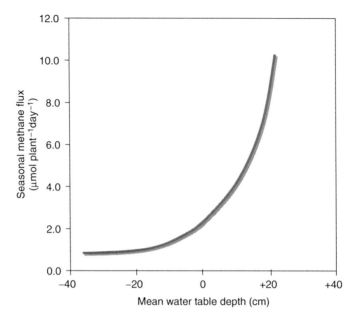

FIGURE 16.4 Relationship between seasonal methane and water table position. (Adapted from Delaune et al., 1983. Data summarized by Moore and Roulet, 1993.)

substrates to methanogens in the form of root exudates and plant litter, thereby increasing production. Oxygen transfer to the rhizosphere within soils by wetland-adapted plants also transports reduced gases such as methane and nitrous oxide to the atmosphere.

Another environmental parameter that is important to methane emissions is soil temperature. Soil temperature affects the metabolic activity of microorganisms that produce substrates for inhibitors of methanogenesis as well as the methanogens themselves. Temperature is directly related to methanogenesis under field conditions (Schutz et al., 1990; Westermann, 1993). Methane emission rates generally parallel soil temperature levels.

Methanogens require neutral pH conditions to function optimally. Methane production occurs only after a sequential reduction of other inorganic electron acceptors (e.g., Fe^{3+}, Mn^{4+}, SO_4^{2-}); it is produced at redox levels below which sulfate in soil is reduced. Methane production is inversely related to sulfate concentration in wetland soils and sediments. High sulfate levels such as those found in seawater can inhibit methane formation (DeLaune et al., 1983).

16.3.5 METHANE SINKS

There are only two important sinks that serve to destroy methane. The first is the oxidation of methane by aerobic bacteria in soils whereas the second and the most important sink is reaction (oxidation) with hydroxyl radicals in the atmosphere. Biological oxidation of methane in soils is responsible for 6–10% of the global source strength. Oxidation due to the reaction of methane with hydroxyl radicals in the atmosphere, however, accounts for the remaining 90% (Cicerone and Oremland, 1988). An estimated 500 Tg year^{-1} is removed from the atmosphere each year; over 95% of the annual emission is removed through these two primary sinks (Khalil et al., 1992).

In general there is a homogenous consumption rate of methane in soils as opposed to a highly variable methane emission rate within specific geographical regions. Both methane production and consumption are microbially mediated. Methane oxidation rates in soil are controlled in part by nonbiological factors, primarily redox status, and soil porosity. Generally, the rate of methane diffusion into the soil controls the rate of consumption (Keller et al., 1990).

Methane oxidation cannot occur in the absence of oxygen; therefore, in wetland soils the process is only important in the surface oxic–anoxic boundary or within the plant root rhizosphere. However, aerated soil regions or oxic layers have been demonstrated to consume as much as 80–90% of methane produced (Bender and Conrad, 1992; Oremland and Culbertson, 1992a). Oxidation within a wetland soil is facilitated by rhizospheres where oxygen is transported from the plant to its roots. Florida Everglade studies have shown that the oxidation of methane by the rhizosphere was small compared to that oxidized at the sediment–water interface (King et al., 1990). Studies performed in rice paddies have shown rhizospheric oxidation of up to 80% of the methane available (Holzapfel-Pschorn and Seiler, 1986; Sass et al., 1990; Schutz et al., 1989).

16.4 NITROUS OXIDE

16.4.1 Wetlands as a Source of Nitrous Oxide

Nitrous oxide, a key compound in the nitrogen cycle, is an intermediate of biological denitrification and a by product of biological nitrification. Gaseous nitrous oxides have multiple influences on atmospheric chemistry. Nitrous oxide is a radiatively active gas that contributes to ozone depletion in the stratosphere where it is photochemically oxidized to NO, which catalyzes a set of reactions resulting in ozone destruction (Rodhe, 1990). Although NO is not radiatively active it governs the concentration of greenhouse gases by regulating the atmospheric oxidants in removal of CO, CH_4, and other gases. NO_x can be oxidized to HNO_3, which is removed from the atmosphere by acid rainfall deposition. Both biotic and abiotic processes are involved in the production of NO_x and nitrous oxide in soil environments. Most microbial processes that involve oxidation or reduction of N through +1 or +2 oxidation state yield trace amounts of NO and nitrous oxide. Nitrification and denitrification are the principal biotic processes responsible for production of nitrous oxide in wetland soils. Abiotic production of nitrous oxide and NO occurs through a set of reactions called chemo-denitrification, the most important being the disproportioning of HNO_2, which can occur in acid soils.

The global emission of nitrous oxide to the atmosphere is estimated to be 17.7 Tg N_2O-N $year^{-1}$ (Kroeze et al., 1999). Although over 20% of the total global nitrous oxide emission and 50% of the total ($N_2O + N_2$) emission may be due to natural terrestrial emissions, the observed increase is essentially due to the increased use of N fertilizers and combustion processes.

Although the estimated source strengths remain uncertain, emissions from soils appear to dominate the nitrous oxide budget (IPCC, 1990). The main biogenic sources of nitrous oxide, NO_x, and N_2 in soils are the microbial processes of denitrification and nitrification.

16.4.2 Nitrous Oxide Production in Wetlands

Soils including wetland soils are important sources of atmospheric nitrous oxide. A wide range of processes may produce nitrous oxide, as well as minor amounts of NO_x, but not all of these seem to be fully understood. The main biological processes of nitrous oxide formation in soils are shown in Figure 16.5. They include nitrification, denitrification, the dissimilatory reduction of nitrate to ammonium, and the assimilatory reduction of nitrate wherein N is incorporated in the cell biomass. Additionally, some NO_x and nitrous oxide may be released due to chemo-denitrification and "pyro-denitrification." Of these processes, nitrification and denitrification are the most important with respect to nitrous oxide production.

Nitrification–denitrification reactions are the major pathway for N loss from wetland ecosystems to the atmosphere. The process involves a series of sequential microbial processes that include mineralization of organic nitrogen to ammonium, oxidation of ammonium to nitrate, and denitrification of nitrate to nitrous oxide or dinitrogen gas. Nitrification–denitrification in wetlands occurs primarily in two zones: the aerobic–anaerobic interface at the surface of the flooded soil or sediment, and the oxidized rhizosphere of wetland plants (Figure 16.6).

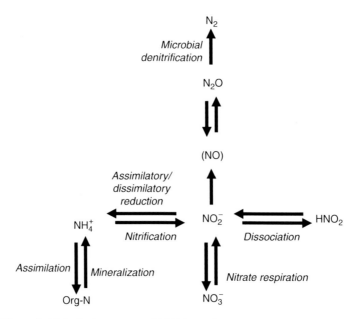

FIGURE 16.5 The main biological processes of N_2O formation.

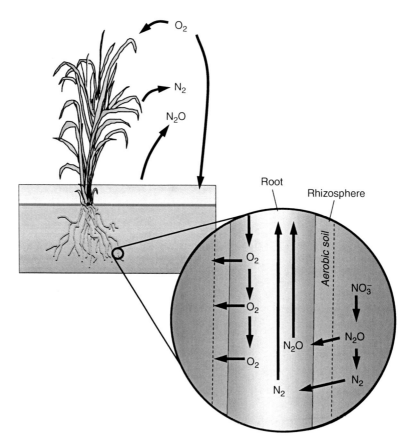

FIGURE 16.6 Zones of nitrification–denitrification reactions in continuously flooded wetland soils.

In both zones, there are two basic requirements: oxygen supply from the atmosphere into the zone where nitrification occurs and ammonium supply through mineralization of organic N or from external inputs of inorganic N. Nitrate concentrations in natural wetland systems are generally low, primarily due to a slow rate of nitrification in relation to the rate of denitrification. However, many natural wetlands receive nitrogen inputs from agricultural runoff. Rice culture includes high nitrogen fertilization rates. In such systems, denitrification and N_2O emissions may be higher than in natural wetlands. Denitrification not only is a major source of N_2O, but also represents the only known biological mechanism to consume N_2O (Bremner et al., 1990; Hutchinson et al., 1993). However, the reduction activity of N_2O greatly depends on the ambient nitrate concentration. The inhibitory effect of nitrate on N_2O reductase is mainly due to electron competition.

In wetlands, N_2O production via denitrification and subsequent emission is governed by the oxygen status of the sediment, supply of nitrifiable nitrogen, moisture content, and temperature. Since the denitrification pathway requires anaerobic conditions, a low partial oxygen pressure during the denitrification process enhances the formation of N_2O.

16.4.3 N_2O Emission from Wetlands

The root rhizosphere and oxidized surface layer of flooded rice soils are zones of active nitrification–denitrification, and hence sources of N_2O production (Figure 16.5). Ammonia-N can be converted to nitrate by nitrification in floodwater and oxidized soil zones. This NO_3^- can then move into the reduced soil zones where it is readily denitrified to N_2O and N_2 (e.g., Reddy and Patrick, 1986). Ammonium in the soil, diffusing to the root, is taken up directly by wetland macrophytes while a part is oxidized to nitrate. The nitrate produced in this reaction either is assimilated by the plant or diffuses back into the adjacent anaerobic zone where it is denitrified to nitrogen gas or nitrous oxide. Wetland plants compete with the denitrifiers for nitrate but because of high electron pressure in the root zone and preferential uptake of ammonium over nitrate by the plant, dentrifying bacteria may outcompete plants for nitrate, thereby enhancing denitrification. The nitrogen gases produced by denitrification can diffuse into the atmosphere through the same aerenchyma passageways that oxygen uses to diffuse down to the root or remain trapped in the saturated soil.

Nitrification–denitrification reactions in flooded soils generally occur at the oxidized–reduced interface of flooded soils or in the root rhizosphere of wetland plants. Oxygen entering wetland soils by diffusing through the stems and roots of wetland plants create an oxidized layer in the reduced soil environment where nitrification–denitrification can occur. Denitrifying activity, measured as a reduction of nitrous oxide to nitrogen gas, is greater in the rhizosphere zone than in bulk soil.

Alternating aerobic–anaerobic cycles in wetland soils promote N_2O emissions to the atmosphere, whereas under continuously flooded conditions most of the N_2O formed will be reduced to N_2. Like methane, appreciable amounts of N_2 and N_2O produced from denitrification remain entrapped in wetland soils and sediments (Lindau and DeLaune, 1991). A low partial oxygen pressure during the denitrification process enhances the formation of N_2O. The critical redox potential for N_2O production in soil is between +100 and +300 mV (Figure 16.7).

The ratio of N_2 to N_2O production in wetlands varies. It ranges from 3 to 25 in a *Spartina* salt marsh soil following nitrogen application (Lindau and DeLaune, 1991). In a freshwater marsh, ratio ranged from 2.5 at day 1 to 80.5 at day 6 following addition of nitrate nitrogen at a rate of 3.85 g N m^{-2} (Yu et al., 2006a). The range can be from 1 to over 400 depending on various factors and conditions. Higher levels of nitrate introduce more competition for electrons from soil organic matter favoring smaller N_2/N_2O ratios (Blackmer and Brenner, 1978). Factors that can influence the N_2/N_2O ratio are listed in Table 16.4.

16.4.4 Production and Emissions from Natural Wetlands

Estimates of the contribution of natural sources of N_2O to the atmosphere have a high degree of uncertainty. Differences in the rate of N cycling among ecosystems account for differences in N_2O

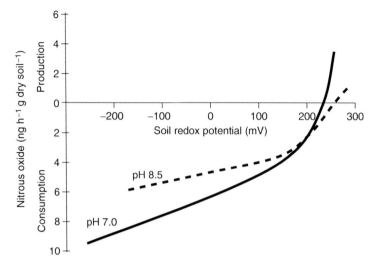

FIGURE 16.7 The critical redox potential for nitrous oxide formation in flooded soil. (From Smith et al., 1983.)

TABLE 16.4
Factors Affecting Relative Proportion of Dinitrogen (N_2) to Nitrous Oxide (N_2O)

Factor	Events	N_2/N_2O Ratio
NO_3, NO_2	Concentration increases	Decreases
Available C	Availability increases	Increases
Anoxicity	O_2 increases	Decreases
pH	Increases	Increases
Temperature	Low	Decreases
Sulfide	High	Decreases
Enzyme status		Decreases
Plants (NO_3, exudates, O_2, H_2O)	Presence increases	Increases
Soil depth	Deeper	Increases
Drying/wetting	Period increases	Increases
Moisture	Content increases	Increases
Marginal aerobic conditions		Decreases
Low rate of denitrification		Decreases

gas emissions. For example, in paddy rice and constructed wetlands used in wastewater treatment, due to heavy nitrogen loading, one would expect larger quantities of N_2O emissions when compared to natural wetlands receiving little nitrogen inputs.

The emissions of N_2O from adjoining salt, brackish, and freshwater marshes within the Louisiana Barataria basin are 31, 48, and 55 mg N m^{-2}, respectively (Smith et al., 1983). The N_2O flux from this study was similar to N_2O from the Narragansett Bay sediment (Seitzinger et al., 1980). The addition of nitrate to the salt marsh significantly increased the rate of N_2O emission (Smith et al., 1983). Appreciable loss of N_2O would only be expected when the marshes receive an extraneous source of nitrate such as sewage or waste water.

Draining and flooding cycles influence N_2O emissions (Figure 16.8). Draining the floodwater from salt marsh cores resulted in increased N_2O emissions. However, the emission of N_2O decreased when the sediment was reflooded. Results demonstrate that the N_2O emission from natural wetlands

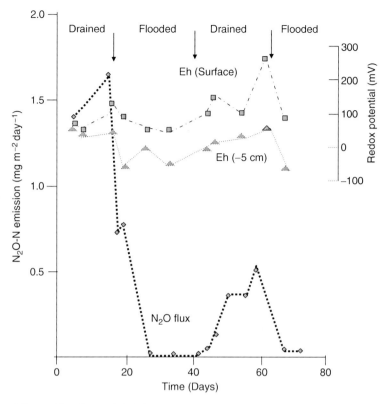

FIGURE 16.8 Effects of flooding and draining cycles on nitrous oxide emission from sediment cores collected from the salt marsh. (Modified from Smith et al., 1983.)

would be greater under the conditions of fluctuating redox potentials than under continuously flooded conditions with low redox potential.

Ammonium nitrogen is the predominant form of inorganic nitrogen in wetland soils and is stable under anaerobic conditions. However, in the presence of both aerobic and anaerobic conditions, ammonium can be nitrified and subsequently denitrified. As a result there is an increase in N_2O emissions.

16.4.5 Regulators of N_2O Production and Emissions

The release of N_2O to the atmosphere from wetlands is enhanced by alternating flooding and drainage cycles. The wetting enhances mineralization of organic matter with resultant pulses in nitrification, and creates (local) anaerobic conditions that favor denitrification. Under these conditions, the production of N_2O may exceed the reduction of N_2O to N_2.

Denitrification generally increases when the oxygen is depleted in the soil. Prolonged flooding can limit denitrification by restricting nitrification that produces the nitrate for denitrification; such conditions may prevail in permanently flooded soils.

Soil pH clearly influences the rate of denitrification as well as the distribution of gaseous end products. The optimal pH range for denitrification is at or near pH 7.0. The redox zone in wetland soils where nitrogen gas and nitrous oxide are formed is generally between +150 and +250 mV (Figure 16.9).

In general, there is a clear separation between zones where methane is formed and where nitrous oxide is formed in wetland soils as depicted in Figure 16.10. When soils are incubated at different pH, the Eh range for CH_4 and N_2O formation shifts according to the inverse relation of pH and Eh described by the Nernst equation. However, the Eh range where both CH_4 and N_2O formations

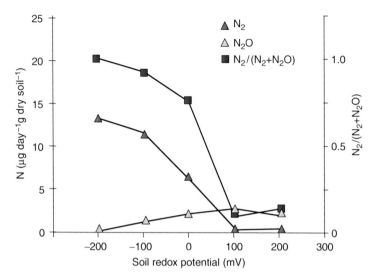

FIGURE 16.9 The effect of redox potential on relative production of nitrous oxide and nitrogen gas during denitrification. (From Kralova et al., 1992.)

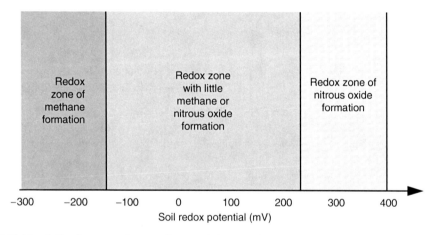

FIGURE 16.10 Soil redox zone where methane and nitrous oxide denitrification occur. (Modified from data by Yu and Patrick, 2003.)

are low remains the same, and is between +180 and −150 mV at pH 7 regardless of different soil characteristics.

The source of nitrogen entering a wetland also governs N_2O production. When nitrate versus ammonium nitrogen was added to a flooded swamp soil, nitrous oxide emissions were recorded in the second day following application of NO_3-N, and lasted for about 2 weeks. It took 15 days before N_2O was evolved following addition of NH_4-N. The lag is attributed to the time that is needed for the ammonium to be nitrified (Table 16.5).

16.4.6 NITROUS OXIDE CONSUMPTION

Soils can both produce and consume nitrous oxide (Blackmer and Bremner, 1978; Letey et al., 1981). The net source of N_2O from wetlands represents the combined production and consumption rates of

TABLE 16.5
Emission of Nitrous Oxide from Flooded Swamp Forests Amended with 10 g N m^{-2} of Ammonium Sulfate and KNO$_3$, Respectively

Sampling Days	N$_2$O Evolved (g N m^{-2} year^{-1}) Equivalent	
	NO$_3^-$-N Source	NH$_4^+$-N Source
0	ND	ND
2	16	ND
5	20	ND
8	4	ND
15	ND	4
21	ND	9
27	ND	9

Note: ND stands for not detectable.
Source: Lindau et al. (1988).

nitrous oxide. The enzyme of N$_2$O reduction can reduce N$_2$O, as it is a free obligatory intermediate in denitrification. Biologically, N$_2$O can only be consumed by reduction to N$_2$ (the last step in denitrification). Wetland soils can serve as a sink for N$_2$O when the redox potential is lower than the critical redox potential for N$_2$O reduction. Such conditions for N$_2$O sink exist under conditions of prolonged and flooding conditions. When reducing conditions in soils are intense enough, complete denitrification is likely to occur with N$_2$ instead of the intermediate product of N$_2$O, as the end product. Letey et al. (1981) indicated that soil redox potential between +200 and +250 mV is critical for N$_2$O production and consumption. Smith et al. (1983) and Yu and Patrick (2003) also reported that the critical redox was +250 mV. Nitrous oxide reduction would occur at redox value below +250 mV. Yu et al. (2006b) in a study of greenhouse gases concentration in soil profile across a hydrological gradient in a swamp forest showed that soil N$_2$O production levels were lower than atmospheric levels in strongly reducing swamp soil. Such finding supports N$_2$O consumption by wetland soils.

16.5 CARBON SEQUESTRATION

Carbon sequestration is the capture and storage of carbon that would otherwise be emitted or remain in the atmosphere. Terrestrial ecosystems, which consist of vegetation and soils containing microbial and invertebrate communities, sequester CO$_2$ directly from the atmosphere. The terrestrial ecosystem is essentially a huge natural biological scrubber for CO$_2$ from all sources of fossil fuel emission, such as automobiles, power plants, and industrial facilities. Terrestrial ecosystems (forests, vegetation, soils, farm crops, pastures, tundras, and wetlands) have a net carbon accumulation of about one-fourth of the amount emitted to the atmosphere from fossil fuels.

Estimates of the carbon stored in soils (1,500 Pg) are roughly three times that stored in the vegetation (Post et al., 1990). Thus, the storage of carbon in soils and the dynamics of soil carbon are of paramount importance to understanding the global carbon budget. Organic soils commonly associated with wetlands contain a large amount of stored C.

Wetlands have the highest carbon density among all terrestrial ecosystems. Because of their low drought stress and high nutrient availability, wetland plants have a large capacity to remove carbon dioxide from the atmosphere. Wetland ecosystems taken as a whole comprise only about 2–6% of the world's land area, but they account for a disproportionately high percentage (14.5%) of the stored soil carbon (Figure 16.11). This is largely due to large amount of carbon sequestered in the organic soils common to wetlands. For example, estimated carbon storage in temperate forests

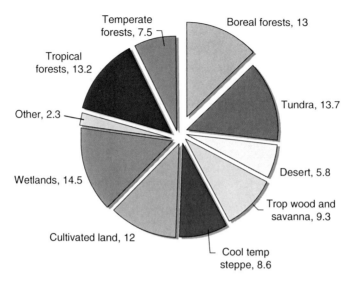

FIGURE 16.11 Estimates of carbon stored in soils of the world as distributed among various ecological zones; total carbon = 1395 Pg (Pg = 10^{15} g). (Redrawn from data by Post et al., 1982.)

is in the range of 7–13 kg C m^{-2}, whereas carbon storage in wetlands is nearly an order of magnitude greater (Post et al., 1982). In North America, approximately 50% of the wetlands are forested.

Wetlands, in addition to being an important carbon sink, are a major source of atmospheric methane. Although soil carbon in wetlands is recognized as being an important component of global carbon budgets and future climate change scenarios, additional studies are needed to quantify the role of coastal wetland ecosystems in carbon sequestration. High rates of organic matter input and reduced rates of decomposition are the main factors resulting in carbon sequestration in wetlands.

16.6 IMPACT OF SEA-LEVEL RISE ON COASTAL WETLANDS

Sea-level rise from the time of glacial retreat until approximately 3000 years ago was quite rapid. From approximately 3,000 years ago to the present, the sea-level rise was relatively slow, which allowed marsh vegetation to grow in tidal areas. During this period the development of a mosaic of marshes in coastal regions was driven by a relative slow rate of sea-level rise. The gradual rise in sea level allows coastlines to accumulate organic carbon as the marsh vertically accretes to keep pace with sea-level changes (Figure 16.12). There is an uncertainty about the ecological and physical changes that will occur in coastal wetlands as a result of global climate change. The resultant increase in sea level associated with global warming is expected to magnify the effect of natural and anthropogenic stress on the coastal marsh areas. Effects will range from increasing stress on coastal marsh vegetation to accelerating the rate of marsh loss. Currently, such conditions exist in coastal Louisiana, where the rapid rate of wetland loss occurs due to subsidence-induced water level and salinity increases. Stresses to marsh plants caused by increases in salinity and water level impacts marsh stability and also causes shift in carbon budget.

The relative elevation of coastal wetlands is a function of a variety of factors, including eustatic sea-level rise, subsidence (which in itself includes many factors), and vertical accretion of mineral and organic sediments (Figure 16.13). Most estimates of eustatic sea-level rise in the last century range from 1 to 2 mm year^{-1} (Gornitz, 1995). Recent estimates of future eustatic sea-level rise vary from 20 to 70 cm by the year 2100 (Houghton et al., 2001). These projected increases are likely to have enormous impacts on coastal wetlands worldwide, inundating low-lying coastal regions. Tidal wetlands developed over a period of relatively low sea-level rise and

Wetlands and Global Climate Change

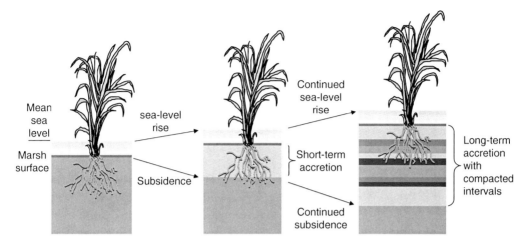

FIGURE 16.12 Factors that affect the accretionary balance of coastal wetlands.

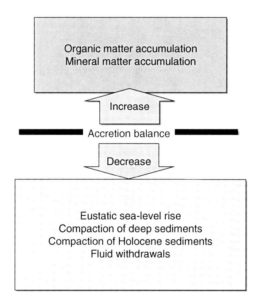

FIGURE 16.13 Processes that affect the accretion balance in response to relative sea-level rise of coastal wetlands.

rapid accumulation of mineral and organic matter. These wetlands can withstand some increases in sea-level rise. There is no clear understanding or prediction of the capacity of coastal wetlands to compensate for increased rates of sea-level rise.

The Mississippi River deltaic plain located along the Louisiana Gulf Coast is currently experiencing an increase in relative sea-level rise mainly due to regional subsidence (see Chapter 19). Changes that occur offer valuable insight into the potential impact of sea-level rise, which will likely occur in other coastal wetlands as a result of global warming. The current rate of relative sea-level rise in the Mississippi River deltaic plain is about 1.2 cm year^{-1} (DeLaune et al., 1978, 1983). Subsidence due to compaction of the Mississippi River sediment accounts for 85–95% of the water-level increase. During the past several thousand years, the Mississippi River deltaic plain expanded considerably, despite the rapid increase in relative sea-level rise. However, in the last century, this accretion trend was reversed with wetland losses of up to 100 km^2 year^{-1}. As a result, marsh

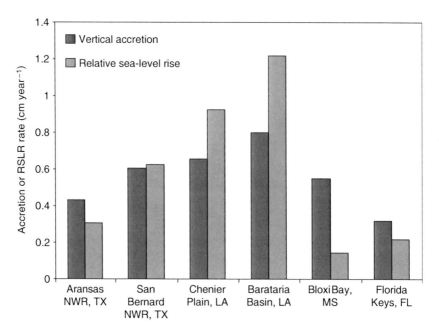

FIGURE 16.14 Vertical marsh accretion and relative sea-level rise at selected marsh locations along the northern shore of the Gulf of Mexico. (Modified from Callaway et al., 1997, which includes data from DeLaune et al., 1983 and Hatton et al., 1983.)

accretion on coastal Louisiana has not kept pace with water-level increases as compared to other Gulf Coast marshes (Figure 16.14). Channeling, leveeing, and dam construction on the Mississippi River have reduced the influx of riverine sediment supply to marshes in coastal Louisiana. As a result, Louisiana wetland marshes that are cut off from their sediment source can no longer vertically accrete at rates high enough to keep pace with relative sea-level rise. Increases in flooding and accompanying salt water intrusion have resulted in the death of marsh vegetation, the source of organic matter to the marsh soil profile.

For coastal marshes to maintain their elevation with respect to projected sea-level increases associated with global warming, sufficient input of sediment and organic matter accumulation will be necessary for both vertical and lateral expansion of the marshes to occur. If the Louisiana Mississippi River deltaic plain serves as an example, such accretion rates will likely be maintained at only a limited number of sites worldwide. The obvious effect of sea-level increases for most coastal marsh regions will be coastal erosion. The nature of the change will vary depending on the nature of the coasts. Intertidal salt marshes and deltaic coasts will be impacted more than steep and cliffed coastlines. There have also been studies of accretion rates in other coastal wetlands outside of Louisiana. Accretion in coastal areas of the United States and Europe show that accretion rates can vary from 0 to 1.5 cm year^{-1}. Globally, most of these wetlands are currently accreting at a rate high enough to compensate for present rates of sea-level rise.

Many factors can affect accretion rates and organic carbon accumulation at a given coastal site. These factors include plant community and density of vegetation, tidal elevation, sediment input from riverine, estuarine, and marine sources, proximity to the sediment source, total organic input from local wetland production, relative sea-level rise, and storm flooding and subsidence. There is also a strong correlation between tidal range and accretionary balance (the difference between accretion and subsidence). Thus, sea-level rise will become an increased threat to coastal marshes. The survival of individual coastal marshes will depend on accretion keeping up with sea-level increases.

16.6.1 MARSH ACCRETION

To counter increases in sea-level rise or other factors that can cause the submergence of coastal marshes, a combination of both mineral and organic matter must accumulate to maintain the marsh surface in the intertidal zone. There is a direct correlation between accretion and organic and mineral accumulation in coastal marshes (Callaway et al., 1997) (Figure 16.15). Otherwise, marsh elevation will become so slow that plant stress or mortality may occur. Mineral matter accumulation varies more than organic matter accumulation. As discussed above, the processes of mineral matter and organic matter are closely related with mineral matter accumulation enhancing organic matter accumulation, and vice versa.

Organic matter accumulation is the main determinant in the vertical growth rates of marshes in the Mississippi River deltaic plain (Hatton et al., 1983). In these marshes, there is less organic matter than mineral matter on a weight basis, but organic matter occupies more volume than mineral matter in fresh and brackish marsh soils (Nyman et al., 1990). Organic matter is believed to contribute greatly to the soil matrix and increase structural strength by forming an interlocking network. Soil mineral matter may represent from 50 to 90% of the dry weight of marsh soils, but mineral matter occupies only 2–7% of soil volume in the Mississippi River deltaic plain marshes (Nyman et al., 1990). Soil organic matter accumulation is the result of *in situ* production by marsh plants, rather than transportation from other areas. Thus factors that regulate plant growth also affect soil organic matter accumulation. Plant growth also promotes mineral matter accumulation. Plant stems reduce water velocities, which promote deposition of mineral matter.

Plant growth, the source of soil organic carbon, is largely controlled by the degree of soil aeration and the amount of reduced sulfur (S^{2-}, HS, and H_2S), which is toxic to plants. The two are

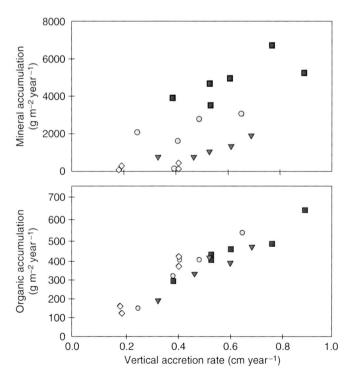

FIGURE 16.15 Correlation of vertical accretion rates with mineral accumulation rates (top) and organic accumulation rates (bottom) for all cores collected in four coastal wetlands along the U.S. Gulf Coast. Accumulation rates are based on accretion since the 1963 peak in bomb-deprived ^{137}Cs. (From Callaway et al., 1997.)

related because sulfate is reduced in waterlogged soils, but only when sulfate is available. Soil water logging is less stressful to marsh vegetation than soil water logging plus sulfides. Because seawater contains more sulfate than freshwater, the importance of sulfide toxicity increases from fresh to brackish to saline marshes. In saline marshes, soil water logging and sulfides combine under some conditions to severely stress vegetation resulting in plant death. Following plant death there is peat collapse and ponding or an increase in open water area (DeLaune et al., 1994).

As shown by these studies, in rapidly subsiding Louisiana Gulf Coast marshes, if marsh vertical accretion in response to sea-level change is inadequate, then the duration of marsh flooding gradually increases. Eventually, flood duration will exceed the tolerance limits of vegetation and plants will die. Following plant death, shallow ponds replace emergent wetlands resulting in coastal marsh deterioration. Evaluation of the future impact of projected sea-level rise on marsh stability in other coastal areas must address marsh accretion–water-level relationships as related to long-term coastal habitat change.

16.7 SUMMARY

- Wetlands are closely connected to global climate change.
- Wetland soils, due to flooded or reducing conditions and high organic matter content, produce methane and nitrous oxide and, as a result, are major anthropogenic sources of these two greenhouse gases.
- The production, emission, and consumption of the greenhouse gases are greatly affected by physical, chemical, and biological properties of wetland soils. The properties include hydrology, organic matter content, nutrients, redox-pH conditions, and salinity.
- Wetland ecosystems consisting of vegetation and soil containing microbial and invertebrate community can sequester a significant amount of CO_2 from the atmosphere.
- The storage of soil carbon in wetland soils is important in the global carbon budget and in predicting the rate of global warming.
- Coastal wetlands are also important in relation to global change but there are concerns regarding how increase in sea levels will impact carbon cycling in these wetlands.
- It is not known whether coastal regions will vertically accrete or store sufficient soil carbon to keep pace with the water-level increases.
- If coastal wetlands do not keep pace with water-level increases, shifts in coastal wetlands carbon budget will occur.

STUDY QUESTIONS

1. What is greenhouse effect? How does change in the greenhouse gas concentrations in the atmosphere affect global temperature?
2. What biological processes contribute to the CH_4 and N_2O production in the wetland soils? Describe redox conditions required for these processes.
3. How do both aerobic and anaerobic conditions exist in soils? Discuss in relation to production and consumption of greenhouse gases.
4. What is the definition of global warming potential? What is the value of this definition?
5. On the molecular basis what is the global warming potential for thermal adsorption of nitrous oxide as compared to carbon dioxide?
6. Is gas transport through wetland plant's aerenchyma system a biological, chemical, or physical process? Why?
7. Discuss the role of sea-level rise on marsh accretion and organic carbon accumulation.

FURTHER READINGS

Bouwman, A. F. 1990. Background. In A. F. Bouwman (ed.) *Soils and the Greenhouse Effect*. Wiley, New York. pp. 25–192.

Eisma, D. (ed.) 1995. *Climate Change: Impact on Coastal Habitation*. CRC Press, Boca Raton, FL.

Lal, R., J. Kimble, E. Levine, and B. A. Stewart (eds.) 1995. *Soils and Global Change*. CRC Press, Boca Raton, FL.

Oremland, R. S. (ed.) 1991. *Biogeochemistry of Global Change*. Selected papers from 10th International Symposium on Environmental Biogeochemists, San Francisco, August 1991. Chapman and Hall, New York.

17 Freshwater Wetlands: The Everglades

17.1 INTRODUCTION

Freshwater wetlands are one of the most productive ecosystems in the world. Biogeochemical cycles in freshwater wetlands are regulated by external material inputs, hydrology, vegetative communities, physicochemical characteristics of soils, and associated internal processes. Marshes, swamps, bogs, and fens constitute broad categories, with each group exhibiting distinct differences with respect to their hydrology, vegetative communities, and soil types (see Chapter 3). For example, marshes (tidal and nontidal) are periodically saturated, flooded, or ponded with water and characterized by herbaceous (nonwoody) vegetation. Soils can be either mineral or organic. Swamps are fed primarily by surface inputs and are dominated by trees and shrubs or mangroves (some examples in the United States include bottomland hardwood forests, Okefenokee Swamp, Georgia; Great Dismal Swamp, Virginia). Soils in swamps can be either mineral or organic. Bogs are freshwater wetlands characterized by spongy peat deposits. Many of these systems receive rainwater as the main input source. Fens are groundwater-fed, peat-forming wetlands covered with grasses, sedges, reeds, and other plants. Biogeochemical differences and relative rates of many of the individual processes measured in these systems are discussed in various chapters of this book.

Many freshwater wetlands are open systems receiving inputs of organic matter and nutrients from upstream portions of the watershed including agricultural and urban areas. Prolonged nutrient (such as nitrogen and phosphorus) loading to wetlands can result in distinct gradients in floodwater and soil. As a result of their position on the landscape, wetlands tend to accumulate organic matter, nutrients, and other contaminants and often exhibit eutrophic conditions. Very few wetlands in the world are oligotrophic. A large body of information is available on productivity and aboveground nutrient cycling in freshwater wetlands (Batzer and Sharitz, 2006); however, little is known about the impact of exogenous nutrients on the diversity and ecology of microbial decomposers and periphyton communities, associated biogeochemical cycling, and long-term storage and bioavailability of nutrients. Nutrient loading to oligotrophic wetlands results in a zone of high nutrient availability or nonlimiting nutrient conditions near the input and low-nutrient availability or nutrient-limiting conditions furthest from the input point. Between these two extremes, there exists a gradient in quality and quantity of organic matter, nutrient accumulation, microbial communities, composition, and biogeochemical cycles.

Low-nutrient systems are characterized by low external loading of nutrients and relatively closed, efficient elemental cycling (Odum, 1969, 1985) (Figure 17.1). Microbial activity and plant productivity are therefore nutrient limited. In response, vascular plants, periphyton, and microbial communities are extremely efficient in utilizing and conserving nutrients through reallocation and uptake of nutrients at very low nutrient concentrations. Plant detritus in these systems generally has high C:N:P ratios and decomposition of this material results in conditions where microbial and periphytic communities outcompete vascular plants for nutrients. Nutrients are held in tight, closed cycles, whose efficiency enables the maintenance of energy flow. The overall turnover rate of organic matter with high C:N:P ratio is usually slow, and long-term decomposition may be both carbon and nutrient limited (Davis, 1991; DeBusk and Reddy, 1998, 2003, 2005).

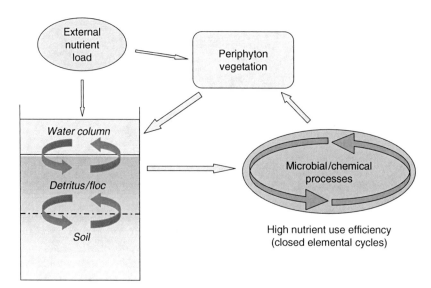

FIGURE 17.1 Schematic representation of a wetland with closed elemental cycles and high nutrient use efficiency (oligotrophic wetlands).

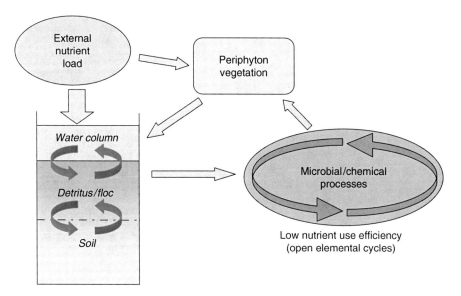

FIGURE 17.2 Schematic representation of a wetland with open elemental cycles and low nutrient use efficiency (eutrophic wetlands).

Environmental factors such as water table fluctuations and fire can result in pulsed release of nutrients, which may provide a significant source of plant-available nutrients in low-nutrient systems (Lodge et al., 1994).

High-nutrient wetland systems are characterized by rapid turnover of carbon and nutrients, and by open elemental cycling, where the nutrient inputs often exceed the demand (Figure 17.2). These systems are low stress systems and contain varying degrees of internal cycling. In response,

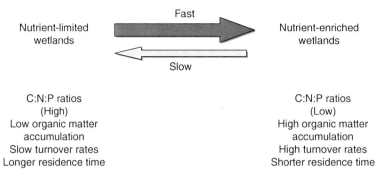

FIGURE 17.3 Schematic showing the transition between oligotrophic and eutrophic wetlands.

vascular plants and microbial/periphyton communities are less efficient in nutrient utilization. Plant detritus in impacted areas generally has low C:N:P ratios, and high net mineralization or release of nutrients during decomposition results in the decreased importance of internal cycling by microbes and plants, as compared with nutrient loading from external sources.

Phosphorus is often the limiting nutrient in oligotrophic wetlands. Increased phosphorus loading to these systems results in a zone with nutrient-nonlimiting conditions near the source, and nutrient-limiting conditions further from the source. Between these two extremes there exists a gradient in quality and quantity of organic matter, nutrient accumulation, microbial communities, and biogeochemical cycles, resulting in diversity of microbial consortia and associated processes (Figure 17.3). These gradients are observed in the detrital layer, and the soil and water columns of many wetlands, most notably in the Everglades.

Individual biogeochemical processes related to carbon, nitrogen, phosphorus, sulfur, and other elemental cycling have been extensively studied in many freshwater and coastal wetlands. For example, individual processes related to organic matter turnover, denitrification, sulfate reduction, methanogenesis, phosphate sorption, metal precipitation, and other reactions have been studied in wetlands. However, only few wetlands are characterized in detail for biogeochemical properties and associated processes. One such wetland where biogeochemical processes (discussed in this book) were measured in detail is the Everglades, a subtropical freshwater wetland. In this chapter, a brief review of the Everglades biogeochemistry is presented.

17.2 EVERGLADES WETLANDS

The Everglades covers approximately 9,000 km^2 (approximately 1.5 million acres) and is the largest subtropical wetland in the United States. The historic Everglades extended from the south shore of Lake Okeechobee to the mangrove estuaries of Florida Bay (Figure 17.4). More than half of the original system has been lost to drainage and development (Davis and Ogden, 1994), including the Everglades Agricultural Area (EAA) located south of Lake Okeechobee. Today's Everglades includes Everglades National Park (ENP), including Florida Bay and the Water Conservation Areas (WCAs) comprising WCA-1, WCA-2A/2B, and WCA-3A/3B. These areas are the primary targets of the Everglades restoration and are grouped as the Everglades Protection Area (EPA).

The Everglades, one of the most unique subtropical wetland ecosystems in the world, evolved in a low-nutrient environment and is unique in that its formation is the result of the accumulation of organic matter over a limestone depression. Nutrient limitation, hydrology, and fire are some of the key factors in the establishment of the endemic Everglades flora, which has adapted to a

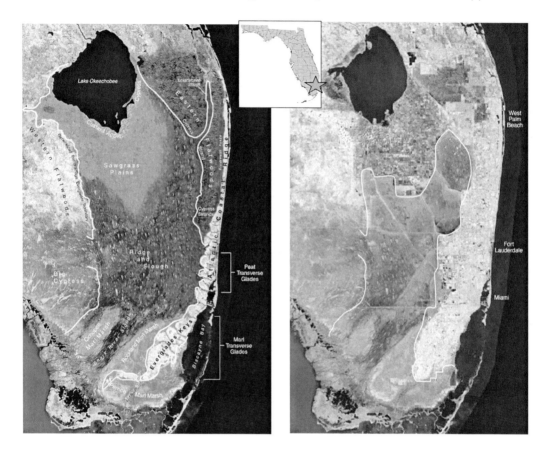

FIGURE 17.4 Historic (map on left side) and current features (map on right side) of the Everglades landscape. (From South Florida Water Management District, West Palm Beach, Florida.)

low-nutrient environment (Davis, 1991). The allochthonous system is the one adapted to low nutrient content (oligotrophic), particularly phosphorus, which in addition to fire and hydrological conditions have resulted in endogenous communities characterized by strands of saw grass (*Cladium jamaicense* Crantz) and open slough areas (Loveless, 1959). Historically, the major source of nutrients to the Everglades has been from atmospheric deposition, with minimum secondary nutrient inputs through infrequent sheet flooding in the northern Everglades from Lake Okeechobee. The subtropical climate of the Everglades can be described by dry (November through May) and wet (June through October) seasons (Duever et al., 1994).

17.2.1 Historical Perspective

The predrainage Everglades consisted of 1.17 million ha of South Florida. The Everglades is part of a larger watershed ranging from Kissimmee River–Lake Okeechobee on the north to Florida Bay on the south, the Gulf of Mexico and Big Cypress Swamp on the west, and the Atlantic Coastal Ridge on the east. The landscape consisted of plains of saw grass, wet prairies, sloughs, tree islands, and marl-forming marshes. Prior to drainage and the installation of water control structures, the natural system was connected. Hydrology of this ecosystem was dominated by pulsed sheet flow from Lake Okeechobee to Florida Bay, with seasonal patterns of rainfall. Figure 17.5 shows a

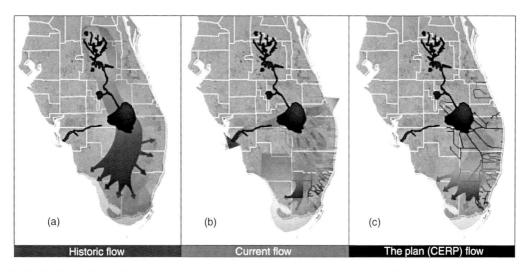

FIGURE 17.5 Water flow in the Everglades under (a) historical conditions, (b) current conditions, and (c) conditions envisioned upon completion of the restoration. (From U.S. Army Corps of Engineers, Jacksonville District, Florida.)

schematic representation of the water flow in the Everglades under (a) historical conditions, (b) current conditions, and (c) conditions envisioned after restoration programs.

17.2.2 Hydrologic Units

Agricultural and urban development reduced the present-day Everglades to only 50% of its original extent, of which approximately 3,500 km² is impounded within the shallow, diked reservoirs known as Water Conservation Areas (WCAs). Postdrainage is extensively managed and the ecosystem is divided into distinct hydrologic units. Approximately, 2,400 km of canals and dikes now manage water in the Everglades ecosystem. The wetland that remains include the WCAs, the Holeyland and Rotenberger Wildlife Management Areas, Big Cypress, and the ENP, and still support unique biotic communities typical to subtropical climate. A brief description of each hydrologic unit is in Table 17.1.

17.2.2.1 Everglades Agricultural Area and C-139 Basin

The EAA and C-139 basins are south of Lake Okeechobee and north of the EPA. The EAA extends south from Lake Okeechobee to the northern levee of WCA-3A, from its eastern boundary at the L-8 canal to the western boundary along the L-1, L-2, and L-3 levees. It incorporates approximately 2,872 km² of highly productive agricultural land containing rich, organic peat or muck soils. Approximately 77% of the EAA is in agricultural production. Runoff discharges from these two basins drain into Stormwater Treatment Areas (STAs) prior to discharge into the EPA. The EAA encompasses 2,800 km² (280,000 ha), with most of these soils being classified as Histosols. The area is intensively farmed with crops such as sugarcane (*Saccharum officinarum* L.), winter vegetables, and turfgrass. Optimal field moisture conditions are maintained through the use of field ditches, water control structures, and pumps. During heavy rainfall events, excess drainage water from the farms is pumped into adjacent WCAs. Because of extensive drainage and subsequent oxidation, these soils are subsiding at a rate of approximately 2.5 cm year^{-1}.

TABLE 17.1
Major Features of the Everglades Hydrologic Units

Hydrologic Units	Area (km²)	Description
Everglades Agricultural Area (EAA)	2,872	Highly productive agricultural land containing rich, organic peat or muck soils; 77% is in agricultural production; recognized as a major contributor to nutrient enrichment of the region; basin is the subject of a water quality monitoring program and a regulatory Best Management Practices program
C-139 Basin	686	Agriculture is the dominant land use; discharges into WCA-3A via structures; basin is the subject of a water quality monitoring program and a regulatory Best Management Practices program
Stormwater Treatment Areas (STAs)	166	Stormwater Treatment Areas (STAs) are built on agricultural lands as a buffer between the EAA and EPA. The STAs are constructed by SFWMD and USACE
Everglades Protection Area (EPA)	9,000	Comprises Water Conservation Areas 1, 2A, 2B, 3A, and 3B; Arthur R. Marshall Loxahatchee National Wildlife Refuge; and Everglades National Park
Water Conservation Area 1 (WCA-1)	566	The Refuge; managed by USFWS, SFWMD, and USACE; saw grass wetland with many tree islands; receives water primarily from STA-1W, STA-1E, and EAA regions
Water Conservation Area 2 (WCA-2)	537	Managed by District with USACE and FWC; smallest WCA divided into WCA-2A and 2B; saw grass wetland with tree islands; receives water primarily from STA-2, STA-3/4, WCA-1, and EAA regions
Water Conservation Area 3 (WCA-3)	2,339	Managed by District with USACE and FWC; largest WCA divided into WCA-3A and 3B; saw grass marsh with tree islands, wet prairies and sloughs; receives water primarily from STA-5, STA-6, WCA-2, Big Cypress National Preserve and EAA region
Holey Land Wildlife Management Area	140	Managed by FWC; lies within the EAA boundaries; heavily used for deer and hog hunting; important for game management, water resource protection, and providing habitat corridors adjacent to the EPA
Rotenberger Wildlife Management Area	96	Managed by FWC; lies within the EAA boundaries; heavily used for deer and hog hunting; important for game management, water resource protection, and providing habitat corridors adjacent to the EPA
Everglades National Park (ENP)	5,560	Second largest national park and one of the nation's 10 most endangered parks; established in 1934 to preserve the unique Everglades ecology; managed by USFWS and NPS with USACE and SFWMD; freshwater sloughs, marl-forming marshes, and mangroves
Big Cypress National Preserve	2,280	Established in 1974 to protect natural and recreational values of the Big Cypress Watershed; land supports hunting, fishing, and oil and gas production; provides an ecological buffer zone and water supply for Everglades National Park
Florida Bay and Florida Keys	2,200	About 80% of the bay lies within Everglades National Park; a broad, shallow expanse of brackish-to-salty water that contains numerous small islands, extensive sandbars and grass flats; mangroves and sea grasses provide valuable habitat for many species; keys watershed consists of a limestone island archipelago of about 800 islands extending southwest for over 320 km
Biscayne Bay	1,100	Subtropical estuary designated as an aquatic preserve and Outstanding Florida Water; bay comprises north, central, and south regions; contains a coral reef system, which is the world's third longest and the only one in the world located in close proximity to a large highly urbanized coastal area; reef is home to more than 200 marine species of fish and is important for fisheries

Source: Redfield et al. (2007) and South Florida Water Management District (2007).

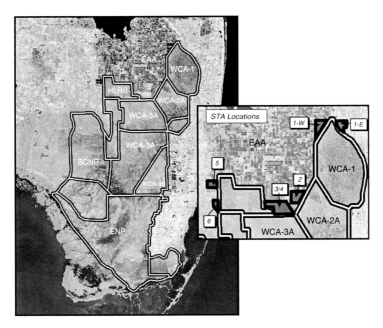

FIGURE 17.6 Hydrologic units of the Everglades landscape, including Everglades Agricultural Area, Water Conservation Areas, Everglades National Park, and Stormwater Treatment Areas. (From South Florida Water Management District (2007).)

17.2.2.2 Stormwater Treatment Areas

STAs are built as additional means of reducing nutrient loads from the EAA into the EPA. Six STAs are currently built and are in full operation. The STAs include STA-1E (2 years) and STA-1W (12 years), STA-2 (6 years), STA-3/4 (3 years), STA-5 (6 years), and STA-6 (9 years) (Figure 17.6). The STA performance, compliance, and optimization are discussed in a recent 2007 report on South Florida Environment (Pietro et al., 2007; Reddy et al., 2006).

17.2.2.3 Water Conservation Areas

The Central and South Florida (C&SF) Project (during 1950s) established three WCAs, which are managed by a set of regulation schedules that determine the water levels in these systems.

- Water Conservation Area-1 (WCA-1) encompasses 572 km^2 of the northern Everglades and is part of the Arthur R. Marshall Loxahatchee National Wildlife Refuge managed by the U.S. Fish and Wildlife Service (USFWS). Rainfall is the primary water input, while other sources include phosphorus-enriched runoff from the EAA. Elevated soil total phosphorus levels have been observed in areas adjacent to water-inflow points. *Typha* dominates in phosphorus-impacted areas whereas *Cladium*, open sloughs, and tree islands are common in unimpacted interior areas.
- Water Conservation Area-2A (WCA-2A) covers an area of 543 km^2 and receives drainage water from WCA-1 in addition to discharge water from the EAA. The P-impacted areas of WCA-2A are much broader and extend much further into the interior than the impacted areas of WCA-1; thus, P impacts on vegetation community structure are more evident in WCA-2A (McCormick et al., 2000). The WCA-2A vegetation consists of *Typha* in P-impacted areas, whereas *Cladium* and open-water sloughs dominate in unimpacted interior areas. This gradient extends from P-impacted, *Typha*-dominated areas near S10-C to unimpacted *Cladium*-dominated areas in the interior.

Water Conservation Area-3 (WCA-3) covers an area of 2,369 km² and receives drainage water from northern hydrologic units, particularly WCA-2A, and the Big Cypress National Preserve through the L-28 gap. In WCA-3, tree islands and wet prairies comprise the vegetation community structure, and most sites adjacent to water-inflow points have been exposed to external nutrient loading.

17.2.2.4 Holeyland and Rotenberger Wildlife Management Areas

The Holeyland Water Management Area (HWMA) is located in the S-8 and S-7 subbasins of the EAA. To restore some of the original functions and values of the Everglades, the Florida Game and Freshwater Fish Commission and the South Florida Water Management District have established water regulation schedules that simulate natural hydroperiods (Surface Water Improvement and Management, 1992). Water leaving the HWMA is discharged into WCA-3A.

17.2.2.5 Everglades National Park

ENP with ridge and slough landscapes includes freshwater sloughs, saw grass ridges, tree islands, marl-forming prairies, mangrove forests, and saline tidal mud flats. The freshwater and mangrove wetlands are located at the southernmost extent of the Florida peninsula and border Florida Bay. Shark River Slough and the wetland area south of the C-111 canal comprise an area of 460 km²; it is a broad shallow basin located in ENP, which serves as a major conduit for freshwater flow to Florida Bay. Shark River Slough receives water from WCA-3A through the L31W and C-111 canals. As surface water from Shark River Slough and C-111 canal flows south toward the bay, it becomes increasingly channelized and flows into five major creek systems that discharge into Florida Bay. Vegetation includes *Cladium*-dominated marshes at northern areas, whereas mangroves are common in southern areas. Taylor Slough (46,000 ha), a broad shallow basin located in ENP, serves as a major water input to Florida Bay. It receives water from WCA-3A through the L31W and C-111 canals. Vegetation includes *Cladium*-dominated marshes in northern areas, whereas mangroves are common in southern areas.

17.3 NUTRIENT LOADS AND ECOLOGICAL ALTERNATIONS

Historically, the major source of nutrients to the Everglades has been from atmospheric deposition, with minimum secondary nutrient inputs through infrequent sheet flooding in the northern Everglades from Lake Okeechobee. At present, approximately two-thirds of the phosphorus loads from Lake Okeechobee is discharged into the St. Lucie and Caloosahatchee estuaries in the east and west, respectively. Approximately one-third of the phosphorus load from Lake Okeechobee enters the EAA and other small basins, which contribute phosphorus loads downstream into WCAs (Redfield and Efron, 2007). Nutrient loading to WCAs of the northern Everglades has not only altered algal and plant communities, but also increased nutrient accumulation (Davis, 1991; DeBusk et al., 1994; Newman et al., 1997). Autochthonous nutrient inputs into the northern areas of the Everglades have resulted in significant alterations to the indigenous system with large incursions of cattail (*Typha domingensis*). Extensive documentation of the temporal and spatial distribution of the nutrients across the northern marshes of the Everglades has established areas of nutrient enrichment (Davis, 1991; Craft and Richardson, 1993a, 1993b; Reddy et al., 1993; DeBusk et al., 1994, 2001; Newman et al., 1997; Corstanje et al., 2006; Bruland et al., 2007; Rivero et al., 2007), associated with shifts in the predominant plant communities (Davis, 1991; Craft and Richardson, 1997; Sklar et al., 2005). Nutrient accumulation rates of 0.11–1.14 g P m^{-2} year^{-1} and 5.4–24.3 g N m^{-2} year^{-1} have been reported for the Everglades (Craft and Richardson, 1993a, 1993b; Reddy et al., 1993). The highest accumulation rates are noted in areas closer to the source of nutrient inputs, and the lowest accumulation rates occurred in areas furthest from the input points. Continual nutrient loading to oligotrophic Everglades wetlands resulted in a zone of high nutrient availability near the input and low nutrient availability in the interior nonimpacted marsh. Unlike carbon and nitrogen, phosphorus added to wetlands accumulates within the system, because there is no significant gaseous loss mechanism in the P cycle. Increased nutrient loading resulted in a zone of high nutrient availability or nutrient-nonlimiting conditions near the source, and low nutrient availability or nutrient-limiting conditions further from the source. This gradient is observed in soils and water columns of WCA-2, WCA-1, HWMA, and WCA-3 (DeBusk et al., 1994, 2001; Newman et al., 1996; Bruland et al., 2006; Corstanje et al., 2006). In this chapter, we will discuss the biogeochemical processes measured along the phosphorus gradient in several hydrologic units of the Everglades.

Carbon, nitrogen, phosphorus, sulfur, and other related elemental cycles regulate many biogeochemical processes in the soil and water column of an ecosystem, and the majority of these biogeochemical processes are mediated by microbial, periphyton, and macrophyte communities. The biogeochemical processes that respond rapidly to nutrient loading and related physicochemical properties of soil, detritus, and water column can be used as indicators to determine nutrient impacts or the recovery after restoration (see Chapter 15). In this chapter we present a summary of the research findings during the past 15 years on the effects of nutrient loading and hydrology on changes in plant, periphyton, and microbial communities and associated biogeochemical processes regulating the nutrient availability and cycling. Publications related to some of the research presented in this chapter can be found at http://wetlands.ifas.ufl.edu.

17.3.1 Surface Water Quality and Loads

The flora and fauna of the Everglades ecosystem are severely phosphorus limited and adapted to nutrient-poor conditions. As a result any small addition of nutrients, especially phosphorus, can have a dramatic effect on growth and productivity of this ecosystem (Childers et al., 2001; Gaiser at al., 2005; Chiang et al., 2000). To determine nutrient impacts on this ecosystem, state and federal agencies have established extensive monitoring networks to determine nutrient concentrations and loads in and out of the hydrologic units of the Everglades. For this reason, major emphasis is placed on monitoring surface water for phosphorus. In addition, surface water is also monitored for various physicochemical parameters including nitrogen, metals, and pesticides. Detailed information on monitoring network and the data can be obtained from the South Florida Water Management District (SFWMD). The Everglades Forever Act (EFA) stipulates surface water total phosphorus concentration limit of 10 μg L^{-1} for the EPA. No numeric limits are set for the surface water total nitrogen.

Nutrients discharged from drainage waters of the EAA and C-139 basins have been identified as major sources impacting downstream EPA. Source control strategies such as BMPs and STAs have been used as the first line of defense to reduce phosphorus loads from these basins to the EPA. The EFA-mandated regulatory program requires BMP implementation and specific phosphorus load limits in discharges from the basins. The EAA basin must achieve a 25% reduction in total phosphorus load in any given water year (since WY 1996) when compared to a pre-BMP baseline period. For the past 11 years, the BMP program has exceeded the 25% requirement (Figure 17.7, Table 17.2) (Adorisio et al., 2007). Similarly, the C-139 basin must also maintain total phosphorus loads leaving the basin in any given year at or below pre-BMP baseline period levels. For the past five years (2002–2006), the C-139 basin did not meet this requirement. Poor performance was attributed to an active hurricane season across South Florida; however, impacts were minimal in most source control program basins, with the exception of C-139.

To reduce phosphorus loads to the EPA, approximately 16,000 ha of the treatment areas, referred to as STAs, are built at strategic locations to intercept the flow of nutrients before discharge into the EPA. Since 1994, these STAs have removed approximately 800 mt of phosphorus from the agricultural drainage waters (Pietro et al., 2007) (Table 17.3).

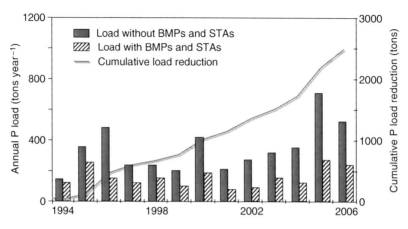

FIGURE 17.7 Cumulative phosphorus loads with and without BMPs and STAs. (From South Florida Water Management District (2007).)

TABLE 17.2
Flow, Total Phosphorus Load, and Total Phosphorus Concentration Comparisons of the Baseline Period (1980–1988) and the Most Recent Five-Year Period (2002–2006) in the EPA

Hydrologic Unit	Parameter	Units	1980–1988	2002–2006	Percent Change
EAA	Flow	$m^3\,year^{-1}$	12.8×10^8	13.0×10^8	2%
	TP Load	$mt\,year^{-1}$	223	120	46%
	TP Conc.	$\mu g\,L^{-1}$	172	119[a]	−31%
C-139	Flow	$m^3\,year^{-1}$	1.6×10^8	2.8×10^8	75%
	TP Load	$mt\,year^{-1}$	37	72	95%
	TP Conc.	$\mu g\,L^{-1}$	227	260[a]	15%

[a] Data represent the year 2006.
Source: Adorisio et al. (2007).

TABLE 17.3
Phosphorus Retention by Stormwater Treatment Areas (STAs) of the Everglades

Period of Record	STA-1E	STA-1W	STA-2	STA-3/4	STA-5	STA-6
Area (ha)	1,630	1,690	2,600	5,760	1,300	350
Start Date	Sept-04	Oct-93	Jun-99	Oct-03	Oct-99	Oct-97
Phosphorus retained to-date (mt)	1.6	314.1	149.5	160.7	151.2	32.2
Phosphorus outflow ($\mu g\,L^{-1}$)	214	52	18	19	100	19

Source: Pietro et al. (2007).

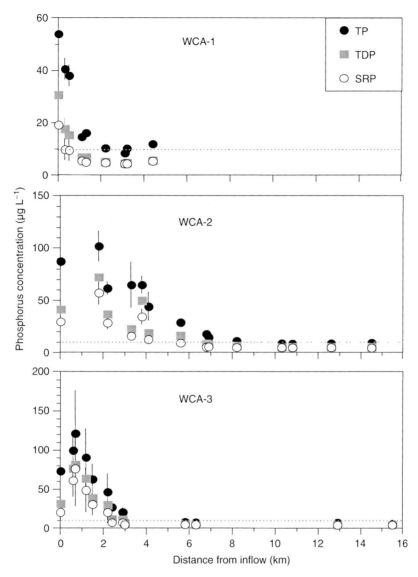

FIGURE 17.8 Surface water phosphorus concentration as a function of distance from inflow in WCA-1, WCA-2, and WCA-3 (SFWMD, 2007).

Distinct gradients in total phosphorus concentrations of water column are reported for WCA-1, WCA-2, and WCA-3 (Figure 17.8). Total phosphorus concentrations are higher near the inflow and decreases significantly with the distance from the inflow to the background levels of less than 10 µg L^{-1}. Total nitrogen concentrations typically do not show similar gradients. Low total phosphorus levels in the interior marsh reflect phosphorus limitation, and high total phosphorus levels near the inflow can create a situation of nitrogen limitation.

17.3.2 Soil Nutrient Distribution and Storage

Nutrient concentrations of the water column are highly variable in time as a result of changes in water depth and hydroperiod. A major portion of the nutrients are stored as organic matter because the nutrient storage in periphyton and vegetation represents only short-term nutrient sink.

Thus, soils serve as long-term sinks for nutrients and as integrators of environmental and ecological conditions of wetlands. In the Everglades, soils are used as one of the key components for assessing the status and recovery after restoration is implemented. For the past two decades, extensive monitoring of soil nutrients has been conducted in select hydrologic units of the Everglades, primarily to determine the zone of nutrient impact. Earlier monitoring studies included soil sampling along specific transects extending from the areas closer to the inflow structures to the areas interior of a hydrologic unit (Koch and Reddy, 1992; Craft and Richardson, 1993a, 1993b, 1998; Reddy et al., 1993; Qualls and Richardson, 1995; DeBusk and Reddy, 1998; Reddy et al., 1998; White and Reddy, 1999; Childers et al., 2003). These transects provided a base line for long-term biogeochemical studies to determine the influence of anthropogenic nutrient loading (Figures 17.9–17.11). A much more intensive spatial sampling was conducted in all hydrologic units using either grid or stratified random sampling. First such a sampling was conducted in 1990 in the WCA-2A, followed by additional sampling in 1998 and 2003 (DeBusk et al., 1994, 2001; Grunwald et al., 2004; Rivero et al., 2007). Similarly, WCA-1 was also sampled in 1992 and 2003 (Newman et al., 1997; Corstanje et al., 2006), WCA-3 in 1992 and 2003 (Bruland et al., 2006, 2007), HWMA and RWMA in 1995 and 2003, and ENP in 2003 (Osborne et al., 2008). During the past two decades, limited sampling was conducted in all other hydrologic units to determine the nutrient impacts. All hydrologic units included in the EPA were sampled in 2003 at approximately 1,400 stations to determine the spatial distribution of soil phosphorus and other associated nutrients (Reddy et al., 2007) (Figure 17.12).

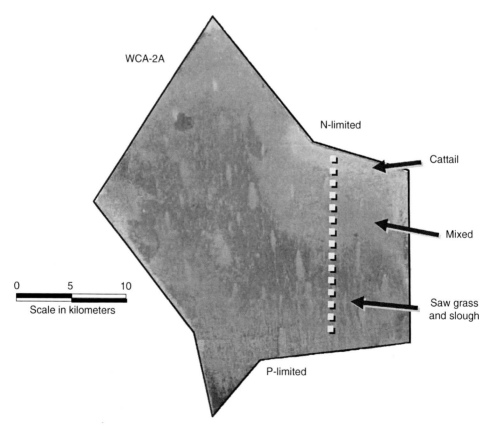

FIGURE 17.9 An experimental transect established along soil phosphorus gradient has been used by a number of investigators to study biogeochemical processes in soil and water column.

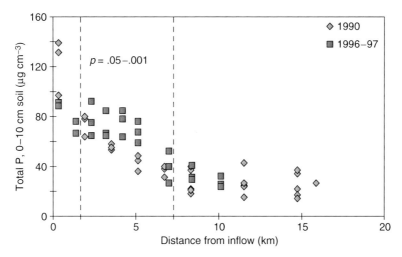

FIGURE 17.10 Soil phosphorus concentration along the transect in WCA-2A of the Everglades (Reddy et al., 1998).

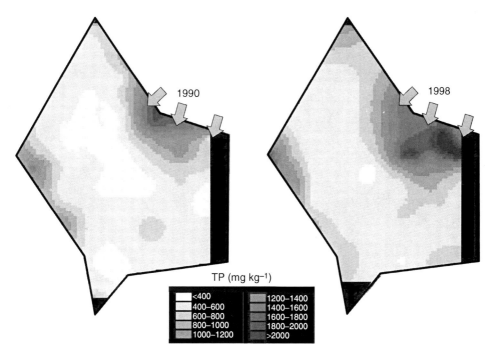

FIGURE 17.11 Spatial distribution of soil phosphorus during 1990 and 1998. (From DeBusk et al., 2001.)

Soil cores obtained from the Everglades show distinct horizons. Surface floc layer typically consists of a well-decomposed organic matter as seen in nutrient-impacted sites (Figure 17.13a). In the interior sites (with the exception of WCA-1), the surface floc is represented by partially developed calcareous benthic periphyton (Figure 17.13b). The WCA-1, a soft water system, is colonized by extensive noncalcareous periphyton mats. The loss of benthic periphyton in the nutrient-impacted sites represents change in vegetation communities (dominated by *Typha*) and accumulation of detrital matter derived from this highly productive macrophyte.

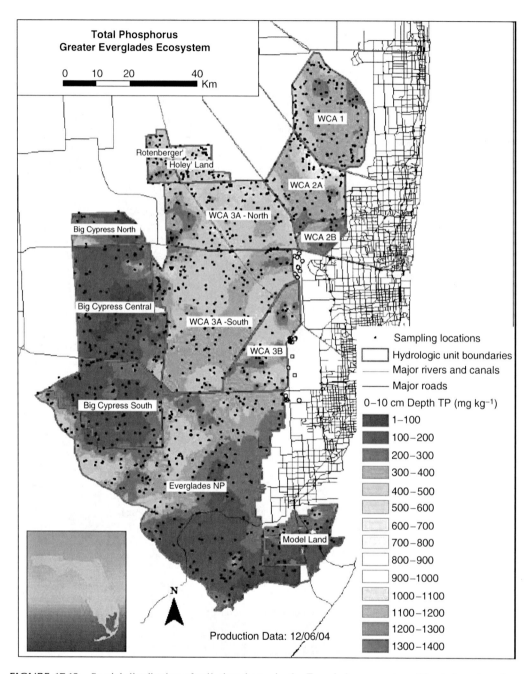

FIGURE 17.12 Spatial distribution of soil phosphorus in the Everglades ecosystem. Map is based on data from 1,400 sampling stations. (Reddy, K. R., Osborne, T. Z., Newman, S., and Grunwald, S., 2007, unpublished results.)

Distinct differences are noted in several physicochemical properties of floc and soils of WCA-1, WCA-2, WCA-3, and ENP (Tables 17.4–17.7). Spatial patterns of total phosphorus in WCA-1, WCA-2A, WCA-3, and ENP showed higher total soil phosphorus in areas closer to inflow structures (predominantly cattail vegetation) than the interior areas with native saw grass vegetations

FIGURE 17.13 Intact soil cores retrieved from impacted and unimpacted sites of the Everglades. (From Newman, S., SFWMD.)

TABLE 17.4
Selected Physicochemical Properties of Surface Floc (Detrital Matter and Benthic Periphyton) and Soils (0–10 cm Depth) in WCA-1 Sampled in the Year 2003

Parameter	Units	Floc	Soil (0–10 cm Depth)
Number of samples		125	131
Bulk density	g cm^{-3}	0.022 (0.001–0.14)	0.073 (0.03–0.40)
LOI	%	92 (73–99)	93
TIP	mg kg^{-1}	158 (38–567)	80 (16–351)
TP	mg kg^{-1}	632 (226–1,462)	405 (113–1,047)
TN	g kg^{-1}	33 (20–44)	33 (17–46)
TC	g kg^{-1}	447 (385–590)	473 (221–544)
Tca	g kg^{-1}	15 (3.6–74)	16 (6–33)

Source: Corstanje et al. (2006).

and open sloughs (DeBusk et al., 1994; Qualls and Richardson, 1995; Newman et al., 1997) (Figure 17.12). Total phosphorus is higher in the floc and decreases with depth (Figure 17.14). Total soil phosphorus profiles showed distinct differences between eutrophic (nutrient-impacted) and oligotrophic (nutrient-nonimpacted) sites up to a depth of approximately 15 cm and below this point there were no significant differences. This suggests that top 10 cm is accreted as a result of nutrient loading during the past few decades.

TABLE 17.5
Selected Physicochemical Properties of Surface Floc (Detrital Matter and Benthic Periphyton) and Soils (0–10 cm Depth) in WCA-2 Sampled in the Year 2003

Parameter	Units	Floc	Soil (0–10 cm Depth)
Number of samples		111	111
Bulk density	$g\,cm^{-3}$	0.046 (0.005–0.3)	0.11 (0.06–0.33)
LOI	%	84 (39–94)	87 (53–94)
TIP	$mg\,kg^{-1}$	249 (82–999)	137 (24–1,277)
TP	$mg\,kg^{-1}$	827 (194–1,865)	551 (155–1,701)
TN	$g\,kg^{-1}$	27 (12–38)	27 (12–40)
TC	$g\,kg^{-1}$	402 (229–468)	427 (247–474)
Tca	$g\,kg^{-1}$	54 (17–277)	39 (17–194)

Note: For the data on 1990 and 1998 sampling, see DeBusk et al. (1994, 2001).

Source: Rivero et al. (2007).

TABLE 17.6
Selected Physicochemical Properties of Surface Floc (Detrital Matter and Benthic Periphyton) and Soils (0–10 cm Depth) in WCA-3 Sampled in the Year 2003

Parameter	Units	Floc	Soil (0–10 cm Depth)
Number of samples		148	389
Bulk density	$g\,cm^{-3}$	0.031 (0.003–0.145)	0.160 (0.031–1.35)
LOI	%	80 (20–96)	77 (4–97)
TIP	$mg\,kg^{-1}$	163 (41–1,420)	97 (4–910)
TP	$mg\,kg^{-1}$	541 (125–1,953)	420 (29–1,169)
TN	$g\,kg^{-1}$	31 (9–46)	28 (2–44)
TC	$g\,kg^{-1}$	401 (176–505)	391 (20–523)
Tca	$g\,kg^{-1}$	67 (17–334)	35 (1–319)

Source: Bruland et al. (2006).

TABLE 17.7
Selected Physicochemical Properties of Surface Floc (Detrital Matter and Benthic Periphyton) and Soils (0–10 cm Depth) in ENP Sampled in the Year 2003

Parameter	Units	Floc	Soil (0–10 cm Depth)
Number of samples		142	310
Bulk density	$g\,cm^{-3}$	0.120 (0.01–0.6)	0.252 (0.060–0.870)
LOI	%	35.9 (10.4–87.8)	47.7 (7.1–94.6)
TIP	$mg\,kg^{-1}$	39.9 (7.7–220.1)	69.5 (2.29–516.0)
TP	$mg\,kg^{-1}$	143.3 (24.5–684.7)	312.1 (37.5–1,316.8)
TN	$g\,kg^{-1}$	13.5 (5.5–37.0)	18.6 (3.0–43.1)
TC	$g\,kg^{-1}$	230 (146–431)	273 (29–495)
Tca	$g\,kg^{-1}$	240 (19–365)	162 (8–374)

Source: Osborne et al. (2008).

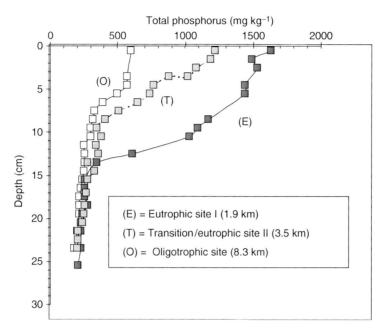

FIGURE 17.14 Vertical distribution of soil phosphorus in the WCA-2A of the Everglades. (From Reddy et al., 1993.)

17.3.3 VEGETATION

Major vegetation communities in the Everglades included saw grass marsh, wet prairies, sloughs/ridge, and tree islands. The saw grass marsh is the dominant plant community and accounts for nearly 70% of vegetative cover in the Everglades (Davis, 1943; Loveless, 1959; Rutchey and Vilchek, 1994; Craft and Richardson, 1997; Miao et al., 2000, 2001; Smith and Newman, 2001; Sklar et al., 2005). Saw grass marshes are of two types: (1) tall dense stands, generally monotypic and (2) sparse stands of short plants mixed with a variety of other sedges, grasses, and floating plants. The distribution of these types is attributed to nutrient availability, water levels, fire, and competition by other macrophytes (Miao and Sklar, 1998). Wet prairies include vegetation such as *Panicum, Eleocharis*, and *Rhynchospora* (Gunderson, 1994). Sloughs are deeper water habitats that include macrophytes such as *Nymphaea, Nuphar*, and *Utricularia*. Historically *Typha* was a minor component of the vegetative communities of the Everglades and was associated largely with areas such as alligator holes and other disturbed deep pockets. Pollen analysis showed no evidence that cattails contributed to any historical peat formation (Wood and Tanner, 1990). Stable isotopic signatures (nitrogen and carbon) support the evidence that saw grass is the main source of historic peat formation (Inglett and Reddy, 2007) (Figure 17.15). Long-term nutrient loading to the northern Everglades has resulted in the replacement of several thousand hectares of saw grass and slough communities with dense stands of cattails. The cattail expansion was associated with phosphorus enrichment and change in hydroperiod (Craft et al., 1995; Craft and Richardson, 1997). The abundance of cattails in WCA-2A was closely correlated with the phosphorus content of soils (Urban et al., 1993; Doran et al., 1997; Miao and DeBusk, 1999; Vaithiyanathan and Richardson, 1997). These studies have shown that cattail invasion occurs in soils with total phosphorus content in the range of 650–1,200 mg kg^{-1}. The mean total phosphorus content of macrophytes in the oligotrophic areas of WCA-2A was found to be in the range of 200–420 mg kg^{-1}. Nitrogen concentrations of the plant tissue were not affected by nutrient loading. Nitrogen concentrations were approximately the same in both enriched and unenriched sites of WCA-2A. The N:P ratio of plant tissue has been used as an indicator to determine nitrogen or phosphorus limitation. The N:P ratio of cattails was found to be around 15:1 for nutrient-enriched site and 33:1 for samples

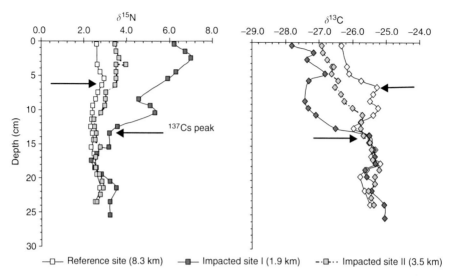

FIGURE 17.15 Carbon and nitrogen stable isotope signatures of soils as a function of depth in the WCA-2A of the Everglades. (Modified from Inglett and Reddy, 2007.)

TABLE 17.8
Relationship between Soil Total Phosphorus and Plant Tissue N:P Ratios of Saw Grass and Cattails

Plant Parts	Regression [$Y = -mX + b$]	Soil TP at Plant Tissue N:P Ratio = 16 (mg kg^{-1})
Cattails		
Leaves	$Y = -0.0124X + 34$; $n = 6$; $R^2 = 0.62$	1,452
Roots	$Y = -0.0447X + 95$; $n = 6$; $R^2 = 0.56$	1,774
Rhizomes	$Y = -0.0235X + 46$; $n = 6$; $R^2 = 0.89$	1,277
Shoot base	$Y = -0.0282X + 46$; $n = 6$; $R^2 = 0.88$	1,064
Saw Grass		
Leaves	$Y = -0.0397X + 78$; $n = 6$; $R^2 = 0.91$	1,572
Roots	$Y = -0.0872X + 158$; $n = 6$; $R^2 = 0.60$	1,623
Rhizomes	$Y = -0.0774X + 115$; $n = 6$; $R^2 = 0.83$	1,276
Shoot base	$Y = -0.0672X + 89$; $n = 6$; $R^2 = 0.64$	1,086
All Plants[a]		
Aboveground tissue	$Y = -0.0588X + 96$; $n = 6$; $R^2 = 0.86$	1,361

Note: Y = mass N:P ratio of the plant tissue; X = soil total phosphorus concentration (mg kg^{-1}); b = N:P ratio of plant tissue at soil total phosphorus concentration equals near zero level.

[a] Includes cattails, saw grass, and macrophytes from slough and wet prairie.

Source: Miao and DeBusk (1999) and Noe et al., (2001a, 2001b).

from unenriched sites of the WCA-2A (Koch and Reddy, 1992). Along the same gradient, the N:P ratio of saw grass increased from 20:1 in the enriched site to 139:1 in the unenriched site. The N:P ratios of various plant parts and the whole are regressed with the soil phosphorus content (Table 17.8). The N:P ratio of 16 or less indicates that the system may be nitrogen limited, whereas higher N:P ratios indicate phosphorus limitation. Regression equations shown in Table 17.8 suggest that total

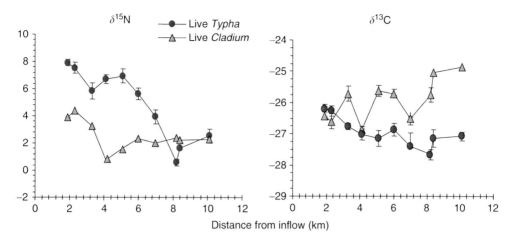

FIGURE 17.16 Carbon and nitrogen stable isotope signatures of live vegetation along soil phosphorus gradient in the WCA-2A of the Everglades. (Redrawn from Inglett and Reddy, 2006.)

soil phosphorus content in the range of 1,064–1,774 mg kg^{-1} can result in nitrogen limitation in the system. These high soil phosphorus values are noted in the nutrient-enriched sites of the Everglades, suggesting sustained phosphorus enrichment may be changing the system from phosphorus limitation to nitrogen limitation. Inglett and Reddy (2006) suggested that macrophyte-stable carbon and nitrogen ratios can serve as indicators of eutrophication and shifts between nitrogen and phosphorus limitation (Figure 17.16). In their study, significant nitrogen isotopic enrichment was observed in live tissue of *Typha* and *Cladium* (four and six parts per thousand) in areas affected by nutrient loading up to 7 km from the inflow. The enrichment of nitrogen isotopes in live tissue also suggests the nitrogen demand by macrophytes.

17.3.4 Periphyton

Periphyton refers to microbial communities attached to substrata surfaces. These surfaces may include plant surfaces below the water level, organic detrital materials in various stages of decomposition, and other nonliving material such as the soil surface. Periphyton is an important ecological component of the Everglades wetlands (McCormick and Stevenson, 1998). In open waters of the Everglades (other than WCA-1), three types of calcareous cyanobacterial mats occur: epipelon (attached to soils), epiphyton (attached to plants), and metaphyton (floating). Each mat type is dominated by filamentous cyanobacteria and diatoms that occur in discrete laminations. Mineral-rich waters, such as those found in WCA-2A and Taylor Slough (ENP), support a periphyton assemblage dominated by a few species of calcium-precipitating cyanobacteria and diatoms (McCormick et al., 2001). This assemblage appears to be favored by waters that are both low in P and at or near saturation with respect to calcium carbonate ($CaCO_3$) (Gleason et al., 1974), the latter condition reflecting the influence of the limestone geology of the region. In contrast, periphyton assemblages consist of green algae and diatoms adapted to the extremely low mineral content of surface waters (McCormick et al., 2002). The species composition and structure of these benthic mats are indicative of pristine Everglades communities (McCormick and O'Dell, 1996; Noe et al., 2001b; McCormick et al., 2002; Gaiser et al., 2006a).

Periphyton is abundant in oligotrophic areas of the Everglades and is responsible for significant primary production and phosphorus storage. In these systems, periphyton productivity and tissue phosphorus concentration are strongly related to phosphorus concentration of the water column.

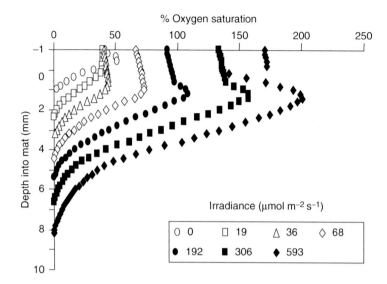

FIGURE 17.17 Dissolved oxygen gradients in benthic periphyton mats in Taylor Slough of the Everglades. (Hagerthey, S., SFWMD.)

Periphyton mats store large amounts of phosphorus with values approaching up to 1 kg P m^{-2} and maintain low phosphorus concentrations in surface waters (McCormick et al., 1998, 2002). The cyanobacterial mats adapted to oligotrophic conditions of the Everglades are very sensitive to phosphorus addition. Controlled phosphorus dosing studies and field monitoring have shown that cyanobacterial mats are replaced by filamentous green algae at surface water phosphorus concentrations between 10 and 20 µg L^{-1} (McCormick and O'Dell, 1996; McCormick et al., 2001; Gaiser et al., 2006a).

The Everglades cyanobacterial mats display distinct vertical stratification of organisms' physical and nutrient conditions. Many of these gradients are the direct result of light intensity and subsequent photosynthetic activity. For example, pH and dissolved oxygen profiles measured in cyanobacterial mats exposed to various levels of irradiance are shown in Figure 17.17 (S. Hagerthey, unpublished results). Results of this and other studies (Grimshaw et al., 1997) have demonstrated that cyanobacteria within the Everglades mats become light saturated at low irradiances relative to those of the mat surface (~500 µmol m^{-2} s^{-1}). Therefore, there is a rapid attenuation of light at the surface of the mat with the peak of photosynthetic activity occurring at approximately 2 mm from the surface (Figure 17.17). Periphyton mats in oligotrophic areas produce more oxygen during photosynthesis than they consume during respiration. Distinct diel patterns are noted in the water column of oligotrophic areas, with high dissolved oxygen of 7 mg L^{-1} during the day to 2 mg L^{-1} during the night (McCormick and Laing, 2003). In nutrient-impacted areas, loss of periphyton communities and increased activity of heterotrophic organisms in the water column (supported by high rate of detrital organic matter accumulation) result in depletion of oxygen. In WCA-2A, the dissolved oxygen concentration of water column decreases exponentially with steady increase in the total phosphorus concentration of the water column (Figure 17.18). Periphyton assemblages are involved in regulating various biogeochemical processes (as discussed in the next section of this chapter) by providing aerobic and anaerobic interfaces with the mat. Many oxidation–reduction reactions regulating carbon, nitrogen, phosphorus, and sulfur cycling are influenced by periphyton mats. Periphyton mats represent an important indicator of wetland condition and loss of these mats may have several implications on the ecosystem restoration (McCormick and O'Dell, 1996; McCormick et al., 2002).

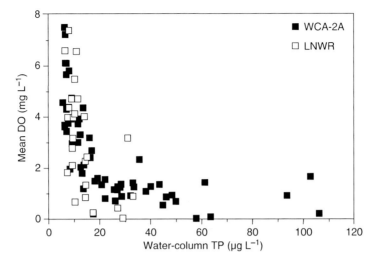

FIGURE 17.18 Relationship between dissolved oxygen and total phosphorus of the water column of LNWR and WCA-2A of the Everglades. (Data from McCormick and Laing, 2003.)

17.3.5 MICROBIAL COMMUNITIES AND BIOMASS

17.3.5.1 Microbial Communities

The unique physicochemical environment of the Everglades wetlands supports a range of microorganisms adapted to oligotrophic conditions. The size and diversity of microbial communities are directly related to the quality and quantity of resources available in their habitat. Many of these communities may respond rapidly to natural disturbances or to external nutrient loading. With external nutrient loading, the nutrient-limited system is converted into a nutrient-enriched system resulting in increased rates of processes conducted by prokaryotes. Microbial populations, biomass, and activity may be low at reference sites that are typically nutrient/resource limited. Any perturbations resulting from addition of small amounts of nutrients can potentially stimulate their growth and activity. The changes in microbial community structure resulting from eutrophication can dramatically influence the biogeochemical cycles they control (Figure 17.19). Several trophic levels of Bacteria and Archaea are connected via the flow of carbon and several oxidation–reduction reactions in wetlands, and the nature and quality of these relationships are affected by the degree of nutrient enrichment they experience. Eutrophication impacts oligotrophic wetlands initially at the microbial level, and this impact is reflected in the quality and magnitude of biogeochemical cycles that microorganisms control. Elevated nutrient levels present in nutrient-impacted marshes may have a variety of effects on the composition and activities of assemblages of microorganisms involved in carbon and associated elemental cycling. Higher levels of nutrients, increased carbon flow through the system due to higher amounts of plant detritus, and lowered redox conditions due to greater electron pressure are some of the factors that may influence the composition of these assemblages (see Chapters 4 and 5).

Biogeochemical cycles are controlled by guilds of microorganisms that are interdependent and sensitive to changes in their environment. The composition of guilds, or assemblages of microorganisms participating in a given function such as fermentation, methanogenesis, and sulfate reduction, is determined by various environmental factors, including quality and quantity of electron donors and acceptors, and concentrations of nutrients. The physiologies of members of the guild, in turn, determine the pathway through which biogeochemical cycles proceed. Changes in the structure of these guilds reflect changes in the environment, and may result in changes in the specific function of the guilds. Linking guild structure and activity with biogeochemical cycles will provide greater insight into the mechanisms that drive these cycles and how they respond to anthropogenic impacts.

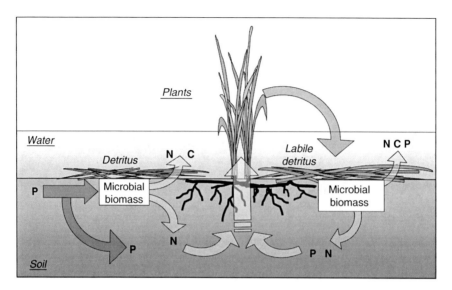

FIGURE 17.19 Schematic showing interactions among carbon, nitrogen, and phosphorus cycling in soil and water column.

Microbial communities respond to environmental impact by changes in size and composition of populations and a decrease in diversity (Drake et al., 1996). Taxonomically distinguishable microbes can replace each other because biogeochemical cycles are altered as a result of increased/decreased availability of substrates for catabolic and anabolic functions and changes in redox conditions. Surface soils from phosphorus-enriched areas yielded 10^3–10^4-fold higher numbers of anaerobes, including methanogens, sulfate reducers, and H_2 and acetate consumers, than did soils from unimpacted sites (Drake et al., 1996). Conversely, similar numbers of aerobes were present at impacted and reference sites. High numbers of anaerobes at the impacted site suggest oxygen limitation and intense reducing conditions, as reflected by low redox potentials at impacted site (Qualls et al., 2001). For example, the population of sulfate-reducing prokaryotes was enhanced by nutrient loading. This is probably due to increased rate of organic matter availability in nutrient-enriched sites. Drake et al. (1996) reported 4×10^{11} sulfate reducing prokaryotes per g (dry weight) for soils in P-enriched areas of the WCA-2A and 5×10^7 sulfate reducing prokaryotes per g in nonimpacted areas of the WCA-3A. They postulated that the enrichment of sulfate reducing prokaryotes in the P-impacted site was due to a high concentration of sulfate coming from EAA runoff. Sulfate reduction rates and most-probable-number enumerations revealed sulfate reducing prokaryotes populations and activities to be greater in eutrophic zones than in more pristine soils. Castro et al. (2002) showed that sulfate reducing prokaryotes populations and activities in eutrophic zones are greater than that in more oligotrophic soils (Table 17.9).

Assemblages of syntrophic bacteria and their methanogenic partners also differ with respect to the position along nutrient gradients. Syntrophs are limited to metabolism of either butyrate or propionate as carbon and energy source, and depend on a hydrogenotrophic methanogen partner to allow the metabolism to proceed. In our studies, the production of methane from butyrate occurs at a much higher rate in eutrophic than in low-nutrient soils, and much more methane is produced from butyrate than from propionate. This is likely due to the fact that more energy is released during fermentation of butyrate than propionate, as is more acetate and H_2 that can be supplied to methanogens. These differences in activities are reflected in corresponding differences in the composition of syntrophic and methanogenic assemblages in eutrophic versus nutrient-poor regions of the marsh.

TABLE 17.9
Microbial Populations Influenced by Nutrient Loading in the WCA-2A Soils of the Everglades

Substrate	Enriched Site (Eutrophic)	Reference (Oligotrophic)
Lactate	9.3×10^5	9.2×10^3
Acetate	2.3×10^5	3.62×10^3
Propionate	4.3×10^5	9.2×10^3
Butyrate	4.3×10^5	$<3.0 \times 10^3$
Formate	2.3×10^5	$<3.0 \times 10^3$

Note: Microbial numbers expressed as MPN g^{-1} soil.

Source: Castro et al. (2002).

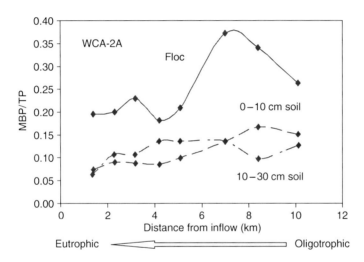

FIGURE 17.20 Ratio of microbial biomass phosphorus to total phosphorus in soils along soil phosphorus gradient in WCA-2A. (From Chua, 2000.)

17.3.5.2 Microbial Biomass

The soil microbial biomass is a key component of the soil ecosystem and is responsible for organic matter decomposition, nutrient cycling, and energy flow (Figure 17.20). The pool size of microbial biomass reflects the microbial population in soils. The ratio of microbial biomass carbon to total organic carbon (MBC/TOC) has been related to the soil carbon availability and the tendency for a soil to accumulate organic carbon. Phosphorus loading increased the microbial biomass pool of detritus and surface soils of WCA-2A (DeBusk and Reddy, 1998; Qualls and Richardson, 2000). MBC is greater in recently accreted detrital plant tissue than in the surface 0–10 cm soil. Microbial biomass decreases with soil depth. Microbial biomass is also influenced by lability and quality of soil organic matter. The assimilation coefficients for carbon (MBC/TOC) are generally higher in the floc or surface soil at the reference site as compared to impacted sites in four key hydrologic units of the Everglades (Table 17.10).

Microbial communities assimilate and recycle nutrients within the biomass pool as needed. Soils that maintain high microbial biomass are not only capable of storing more nutrients but also greater cycling within the system. Microbial biomass nitrogen (MBN) was higher in detrital and surface soil layers of the impacted site than the reference site (Reddy et al., 1999; White and Reddy,

TABLE 17.10
Carbon, Nitrogen, and Phosphorus Assimilation Coefficients in Selected Hydrologic Units of the Everglades

	MBC/TC		MBN/TN		MBP/TP	
Hydrologic Unit	Impacted	Reference	Impacted	Reference	Impacted	Reference
Floc						
WCA-1	0.06	0.12	0.13	0.17	0.41	0.65
WCA-2A	0.03	0.06	0.06	0.09	0.23	0.32
WCA-3	0.10	0.17	0.11	0.10	0.30	0.37
Taylor Slough	0.06	0.06	0.08	0.03	0.27	0.33
0–3 cm Soil						
WCA-1	0.03	0.02	0.07	0.04	0.19	0.40
WCA-2A	0.01	0.02	0.03	0.03	0.13	0.27
WCA-3	0.02	0.01	0.04	0.01	0.21	0.16
Taylor Slough	0.03	0.02	0.05	0.01	0.20	0.26

Note: MBC = microbial biomass carbon; MBN = microbial biomass nitrogen; MBP = microbial biomass phosphorus. MBP values are corrected for fumigation extraction efficiency.

Source: Wright et al. (2007).

TABLE 17.11
Molar Nitrogen to Phosphorus Ratios of Soils and Microbial Biomass in Select Hydrologic Units of the Everglades

	Soil TN/TP		Microbial Biomass MBN/MBP	
Hydrologic Unit	Enriched (Eutrophic)	Reference (Oligotrophic)	Enriched (Eutrophic)	Reference (Oligotrophic)
WCA-1	46	132	16	34
WCA-2A	47	125	12	36
WCA-3A	50	171	18	44
Taylor Slough	112	346	34	34

2000; Wright et al., 2007) (Table 17.10). The nitrogen assimilation coefficient (the ratio of MBN to total Kjeldahl N, MBN/TKN) was higher for detritus than 0–10 cm soil for both the impacted and reference sites. High N assimilation coefficient at impacted sites suggests high N demand by microbes, and possible N limitation. Low detrital or soil N:P ratio at the impacted sites suggests high nitrogen demand in relation to phosphorus availability. The N:P ratios of the microbial biomass suggest possible nitrogen limitation at eutrophic sites in WCAs (Table 17.11).

Nutrient limitations affect the overall nutrient status of the wetland ecosystem. The most limiting nutrient will be held in the closed microbial compartment and will be unavailable to higher plants. Phosphorus has historically been the limiting nutrient to the microbial biomass in the Everglades, and hence has controlled the size and activity of the microbial pool. In turn, phosphorus has been limiting for macrophytes, and the Everglades ecosystem evolved and sustained itself under these historically low nutrient conditions. The phosphorus assimilation coefficient (MBP/TP) for the impacted sites ranged from 0.13 to 0.40, as compared with 0.16 to 0.65 for the reference site (Reddy et al., 1999; Wright et al., 2008). This comparison suggests that phosphorus limitation to the microbial communities is less at the inflow points and consequently more phosphorus is released into the soil pore water. The C:P ratio of detritus and surface soils was 270 and 291, respectively, for the impacted site, as compared with 850 and 915 for the reference site (Reddy et al., 1999). The MBP:TP ratio can serve as an indicator of the phosphorus assimilation efficiency of microbes. In WCA-2A, the phosphorus assimilation coefficient increased with distance from inflow (Figure 17.20). A high ratio implies phosphorus limitation and rapid assimilation and storage of bioavailable phosphorus in the microbial biomass, as was observed in unimpacted sites of the Everglades (Chua, 2000; Wright et al., 2008). A highly significant correlation was observed between alkaline phosphatase (APA) and the phosphorus assimilation coefficient, suggesting phosphorus limitation in the system (Chua, 2000).

17.4 BIOGEOCHEMICAL CYCLES

Many biogeochemical processes that affect plant productivity and water quality are accelerated by the addition of limiting nutrients. When this situation arises, the vegetation and algal/microbial communities essentially respond to nutrient availability. Although loading of anthropogenic nutrients stimulates the growth of aquatic vegetation in wetlands, a significant portion of the nutrient requirements may be met through remineralization during the decomposition of organic matter. The rate of organic matter turnover and nutrient regeneration is influenced by hydroperiod, characteristics of organic substrates, supply of electron acceptors, and addition of growth-limiting nutrients (see Chapters 5, 8, and 9). In the Everglades ecosystem, addition of limiting nutrient such as phosphorus stimulated many biogeochemical processes regulating carbon, nitrogen, phosphorus, sulfur, and mercury cycling. In this section, we will provide a brief review of recent research on biogeochemical cycling of elements in the Everglades.

17.4.1 Enzymes

During the initial stages of microbial decomposition, complex polymers of plant detritus and soil organic matter are hydrolyzed through the activity of extracellular enzymes into simple organic molecules. These low-molecular-weight compounds are directly transferred into microbial cells, oxidized, and used as an energy source. Nutrient enrichment can significantly influence the production of extracellular enzymes as a result of increased activity of microbes and plant productivity. Measurement of enzyme activity can be linked to detritus characteristics and decomposition rates. Thus, changes in extracellular enzyme activity may prove to be a sensitive indicator of wetland eutrophication (Wright and Reddy, 2001a; Newman et al., 2003).

Soils, plant detritus, and periphyton mats support a wide range of extracellular enzymes (supplied by microbes and plants) that catalyze the decomposition of organic matter. Measurement of enzymes in detritus and soil samples obtained from WCA-2A showed that APA decreased with phosphorus enrichment, whereas B-D glucosidase increased with P enrichment (Table 17.12) (Wright and Reddy, 2001a). High B-D glucosidase activity reflects the accumulation of organic substrates derived from phosphorus-enriched plant detritus at the impacted site. High levels of inorganic phosphorus at the impacted site apparently inhibited APA production. Arylsulfatase, phenol oxidase, and protease activities were unaffected by phosphorus loading. Activities of all enzymes were higher in the detrital layer than in the 0–10 cm soil layer. Phosphatase activity in periphyton mats was affected within 2–3 weeks after the initiation of phosphorus loading to the unimpacted Everglades soil (Newman et al., 2003). Phosphatase activity is inversely related to bioavailable phosphorus in soils (Figure 17.21). Research conducted indicates that enzymes can be used as useful indicators to determine nutrient limitation or nutrient impact.

The phosphorus limitation of periphyton mats is also evidenced by biomass N:P ratios >200 (McCormick and O'Dell, 1996) and increased activity of phosphatase in non-P-impacted areas (Newman et al., 2003; Sharma et al., 2005). As both algae and bacteria are known to produce phosphatases, it is uncertain which groups are dominantly responsible for the observed patterns of phosphatase. Recent work with floating mats from P-limited regions of the Everglades has shown

TABLE 17.12
Extracellular Enzyme Activity in Litter and 0–10 cm Soil Layer of Samples Collected along the Gradient in WCA-2A of the Everglades

Extracellular Enzyme Activity	Units	Distance from Inflow (km)	
		2.3 (Impacted)	10.1 (Reference)
Litter Layer			
B-D glucosidase	mg p-nitrophenol kg^{-1}h^{-1}	1,515	370
Protease	mg tyrosine kg^{-1}h^{-1}	19,170	4,800
Alkaline phosphatase	mg p-nitrophenol kg^{-1}h^{-1}	1,430	8,340
Arylsulfatase	mg p-nitrophenol kg^{-1}h^{-1}	2,270	2,490
Phenol oxidase	mmol dicq kg^{-1}h^{-1}	70	44
0–10 cm Soil			
B-D-glucosidase	mg p-nitrophenol kg^{-1}h^{-1}	760	310
Protease	mg tyrosine kg^{-1}h^{-1}	9,770	5,960
Alkaline phosphatase	mg p-nitrophenol kg^{-1}h^{-1}	860	2,160
Arylsulfatase	mg p-nitrophenol kg^{-1}h^{-1}	1,720	1,500
Phenol oxidase	mmol dicq kg^{-1}h^{-1}	13	14

Source: Wright and Reddy (2001a).

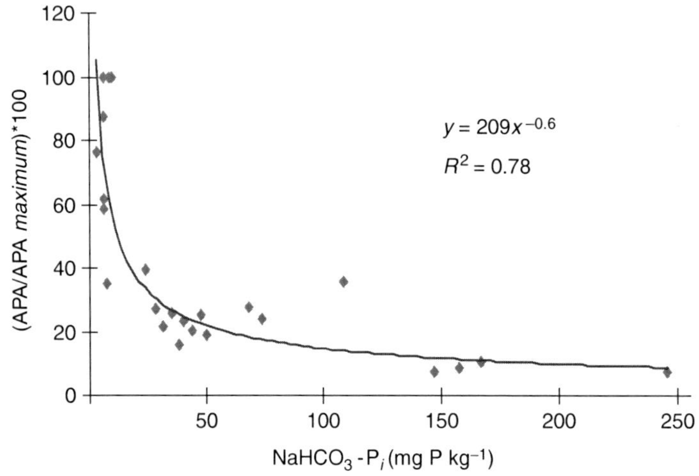

FIGURE 17.21 Relationship between alkaline phosphatase activity and bioavailable phosphorus in soils along soil phosphorus gradient in the WCA-2A. (From Wright and Reddy, 2001a.)

that APA activity is mainly concentrated in the lower sections of the mat and that phosphatase may be associated with organisms other than cyanobacterial and algal cells. This conclusion was based on microscopic examinations using the fluorescent phosphatase substrate ELF, which indicated that phosphatase enzyme is concentrated in the lower sections of the mat and phosphatase may be associated with heterotrophic bacterial cells in close proximity to (i.e., on the surface) cyanobacterial and algal cells (Sharma et al., 2005). For phosphorus-limited system such as the Everglades, phosphatase activity in periphyton, detrital matter, soils, and plant roots can provide an early warning signal for nutrient impacts and signs of recovery after restoration (Reddy et al., 1999; Wright and Reddy, 2001a; Kuhn et al., 2002; Newman et al., 2003; Sharma et al., 2005; Turner and Newman, 2005; Corstanje et al., 2006; Penton and Newman, 2007; Chua et al., 2007).

17.4.2 CARBON CYCLING

Nutrient loading to portions of the northern Everglades has resulted in a shift in saw grass communities to cattails communities. Cattails are very productive and respond to nutrient loading and change in hydrology. Increased primary productivity of this plant community has resulted in the increased rate of organic matter accumulation (Davis, 1991). The impacted zone has accumulated organic matter at a rate exceeding 1 cm year^{-1} (Reddy et al., 1993). Much of this accumulation is from the deposition of detrital matter on the soil surface, which supports a range of microbial communities.

17.4.2.1 Decomposition of Organic Matter

The combination of nutrient availability and changes in the litter source has resulted in a shift in litter quality and quantity (Davis, 1991; DeBusk and Reddy, 1998), with concomitant increases in organic matter mineralization rates (Davis, 1991; Qualls and Richardson, 2000) and significant shifts in carbon (DeBusk and Reddy, 1998). Differences in organic matter decomposition in WCA-2A were attributed to the differences in lignin and cellulose content (expressed as lignocellulose index or LCI) associated with phosphorus content of the tissue (DeBusk and Reddy, 1998). The LCI (ratio of lignin content to lignin + cellulose content) of plant litter was 0.2, which increased to 0.8 as the material was decomposed and incorporated into soil organic matter. Estimated organic matter turnover time for detrital pools under aerobic conditions ranged from 1.2 years for the standing dead material and litter to 9.2 years for the organic matter. Mean turnover time under anaerobic conditions ranged from 3.6 years for litter to 24 years for the organic matter (DeBusk and Reddy, 1998). Total phosphorus and lignocellulose content of the detrital tissue and soil organic matter were key variables regulating organic C mineralization in WCA-2A soils (DeBusk and Reddy, 1998). In the WCA-1, Newman et al. (2001) measured decomposition of litter material in the range of 0.00027–0.00041 day^{-1} or approximately 30% of the litter weight was lost over the 3-year period. In the STAs, Chimney and Pietro (2006) reported decomposition rates in the order of *Najas/Ceratophyllum* (0.0568 day^{-1}) > *Pistia* (0.0508 day^{-1}) > *Eichhornia* (0.0191 day^{-1}) > submerged *Typha* (0.0059 day^{-1}) > aerial *Typha* (0.0008 day^{-1}). Decomposition of detrital plant tissue was enhanced by high carbon quality (high cellulose content relative to lignin content) in P-enriched areas of WCA-2A, limited by low nutrient availability in the interior marsh (DeBusk and Reddy, 1998).

17.4.2.2 Microbial Respiration

Microbial respiration reflects the activity of microorganisms under ambient substrate and nutrient concentrations. The addition of low levels of phosphorus to a P-limited system can increase microbial activity (Amador and Jones, 1995). Thus, measurement of microbial activity can provide a sensitive indicator of nutrient loading to such an ecosystem. Both aerobic and anaerobic microbial respiration rates were greater in the detrital layer than in the 0–10 cm soil layer (DeBusk and Reddy, 1998; Wright and Reddy, 2001b). Respiration rates measured within 2 h on freshly collected detritus and surface soil samples were greater under drained conditions than flooded conditions (Wright and Reddy, 2001b). Approximately a twofold increase in respiration rates was observed in the detritus layer of the impacted site, as compared to the reference site. Anaerobic respiration represented approximately one-third of aerobic respiration in detritus and soil samples of the Everglades (DeBusk and Reddy, 1998). Addition of P accelerated the rate of microbial respiration, apparently by an increased supply of electron donors from labile detrital plant tissue.

The metabolic quotient, or specific respiration rate (qCO_2), is the ratio of the basal respiration rate (as CO_2-C) per unit MBC, and provides an indication of efficiency of microbial in utilization of available substrates (Anderson and Domsch, 1993). Low qCO_2 in detritus of the impacted site suggests poor efficiency of microbes in utilization of available substrates (DeBusk and Reddy, 1998; Wright and Reddy, 2001b). Minimal differences in qCO_2 were observed for 0–10 cm soil of impacted and reference sites. Consumption of electron acceptors such as oxygen, nitrate, and sulfate during microbial respiration was higher in soils collected from P-enriched area of WCA-2A compared with interior unenriched locations (Fisher and Reddy, 2001). Microbial respiration expressed as soil oxygen demand was distinctly different in four hydrologic units in the order of TS > WCA-1 = WCA-2A > WCA-3, and correlated with the MBC (Wright et al., 2007) (Figure 17.22).

Surface water draining the EAA and the marshes in the EPA contain high concentration of DOM, which transports approximately 90% of dissolved organic carbon and nitrogen, and 25% of dissolved organic phosphorus (Qualls and Richardson, 2003). Microbial degradation of the DOM is very slow with less than 10% decomposed in 6 months, and exposing the material to solar radiation enhanced the decomposition by 25% (Qualls and Richardson, 2003). Soil pore water dissolved organic carbon concentrations were significantly higher in soils impacted by nutrient loading (cattail areas) than the interior oligotrophic saw grass marsh (Figure 17.23). In a recent study,

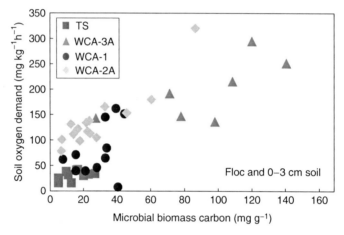

FIGURE 17.22 Relationship between soil oxygen demand and microbial biomass carbon of soils from select sites in four hydrologic units of the Everglades. (From Wright et al., 2007.)

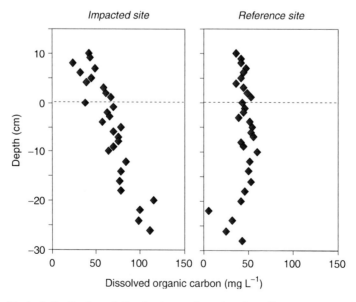

FIGURE 17.23 Vertical distribution of dissolved organic carbon in soil pore water and water column at impacted and unimpacted sites in the WCA-2A of the Everglades. (Reddy, K. R., Unpublished Results, University of Florida.)

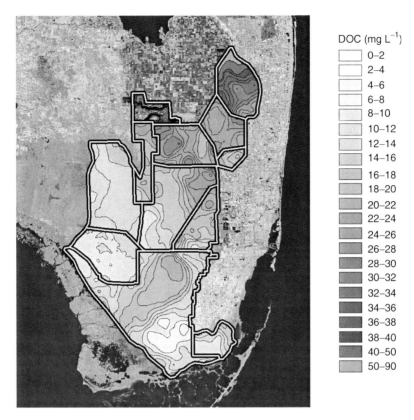

FIGURE 17.24 Spatial distribution of dissolved organic carbon of surface water in the Everglades. (From Osborne, 2005.)

Osborne et al. (2007) demonstrated that individual species of wetland vegetation commonly found in the Everglades ecosystem produce characteristically different DOM products on the basis of major nutrient content (TC, TN, TP), and total carbohydrate and phenolic content. Because hydrologic and nutrient impacts on wetlands often result in changes to dominant vegetation patterns, as is the case in the Everglades, these findings suggest a need to better understand the role of individual plant species as contributors to the DOM pool and possible modulators of DOM cycling (Osborne, 2005). Surface water DOM was shown to be different in various hydrologic units of the Everglades (Osborne, 2005). The data shown in Figure 17.24 show high DOM in WCA-1, and DOM decreased southward toward ENP and Florida Bay.

17.4.2.3 Methane Emissions

Phosphorus enrichment resulted in a significant increase in methane emissions from WCA-2A in areas with cattails. Methane production was significantly different among the soil components of WCA-2A, in the order of litter > surface peat > subsurface peat (DeBusk and Reddy, 1998). Methane production in nutrient-enriched sites was limited by available carbon. The addition of organic substrate significantly increased methane production in soils from the nutrient-enriched sites of the Everglades (Bachoon and Jones, 1992; Wright and Reddy, 2007) (Figures 17.25 and 17.26). Methane production potential of floc and soils was generally higher in impacted sites than in the interior reference sites of WCA-3A and Taylor Slough. These results suggest that phosphorus enrichment can increase the methane production potential of oligotrophic wetlands (Table 17.13).

Methane concentrations in soil pore water increased exponentially with depth and were significantly higher at eutrophic sites than oligotrophic sites of WCA-2A (Figure 17.27), and pore

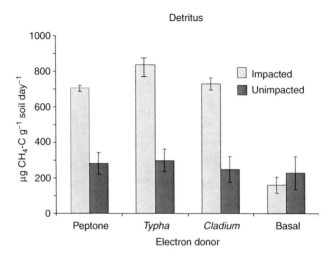

FIGURE 17.25 Influence of electron donors on methane production from detrital layer of WCA-2A. (From Wright and Reddy, 2007.)

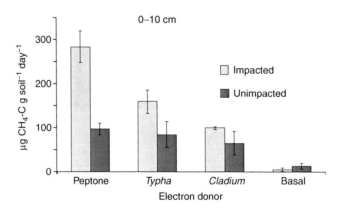

FIGURE 17.26 Influence of electron donors on methane production from 0–10 cm soil layer of WCA-2A. (From Wright and Reddy, 2007.)

TABLE 17.13
Potential Methane Production Rates in the Floc and Surface Soils from Select Hydrologic Units of the Everglades

	Methane Production (mg kg^{-1} day^{-1})	
Hydrologic Unit	Impacted	Reference
Floc		
WCA-3A	710	454 [S]
Taylor Slough	133	216 [NS]
0–3 cm Soil		
WCA-3A	258	165 [S]
Taylor Slough	158	57 [S]

Note: NS = not significant and S = significant at $p < .05$.

Source: Wright et al. (2008).

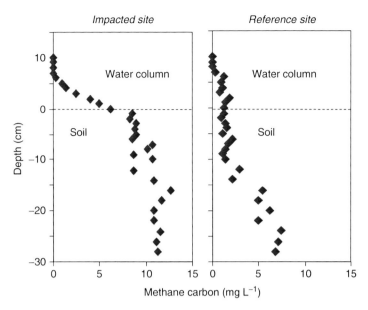

FIGURE 17.27 Vertical distribution of dissolved methane in soil pore water and water column at impacted and unimpacted sites in the WCA-2A. (Reddy, K. R., Unpublished Results, University of Florida.)

water methane concentrations were significantly correlated with the pore water ammonium and soluble reactive phosphorus (Koch-Rose et al., 1994). Methane within surface water contained ^{13}C enriched by six parts per thousand as compared to methane bubbles collected from soil pore water, consistent with the activity of methane-oxidizing bacteria consuming methane as it diffused across soil–floodwater interface. Methane oxidation consumed 91% of the methane diffusing across the soil–water interface in the ENP peat soils (Happell et al., 1993). Similar results were also observed in WCA-2A soils (Koch-Rose et al., 1994).

Detailed information on organic carbon cycling can be found in the following papers: Davis (1991); Koch and Reddy (1992); Reddy et al. (1993, 1998, 1999); Koch et al. (1994); DeBusk and Reddy (1998, 2003, 2005); Newman et al. (2001); Qualls and Richardson (2000, 2003); Qualls et al. (2001); Wright and Reddy (2001a); Wright and Reddy (2001b); Chimney and Pietro (2006); Osborne (2005); Osborne et al. (2007); Wright et al. (2008). Some examples of research conducted on carbon cycling in the Everglades are also presented in Chapter 5 of this book.

17.4.3 Nitrogen Cycling

17.4.3.1 Organic Nitrogen Mineralization

Organic nitrogen mineralization occurs through (i) hydrolytic deamination of amino acids and peptides, (ii) degradation of nucleotides, and (iii) metabolism of methylamines by methanogenic bacteria (see Chapter 8). Both potentially mineralizable N (PMN) and substrate-induced N mineralization (SINM) in detritus and soil layers were higher at the impacted site than at the reference site (White and Reddy, 2000). PMN rates in floc and soils from four key hydrologic units significantly correlated with the MBN (Figure 17.28). SINM indicates the activity of microbes and associated enzymes involved in organic nitrogen mineralization. This concept is similar to substrate-induced respiration (SIR), typically used to estimate the size of microbial populations. Low C:N ratio and phosphorus-nonlimiting conditions at the impacted site resulted in higher rates of PMN. Phosphorus loading increased the rates of organic nitrogen mineralization. As the microbial phosphorus needs are satisfied, a significant portion of nitrogen needs of macrophytes and microbial communities are met through the mineralization of organic nitrogen. Increased rates of organic nitrogen

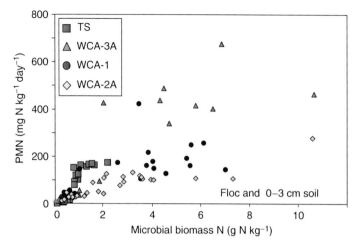

FIGURE 17.28 Relationship between potentially mineralizable nitrogen (PMN) and microbial biomass nitrogen of soil in four hydrologic units of the Everglades. (Wright, A., and Reddy, K. R., Unpublished Results, University of Florida.)

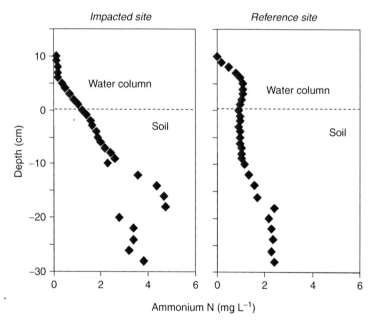

FIGURE 17.29 Vertical distribution of dissolved ammonium in soil pore water and water column at impacted and unimpacted sites in the WCA-2A. (Reddy, K. R., Unpublished Results, University of Florida.)

mineralization resulted in elevated levels of ammonium nitrogen in the soil pore water (Figure 17.29). Since phosphorus requirements of macrophytes are met in the impacted area, growth of macrophytes may be regulated to some extent by the supply of inorganic nitrogen, and thus the rate of organic nitrogen mineralization. The (NH_4-N:SRP) ratio in the soil pore water of the impacted site is in the range of 2–4, as compared to 200–400 for the unimpacted reference site (Koch-Rose et al., 1994). The high NH_4-N:SRP ratio at the reference site reflects P limitation in the system. Extractable ammonium nitrogen was significantly correlated with both MBC and MBN in WCA-2A soils (White and Reddy, 2000).

In a recent study, Inglett and Reddy (2006) showed distinct differences in stable isotope ^{15}N signatures, with high $\delta^{15}N$ shown in plants closer to inflow areas as compared to interior marsh

Freshwater Wetlands: The Everglades

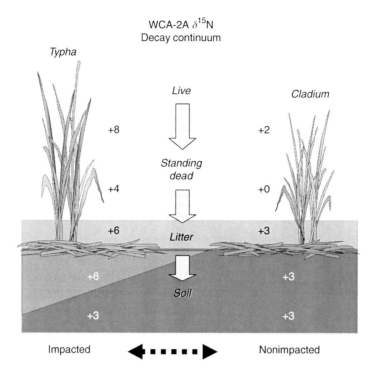

FIGURE 17.30 Nitrogen stable isotope signatures in various vegetation and soil components in impacted and nonimpacted sites of WCA-2A. (From Inglett and Reddy, 2006.)

of WCA-1A (Figure 17.30). Similarly, surface litter material and soils on the WCA-2A transect also showed increased $\delta^{15}N$ (Figure 17.30) with values of approximately 4‰ higher near the inflow sources relative to the interior marsh (Inglett et al., 2007). The values were also nearly identical for both litter and surface soil at all transect sites. Since litter is the main source of organic matter, it is not surprising that the similarity exists between the litter and the soil. However, this contrasts with other studies showing gradual alteration of original plant $\delta^{15}N$ during diagenesis. In this case, the similarity between the litter and soil at all WCA-2A transect sites indicates that the immobilization and mineralization processes affecting $\delta^{15}N$ are rapid and largely completed during the early phases of litter diagenesis (0.5–1 year) (Inglett et al., 2007). For this reason, Inglett et al. (2007) suggest that the $\delta^{15}N$ of peat could serve as an accurate record of conditions at the time of litterfall, and thus is also a sensitive temporal indicator of nutrient enrichment.

17.4.3.2 Nitrification–Denitrification

Nitrification rates at the impacted site were higher than that at the reference site for both detrital and surface soil layers of WCA-2A soils (White and Reddy, 2003) (Figure 17.31). Since ammonium and alkalinity were not limiting at either site, low rates of nitrification at the reference site were probably due to P limitation of the microbial pool. The denitrification enzyme activity (DEA) of detrital and soil layers at the impacted site was approximately threefold higher than that at the reference site. Much higher DEA values were observed at impacted sites <1 km from the inflow (White and Reddy, 1999). The denitrifying population increases with nitrate loading, and high DEA values at the impacted site reflect nitrate loading from drainage water. It is likely that the DEA in the Everglades soils is limited by nitrate concentration rather than phosphorus concentration. Potential denitrification rates increased with increase in soil phosphorus enrichment (Figure 17.32). This may be an indirect effect because the areas with soil phosphorus enrichment are dominated with cattails and highly labile organic matter that may fuel the denitrification rates.

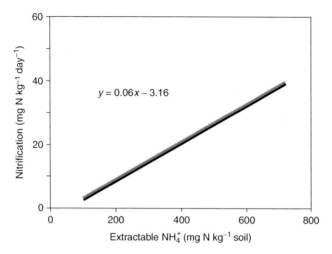

FIGURE 17.31 Relationship between extractable ammonium nitrogen and nitrification rates in surface soils along phosphorus gradient in the WCA-2A. (From White and Reddy, 2003.)

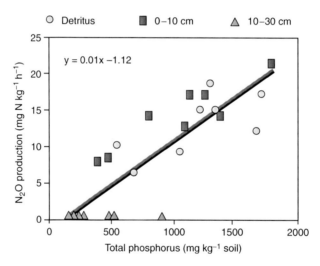

FIGURE 17.32 Relationship between total phosphorus and denitrification rates in surface soils along phosphorus gradient in the WCA-2A. (From White and Reddy, 2003.)

Both potential nitrification and denitrification rates decreased with distance from the inflow along the eutrophic gradient in WCA-2A and the rates were positively correlated to total phosphorus (White and Reddy, 2003). Higher nitrification rates in soils from eutrophic areas were due to high ammonium levels, whereas higher denitrification rates were due to greater availability of bioavailable carbon. Both nitrifiers and denitrifiers were limited by phosphorus in the oligotrophic areas of WCA-2A (White and Reddy, 1993, 2003).

17.4.3.3 Biological Nitrogen Fixation

In eutrophic areas of the Everglades (e.g., in areas receiving discharges of agricultural drainage), phosphorus inputs have increased nitrogen limitation of the system as evidenced by a lowered water column N:P (TN:TP < 30) and algal nutrient limitation assays (McCormick et al., 1996; Inglett et al., 2004). In these more eutrophic marsh areas, periphytic mats are much less abundant, lack a visible calcareous structure, and comprise filamentous algal assemblages. The presence of cyanobacteria (and other diazotrophic microorganisms) indicates the potential for nitrogenase activity within the oligotrophic Everglades periphyton mats. Additionally, the lowered N:P ratio coupled with the presence of cyanobacteria in the periphyton of nutrient-impacted marsh areas could lead to enhanced rates of nitrogen fixation in the periphyton near the points of agricultural drainage discharges. As such, periphyton nitrogenase activity could serve as an indicator of nutrient impacts (in particular those of phosphorus) on the Everglades ecosystem (Inglett et al., 2004). Biological N_2 fixation in periphyton mats obtained from the impacted site (2 km from the inflow) was higher than that from the reference site (11 km from the inflow) (Inglett et al., 2004). Potential biological nitrogen fixation was estimated by using the theoretical ratio of three moles of C_2H_2 reduced per mole of N_2 fixed. The estimated seasonal rates of unimpacted WCA-2A slough N_2 fixation range from 0.21 to 2 mg N m^{-2} h^{-1} or 1.8 to 18 g N m^{-2} year^{-1}, with a mean annual fixation rate of 9.7 g N m^{-2} year^{-1} (Inglett et al., 2004). It should be noted that Inglett et al. (2004) included all periphyton forms (benthic, epiphytic, and floating mat) in estimating nitrogen fixation rates. Of this annual 9.7 g N m^{-2} year^{-1} fixation by all slough periphyton, floating mats represent approximately between 8 and 30% (based on dry/wet season standing-crop biomass as reported by McCormick et al., 1998; Inglett et al., 2004).

Detailed information on nitrogen cycling can be found in the following papers: Reddy et al. (1993, 1999); Koch et al. (1994); Qualls and Richardson (2003); White and Reddy (1999, 2000, 2001, 2003); Wright and Reddy (2001a); Inglett et al. (2004); Inglett and Reddy (2006). Some examples of the research conducted on nitrogen cycling in the Everglades are also presented in Chapter 8 of this book.

17.4.4 Phosphorus Cycling

Phosphorus is an essential nutrient frequently limiting the productivity of oligotrophic Everglades wetland ecosystems. The phosphorus demand by periphyton and macrophyte communities in extremely oligotrophic wetlands often exceeds supply and maintains water column chemistry in a severely phosphorus-limited condition. Phosphorus in water, soil, and periphyton components is present in both organic and inorganic forms, with organic forms present as the dominant pool in many hydrologic units of the Everglades wetlands. Using 1M HCl extraction of the Everglades soils, Reddy et al. (1998) developed the following empirical relation to estimate total inorganic phosphorus in the Everglades soils.

$$TP_i = 0.01\ TP^{1.54};\quad r^2 = 0.897,\ n = 390$$

Steady loading of phosphorus into the Everglades wetlands has increased the relative proportion of all forms of soil phosphorus, with the largest proportion stored as refractory forms of organic phosphorus. Qualls and Richardson (1995) and Reddy et al. (1998) evaluated phosphorus forms of soils collected from several hydrologic units of the Everglades and presented the following conclusions. Significant gradients are observed as a function of both distance from inflow and soil depth. Phosphorus loading increased the proportion of phosphorus stored as inorganic phosphorus compared with unimpacted areas. The ratio of $TP_o:TP_i$ was lower in near-surface soils and increased with depth, suggesting accumulation of inorganic phosphorus in surface soil layers. This observation was substantiated by a positive correlation between Pi and calcium and magnesium. These relationships suggest that inorganic phosphorus dynamics in the Everglades soils may be governed by calcium and magnesium. Phosphorus fractions associated with calcium and magnesium and organic phosphorus are found to be the dominant forms for long-term phosphorus storage in the Everglades soils.

Total phosphorus accretion in the WCA-2A was estimated to be 0.06–1.1 g m^2 year^{-1}, as compared to 1.0–2.2 g m^{-2} year^{-1} in STA-1W (Craft and Richardson, 1993a, 1993b; Reddy et al., 1993; Turner et al., 2006). Although phosphorus loading increased the storage of phosphorus through organic soil accretion, it also resulted in increased levels of dissolved phosphorus and other bioavailable forms. Distinct gradients with distance are noted in areas adjacent to canals and inflow structures, suggesting the influence of hydrology and nutrient loading. Much of the nutrient-loading effects were confined to shallow soil layers. Oxidation of organic matter, resulting from biological processes or fire, increased inorganic phosphorus, as observed for the WCA-3A, HWMA, and EAA soils. Both biotic and abiotic mechanisms regulate relative pool sizes and transformations of phosphorus compounds within the water column and soil. Alterations in these fractions can occur during flow in wetlands that depend on the physical, chemical, and biological characteristics of the systems. Thus, when evaluating phosphorus retention capacities of the Everglades wetlands, both biotic and abiotic processes must be considered. Biotic processes in the Everglades wetlands include assimilation by vegetation, periphyton, and microorganisms, and abiotic processes include sedimentation, adsorption by soils, precipitation, and exchange processes between soil and the overlying water column (see Chapter 9 for details).

17.4.4.1 Biotic Processes

Phosphorus added through external loads in the Everglades wetlands is readily utilized near the inflow, resulting in increased productivity of macrophytes and loss of calcareous periphyton mats in sloughs, as noted in the WCA-2A. It is likely that addition of phosphorus to oligotrophic system can potentially increase the heterotrophic activity of bacteria, resulting in decreased oxygen levels and pH. These conditions potentially decrease calcification process and shift in cyanobacterial communities. Unlike macrophytes, rapid growth rates and turnover of these organisms within the mats allow them to efficiently process the available nutrients. Periphyton can assimilate both organic and inorganic forms of phosphorus (Scinto and Reddy, 2003), and can induce marked changes in dissolved oxygen concentration of water column and soil–floodwater interface (McCormick and Laing, 2003; Hagerthey, 2007). These changes can potentially influence the solubility of inorganic phosphorus in wetlands. Periphyton is known to mediate the precipitation of $CaCO_3$ (Gleason, 1972) and to coprecipitate P (Scinto and Reddy, 2003). Calcite formed in the mats is eventually incorporated into the surface soil and can be important to long-term phosphorus storage (Koch and Reddy, 1992; Reddy et al., 1993, 1998).

Phosphorus added to a wetland or released during decomposition of soil organic matter is usually retained in the system through sorption and precipitation reactions. Phosphorus stored in vegetation during the active growth period is potentially released during winter months as a result of leaching and decomposition. Potentially mineralizable phosphorus (PMP) was higher at the impacted site for both litter and surface 0–10 cm soil layer of WCA-2A as compared to the interior site (Chua, 2000) (Figure 17.33). Similar observations were made in other hydrologic units of the Everglades (Table 17.14).

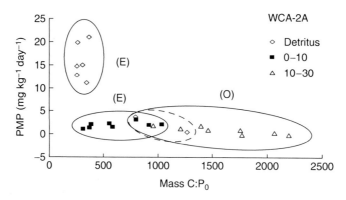

FIGURE 17.33 Relationship between mass $C:P_0$ ratio and potentially mineralizable organic phosphorus in soils along phosphorus gradient in the WCA-2A. (From Chua, 2000.)

TABLE 17.14
Potentially Mineralizable Phosphorus Rates in the Floc and Surface Soils from Select Hydrologic Units of the Everglades

Hydrologic Unit	Potential Mineralizable Phosphorus (mg kg^{-1} day^{-1})	
	Impacted	Reference
Floc		
WCA-1	45	11 [S]
WCA-2A	14	6 [S]
WCA-3	144	8 [S]
Taylor Slough	10	2 [S]
0–3 cm Soil		
WCA-1	24	15 [NS]
WCA-2A	10	6 [S]
WCA-3	24	4 [S]
Taylor Slough	3	1 [S]

Note: NS = not significant and S = significant at $p < .05$.
Source: Wright et al. (2008).

Substrate-induced phosphorus mineralization (SIPM) can indicate activity of microbes and associated enzymes involved in organic phosphorus mineralization. Hydrolysis of added substrates such as glucose-6-phosphate was not influenced by P enrichment in the litter layer (Chua, 2000). In the 0–10 cm soil layer, substrate hydrolysis decreased by approximately 67% at the impacted site. The mineralization coefficient (SIPM/MBP) also decreased by 54 and 74% in the litter and 0–10 cm soils, respectively, at the impacted site. These decreases suggest that phosphorus is not limiting to microbes. High substrate phosphorus hydrolysis at the reference site was attributed to limitation of labile inorganic phosphorus and high phosphatase activity.

17.4.4.2 Abiotic Processes

Abiotic phosphorus retention by Everglades soils is regulated by various physicochemical properties including pH; redox potential; Fe, Al, and Ca content of soils; organic matter content; phosphorus loading; and ambient phosphorus content of soils. Phosphorus sorption capacity of Everglades soils decreased with loading, as reported for WCA-2A soils (Richardson and Vaithiyanathan, 1995; Clark et al., 2007). Soils from the eutrophic sites of WCA-2A exhibited high EPC_0 (equilibrium phosphorus concentration where adsorption equals desorption, see Chapter 5) as compared to soils from oligotrophic areas. High EPC_0 in the soil suggest that soil's capacity to buffer dissolved phosphorus concentrations in the pore waters is decreasing as a result of anthropogenic loading. Distinct gradients in pore water phosphorus concentrations are observed in impacted sites of WCA-2A (Figure 17.34), suggesting flux of phosphorus from soil to the overlying water column.

Photosynthesis and respiration can initiate significant changes in water column pH on a diurnal basis. These processes can increase pH to as high as 10, depending on the buffering capacity of the water column, can precipitate a significant portion of water column phosphorus, and can coprecipitate with $CaCO_3$ where dissolved calcium is available. The surface waters in the Everglades (with the exception of WCA-1) typically have high alkalinity that may buffer the pH around 8.5. However, Diaz et al. (1994) noted that about 75–90% of the phosphorus precipitated was solubilized when pH levels decreased to below 8 as a result of an increase in carbon dioxide levels. Retention of inorganic phosphorus by precipitation will be significant in waters with high calcium ions and alkalinity. The mechanism regulating abiotic phosphorus retention in the Everglades soils is not

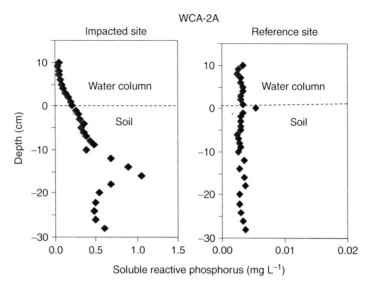

FIGURE 17.34 Vertical distribution of dissolved reactive phosphorus in soil pore water and water column at impacted and unimpacted sites in the WCA-2A of the Everglades. (Reddy, K. R., Unpublished Results, University of Florida.)

clearly understood, especially with respect to microgradients that exist within the periphyton mats and the organic matter. The dominance of calcium in the Everglades system suggests that it may be one of the main regulators. However, the role of iron and aluminum in inorganic phosphorus retention should not be ruled out because these metals can play a significant role in the soils of some of the hydrologic units (HWMA, WCA-1, and WCA-3). Chemical amendments containing these metals are considered as one of the options for *in situ* immobilization of phosphorus in constructed wetlands. At present these options have not been evaluated for Everglades soils.

Detailed information on phosphorus cycling can be found in the following papers: Davis (1991); Koch and Reddy (1992); Reddy et al. (1993, 1998, 1999); Koch et al. (1994); Richardson and Vaithiyanathan (1995); Qualls and Richardson (1995, 2000, 2003); Newman et al. (1998); Noe et al. (2001a,b); McCormick et al. (1996, 2002); Scinto and Reddy (2003); Wright and Reddy (2001b); Clark et al. (2007); Chua (2000); Wright et al. (2007). Some examples of research conducted on phosphorus cycling in the Everglades are also presented in Chapter 9 of this book.

17.4.5 Sulfur Cycling

Sulfur cycling in the Everglades is gaining importance because of its role in organic matter mineralization, phosphorus mobilization, and linkage of sulfate-reducing bacteria to the formation of toxic methyl mercury (MeHg) in wetlands. Sulfur cycle in wetland ecosystems is discussed in detail in Chapter 10 of this book. Sources of sulfur in the Everglades soils include natural sulfur content of soils and sulfur export from the EAA into canals and waterways and eventually into the Everglades wetlands (Bates et al., 2002). Bates et al. (2002) showed that spatial patterns in the range of concentrations and ^{34}S values of sulfate in surface water indicate that the major source of sulfate in sulfur-contaminated marshes is water from canals draining the EAA. Shallow groundwater underlying the Everglades and rainwater samples had much lower sulfate concentrations and ^{34}S values distinct from those found in surface water.

Surface water sulfate concentrations are generally higher in the northern Everglades and tend to decrease along north–south gradient. Within each hydrologic units, sulfate concentrations are higher in canals and in areas closer to inflow structures (Table 17.15). Sulfate concentrations are expressed in various units including mmol L^{-1} or mg SO$_4^{2-}$ L^{-1} or mg SO$_4^{2-}$-S L^{-1} (sulfate concentration = 1 mmol L^{-1} = 96 mg SO$_4^{2-}$ L^{-1} = 32 mg SO$_4^{2-}$-S L^{-1}).

High sulfate concentrations and low δ^{34}S are reported for surface waters of the canals in the EAA, as compared to other hydrologic units of the Everglades. Distinct vertical gradients between soil pore water and surface water were observed in both eutrophic and oligotrophic areas of the WCA-2A (Figure 17.35). The lowest sulfate concentrations were reported in WCA-1 surface waters. The δ^{34}S values suggest that the EAA and groundwater are major sources of sulfate in canal waters.

Total sulfur concentrations of WCA-2A soils were higher in surface layers (0–25 cm depth) and decreased with depth (Bates et al., 1998) (Figure 17.36). These profiles were similar to those measured for phosphorus by Craft and Richardson (1993a, 1993b) and Reddy et al. (1993). Total carbon content showed no vertical stratification. Molar ratios of carbon to sulfur increased with depth, with

TABLE 17.15
Surface Sulfate Concentration (1995–1999) and δ^{34}S Values in the Everglades Agricultural Area (EAA), Stormwater Treatment Area-1W, and the Water Conservation Areas (WCAs) in the Northern Everglades

Hydrologic Unit	Sulfate-S (mg L^{-1})	δ^{34}S
Hillsboro, Miami, and N. New river canals in the EAA	22.7 (13.8)	17.79 (1.49)
WCA-1	1.3 (1.5)	17.96 (2.47)
Hillsboro Canal at spillways bordering WCA-2A	23.0 (10.2)	20.21 (1.49)
WCA-2A	16.6 (6.4)	23.12 (2.52)
WCA-2B	5.8 (4.5)	25.99 (3.72)
WCA-3A and canals	1.6 (2.2)	24.03 (2.70)
STA-1W	18.4 (8.6)	22.85 (3.98)

Source: Bates et al., (2002).

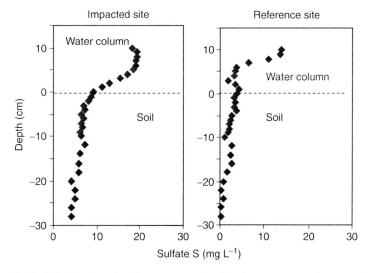

FIGURE 17.35 Vertical distribution of sulfate concentration in soil pore water and water column at impacted and unimpacted sites in the WCA-2A. (Reddy, K. R., Unpublished Results, University of Florida.)

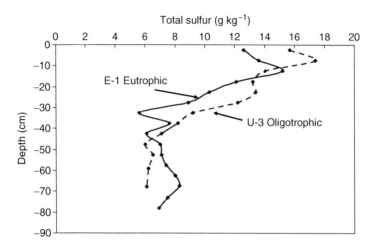

FIGURE 17.36 Vertical distribution of total sulfur in soils at impacted and unimpacted sites in the WCA-2A (Bates et al., 1998.).

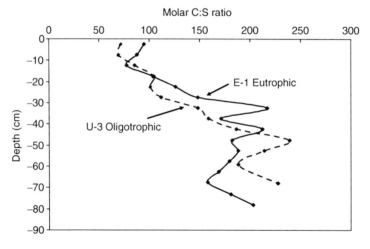

FIGURE 17.37 Vertical distribution of carbon:sulfur ratio (C:S) in soils at impacted and unimpacted sites in the WCA-2A (Bates et al., 1998.).

values in the range of 50–60 in surface layers and 150–200 in subsurface layers (Bates et al., 1998) (Figure 17.37). Increase in sulfur concentrations in surface layers can be due to the assimilation of added sulfur by periphyton and macrophytes, hydroperiod, and remobilization of reduced sulfur from subsurface to surface layers.

Organic matter mineralization rates due to sulfate reduction accounted for approximately one-third of the aerobic mineralization rates in the WCA-2A soils (Wright and Reddy, 2001a, 2001b). Sulfate reduction supports organic matter mineralization in wetland soils. For example, under sulfate-reducing conditions (with sulfate as the dominant electron acceptor), organic nitrogen mineralization rates in WCA-2A soils were 36, 9, and 3 mg N kg^{-1} day^{-1} for detritus, 0–10, and 10–30 cm soil depths (White and Reddy, 2001). Corresponding rates under aerobic conditions were 237, 143, and 75 mg N kg^{-1} day^{-1} for detritus, 0–10, and 10–30 cm soil depths, respectively.

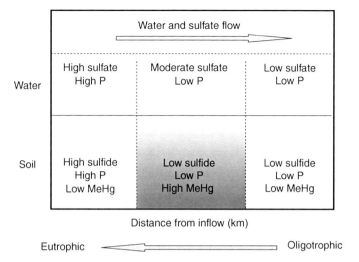

FIGURE 17.38 The methylmercury–sulfate connection in the Everglades. Hg, mercury; P, phosphorus; MeHg, methylmercury. (From Orem and Krabbenhoft, 2004.)

17.4.5.1 The Methylmercury–Sulfate Link

Natural and anthropogenic sulfate inputs to the Everglades have resulted in relatively high levels of sulfate, the primary electron acceptor of SRPs (Gilmour et al., 1998). Sulfate loading from the EAA canals into northern Everglades was attributed to be a key factor in controlling the distribution and extent of methylmercury production in the Everglades. The processes regulating the link between sulfate reduction and methylmercury production are presented in Figure 17.38. Sulfate entering the northern Everglades wetlands stimulates sulfate reduction and promotes production of methylmercury. Sulfate-reducing bacteria are important methylators of inorganic mercury; however, the product of their metabolism, sulfide, can potentially inhibit the methylation reaction, but the mechanisms whereby sulfide inhibits Hg methylation are not well understood (Gilmour et al., 1998; Orem and Krabbenhoft, 2004). The methylmercury concentrations and production were inversely related to sulfate reduction rate and pore water sulfide (Gilmour et al., 1998).

The linkage between sulfate reduction and methlymercury formation explained by Gilmour et al. (1998) and Orem and Krabbenhoft (2004) can be summarized as follows:

- Areas with high sulfate concentrations have low-to-moderate methylmercury production stimulated by sulfate reduction but inhibited by sulfide accumulation.
- Interior oligotrophic areas have low methylmercury production because of low sulfate concentrations that limit sulfate reduction.
- Highest rates of methylmercury production can potentially occur in areas where sulfate concentrations are moderate (2–10 mg L^{-1}) and sulfide levels are not high enough to inhibit methylmercury production.
- In the Everglades, highest production of methylmercury was observed in the center of WCA-3, where sulfate concentrations are low.

17.5 RESTORATION AND RECOVERY

"Get the water right" is one of the commonly stated goals of the Everglades restoration plan, which has largely meant to restoring the timing and duration of the water levels and the water quality in various hydrologic units of this ecosystem. The flow of water affects several physical, chemical, and biological processes in the Everglades. Flow can generally enhance mixing and transport of materials including particulate matter and nutrients and significantly affect various biogeochemical processes involved in transformation of these materials. The Everglades ecosystem is historically characterized by the peat-based ridge and slough landscape, with ridges dominated by saw grass and sloughs by submersed aquatic vegetation. In pristine state, the soil surface of ridges is approximately 60–90 cm higher than the adjacent sloughs (SCT, 2003; CROGEE, 2003). This microtopography can result in sharp redox gradients that may influence organic matter degradation and peat accumulation rates. For example, during dry season, the ridge soils may support biogeochemical processes regulated by aerobic microorganisms, such as enhanced decomposition rates of organic matter, mineralization of organic nitrogen and phosphorus, nitrification of ammonium nitrogen, oxidation of sulfides, and reduced emissions of methane. During wet season, saturated soil conditions may inhibit aerobic processes and enhance anaerobic processes such as denitrification, sulfate reduction, and methanogenesis.

The concept of utilizing microbial ecophysiological measures as indicators of disturbance is particularly reinforced by the spiked microbial response characteristic of the intermediate site (Corstanje et al., 2007). Soil biogeochemical characteristics changed gradually as a result of the influx of nutrients; soil microbial measures exhibited a threshold-type (step) response. These distinct, abrupt changes in microbial measures compared to the more progressive change in the soil chemical characteristics indicate that microbial indicators function effectively as early warning signals (Corstanje et al., 2007). A composite analysis using all the measures, biotic and abiotic, suggests a clear separation of impacted, intermediate, and unimpacted sites of the Everglades (Figures 17.39 and 17.40). In summary, at the three sites in this system, Corstanje et al. (2007) were able to effectively describe microbial ecophysiology in response to the environmental conditions present. Profiles of the relative enzyme activity provided insights into the initial microbial response to the changes in environmental conditions. Subsequent analysis of the measures associated with microbial biomass and associated nutrient turnover rates illustrated the changes in microbial physiological responses to an altered physicochemical environment.

Soils and the plant detrital matter are important nutrient storage components of the Everglades ecosystem. Cycling of stored nutrients through various biogeochemical processes can have significant effect on the release or retention of nutrients, thus affecting the downstream surface water

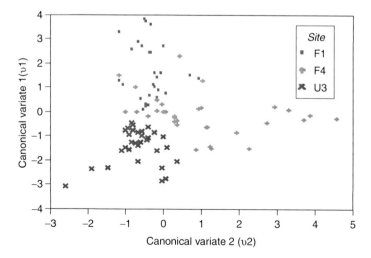

FIGURE 17.39 Stepwise canonical discriminant analysis of biotic indicators measured along phosphorus gradient in WCA-2A. (From Corstanje et al., 2007.)

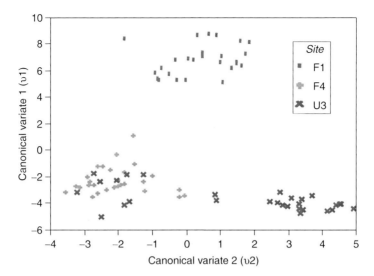

FIGURE 17.40 Stepwise canonical discriminant analysis of abiotic indicators measured along phosphorus gradient in WCA-2A. (From Corstanje et al., 2007.)

quality and productivity of microbes, periphyton, and vegetation. This internal cycling of nutrients in soil and water column can influence the time required for ecosystem recovery after restoration. For example, in the impacted sites of the Everglades (WCA-2A), after external nutrient loads are curtailed, internal cycling of nutrients can maintain the flux of nutrients from soil to water column and maintain eutrophic conditions. This internal memory of nutrients can last for a long period of time in the order of several years. The lag time in recovery process can be shortened by *in situ* remediation of sites.

Constructed wetlands such as STAs are used as buffers to retain nutrients and other contaminants are usually managed to improve their overall performance, and to maintain expected water quality. The extent of management required depends on the nutrient/contaminant retention capacity of the wetlands and the desired effluent quality. Management scenarios can vary, depending on wetland type and hydraulic loading rate. For example, small-scale wetlands can be managed effectively by altering hydraulic loading rates or integrating them with a conventional treatment system, while large-scale systems can be managed by controlling nutrient/contaminant loads.

Phosphorus retention by soils and sediments, which store most of the phosphorus relative to other ecosystem components, is regulated by surface adsorption on minerals, precipitation, microbial activities, and immobilization. In STAs these processes may be combined into two distinct P retention pathways: sorption and burial. When plants and microbes die off, the P contained in cellular tissue may either recycle within the wetland, or may be buried with refractory organic compounds. Accretion of organic matter has been reported as a major mechanistic long-term sink for nutrients in wetlands. Wetland soils tend to accumulate organic matter due to the production of detrital material from biota and the suppressed rates of decomposition. Soil accretion rates for STAs can range between a few millimeters and more than one centimeter per year. Accretion rates in productive natural wetland systems such as the Everglades have been reported as high as one centimeter or more per year. The genesis of this new material is a relatively slow process, which may affect the nutrient retention characteristics of the wetland. With time, productive wetland systems will accumulate organic matter (which ultimately forms peat) that has different physical and biological characteristics than the underlying soil. Eventually, this new material settles and compacts to form new soil with perhaps different nutrient retention characteristics than the original soil. Management of newly accreted material by consolidation, hydrologic manipulation (water-level drawdown), and soil removal can improve the overall life span of STAs.

Phosphorus management in STAs has been the major focus of the state and federal agencies involved in restoring the Everglades. As discussed in this book, phosphorus cycling is tightly coupled to organic matter turnover and cycling of other nutrients such as nitrogen and sulfur. A recent series of papers on STA-1W published in a special issue of *Ecological Engineering* (2006) clearly indicates the importance of hydrology, vegetation, periphyton, and biogeochemical processes regulating the long-term retention of phosphorus in wetlands. However, to understand the long-term performance of STAs and for effective management of phosphorus retention on a long-term basis, it is critical for governmental agencies to determine the fundamental processes and the linkages between these processes. The relative rates of these coupled processes affect the long-term accretion of phosphorus and vary across time and space as affected by hydraulic and nutrient-loading rates. For example, the types of vegetation communities (emergent macrophytes vs. SAV) and periphyton can influence the characteristics and stability of accreted organic matter and alter water turbidity thereby causing a potential export of dissolved and particulate organic matter from STAs.

Sustainable nutrient removal by STAs depends on the creation and burial of refractory residuals created by microbial, periphyton, and vegetation communities. As a result, the process of phosphorus retention in wetland systems is also tied or coupled to other nutrients that affect vegetation growth, microbial activity, or other chemical reactions determining phosphorus availability and cycling. Among these other potential elements affecting phosphorus stability in wetlands are nitrogen and sulfur. It has long been known that in highly P-affected areas, biotic growth/productivity is most frequently limited by nitrogen. Also, the process of litter decomposition is known to have a high demand for nitrogen. In this way, nitrogen availability could be a primary factor regulating phosphorus storage and stability in wetland systems where a significant uptake of nutrients by biota can be counterbalanced by release of nutrients during decomposition, resulting in the retention of a small fraction of the total material produced in accreted material. Similarly, sulfur can interact with soils to affect phosphorus retention and stability. In highly anaerobic wetland conditions, excess sulfate can stimulate organic matter decomposition, thus promoting organic phosphorus mineralization as well as increasing the soluble phosphorus. Sulfur can also act through oxidation-reduction reactions to affect soil-binding capacity for phosphorus, particularly in calcareous systems where pH directly determines mineral stability and phosphorus retention.

17.6 SUMMARY

- Organic matter accumulation in wetlands is a result of greater net primary productivity and slower rates of decomposition than terrestrial ecosystems. Consequently, organic matter accretion provides a long-term storage compartment for nutrients including carbon, nitrogen, phosphorus, and sulfur.
- The Everglades ecosystem has adapted to low external nutrient loads, relying primarily on low P-limited production of plant biomass and slow recycling of organic substrates by a small nutrient-limited microbial pool.
- The once-nutrient-limited microbial pool has increased in size due to the availability of nutrients, particularly phosphorus, in the inflow of water and currently mediates the large-scale release of nutrients from the deposited organic matter.
- Several microbial processes are affected by the phosphorus enrichment of the soil and litter. Recently, accreted plant detritus and floc showed a high degree of positive response to phosphorus loading on various microbial processes.
- Increases in phosphorus loading in the impacted areas decreased C:P ratio of litter and soil, resulting in an increased rate of organic nitrogen and phosphorus mineralization.

- Although phosphorus may no longer be limited to biota in the impacted zone, a high degree of phosphorus enrichment has induced nitrogen limitation.
- The spread and growth of *Typha* in the impacted zone may be directly linked to the increased supply of bioavailable nutrients liberated through an enhanced decomposition of P-enriched litter and organic matter.

STUDY QUESTIONS

1. List the unique characteristics of the Everglades ecosystem. What is an oligotrophic wetland?
2. Historically what was the major source of nutrients to the Everglades? What changes have occurred in the Everglades in the past century?
3. Discuss the difference between high nutrient and low nutrient wetland systems. Explain the relationship of closed to open elemental cycling in these systems.
4. Explain why water conservation areas and storm water treatment areas are established. Discuss briefly pros and cons of these systems in meeting long-term goals of the Everglades restoration.
5. How has increased phosphorus loading affected the flora and fauna of the Everglades ecosystem?
6. What are some of the major vegetation communities and how have they changed during the past few decades and why?
7. Discuss external and internal factors governing competition between cattails and saw grass.
8. How have periphyton communities been impacted by nutrient enrichment? What is the overall role of periphyton in regulating surface water phosphorus concentrations?
9. Explain the impact of nutrient enrichment on microbial community structure. What impact has the change had on biogeochemical cycles?
10. What is the role of extracellular enzymes in the Everglades. Specifically discuss the importance of alkaline phosphatase activity.
11. Explain aspects of carbon, nitrogen, and sulfur cycles impacted by nutrient enrichment to the Everglades.
12. Explain how methane emissions are influenced by phosphorus enrichment.
13. What is the relationship between sulfur cycling and methylmercury formation? In what areas of the Everglades methylmercury formation would be significant?
14. Discuss how hydrology, vegetation, periphyton, and microbial communities, and associated biogeochemical processes regulate long-term retention of phosphorus in the Everglades wetlands.
15. What is an internal phosphorus load and how it affects the overall restoration and recovery?
16. Overall, based on your understanding of the Everglades biogeochemistry, is the Everglades a carbon sink or source?

FURTHER READINGS

Committee on Independent Scientific Review of Everglades Restoration Progress (CISRERP). 2006. Progress Toward Restoring the Everglades, National Research Council. National Academy of Sciences. Washington DC. 234 pp.

McCormick, P. V., S. Newman, S. L. Miao, D. E. Gawlik, D. Marley, K. R. Reddy, and T. D. Fontaine. 2002. Effects of anthropogenic P inputs on the Everglades. In K. G. Porter and J. Porter (eds.) *The Everglades, Florida Bay, and Coral Reefs of the Florida Keys: An Ecosystem Sourcebook*. (CRC Press, Boca Raton, FL. pp. 83–126.

Noe, G. B., D. L. Childers, and R. D. Jones. 2001. Phosphorus biogeochemistry and the impact of phosphorus enrichment: why is the Everglades so unique? *Ecosystems* 4:603–624.

Reddy, K. R., G. A. O'Connor, and C. L. Schelske. 1999. *Phosphorus Biogeochemistry in Sub-tropical Ecosystems: Florida as a Case Example.* CRC/Lewis Publishers.

Reddy, K. R., R. H. Kadlec, M. J. Chimney, and W. J. Mitsch. 2006. The Everglades Nutrient Removal Project. Special Issue. *Ecol. Eng.* 27:1–379.

Sklar, F. H. and A. van der Valk (eds). 2003. *Tree Islands of the Everglades.* Kluwer Academic Publishers, Boston, MA.

South Florida Water Management District. 2007. Phosphorus source controls for the basins tributary to the Everglades Protection Area. In *South Florida Environmental Report.* South Florida Water Management District, West Palm Beach, FL. http://www.sfwmd.gov/sfer/SFER_2007/index_draft_07.html

18 Coastal Wetlands: Mississippi River Deltaic Plain Coastal Marshes, Louisiana

18.1 INTRODUCTION

Riverine-dominated coastal systems are complex ecosystems that are in dynamic equilibrium with their surroundings. The important variables that determine the structure and function of coastal deltaic wetlands are pulses from the river during high discharge or flood events, which supply nutrients and sediments to the coastal marshes. These marshes maintain the existence when the accumulation of sediment is equal to or greater than the rate of subsidence. Coastal marshes are among the most productive ecosystems in the world. Productivity is related to nutrient and sediment input from rivers feeding the system. Louisiana coastal marshes located in the Mississippi River deltaic plain are outwelling systems that export organic energy (detritus) to the estuary and the Gulf of Mexico.

Louisiana coastal marshes represent a complex assortment of plants, animals, and microbes that are impacted by daily and seasonal fluctuations in temperature, water levels, and salinity. The stress factors governing the major biogeochemical processes include carbon, nitrogen, and sulfur cycling. Because of leveeing of the Mississippi River, Louisiana coastal wetlands do not receive normal pulses of freshwater and sediment historically associated with spring floods. This fact along with rapid subsidence has altered the processes governing the normal biogeochemical functioning of these wetlands. As a result Louisiana Gulf Coast wetlands are experiencing rapid rate of wetlands loss. Louisiana coastal wetlands are a case example of what may occur in other coastal areas in response to global warming and the accompanying increase in sea level.

18.2 BIOGEOGRAPHY AND GEOLOGY OF LOUISIANA COASTAL WETLANDS

Louisiana Mississippi deltaic plain wetlands were created over the past several thousand years through a series of delta lobes (Figure 18.1). The Mississippi River deposited sediment along the main channel and distributaries, switching to a shorter route and subsequently building a new delta as this natural phenomenon continued. Simultaneously, wetlands resulting from the previous route began to deteriorate to open water while new wetlands were created along the new route, which compensated for wetland loss. The present deltaic landscape in coastal Louisiana represents a relatively thin layer of Holocene sediment deposits overlying earlier deposits. Rapid delta switching of the Mississippi River during the past 7,000 years has resulted in the formation of a vast deltaic plain. At the same time parallel progradation of Louisiana coastline occurred. The process was maximized during this period with the Mississippi River utilizing its sediment load for delta building, resulting in sediment deposits approximately 300 km wide and nearly 100 km inland, which is referred to as the deltaic plain. The delta building represented a relatively rapid process leading to the basinwide extension of marshlands on the subaerial to near-subaerial parts of a newly formed deltaic land mass. When the river abandoned its delta, the fluvial process forced by abundant sediment and freshwater was replaced by submergence and wetland deterioration.

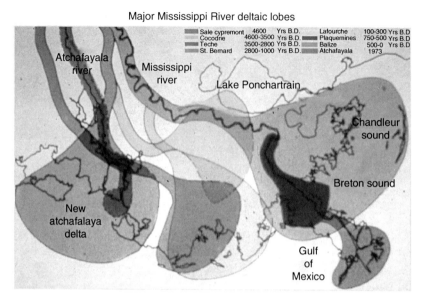

FIGURE 18.1 The deltaic plain landmass was built by a sequence of overlapping deltaic lobes that developed during the last 5,000 years. (Modified from Kolb and van Lopik, 1958.)

18.3 COASTAL WETLAND LOSS

Currently there is rapid wetland loss occurring in Louisiana's Mississippi River deltaic plain, which is attributed to a combination of natural processes and human activities. Massive coastal loss that began around 1890 and peaked during the 1950s and 1960s has resulted in the loss and deterioration of coastal wetlands. During the last 50 years, land loss rates have exceeded 100 km^2 year^{-1}, and in the 1990s the rate has been estimated to be between 65 and 91 km^2 year^{-1} (Figure 18.2). This accounts for approximately 80% of the coastal wetland loss in the continental United States. During a period of little more than 100 years, more than 404 ha or about 20% of coastal wetlands have deteriorated.

Deterioration of Louisiana coastal wetlands began in the early nineteenth century at approximately the same period that the Mississippi River was leveed. Submergence resulting from subsidence is the major factor contributing to wetland loss in coastal Louisiana. Adequate sediment input is necessary for maintaining marsh surface elevation in response to an increasing water level. During the last decades, sediment load has decreased dramatically as a result of human management of the Mississippi River by forcing the river down its present channel. This has deprived wetlands and coastal areas of mineral sediment critical to plant productivity and deltaic expansion. Maintaining the Mississippi River in its present channel halted the delta switching process, causing sediment to be deposited off the continental shelf, and cut off freshwater, sediment, and nutrient supply to vast areas of coastal marsh. The compaction of existing sediment, absence of sediment inputs, and resultant submergence and saltwater intrusion contribute to the wetland loss.

The relative rates of vertical marsh accretion and submergence determine the long-term stability of Louisiana coastal marshes. Coastal marshes are highly susceptible to submergence associated with a rise in relative sea level (Penland and Ramsey, 1989). Louisiana coastal marshes are undergoing rapid subsidence and currently experiencing rapid increases in the water level. Research conducted over the past quarter century in coastal Louisiana has shown that marsh accretion at many

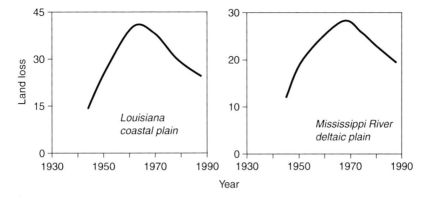

FIGURE 18.2 Wetland loss rates over the entire Louisiana coastal area and in the Mississippi River deltaic plain.

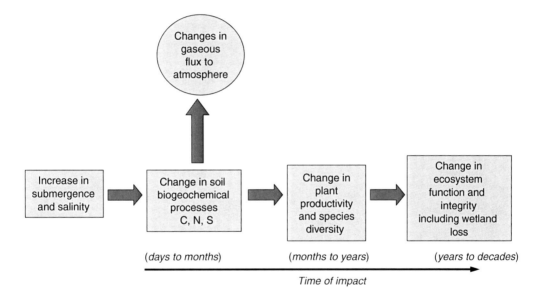

FIGURE 18.3 Impact of increase in submergence on wetland ecosystem process along the Louisiana Gulf Coast.

locations has not kept pace with submergence. A schematic of ecosystem processes impacted by an increase in submergence and salinity is shown should read: "A schematic of …" in Figure 18.3.

Louisiana's coastal wetland is characterized by dominant wetland plant species that respond well to salinity and flood duration. Physical processes (flood duration, salinity, and inorganic sediment availability) determine the productivity of emergent vegetation (above- and belowground biomass), and over a period of time ecosystem function. These same variables are also important factors governing not only plant growth but also biogeochemical cycling of carbon, nitrogen, sulfur, and other elements. The factors governing carbon fixation (plant biomass), organic matter accumulation, decomposition, and flux are depicted in Figure 18.4. In the following sections we describe case studies of processes such as sedimentation, plant growth, and elemental cycles that are strongly influenced by subsidence, and saltwater infusions that are unique in the Louisiana Gulf Coast but may serve to show what might occur in other coastal regions with the predicted sea-level increase associated with global warming.

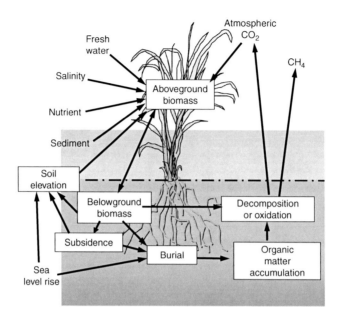

FIGURE 18.4 Factors influencing carbon cycling in the rapidly subsiding Mississippi River deltaic plain.

18.4 CASE STUDIES

18.4.1 Processes Governing Coastal Marsh Stability

The major processes involved in marsh stability dynamics within Louisiana coastal wetlands are listed as follows:

1. *Eustatic sea-level rise*, or global changes in the sea level because of changes in the mass of water in oceans, ocean volumes, and temperature changes.
2. *Compaction of deep sediments* (except Holocene sediments) includes both tectonic downwarping of basement sediments and compaction of Tertiary, Pleistocene, and older sediments. Tectonic downwarping is probably the only factor in major depositional areas such as large deltas, where the loads from massive amounts of sediments cause the bottom of the sedimentary basin to warp. In most studies of modern sediments, very little is known about this factor, so it is often treated as an unknown or something outside of the system.
3. *Compaction of Holocene sediments* includes dewatering of sediments (primary consolidation), rearrangement of mineral structure of the sediment and subsequent loss of volume (secondary consolidation), and the decomposition of organic matter in the sediment. This is the most important process affecting subsidence in Louisiana coastal areas; however, it is a very difficult process to evaluate.
4. *Marsh accretion* is the vertical accumulation of material (organic matter and mineral sediment) on the surface of a marsh.
5. *Subsidence* is the decrease in land elevation relative to some elevation (benchmark or water level).
6. *Accretion balance* refers to the net change in sea level relative to marsh surface elevation at a particular site over time. For a coastal wetland to remain at the same elevation over time, accretion must be equal to subsidence plus eustatic sea-level rise.
7. *Submergence* refers to negative accretion balance, that is, a marsh surface being flooded to greater and greater depths with time.

These processes govern marsh stability along the Louisiana Coast as depicted in Figure 18.5.

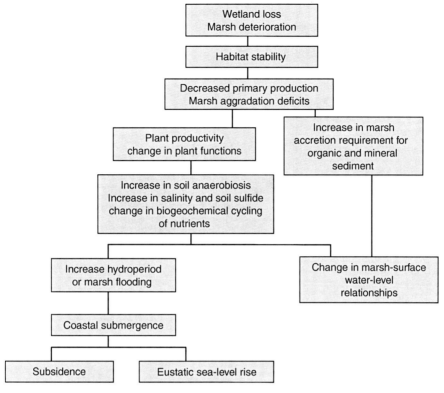

FIGURE 18.5 Schematic of how the increase in coastal submergence impacts Louisiana coastal marsh stability.

18.4.2 COMPARISON OF VERTICAL ACCRETION OF LOUISIANA MARSH TO OTHER GULF COAST MARSH

The relationship between vertical marsh accretion and marsh surface submergence in various Louisiana coastal regions is shown in Figure 18.6. It is apparent that in many geomorphic regions in coastal Louisiana vertical accretion is not keeping pace with submergence. This is especially true in the Teche, Terrebonne, Barataria, and St. Bernard regions. This is not a function of low rates of vertical accretion, but rather a function of the relatively high subsidence rates in these geomorphic regions. Comparing the differences between submergence rates and vertical accretion rates shows that inadequate vertical accretion is an important factor in wetland loss in much of Louisiana coastal region as reported by Britsch and Dunbar (1993).

In addition to mineral sediment, organic matter accumulation is also a major determinant in the vertical growth rate of marshes in the Mississippi River deltaic plain (Hatton et al., 1983). There is less organic matter than mineral matter on a weight basis, but the organic matter occupies more volume than the mineral matter in fresh and brackish marsh soil. The organic matter contributes greatly to the soil matrix and increases structural strength by forming an interlocking network. The amount of soil organic matter varies from 3% by volume in fresh marsh to 5% in saline marshes, and the majority of the soil volume is made up of pore space (either air or water depending on tidal conditions) (Nyman et al., 1990).

The majority of the soil organic matter accumulation results from *in situ* production by marsh plants, rather than transportation of particulate organic matter from other areas. Factors that regulate plant growth also affect soil organic matter accumulation. Aboveground plant growth also promotes mineral matter accumulation by slowing down water velocity and increasing sediment deposition.

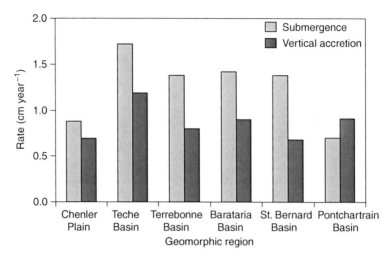

FIGURE 18.6 Relationship between vertical accretion and submergence of the marsh surface in different geomorphic regions of the Louisiana Gulf Coast; estimated in part from tide gauge data, and accretion rates. (Redrawn from Penland and Ramsey, 1989; Gornitz and Lebedeff, 1987; DeLaune et al., 1978, 1983a, 1986, 1987; Hatton et al., 1983.)

TABLE 18.1
Rates ($g\,m^{-2}\,year^{-1}$) of Mineral and Organic Matter Accumulation in Inland Marshes of the Mississippi River Deltaic Plain

Organic Matter	Mineral Matter	Source
Saline marsh		
435	1,740	Hatton et al. (1983)
315		Smith et al. (1983b)
Brackish marsh		
348	478	Hatton et al. (1983)
414		Smith et al. (1983b)
262		DeLaune et al. (1987)
Fresh marsh		
306	280	Hatton et al. (1983)
386		Smith et al. (1983b)
312		DeLaune et al. (1987)

The rates of organic matter accumulation and decomposition in fresh, brackish, and saline marsh soils of the Mississippi River deltaic plain (Table 18.1) have been used to estimate the amount of soil organic matter production required to maintain a positive accretion balance (Nyman et al., 1990).

Soil mineral matter consists of 50–90% of the dry weight of marsh soils (Nyman et al., 1990). Storm passage including hurricanes is a primary mechanism that delivers mineral sediments to Louisiana marshes (Reed, 1989; Roberts et al., 1989; Turner et al., 2006). Increased water levels and wave action in bays and lakes prior to frontal passage transport sediments from water bottoms to

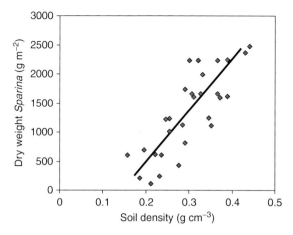

FIGURE 18.7 Relationship between soil bulk density and aboveground biomass of the Louisiana saline coastal marsh supporting the species *Spartina alterniflora*. (Adapted from DeLaune et al., 1990b.)

marsh surfaces. Hurricanes Katrina and Rita, which crossed the Louisiana Coast in 2005, deposited several centimeters of sediment on the marsh surface (Turner et al., 2006). It was estimated that the amount of storm-transported material was greater than the amount introduced into Louisiana wetlands via historical overbank flow. However, the hurricane also converted several hundred km² of coastal marshes to open water. Some of the hurricane-deposited sediment may be reworked sediment coming from the destroyed marsh soil profiles.

Soil mineral content, in addition to contributing to soil volume and marsh surface elevation, promotes vigorous plant growth in several ways. First, it provides mineral nutrients for plant growth. Marsh soils with low bulk density support less robust vegetation because the nutrient content of the vegetation is related to that of the soil (DeLaune and Pezeshki, 1988). The iron content of marsh soil also promotes growth of marsh vegetation because reduced iron in wetland soil precipitates sulfide, a plant toxin (Pezeshki et al., 1988) produced through sulfate reduction by bacteria. Bulk densities less than 0.20 g cm^{-3} will not support growth of *Spartina alterniflora* (Figure 18.7). Soil with bulk density greater than 0.20 g cm^{-3} often contains sufficient mineral sediment (i.e., iron) to precipitate toxic sulfides.

Fresh, brackish, and saline marshes have different mineral matter requirements for maintaining accretion. It is estimated that to vertically accrete at 1 cm year^{-1} fresh marsh required 424 g m^{-2} year^{-1}, brackish marsh required 1,052 g m^{-2} year^{-1}, and saline marsh required 1,789 g m^{-2} year^{-1} (Nyman et al., 1990). Marsh loss may occur if low mineral and organic matter accumulation result in inadequate rates of vertical accretion to keep pace with submergence or if the composition in the soil substrate is insufficient to support growth of marsh vegetation in individual marsh types.

As described earlier, vertical marsh accretion rates in the Louisiana Mississippi deltaic plain is not sufficient to keep pace with submergence. A study of accretion from Biloxi Bay, Mississippi, and the Florida Keys indicated low gross rates of accretion but very low rates of subsidence (Callaway et al., 1997). Accretion rates were greater in Louisiana marshes when compared to other marsh sites along the Gulf of Mexico (Figure 18.8). However, the marsh locations along the Gulf of Mexico outside of Louisiana show that accretion is keeping pace with relative sea-level rise. Even though there are regional differences in accretion rates in other U.S. Gulf Coast marshes, in contrast to Louisiana's Gulf Coast marsh vertical accretion has in general been sufficient to compensate for the sea-level rise.

Along the east coast of the United States, eustatic sea level rise is generally the primary factor governing marsh accretion and stability. In comparison, along the Louisiana Gulf Coast, subsidence is of far greater significance than global sea-level rise in governing the coastal marsh stability.

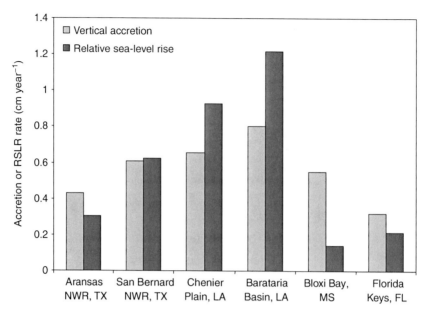

FIGURE 18.8 Vertical marsh accretion and relative sea-level rise at select marsh locations along the northern shore of the Gulf of Mexico. (Modified from Callaway et al., 1997, which includes data from DeLaune et al., 1983a and Hatton et al., 1983.)

18.4.3 INFLUENCE OF SEDIMENT ADDITION TO A DETERIORATING LOUISIANA SALT MARSH

Sediment accretion or additions can improve primary productivity of marsh vegetation. For example, sediment addition to a deteriorating *Spartina alterniflora* salt marsh at a rate equivalent to 94 kg m^{-2} increased the aboveground biomass production by 100% (DeLaune et al., 1990a) (Table 18.2a). The sediment addition increased marsh surface by approximately 10 cm. Similarly, the number of regenerating shoots increased substantially compared to control plots.

Vertical marsh accretion determined from ^{137}Cs dating indicated the deteriorating inland salt marsh to be accreting at the rate of 0.44 cm year^{-1}. This rate is approximately 0.8–1.0 cm year^{-1}, less than the accretion rate needed to maintain a productive salt marsh typical of Louisiana Barataria Basin.

Sediment addition also resulted in an increase in nutrient and mineral content of *Spartina alterniflora* plant tissue (Table 18.2b). There was a significant increase in Fe and Mn content of plant tissue in plots receiving sediment input. The plant tissue content of Fe and Mn essentially doubled from the 94 kg sediment m^{-2} input as compared to plots with no sediment additions. Phosphorus content of the plant tissue also increased as a result of sediment addition to the inland marsh. Although sediment addition did not increase nitrogen content of plant tissue, there was an increase in total uptake of nitrogen as a result of plant biomass increase. Calcium content of plant tissue was enhanced only at the highest level of sediment input to the marsh. The study clearly demonstrates that if the Mississippi River sediment could be reintroduced and distributed into these rapidly deteriorating salt marshes, plant regeneration and productivity would be enhanced, thus reducing the rate of wetland loss.

18.4.4 IMPACT OF MISSISSIPPI RIVER DIVERSION ON ENHANCING MARSH ACCRETION

Currently, diversions are the primary focus of major restoration efforts in coastal Louisiana. Reintroduction of freshwater and sediment will play a determining role in slowing and perhaps reversing

TABLE 18.2a
Plant Biomass Sampled from Control and Sediment Treated Plots

Sediment Addition ($kg\,m^{-2}$)	Total Aboveground Biomass ($g\,m^{-2}$)	Live Plant Biomass ($g\,m^{-2}$)	Number of Plant Shoots	Plant Height (cm)
Control	1,152a	736a	188a	43a
47	1,624a,b	1,072a,b	276b	41a
94	2,108b	1,472b	300b	41a

Note: Values with the same letter (a or b) within the same column are not significant at the 5% level.
Source: DeLaune et al. (1990a).

TABLE 18.2b
Plant Tissue Nutrient and Mineral Content of *Spartina alterniflora* Sampled from Control Plots and Plots with Sediment

Sediment Addition ($kg\,m^{-2}$)	Tissue Nutrient Content							
	N ($g\,kg^{-1}$)	Fe ($mg\,kg^{-1}$)	Mn ($mg\,kg^{-1}$)	P ($mg\,kg^{-1}$)	Al ($mg\,kg^{-1}$)	Ca ($mg\,kg^{-1}$)	Mg ($mg\,kg^{-1}$)	K ($mg\,kg^{-1}$)
Control	7.6a	222a	19.5a	1,256a	279a	1,588a	2,340a	12,890a
47	7.1a	346b	40b	1,544a,b	364a	1,815a,b	2,380a	14,620a
94	7.4a	397b	47b	1,631b	420b	2,266b	2,270a	12,490a

Note: Values with the same letter (a or b) within the same column are not significant at the 5% level.
Source: DeLaune et al. (1990a).

the loss of wetlands in Louisiana. The diversions will in part replicate the action of the Mississippi River prior to the existence of the levee system. By siphoning, pumping, or cutting through existing levee system, diversion projects can move millions of liters of water and sediment from the Mississippi River into targeted wetlands (Lane et al., 1999). The infusion of freshwater and sediment into Louisiana estuaries will help offset submergence and reduce salinity levels, supporting the growth of vegetation and allowing for marsh accretion to proceed at rates sufficient to keep pace with water level increase as a result of subsidence and sea-level rise.

A series of diversion projects have been implemented to reintroduce freshwater and sediment from the Mississippi River into Louisiana coastal wetlands (Figure 18.9). A recent study examined the impact of Mississippi River freshwater diversion on enhancing vertical marsh accretion (mineral and organic matter accumulation) in Breton Sound estuary, a coastal wetland experiencing marsh deterioration as a result of subsidence and saltwater intrusion (DeLaune et al., 2003). The Caernarvon diversion has positively impacted marsh accretion in Brenton Sound estuary helping to slow or reverse wetland loss. Several hundred hectares of new marsh have been created by the introduction of Mississippi River water into the system (Villarrubia, 1998).

Sedimentation and vertical marsh accretion, soil mineral and organic matter accumulation, nutrient status, and plant biomass were measured at 20 marsh sites with distance (15–20 km) into estuary at the Caernarvon freshwater diversion (DeLaune et al., 2003). Vertical accretion and

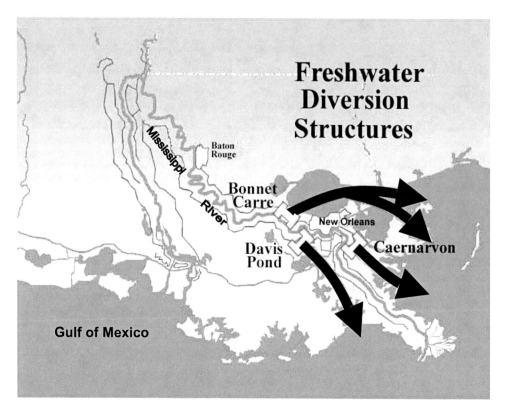

FIGURE 18.9 Location of Caernarvon Diversion and other freshwater sites.

accumulation of mineral sediment organic matter and nutrients in the marsh soil profile increased at marsh sites receiving freshwater and sediment input (Figure 18.10).

Vertical marsh accretion rates were determined from ^{137}Cs dating of soil profiles at 20 sites, ranging from 0.55 to 1.11 cm year^{-1}. Marsh sites nearest to the river water input had significantly higher rates ($p < 0.05$) of vertical accretion compared to accretion at sites located at a greater distance from the freshwater input.

The bulk density of the marsh soil profiles showed a significant increase ($p < 0.05$) in sediment in the surface (0–9 cm) at the sites nearest to the freshwater input. The bulk density in the surface (0–9 cm) averaged over 0.30 g cm^{-3} at sites nearest to the diversion. The higher bulk density represented a sediment spike or bulk density increase in the surface profile as compared to the lower bulk density observed with depth in the marsh profile (DeLaune et al., 2003).

The accretion rates determined using feldspar marker horizon were also greater at sites nearest to the diversion as compared to marsh sites at a distance (Figure 18.11). The accretion rates determined from feldspar marker horizon were significantly greater than that determined from ^{137}Cs dating, which includes rates prior to the diversion. Mineral accumulation determined from the feldspar (Figure 18.12) measurement, mineral and organic matter percentage of deposited material, and bulk density of marsh soil profile shows a significant increase ($p < 0.05$) in mineral sediment deposition at sites nearest to the freshwater input. The Caernarvon diversion confirms that the inputs of freshwater and sediment into the area will increase marsh elevation and enhance marsh stability.

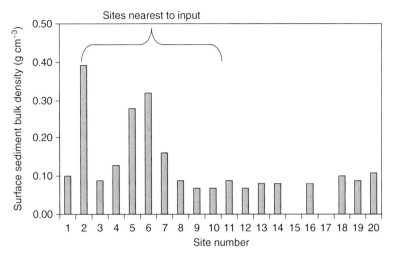

FIGURE 18.10 Bulk density of the surface (0–9 cm) marsh soil at the sampling site. (Modified from DeLaune et al., 2003.)

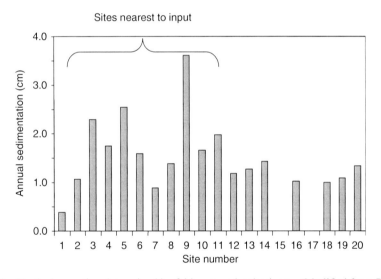

FIGURE 18.11 Vertical accretion determined by feldspar marker horizons. (Modified from DeLaune et al., 2003.)

This case study demonstrated that freshwater diversion projects such as Caernarvon serve an important role in the reintroduction of freshwater and limited amount of mineral sediment necessary for promoting growth of marsh vegetation (the source of organic matter in marsh soil) in subsiding delta environments. In addition to sediment input, an important influence of the diversion on enhancing marsh stability is associated with the lowering of salinity, which reduces the sediment requirement for marsh maintenance (freshwater marshes have lower mineral sediment requirement than saline marshes).

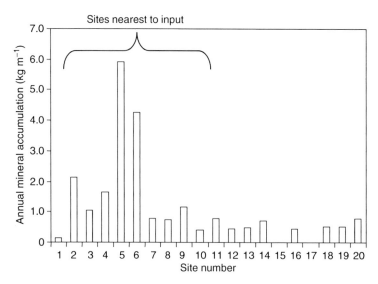

FIGURE 18.12 Mineral accumulation determined using feldspar marker horizons. (Modified from DeLaune et al., 2003.)

18.5 IMPACT OF FLOODING AND SALTWATER INTRUSION ON LOUISIANA COASTAL VEGETATION

A typical wetland transition along the Louisiana Gulf Coast extending inland consists of a remarkably distinct zonation of salt marsh, brackish marsh, and freshwater habitats. Further inland along the transitional gradient, the vegetation is composed of bottomland forest species changing into upland forest communities (Figure 18.13). The transition occurs primarily as a result of water depth and salinity gradients and the corresponding differences in the soil physicochemical properties. The increase in anaerobic soil conditions and salinity as a result of submergence and saltwater intrusion can adversely affect plants from various habitat types.

Numerous field studies have concluded that flooding and saltwater intrusion was a major cause for vegetation dieback in Louisiana Gulf coastal marshes. The frequency, duration, depth of flooding, and concentration of salt are important factors that affect plant distribution, productivity, and coastal marsh stability. Sediment physiochemical characteristics including pH, redox potential, and ionic content become critical to plant productivity once the soil is flooded. Under reduced soil conditions (low redox potential) organic substrates are not decomposed and phytotoxic compounds such as sulfides are formed that create additional stress on vegetation. This is well demonstrated, because the zonation of plants in the U.S. Gulf coastal zone is a classic example of how plant zonation is closely associated with dominant environmental gradients, in this case, governed by water level and salinity.

Plant stress factors in rapidly subsiding coastal marsh are the result of an

1. Increase in submergence (low redox potential and sulfide accumulation)
2. Increase in salinity (salt stress to fresh and brackish marsh plant species)

Saltwater intrusion is a dominant edaphic factor that influences the survival, growth, and species composition in coastal wetlands along the Louisiana Gulf Coast. The net effect of sublethal salt stress is the reduction in growth especially in marsh soils with low bulk density or mineral content in the soil profile. For most species growing in salt and brackish marshes, elevated salinity does not create a favorable environment for growth and productivity (Figure 18.14). In fact, numerous

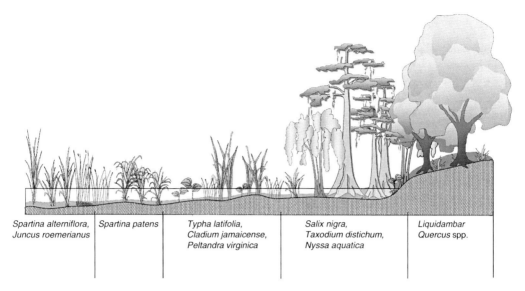

FIGURE 18.13 Plant distribution along flooding and salinity gradient extending inland to the coast.

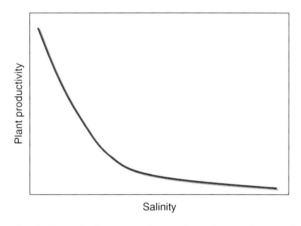

FIGURE 18.14 Schematic of effect of salinity or saltwater intrusion on plant productivity (freshwater and brackish plant species).

studies under laboratory and greenhouse conditions have shown that salinity in the range encountered by these species in the U.S. Gulf coastal marshes reduces the plant carbon fixation capacity. For instance, marshes with soil salinities in the range of 170–850 mol m^{-3} resulted in reduction of net photosynthesis of *Spartina alterniflora* by approximately 28–52% (Pezeshki et al., 1989a). The growth potential of major species from Louisiana Gulf Coast brackish and salt marsh communities is also adversely affected by salinity. As it is evident from the literature (Pezeshki et al., 1989a), saltwater intrusion can significantly impact survival and growth of wetland species from various habitats in the U.S. Gulf Coast. For example, in salt marsh species such as *Distichlis spicata*, plant dry weight was reduced by 35% in response to elevated salinities. In *S. cynosuroides* and *S. alterniflora*, plant dry weight reduced by as much as 80% in response to elevated salinities. *S. alterniflora* plants from different U.S. Gulf Coast populations responded to excess soil salinities ranging from 170 to 850 mol m^{-3} by reduced biomass production in all populations (Pezeshki et al., 1989a). In fresh and brackish marsh communities, saltwater intrusion is expected to cause substantial greater stress on plants. For instance, salinities ranging between 170 and 200 mol m^{-3} resulted in severe injury and death of *Panicum hemitomon*, a dominant fresh marsh species. Populations of *S. patens*, a dominant brackish marsh species, showed reduced growth in response to elevated floodwater salinity from 85 to 430 mol m^{-3}. Various biomass components including foliage and roots were reduced significantly.

Bottomland hardwood swamps and coastal forests are important components of forested systems along the Louisiana Gulf Coast. Portions of these forests are under increasing stress associated with deep flooding and saltwater intrusion. In this region, cypress–tupelo swamps are composed of bald cypress (*Taxodium distichum* (L) Rich.) and water tupelo (*Nyssa aquatica* L.), as two major species that cover between 1.2 and 2 million ha. In Louisiana alone, these forests cover approximately 260,000 ha. Degradation of coastal forests and associated wetland habitats by excessively deep flooding and saltwater intrusion is dramatically high especially in the Mississippi River Delta. In addition to anthropogenic factors, both land subsidence and sea-level rise have contributed to degradation and death of trees in coastal forests. Further rise in the sea level and a consequent increase in inundation and salinity along coastal regions can be expected. As water level increases, deep flooding and saltwater intrusion may occur further upstream and inland. Along the Louisiana Gulf Coast, saltwater intrusion has been reported to cause substantial degradation of forest species and the consequences such as heavy tree mortality in some forested areas are already evident. The decline of cypress forests in coastal Louisiana is an example of how these environmental factors may adversely affect plant communities and how vulnerable these communities may become. These swamp forests are presently being converted to open waters at an alarming rate. In a reported study, bald cypress (*T. distichum*) seedlings from different sources in coastal Louisiana were tested for possible difference in the level of salt tolerance. Seedlings were subjected to a different combination of flooding and salinity treatments at moderate levels. However, plant biomass production was adversely affected as salinity of floodwater increased from 0 to 136 mol m^{-3} (Pezeshki and DeLaune, 1990). The data indicated that bald cypress seedlings from different sources were sensitive to saltwater intrusion. Thus, sea-level rise and saltwater intrusion will continue to be important factors threatening the health and the very existence of these forests.

Deep, prolonged coastal submergence and the resultant increase in the intensity and duration of flooding result in substantial changes in marsh soil redox conditions, which also affect wetland vegetation. Stressed wetland plants have limited ability to transport oxygen to the root zone to abate soil anaerobic conditions. Under these conditions, plant root respiration to some extent is regulated by aerobic pathways (see Chapter 7). Such reducing conditions influence wetland macrophyte productivity and distribution along the Louisiana Gulf Coast.

Increased flooding from subsidence affects carbon fixation by marsh vegetation growing along the Gulf Coast. The growth and productivity of *S. alterniflora* was seriously impacted by low sediment redox conditioning (Pezeshki et al., 1989a). Carbon assimilation rates declined significantly (15–20%) when soil redox level decreased rapidly to less than –200 mV (Figure 18.15). The results indicated the adverse effects of extreme anoxia on carbon assimilation by *S. alterniflora*, a possible reflection of the limited ability of this species to maintain root oxygenation during a rapid intense reduction in soil redox potential.

Potential toxic substances produced at low soil redox potentials can also influence wetland plant growth. Such toxins are of importance to roots that have not been killed by oxygen depletion itself. After prolonged periods of flooding, hydrogen sulfide is produced by the bacterial reduction of sulfates and the dissimilation of sulfur-containing amino acids. Sulfide has been implicated as a cause of reduced plant growth in Louisiana Gulf Coast salt marsh (Figure 18.16). Other studies have also examined various responses of selected populations of *S. patens* to changing rhizosphere redox potential (Pezeshki et al., 1988). Rhizotrons were utilized to control redox potential using various gas mixtures. The results indicated that, despite high levels of flood tolerance, anaerobiosis or soil redox conditions inhibited root elongation in some populations (Figure 18.17).

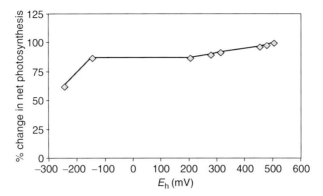

FIGURE 18.15 Effect of low soil redox conditions on the photosynthesis activity of *Spartina alterniflora*. Plants were grown in soil suspension in which soil redox level was controlled using techniques described by DeLaune et al. (1984). (Redrawn from DeLaune et al., 1984.)

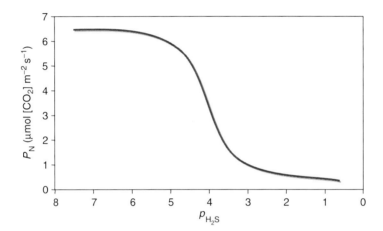

FIGURE 18.16 Relationship between net photosynthetic rate (P_N) and hydrogen sulfide concentrations (p_{H_2S}) for *S. alterniflora*. (Modified from Pezeshki et al., 1989b.)

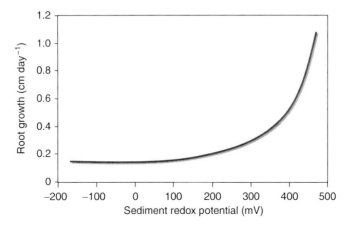

FIGURE 18.17 Schematic relationship between root growth of *Spartina patens* and sediment redox conditions. (Modified from Pezeshki and DeLaune, 1990.)

18.6 CARBON CYCLING

Research conducted in Barataria Basin, Louisiana, has quantified the major carbon fluxes occurring in fresh, brackish, and salt marshes of the Mississippi River deltaic plain (Feijtel et al., 1985). The results are summarized as follows:

1. High primary productivity or carbon fixation
2. Large sedimentation or accretion sinks
3. High greenhouse gas (carbon dioxide and methane) emissions

18.6.1 Primary Production

Louisiana coastal marsh vegetation is highly productive, fixing considerable amount of carbon dioxide through photosynthesis. Reported standing crop and aboveground primary productivity estimates for dominant marsh macrophytes (fresh, brackish, and salt) in the Louisiana Mississippi River deltaic plain are shown in Table 18.3. Aboveground production averages from 1,410–2,895 g organic matter m^{-2} $year^{-1}$ for salt marsh habitats. Brackish marshes had the highest range in productivity with values of 1,342–5,812 g organic matter m^{-2} $year^{-1}$. Freshwater marshes produced

TABLE 18.3
Production of Major Marsh Macrophytes in the Louisiana Mississippi River Deltaic Plain Marshes

Species	Peak Biomass (g Organic Matter m^{-2} $year^{-1}$)	Productivity (g Organic Matter m^{-2} $year^{-1}$)	Method/Source	
Spartina alterniflora	1,018	1,410	a	(1)
		2,645	b	"
	754	2,658	b	(2)
	1,070	1,527	a	(3)
		2,895	b	"
	1,080	2,160	c	"
Spartina patens	1,375	2,000	a	(2)
		5,812	b	"
	1,350	1,342	a	(3)
		1,428	b	"
	1,248–2,466	2,605–1,411	a	(4)
		3,056–3,464	b	"
	1,100	1,350	c	(5)
Panicum hemitomon	1,160	1,700	d	(6)
	1,200	1,500	c	(5)

[a] Smalley (1958)
[b] Wiegert and Evans (1964)
[c] Bartaria Basin study-site estimate
[d] Lomnincki et al. (1968)

(1) Kirby and Gosselink (1976)
(2) Hopkinson et al. (1978)
(3) White et al. (1978)
(4) Cramer and Day (1980)
(5) Delaune and Smith (1984)
(6) Sasser and Gosselink (1984)

between 1,500 and 1,700 g organic matter m^{-2} $year^{-1}$. These productivity estimates suggest higher turnover rates as shown by comparing the peak standing crop estimates to productivity estimates. There is currently a lack of reliable data for belowground productivity estimates.

18.6.2 Methane and Carbon Dioxide Emission along a Salinity Gradient in Louisiana Coastal Marshes

A seasonal study of methane emissions from adjoining salt, brackish, and fresh marsh soils in Louisiana's Barataria Basin demonstrated that methane emission is a significant process in the carbon and energy flow of Louisiana coastal ecosystem (DeLaune et al., 1983b). Methane emission was inversely related to salinity and sulfate concentration, with methane increasing and salinity and sulfate decreasing with increasing distance from the coast. The annual amounts of methane evolved were 4.3, 7.3, and 160 g C m^{-2} $year^{-1}$ for salt, brackish, and freshwater marshes, respectively (Figure 18.18). *In vitro* experiments show that methane production is sensitive to the addition of sulfate, high concentrations (10 mM SO_4) inhibiting methane evolution. Annual carbon dioxide fluxes were 418, 180, and 618 g C m^{-2} $year^{-1}$ from the salt, brackish, and freshwater marshes, respectively (Smith et al., 1983b) (Figure 18.19).

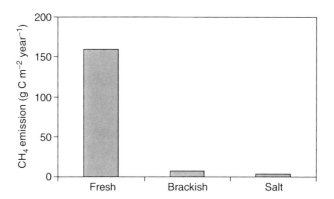

FIGURE 18.18 Methane emissions from fresh, brackish, and salt marshes of the Mississippi River deltaic plain. (Modified from DeLaune et al., 1983a.)

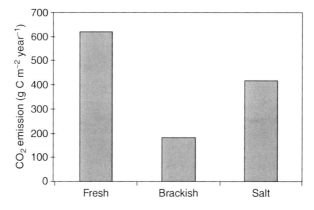

FIGURE 18.19 Carbon dioxide emissions from fresh, brackish, and salt marshes of the Mississippi River deltaic plain. (Modified from Smith et al., 1983b.)

18.6.3 Carbon Sinks

The role of these marshes as a major carbon sink has been determined from the carbon content of the sediment, vertical accretion rates, and the bulk density of the sediment (Smith et al., 1983b). Accretion rates were calculated from depth in sediment of the horizon for 1963, the year of peak ^{137}Cs fallout (DeLaune et al., 1978). Net carbon accumulation was essentially the same in all three marshes: 183, 296, and 224 g C m^{-2} year^{-1} from the saltwater, brackish, and freshwater marshes, respectively. A large percentage of fixed carbon, immobilized in accretionary processes, remained on the marshes (Table 18.4). Hatton et al. (1983) found similar carbon accumulation rates in these marshes.

18.6.4 Decomposition of Surface Peat

The oxidation and decomposition process in surface peats can annually remove several hundred grams of organic carbon per square meter through carbon dioxide and methane emission to the atmosphere (Smith et al., 1983b; DeLaune et al., 1983c). Organic matter (dry weight %) is greater in fresh and brackish than in salt marshes. Organic matter on a dry weight basis constitutes an increasing fraction of soil solids as the marine influence diminishes inland from the coast. Organic matter is of greatest structural significance in low density, fresh, and brackish marsh environments (Nyman et al., 1990). However, on a unit volume basis, the organic matter occupies the same volumes in fresh, brackish, and salt marshes (Hatton et al., 1983). Oxidation of the surface peats can contribute to negative accretion balances (marsh surface not keeping pace with level increases). Methane and carbon dioxide emissions from Louisiana fresh, brackish, and salt marshes are equivalent to 760, 250, and 420 g C m^{-2} year^{-1}, respectively (DeLaune et al., 1990b, Smith et al., 1983b). Converting to organic matter using the conversion factor 1.724, the equivalent of 1,301, 431, and 724 g organic matter m^{-2} year^{-1} is removed annually by oxidation or decomposition of the surface peats. The major carbon flux occurring in Mississippi River deltaic plain marshes is summarized in Table 18.5.

18.6.5 Carbon Losses Resulting from Wetland Deterioration

Much of the wetland loss occurring in coastal Louisiana is due to the deterioration of highly organic marsh soil. As discussed earlier, conversion of coastal marshes to inland open water is associated with plant stresses such as saltwater intrusion into nonsaline marshes and increased soil waterlogging as a result of subsidence. Marsh elevation decreases rapidly following plant mortality because of the structural collapse of the living root networks (DeLaune et al., 1994). The peat collapse and the associated erosion result in the conversion of marsh into open water. Conversions to open water system releases a considerable amount of carbon into the estuary where it is either decomposed or

TABLE 18.4
Accumulation Rates of Carbon and Net Accretion Rates

Marsh Type	Site	Accretion Rate (cm year^{-1})	Bulk Density (g cm^{-3})	Organic Carbon (mg g^{-1})	Carbon Accumulation (g C m^{-2} year^{-1})
Spartina alterniflora (salt)	Marsh surface	0.76 ± 0.12	0.28 ± 0.06	86 ± 16	183
Spartina patens (brackish)	Marsh surface	0.95 ± 0.05	0.13 ± 0.02	240 ± 32	296
Panicum hemitomon (fresh)	Marsh surface	0.85 ± 0.02	0.10 ± 0.04	263 ± 14	224

TABLE 18.5
Carbon Flux Occurring in Mississippi Deltaic Plain Marshes

Marsh	Fresh (g m^{-2})	Brackish (g m^{-2})	Salt (g m^{-2})
Net primary production	600[a]	550[a]	860[a]
CO_2 emission	618[b]	180[b]	418[b]
CH_4 production	160[c]	73[c]	4.3[c]
Sediment C accumulation	224[d]	296[d]	183[d]

[a] Estimation based on study sites where flux determinations were made (DeLaune et al., 1990a) (Table 18.1).
[b] Smith et al. (1983b): triplicate determinations, 12 times during 1980–1981.
[c] DeLaune et al. (1983b): triplicate determination in fresh and salt marshes; duplicate in brackish marsh; 13 sampling dates during 1980–1981.
[d] Hatton et al. (1983): ^{137}Cs dating on 10 cores from streamside on lateral transects.

TABLE 18.6
Amount of Organic Carbon Loss in a 100 cm Depth Section Eroded

Marsh Types	Bulk Density (g cm^{-3})	Organic Soil Carbon (mg g^{-1})	Carbon Loss to Erosion (g m^{-2})
Salt	0.28	86	24.1
Brackish	0.13	240	31.2
Fresh	0.10	263	23.6

accumulated in bottom sediment of estuary or exported from the system. With the loss of 25–35 square mile of wetland and the majority of the loss being highly organic marsh soil, a large amount of soil organic carbon is lost through marsh deterioration.

Assuming the marsh soil depth of 100 cm with the bulk density and organic carbon content as reported earlier for carbon sink, the following amount of organic carbon (Table 18.6) would be lost per m² (depth of 100 m) from fresh, brackish, and salt marsh. The data reflect the amount of organic carbon loss in a 100 cm depth section eroded. For marsh soil with deeper peat layers, there would be a larger amount of loss. If extrapolated to amount of loss per ha, the equivalent (for 100 cm depth) of 241, 312, and 24 metric tons of carbon would be lost, respectively, from the salt, brackish, and fresh marsh as a result of marsh deterioration. The actual amount for any particular location may vary depending on the difference in bulk density, soil depth, and organic carbon content. However, Table 18.6 demonstrates that a considerable amount of organic carbon can be lost through the marsh deterioration occurring in coastal Louisiana.

18.7 NITROGEN CYCLING

Nitrogen is the most important nutrient regulating Mississippi River deltaic plain wetland macrophyte production. The nitrogen cycle in coastal Louisiana wetlands is extremely complex. The nitrogen cycle relies heavily on both biological and physical processes. Microorganisms are responsible for the fixation of atmospheric nitrogen, which can be transformed into forms usable by other organisms and marsh plants. Microorganisms also denitrify, that is, convert nitrogen back into atmospheric nitrogen. Sedimentation and tidal exchanges also influence the biochemical nitrogen cycle in marshes.

18.7.1 Nitrogen Inputs

A large amount of nitrogen is required to produce the biomass of marsh vegetation. Nitrogen enters the marsh estuarine ecosystem by several sources:

1. Fixation
2. Sediment input
3. Wet deposition
4. Tidal exchanges
5. Agricultural and urban runoff

Molecular nitrogen is fixed by various microbial inhabitants in marshes and estuaries. Nitrogen fixation provides the second largest input of nitrogen to Louisiana salt marsh. Seasonally, the greatest measured fixation was in the streamside marsh soil where fixation rates equivalent to 15.4 g N m^{-2} $year^{-1}$ were measured (Casselman et al., 1981). Fixation was considerably less (4.5 g N m^{-2} $year^{-1}$) in the adjoining inland marsh. Reported fixation in Louisiana salt marsh closely paralleled the root distribution in the soil profile. A highly significant negative relationship between nitrogenase activity and extractable ammonium nitrogen in the soil profile was observed. For Louisiana salt marsh as a whole it is estimated that 10 g N m^{-2} $year^{-1}$ is fixed. There is a need for additional information on the amount of nitrogen fixed in other locations and habitats within the Louisiana coastal system.

Another mechanism by which nitrogen enters the Louisiana salt marsh is sediment import in tidal waters. Particulate nitrogenous compounds or other nitrogenous materials sorbed to the various suspended sediments may be carried into the estuary and may then be deposited on the marsh surface by tides. Using sediment traps, DeLaune et al. (1981a,b) found that an average of 17 g N m^{-2} $year^{-1}$ imported to the salt marsh surface in sediment. The sediment import figure represents an average amount for streamside and inland locations. Deposition of particulate-bound nitrogen is an important fertilization mechanism for the marsh grasses.

Nitrogen may also be brought into the estuary in solution by the incoming streams and imported into the marsh by the tides and by wet deposition. Nitrogen entering from adjacent uplands is an important source in the upper Barataria Basin (Day et al., 1977; Hopkinson and Day, 1979). Total nitrogen loading into Lac des Allemands is in excess of 30 g N m^{-2} $year^{-1}$. However, this nitrogen is adsorbed or removed before reaching the salt marsh in the lower estuary. There is no information on nitrogen's input by wet deposition.

18.7.2 Nitrogen Regeneration and Uptake

Mineralization of soil organic nitrogen to ammonium form is the primary source of inorganic nitrogen for *S. alterniflora*. Very little nitrate is present in flooded marsh soils because there is a lack of oxygen for appreciable nitrification and also any nitrate formed is rapidly denitrified (DeLaune et al., 1976, 1983c). Laboratory studies have shown that streamside soils are capable of mineralizing approximately 9.0 µg nitrogen per gram of soil per week (DeLaune and Patrick, 1979).

Even though the mineralization and release of nitrogen in the sediment fulfill the plant requirement to a great extent, Louisiana salt marshes are still nitrogen limited. Supplemental labeled inorganic nitrogen applied in the spring increased the aboveground biomass *S. alterniflora* at the streamside marsh (DeLaune et al., 1983c; DeLaune and Patrick, 1979). The added nitrogen caused a yield increase equivalent to 250 g m^{-2}. The use of labeled nitrogen made it possible to distinguish between plant nitrogen derived from the sediment and the added fertilizer nitrogen. The amount of plant nitrogen derived from the sediment was about 59% during June and July and increased to about 69% by September. Nitrogen balance studies have shown that N-loss was less in successive years when nitrogen entered the organic pool (DeLaune et al., 1983c). The relatively large recovery from a single addition of large quantities of fertilizer ammonium nitrogen indicates that *S. alterniflora* has a high capacity to assimilate the soil inorganic nitrogen forms.

18.7.3 NITROGEN LOSSES

The main nitrogen losses from marshes are caused by denitrification and detrital export and sedimentation sinks. Denitrification has been reported to be the major cause of nitrogen loss from marshes. Although the potential for denitrification is very high in the reduced marsh soil, the level of *in situ* denitrification depends on the availability of nitrate. Nitrate must be either formed through nitrification of ammonium or supplied from an extrinsic source such as tidal input, wet deposition, or groundwater flow.

Studies indicate that under natural conditions the gaseous N losses from salt and brackish marshes of the Louisiana Gulf Coast are minimal (Smith and DeLaune, 1983; Smith et al., 1983a) (Table 18.7). A 2-year study of N_2O emission from Barataria Basin salt marshes indicated the annual emission of 31 N m^{-2} year^{-1} (Smith et al., 1983a).

These reported N_2O emissions provide the evidence that nitrification–denitrification occurs in coastal wetlands that are generally considered anaerobic. The low N_2O emission rates also indicate that a relatively small amount of N_2 is lost as a result of nitrification–denitrification reactions. High denitrification rates would be expected in marshes that receive large inputs of nitrate from agricultural runoff or other sources.

Nitrogen mass balance studies have also demonstrated that added nitrogen is not lost to any extent during the first growing season (DeLaune et al., 1983c). Loss of ^{15}N-labeled nitrogen in a *Spartina alterniflora* salt marsh was measured over three growing seasons. Labeled ammonium nitrogen equivalent to 100 μg ^{15}N g^{-1} of dry soil was added in four installments over an 8-week period. Recovery of the added nitrogen ranged from 93%, 5 months after the addition of the NH_4^+-N, to 52% at the end of the third growing season, which represented a nitrogen loss equivalent to 3.4 g N m^{-2}. A significant portion of this loss attributed to transport of nitrogen-enriched plant material from the marsh. The availability of the labeled NH_4^+-N incorporated into the organic fraction was estimated by calculating the rate of mineralization. The time required for mineralization of 15% of the tagged organic N increases progressively with successive cuttings of *S. alterniflora* and ranged from 152 to 299 days. The N remaining in the system was in the organic N pool and only slowly mineralized and released.

Nitrogen losses through sedimentation below the root zone are also appreciable in these rapidly accreting Louisiana salt marshes. Sedimentation rates from ^{137}Cs dating show that these marshes are vertically accreting at a rate in the order of 1 cm year^{-1}. The range varies depending on the marsh site, and measured rates were 0.75–1.35 cm year^{-1} (DeLaune et al., 1978). Calculations using

TABLE 18.7
Estimates of Nitrogen Loss as Nitrous Oxide from the Three Predominant Marsh Environments Found within Barataria Basin

Location	N_2O Evolution (Annual mg N m^{-2} year^{-1})
Salt marsh	31
Brackish marsh	48
Fresh marsh	55

Note: Measurement period: 730 days.

Source: Smith et al. (1983a).

TABLE 18.8
Sedimentation as a Nitrogen Sink

	N Content of Marsh Sediment (mg g^{-1})	Sedimentation Rate ^{137}Cs (cm year^{-1})	N Accumulation Rate (g m^{-2} year^{-1})
Streamside	6.2	1.35	21.0
Inland	8.7	0.75	13.4

Source: DeLaune et al. (1989).

TABLE 18.9
Local Nitrogen Budget (Average Streamside and Inland Site for Louisiana *Spartina alterniflora* Salt Marsh) Representing Gains and Losses from Plant and Plant Root Zone (Not Including Marsh Deterioration)

	Gains (g N m^{-2} year^{-1})		Losses (g N m^{-2} year^{-1})
Nitrogen fixation	10.0	Sedimentation	17.0
Import	17.0	Denitrification	4.0
Wet deposition	<1.0	Import	5.0
		Volatilization	<1.0
Total	~27.0		~26.0

Source: DeLaune et al. (1989).

accretion rates, bulk density, and nitrogen content of the marsh soil show that the salt marsh is undoubtedly a large sink for nitrogen. For a streamside and inland marsh location sedimentation would remove 21.0 and 13.4 N m^{-2} year^{-1} (Table 18.8). Using an average for streamside and inland marshes we estimated that nitrogen is accumulating at a rate of 17 g N m^{-2} year^{-1}. These sedimentation losses of nitrogen represent an average for the past several decades. These losses are considerably greater than that for Atlantic Coast marshes, which are accreting at a slower rate. For instance, Valiela and Teal (1974) estimated sedimentation losses to be in the order of 4 g m^{-2} year^{-1}.

The relationship among land sinking, vertical marsh accretion, and marsh deterioration is important in understanding regional influences on salt marsh nitrogen budgets. It has been estimated that the salt marshes are deteriorating at a rate of 2% per year (Leibowitz and Hill, 1987). An appreciable amount of nitrogen is stored in these highly organic soils. As the marsh breaks up and erodes, nitrogen is removed. Calculation, using average soil depth (50 cm), bulk density (0.25 g cm^{-3}), and total nitrogen content of soil (5 mg N g), shows that on a regional basis the equivalent of 12.5 g N m^{-2} year^{-1} is lost from the salt marshes due to deterioration of Barataria Basin salt marshes.

18.7.4 Nitrogen Budget

For stable soil–plant system, nitrogen gains and losses are nearly balanced in the Louisiana Gulf Coast salt marshes (Table 18.9). Nitrogen-rich sediment being deposited on the marsh is the primary source of nitrogen, supplying 17 g N m^{-2} year^{-1}. Nitrogen fixation provides approximately 10 g N m^{-2} year^{-1}, whereas rainfall was estimated to be less than 1 g N m^{-2} year^{-1}. For a stable marsh the primary losses are net sedimentation 17 g N m^{-2} year^{-1}, detrital export 5 g N m^{-2} year^{-1}, and denitrification 4 g N m^{-2} year^{-1}. The loss due to ammonia volatilization is insignificant. On a site-specific basis these losses balance the nitrogen inputs when developing a budget for a soil–plant system with limited rooting depth. However, per unit surface (e.g., m^2), marsh is a major sink due to the rapid accretion that removes 17 g N m^{-2} year^{-1} below the plant root zone. In contrast, when nitrogen loss through marsh deterioration on a regional basis is considered the marshes of the rapidly subsiding Mississippi River deltaic plain may not serve as a significant sink with approximately 5 g N m^{-2} year^{-1} accumulation.

18.7.5 Processing Capacity of Added Nitrogen Entering Louisiana Wetland

Agricultural runoff and municipal discharges are major contributors to nutrient enrichment in the Louisiana coastal wetlands. One primary agricultural nutrient source is the sugarcane grown on the natural levees of the Mississippi River and its distributaries. The elevated natural levees border coastal drainage basins with a slope extending into cypress and tupelo gum swamps. The swamps grade into the freshwater marshes and lakes. The northern portion of these coastal drainage basins, which receive agricultural runoff and nutrients from municipal discharges, causes water quality problems in streams and other water bodies (Stow et al., 1985).

The capacity of the swamp forest to remove large quantities of added nitrogen via nitrification–denitrification processes has been demonstrated (Lindau et al., 1988). Added ^{15}N-labeled inorganic nitrogen was used to determine the significance of nitrification–denitrification in flooded swamp soil in removing nitrogen. Nitrogen-15 labeled $(NH_2)_2SO_4$ and KNO_3 were added to replicate plots in a swamp forest receiving agricultural runoff. Nitrous oxide and N_2 fluxes (Figures 18.20 and 18.21) were measured and maximum fluxes were estimated at 24 and 110 g N m^{-2} year^{-1}, respectively.

Denitrification was shown to be a major removal mechanism of applied nitrogen in a Louisiana Gulf Coast swamp forest. Applied nitrate underwent denitrification and produced nitrous oxide and dinitrogen that escaped into the atmosphere. Ammonium was nitrified and nitrate formed was also denitrified to nitrous oxide and nitrogen gas. Initial detection of denitrification products after ammonium addition lagged several days behind the applied nitrate. In addition, nitrous oxide emissions were observed before dinitrogen detection. Ratios of evolved N_2 and N_2O varied over the length of the study. The ratios could be used to estimate future swamp forest denitrification fluxes if only one gaseous product is measured.

Results demonstrate that Louisiana swamp forests play an important role in controlling eutrophication and improving the deteriorating water quality caused by nutrient inputs from agricultural and municipal discharges. Nitrification–denitrification reactions are important as a removal mechanism for nitrogen entering the coastal Louisiana wetlands.

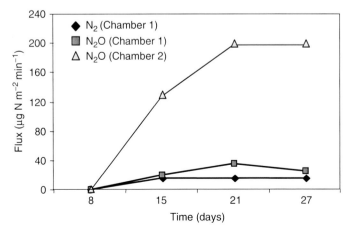

FIGURE 18.20 Nitrous oxide and dinitrogen fluxes after addition of nitrate (KNO_3 10 g N m^{-2}). (Modified from Lindau et al., 1988.)

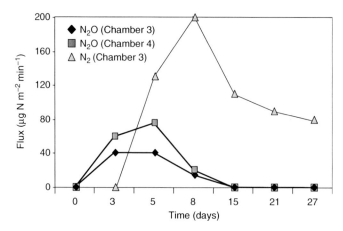

FIGURE 18.21 Nitrous oxide and dinitrogen fluxes after addition of ammonium (($NH_4)_2SO_4$ 10 g N m^{-2}). (Modified from Lindau et al., 1988.)

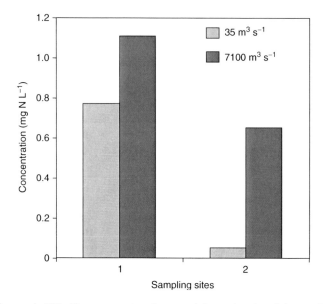

FIGURE 18.22 Changes in NO_3-N concentrations between inlet and outlet of the ponded freshwater marsh receiving diverted Mississippi River water during periods of high pulse and medium pulse events. (Modified from DeLaune et al., 2005.)

18.7.6 Capacity of Freshwater Marsh to Process Nitrate in Diverted Mississippi River Water

Mississippi River water containing nitrate is currently being diverted into Louisiana coastal wetlands for slowing or reversing marsh deterioration caused by the rapid subsidence and accompanying saltwater intrusion. It been shown that in a 3,700 ha ponded freshwater wetland in Northern Barataria Basin through which the diverted Mississippi River water enters Louisiana Barataria Basin estuary can effect the process and remove nitrate in the diverted Mississippi River water (DeLaune et al., 2005). Nitrate removal rates using mass balance approach (measuring changes in nitrate content between the inlet and the outlets) showed that the freshwater wetland removed practically all the nitrate in the diverted river water ($\approx$35 m^3 s^{-1}) discharge event (Figure 18.22).

Denitrification was shown to be a major process in removing nitrate in diverted river water (Yu et al., 2006). The capacity of freshwater marsh to remove all nitrate in the diverted river water was strongly influenced by the flow or discharge rate (DeLaune et al., 2005). The ponded wetland did not completely remove the nitrate in river water at high flow or discharge (>100 m^3 s^{-1}). On the basis of mass balance calculation using concentration difference between inlet and outlet, the system removed the nitrate-N at the rate of approximately 60 mg N m^{-2} day^{-1} or approximately 2,000 kg nitrate-N day^{-1} for the 3,700 ha marsh. The study clearly showed that freshwater marsh can effectively remove nitrate in river water. However, the discharge or pulsing rate is important in determining removal rate in freshwater marsh or amount of nitrate passing through the system entering lakes and other water bodies in Barataria Basin.

18.8 SULFUR CYCLING

18.8.1 Forms of Sulfur in Louisiana Marsh Soil

The amounts and profile distribution of various sulfur forms in Louisiana coastal marshes are important in understanding sulfur cycling as related to the origin and type of tidal wetland marshes. Sulfur forms and distribution were determined in *P. hemitomon* freshwater marsh, a *S. patens* brackish marsh, and a *S. alterniflora* salt marsh along a salinity gradient in Barataria Basin, Louisiana. Soil samples were fractionated into acid volatile sulfur (AVS), elemental sulfur, HCl-soluble sulfur, pyrite sulfur, ester sulfate sulfur, carbon-bonded sulfur, and total sulfur (see Chapter 11 for details).

In Louisiana, extending along a salinity gradient from the Gulf of Mexico, organic S in the forms of ester sulfate S and C-bonded S is the predominant form of all marsh types representing 76–87% of the total S present in fresh, brackish, and salt marshes (Krairapanond et al., 1992). The contribution of inorganic S to the total S pool was greater in salt marsh (24%) than in brackish (13%) and freshwater (22%) marshes. HCl-soluble S was a major component of the inorganic S fraction. Pyrite S comprised <3% of the total S present in individual marsh site (Table 18.10).

TABLE 18.10
(a) Distribution of Sulfur Fractions Average over Depth (0–50 cm) as Percent of Total Sulfur, Pore Water Sulfur and (b) Interstitial Iron Concentrations (Mean ± S, N = 15) in Soils Collected from the Three Marsh Types of Louisiana (% of Total Sulfur)

(a)

Sampling Site	AVS-S	HCL-Soluble S	Elemental S (% Total S)	Pyrite S	Ester Sulfate S
Freshwater	0.19b ± 0.07	19.0b ± 5.2	1.59b ± 0.89	1.76b ± 0.24	20.7a ± 3.2
Brackish	0.19b ± 0.03	9.9a ± 2.7	0.54a ± 0.08	2.01c ± 0.32	23.2b ± 5.4
Salt	0.15a ± 0.07	20.7b ± 5.8	1.73b ± 0.88	1.56a ± 0.20	21.1b ± 4.3

(b)

Sampling Site	C-bonded S	Total S	Pore Water SO$_4$-S (g S kg^{-1})	Interstitial Fe (mg Fe kg^{-1})
Freshwater	57.0b ± 5.9	5.7a ± 0.9	0.55a ± 0.14	2.09c ± 0.63
Brackish	64.1c ± 6.6	15.6b ± 1.1	1.15a ± 0.43	0.74a ± 0.14
Salt	54.7a ± 9.9	15.7b ± 4.9	2.66b ± 0.44	1.13b ± 0.60

Note: Any two values having a common letter are not significantly different at the 5% level of significance on the basis of Duncan's Multiple Range Test (DMRT).

Source: Krairapanond et al. (1992).

TABLE 18.11
Rates of S Accumulation in the Barataria Basin Tidal Marsh Soils (mean + SE, $n = 15$)

Sampling Site	Sedimentation Rate (mm year^{-1})	AVS-S	HCL-Soluble S (mg S m^2 year^{-1})	Elemental S	Pyrite S	Ester Sulfate S	C-bonded S (g S m^2 year^{-1})	Total S
Freshwater	8.5	11.1a ± 2.6	1,077a ± 228	86a ± 31	104a ± 41	1.27a ± 0.46	3.5a ± 1.3	6.1a ± 1.9
Brackish	9.5	45.6b ± 6.7	2,309b ± 654	126b ± 19	479 ± 96	5.37b ± 0.83	15.3b ± 3.4	23.7b ± 3.4
Salt	10.5	45.6b ± 29.4	9,030c ± 1043	726 ± 236	713c ± 185	9.57c ± 2.44	25.9c ± 8.8	46.0c ± 10.3

Note: Any two values having a common letter are not significantly different at the 5% level of significance on the basis of Duncan's Multiple Range Test (DMRT).

The AVS-S and fractions together comprised <2% of the sulfur fractions. Sinks of reduced sulfur compounds followed the inorganic and organic sedimentation patterns found along a hydraulic gradient extending from freshwater to salt marshes. The accumulation of reduced S forms followed seaward increase (Table 18.11). The accumulation of reduced S species in freshwater, brackish, and salt marshes was a function of marsh sedimentation patterns. In Louisiana's rapidly accreting coastal marsh a significant portion of marsh energy flow enters this sedimentation pool. Since during dissimilatory sulfate reduction only a fraction of the energy from marsh soil organic matter is lost, most of the free energy is conserved in reduced S compounds.

18.8.2 Sulfate Reduction Rates in Louisiana Marsh Soils

Potential sulfate reduction rates for three Louisiana marsh soils of varying salinities (salt, brackish, and freshwater) have been determined (DeLaune et al., 2002a). The three Louisiana marshes represent different physiochemical environments. The fresh and brackish marsh soils are composed predominantly of organic matter, whereas the salt marsh soils are higher in mineral matter. Sulfate content is higher in the salt marsh.

The brackish and fresh marshes contain similar sulfate contents, although means were higher in the brackish marsh. Average potential sulfate reduction rates over three sampling periods (August, December, and April) were 45.9, 40, and 33.4 mol S m^{-2} year^{-1}, respectively, for fresh, brackish, and salt marsh (Table 18.12). Turnover times are consistently longer in the salt marsh (approximately 40 days, Figure 18.23) than in either the fresh or the brackish marshes (3–8 days).

The lowest potential sulfate reduction rates for all marshes occurred during the winter. The sulfate reduction rates in Louisiana marshes were controlled by an interaction between the availability of sulfate, carbon as an energy source, and microsites for sulfate-reducing microorganisms. Even though salt marshes had the greatest sulfate content, potential sulfate reduction rates were not always high.

18.8.3 Flux of Reduced Sulfur Gases

Seasonal field sulfur emission measurements (DeLaune et al., 2002b) were determined in a *Spartina alterniflora* salt marsh (10–12 ppt salinity), a *Spartina patens* brackish marsh (5–8 ppt), and a *Sagittaria lancifolia* freshwater marsh (0 ppt salinity), along a salinity gradient extending inland from the coast in the Mississippi River deltaic plain region of the coastal Louisiana. Results

TABLE 18.12
Comparison of Potential Sulfate Reduction Rates for Louisiana Marsh Soil to Other Wetland Sites

Location	Sulfate Reduction (mol S m^{-2})	Source
Freshwater		
Buckles Bog, Maryland	10.5	Weider et al. (1990)
Big Run Bog, West Virginia	17.2	Weider et al. (1990)
Cedar Swamp, New Jersey	2.7	Spratt and Morgan (1990)
Louisiana Freshwater Marsh	45.9	DeLaune et al. (2002)
Saline	75	Howarth and Teal (1979)
Great Salt Marsh, Massachusetts	40	Howes et al. (1984)
Belle Baruch, Goat Island, North Carolina		King (1988)
Short form	13.3	
Long form	5.9	
Sapelo Island, Georgia	40	Howarth and Giblin (1983)
Chapmans Marsh, New Hampshire		Hines et al. (1989)
Tall form	19	
S. patens	8	
Louisiana Brackish Marsh	40	DeLaune et al. (2002)
Louisiana Salt Marsh	33.4	DeLaune et al. (2002)

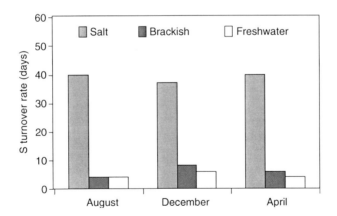

FIGURE 18.23 Sulfate turnover rates. (Redrawn from DeLaune et al., 2002b.)

of these studies show that salinity and flooding are important variables governing sulfur flux from coastal wetlands (see Chapter 11).

Seasonal and diurnal reduced sulfur gas emission measured along a salinity gradient in the Louisiana Gulf Coast salt, brackish, and freshwater marshes was strongly associated with habitat and salinity gradient. The dominant emission component was dimethyl sulfide (average 57.3 µg S m^{-2} h^{-1}) in salt marsh with considerable seasonal (max: 144.03 µg S m^{-2} h^{-1}, min: 1.47 µg S m^{-2} h^{-1}) and diurnal (max: 83.58 µg S m^{-2} h^{-1}; min: 69.6 µg S m^{-2} h^{-1}) changes in flux. Hydrogen sulfide was the dominant (average: 21.2 µg S m^{-2} h^{-1}; max: 79.2 µg S m^{-2} h^{-1}; min: 5.29 µg S m^{-2} h^{-1}) form in brackish marsh, and carbonyl sulfide (average: 1.09 µg S m^{-2} h^{-1}; max: 3.42 µg S m^{-2} h^{-1}; min: 0.32 µg S m^{-2} h^{-1}) was the dominant form in freshwater marsh. A greater amount of H_2S was evolved from brackish (21.22 µg S m^{-2} h^{-1}) as compared to the salt marsh (2.46 µg S m^{-2} h^{-1}) and freshwater marsh (0.30 µg S m^{-2} h^{-1}). Emission of total reduced sulfur gases decreases with decrease in salinity and distance inland from the coast (DeLaune et al., 2002b).

18.8.3.1 Salt Marsh

Dimethyl sulfide (DMS) was the major gas emitted from the *S. alterniflora* salt marsh (Table 18.13a). Average emission was 57.3 µg S m^{-2-2} h^{-1} over the 2 year sampling period. The range in emission of DMS was from 1.5 to 144 µg S m^{-2} h^{-1}. The lowest emission rate was measured in February and the highest in August. Carbonyl sulfide was the second most prevalent reduced sulfur gas evolved with an average emission of 6.63 µg S m^{-2} h^{-1} (range 0.83–10.98 µg S m^{-2} h^{-1}). Methanethiol was emitted to the atmosphere at an average of 4.48 µg S m^{-2} h^{-1}. Hydrogen sulfide and carbon disulfide were evolved in smaller concentrations, 2.40 and 2.44 µg S m^{-2} h^{-1}, respectively (DeLaune et al., 2002b).

18.8.3.2 Brackish Marsh

Hydrogen sulfide was the predominant sulfur gas emitted from the brackish marsh; an average emission was 21.22 µg S m^{-2} h^{-1} (Table 18.13b). H_2S emission ranged from 5.29 to 79.20 µg S m^{-2} h^{-1}.

TABLE 18.13a
Spartina elterniflora Louisiana Salt Marsh

	Average Emission (µg S m^{-2} h^{-1})
Hydrogen sulfide	2.46
Dimethyl sulfide	57.30
Methanethiol	4.48
Carbonyl sulfide	6.63
Carbon disulfide	2.44
Total reduced sulfur compound	73.30

TABLE 18.13b
Spartina patens Louisiana Brackish Marsh

	Average Emission (µg S m^{-2} h^{-1})
Hydrogen sulfide	21.22
Dimethyl sulfide	0.84
Methanethiol	2.26
Carbonyl sulfide	6.05
Carbon disulfide	1.74
Total reduced sulfur compound	32.11

TABLE 18.13c
Sagittaria lancifolia Louisiana Freshwater Marsh

	Average Emission ($\mu g\, S\, m^{-2}\, h^{-1}$)
Hydrogen sulfide	0.30
Dimethyl sulfide	0.27
Methanethiol	0.46
Carbonyl sulfide	1.09
Carbon disulfide	0.64
Total reduced sulfur compound	2.76

18.8.3.3 Freshwater Marsh

Measured emission of total reduced sulfur was considerably less in the freshwater marsh as compared to emissions from the salt and brackish marsh sites (Table 18.13c). Carbonyl sulfide emitted at a rate of 1.09 $\mu g\, S\, m^{-2}\, h^{-1}$ was the dominant form evolved from the freshwater marsh site. Emissions of H_2S, DMS, MeSH, and CS_2 to the atmosphere were less and averaged 0.30, 0.27, 0.46, and 0.64 $\mu g\, S\, m^{-2}\, h^{-1}$, respectively (DeLaune et al., 2002b).

Louisiana coastal wetland reduced sulfur gas emission was strongly associated with changes in marsh habitat along a salinity gradient extending inland from the coast. Dimethyl sulfide (57.3 $\mu g\, S\, m^{-2}\, h^{-1}$) was the dominant reduced sulfur gas evolved from the salt marsh near the coast. Hydrogen sulfide (21.2 $\mu g\, S\, m^{-2}\, h^{-1}$) was the dominant reduced sulfur gas emitted from the brackish marsh. Carbonyl sulfide (1.09 $\mu g\, S\, m^{-2}\, h^{-1}$) was the dominant gas evolved from the freshwater marsh studied. Emission of total reduced sulfur gases averaged 73.3 $\mu g\, S\, m^{-2}\, h^{-1}$ in salt marsh, 32.1 $\mu g\, S\, m^{-2}\, h^{-1}$ in brackish marsh, and less than 3 $\mu g\, S\, m^{-2}\, h^{-1}$ in freshwater marsh.

18.9 CASE STUDIES OF FACTORS GOVERNING THE FATE OF TOXIC ORGANIC COMPOUNDS AND POLLUTANTS IN THE LOUISIANA COASTAL WETLAND

In Louisiana, numerous potentially toxic compounds are entering the inshore and nearshore coastal environment. The Types of anthropogenic compounds entering this coastal region are diverse and range from the common herbicides and pesticides to the toxic hydrocarbons associated with petroleum exploration and refining and the chemical manufacturing industries. The duration of toxic organic compounds remaining in aquatic environments is largely dependent on the rate of their degradation.

18.9.1 TOXIC ORGANIC COMPOUNDS

The effectiveness of microbial degradation of chemicals in the environment is dictated primarily by environmental physicochemical conditions. Like other organisms, toxic organic-degrading microbes have certain requirements for growth that vary considerably from species to species.

Louisiana's coastal environment, particularly the sediment–water systems, provides a wide range of microhabitats for microbial activity. Sediments range from saline to fresh, well oxidized to strongly reduced, and nutrient poor to nutrient rich. Oxygen availability, pH, temperature, salinity, and nutrient status are a few parameters, which can vary considerably within a coastal sediment system.

One of the most significant environmental factors affecting microbial degradation is the oxygen status of the sediment, often indirectly measured as the oxidation–reduction potential. Bottom sediments are viewed as the ultimate "sink" for many toxic organics. The oxygen status

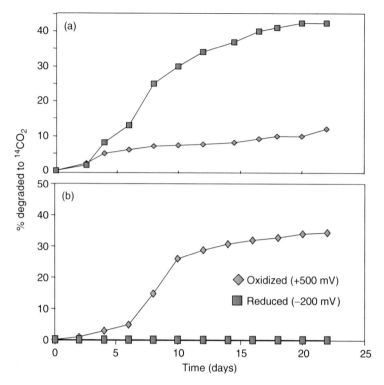

FIGURE 18.24 Degraded rates of ^{14}C-labeled octadecane (a) and naphthalene (b) in oxidized and reduced sediment from Barataria Bay, Louisiana. (Drawn using data from Hambrick et al., 1980.)

of coastal region sediments is variable and supports a wide range of aerobic, facultative aerobic, and anaerobic microbial communities. Bottom sediment redox potentials range from +700 mV in highly aerobic sediment to −200 mV in the bulk sediment profile. Sediments are usually characterized by a distinct thin surface-oxidized layer.

Sediment oxidation–reduction potential and pH were important factors governing the activity of petroleum hydrocarbon-degrading microorganisms and subsequent mineralization rates (Figure 18.24). The highest mineralization occurred at pH 8.0 and the lowest at pH 5.0. At all pH levels, mineralization decreased with decreasing oxidation–reduction potential (increasing sediment anaerobiosis). Aerobic microorganisms in the oxidized sediment were more capable of degrading hydrocarbons than anaerobic microorganisms in reduced sediment of the same pH. Likewise degradation of benzo(a)pyrene increased with increasing redox potential (DeLaune et al., 1981a, 1981b) (Figure 18.25). Up to a 100-fold difference in benzo(a)pyrene degradation rates could be attributed to changes in sediment pH and redox potential.

Herbicides are mineralized or degraded at slower rates under anaerobic conditions. Pronounced differences in degradation rates of added, labeled 2,4-D as affected by sediment redox potential were observed (Figure 18.26). Degradation of 2,4-D was approximately six times faster under aerobic conditions (+500 mV) as compared to anaerobic conditions (−200 mV) (DeLaune et al., 1997). Degradation of pentachlorophenol (PCP) (a wood-preserving compound) entering Louisiana's coastal environment following a major spill showed that degradation was strongly influenced by sediment pH and oxidation–reduction potential conditions. Mineralization rates decreased with decreasing sediment oxidation–reduction potential.

Sediment pH and redox potential also determined the rate of polychlorobiphenyl (PCB) mineralization (Pardue et al., 1988). In contrast to PCP, which degrades faster under strongly aerobic

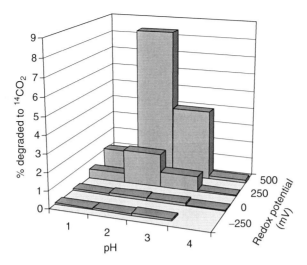

FIGURE 18.25 Effect of pH and redox potential on degradation of ^{14}C-labeled benzo(a)pyrene in Airplane lake, Louisiana sediment after 37 days incubation period. (Modified from DeLaune et al., 1981.)

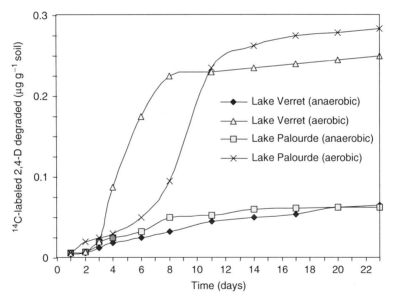

FIGURE 18.26 Degradation of ^{14}C-labeled 2,4-D under aerobic and anaerobic conditions in bottom sediment from Lake Verret and Lake Palourde, Louisiana. (Adapted from DeLaune and Salinas, 1985.)

conditions, mineralization rates of PCB were higher under moderately aerobic conditions (+250 mV) than under strongly aerobic conditions (+500 mV) or anaerobic conditions (0 and –200 mV). PCB mineralization rates in moderately aerobic sediment were 30–40 fold higher than those in anaerobic sediment.

With exceptions of chlorinated organics such as DDT and PCBs, most toxic organics are degraded more rapidly under highly aerobic conditions. Therefore, the fate of any toxic organic entering Louisiana's estuarine system is influenced by the thickness of the oxidized surface layer of the marsh soils and sediments. Hydrocarbons introduced into a wetland ecosystem will eventually enter the sediment column through the surface-oxidized layer. This thin oxidized surface layer may be very important in microbial degradation.

There are conflicting reports on the ecological effect of toxic organics in Louisiana coastal areas. In a recent study in which petroleum was applied in doses of 0, 1, 2, 4, and 8 L m^{-2} to replicated plots in a *S. alterniflora* salt marsh, it was found that oiling the marsh caused no reduction in macrophyte production as compared to the nonoiled plots (DeLaune et al., 1984). Likewise, there was no oil-induced mortality for the marsh microfauna or meiofauna. This is an example of contaminant level, which was considered to be sufficiently high to adversely affect microbial activity and subsequently the physicochemical sediment environment.

The biotransformation of atrazine as affected by soil redox conditions was investigated in soil collected from a *T. distichum/N. aquatica* (bald cypress/water tupelo) swamp-forest receiving runoff from sugarcane growing on adjacent natural levees of the Mississippi River deltaic plain, Louisiana (DeLaune et al., 1997). The soil was incubated under controlled redox conditions over a range representing both reducing and oxidizing conditions (−164, +169, +392, and +584 mV). The atrazine biotransformation rate was extremely rapid in oxidized soil (+392 and +584 mV). The concentration of atrazine dropped from approximately 70 µg g^{-1} soil to nondetectable levels after only 2 weeks of incubation under both oxidized redox treatments. Biotransformation of atrazine was considerably slower in soils maintained under reducing or anaerobic redox conditions (−164 and +169 mV). From an initial atrazine concentration of 70 mg g^{-1} soil, 9 mg g^{-1} and 3 mg g^{-1} atrazine, respectively, remained after 99 d for soil incubated at −164 mV and +169 mV. This study showed that there is a very clear demarcation in atrazine biotransformation between soil redox levels representing aerobic and anaerobic conditions.

18.9.2 Mercury

The microbial-mediated processes of methylation and demethylation of mercury are strongly influenced by redox potential, pH, and microbial activity of the sediment water system (DeLaune et al., 2004). The effect of sediment redox condition on methyl Hg formation in surface sediment of selected Louisiana Lakes has recently been examined. Methylation of added Hg in sediment was higher under reduced conditions as compared to oxidized conditions (Figure 18.27). Mercury methylation in sediment has been attributed to anaerobic sulfate-reducing bacteria such as

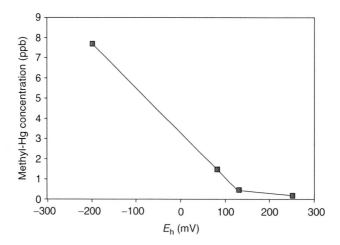

FIGURE 18.27 Effect of sediment redox potential (mV) on methyl-Hg content of sediment in a freshwater lake. (Modified from DeLaune et al., 2004.)

Desulfovibrio spp. High sulfate levels inhibit methylation while methylation is increased when sulfate levels are low and concentrations of organic fermentation products are high. Sulfide generated by sulfate-reducing bacteria inhibits mercury methylation by rendering mercury less available for methylation.

In a controlled redox study in which redox potential levels of sediment from a Louisiana lake (False River) were increased, demethylation of methyl-Hg was found to be strongly influenced by sediment redox potentials (DeLaune et al., 2003). As redox potential increased, methyl-Hg content of sediment decreased. Increasing redox potential from approximately –200 mV (strong reducing) to +245 mV (moderate reducing) reduced methyl-Hg content by 95% in False River sediment. Methyl-Hg content during the period of incubation was reduced from 7.8 to less than 0.3 ppb.

18.10 SUMMARY

- The Louisiana Mississippi River deltaic plain is undergoing rapid subsidence with parallel increases in water level and saltwater intrusion. As a result there is rapid wetland loss.
- The coastal wetlands are currently experiencing increases in water level projected in the near future for other coastal areas as a result of global warming and associated increase in global sea level.
- As shown in this chapter, increases in water level can impact many wetland processes. These include biogeochemical cycles of carbon, nitrogen, and sulfur.
- Increases in soil anaerobiosis and salinity will affect wetland plant growth functions resulting in shifts in plant distribution pattern including marsh loss. This will result in the decrease in carbon storage.
- Changes in redox/pH and salinity associated with increased coastal submergence and accompanying saltwater intrusion will also impact environmental issues such as degradation of petroleum hydrocarbon and heavy metal transformation.

STUDY QUESTIONS

1. Explain how wetlands in the Mississippi River deltaic plain were formed.
2. How did leveeing of the Mississippi River influence coastal marsh stability?
3. What are the main factors affecting marsh loss in coastal Louisiana?
4. What influence does sediment addition have on marsh stability?
5. Explain and discuss the benefit of diverting Mississippi River water into Louisiana Coastal wetland.
6. How will coastal submergence impact coastal marsh vegetation?
7. Compare carbon dioxide and methane flux in fresh, brackish, and salt marshes.
8. What are the major carbon flux processes? Compare rates.
9. List the major nitrogen inputs into Louisiana coastal wetlands.
10. List the factors influencing N_2O emissions from coastal marshes.
11. What are the dominant sulfur forms found in Louisiana marsh soils?
12. List the dominant form of reduced sulfur gas emissions from fresh, brackish, and salt marshes.
13. What influence does soil–sediment redox have on mercury methylation and toxic organic degradation?
14. Why does Louisiana Gulf Coast serve as a case example of what may occur globally in other coastal regions as a result of global increase in sea level?
15. List the main processes governing coastal marsh stability.
16. How is the carbon cycle impacted by a subsiding coastal environment?

17. Discuss the role of sediment and organic matter in maintaining the marsh surface water level relationships.
18. How will coastal submergence impact coastal vegetation?
19. Will increase in salinity impact carbon assimilation? Explain.
20. Explain how salt water intrusion can reduce methane emissions.
21. What is the primary source of inorganic nitrogen for plant growth in Louisiana coastal marsh?
22. Explain the importance of sediment as a source of nitrogen to Louisiana coastal marsh.
23. Does Louisiana wetlands have the ability to process and remove a large quantity of nitrogen? Explain the process involved.
24. Briefly describe the rates of sulfate reduction and the form of reduced sulfur gas emitted from Louisiana fresh, brackish, and salt marshes.
25. What are the primary factors governing the degradation of toxic organics in Louisiana coastal sediment?
26. How can sulfate influence mercury methylation in sediments? In which environment would mercury methylation be restricted?

FURTHER READINGS

Coast 2050: Towards a Sustainable Coastal Louisiana. http://www.coast2050.gov/2050reports.htm

Day, J. W. and P. W. Templet, 1989. Consequence of sea level rise: implication for Mississippi delta. *Coast Manage.* 17:241–257.

Day, J. W., D. Pont, P. E. Henzel, and C. Ibanez, 1995. The Impacts of sea level rise on the Deltas in the Gulf of Mexico and the Mediterranean: the importance of pulsing events to sustainability. *Estuaries* 18: 637–647.

Reyes, E., M. L. White, J. F. Martin, G. P. Kemp, J. W. Day, and V. Aravamuthan. 2000. Landscape modeling of coastal habitat change in the Mississippi Delta. *Ecology* 81:2331–2349.

Roberts, H. H. 1997. Dynamic changes of the Holocene Mississippi River delta plain: the delta cycle. *J. Coast. Res.* 13:605–627.

The coastal zone: Papers in honor of H. Jesse Walker, edited by Donald W. Davis and Miles Richardson, 2004. Geoscience Publications, Department of Geography and Anthropology Louisiana State University, Baton Rouge, LA, 70803-4105.

Wells, J. T. 1996. Subsidence, sea-level rise, and wetland loss in the lower Mississippi River delta. In J. D. Milliman and B. U. Haq (eds.) *Sea-Level Rise and Coastal Subsidence*. Kluwer Academic Publishers, Dordrecht, The Netherlands. pp. 281–311.

19 Advances in Biogeochemistry

19.1 INTRODUCTION

By now, the reader should realize that wetlands are critical components of landscape. In previous chapters (3–18), we presented an overview of basic elemental cycles and their role in regulating larger wetland functions including water quality, production of greenhouse gases, carbon sequestration, and long-term burial of various elements into soil organic and mineral matter. Despite this discussion and better understanding of biogeochemical cycles, the roles of many of these processes in global elemental cycles are not well understood. In this chapter, we briefly review some of the recent advances in wetland biogeochemistry and attempt to direct readers to some promising areas for future research.

The basic concepts currently used and taken for granted in wetland biogeochemistry were historically developed by soil scientists studying chemistry and biochemistry of submerged soils and paddy soils, geochemists working in freshwater and marine ecosystems, and microbiologists working with anaerobic bacteria and their activities in soils and sediments. Some of this research was initiated during the early twentieth century in Europe, Asia, and North America. This early research primarily focused on measuring processes such as organic matter decomposition, mineralization, nitrogen fixation, denitrification, reduction of iron and manganese oxides, sulfate reduction, methanogenesis, and several other oxidation–reduction reactions of various metals. In addition, research also included microbial physiology and ecology. In soil science, much of the early research was centered on how these processes influenced the productivity and yield of rice. Some of this early research in submerged soils was summarized in review articles by Patrick and Mahapatra (1968), Ponnamperuma (1972), and several others. However, during the past three decades the value and the importance of wetlands as a key ecosystem of the biosphere have been recognized. This has resulted in studying many of these biogeochemical processes identified in early research on wetland ecosystems. Much of this research is done in two areas: biogeochemical processes (rates and controls) and more recently microbial community structure and populations and their interactions with plant and algal communities. Some of the biogeochemical measurements have been made at multiple scales (microbial to landscape) using both experimental and modeling approaches.

Introduction of new tools and techniques in recent years has made it possible to study many of the well-established biogeochemical processes in greater detail and at different scales. A number of advances have been made in recent years and never before has there been such interest (or capability) to study wetland processes at both extremes of scale (microbial and landscape). In recent years several texts have reviewed the usefulness of these new tools in studying biogeochemical processes in several ecosystems (Fenchel et al., 1998; Sterner and Elser, 2002; Melillo et al., 2003; Schlesinger, 2005). Some of the new tools that have aided in advancements in wetland biogeochemistry are as follows:

- Stable isotopes (tracers at molecular scale, cost of measurements allows massive sample numbers, models for ecosystem cycling)
- Molecular tools (culture-independent methods [new organisms/previously unknown diversity])
- Microscopy with molecular tools (fluorescence *in situ* hybridization [FISH] and enzyme-fluorescent substrates)
- *In situ* probes and real-time analytical sensors

- Nondestructive chemical (nuclear magnetic resonance [NMR]) and physical methods (e.g., diffuse reflectance spectroscopy [DRS])
- Mechanistic simulation models
- Statistical/geostatistical models and techniques
- Geographic information systems (GIS) and remote sensing tools for spatial scaling of measurements

With the availability of modern tools, there is now considerable interest in studying these processes at molecular level and identifying specific microbial communities involved in performing biogeochemical functions. However, a significant amount of current research is driven by new tools rather than new concepts. In many situations, the old established concepts are studied with the use of new tools, with very little or no advancement in understanding the actual process. At times it gives an impression that it is like putting "old wine in a new and a more sophisticated looking bottle." Yet advanced modern stochastic and mechanistic models and computational power have facilitated to synergize biogeochemical properties and processes into coherent model structures to describe the complexity of interfacing pedosphere, hydrosphere, biosphere, and atmosphere as well as internal and external stresses. Some of these new tools have improved our understanding of carbon, nitrogen, phosphorus, sulfur, and other elemental cycles in wetlands and aquatic systems. Some examples are (1) new pathways (Anammox, anaerobic methane oxidization, etc.); (2) new consortia ecological theory (e.g., syntrophs, anaerobic methane oxidizers); (3) diverse patterns of biogeochemical events (diurnal/seasonal heterogeneity) hot spots and hot moments; (4) finger printing molecular structures of organic compounds typically not measured using traditional methods; (5) isotopic signatures to identify sources, sinks, and transformations; (6) upscaling of ecosystem processes to the landscape scale; and (7) historical analyses of paleo systems/conditions. These advancements have led to better estimates of ecosystem-level fluxes, providing glimpses inside the black box of process-level controls.

Although a significant amount of literature is now emerging on both biogeochemical processes and microbial communities by independent researchers, we still lack a fundamental understanding of the linkage between measured processes and microbial community structure and their functions (see select reviews: Lovley et al., 2004; Megonigal et al., 2004; Reddy et al., 2005; Gutknecht et al., 2006; Ogram et al., 2006; Weber et al., 2006). Differences in methodological approaches and measurement scales between biogeochemists and molecular/microbial ecologists and lack of interdisciplinary approaches are probably one of the main reasons why the linkage between biogeochemical processes and microbial community structure has not been made. Studying biogeochemical cycles at different scales requires the expertise from different disciplines.

Biogeochemical cycles of elements involve exchange of material between abiotic (reservoir pool, long term) and biotic (exchange pools, short term) pools and are regulated by physical, chemical, and biological processes occurring within ecosystem components of a wetland. Since organic matter consists of essentially all major elements, cycling or transformation of one element can have profound influence on cycling of other associated elements in both biotic and abiotic pools. For example, decomposition of organic matter (as a result of heterotrophic microbial activity) can not only result in production of greenhouse gases such as carbon dioxide and methane, but also promote mineralization of organic nitrogen, sulfur, and phosphorus. Several examples can be cited on interrelationship between cycles and mutual dependency of one cycle over another cycle (see Chapters 3–18). These relationships are often referred to as coupled biogeochemical cycles and defined as regulation or influence of one cycle (or associated processes within the cycle) over another cycle through feedbacks and controls by another cycle or associated abiotic and biotic processes within that cycle. This adds even more complexity to the advancements made in wetland biogeochemistry and to our understanding of elemental cycles. Improving our understanding of coupled biogeochemical cycles across different spatial and temporal scales will provide much research opportunities in the future.

Anthropogenic activities often decouple elements from their natural stoichiometry through selective release or retention of elements such as carbon, nitrogen, sulfur, and phosphorus from their long-term reservoir stores (Sterner and Elser, 2002; Molden et al., 2003). The concept of decoupling is demonstrated in the Everglades ecosystem, where anthropogenic loading of phosphorus has resulted in alternation of stoichiometry and altered cycles of carbon, nitrogen, and sulfur (see Chapter 17). For example, in areas closer to nutrient loading, a phosphorus-limited Everglades system is now converted into a nitrogen-limited system, as a result of alternation of natural stoichiometry. Molden et al. (2003) also suggested that recoupling of elements can be a successful strategy for restoring degraded ecosystem. In the case of Everglades, recoupling can occur through reduction of external phosphorus loading and stabilizing internal phosphorus (see Chapter 17).

Elemental cycles in wetlands are influenced by biotic communities (vegetation, periphyton, and microbes). These organisms may influence a given biogeochemical cycle or a process directly by transforming organic forms to inorganic forms and vice versa or indirectly by providing substrates to microbes or by altering their physicochemical environment and creating conditions for a given biogeochemical reaction to occur. Many examples of the role of microbes, algae, and vegetation in regulating various biogeochemical processes are discussed in several chapters of this book.

19.2 BIOGEOCHEMICAL PROCESSES

In recent years, biogeochemists, microbiologists, limnologists, and soil scientists have extensively measured many biogeochemical processes regulating elemental cycling in wetland soils and aquatic sediments. The reader is directed to a *Treatise on Biogeochemistry*, which presents reviews by several authors on some of the recent advances on biogeochemical cycling of elements in various ecosystems and environments, including wetlands and aquatic systems (Schlesinger, 2005). It is beyond the scope of this chapter to review all recent advances made in this topical area. However, we will present few examples of key biogeochemical processes pertinent to wetland ecosystems.

Resource limitations of productivity in wetlands and aquatic ecosystems have focused on inorganic nutrient availability and loading rates from drainage basins. Recently, research has shown that much of the nutrient bioavailability in the soil and water column is associated with nutrients bound in or with naturally dissolved organic compounds. Many of the nutrient and organic substrate limitations are focused on hydrolytic activity of enzymes. The role of enzymes in regulating elemental cycles is addressed in various chapters of this book.

Enzymes represent a functional response of the microorganisms to the environmental variables they are exposed to. Several studies have shown that small shifts in enzyme activities can significantly alter organic matter decomposition (Wright and Reddy, 2001a; Sinsabaugh et al., 2002). A slow rate of organic matter decomposition in wetlands and peatlands, resulting from lack of oxygen during periods of flooding, can potentially inhibit the activity of phenol oxidase, causing accumulation of phenolic compounds (Freeman et al., 2001). A multitude of evidence points to complexing of humic substances with enzymes as a major impediment to the hydrolytic availability of these nutrients to microbiota and dependent higher trophic levels (Wetzel, 2001). Under conditions of reduced inorganic phosphorus availability, microbial synthesis of phosphatases, nucleotidases, and other enzymes increases markedly in order to exploit an abundance of organic phosphorus moieties in natural waters. However, in the presence of natural humic substances, nearly all associated with decomposition products of higher plant structural tissues, membrane-bound and free enzymes are complexed and inactivated by noncompetitive inhibition. As a result, prodigious amounts of energy are diverted from growth and productivity to compensate for these losses by increased production of enzymes. Recent studies have shown the influence of photolysis on decoupling or systematic fragmentation of humic-enzyme complexes (Wetzel, 2001, 2002, 2003).

Water-level drawdown and increase in temperature were shown to enhance the decomposition process and reduce the carbon sequestration potential of wetlands (DeBusk and Reddy, 1998; Bridgham et al., 2001; Davidson and Janssens, 2006) (see Chapter 5). In addition, decomposition of organic matter in nutrient-poor environments increases by anthropogenic nutrient loading to wetlands (Reddy et al., 1999). This increase in the rate of decomposition is due to increased enzyme and microbial activities. Carreiro et al. (2000) showed that nitrogen additions to low-lignin organic matter stimulated the decomposition, while addition of nitrogen to high-lignin organic matter inhibited the decomposition due to suppression of phenol oxidase activity. Enzyme activities are also influenced by soil redox conditions, hydrology, and nutrient enrichment (McLatchey and Reddy, 1998; Freeman et al., 2001; Wright and Reddy, 2001a). Phosphorus metabolism of the microbial community is often linked to the activity of phosphomonoesterases that are induced in response to phosphorus limitation and inhibited under high concentrations of inorganic phosphorus. Phosphatase activity is now used as an indicator of nutrient impacts in phosphorus-limited wetlands (Wright and Reddy, 2001a; Newman et al., 2003; Sharma et al., 2005).

Decomposition of organic matter is influenced by the availability of electron acceptors (Lovley et al., 1996; Conrad, 1996; McLatchey and Reddy, 1998; Roden and Wetzel, 1996; Wright and Reddy, 2001b). In addition to traditional inorganic electron acceptors, research also suggests that humic acids can be an important electron acceptor in wetland environments (Lovley et al., 1996). For example, electrons produced during fermentation may be shuttled via humics to ferric iron, which is reduced to ferrous iron (Lovley et al., 1996). This mechanism has important implications in wetland biogeochemistry because humic substances may be involved in shuttling electrons to insoluble/immobile iron and manganese oxides and enhance their reduction.

Wetlands are known to be a major source of methane. Recent interest has been on determining different anaerobic pathways that regulate the ratios of methane and carbon dioxide produced in wetlands. Regulation of methane production by electron acceptors (such as iron oxides and sulfates) with higher reduction potentials has been demonstrated. Although, this concept has been recognized about more than four decades ago, only recently its significance with respect to methane emissions has been quantified (Roden and Wetzel, 1996). Approximately 70% methane emissions from a wetland were reduced due to reduction of iron oxides (Roden and Wetzel, 1996). The relative role of methanogens, homoacetogens, and sulfate reducing prokaryotes in regulating methane production is now better understood (Castro et al., 2002, 2005; Drake et al., 1996). Methane production from H_2–CO_2 and acetate has been quantified by various tools including stable isotopes (Ogram et al., 2006; Penning et al., 2006). For example, Hines and Duddleston (2001) found that acetoclastic methanogenesis was insignificant in peatlands when the pH was <4.6 or temperature was <15°C and all methanogenesis in bogs and northern fens proceeded via hydrogenotrophic methanogenesis. In another study, Castro et al. (2005) observed that hydrogenotrophic methanogenesis was dominant in eutrophic wetlands than in oligotrophic wetlands. Given the importance of methane emissions from wetlands, a better understanding of controls and regulators of anaerobic catabolism is needed. Coupling plant- and algal-derived carbon substrates to methanogenesis needs to be better understood to determine the role of wetlands as a source of methane to the atmosphere.

The global carbon cycle is driven by microorganisms feeding on one of the carbon compounds such as methane or carbon dioxide. Wetlands are known to be a major source of methane to the atmosphere estimated at 20–25% to global methane emissions (Chappalaz et al., 1993). A large proportion of methane produced in wetlands can be oxidized anaerobically using alternate electron acceptors such as nitrate, iron and manganese oxides, and sulfates. Now, there is evidence that methane oxidation is coupled to denitrification and sulfate reduction. Wetlands located in agricultural watersheds are known to receive significant loadings of nitrate. Boetius et al. (2000) have documented the coupling of methane oxidation to sulfate reduction in marine sediments. Raghoebarsing et al. (2006) have identified microorganisms that can oxidize methane to carbon dioxide using nitrate as an electron acceptor, showing coupling of methane oxidation to denitrification.

As discussed in Chapters 4 and 5, it is well known that wetlands are not limited by electron donors, and anaerobic conditions in these systems provide high electron pressure. Some of the end products (ethanol, methane, and hydrogen) of organic matter decomposition are also known to produce electricity. However, capturing these electrons into fuel cells is a relatively new and novel idea. Lovley (2006) presented a review on how these electrons can be captured to produce electricity. Recent studies presented in this review show that during microbial respiration some of the microorganisms conserve energy to support growth by oxidizing organic compounds to carbon dioxide with direct quantitative electron transfer to electrodes. The review by Lovley (2006) provides a discussion on practical application of this novel concept, first described in wetland soils and aquatic sediments, to power electronic devices, such as monitoring equipment.

Although we have a good understanding of potential rates of various nitrogen transformations in wetlands, we have not advanced in determining *in situ* rates and assess the importance of these processes in regulating the global nitrogen cycle, especially the nitrogen balance (nitrogen inputs including fixation versus nitrogen losses including denitrification). New information is now emerging on coupling of nitrification–denitrification, dissimilatory nitrate reduction to ammonia (DNRA) versus denitrification, and alternative pathways to nitrogen production (Galloway, 2005; Burgin and Hamilton, 2007).

Oxidation of ammonium to nitrogen gas (ANAMOX) is now considered as a significant process in marine, coastal, and estuarine sediments (see Chapter 8). Several studies have now provided evidence using a combination of ^{15}N-based tracer studies, lipid biomarkers, FISH, and phylogenetic and quantitative PCR analysis (Francis et al., 2007).

A vast amount of literature has emerged during the last two decades on the importance of dissimilatory Mn(IV) and Fe(III) reduction in various environments including wetlands. Specific microbial groups involved in this process have been identified and well studied (Lovley et al., 2004). Unlike methanogens and sulfate reducers, these organisms can use a wide range of substrates including various end products of fermentation. Reduced end products are known to serve as electron donors for iron- and manganese-oxidizing bacteria (Lovley et al., 2004; Weber et al., 2006). Oxidation of Fe(II) can occur in both aerobic and anaerobic environments, whereas reduction is restricted to anaerobic environments (see Chapter 10). Both biotic and abiotic reactions are involved in the oxidation of these reduced compounds. Roden and Wetzel (2002) showed a significant relationship between labile organic carbon and Fe(III) reduction rates, linking the role of this electron acceptor during organic matter decomposition (see Chapter 10 for details). Although Fe(III) and Mn(IV) reductions have been recognized for a long time, recent advances revealed specific groups of organisms involved in conducting this process (Lovley et al., 2004).

Sulfur transformations in wetlands have been well studied and the importance of oxidation–reduction reactions has been well established, especially for coastal systems (see Chapter 11). In general, sulfate when available is subject to rapid utilization by dissimilatory sulfate reducing prokaryotes, and in freshwater wetlands sulfate availability is often limiting. New information is now emerging on phylogenetic groups of sulfate reducing prokaryotes and their functional role in carbon metabolism (Megonigal et al., 2004; Campbell et al., 2006; Ogram et al., 2006).

Microbial sulfate reduction is the process by which mercury (Hg) is converted to the methylated forms. The methylated forms are bioaccumulated and are potent neurotoxins. Bioaccumulated methyl-Hg poses a threat to the health of both humans and wildlife. Recent studies have shown that methyl-Hg concentrations are the highest at moderate sulfate concentrations in freshwater wetlands (Gilmour et al., 1998), whereas at high or low sulfate concentrations this process is inhibited (see Chapters 11 and 17). Sulfur isotope studies are needed to track the sources of sulfur that enter wetlands, and to determine the range of sulfate concentrations over which maximum methyl-Hg production occurs. Mercury occurs naturally in moderate abundance in wetlands, yet is accumulated to very high levels in fish. Biogeochemical conditions regulating mercury methylation and mechanisms for bioaccumulation are critical factors in determining mercury impact on wetland ecosystems. Regional and historical variations in mercury concentration and accumulation including atmospheric inputs of mercury and other metals of environmental concern in sediments are important factors that require further research and studies.

19.3 ALGAL AND MICROBIAL INTERACTIONS

Periphyton mats are abundant in freshwater wetlands and are ecologically important communities because they play a critical role in elemental cycling. The pathways regulating diversity and activity of periphyton communities are exclusive (or competing processes) and are frequently separated within the mat structure. This separation can occur spatially and temporally where distinct zones of processes are structured vertically, but at the same time, can occur in close proximity (ca., millimeters). These zones can be represented by redox, pH, or nutrient gradients, which result in flux-driven cycles that may occur entirely within the mat structure. The phototrophic cyanobacteria and heterotrophs orient along these gradients to optimize biogeochemical processes (carbon, nitrogen, phosphorus, and sulfur cycling) essential for growth, survival, and maintenance of genetic diversity (Pearl et al., 2000). For example, ammonium produced by mineralization of organic substrates in deeper anaerobic mat layers can diffuse upward and be utilized as an energy source by chemoautotrophic nitrifying organisms in aerobic zones and converted to nitrate. Nitrate-nitrogen can then diffuse from aerobic zones to anaerobic zones, where it can be used as an electron acceptor (see Chapters 6 and 8). Similarly, other dissolved reduced compounds can flux from anaerobic zones to aerobic zones, whereas the dissolved oxidized compounds can flux from aerobic zones to anaerobic zones. Separation of competing processes can also occur temporally; for example, in many mats of nonheterocystous cyanobacteria, dinitrogen fixation is primarily conducted at night to avoid competition with oxygen produced during daytime photosynthesis. Carbon is frequently the shared resource within the mat structure and is transferred and competed for by the mat organisms. For this reason, photosynthetic production, and subsequent decomposition are often the driving forces of mat metabolism. The presence of highly productive periphyton communities in nutrient-limited systems has been linked to increased efficiency of uptake, and rapid and efficient release and remineralization of limiting nutrients due to the close association of autotrophic and heterotrophic microbial components in these communities (Wetzel, 2002). The efficiency by which these communities regulate coupled biogeochemical cycles depends on mutual dependency of different types of organisms and associated individual processes regulating individual elemental cycle and feedback mechanisms within and between elemental cycles.

The stability of periphyton mat communities, and likewise their ability to maintain their functional role in an ecosystem, is largely based on the idea of mutual dependency between groups of organisms. In this manner, organisms perform one functional role in the mat and, at the same time, are dependent on the functional roles of other representative mat groups. For example, heterotrophic bacteria are dependent on the release of carbon (e.g., exopolymeric substances) from photosynthetic cyanobacteria or diatoms, and in return, these bacteria provide a vitamin or nutrient source for the photosynthesizers (Marshall, 1989). In this way, mat communities not only are utilizing the most advantageous features of each member, so that the whole is better than the sum of the parts, but also are conserving nutrient resources through efficient use and recycling within the mat structure (Wetzel, 2001). This cooperative existence of autotrophs and heterotrophs is potentially the reason why such microbial mats can colonize and thrive in nutrient-poor environments. Several external and internal factors control the stability of periphyton in wetlands. Internal factors include autotrophic and heterotrophic processes within the boundary layer of the mat that result in developing gradients in redox potential, dissolved oxygen, pH, and nutrients. These sharp gradients can support high rates of oxidation and reduction reactions over short distances. As photosynthesis ceases in surface layers of periphyton mats, they may rapidly turn into anaerobic zones. In the dark, many cyanobacteria were shown to ferment endogenous glycogen (Stal, 2000). Excretion of fermentation products can be a source of labile carbon for heterotrophic bacteria such as nitrate or sulfate-reducing bacteria. To what extent these gradients are altered depends on the rate of exchange of solutes between periphyton mats and the water column. Exchange processes within these gradients occur through diffusion in response to gradients, in contrast to mixing and mass flow in the free-water phase. These gradients can also affect mat community composition and associated functions. External regulators include anthropogenic nutrient loads, physicochemical characteristics of the water column, flux of nutrients from the soil, and solar radiation. Both internal and external factors can alter the relative rates of various biogeochemical processes regulating carbon, nitrogen, and phosphorus cycling in wetlands, as related to water quality and long-term storage of nutrients.

19.4 VEGETATION AND MICROBIAL INTERACTIONS

Historically, research in wetland macrophytes has focused on the physiological and ecological mechanisms enabling plants to survive under flooded/anaerobic soil conditions. Many of these mechanisms/adaptations are presented in Chapter 7, and mainly deal with oxygenation processes and the oxygen release from plant roots into the rhizosphere. Identification of internal gas transport systems also led to the measurement of gaseous fluxes from the root zone and the knowledge that these are very significant pathways of greenhouse gas emissions.

More recent advances in wetland vegetation research involve the role of vegetation in nutrient cycling and carbon storage within wetland ecosystems. Much of this work has also identified the association of plant species composition and the development of soil microbial communities and processes that determine rates of litter decomposition, and nitrogen and carbon availability for microbial processes like denitrification or methanogenesis among many others (Gutknecht et al., 2006). These studies have increased our understanding of the linkages between macro- and microscale processes in wetlands.

Recently, the idea of competition between plants and bacteria for nutrients has received some attention (e.g., Sundareshwar et al., 2003). These and other novel studies have demonstrated that bacteria and plants in the same system may be limited by different nutrients (in particular, nitrogen versus phosphorus). In this context, these new findings are helping to refine our simplistic models of nutrient limitation in wetlands. While this concept is not entirely new, having been proposed for algae as far back as the 1980s (Cole, 1982), it has important implications for both macrophyte composition/productivity and nutrient cycling in wetland systems and places new emphasis on the role of carbon (e.g., through exudates of roots) as the currency of these interactions.

We are only beginning to focus on the importance of plant species composition in determining not only microbial composition but also community function. One of the largest areas of research in this area involves the interactions of bacteria with plants in regulating biogeochemical processes in the rhizosphere. This is one of the fastest growing areas of plant and soil ecology with huge implications for ecosystem function in the light of effects of biodiversity, climate change, carbon cycling/storage, and accelerated nutrient inputs (Zak et al., 2003).

In a recent study on terrestrial ecosystems, Wassen et al. (2005) demonstrated that endangered plant species persist under phosphorus- and nitrogen-limited conditions, and concluded that enhanced phosphorus is more likely to be the cause of species loss than nitrogen enrichment. In the same manner as for upland plant ecology, new tools have greatly increased our understanding of processes in the rhizosphere (Killham and Yeomans, 2001). Of these tools, stable isotopic approaches are proving very useful in wetland systems. For example, a recent study used stable isotopic techniques to trace carbon from recently fixed photosynthates into methane production in the rice rhizosphere (Kimura et al., 2004). Similar studies have also demonstrated the potential of the tracing ability of isotopes in combination with molecular techniques for identifying involved bacterial populations (Lu et al., 2005, 2007). The use of these techniques also extends into the phyllosphere, where bacteria on (and inside) the leaf surfaces of plants have been shown to alter elemental cycling and associated feedbacks with root systems. One example of this is found in peat bogs where methanotrophic symbionts in the leaves of *Sphagnum* sp. can provide carbon dioxide for photosynthesis through oxidation of methane (Raghoerbarsing et al., 2006). With expected developments and increased sensitivity in molecular techniques, these types of approaches offer great potential to elucidate the complexities of these plant–microbe interactions.

19.5 MODERN TOOLS TO STUDY BIOGEOCHEMICAL CYCLES

19.5.1 Microbial Communities and Diversity

Microbial communities are pivotal to many biogeochemical processes functioning in wetlands. The use of molecular approaches over the past two decades has shown that only a small fraction of the microbial community has been cultivated from different ecosystems. The functional role of many of these microbes in wetland ecosystems is unclear. Molecular methods including FISH, metabolic gene presence/expression quantification, and genome sequencing hold significant promise for understanding the biogeochemical roles of various groups of bacteria in wetland environments. Some of the modern methods currently used are summarized in the following paragraphs.

Phospholipid fatty acid analysis (PLFA) is based on the determination of "signature" lipid biomarkers from the cell membranes and walls of microorganisms. Phospholipids are an essential part of intact cell membranes, and information from the lipid analysis provides quantitative insight into three important attributes of microbial communities: viable biomass, community structure, and nutritional status. Phospholipid fatty acid profiles have been used to show the response of the microbial community to phosphorus availability (Keinanen et al., 2002). Signature lipid biomarker analysis may not detect every species of microorganism in an environmental sample accurately, because many species have similar PLFA patterns.

Aerobic metabolic activities have been used for comparing bacterial community composition in soils and sediments. Garland and Mills (1991) used a redox technology developed by BIOLOG, Inc., to determine microbial community structure. This technique uses tetrazolium dye reduction as an indicator of sole-carbon-source utilization to characterize and classify heterotrophic microbial communities. Whole environmental samples (water, soil, and rhizosphere) are incubated in BIOLOG plates containing 95 separate carbon sources. Correlation of the original carbon source variables to the principal components provided a functional basis to distinctions among communities.

Availability of molecular approaches including DNA sequencing analysis has revolutionized the current understanding of microbial communities and ecology.

Microbiologists used nucleic acids (DNA or RNA) extracted from soils to determine community-level changes, measure microbial diversity, identify key organisms, and discover new microbial processes (Zwolinski, 2007). The 16S ribosomal RNA (rRNA) gene sequences of Bacteria and Archaea are now routinely used to determine the soil microbial populations and for monitoring microbial community dynamics following natural or anthropogenic environmental changes (Zwolinski, 2007). Molecular methods currently used to determine soil microbial communities include density gradient gel electrophoresis (DGGE), length heterogeneity-PCR (LH-PCR), terminal restriction fragment length polymorphism (TRFLP), clone libraries and sequences, FISH, stable isotope probing (SIP), and molecular analysis of ectomycorrhizal communities (Entry et al., 2007). Recently, these methods have been reviewed in a series of articles by Nakatsu (2007), Thies (2007), and Zwolinski (2007).

Terminal restriction fragment length polymorphism (T-RFLP) (Liu et al., 1997; Thies, 2007) is used for rapid analysis of microbial community diversity in various environments. The technique employs PCR in which one of the two primers used are fluorescently labeled and used to amplify a selected region of bacterial genes encoding 16S rRNA from total community DNA (Liu et al., 1997). Castro et al. (2005) used this method for analysis of *dsrAB* (dissimilatory sulfite reductase) and *mcrA* (methyl coenzyme M reductase) distribution in the Everglades wetland soils. T-RFLP profiles reflect the distribution of the numerically dominant genotypes in each sample based on the analysis of the terminal end of the target gene. T-RFLP analyses were conducted on DNA extracted from soil samples taken monthly from the oligotrophic, transition, and eutrophic regions. Individual T-RFLP profiles (representing assemblages of the respective genotypes from each sample) were analyzed by principal component analysis (PCA). Assemblages of both *dsrAB* and *mcrA* genotypes were relatively stable and the compositions of these assemblages were clearly characteristic of their position along the nutrient gradient (Castro et al., 2002). Assemblages of SRP from the three sites separated more clearly by PCA than did methanogens, which may be a function of the greater metabolic diversity of SRP than of methanogens (Castro et al., 2002; Ogram et al., 2006).

19.5.2 Nuclear Magnetic Resonance Spectroscopy

Nuclear magnetic resonance spectroscopy (NMR) commonly known as NMR has been used by biogeochemists for determining the structure of organic compounds in soils and sediments. Solution ^{31}P NMR spectroscopy was first used to determine the phosphorus composition of soil extracts by Newman and Tate (1980), and has since become the method of choice for such studies. Using ^{31}P NMR, several researchers identified organic phosphorus compounds in wetland soils. For example, organic phosphorus compounds such as (1) inositol phosphates (IHP), (2) nucleic acids, and (3) phospholipids can be now identified and quantified using these techniques. Turner and Newman (2005) analyzed organic phosphorus compounds in surficial soils in the Florida Everglades and noted that diester to monoester ratio was >1. Further, in a constructed wetland designed for phosphorus removal, >70% of the soluble organic phosphorus in the surface water was hydrolyzed by phosphodiesterase, with <10% hydrolyzed by phosphomonoesterase (Pant et al., 2002). Recent advances in high-mass, high-resolution Fourier transform ion cyclotron resonance mass spectrometry (FT-ICR MS; Marshall and Guan, 1996) combined with "soft" ionization techniques such as electrospray ionization (ESI; Senko et al., 1996) now makes it feasible to characterize individual organic phosphorus molecules.

19.5.3 Diffuse Reflectance Spectroscopy

DRS is a technology used for nondestructive characterization of the physical, chemical, and biological properties of soil based on the interaction of visible-infrared light (electromagnetic energy) with matter (Shepherd and Walsh, 2002). Typically, air-dry samples are illuminated with an artificial light source, and reflected light is collected and passed through fiber-optic cables to an array of light detectors and measured in narrow wavelength (3–10 nm) in the visible-infrared range as an electrical signal. For African top soils (total number of soils in the range of 680–1,100), highly significant correlations were observed between spectral signatures and various soil physicochemical properties including pH, organic carbon, percent sand and clay, conductivity, and exchangeable calcium and magnesium (Shepherd and Walsh, 2002). Recently, this technique has been used in wetland soils, where several biogeochemical properties have been significantly correlated to spectral signatures (Cohen et al., 2005).

19.5.4 Stable Isotopes

Biogeochemists have used stable isotope ratios of carbon, nitrogen, oxygen, and sulfur to determine sources and track these elements as they cycle in the environment. Interactions within biological systems result in fractionation of stable isotopes. Enzyme-catalyzed reactions favor the transformation of lighter isotopes, resulting in enriching the source material and producing products with lighter isotopes (Nealson and Rye, 2005). In recent years, there has been an explosion in the use of stable isotope ratios to address research problems over a broad range of biogeochemical studies in wetlands and aquatic systems. Both natural abundance and enriched isotopes are used to study biogeochemical processes. In natural abundance studies, isotope ratios are expressed in relation to international standards as delta (δ) values in parts per thousand. In enrichment studies, isotope ratios are often expressed as atom percent, which expresses the number of atoms of the isotope for every one hundred atoms of that element in the sample.

Such isotope techniques, once the realm of specialized analytical chemists and largely unavailable to many biogeochemists, are now becoming common tools to address sophisticated questions on biogeochemical cycling, interaction among microbial, algae, and plant communities, and food–web interactions. One of the recent developments in isotope ratio analysis involves linking of gas chromatography (GC) and isotope ratio mass spectrometer (IRMS). A gas chromatograph–combustion furnace (GC-C) interface was developed in which the effluent stream from a GC column was fed into a combustion furnace (Kreuzer-Martin, 2007). Stable isotope techniques have been used to link organisms or groups of organisms to specific functions within natural environments. SIP involves introducing a stable isotope-labeled substrate into a microbial community and following the fate of the substrate by extracting diagnostic molecular species such as fatty acids and nucleic acids from the community and determining which specific molecules have incorporated the isotope (Kreuzer-Martin, 2007).

19.6 SYNTHESIS: MECHANISTIC AND STATISTICAL MODELS

In the increasingly diverse body of literature concerned with wetland biogeochemistry, the common thread is understanding the processes that occur in wetlands and identifying the constituent elements that take part in these processes. Examples can be found throughout the book, the description of our understanding of carbon, nitrogen, phosphorus, and sulfur cycles, all constitute a narrative of biotic and abiotic processes, storages and pool sizes, and microbial and plant communities involved in these cycles at a range of spatial and temporal scales.

The rationale behind much of this research has been extensively discussed in chapters throughout the book; however, it is important to note that the end point is often to understand and possibly to manipulate these processes to maintain or obtain some desired system status. For example, much of the research presented for the two case studies (Chapters 17 and 18) is oriented at understanding how the biogeochemical processes have changed as a result of anthropogenic inputs of nutrients and sediments, where they have changed, and how to mitigate for these changes. In order for managers to identify the appropriate management strategies and then monitor their effect, management tools such as predictive models are used to describe processes or empirical relationships among select properties.

19.6.1 MECHANISTIC MODELS

Mechanistic models synergize our understanding of properties (state variables) and ecosystem processes. These models are built on the premise that fundamental knowledge of processes and their interactions is available and numerical methods are then used to describe the underlying physics and chemistry governing processes. Such process-based mechanistic models are spatially and temporally explicit and allow modeling of the components of wetland ecosystems over longer periods of time. A major limitation of such models is the vast amount of input data to run (test, develop) and evaluate (validate) model outputs. Because of our limited understanding of some ecosystem processes the deterministic approach has been relaxed using empirical algorithms, meaning that a simplified representation of the system or phenomenon is used based on the experience or experimentation. These quasi physically based models have been used extensively to model water flux, nutrient transport, and transformations (Rousseau et al., 2004), often containing simple kinetics to describe nutrient transformations. More challenging has been to model microbial-induced biogeochemical processes that are highly dynamic and nonlinear and occur at fine spatial scales.

Various mechanistic ecohydrological simulation models for wetlands have been presented including the comprehensive biogeochemical model, Wetland-DNDC, for modeling of carbon cycle and hydrology (Jianbo et al., 2005); the Everglades Landscape Model (ELM) that models water flux, biomass production, decomposition, phosphorus transport, and other processes (Fitz et al., 2004); the Ecological Dynamic Simulation Model (EDYS) that describes soil, water, and nutrient dynamics among many other processes (Childress et al., 2002); and a spatially heterogeneous ecosystem model covering different freshwater habitat types in the Everglades that models patterns of biomass, energy flux, and species composition (Diffendorfer et al., 2001). Watershed-based simulation tools model wetlands and uplands, thus are ideally suited to assess the impact of land use management practices (e.g., nutrient input) and other stresses on wetlands. Examples for mechanistic watershed simulation models are given by Kort et al. (2007) who presented the watershed ecosystem nutrient dynamics-phosphorus (WEND-P) model; by Grunwald and Qi (2006) and Chaubey et al. (2007) who utilized the soil and water assessment tool to model nutrient transport processes across multiple ecosystem types; and by Refsgaard and Storm (1995) who introduced MIKE SHE that is a comprehensive deterministic, spatially distributed, physically based modeling system for the simulation of all major hydrological processes.

Mechanistic model applications are hampered by (i) lack in input data to characterize a wetland ecosystem; (ii) available datasets on level II and III biogeochemical indicators (see Chapter 15); (iii) spatially and temporally explicit descriptions of all wetland components (soils, geology, hydrology, vegetation/fauna, bathymetry, and anthropogenic factors); (iv) the scale of measurements of biogeochemical variables to describe processes (e.g., transfer of knowledge from lab experiments under controlled conditions to field/landscape scale); (v) matching the scale of modeling and scale of processes (e.g., nitrification, methanogenesis, and other processes occur at microscale but often coarse model units of tens, hundreds, or thousands of meters are used to discretize a wetland/watershed); (vi) lack of complete understanding of the processes, interactions, and feedbacks in quantitative form; (vii) lack in explicit incorporation of uncertainty into the modeling process; and (viii) quasi physically based nature of models that incorporate simplifications of fundamental processes and use empirical coefficients (fudge factors) in place of physical/chemical/biological parameters.

As Albert Einstein put it "a model should be as simple as possible but not too simple" may serve as an underlying conceptual framework for process-based modeling of complex, coupled biogeochemical cycles in wetlands. Research scientists strive to understand all bits and pieces of every single process and how it is controlled by environmental factors and other processes at various spatial and temporal scales resulting in the most complex mechanistic models. Yet, watershed managers, regulators, and decision makers require tools that facilitate to understand major outcomes of dominant ecosystem processes that have caused imbalances within a wetland ecosystem (e.g., phosphorus enrichment). Such models may be less complex (i.e., more empirical in nature), but serve the purpose to guide restoration of impaired wetlands or provide recommendations for conservation management. Over the last decades major advances have been made to translate conceptual models into quantitative form generating complex mechanistic models. With the advent of novel measurement techniques and rapid, nondestructive methods, as outlined in this chapter, there are exciting advances in mechanistic modeling expected to follow in the near future. Major challenges are to better understand scaling behavior of biogeochemical processes, to quantify controlling environmental factors under variable field conditions, and to better understand how processes behave across space and through time.

19.6.2 Stochastic (Statistical) Models

Placing our understanding of wetland processes in a model of some form, whether it is stochastic or deterministic, forces to synthesize knowledge in an unambiguous, coherent manner. In practice, the two approaches do not exist in separate realms. Addiscott and Mirza (1998) noted that many equations in process models can be approximated stochastically and vice versa. Meaningful stochastic relationships must bear some resemblance to our understanding of the mechanisms that operate in the system. Stochastic models aim to identify functional relationships (i.e., deterministic model structure) and include some sort of random forcing (i.e., random error or variation). In many cases, stochastic models are used to simulate deterministic systems that include fine-scale phenomena that cannot be accurately observed or modeled. As such, these fine-scale phenomena are effectively unpredictable. A good stochastic model manages to represent the average effect of unresolved phenomena on larger-scale phenomena in terms of a random forcing. In essence, stochastic approaches state that we cannot know all factors and processes that influence the values we observe for a certain wetland biogeochemical property. Simple statistical methods as a continuous response (regression analysis) or as a set of responses to discrete factors (analysis of variance, ANOVA) have been used extensively in wetland research, in order to relate elements in biogeochemical processes or determine the extent to which a particular factor may influence a process.

Modern observation techniques have allowed to collect comprehensive datasets on multiple biogeochemical variables that are hypothesized to be interlinked. These data are complemented by dense datasets on ancillary environmental properties (e.g., remote sensing-derived vegetation patterns at high spatial resolution, water levels, and bathymetry) and facilitate to employ stochastic models. Multivariate techniques allow to model complex behavior and interaction among multiple biogeochemical variables. Ordination and classification (cluster) techniques describe the structure within the multivariate dataset by reducing the dimensionality of the set of properties to an underlying smaller set of variables, with a minimum loss of information. The most extensively used technique belonging to this group is PCA. The uncorrelated factors extracted through this form of ordination are linear combinations of the variables, and have been used to examine and compare the structure of biological communities (Kreutzweiser and Faber, 1999), microbial community composition (Castro et al., 2002), and variability in physicochemical properties (Grunwald et al., 2007a) in various wetlands. Cluster analysis is generally applied to larger landscapes and has been used to identify the underlying landscape arrangements such as habitat types in wetlands (e.g., Knight and Morris, 1996) or impairment by nutrient influx (e.g., Grunwald et al., 2006a).

Often the aim of multivariate data analysis is to elucidate the relationships between the abiotic environment (independent variables) and biological responses (dependent variables) and multiple biogeochemical properties. Multivariate correlation and regression analysis has been used extensively (White and Reddy, 2000; Fisher and Reddy, 2001; and others) to characterize the relationships between biogeochemical variables in various wetlands. However, it has severe limitations because it cannot identify spatial dependencies and autocorrelation structures and complex (high-order) relationships among multiple interacting states and response variables. Another limitation is that regression and correlation analysis assumes independence of observations (i.e., no spatial autocorrelation). But as almost all ecosystem properties are spatially autocorrelated it is imperative to map spatial autocorrelations among biogeochemical properties and models (Grunwald, 2006a, 2006b). To overcome these limitations of classical statistics, innovative research has been presented by King et al. (2004) who used Mantel tests to remove spatial autocorrelation structures (spatial dependencies) before testing for unmasked (true) relationships between environmental properties and vegetation in an anthropogenically influenced wetland ecosystem. Grunwald et al. (2006a) used a spatial moving correlation analysis to reveal that local spatial relationships among total, organic, inorganic, and microbial biomass phosphorus differed geographically across a subtropical wetland rendering global (traditional) correlation coefficients meaningless. Corstanje et al. (2007) demonstrated that canonical discriminant analysis (CDA) can identify the set of properties that best distinguish between eutrophication status among sites.

Modern regression variants, such as classification and regression trees (CART) (Breiman et al., 1984), generalized linear mixed models (Jiang, 2007), and artificial neural networks models (Haykin, 1999), have been employed to model complex relationships among environmental properties. CART has been adopted widely in other disciplines to model soil properties across large watersheds (Grunwald et al., 2006b) but less so to analyze relationships of biogeochemical variables in wetlands. For example, De'Ath and Fabricius (2000) used CART to analyze ecological data of the Great Barrier Reef and Cohen et al. (2005) to analyze wetland soil quality.

Challenges in the future remain to model the high-order interactions among biogeochemical variables, to model complex linear and nonlinear processes, and to incorporate spatial and temporal autocorrelations into models. Scale is one of the most confounding factors when synthesizing and modeling wetland processes. Process-based models implicitly assume a specific scale at which processes operate (e.g., rhizosphere). Yet biogeochemical processes operate at escalating spatial and temporal scales ranging from molecular to landscape scales. How these processes operate across scales and how the relationship between soil, water, and vegetative properties is affected by scale in wetlands is still poorly understood. When a model describing a biogeochemical process is constructed, whether this model is stochastic or process-based is scale dependent. More research on scaling of processes from the laboratory/site-specific to large landscape scales will be critical to translate biogeochemical processes into models that address natural and anthropogenic stresses imposed on wetland ecosystem.

For a more extensive overview of descriptive multivariate stochastic techniques it is referred to Hair et al. (1995), Berthouex and Brown (2002), and Schabenberger and Pierce (2002). For scaling methods in aquatic ecology it is referred to Seurant and Strutton (2004).

19.6.3 Geospatial Models

Site-specific biogeochemical wetland studies have generated extensive knowledge of how individual parts and processes of the ecosystems function. Understanding of how various components interact as a whole requires a holistic perspective that considers the spatial variability, distribution, and interaction of all components of a landscape (Grunwald et al., 2007b). GIS provide the framework to synthesize biogeochemical data and scale up the information to landscape level. To develop realistic geospatial biogeochemical wetland models it is imperative to map the underlying spatial variability and distribution of properties and monitor their change (see Chapter 15). Although biogeochemical properties are measured at various spatial and temporal scales, observations are confined to small samples (or small sample support) at select (point) locations to analyze individual processes. Transforming these process-level data into spatially explicit context has been hampered by ecosystem complexity, multiple nested levels of interrelated physical, biological, and chemical processes, and the lack of sufficient quantitative data (Grunwald et al., 2007b). Only recently, wetland science and geoscience have intersected to investigate wetland characteristics in a spatially explicit framework although geostatistical theory was developed in the early 1960s (Matheron, 1963) and GIS science boomed in the early 1980s. Geostatistics allows quantification of spatial autocorrelation patterns of biogeochemical variables and their spatial covariations with environmental landscape properties. An overview of geostatistical methods is provided by Goovaerts (1997), Chilès and Delfiner (1999), Webster and Oliver (2001), and Grunwald (2006b).

True spatial independence of properties in aquatic, wetland, and terrestrial landscapes does not really exist due to the interconnectedness of biogeochemical processes of nitrogen, carbon, phosphorus, sulfur, and other elements, and transport processes that move water, material, and energy through the ecosystem (Grunwald et al., 2007b). But the sampling density and spacing between observations determine whether the underlying spatial autocorrelation can be captured by a model or not. Each soil biogeochemical property exhibits specific behavior of spatial autocorrelation that is related to environmental factors such as internal nutrient cycling, microbial-induced processes, hydrology, anthropogenic-induced nutrient inputs and disturbances, and vegetation types.

The knowledge on spatial autocorrelation structures is important for the interpolation of biogeochemical properties across a wetland. Kriging, which is a weighted interpolation technique that accounts for the spatial structure and variability of properties, and kriging variants (e.g., co-kriging, regression kriging, kriging with external drift) have been employed in various wetland/aquatic studies. Extensive work was presented by DeBusk et al. (2001) and Bruland et al. (2007) to quantify the spatial patterns of various physicochemical properties in the Everglades and how they relate with other landscape properties (e.g., vegetation, hydrology, nutrient influx points). A novel geostatistical method was used by Grunwald et al. (2007a) to identify the magnitude and scale at which multiple biogeochemical properties accounted for variability within the subtropical wetland and to map the distribution of this variability using PCA-kriging. A comparison of multivariate regression analysis and three different geostatistical methods (ordinary kriging, co-kriging, and regression kriging) demonstrated the success to predict floc/soil total phosphorus by incorporating remote-sensing imagery (Landsat Enhanced Thematic Mapper [ETM] and Advanced Spaceborne Thermal Emission and Reflection Radiometer [ASTER]) into geospatial predictions in WCA-2A, Everglades (Rivero et al., 2007). The integration of sparse point observations of biogeochemical properties and dense ancillary environmental datasets (e.g., derived from satellite imagery) has the potential to improve spatially explicit predictions of properties throughout aquatic/wetland ecosystems. This geospatial research is still in its infancy and will allow tackling issues on spatial and temporal scaling of biogeochemical properties and integration of suites of biogeochemical variables into complex geospatial models. In particular, remote sensing has enormous potential to provide dense geospatial information layers on biophysical landscape properties across large wetlands or whole watersheds. Besides remote-sensing-based mapping of wetland vegetation (Schmidt and Skidmore, 2003) and wetland types using various active and passive sensors (Landsat ETM, SPOT, AVHRR, JERS-1, ERS-1, and RADARSAT) (Ozesmi and Bauer, 2002), spectral indices can be derived to analyze biophysical properties characterizing wetland surfaces (water, ridges, floc, detritus, periphyton, and vegetation). Indices such as the Normalized Difference Vegetation Index (NDVI) or Leaf Area Index (Jensen, 2000) provide spectral signatures that allow direct sensing of biophysical properties and (empirical) inference on floc/soil properties.

Because of the complexity of wetland ecosystems a synergistic approach is required that integrates knowledge from different disciplines including biogeochemistry, geography, statistics/geostatistics, ecology, hydrology, and others. Interdisciplinary collaboration will be key to reconcile deductive and inductive science and allow us to better understand the linkages between biogeochemical properties and processes across different spatial and temporal scales.

19.7 FUTURE DIRECTIONS AND PERSPECTIVES

Wetlands provide a range of ecosystem services. Those pertinent to biogeochemistry include water quality, accretion of organic matter and associated elements, and carbon sequestration. Wetlands also serve as sources of methane and sources/sinks for nitrous oxide.

Retention of elements such as carbon, nitrogen, phosphorus, sulfur, and metals and toxic organic compounds in wetlands is an important function in watershed biogeochemistry particularly in drainage basins with significant nonpoint nutrient contributions from agriculture and urban sources. Thus, retention of these compounds in wetlands involves complex intercoupled physical, chemical, and biological processes that ultimately retain them either in organic or inorganic forms. Many of the biogeochemical processes interact with the elements loaded to the wetland ecosystem within various internal storage compartments (soils, vegetation, detritus, microorganisms, and fauna). When evaluating elemental retention by wetland ecosystems, both short-term storage (assimilation into vegetation, translocation within above- and below-ground plant tissues, microorganisms, periphyton, and detritus) and long-term storage components (retention by inorganic and organic soil particles and net accretion of organic matter) have to be taken into account.

Future research directions should focus on enhanced basic understanding of biogeochemical cycles and the roles of key players (biotic communities) at both fine and coarse spatial and temporal scales, and integration of information across scales. Some of the critical research areas may include

- Identification of organic and inorganic forms of elements in the water column and soils, with emphasis on their stability under various hydrological and redox conditions, particularly as mediated by microbial and plant metabolism.
- Multiple, dynamic roles of microbial and plant rhizosphere biotic processes on abiotic retention of elements in wetlands.
- *In situ* rates of various biogeochemical processes and associated controls and regulators.
- Biogeochemical hot spots and significance of these zones in regulating kinetics of various reactions.

- Mutual dependency of elemental cycles and roles in regulating diversity of biotic communities.
- Identification and functions of key microbial species involved in biogeochemical reactions. Determination of linkages between microbial community structure and functions.
- Simultaneous measurements of "processes" regulating coupled cycles at various spatial and temporal scales.
- Effective nonlinear predictive models based on mechanistic understanding of processes regulating retention of elements.
- Development of simple indices that can extrapolate biogeochemical dynamics information to regional and global scales by GIS-based modeling.
- Integration of biogeochemical measurements across scales using statistical, geospatial, and process-based models.
- Role of wetlands in regulating global elemental cycles including eutrophication, carbon sequestration, and regulation of greenhouse gases and climate change.

FURTHER READINGS

Burgin, A. J. and S. K. Hamilton. 2007. Have we overemphasized the role of denitrification in aquatic ecosystems? A review of nitrate removal pathways. *Front. Ecol. Environ.* 5:89–96.

Fenchel, T., G. M. King, and T. H. Blackburn. 1998. *Bacterial Biogeochemistry*. Academic Press, New York. 307 pp.

Francis, C. A., J. M. Beman, and M. M. M. Kuypers. 2007. New processes and players in the nitrogen cycle: the microbial ecology of anaerobic and archaeal ammonia oxidation. *Int. Soc. Microbial Ecol. J.* 1:19–27.

Gutknecht, J. L. M. and R. M. Goodman. 2006. Linking soil process and microbial ecology in freshwater wetland ecosystems. *Plant Soil* 289:17–34.

Lovley, D. R. 2006. Bug juice: harvesting electricity with microorganisms. *Nature Rev.* 4:497–508.

Lovley, D. R., D. E. Holmes, and K. P. Nevin. 2004. Dissimilatory Fe(III) and Mn(IV) reduction. *Adv. Microbial Physiol.* 49:220–286.

Megonigal, J., M. Hines, and P. Visscher. 2004. Anaerobic metabolism: linkages to trace gases and aerobic processes. In W. H. Schlesinger (ed.) *Biogeochemistry*. Elsevier-Pergamon, Oxford, UK.

Melillo, J. M., C. B. Field, and B. Moldan. 2003. *Interaction of the Major Biogeochemical Cycles*. SCOPE 61. Island Press, Washington, DC. 357 pp.

Ogram, A., S. Bridgham, R. Corstanje, H. Drake, K. Kosel, A. Mills, S. Newman, K. Portier, and R. Wetzel. 2006. Linkages between microbial community composition and biogeochemical processes across scales. *Ecol. Studies* 190:239–268.

Pearl, H. W., J. L. Pinckeny, and T. F. Steppe. 2000. Cyanobacterial-bacterial mat consortia: examining the functional unit of microbial survival and growth in extreme environments. *Environ. Microbiol.* 2:11–26.

Reddy, K. R., R. G. Wetzel, and R. Kadlec. 2005. Biogeochemistry of phosphorus in wetlands. In J. T. Sims and A. N. Sharpley (eds.) *Phosphorus: Agriculture and the Environment* Soil Science Society of America. pp. 263–316.

Schlesinger, W. H. 2005. Biogeochemistry. In H. D. Holland and K. K. Turekian (eds.) *Treatise in Geochemistry*. Vol. 8. Elsevier-Pergamon, Oxford. 702 pp.

Sterner, R. W. and J. L. Elser. 2002. *Ecological Stoichiometry*. Princeton University Press, New Jersey. 439 pp.

Weber, K. A., L. A. Achenbach, and J. D. Coates. 2006. Microorganisms pumping iron: anaerobic microbial iron oxidation and reduction. *Nature Rev.* 4:732–764.

References

Adams, D. F., S. O. Farewell, M. R. Pack, and W. L. Bamesberger. 1979. Preliminary measurements of biogenic sulfur containing gas emissions from soils. *Air Pollut. Control Assoc. J.* 29:380–383.

Adams, D. F., S. O. Farewell, M. R. Pack, and E. Robinson. 1981. Biogenic sulfur gas emission from soils in eastern and south-eastern United States. *J. Air Pollut. Control. Assoc.* 31:1035–1140.

Addiscott, T. M. and N. A. Mirza. 1998. New paradigms for modelling mass transfers in soils. *Soil Tillage Res.* 1998:105–109.

Adorisio, C., C. Bedregal, J. Gomez, J. Madden, C. Miessau, D. Pescatore, P. Sievers, S. Van Horn, T. van Veen, and J. Vega. 2007. Phosphorus source controls for the basins tributary to the Everglades Protection Area. Volume 1. In South Florida Environmental Report, South Florida Water Management District, West Palm Beach, FL. http://www.sfwmd.gov/sfer/SFER_2007/index_draft_07.html.

Aelion, M. C. and D. C. Cresi. 1999. Impact of land use on microbial transformation of atrazine. *J. Environ. Qual.* 28:683–691.

Aitkenhead, J. A. and W. H. McDowell. 2000. Soil C:N ratio as a predictor of annual riverine DOC flux at local and global scale. *Global Biogeochem. Cycles* 14:127–138.

Aftring, R. P., B. E. Chalker, and B. F. Taylor. 1981. Degradation of phthalic acids by denitrifying mixed cultures of bacteria. *Appl. Environ. Microbiol.* 41:1177–1183.

Aleem, M. I. H., H. Lees, and D. J. D. Nicholas. 1963. Adenosine triphosphate-dependent reduction of nicotinamide adenine dinucleotide by ferro-cytochrome c in chemoautotrophic bacteria. *Nature* 200:759–761.

Alexander, M. 1994. *Biodegradation and Bioremediation*. Academic Press, New York. 302 pp.

Aller, R. C. 1982. The effects of macrobenthos on chemical properties of marine sediment and overlying water. In P. L. McCall and M. J. S. Tevesz (eds.) *Animal-Sediment Relations: The Biogenic Alteration of Sediments*. Plenum Press, New York. pp. 53–102.

American Public Health Association (APHA). 1999. Standard Methods for Examination of Water and Wastewater. In L. Clescerl, A. Greenberg, and A. Eaton (eds.), 20th Edition. New York.

Anderson, T. H. and K. H. Domsch. 1989. Ratios of microbial biomass carbon to total organic carbon in arable soils. *Soil Biol. Biochem.* 21:471–479.

Aller, R. C. 1983. The importance of the diffusive permeability of animal burrow linings in determining marine sediment chemistry. *J. Mar. Res.* 41:299–322.

Aller, R. C. and J. Y. Aller. 1992. Meiofauna and solute transport in marine muds. *Limnol. Oceanogr.* 37:1018–1033.

Amador, J. A. and R. D. Jones. 1995. Carbon mineralization in pristine and phosphorus-enriched peat soils of the Florida Everglades. *Soil Sci.* 159:129–141.

Anderson, T. H. and K. H. Domsch. 1993. The metabolic quotient for CO_2 (Q_{CO_2}) as a specific activity parameter to assess the effects of environmental-conditions, such as pH, on the microbial biomass of forest soils. *Soil Biol. Biochem.* 25:393–395.

Anisfeld, S. C., M. Tobin, and G. Benoit. 1999. Sedimentation rates in flow-restricted and restored salt marshes in Long Island Sound. *Estuaries* 22:231–244.

Anschutz, P., K. Dedieu, F. Desmazes, and G. Chaillou. 2005. Speciation, oxidation state and reactivity of particulate manganese in marine sediments. *Chem. Geol,* 218(3–4):265–279.

Ann, Y., K. R. Reddy, and J. J. Delfino. 2000a. Influence of chemical amendments on P retention in a constructed wetland. *Ecol. Eng.* 14:157–167.

Ann, Y., K. R. Reddy, and J. J. Delfino. 2000b. Influence of redox potential on P solubility in chemically amended wetland soils. *Ecol. Eng.* 14:169–180.

Antweiler, R. C., D. A. Goolsby, and H. E. Taylor. 1995. Nutrients in the Mississippi River. In R. H. Meade (ed.) *Contaminants in the Mississippi River: 1987–1992*. U.S. Geology Survey Circular 1133, 140 pp.

APHA, 1992. In A. E. Greenburg, L. S. Clesceri, and A. D. Eaton (eds.) *Standard Methods for the Examination of Water and Wastewater*. American Public Health Association, Washington, DC.

APHA, 1999. *Standard Methods for the Examination of Water and Wastewater*. American Public Health Association.

Armentano, T. M. and G. M. Woodwell. 1975. Sedimentation rates in a Long Island marsh determined by ^{210}Pb dating. *Limnol. Oceanogr.* 20:452–456.

Armentano, T. V. and E. S. Menges. 1986. Patterns of change in the carbon balance of organic soil wetlands of the temperate zone. *J. Ecol.* 74:755–774.

Armstrong, W. and P. M. Beckett. 1987. Internal aeration and the development of stellar anoxia in submerged roots. *New Phytol.* 105:221–245.

Armstrong, W. and P. M. Beckett. 1996. Pressurized aeration in wetland macrophytes: some theoretical aspects of humidity-induced convection and thermal transpiration. *Folia Geobot. Phytotaxinom.* 31:25–36.

Armstrong, J. and W. Armstrong. 1991. A convective through-flow of gases in *Phragmites australis* (Cav.) Trin ex Steud. *Aquat. Bot.* 39:75–88.

Armstrong, J., W. Armstrong, and P. M. Beckett. 1992. *Phragmites australis*: Venturi- and humidity-induced pressure flows enhance rhizome aeration and rhizosphere oxidation. *New Phytol.* 114:197–207.

Armstrong, W. 1967. The use of polarography in the assay of oxygen from roots in anaerobic medium. *Physiol. Plants* 20:540–553.

Armstrong, W. 1967. The oxidizing activity of roots in waterlogged soils. *Physiol. Plant.* 20:920–926.

Armstrong, W. 1968. Oxygen diffusion from roots of woody species. *Physiol. Plant.* 21:539–543.

Armstrong, W. 1971. Radial oxygen losses from intact rice roots as affected by distance from the apex, respiration and water logging. *Physiol. Plant.* 25:192–197.

Armstrong, W. 1978. Root aeration in wetland conditions. In D. D. Hook and R. M. M. Crawford (eds.) *Plant Life in Anaerobic Environments.* Ann Arbor Science, Ann Arbor, MI. pp. 269–297.

Armstrong, W. 1979. Aeration in higher plants. *Adv. Bot. Res.* 7:225–329.

Armstrong, W., J. Armstrong, P. M. Beckett, and S. H. F. W. Justin. 1991. Convective gas-flows in wetland plant aeration. In M. B. Jackson, D. D. Davies, and H. Lambers (eds.) *Plant Life Under Oxygen Deprivation.* SPB Academic Publishing, The Hague, The Netherlands. pp. 283–302.

Armstrong, W., P. M. Beckett, S. H. F. W. Justin, and S. Lythe. 1991. Modelling and other aspects of root aeration by diffusion. In M. B. Jackson, D. D. Davies, and H. Lambers (eds.) *Plant Life Under Oxygen Deprivation.* SPB Academic Publishing, The Hague, The Netherlands. pp. 267–282.

Armstrong, W. and D. J. Boatman. 1967. Some field observations relating to growth of bog plants to conditions of soil aeration. *J. Ecol.* 55:101–110.

Armstrong, W., R. Brandle, and M. B. Jackson. 1994. Mechanisms of flood tolerance in plants. *Acta Bot. Neerland.* 43:307–358.

Armstrong, W. and D. J. Gaynard. 1976. The critical oxygen pressures for respiration in intact plants. *Physiol. Plant.* 37:200–206.

Arnosti, C. 1998. Rapid potential rates of extracellular enzymatic hydrolysis in Arctic sediments. *Limnol. Oceanogr.* 43:315–324.

Aselman, I. and P. J. Crutzen. 1989. Global distribution of natural freshwater wetlands and rice paddies: their net primary productivity, seasonality and possible methane emissions. *J. Atmos. Chem.* 8:307–358.

Avnimelech, Y. 1980. Calcium-carbonate-phosphate surface complex in calcareous systems. *Nature* 266: 255–257.

Axt, J. R. and M. R. Walbridge. 1999. Phosphate removal capacity of palustrine forested wetlands and adjacent uplands in Virginia. *Soil Sci. Soc. Am. J.* 63:1019–1031.

Baas Becking, L. G. M., I. R. Kaplan, and D. Moore. 1960. Limits of the natural environment in terms of pH and oxidation–reduction potentials. *J. Geol.* 68:243–284.

Bacha, R. E. and L. R. Hossner. 1977. Characteristics of coatings formed on rice roots as affected by iron and manganese additions. *Soil Sci. Soc. Am. J.* 41:931–935.

Bache, B. W. and E. G. Williams. 1971. A phosphate sorption index for soils. *J. Soil Sci.* 22:289–301.

Bagley, D. and J. Gosset. 1990. Tetrachloroethylene transformation to trichloroethylene and cis-1,2-dichloroethylene by sulfate-reducing enrichment cultures. *Appl. Environ. Microbiol.* 56:2511–2516.

Baldwin, D. S. 1996. Effects of exposure to air and subsequent drying on the phosphate sorption characteristics of sediments from a eutrophic reservoir. *Limnol. Oceanogr.* 41:1725–1732.

Balistrieri, L. S., J. W. Murray, and B. Paul. 1992. The biogeochemical cycling of trace metals in the water column of Lake Sammamish, Washington: response to seasonally anoxic conditions. *Limnol. Oceanogr.* 37:529–548.

Banerjee, M. R. and S. J. Chapman. 1996. The significance of microbial biomass sulfur in soil. *Biol. Fert. Soils* 22:116–125.

Barrow, N. J. and T. C. Shaw. 1980. Effect of drying on the measurement of phosphate adsorption. *Commun. Soil Sci. Plant Anal.* 11:347–353.

Bartlett, R. J. and B. R. James. 1991. Redox chemistry of soils. *Adv. Agron.* 50:151–208.

Bartlett, K. B. and R. C. Harriss. 1993. Review and assessment of methane emissions from wetlands. *Chemosphere* 26(1–4):261–320.

Bartlett, K. B., R. C. Harriss, and D. I. Sebacher. 1985. Methane flux from coastal salt marshes. *J. Geophys. Res.* 90:5710–5720.

Bates, A. L., E. C. Spiker, and C. W. Holmes. 1998. Speciation and isotopic composition of sedimentary sulfur in the Everglades, Florida, USA. *Chem. Geol.* 146:155–170.

Bates, A. L., W. H. Orem, J. W. Harvey, and E. C. Spiker. 2002. Tracing sources of sulfur in the Florida Everglades. *J. Environ. Qual.* 31:287–299.

Batjes, N. H. and E. M. Bridges. 1992. A review of soil factors and processes that control fluxes of heat, moisture and greenhouse gases. Technical Paper 23, International Soil Reference and Information Centre, Wageningen. 1–4:37–66.

Batzer, D. P., R. B. Rader, and S. A. Wissinger. 1999. *Invertebrates in Freshwater Wetlands of North America: Ecology and Management.* Wiley, New York.

Batzer, D. P. and R. R. Sharitz (eds.). 2006. *Ecology of Freshwater and Estuarine Wetlands.* University of California Press, Berkley and Los Angeles, CA. 568 pp.

Beare, M. H., C. L. Neely, D. C. Coleman, and W. L. Hargrove. 1990. A substrate-induced respiration (SIR) method for measurement of fungal and bacterial biomass on plant residues. *Soil Biol. Biochem.* 22:585–594.

Beckett, P. M., W. Armstrong, F. W. Justin, and J. Armstrong. 1988. On the relative importance of convective and diffusive gas-flows in plant aeration. *New Phytol.* 110:463–469.

Bedford, B. L., D. R. Bouldin, and B. D. Beliveau. 1991. Net oxygen and carbon dioxide balances in solution bathing roots of wetland plants. *J. Ecol.* 79:943–959.

Bedford, B. L., M. R. Walbridge, and A. Aldous. 1999. Patterns in nutrient availability and plant diversity of temperate North American wetlands. *Ecology* 80:2170–2181.

Begg, C. B. M., G. J. D. Kirk, A. F. Mackenzie, and H. U. Neue. 1994. Root-induced iron oxidation and pH changes in the lowland rice rhizosphere. *New Phytol.* 128:469–477.

Bender, M. and R. Conrad. 1992. Kinetics of CH_4 oxidation in oxic soils exposed to ambient air or high CH_4 mixing ratios. *FEMS Microbiol. Ecol.* 101:261–270.

Benedetti-Cecchi, L. 2001. Beyond BACI: optimization of environmental sampling designs through monitoring and simulation. *Ecol. Appl.* 11:783–799.

Benner, R., M. A. Moran, and R. E. Hodson. 1985. Effect of pH and plant source on lignocellulose biodegradation rates in two wetland ecosystems, the Okefenokee Swamp and Georgia salt marsh. *Limnol. Oceanogr.* 30:489–499.

Benz, M., A. Brune, and B. Schink. 1998. Anaerobic and aerobic oxidation of ferrous iron at neutral pH by chemoheterotrophic nitrate-reducing bacteria. *Arch. Microbiol.* 169:159–165.

Berkheiser, V. E., J. J. Street, P. S. C. Rao, and T. L. Yuan. 1980. Partitioning of inorganic orthophosphate in soil–water systems. *Crit. Rev. Environ. Sci. Technol.* 10:179–224.

Berner, R. A. 1970. Sedimentary pyrite formation. *Am. J. Sci.* 268:1–23.

Berner, R. A. 1971. *Principles of Chemical Sedimentology.* McGraw-Hill, New York. 240 pp.

Berner, R. A. 1980. *Early Diagenesis. A Theoretical Approach.* Princeton University Press, Princeton, NJ. 481 pp.

Berner, R., A. E. Maccubbin, and R. E. Hodson. 1984. Anaerobic biodegradation of the lignin polysaccharide components of lignocellulose and synthetic lignin by sediment microflora. *Appl. Environ. Microbiol.* 47:998–1004.

Berthouex, P. M. and L. C. Brown. 2002. *Statistics for Environmental Engineers.* Lewis Publishers, Boca Raton, FL.

Billings, W. D. 1987. Carbon balance of Alaskan tundra and taiga ecosystems—past, present and future. *Quat. Sci. Rev.* 6:165–177.

Blackburn, T. H. 1979. Methods for measuring rates of NH_4^+ turnover in anoxic marine sediments using a ^{15}N-NH_4^+ dilution technique. *Appl. Environ. Microbiol.* 37:760–765.

Blackmer, A. M. and J. M. Bremner. 1978. Inhibitory effect of nitrate on reduction of N_2O to N_2 by soil microorganisms. *Soil Biol. Biochem.* 10:187–191.

Blackmer, A. M. and J. M. Bremner. 1981. Terrestrial nitrification as a source of atmospheric nitrous oxide. In C. C. Delwiche (ed.) *Denitrification, Nitrification, and Atmospheric Nitrous Oxide.* Wiley, New York. pp. 151–170.

Blackmer, A. M., J. M. Bremner, and E. L. Schmidt. 1980. Production of nitrous oxide by ammonia-oxidizing chemoautotrophic microorganisms in soil. *Appl. Environ. Microbiol.* 40:1060–1066.

Boetius, A., K. Ravenschlag, C. J. Schubert, D. Rickert, F. Widdel, A. Gieseke, R. Amann, B. B. Jùrgensen, U. Witte, and O. Pfannkuche. 2000. A marine microbial consortium apparently mediating anaerobic oxidation of methane. *Nature* 407:623–626.

Bohn, H. L. 1976. Arsenic Eh–pH diagram and comparison to soil chemistry of phosphorus. *Soil Sci.* 121:125—127.

Bohn, H. L., B. L. McNeal, and G. A. O'Connor. 1985. *Soil Chemistry*. Wiley, New York. 341 pp.

Bohn, H. L., B. L. McNeal, and G. A. O'Connor. 2001. *Soil Chemistry*. 3rd Edition. Wiley, New York. 306 pp.

Bol, R., T. Bolger, R. Cully, and D. Little. 2003. Recalcitrant soil organic materials mineralize more efficiently at higher temperatures. *J. Plant Nutr. Soil Sci.* 166:300–307.

Bolin, B. 1977. Changes of land biota and their importance for the carbon cycle. *Science* 196:613–615.

Bolin, B., E. T. Degens, P. Duvigneau, and S. Kempe. 1979. The global biogeochemical carbon cycle. In B. Bolin (ed.) *The Global Carbon Cycle*. Wiley, Chichester. pp. 1–56.

Bollag, J. M. and S. Liu. 1990. Biological transformation process of pesticides. In *Pesticides in the Soil Environment*. SSSA Book Series No. 2. Soil Science Society of America, Madison, WI. pp. 169–211.

Boudreau, B. P. 1997. *Diagenic Models and their Implementation*. Springer-Verlag, Berlin, Heidelberg, New York. 414 pp.

Boulegue, J., T. M. Church, and C. J. Lord. 1982. Sulfur speciation and associated trace metals (Fe, Cu) in the pore waters of Great Marsh, Delaware. *Geochim. Cosmochim. Acta* 46:453–464.

Boultin, A. J. and P. I. Boon. 1991. A review of methodology used to measure leaf decomposition in lotic environments: time to turn over an old leaf? *Aust. J. Mar. Freshw. Res.* 42:1–43.

Boumans, R. M. J. and J. W. Day, Jr. 1993. High precision measurements of sedimentation in shallow coastal areas using a sedimentation–erosion table. *Estuaries* 16:375–380.

Boyd, C. E. 1970. Production, mineral accumulation and pigment concentration in *Typha latifolia* and *Scirpus americanus*. *Ecology* 51:285–290.

Bradford, M. A., P. Ineson, P. A. Wookey, and H. M. Lappin-Scott. 2001. Role of CH_4 oxidation, production, and transportation in forest soil CH_4 flux. *Soil Biol. Biochem.* 33:1625–1631.

Brady, N. C. and R. R. Weil. 2003. *The Nature and Properties of Soils*. Prentice Hall, New Jersey. 960 pp.

Breiman, L., J. H. Friedman, R. A. Olshen, and C. J. Stone. 1984. *Classification and Regression Trees*. Chapman & Hall, CRC Press, Boca Raton, FL.

Bremner, J. M., S. G. Robbins, and A. M. Blackmer. 1990. Seasonal variability in emission of nitrous oxide from soil. *Soil Biol. Biochem.* 7:641–644.

Brenner, M., C. L. Schelske, and L. W. Keenan. 2001. Historical rates of sediment and nutrient accumulation in marshes of the Upper St. Johns River Basin, Florida, USA. *J. Paleolimnol.* 26:241–257.

Bricker-Urso, S., S. W. Nixon, J. K. Cochran, D. J. Hirschberg, and C. Hunt. 1989. Accretion rates and sediment accumulation in Rhode Island salt marshes. *Estuaries* 12(4):300–317.

Bridgham, S. D. and C. J. Richardson. 1992. Mechanisms controlling soil respiration (CO_2 and CH_4) in southern peatlands. *Soil Biol. Biochem.* 24:1089–1099.

Bridgham, S. D., K. Updegraff, and J. Pastor. 1998. Carbon, nitrogen, and phosphorus mineralization in northern wetlands. *Ecology* 79:545–561.

Bridgham, S. D., K. Updegraff, and J. Pastor. 2001. A comparison of nutrient availability indices along an ombrotrophic-minerotrophic gradient in Minnesota wetlands. *Soil Sci. Soc. Am. J.* 65:259–269.

Brinkman, R. 1970. Ferrolysis, a hydromorphic soil forming process. *Geoderma* 3:199–206.

Brinson, M. M. 1993. *A Hydrogeomorphic Classification of Wetlands*. Technical Report; WRP-DE-4, U.S. Army Corps. 79 pp.

Britsch, L. D. and J. B. Dunbar. 1993. Land loss rates: Louisiana coastal plain. *J. Coastal Res.* 9:324–338.

Brix, H., B. K. Sorrell, and P. T. Orr. 1992. Internal pressurization and convective gas flow in some emergent freshwater macrophytes. *Limnol. Oceanogr.* 37:1420–1433.

Brock T. D. and M. T. Madigan. 1991. *Biology of microorganisms*. Toronto: Prentice-Hall. 874 pp.

Brookins, D. G. 1988. *Eh–pH Diagrams for Geochemistry*. Springer-Verlag, Berlin. 176 pp.

Brooks, B. W., C. M. Foran, S. M. Richards, P. K. Westonner, J. K. Stanley, K. R. Solomon, M. Slattery, and T. W. La point. 2003. Aquatic ecotoxicology of fluoxetine. *Toxicol. Lett.* 142:169–183.

Brown, B. R. 1967. Biochemical aspects of oxidative coupling of phenols. In W. I. Taylor and A. R. Battersby (eds.) *Oxidative Coupling of Phenols*. Marcel Dekker, New York. 167 pp.

Brown, C. E. and S. R. Pezeshki. 2000. A study on waterlogging as a potential tool to control *Ligustrum sinense* populations in western Tennessee. *Wetlands* 20:429–437.

Brown, K. A. 1982. Sulfur in the environment. A review. *Environ. Pollut. Ser. B.* 3:47–80.

Bruland, G. L., S. Grunwald, T. Z. Osborne, K. R. Reddy, and S. Newman. 2006. Spatial distribution of soil properties in Water Conservation Area 3 of the Everglades. *Soil Sci. Soc. Am J.*, 70:1662–1676.

Bruland, G. L., T. Z. Osborne, K. R. Reddy, S. Grunwald, S. Newman, and W. F. DeBusk. 2007. Recent changes in soil total phosphorus in the Everglades: Water Conservation Area 3. *Environ. Monit. Assess.* 129:379–395.

Brus, D. J. and J. J. de Gruijter. 1997. Random sampling or geostatistical modelling? Choosing between design-based and model-based sampling strategies for soil. *Geoderma* 80:1–59.

Bruun, S., J. Luxhøi, J. Magid, A. de Neergaard, and L. S. Jensen, 2006. A nitrogen mineralization model based on relationships for gross mineralization and immobilization. *Soil Biol. Biochem.* 38:2712–2721.

Bruun, S., B. Stenberg, T. A. Breland, J. Gudmundsson, T. Henriksen, L. S. Jensen, A. Korsaeth, J. Luxhøi, F. Palmason, A. Pedersen, and T. Salo, 2005. Empirical predictions of plant material C and N mineralization patterns from near infrared spectroscopy, stepwise chemical digestion and C/N ratios. *Soil Biol. Biochem.* 37:2283–2296.

Buford, J. R. and J. M. Bremner. 1975. Relationships between the denitrification capacities of soils and total water-soluble and readily decomposable soil organic matter. *Soil Biol. Biochem.* 7:389–394.

Burdt, A. C., J. M. Galbraith, and W. L. Daniels. 2005. Season length indicators and land-use effects in southeast Virginia wet flats. *Soil Sci. Soc. Am. J.* 69:1551–1558.

Buresh, R. J. and W. H. Patrick, Jr. 1981. Nitrate reduction to ammonium and organic nitrogen in an estuarine sediment. *Soil Biol. Biochem.* 13:279–283.

Buresh, R. J., K. R. Reddy, and C. van Kessel. 2008. Nitrogen transformations in submerged soils. Nitrogen in Agricultural System.

Buresh, R. J., M. E. Casselman, and W. H. Patrick. 1980. Nitrogen fixation in flooded soil systems, a review. *Adv. Agron.* 33:149–192.

Burgin, A. J. and S. K. Hamilton. 2007. Have we overemphasized the role of denitrification in aquatic ecosystems? A review of nitrate removal pathways. *Front. Ecol. Environ.* 5:89–96.

Burgoon, P. S. and K. R. Reddy. 1996. Influence of temperature on biogeochemical processes in constructed wetlands. *J. Ecol.* 56:77–87.

Burkett, V. and J. Kusler. 2000. Climate change: potential impacts and interactions in wetlands of the United States. *J. Am. Water Res. Assoc.* 32(2):313–320.

Burnett, W. C. and H. Dulaiova. 2003. Estimating the dynamics of ground water input into the coastal zone via continuous radon-222 measurements. *J. Environ. Radioactivity* 69(1–2):21–35.

Burnett, W. E. C., H. Bokuniewicz, M. Huettel, W. S. Moore, and M. Taniguchi. 2003. Ground water and pore water inputs to the coastal zone. *Biogeochemistry* 66(1–2):3–33.

Burrough, P. A. and R. A. McDonnell. 1998. *Principles of Geographical Information Systems: Spatial Information Systems and Geostatistics.* Oxford University Press, Oxford.

Byrnes, B. H., E. R. Austin, and B. K. Tays. 1995. Methane emissions from flooded rice soils and plants under controlled conditions. *Soil Biol. Biochem.* 27:331–339.

Cahoon, D. R., D. J. Reed, and J. W. Day, Jr. 1995. Estimating shallow subsidence in microtidal salt marshes of the southeastern United States: Kaye and Barghoorn revisited. *Mar. Geol.* 128:1–9.

Cahoon, D. R., J. C. Lynch, and A. N. Powell. 1996. Marsh vertical accretion in a southern California estuary, U.S.A. *Estuar. Coast. Shelf Sci.* 43:19–32.

Callahan, M. A., M. W. Slimak, and N. W. Gabel. 1979. *Water-Related Environmental Fate of 129 Priority Pollutants. Volume 1. Introduction and Technical Background, Metals and Inorganics, Pesticides and PCBs.* Office of Water Planning and Standards, Washington, DC, by Versar Incorporated, Springfield, VA. EPA/440/4-79/029a.

Callaway, J. C., R. D. DeLaune, and W. H. Patrick. 1997. Sediment accretion rates from four coastal wetlands along the Gulf of Mexico. *J. Coast. Res.* 13(1):181–191.

Callaway, J. C. 2001. Hydrology and substrate. In J. B. Zedler (ed.) *Handbook for Restoring Tidal Wetlands.* CRC Press, Boca Raton, FL. pp. 89–117.

Callaway, J. C., R. D. DeLaune, and W. H. Patrick, Jr. 1996. Chernobyl ^{137}Cs used to determine sediment accretion rates at selected northern European coastal wetlands. *Limnol. Oceanogr.* 41(3):444–450.

Callaway, J. C., J. A. Nyman, and R. D. DeLaune. 1996. Sediment accretion in coastal marshes: a review and a simulation model of processes. *Curr. Top. Wetland Biochem.* 2:1–9.

Callaway, J. C., R. D. DeLaune, and W. H. Patrick. 1997. Sediment accretion rates from four coastal wetlands along the Gulf of Mexico. *J. Coastal Res.* 13(1):181–191.

Campbell, B. J., A. S. Engel, M. L. Porter, and K. Takai. 2006. The versatile e-proteobacteria: key players in sulphidic habitats. *Nat. Rev.* 4:458–468.

Capone, D. G. and R. P. Kiene. 1988. Comparison of microbial dynamics in marine and freshwater sediments: contrasts in anaerobic carbon metabolism. *Limnol. Oceanogr.* 33:725–749.

Caraco, N., J. J. Cole, and G. E. Likens. 1991. A cross-system study of phosphorus release from lake sediments. In J. Cole, G. Lovett, and S. Findlay (eds.) *Comparative Analyses of Ecosystems: Patterns, Mechanisms and Theories.* Springer-Verlag, New York. pp. 241–258.

Carlson, P. R. and J. Forrest. 1982. Uptake of dissolved sulfide by *Spartina alterniflora*: evidence from natural sulfur isotope abundance ratios. *Science* 216:633–635.

Carlton, R. G. and R. G. Wetzel. 1987. Distributions and fates of oxygen in periphyton communities. *Can. J. Bot.* 65:1031–1037.

Carlton, R. G. and R. G. Wetzel. 1988. Phosphorus flux from lake sediments: effects of epipelic algal oxygen production. *Limnol. Oceanogr.* 33:562–570.

Carreiro, M. M., R. L. Sinsabaugh, D. A. Repert, and D. F. Parkhurst. 2000. Microbial enzyme shifts explain litter decay responses to simulated nitrogen deposition. *Ecology* 81:2359–2365.

Casselman, M. E., W. H. Patrick, Jr., and R. D. DeLaune. 1981. Nitrogen fixation in a Gulf Coast Salt Marsh. *Soil Sci. Soc. Am. J.* 45:51–56.

Castro, H., K. R. Reddy, and A. Ogram. 2002. Composition and function of sulfate reducing prokaryotes in eutrophic and pristine areas of the Florida Everglades. *Appl. Environ. Microbiol.* 68:6129–6137.

Castro, H., S. Newman, K. R. Reddy, and A. Ogram. 2005. Distribution and stability of sulfate-reducing prokaryotic and hydrogenotrophic methanogenic assemblages in nutrient-impacted regions of the Florida Everglades. *Appl. Environ. Microbiol.* 71:2695–2704.

Chan, A. S. K. and T. B. Parkin. 2000. Evaluation of potential inhibitors of methanogenesis and methane oxidation in a landfill cover soil. *Soil Biol. Biochem.* 32:1581–1590.

Chang, S. C. and M. L. Jackson. 1957. Fractionation of soil phosphorus. *J. Soil Sci.* 84:133–144.

Chappalaz, J. A., I. Y. Fung, and A. M. Thompson. 1993. The atmospheric CH_4 increase since the last glacial maximum. 1. Source Estimates. *Tellus* 45B:228–241.

Chaubey, I., K. W. Migliaccio, C. H. Green, J. G. Arnold, and R. Srinivasan. 2007. Phosphorus modeling in Soil and Water Assessment Tool (SWAT) model. In D. E. Radcliffe and M. L. Cabrera (eds.) *Modeling Phosphorus in the Environment.* CRC Press, New York. pp. 161–189.

Cheng, H. H. and L. T. Kurtz. 1963. Chemical distribution of added nitrogen in soils. *Soil Sci. Soc. Am. Proc.* 16:276–292.

Childers, D. L., R. D. Jones, J. C. Trexler, C. Buzzelli, J. Boyer, A. L. Edwards, E. E. Gaiser, K. Jayachandaran, D. Lee, J. F. Meeder, J. Pechmann, J. H. Richards, and L. J. Scinto. 2001. Quantifying the effects of low level phosphorus enrichment on unimpacted Everglades wetlands with in situ flumes and phosphorus dosing. In Porter, J. and K. Porter (eds.) *The Everglades, Florida Bay, and Coral Reefs of the Florida Keys.* CRC Press, Boca Raton, FL. pp. 127–152.

Childers, D. L., R. F. Doran, R. Jones, G. B. Noe, M. Rugge, and L. J. Scinto. 2003. Decadal changes in vegetation and soil phosphorus patterns across the Everglades landscape. *J. Environ. Qual.* 32:344–362.

Childress, W. M., C. L. Coldren, and T. McLendon. 2002. Applying a complex, general ecosystem model (EDYS) in large-scale land management. *Ecol. Model.* 153:97–108.

Chilès, J.-P. and P. Delfiner. 1999. *Geostatistics: Modeling Spatial Uncertainty.* Wiley, New York.

Chilvers, D. C. and P. J. Peterson. 1987. Global cycling of arsenic. In T. C. Hutchinson and K. M. Meema (eds.) *Lead, Mercury, Cadmium and Arsenic in the Environment.* Scientific Committee on Problems of the Environment (SCOPE) 31. Wiley, New York.

Chimney, M. J. and K. C. Pietro. 2006. Decomposition of macrophyte litter in a subtropical constructed wetland in south Florida (USA). *Ecol. Eng.* 27:301–321.

Chiou, C. T., V. H. Freed, D. Schmedding, and R. L. W. Kohnert. 1977. Partition coefficient and bioaccumulation of selected organic chemicals. *Environ. Sci. Technol.* 11:475–478.

Choi, S. C. and R. Bartha. 1993. Cobalamin-mediated mercury methylation by *Desulfovibrio desulfuricans* Ls. *Appl. Environ. Microbiol.* 59:290–295.

Christensen, P. B., N. P. Revsbech, and K. Sand-Jensen. 1994. Microsensor analysis of oxygen in the rhizosphere of the aquatic macrophyte *Littorella uniflora* (L) Ascherson. *Plant. Physiol.* 105:847–852.

Chrost, R. and W. Siuda. 2002. Ecology of microbial enzymes in lake ecosystems. In R. G. Burns and R. P. Dick (eds.) *Enzymes in Environment.* Marcel Dekker, New York. pp. 35–72.

Chrost, R. J. 1991. Environmental control of the synthesis and activity of aquatic microbial ectoenzymes. In R. J. Chrost (ed.) *Microbial Enzymes in Aquatic Environments.* Springer-Verlag, New York. pp. 29–59.

Chua, T. 2000. Mineralization of organic phosphorus in a subtropical freshwater wetland. Ph.D. Dissertation, University of Florida, Gainesville, FL.

Church, T. M., C. J. Lord, III, and B. L. K. Somayajulu. 1981. Uranium, thorium, and lead nuclides in a Delaware salt marsh sediment. *Estuar. Coast. Shelf Sci.* 13:267–285.

Cicerone, R. J. and R. S. Oremland. 1988. Biogeochemical aspects of atmospheric methane. *Global Biogeochem. Cycles* 2:299–327.

Clair, T. A., B. G. Warner, R. Roberts, H. Murkin, J. Lilley, L. Mortsch, and C. Rubec. 1995. Canadian inland wetlands and climate change. In *Canadian Country Study: Climate Impacts and Adaptations*. Environmental Canada, Ottawa. pp. 189–218.

Clark, Q. 2002. Phosphorus retention characteristics of newly accreted soils in treatment wetlands of the Florida Everglades. M.S. Thesis. Gainesville, FL University of Florida.

Clark, I. D. and P. Fritz. 1997. *Environmental Isotopes in Hydrogeology*. Lewis Publishers, Boca Raton, FL. 328 pp.

Clesceri, L. S., A. E. Greenberg, and A. D. Eaton. 1998. *Standard Methods for the Examination of Water and Wastewater*. 20th Edition. American Public Health Association, the American Water Works Association and the Water Environment Federation.

Clymo, R. C. 1983. Peat. In: Gore, A. J. P. (ed.) Mires: Swamp, Bog, Fen, and Moor. Elsevier, Amsterdam, pp. 159–224.

Cohen, A. D. 1974. Petrography and paleoecology of Holocene peats from the Okefenokee swamp- marsh complex of Georgia. *Journal of Sedimentary Research*. 44:716–726.

Cohen, M. J., J. P. Prenger, and W. F. DeBusk. 2005. Visible-near infrared reflectance spectroscopy for rapid, nondestructive assessment of wetland soil quality. *J. Environ. Qual.* 34:1422–1434.

Colberg, P. J. 1988. Anaerobic microbial degradation of cellulose, lignin, oligolignols, and monoaromatic lignin derivatives. In A. J. Zehnder (ed.) *Biology of Microorganisms*. Wiley, New York. pp. 333–372.

Cole, J. J. 1982. Interactions between bacteria and algae in aquatic ecosystems. *Annu. Rev. Ecol. Syst.* 13:291–314.

Compeau, G. C. and R. Bartha. 1984. Methylation and demethylation of mercury under controlled redox, pH and salinity conditions. *Appl. Environ. Microbiol.* 48:1203.

Compeau, G. C. and R. Bartha. 1985. Sulfate-reducing bacteria—principal methylators of mercury in anoxic estuarine sediment. *Appl. Environ. Microbiol.* 50:498–502.

Compeau, G. C. and R. Bartha. 1987. Effect of salinity on mercury-methylating activity of sulfate-reducing bacteria in estuarine sediments. *Appl. Environ. Microbiol.* 53:261—265.

Connell, W. E. and W. H. Patrick, Jr. 1968. Reduction of sulfate to sulfide in waterlogged soil. *Soil Sci. Soc. Am. Proc.* 33:711–715.

Conrad, R. 1989. Control of methane production in terrestrial ecosystems. In M. O. Andreae and D. S. Schimel (eds.) *Exchange of Trace Gases between Terrestrial Ecosystems and The Atmosphere*. Dahlem Konferenzen, Wiley, Chichester, England. pp. 39–58.

Conrad, R. 1996. Soil microorganisms as controllers of atmospheric trace gases (H_2, CO, CH_4, OCS, N_2O, and NO). *Microbiol. Rev.* 60:609–640.

Constanza, R., R. d'Arge, R. deGroot, S. Farber, M. Grasso, B. Hannon, K. Limburg, S. Naeem, R. V. O'Neill, J. Paruelo, R. G. Raskin, P. Sutton, and M. van den Belt. 1997. The value of the worlds ecosystem services and natural capital. *Nature* 387:253–260.

Cooke, J. G. 1992. Phosphorus removal processes in a wetland after a decade of receiving a sewage effluent. *J. Environ. Qual.* 21:733–739.

Cooper, A. B. 1990. Nitrate depletion in the riparian zone and stream channel of a small headwater catchment. *Hydrobiologia* 202:13–26.

Corbet, A. S. 1935. The formation of hyponitrous acid as an intermediate compound in the biological or photochemical oxidation of ammonia to nitrous acid: microbiological oxidation. *Biochem. J.* 29(5):1086–1096.

Corbridge, D. E. C. 2000. *Phosphorus 2000: Chemistry, Biochemistry & Technology*. Elsevier, Amsterdam. 1267 pp.

Cornu, S., V. Deschatrettes, S. Salvador-Blanes, B. Clozel, M. Hardy, S. Branchut, and L. Le Forestier. 2005. Trace element accumulation in Mn–Fe-oxide nodules of a planosolic horizon. *Geoderma* 125:11–24.

Corstanje, R., S. Grunwald, K. R. Reddy, T. Z. Osborne, and S. Newman. 2006. Assessment of the spatial distribution of soil properties in a northern Everglades marsh. *J. Environ. Qual.* 35:938–949.

Corstanje, R., K. R. Reddy, J. P. Prenger, S. Newman, and A. V. Ogram. 2007. Soil microbial eco-physiological response to nutrient enrichment in a sub-tropical wetland. *Ecol. Indicators* 7:277–289.

Cowardin, L. M., M. V. Carter, F. C. Golet, and E. T. LaRoe. 1979. *Classification of Wetlands and Deepwater Habitats of the United States*. U.S. Department of Interior, Fish and Wildlife Service, Washington, DC, FWS/OBS-79/31.

Coyne, M. S. and J. A. Thompson. 2006. *Math for Soil Scientists*. Thomson Delmar Learning, New York. 285 pp.

Craft, C. B., S. W. Broome, and E. D. Seneca. 1989. Exchange of nitrogen, phosphorous, and organic-carbon between transplanted marshes and estuarine waters. *J. Environ. Qual.* 18(2):206–211.

Craft, C. B. and C. J. Richardson. 1993. Peat accretion and nutrient accumulation in nutrient enriched and unenriched Everglades peatlands. *Ecol. Appl.* 3:446–458.

Craft, C. B. and C. J. Richardson. 1993. Peat accretion, nutrient accumulation and phosphorus storage efficiency along an eutrophication gradient in the northern Everglades. *Biogeochemistry* 22:133–156.

Craft, C. B. and C. J. Richardson. 1997. Relationships between soil nutrients and plant species composition in Everglades peatlands. *J. Environ. Qual.* 26:224–232.

Craft, C. B. and C. J. Richardson. 1998. Recent and long-term organic soil accretion and nutrient accumulation in the Everglades. *Soil Sci. Soc. Am. J.* 62:834–843.

Craft, C. B., E. D. Seneca, and S. W. Broome. 1993. Vertical accretion in microtidal regularly and irregularly flooded estuarine marshes. *Estuar. Coast. Shelf Sci.* 37:371–386.

Craft, C. B., J. Vymazal, and C. J. Richardson. 1995. Response of Everglades plant communities to nitrogen and phosphorus additions. *Wetlands* 15:258–271.

Cramer, G. W. and J. W. Day, Jr. 1980. Productivity of the swamps and marshes surrounding Lake Pontchartrain, La. In J. H. Stone (ed.) *Environmental Analyses of Lake Pontchartrain, Louisiana, Its Surrounding Wetlands and Selected Land Uses*. Vol. 2. Coastal Ecology Laboratory, Center for Wetland Resources, Louisiana State University, pp. 593–645.

Crawford, R. M. M. 1989. The anaerobic retreat. In *Studies in Plant Survival. Studies in Ecology*. Vol. 11. Blackwell Scientific Publications, London. pp. 105–129.

Crawford, R. M. M. 1992. Oxygen availability as an ecological limit to plant distribution. *Adv. Ecol. Res.* 23:93–285.

CROGEE, 2003. Does the water flow influence Everglades landscape patterns. National Research Council Report, The National Academy of Sciences, Washington, DC. 41 pp.

Crosby, D. G. 1976. Nonbiological degradation of herbicides in the soil. In L. J. Audus (ed.) *Herbicides, Physiology, Biochemistry, Ecology*. Vol. 2. 2nd Edition. Academic Press, London. 65 pp.

Curran, M., M. Bole, and W. G. Allaway. 1986. Root aeration and respiration in young mangrove plants (*Avicennia marina* (Forsk) Vierh.). *J. Exp. Bot.* 37:1225–1233.

Davis, J. H. 1943. The natural features of southern Florida, FL. Geol. Surv. Biol. Bull. No. 25. Tallahassee, FL.

Davis, S. M. 1994. Phosphorus inputs and vegetation sensitivity in the Everglades. In S. M. Davis and J. C. Ogden (eds.) *Everglades: The Ecosystem and its Restoration*. St. Lucie Press, Delray Beach, Fl. pp. 357–378.

D'Amore, D. V., S. R. Stewart, and J. H. Huddleston. 2004. Saturation, reduction, and the formation of iron–manganese concretions in the Jackson-Frazier wetland. *Oregon Soil Sci. Soc. Am. J.* 68:1012–1022.

D'Angelo, E. M. 1998. Soil processes regulating the fate of chlorophenols in wetlands. Ph.D. Dissertation. University of Florida, Gainesville, FL.

D'Angelo, E. M. 2002. Wetlands: biodegradation of organic pollutants. In G. Bitton (ed.) *Encyclopedia of Environmental Microbiology*. Wiley, New York. pp. 3401–3417.

D'Angelo, E. M. and K. R. Reddy. 1993. Ammonium oxidation and nitrate reduction in a hypereutrophic lake. *Soil Sci. Soc. Am. J.* 57:1156–1163.

D'Angelo, E. M. and K. R. Reddy. 1994a. Diagenesis of organic matter in a wetland receiving hypereutrophic lake water. I. Distribution of dissolved nutrients in the soil and water column. *J. Environ. Qual.* 23:937–943.

D'Angelo, E. M. and K. R. Reddy. 1994b. Diagenesis of organic matter in a wetland receiving hypereutrophic lake water. II. Role of inorganic electron acceptors in nutrient release. *J. Environ. Qual.* 23:928–936.

D'Angelo, E. M. and K. R. Reddy. 1999. Regulators of heterotrophic microbial potentials in wetland soils. *Soil Biol. Biochem.* 31:815–830.

D'Angelo, E. M. and K. R. Reddy. 2000. Aerobic and anaerobic transformations of pentachlorophenol in wetland soils. *Soil Sci. Soc. Am. J.* 64:933–943.

D'Angelo, E. M. and K. R. Reddy. 2003. Effect of aerobic and anaerobic conditions on chlorophenol sorption in wetland soils. *Soil Sci. Soc. Am. J.* 67:787–794.

Dacey, J. W. H. 1981. Pressurized ventilation in the yellow water lily. *Ecology* 62:1137–1147.

Dacey, J. W. H. 1987. Knudsen-transitional flow and gas pressurization in leaves of *Nelumbo*. *Plant. Physiol.* 85:199–203.

References

Dacey, J. W. H. 1980. Internal winds in water lilies: an adaptation for life in anaerobic sediments. *Science* 210:1017–1019.

Dacey, J. W. H. 1984. Water uptake by roots control water table movement and sediment oxidation in short *Spartina* marsh. *Science* 227:487–489.

Dacey, J. W. H. and M. J. Klug. 1979. Methane efflux from lake sediments through water lilies. *Science* 203:1253–1254.

Dalias, P., J. M. Anderson, P. Bottner, and M. M. Couteaux. 2001. Long-term effects of temperature on carbon mineralization process. *Soil Biol. Biochem.* 33:1049–1057.

Dalsgaard, T. and B. Thamdrup. 2002. Factors controlling anaerobic ammonium oxidation with nitrite in marine sediments. *Appl. Environ. Microbiol.* 68:3802–3808.

Dalsgaard, T., B. Thamdrup, and D. E. Canfield. 2005. Anaerobic ammonium oxidation (anammox) in the marine environment. *Res. Microbiol.* 156:457–464.

Dane, J. H. and G. C. Topp (eds.). 2002. *Methods of Soil Analysis: Part 4, Physical Method*. SSSA Book Series No. 5. Soil Science Society of America, Madison, WI. 1692 pp.

Darke, A. K. and M. R. Walbridge. 2000. Al and Fe biogeochemistry in a floodplain forest: implications for P retention. *Biogeochemistry* 51:1–32.

Davidson, E. A. and I. A. Janssens. 2006. Temperature sensitivity of soil carbon decomposition and feedbacks to climate change. *Nature* 440:165–173.

Davidson, E. A., E. Belk, and R. D. Boone. 1998. Soil water content and temperature as independent or confounded factors controlling soil respiration in a temperate mixed hardwood forest. *Global Change Biol.* 4:217–227.

Davis, S. M. 1991. Growth, decomposition, and nutrient retention of *Cladium jamaicense* Crantz and *Typha domingensis* Pers in the Florida Everglades. *Aquat. Bot.* 40:203–224.

Davis, S. M. and J. C. Ogden (eds.). 1994. *Everglades: The Ecosystem and Its Restoration*. St. Lucie Press, Delery Beach.

Day, J. W. Jr., T. J. Butler, and W. H. Conner. 1977. Productivity and export studies in a cypress swamp and lake system in Louisiana. In M. L. Wiley (ed.) *Estuarine Processes*. Vol. II. Academic Press, New York, pp. 255–269.

De'ath, G. and K. E. Fabricius. 2000. Classification and regression trees: a powerful yet simple technique for ecological data analysis. *Ecology* 81:3178–3192.

de Gruijter, J., D. Brus, M. Bierkens, and M. Knotters. 2006. *Sampling for Natural Resource Monitoring*. Springer, Berlin. 332 pp.

DeBusk, T. A., K. A. Grace, F. E. Dierberg, S. D. Jackson, M. J. Chimney, and B. Gu. 2004. An investigation of the limits of phosphorous removal in wetlands: a mesocosm study of a shallow periphyton-dominated treatment system. *Ecol. Eng.* 23:1–14.

DeBusk, W. F. 1996. Organic matter turnover along a nutrient gradient in the Everglades. Ph.D. Dissertation. University of Florida, Gainesville, FL.

DeBusk, W. F., S. Newman, and K. R. Reddy. 2001. Spatio-temporal patterns of soil phosphorus enrichment in Everglades Water Conservation Area 2A. *J. Environ. Qual.* 30:1438–1446.

DeBusk, W. F. and K. R. Reddy. 1987. Removal of floodwater nitrogen in a cypress-mixed hardwood swamp receiving primary sewage effluent. *Hydrobiologia* 153:79–86.

DeBusk, W. F. and K. R. Reddy. 1998. Turnover of detrital organic carbon in a nutrient-impacted Everglades marsh. *Soil Sci. Soc. Am. J.* 62(5):1460–1468.

DeBusk, W. F. and K. R. Reddy. 2003. Nutrient and hydrology effects on soil respiration in a northern everglades marsh. *J. Environ. Qual.* 32(2):702–710.

DeBusk, W. F. and K. R. Reddy. 2005. Litter decomposition and nutrient dynamics in a phosphorus enriched Everglades marsh. *Biogeochemistry* 75:217–240.

DeBusk, W. F., K. R. Reddy, M. S. Koch, and Y. Wang. 1994. Spatial distribution of soil nutrients in a Northern Everglades marsh: water conservation area 2A. *Soil Sci. Soc. Am. J.* 58:543–552.

DeBusk, W. F., J. R. White, and K. R. Reddy. 2001. Carbon and nitrogen dynamics in wetland soils. In M. J. Shaffer, L. Ma, and S. Hansen. (eds.) *Modeling Carbon and Nitrogen Dynamics for Soil Management*. Lewis Publishers, Boca Raton, FL. pp. 27–53.

DeLaune, R. D., R. H. Baumann, and J. G. Gosselink. 1983. Relationship among vertical accretion, coastal submergence, and erosion in a Louisiana Gulf Coast marsh. *J. Sediment. Petrol.* 53(1): 147–157.

DeLaune, R. D., R. J. Buresh, and W. H. Patrick, Jr. 1978. Sedimentation rates determined by ^{137}Cs dating in a rapidly accreting salt marsh. *Nature* 275:532–533.

DeLaune, R. D., J. C. Callaway, W. H. Patrick, Jr., and J. A. Nyman. 2004. Analysis of marsh accretionary process in Louisiana coastal wetlands. In D. W. Davis and M. Richardson (eds.) *The Coastal Zone: Papers in Honor of H. Jesse Walker*. Geoscience Publications, Department of Geography and Anthropology, Louisiana State University, Baton Rouge, LA. pp. 113–130.

DeLaune, R. D., I. Devai, C. R. Crozier, and P. Kelle. 2002a. Sulfate reduction in Louisiana marsh soils of varying salinities. *Commun. Soil Sci. Plant Anal.* 33:79–94.

DeLaune, R. D., I. Devai, and C. W. Lindau. 2002b. Flux of reduced sulfur gases along a salinity gradient in Louisana coastal marshes. *Estuar. Coast. Shelf Sci.* 54:1003–1011.

DeLaune, R. D., I. Devai, C. Mulbar, C. Crozier, and C. W. Lindau. 1997. The influence of redox condition on atrazine degradation in wetlands. *Agric. Environ.* 66:41–46.

DeLaune, R. D., T. C. Feijtel, and W. H. Patrick, Jr. 1989. Nitrogen flows in a La Gulf Coast salt marsh. *Biogeochemistry* 8:25–37.

DeLaune, R. D., R. P. Gambrell, J. H. Pardue, and W. H. Patrick, Jr. 1990a. Fate and effect of petroleum hydrocarbons and toxic organics in Louisiana coastal environments. *Estuaries* 13(1):72–80.

DeLaune, R. D., R. P. Gambrell, and K. S. Reddy. 1983. Fate of pentachlorophenol in estuarine sediment. *Environ. Pollut. Series B* 6:297–308.

DeLaune, R. D., G. A. Hambrick, and W. H. Patrick, Jr. 1980. Degradation of hydrocarbons in oxidized and reduced sediments. *Mar. Pollut. Bull.* 11:103–106.

DeLaune, R. D., A. Jugsujinda, I. Devai, and W. H. Patrick, Jr. 2004. Relationship of sediment redox conditions to methylmercury in surface sediment of Louisiana Lakes. *J. Environ. Sci. Health.* A39(8):1925–1935.

DeLaune, R. D., A. Jugsujinda, G. W. Peterson, and W. H. Patrick, Jr. 2003. Impact of Mississippi River freshwater reintroduction on enhancing marsh accretionary processes in a Louisiana estuary. *Est. Coast. Shelf Sci.* 58:653–662.

DeLaune, R. D., A. Jugsujinda, J. L. West, C. B. Johnson, and M. Kongehum. 2005. A screening of the capacity of Louisiana wetlands to process nitrate in diverted Mississippi River Water. *Ecol. Eng.* 25:315–321.

DeLaune, R. D., A. Jugsujnda, and K. R. Reddy. 1999. Effect of root oxygen stress on phosphorus uptake by cattail. *J. Plant Nutr.* 22:459–466.

DeLaune, R. D., J. A. Nyman, and W. H. Patrick, Jr. 1994. Peat collapse, ponging and wetland loss in a rapidly submerging coastal marsh. *J. Coast. Res.* 10(4):1021–1030.

DeLaune, R. D. and W. H. Patrick, Jr. 1979. Nitrogen and phosphorous cycling in a Gulf Coast salt marsh. *Proceedings of the 5th Biennial International Estuaries Research Conference*, Jekyll Island, Georgia. October 7–12, 1979.

DeLaune, R. D., W. H. Patrick, Jr., and J. Brannon. 1976. *Nutrient Transformation in Louisiana Salt Marsh Soils*. Louisiana State University, Sea Grant Publication. LSU-T-76-009, Center for Wetland Resources, Louisiana State University, Baton Rouge, LA.

DeLaune, R. D., W. H. Patrick, Jr., and M. Casselman. 1981. Effect of sediment pH and redox condition on degradation of benzo(a)pyrene. *Mar. Pollut. Bull.* 12(7):251–253.

DeLaune, R. D., W. H. Patrick, and T. Gao. 1998. The redox–pH chemistry of chromium in water and sediments. In H. E. Allen, A. W. Garrison, and G. W. Luther III (eds.) *Metals in Surface Waters*. Ann Arbor Press, Chelsea, MI. pp. 241–253.

DeLaune, R. D. and S. R. Pezeshki. 1988. Relationship of mineral nutrients to growth of *Spartina alterniflora* in Louisiana salt marshes. *Northeast Gulf Sci.* 10(1):55–60.

DeLaune, R. D., S. R. Pezeshki, and C. W. Lindau. 1998. Influence of soil redox potential on nitrogen uptake and growth of wetland oak seedlings. *J. Plant Nutr.* 21:757–768.

DeLaune, R. D., S. R. Pezeshki, and J. H. Pardue. 1990. An oxidation-reduction buffer for evaluating physiological response of plants to root oxygen stress. *Environ. Exp. Bot.* 30(2):243–247.

DeLaune, R. D., S. R. Pezeshki, J. H. Whitcomb, and W. H. Patrick, Jr. 1990. Some influence of sediment addition to a deteriorating salt marsh in the Mississippi River deltaic plain. A pilot study. *J. Coast. Res.* 4:181–190.

DeLaune, R. D., C. N. Reddy, and W. H. Patrick, Jr. 1981. Accumulation of plant nutrients and heavy metals through sedimentation processes and accretion in a Louisiana salt marsh. *Estuaries* 4:328–334.

DeLaune, R. D. and L. M. Salinas. 1985. Fate of 2,4-D entering a freshwater aquatic environment. *Bull. Environ. Contam. Toxicol.* 35:564–568.

DeLaune, R. D. and C. J. Smith. 1984. The carbon cycle and the rate of vertical accumulation of peat in the Mississippi River Deltaic Plain. *Southeast. Geol.* 25:61–68.

DeLaune, R. D., C. J. Smith, and W. H. Patrick, Jr. 1983. Methane release from Gulf coast wetlands. *Tellus* 35B:8–15.

DeLaune, R. D., C. J. Smith, and W. H. Patrick, Jr. 1983. Relationship of marsh elevation, redox potential, and sulfide to *Spartina alterniflora* productivity. *Soil Sci. Soc. Am. J.* 47:930–935.

DeLaune, R. D., C. J. Smith, and W. H. Patrick, Jr. 1983. Methane release from Gulf coast wetlands. *Tellus* 35:8–15.

DeLaune, R. D., C. J. Smith, and W. H. Patrick, Jr. 1983. Nitrogen losses from a Louisiana Gulf Coast salt marsh. *Est. Coast. Shelf Sci.* 17:133–141.

DeLaune, R. D., C. J. Smith, and W. H. Patrick, Jr. 1986. Land Loss in coastal Louisiana: effect of sea level rise and marsh accretion. Final Report to the L.S.U. Board of Regents, Research and Development Program.

DeLaune, R. D., C. J. Smith, W. H. Patrick, Jr., J. D. Fleeger, and M. D. Tolley. 1984. Effect of oil on saltmarsh biota: methods for restoration. *Environ. Pollut. (Series A)* 36:207–227.

DeLaune, R. D., C. J. Smith, W. H. Patrick, Jr., and H. H. Roberts. 1987. Rejuvenated marsh and bay-bottom accretion on the rapidly subsiding costal plain of the U.S. Gulf Coast: a second order effect of the emerging Atchafalaya delta. *Shelf Sci.* 25:381–389.

DeLaune, R. D., N. Van Breeman, and W. H. Patrick, Jr., 1990b. Processes governing marsh formation in a rapidly subsiding coastal environment. *Catena* 17:277–288.

Delwiche, C. C. 1981. *Denitrification, Nitrification, and Atmospheric Nitrous Oxide.* Wiley, New York. 286 pp.

Demas, G. P. and M. C. Rabenhorst. 1999. Subaqueous soils: pedogenesis in a submersed environment. *Soil Sci. Soc. Am. J.* 63:1250–1257.

Demas, G. P., M. C. Rabenhorst, and J. C. Stevenson. 1996. Subaqueous soils: a pedological approach to the study of shallow-water habitats. *Estuaries* 19:229–237.

Dévai, I. and R. D. DeLaune. 1995. Evidence for phosphine production and emission from Louisiana and Florida marsh soils. *Org. Geochem.* 23:277–279.

Dévai, I., L. Felföldy, I. Wittner, and S. Plösz. 1988. Detection of phosphine: new aspects of the phosphorus cycle in the hydrosphere. *Nature* 333:343–345.

Di Toro, D. M., J. D. Mahony, P. R. Krichgraber, A. L. O'Byrne, and L. R. Pasquale. 1986. Effect of nonreversibility, particle concentration and ionic strength on heavy metal sorption. *Environ. Sci. Technol.* 20:55–56.

Diaz, O. A., K. R. Reddy, and P. A. Moore, Jr. 1994. Solubility of inorganic P in stream water as influenced by pH and Ca concentration. *Water Res.* 28:1755–1763.

Dick, W. A. 1992. Sulfur cycle. In *Encyclopedia of Microbiology*, Vol. 4. Academic Press, New York. pp. 123–133.

Diffendorfer, J. E., P. M. Richards, G. H. Dalrymple, and D. L. DeAngelis. 2001. Applying linear programming to estimate fluxes in ecosystems or food webs: an example from the herpetological assemblage of the freshwater Everglades. *Ecol. Model.* 144:99–120.

Ding, W., Z. Cai, and H. Tsuruta. 2005. Plant species effects on methane emissions from freshwater marshes. *Atmos. Environ.* 39:3199–3207.

Dixon, W. and B. Chiswell. 1996. Review of aquatic monitoring program design. *Water Res.* 30:1935–1948.

Doran, R. F., T. V. Armentano, L. D. Whiteaker, and R. D. Jones. 1997. Marsh vegetation patterns and soil phosphorus gradients in the Everglades ecosystem. *Aquat. Bot.* 30:1–19.

Dove, P. N. and M. F. Hochella. 1993. Calcite precipitation mechanisms and inhibition by orthophosphate: *In situ* observations by scanning force microscopy. *Geochim. et Cosmochim. Acta* 57:705–714.

Drake, H. L., N. G. Aumen, C. Kuhner, C. Wagner, A. Grießhammer, and M. Schmittroth. 1996. Anaerobic microflora of Everglades sediments: effects of nutrients on population profiles and activities. *Appl. Environ. Microbiol.* 62:486–493.

Drever, J. I. 1982. *The Geochemistry of Natural Waters.* Prentice Hall, New Jersey. pp. 250–278.

Duever, M. J., J. F. Meeder, L. C. Meeder, and J. M. McCollom. 1994. The climate of south Florida and its role in shaping the Everglades ecosystem. In S. M. Davis and J. C. Ogden (eds.) *Everglades: The Ecosystem and its Restoration*. St. Lucie Press, Delray Beach, FL. pp. 225–248.

Duncan, C. P. and P. M. Groffman. 1994. Comparing microbial parameters in natural and constructed wetlands. *J. Environ. Qual.* 23:298–305.

Ehrlich, H. L. 2002. *Geomicrobiology.* Marcel Dekker, New York (Chapters 15 and 16).

Eivazi, F. and M. A. Tabatabai. 1988. Glucosidases and galactosidases in soils. *Soil Biol. Biochem.* 20:601–606.

Emerson, D. 2000. Microbial oxidation of Fe(II) and Mn(II) at circumneutral pH. In D. R. Lovley (ed.) *Environmental Metal–Microbe Interactions*. ASM Press, Washington, DC. pp. 31–52.

Engler, R. M., D. A. Antie, and W. H. Patrick, Jr. 1976. Effect of dissolved oxygen on redox potential and nitrate removal in flooded swamp and marsh soils. *J. Environ. Qual.* 5:230–235.

Engler, R. M. and W. H. Patrick. 1974. Nitrate removal from floodwater overlaying flooded soils and sediments. *J. Environ. Qual.* 3:409–413.

Engler, R. M. and W. H. Patrick, Jr. 1975. Stability of sulfides of manganese, iron, zinc, copper, and mercury in flooded and non-flooded soil. *Soil Sci.* 119:217–221.

Entry, J. A., D. Mills, K. Jayachandran, and T. B. Moorman. 2007. Symposium: molecular-based approaches to soil microbiology. *Soil Sci. Soc. Am. J.* 71:561.

Fairchild, J. F., D. N. Ruessler, and R. Carlson. 1998. Comparative sensitivity of fine species of macrophytes and six species of algae to atrazine, mutribuzin alachlor and metolachlor. *Environ. Toxicol. Chem.* 17(9):1830–1834.

Fang, C., P. Smith, J. B. Moncriett, and J. U. Smith. 2005. Similar response of labile and resistant soil organic matter pools to changes in temperature. *Nature* 433:57–59.

Faulkner, S. P. and W. H. Patrick, Jr. 1992. Redox processes and diagnostic wetland Soil indicators in bottomland hardwood forests. *Soil Sci. Soc. Am. J.* 56:856–865.

Faulkner, S. P., W. H. Patrick, Jr., and R. P. Gambrell. 1989. Field techniques for measuring wetland soil parameters. *Soil Sci. Soc. Am. J.* 53:883–890.

Feijtel, T. C., R. D. DeLaune, and W. H. Patrick, Jr. 1985. Carbon flux in Coastal Louisiana. *Mar. Ecol. Prog. Ser.* 24:255–260.

Feijtel, T. C., R. D. DeLaune, and W. H. Patrick, Jr. 1988. Biogeochemical control on metals distribution and accumulation in Louisiana sediment. *J. Environ. Qual.* 17:88–89.

Fenchel, T., G. M. King, and T. H. Blackburn. 1998. *Bacterial Biogeochemistry*. Academic Press, New York. 307 pp.

Findlay, R. H., M. B. Trexler, J. B. Guckert, and D. C. White. 1990. Laboratory study of disturbance in marine sediments: response of a microbial community. *Mar. Ecol. Prog. Ser.* 62:121–133.

Firestone, M. K. 1982. Biological denitrification. *Nitrogen in Agricultural Soils*. Agronomy Monograph 22. American Society of Agronomy, Madison, WI. pp. 289–326.

Fisher, M. M. and K. R. Reddy. 2001. Phosphorus flux from wetland soils affected by long-term nutrient loading. *J. Environ. Qual.* 30:261–271.

Fitz, C., F. Sklar, T. Waring, A. Voinov, R. Costanza, and T. Maxwell. 2004. Development and application of the Everglades Landscape Model. In R. Costanza and A. Voinov (eds.) *Landscape Simulation Modeling—A Spatially Explicit, Dynamic Approach*. Springer, New York. pp. 143–172.

Flessa, H. 1994. Plant induced changes in the redox potentials of the rhizosphere of the submerged vascular macrophytes *Myriophyllum verticillatum* L. and *Ranunculus circinatus* L. *Aquat. Bot.* 47:119–129.

Fossing, H. and B. B. Jørgensen. 1989. Measurements of bacterial sulfate reduction in sediments: evaluation of a single-step chromium reduction method. *Biogeochemistry* 8:205–222.

Francis, C. A., J. M. Beman, and M. M. M. Kuypers. 2007. New processes and players in the nitrogen cycle: the microbial ecology of anaerobic and archaeal ammonia oxidation. *Int. Soc. Microbiol. Ecol. J.* 1:19–27.

Frankenburger, W. T. (eds.). 2001. *Environmental Chemistry of Arsenic*. Marcel Dekker, New York.

Freedman, D. L. and J. M. Gossett. 1989. Biological reductive dechlorination of tetrachloroethylene and trichloroethylene to ethylene under methanogenic conditions. *Appl. Environ. Microbiol.* 55:2144–2151.

Freeman, C., C. D. Evans, D. T. Monteith, B. Reynolds, and N. Fenner. 2001. Export of organic carbon from peat soils. *Nature* 412:785.

Freeman, C., N. Ostle, and H. Kang. 2001. An enzymatic 'latch' on a global carbon store. *Nature* 409:149.

French, J. R. and T. Spencer. 1993. Dynamics of sedimentation in a tide-dominated backbarrier salt marsh, Norfolk, UK. *Mar. Geol.* 110:315–331.

French, P. W., J. R. L. Allen, and P. G. Appleby. 1994. 210-Lead dating of a modern period saltmarsh deposit from the Severn Estuary (Southwest Britain), and its implications. *Mar. Geol.* 118:327–334.

Freney, J. R., O. T. Denmead, I. Watanabe, and E. T. Craswell. 1981. Ammonia and nitrous oxide losses following applications of ammonium sulfate to flooded rice. *Aust. J. Agric. Res.* 32:37–45.

Frenzel, P., U. Bosse, and P. H. Janssen. 1999. Rice roots and methanogenesis in a paddy soil: ferric iron as an alternate electron acceptor in the rooted soil. *Soil Biol. Biochem.* 31:421–430.

Froelich, P. N. 1988. Kinetic control of dissolved phosphate in natural rivers and estuaries: a primer on the phosphate buffer mechanism. *Limnol. Oceanogr.* 33(4):649–668.

Frostegard, A., E. Baath, and A. Tunlid. 1993. Shifts in the structure of soil microbial communities in limed forests as revealed by phospholipid fatty acid analysis. *Soil Biol. Biochem.* 25:723–730.

Gaiser, E. E., J. H. Richards, J. C. Trexler, R. D. Jones, and D. L. Childers. 2006. Periphyton responses to eutrophication in the Florida Everglades: cross-system patterns of structural and compositional change. *Limnol. Oceanogr.* 51:617–630.

Gaiser, E. E., J. C. Trexler, J. H. Richards, D. L. Childers, D. Lee, A. L. Edwards, L. J. Scinto, K. Jayachandran, G. B. Noe, and R. D. Jones. 2005. Cascading ecological effects of low-level phosphorus enrichment in the Florida Everglades. *J. Environ. Qual.* 34:717–723.

Gale, P. M., I. Devai, K. R. Reddy, and D. A. Graetz. 1993. Denitrification potential of soils from constructed and natural wetlands. *Ecol. Eng.* 2:119–130.

Gale, P. M., K. R. Reddy, and D. A. Graetz. 1994. Phosphorus retention by wetland soils used for treated wastewater disposal. *J. Environ. Qual.* 23:370–377.

Galloway, J. N. 2005. The global nitrogen cycle. In W. H. Schlesinger (ed.). *Biogeochemistry.* Elsevier-Pergamon, Oxford, UK. pp. 585–644.

Galloway, J. N., F. J. Dentener, D. G. Capone, E. W. Boyer, R. W. Howarth, S. P. Seitzinger, G. P. Asner, C. C. Cleveland, P. A. Green, E. A. Holland, D. M. Karl, A. F. Michaels, J. H. Porter, A. R. Townsend, and C. J. Vorosmarty. 2004. Nitrogen cycles: past, present, and future. *Biogeochemistry* 70:153–226.

Gambrell, R. P. 1994. Trace and toxic metals in wetlands—a review. *J. Environ. Qual.* 23:883–891.

Gambrell, R. P., R. A. Khalid, M. G. Verloo, and W. H. Patrick, Jr. 1975. "Transformation of Heavy Metals and Plant Nutrients in Dredged Sediments as Affected by Oxidation-Reduction and pH. Volume II. Materials and Methods, Results and Discussion," report submitted to Office of Dredged Material Research, U.S. Army Engineer Waterways Experiment Station, Vicksburg, Mississippi, U.S.A.

Gambrell, R. P., R. A. Khalid, and W. H. Patrick, Jr. 1980. Chemical availability of mercury, lead, and zinc in Mobile Bay sediment suspension as affected by pH and oxidation–reduction conditions. *Environ. Sci. Technol.* 14:431–436.

Gambrell, R. P. and W. H. Patrick, Jr. 1988. The influence of redox potential on the environmental chemistry of contaminants in soils and sediments. In D. D. Hook (ed.) *The Ecology and Management of Wetlands.* Vol. 1. Timber Press, Portland, OR. pp. 319–333.

Gambrell, R. P. and W. H. Patrick, Jr. 1991. *Handbook—Remediation of Contaminated Sediments.* EPA/625/6-91/028.

Gambrell, R. P., C. N. Reddy, V. Collard, G. Green, and W. H. Patrick, Jr. 1984. The recovery of DDT, Kepone and permethrin added to soil and sediment suspensions incubated under controlled redox potential and pH conditions. *J. Water Poll. Control Fed.* 52:174–182.

Gambrell, R. P., J. B. Wiespape, W. H. Patrick, Jr., and M. C. Duff. 1991. The effects of pH, redox and salinity on metal release from a contaminated sediment. *Water Air Soil Pollut.* 57–58:359–367.

Garcia, J. L. 1990. Taxonomy and ecology of methanogens. *FEMS Microbial. Rev.* 87:297–308.

Garland, J. L. and A. L. Mills. 1991. Classification and characterization of heterotrophic microbial communities on the basis of patterns of community-level sole-carbon-source utilization. *Appl. Environ. Microbiol.* 57:2351–2359.

Garrels, R. M. and C. L. Christ. 1965. *Solutions, Minerals, and Equilibria.* Harper & Row, New York. 450 pp.

Geta, R., C. Postolache, and A. Vadineanu. 2004. Ecological significance of nitrogen cycling by Tubificid communities in shallow eutrophic lakes of the Danube delta. *Hydrobiologia* 524(1):193–202.

Ghiorse, W. C. 1988. Microbial reduction of manganese and iron. In A. J. B. Zehnder (ed.) *Biology of Anaerobic Microorganisms.* Wiley, New York. pp. 305–331.

Giannimaras, E. K. and P. G. Koutsoukos. 1987. The crystallization of calcite in the presence of orthophosphate. *J. Colloidal Interface Sci.* 116:423–430.

Gillespie, L. J. 1920. Reduction potentials of bacterial cultures and water-logged soils. *Soil Sci.* 9:199–216.

Gilmour, C. C., G. S. Riedel, M. C. Ederington, J. T. Bell, J. M. Benoit, G. A. Gill, and M. C. Stordal. 1998. Methylmercury concentrations and production rates across a trophic gradient in the northern Everglades. *Biogeochemistry* 40:327–345.

Gleason, P. J., A. D. Cohen, P. Stone, W. G. Smith, H. K. Brooks, R. Goodrick, and W. Spackman, Jr. 1974. The environmental significance of Holocene sediments from the Everglades and saline tidal plains. In P. J. Gleason (ed.) *Environments of South Florida, Present, and Past.* Miami Geol. Soc., Coral Gables, FL. pp. 297–351.

Goldberg, E. D., E. Gamble, J. J. Griffin, and M. Koide. 1997. Pollution history of Narragansett Bay as recorded in its sediments. *Estuar. Coast. Mar. Sci.* 5(4):549.

Goldberg, S. and R. A. Glaubig. 1988. Anion adsorption on calcerious, montmorollonitic soil—arsenic. *Soil Sci. Soc. Am. J.* 52:1297–1300.

Goolsby, D. A., W. A. Battaglin, and E. M. Thurman. 1993. Occurrence and transport of agricultural chemicals in the Mississippi River, July through August 1993. U.S. Geological Survey Circular. 1120-C, 22 pp.

Goovaerts, P. 1994. Study of spatial relationships between two sets of variables using multivariate geostatistics. *Geoderma* 62:93–107.

Goovaerts, P. 1997. *Geostatistics for Natural Resources Evaluation*. Oxford University Press, New York.

Goovaerts, P. 1999. Geostatistics in soil science: state-of-the-art and perspectives. *Geoderma* 89:1–45.

Gopal, B. and V. Masing. 1990. Biology and ecology. In B. C. Pattan et al. (eds.) *Wetlands and Shallow Continental Water Bodies*. Vol. 1. SPB Academic Publishing, the Netherlands. pp. 91–239.

Gorham, E., 1995. The biogeochemistry of Northern Peatlands and its possible response to global warming. In G. M. Woodwell and F. T. MacKenzie (eds.) *Biotic Feedback in the Global Climate System. Will the Warming Feed Warming?* Oxford University Press, New York, NY. pp. 169–187.

Gornitz, V. 1995. Sea-level rise: a review of recent past and near-future trends. *Earth Surf. Proc. Landforms* 20:7–20.

Gornitz, V. and S. Lebedeff. 1987. Global sea-level changes during the past century. In D. Nummedal, O. Pilkey, and J. D. Howard (eds.) *Sea Level Fluctuation and Coastal Erosion*. Vol. 4. SEPM Special Publication. pp. 3–16.

Gotoh, S. and W. H. Patrick, Jr. 1972. Transformation of manganese in a water logged soil as affected by redox potential and pH. *Soil Sci. Soc. Am. Proc.* 36:738–742.

Gotoh, S. and W. H. Patrick, Jr. 1974. Transformation of iron in a water logged soil as influenced by redox potential and pH. *Soil Sci. Soc. Am. Proc.* 38:66–71.

Gottschalk, G. 1996. *Bacterial Metabolism*. Springer Verlag, New York.

Green, R. H. 1979. *Sampling Design and Statistical Methods for Environmental Biologists*. Wiley.

Griffin, T. M. and M. C. Rabenhorst. 1989. Processes and rates of pedogenesis in some Maryland tidal marsh soils. *Soil Sci. Soc. Am. J.* 53(3):862–870.

Grimshaw, H. J., R. G. Wetzel, M. Brandenburg, M. Segerblom, L. J. Wenkert, G. A. Marsh, W. Charnetzky, J. E. Haky, and C. Carraher. 1997. Shading of periphyton communities by wetland emergent macrophytes: decoupling of algal photosynthesis from microbial nutrient retention. *Arch. Hydrobiol.* 139:17–27.

Groffman, P. M., M. A. Altabet, J. K. Bohlke, K. Butterbach-Bahl, M. B. David, M. K. Firestone, A. E. Giblin, T. M. Kana, L. P. Nielsen, and M. A. Voytek. 2006. Methods for measuring denitrification: diverse approaches to a difficult problem. *Ecol. Appl.* 16:2091–2122.

Grunwald, S. (ed.) 2006a. *Environmental Soil-Landscape Modeling: Geographic Information Technologies and Pedometrics*. CRC Press, New York. 488 pp.

Grunwald, S. 2006b. What do we really know about the space-time continuum of soil-landscapes. In S. Grunwald (ed.) *Environmental Soil-Landscape Modeling—Geographic Information Technologies and Pedometrics*. CRC Press, New York. pp. 3–36.

Grunwald, S., R. Corstanje, B. E. Weinrich, and K. R. Reddy. 2006a. Spatial patterns of labile forms of phosphorus in a subtropical wetland 10 years after a sustained nutrient impact. *J. Environ. Qual.* 35:378–389.

Grunwald, S., P. Goovaerts, C. M. Bliss, N. B. Comerford, and S. Lamsal. 2006b. Incorporation of auxiliary information in the geostatistical simulation of soil nitrate-nitrogen. *Vadose Zone J.* 5:391–404.

Grunwald, S. and S. Lamsal. 2006. Emerging geographic information technologies and soil information systems. In S. Grunwald (ed.) *Environmental Soil-Landscape Modeling: Geographic Information Technologies and Pedometrics*. CRC Press, New York. pp. 127–154.

Grunwald, S. and C. Qi. 2006. GIS-based water quality modeling in the Sandusky Watershed. *J. Am. Water Res. Assoc.* 42(4):957–973.

Grunwald, S., K. R. Reddy, S. Newman, and W. F. DeBusk. 2004. Spatial variability, distribution, and uncertainty assessment of soil phosphorus in a south Florida wetland. *Environmetrics* 15:811–825.

Grunwald, S., K. R. Reddy, J. P. Prenger, and M. M. Fisher. 2007a. Modeling of the spatial variability of biogeochemical soil properties in a freshwater ecosystem. *Ecol. Model.* 210:521–535.

Grunwald, S., R. G. Rivero, and K. R. Reddy. 2007b. Understanding spatial variability and its application to biogeochemistry analysis. In D. Sarkar, R. Datta, and R. Hannigan (eds.) *Environmental Biogeochemistry—Concepts and Case Studies*. Elsevier, Amsterdam, The Netherlands, pp. 435–462 (chapter 20).

Gunderson, L. H. 1994. Vegetation of the Everglades: determinants of community composition. In S. M. Davis and J. C. Ogden (eds.) *Everglades: The Ecosystem and its Restoration*. St. Lucie Press, Delray Beach, FL. pp. 323–340.

Gusewell, S. and W. Koerselman. 2002. Variation in nitrogen and phosphorus concentrations of wetland plants. *Perspect. Plant Ecol. Evol. Syst.* 5:37–61.

Gutknecht, J. L. M., R. M. Goodman, and T. C. Balser. 2006. Linking soil process and microbial ecology in freshwater wetland ecosystems. *Plant Soil* 289:17–34.

Hair, Jr., J. F., R. E. Anderson, R. L. Tatham, and W. C. Black. 1995. *Multivariate Data Analysis: With Readings.* Prentice-Hall, Upper Saddle River, NJ.

Hambrick, G. A. III, R. D. DeLaune, and W. H. Patrick, Jr. 1980. Effect of estuarine sediment pH and oxidation–reduction potential on microbial hydrocarbon degradation. *Appl. Environ. Microbiol.* 40(2):365–369.

Happell, J. D., J. P. Chanton, G. W. Whiting, and W. S. Showers. 1993. Stable isotope tracing of methane dynamics in Everglades marshes with and without active populations of methane oxidizing bacteria. *J. Geophys. Res.* 98:14771–14782.

Harris, W. G. 2002. Phosphate minerals. In *Soil Mineralogy with Environmental Applications.* SSSA Book Series No. 7, Soil Science Society of America, Madison, WI. pp. 637–665.

Hartley, A. M., W. A. House, M. E. Callow, and B. S. C. Leadbeater. 1997. Coprecipitation of phosphate with calcite in the presence of photosynthesizing green algae. *Water Res.* 9:2261–2268.

Hassett, J. P. and E. Milicic. 1985. *Environ. Sci. Technol.* 19:638–643.

Hatton, R. S., R. D. DeLaune, and W. H. Patrick, Jr. 1983. Sedimentation, accretion and subsidence in marshes of Barataria Basin, Louisiana. *Limnol. Oceanogr.* 28(3):494–502.

Haykin, S. 1999. *Neural Networks.* Prentice-Hall, Upper Saddle River, NJ.

Haynes, R. J. and R. S. Swift. 1985. Effect of air-drying on the adsorption and desorption of phosphate adsorption and desorption of phosphate and levels of extractable phosphate in a group of New Zealand acid soils. *Geoderma* 35:145–157.

Haynes, R. J. and R. S. Swift. 1986. Effects of soil acidification and subsequent leaching on levels of extractable nutrients in a soil. *Plant Soil* 95:327–336.

Haynes, R. J. and R. S. Swift. 1989. The effects of pH and drying on adsorption of phosphate by aluminum-organic matter associations. *Eur. J. Soil. Sci.* 40(4):773–781.

Hedely, M. J. and J. W. B. Stewart. 1982. Method to measure microbial biomass phosphorus in soils. *Soil Biol. Biochem.* 14:377–385.

Hemond, H. F. and E. J. Fechner. 1994. *Chemical Fate and Transport in the Environment.* Academic Press, New York. 338 pp.

Henriksen, K. and W. M. Kemp. 1988. Nitrification in estuarine and coastal marine sediments. In T. H. Blackburn and J. Sorensen (eds.) *Nitrogen Cycling in Coastal Marine Environments.* Wiley, New York. pp. 207–249.

Hensel, P. F., J. W. Day, Jr., and D. Pont. 1999. Wetland vertical accretion and soil elevation change in the Rhone River Delta, France: the importance of riverine flooding. *J. Coast. Res.* 15(3):668–681.

Herbert, R. A. 1999. Nitrogen cycling in coastal marine sediments. *FEMS Microbiol. Rev.* 23:563–590.

Herman, W. A., McGill, W. B., and Dormaar, J. F., 1977. Effects of initial chemical composition on decomposition of roots of three grass species. *Can. J. Soil Sci.* 57:205–215.

Hesslein, R. H. 1976. An in situ sampler for close interval porewater studies. *Limnol. Oceanogr.* 21:912–914.

Hieltjes, H. M. and L. Lijklema. 1980. Fractionation of inorganic phosphates in calcareous sediments. *J. Environ. Qual.* 9:405–407.

Hines, M. E. and K. N. Duddleston. 2001. Carbon flow to acetate and C_1 compounds in northern wetlands. *Geophys. Res. Lett.* 28:4251–4254.

Hines, M. E., S. L. Knollmeyer, and J. B. Tugel. 1989. Sulfate reduction and other sedimentary biogeochemistry in a northern New England salt marsh. *Limnol. Oceanogr.* 34:578–590.

Hingston, F. J., R. J. Atkinson, A. M. Posner, and J. P. Quirk. 1967. Specific adsorption of anions. *Nature (London)* 215:1459–1461.

Hlawatsch, S., C. D. Garbe-Schonberg, F. Lechtenberg, A. Manceau, N. Tamura, D. A. Kulik, and M. Kersten. 2002. Trace metal fluxes to ferromanganese nodules from the western Baltic Sea as a record for long-term environmental changes. *Chem. Geol.* 182:697–709.

Ho, D. T., P. Schlosser, R. W. Houghton, and T. Caplow. 2006. Comparison of SF6 and fluorescence as traces for measuring transport in a large tidal river. *J. Environ. Eng.* 132(2):1664–1669.

Holzapfel-Pschorn, A. and W. Seiler. 1986. Methane emissions during a cultivation period from an Italian rice paddy. *J. Geophys. Res.* 91:11803–11814.

Houghton, J. T., B. A. Callander, and S. K. Varney. Climate Change 1992. The Supplementary Report to the IPCC Scientific Assessment. Cambridge University Press.

Hook, D. D. 1984. Adaptations to flooding with freshwater. In T. T. Kozlowski and A. J. Ricker (eds.) *Flooding and Plant Growth.* Academic Press, New York. pp. 265–294.

Hook, D. D. and J. R. Scholtens. 1978. Adaptations and flood tolerance of tree species. In D. D. Hook and R. M. M. Crawford (eds.) *Plant Life in Anaerobic Environments.* Ann Arbor Science Publishers, Ann Arbor, MI. pp. 299–231.

Hoosbeek, M. R., R. G. Amundson, and R. B. Bryant. 2000. Pedological modeling. In M. E. Sumner (ed.) *Handbook of Soil Science.* CRC Press, New York.

Hopkinson, C. S. and J. W. Day, Jr. 1979. Aquatic productivity and water quality at the upland-estuary interface in Barataria Basin, Louisiana. In R. Livingston (ed.) *Ecological Processes in Coastal and Marine Systems.* Plenum Press, New York, pp. 291–314.

Hopkinson, C. S., J. G. Gosselink, and R. T. Parrondo. 1978. Aboveground production of seven marsh plant species in coastal Louisiana. *Ecology* 59:750–769.

Hopkinson, C. S. and J. P. Schubauer. 1984. Static and dynamic aspects of nitrogen cycling in the salt marsh graminoid *Spartina alterniflora. Ecology* 65:961–969.

Hoppe, H. G. 2003. Phosphatase activity in the sea. *Hydrobiologia* 493:187–200.

Houghton, J. T., Y. Ding, D. J. Griggs, M. Noguer, P. J. van der Linden, and D. Xiaosu (eds.). 2001. *Climate Change 2001: The Scientific Basis.* Cambridge University Press, Cambridge, UK.

Houghton, R. A. and D. L. Skole. 1990. Carbon. In B. L. Turner, W. C. Clark, R. W. Kates, J. F. Richards, J. T. Matthews, and W. B. Meyers (eds.) *The Earth as Transformed by Human Action.* Cambridge University Press, Cambridge. pp. 393–408.

House, W. A. 1990. The prediction of phosphate co-precipitation with calcite in freshwaters. *Water Res.* 8:1017–1023.

House, W. A. and L. Donaldson. 1986. Adsorption and coprecipitation of phosphate on calcite. *J. Colloidal Interface Sci.* 112:309–324.

Howarth, R. W. 1979. Pyrite: its rapid formation in a salt marsh and its importance in ecosystem metabolism. *Science* 203:49–51.

Howarth, R. W. 1984. The ecological significance of sulfur in the energy dynamics of salt marsh and coastal marine sediments. *Biogeochemistry* 1:5–27.

Howarth, R. W. 1993. Microbial processes in salt-marsh sediments. In T. E. Ford (ed.) *Aquatic Microbiology.* Blackwell Scientific Publications, Boston, MA. Oxford. pp. 239–259.

Howarth, R. W. and A. E. Giblin. 1983. Sulfate reduction in the salt at Sapelo island, Georgia, *Limnol. Oceanogr.* 28:70–82.

Howarth, R. W., R. Marino, J. Lane, and J. J. Cole. 1988. Nitrogen fixation in freshwater, estuarine, and marine ecosystems, 1: Rates and importance. *Limnol. Oceanogr.* 33:669–687.

Howarth, R. W. and S. Merkel. 1994. Pyrite formation and the measurement of sulfate reduction in salt marsh sediments. *Limnol. Oceanogr.* 29:598–600.

Howarth, R. W. and J. M. Teal. 1979. Sulfate reduction in a New England salt marsh. *Limnol. Oceanogr.* 24:999–1012.

Howarth, R. W. and J. M. Teal. 1980. Energy flow in a salt marsh ecosystem: the role of reduced inorganic sulfur compounds. *Am. Nat.* 116:862–872.

Howeler, R. H. 1972. The oxygen status of lake sediments. *J. Environ. Qual.* 1:366–371.

Howeler, R. H. and D. R. Bouldin. 1971. Diffusion and consumption of oxygen in submerged soils. *Soil Sci. Soc. Am. Proc.* 35:202–208.

Howes, B. L., J. W. H. Dacey, and G. M. King. 1984. Carbon flow through oxygen and sulfate reduction pathways in salt marsh sediments. *Limnol. Oceanogr.* 29:1037–1051.

Howes, B. L., R. W. Howarth, J. M. Teal, and J. Valiela. 1981. Oxidation–reduction potentials in a salt marsh: spatial patterns and interactions with primary production. *Limnol. Oceanogr.* 26:350–360.

Howes, B. L. and J. M. Teal. 1994. Oxygen loss from *Spartina alterniflora* and its relationship to salt marsh oxygen balance. *Oecologia.* 97:431–438.

Howler, R. H. and D. R. Bouldin. 1971. The diffusion and consumption of oxygen in submerged soils. *Soil Sci. Soc. Am. Proc.* 35:202–208.

Hugget, D. B., B. W. Brooks, B. Peterson, C. M. Foram, and D. Schlenk. 2002. Toxicity of selected beta adrenergic receptor-blocking pharmaceuticals (B-blockings) on aquatic organisms. *Arch. Environ. Contam. Toxicol.* 43:229–235.

Hutchinson, G. L., W. D. Guenzi, and G. P. Livingston. 1993. Soil water controls on aerobic soil emission of gaseous N oxides. *Soil Biol. Biochem.* 25:1–9.

Inglett, P. W. and K. R. Reddy. 2006. Investigating the use of macrophyte stable C and N isotopic ratios as indicators of wetland eutrophication: patterns in the P-affected Everglades. *Limnol. Oceanogr.* 51:2380–2387.

Inglett, P. W. and K. R. Reddy. 2006. Stable C and N isotopic ratios of macrophytes as an indicator of wetland eutrophication: patterns in the P-affected Everglades. *Limnol. Oceanogr.* 51:2380–2387.

Inglett, P. W, K. R. Reddy, and P. V. McCormick. 2004. Periphyton tissue chemistry and nitrogenase activity in a nutrient impacted Everglades ecosystem. *Biogeochemistry* 67:213–233.

Inglett, P. W., K. R. Reddy, S. Newman, and B. Lorenzen. 2007. Increased soil stable nitrogen isotopic ratio following phosphorus enrichment: historical patterns and tests of two hypotheses in a phosphorus-limited wetland. *Oecologia*. Online reference 10.1007/s00442-007-0711-5.

Ingraham, J. L. 1981. Microbiology and genetics of denitrifiers. In C. C. Delwiche (ed.) *Denitrification, Nitrification, and Atmospheric Nitrous Oxide*. Wiley, New York.

Intergovernmental Panel on Climate Change (IPCC). 1990. In J. T. Houghton, G. J. Jenkins, and J. J. Ephraums (eds.) *Climate Change*. Cambridge University Press, Cambridge. pp. 11043–11051.

Intergovernmental Panel on Climate Change (IPCC). 1996. In R. Watson, M. Zinyowera, and R. Moss (eds.) *Climate Change 1995: Impacts, Adaptations and Mitigation of Climate Change: Scientific-Technical Analysis*. Cambridge University Press, New York. 879 pp.

Intergovernmental Panel on Climate Change (IPCC). 1998. *Regional Impacts of Climate Change: An Assessment of Vulnerability*. Cambridge University Press, New York. 517 pp.

Intergovernmental Panel on Climate Change (IPCC). 2001. *The Third Assessment Report, Climate Change 2001*. Cambridge University Press, Cambridge.

Ise, T. and P. R. Moorcroft. 2006. The global-scale temperature and moisture dependencies of soil organic carbon decomposition: an analysis using a mechanistic decomposition model. *Biogeochemistry* 80:217–231.

Ishikawa, M. and Nishimura, H. 1989. Mathematical model of phosphate release rate from sediments considering the effect of dissolved oxygen in overlying water. *Water Res.* 23(3):351–359.

Ivanoff, D. B., K. R. Reddy, and S. Robinson. 1998. Chemical fractionation of organic P in histosols. *Soil Sci.* 163:36–45.

Iversen, N. and B. B. Jorgensen. 1985. Anaerobic methane oxidation rates at the sulphate-methane transition in marine sediments from Kattegat and Skagerak (Denmark). *Limnol. Oceanogr.* 30:944–955.

Jackson, L. E., J. C. Kurtz, and W. S. Fisher (eds.). 2000. *Evaluation Guidelines for Ecological Indicators*. EPA/620/R-99/005. U.S. Environmental Protection Agency, Office of Research and Development, Research Triangle Park, NC. 107 pp.

Jackson, M. B., T. M. Fenning, and W. Jenkins. 1985. Aerenchyma (gas space) formation in adventitious root of rice (*Oryza sativa* L.) is not controlled by ethylene or small partial pressure of oxygen. *J. Exp. Bot.* 36:1566–1570.

James, W. F., J. W. Barko, and H. L. Eakin. 1995. Internal phosphorus loading in Lake Pepin, upper Mississippi River. *J. Freshwater Ecol.* 10(3):269–276.

James, B. R. and R. J. Bartlett. 1983. Behavior of chromium in soils, interchange between oxidation–reduction and organic complexation. *J. Environ. Qual.* 12(2):173–176.

Jaynes, M. L. and S. R. Carpenter. 1986. Effects of vascular and nonvascular macrophytes on sediment redox and solute dynamics. *Ecology* 67:875–882.

Jensen, J. R. 2000. *Remote Sensing of the Environment*. Pearson Education, India.

Jeppesen, E., J. P. Jensen, M. Sondergaard, K. S. Hansen, P. H. Moller, H. U. Rasmussen, V. Norby, and S. E. Larsen. 2003. Does resuspension prevent a shift to a clear state in shallow lakes during reoligotrophication. *Limnol. Oceangr.* 48:1913–1919.

Jianbo, C., C. Li, and C. Trettin. 2005. Analyzing the ecosystem carbon and hydrologic characteristics of forested wetland using a biogeochemical process model. *Global Change Biol.* 11:278–289.

Jiang, J. 2007. *Linear and Generalized Linear Mixed Models and their Applications*. Springer, Berlin.

Johnson, K. S. 2006. Manganese redox chemistry revisited. *Science* 313:1896–1897.

Johnston, C. A. 1991. Sediment and nutrient retention by freshwater wetlands: effects on surface water quality. *Crit. Rev. Environ. Control* 21:491–565.

Jorgensen, N. O. G., N. Kroer, R. B. Coffin, and M. P. Hoch. 1999. Relationship between bacterial nitrogen metabolism and growth efficiency in an estuarine and an open-water ecosystem. *Aquat. Microb. Ecol.* 18:247–261.

Joshi, M. M. and J. P. Hollis. 1976. Interaction of *Beggiatoa* and rice plant: detoxification of hydrogen sulfide in the rice rhizosphere. *Science* 195:179–180.

Kadlec, R. H. 1993. Natural wetland treatment at Houghton Lake: the first fifteen years, In *Proceedings of the WEF 66th Annual Conference*, Anaheim, CA. WEF, Alexandria, VA. pp. 73–84.

Kadlec, R. H. and R. L. Knight. 1996. *Treatment Wetlands*. Lewis Publishers, Boca Raton, FL. 893 pp.

Kadlec, R. H. and K. R. Reddy. 2001. Temperature effects in treatment wetlands. *Water Environ. Res.* 73:543–557.

Kamp-Nielsen, L. 1975. A kinetic approach to the aerobic sediment–water exchange of phosphorus in Lake Esrom. *Ecol. Model.* 1:153–160.

Kappler, A., M. Benz, B. Schink, and A. Brune. 2004. Electron shuttling via humic acids in microbial iron(III) reduction in a freshwater sediment. *Microb. Ecol.* 47:85–92.

Kastler, J. A. and P. L. Wiberg. 1996. Sedimentation and boundary changes of Virginia salt marshes. *Estuar. Coast. Shelf Sci.* 42:683–700.

Kawase, M. 1981. Effect of ethylene on aerenchyma development. *Am. J. Bot.* 68:651–658.

Kearney, M. S. and L. G. Ward. 1986. Accretion rates in brackish marshes of a Chesapeake Bay estuarine tributary. *Geo Marine Lett.* 6(1):41–49.

Kearney, M. S. and J. C. Stevenson. 1991. Island land loss and marsh vertical accretion rate evidence for historical sea-level changes in Chesapeake Bay. *J Coast. Res.* 7:403–415.

Kearney, M. S., J. C. Stevenson, and L. G. Ward. 1994. Spatial and temporal changes in marsh vertical accretion rates at Monie Bay: implications for sea-level rise. *J. Coast. Res.* 10:1010–1020.

Keinänen, M. M., L. K. Korhonen, M. J. Lehtola, I. T. Miettinen, P. J. Martikainen, T. Vartiainen, and M. H. Suutari. 2002. The microbial community structure of drinking water biofilms can be affected by phosphorus availability. *Appl. Environ. Microbiol.* 68:434–439.

Keller, A. E. and S. G. Zam. 1991. The acute toxicity of selected metals to the freshwater mussel. *Anodonta imbecilis. Environ. Toxicol. Chem.* 10:539–546.

Keller, M., M. E. Mitre, and R. F. Stallard. 1990. Consumption of atmospheric methane in soils of central Panama: effects of agricultural development. *Global Biogeochem. Cycles* 4:21–27.

Kelley, K. R. and J. F. Stevenson. 1995. Forms and nature of organic N in soil. *Fert. Res.* 42:1–11.

Kellog, L. E. and S. Bridgham. 2003. Phosphorus retention and movement across an ombrotrophic–minerotrophic peatland gradient. *Biogeochemistry* 63:299–315.

Kemp, W. M. 1989. Estuarine chemistry. In J. W. Day, Jr., C. A. S. Hall, W. M. Kemp, and A. Yanez-Arancibia (eds.) *Estuarine Ecology*. Wiley, New York. pp. 79–143.

Kemp, W. M., W. R. Boynton, J. Cunningham, J. C. Stevenson, T. Jones, and J. C. Means. 1985. Effects of atrazine and linuron on photosynthesis and growth of the macrophytes, *Potamogeton perfoliatus* L. and *Myriophyllum spicatum* L. in an estuarine environment. *Mar. Environ. Res.* 16:225–280.

Kennedy, M. P., K. J. Murphy, and D. J. Gilvear. 2006. Predicting interactions between wetland vegetation and the soil-water and surface environment using diversity, abundance and attribute values. *Hydrobiologia* 570(1):189–196.

Kemp, W. M. and L. Murray. 1986. Oxygen release from roots of the submersed macrophyte *Potamogeto perfoliatus* L.: regulating factors and ecological implications. *Aquat. Bot.* 26(3–4):271–283.

Keppler, F. J., T. Hamilton, M. Brab, and T. Ruckman. 2006. Methane emissions from terrestrial plants under aerobic conditions. *Nature* 439:187–191.

Khalid, R. A., R. P. Gambrell, and W. H. Patrick, Jr. 1978. Chemical transformations of cadmium and zinc in Mississippi River sediments as influenced by pH and redox potential. In D. C. Adriano and I. L. Brisbin, Jr. (eds.) *Environmental Chemistry and Cycling Processes*. Department of Energy, Symp. Ser. 45:417–433. Tech. Info. Center, U.S. Department of Energy. Proc. 2nd Mineral Cycling Symp., May 1, 1976.

Khalid, R. A., R. P. Gambrell, and W. H. Patrick, Jr. 1981. Chemical availability of cadmium in Mississippi River sediment. *J. Environ. Qual.* 10:523–529.

Khalid, R. A., W. H. Patrick, Jr., and R. D. DeLaune. 1977. Phosphorus sorption characteristics of flooded soils. *Soil Sci. Soc. Am. J.* 41:305–310.

Khalil, M. A. K., M. J. Shearer, and R. A. Rasmussen. 1992. Methane sinks and distribution. In M. A. K. Khalil (ed.) *Atmospheric Methane: Sources, Sinks, and Role in Global Change*. NATO ASI Series 1, Vol. 13. Springer-Verlag, Berlin. pp. 168–180.

Killham, K. and C. Yeomans. 2001. Rhizosphere carbon flow measurement and implications: from isotopes to reporter genes. *Plant Soil* 232 (1–2):91–96.

Killingbeck, K. Y. 1996. Nutrients in senesced leaves: keys to the search for potential resorption and resorption proficiency. *Ecology* 77:1716–1727.

Kim, J. G., J. B. Dixon, C. C. Chusuei, and Y. Deng. 2002. Oxidation of chromium (III) to (VI) by manganese oxide. *Soil Sci. Soc. Am. J.* 66:306–315.

Kimura, M., J. Murase, and Y. H. Lu. 2004. Carbon cycling in rice field ecosystems in the context of input, decomposition and translocation of organic materials and the fates of their end products (CO_2 and CH_4). *Soil Biol. Biochem.* 36(9):1399–1416.

King, G. 1992. Ecological aspects of methane oxidation, a key determinant of global methane dynamics. In K. C. Marshall (ed.) *Advances in Microbial Ecology.* Vol. 12. Plenum Press, New York. pp. 431–468.

King, G. M. 1986. Characterization of β-glucosidase activity in intertidal marine sediments. *Appl. Environ. Microbiol.* 51:373–380.

King, G. M. 1988. Patterns of sulfate reduction and sulfur in a South Carolina salt Marsh. *Limnol. Oceanogr.* 33:376–396.

King, G. M., P. Roslev, and H. Skovgaard. 1990. Distribution and rate of methane oxidation in sediments of the Florida Everglades. *Appl. Environ. Microbiol.* 56:2902–2911.

King, J. K., J. E. Kotska, M. E. Frischer, F. M. Saunders, and R. A. Jahnke. 2001. A quantitative relationship that demonstrates mercury methylation rates in marine sediments is based on the community composition and activity of sulfate reducing bacteria. *Environ. Sci. Technol.* 35(12):2491–2496.

King, R. S., C. J. Richardson, D. L. Urban, and E. A. Romanowicz. 2004. Spatial dependency of vegetation-environment linkages in an anthropogenically influenced wetland ecosystem. *Ecosystems* 7:75–97.

Kirby, C. J. and J. G. Gosselink. 1976. Primary production in a Louisiana Gulf coast *Spartina alterniflora* marsh. *Ecology* 57:1052–1059.

Kirchner, G. and H. Ehlers. 1998. Sediment geochronology in changing coastal environments: potentials and limitations of the Cs-137 and Pb-210 methods. *J. Coast. Res.* 14:483–492.

Kirschbaum, M. U. F. 1995. The temperature dependence of soil organic matter decomposition, and the effect of global warming on soil organic C storage. *Soil Biol. Biochem.* 27:753–760.

Kleiner, J. 1988. Coprecipitation of phosphate with calcite in lake water: a laboratory experiment modeling phosphorus removal with calcite in Lake Constance. *Water Res.* 22:1259–1265.

Kludze, H. K. and R. D. DeLaune. 1994. Methane emission and growth of *Spartina patens* in response to soil redox intensity. *Sci. Soc. Am. J.* 58:1838–1845.

Kludze, H. K. and R. D. DeLaune. 1995. Gaseous exchange and wetland response to soil redox intensity and capacity. *Soil Sci. Soc. Am. J.* 59:939–949.

Kludze, H. K. and R. D. DeLaune. 1995. Straw application effects on methane and oxygen exchange and growth in rice. *Sci. Soc. Am. J.* 59:824–830.

Kludze, H. K., R. D. DeLaune, and W. H. Patrick, Jr. 1993. Aerenchyma formation and methane and oxygen exchange in rice. *Soil Sci. Soc. Am. J.* 57:386–391.

Kludze, H. K., R. D. DeLaune, and W. H. Patrick. 1994. A colorimetric method for assaying dissolved oxygen loss from container-grown rice roots. *Agron. J.* 86:483–487.

Kludze, H. K., S. R. Pezeshki, and R. D. Delaune. 1994. Evaluation of root oxygenation and growth in bald cypress in response to short-term soil hypoxia. *Can. J. Forest. Res.* 24:804–809.

Klute, A. 1986. *Methods of Soil Analysis: Part 1, Physical and Mineralogical Methods.* ASA and SSSA Series No. 5. Soil Science Society of America, Madison, WI. 1188 pp.

Knight, T. W. and D. W. Morris. 1996. How many habitats do landscapes contain? *Ecology* 77:1756–1764.

Koch, M. S. and K. R. Reddy. 1992. Distribution of soil and plant nutrients along a trophic gradient in the Florida Everglades. *Soil Sci. Soc. Am. J.* 56:1492–1499.

Koch, M. S., K. R. Reddy, and J. P. Chanton. 1994. Factors controlling seasonal nutrient profiles in a subtropical peatland: Florida Everglades. *J. Environ. Qual.* 23:526–533.

Kolb, C. R. and J. R. van Lopik. 1958. Geology of Mississippi River Deltaic Plain, Southeastern Louisiana, U.S. Corps of Engineers, Waterways Expt. Sta., Tech Repts. 3-483 and 3-484, 2 vols.

Kolpin, D. W., E. T. Furlong, M. T. Meyer, E. M. Thurman, S. D. Zaugg, L. B. Barber, and H. T. Buxton. 2002. Pharmaceuticals, hormones, and other organic wastewater contaminants in U.S. streams 1999–2000: a national reconnaissance. *Environ. Sci. Technol.* 36:1202–1211.

Koncalova, H., J. Pokorny, and J. Kvet. 1988. Root ventilation in *Carex gracilis* Curt: diffusion or mass flow? *Aquat. Bot.* 30:149–155.

Konings, H. and H. Lambers. 1991. Respiratory metabolism oxygen transport and the induction of aerenchyma in roots. In M. B. Jackson, D. D. Davies, and H. Lambers (eds.) *Plant Life under Oxygen Deprivation.* SPB Academic Publishing, The Hague, The Netherlands. pp. 247–265.

Kort, R. L., E. A. Cassell, and S. G. Aschmann. 2007. Watershed Ecosystem Dynamics-Phosphorus (WEND-P Model). In D. E. Radcliffe and M. L. Cabrera (eds.) *Modeling Phosphorus in the Environment.* CRC Press, New York. pp. 261–276.

Kostka, J. E., A. N. Roychoudhury, and P. Van Cappellen. 2002. Rates and controls of anaerobic microbial respiration across spatial and temporal gradients in salt marsh sediments. *Biogeochemistry* 60:49–76.

Krabbenhoft, D. P., J. P. Hurley, G. Aiken, C. Gilmour, M. Marvin-DiPasquale, W. H. Orem, and R. Harris. 2000. Mercury cycling in the Florida Everglades: a mechanistic field study. *Verh. Internat. Verein. Limnol.* 27:1657–1660.

Krairapanond, N., R. D. DeLaune, and W. H. Patrick, Jr. 1992. Distribution of organic and reduced sulfur forms in marsh soils of coastal Louisiana. *Org. Geochem.* 18(4):489–500.

Kralova, M., P. H. Masscheleyn, C. W. Lindau, and W. H. Patrick, Jr. 1992. Production of dinitrogen and nitrous oxide in soil suspensions as affected by redox potential. *Water, Air, Soil Pollut.* 61:37–45.

Kratz, T. K. and C. B. Dewitt. 1986. Internal factors controlling peatland-lake ecosystem development. *Ecology* 67:100–107.

Krauskopf, C. 1979. *Introduction to Geochemistry.* McGraw-Hill, London.

Krauskopf, K. B. 1979. *Introduction to Geochemistry.* 2nd Edition. McGraw-Hill, New York. 617 pp.

Kreutzweiser, D. P. and M. J. Faber. 1999. Ordination of zooplankton community data to detect pesticide effects in pond enclosures. *Arch. Environ. Contam. Toxicol.* 36:392–398.

Kreuzer-Martin, H. W. 2007. Stable isotope probing: linking functional activity to specific members of microbial communities. *Soil Sci. Soc. Am. J.* 71:611–619.

Kristensen, E. 2000. Organic matter diagenesis at the oxic/anoxic interface in coastal marine sediments, with emphasis on the role of burrowing animals. *Hydrobiologia* 426:1–24.

Kroeze, C., A. Mosier, and L. Bouwman. 1999. Closing the global N_2O budget: a retrospective analysis 1500–1994. *Global Biogeochem. Cycles* 13:1–8.

Krumbein, W. C. and R. M. Garrels. 1952. Origin and classification of chemical sediments in terms of pH and redox potentials. *J. Geol.* 60:1–33.

Kuenen, J. G. and L. A. Robertson. 1994. Combined nitrification–denitrification processes. *FEMS Microbiol. Rev.* 15:109–117.

Kuypers, M. M. M., A. O. Sliekers, G. Lavik, M. Schmid, B. B. Jørgensen, J. G. Kuenen, J. S. Sinninghe Damste, M. Strous, and M. S. M. Jetten. 2002. Anaerobic ammonium oxidation by anammox bacteria in the Black Sea. *Nature* 422:608–611.

Lafleur, P. M., T. R. Moore, N. T. Roulet, and S. Frolking. 2005. Ecosystem respiration in a cool temperate bog depends on peat temperature but not water table. *Ecosystems* 8:619–629.

Lane, R. L., J. W. Day, Jr., and B. Thibodeaux. 1999. Water quality analysis of freshwater diversion at Caernarvon, Louisiana. *Estuaries* 22:327–336.

Lavery, P. S., C. E. Oldham, and M. Ghisalberti. 2001. The use of Fick's First Law for predicting porewater nutrient fluxes under diffusive conditions. *Hydrolog. Process.* 15(13):2435–2451.

Lee, D. R. 1977. A device for measuring seepage flux in lakes and estuaries. *Limnol. Oceanogr.* 22:140–147.

Leibowitz, S. G. and J. M. Hill. 1987. Spatial analysis of Louisiana coastal land loss. In R. E. Turner and D. R. Cahoon (eds.) *Causes of Wetland Loss in the Coastal Central Gulf of Mexico. Volume II: Technical Narrative.* Final report submitted to minerals Management Srvice, New Orleans, LA. Contract No. 14-12-001-30252. OCS Study/MMS 87-0120. 400 pp.

Lemon, E. R. and A. R. Erickson. 1955. Principle of platinum microelectrode as a method of characterizing soil aeration. *Soil Sci.* 79:383–392.

Lerman, A. 1979. *Geochemical Processes; Water and Sediment Environments.* Wiley, New York. 481 pp.

Letey, J., N. Valoras, D. D. Focht, and J. C. Ryden. 1981. Nitrous-oxide production and reduction during denitrification as affected by redox potential. *Soil Sci. Soc. Am. J.* 45:727–730.

Letzsch, W. S. 1983. Seven year's measurement of deposition and erosion, Holocene salt marsh, Sapelo Island, Georgia. *Senckenbergiana Maritima* 15:157–165.

Lewis, W. M. 1995. *Wetlands—Characteristics and Boundaries.* National Research Council, National Academy Press, Washington, DC. 306 pp.

Li, M. R. and M. B. Jones. 1995. CO_2 and O_2 transport in aerenchyma of *Cyperus papyrus* L. *Aquat. Bot.* 52:93–106.

Li, Y. and S. Gregory. 1974. Diffusion of ions in sea water and in deep-sea sediments. *Geochim. Cosmochim. Acta* 38:703–714.

Li, Y., A. J. Mehta, K. Hatfield, and M. S. Dortch. 1997. Modulation of constituent release across the mud–water interface by water waves. *Water Resour. Res.* 33:1409–1418.

Lindau, C. W. and R. D. DeLaune. 1991. Dinitrogen and nitrous oxide emission and entrapment in *Spartina alterniflora* salt marsh soils following addition of N-15 labeled ammonium nitrate. *Estuarine, Coastal Shelf Sci.* 32:161–172.

Lindau, C. N., R. D. DeLaune, and G. L. Jones. 1988. Fate of added nitrate and ammonium: nitrogen entering a Louisiana Gulf Coast Swamp Forest. *J. Water Pollut. Control Fed.* 60:386–388.

Lindsay, W. L. 1979. *Chemical Equilibria in Soils*. Wiley, New York. 449 pp.

Linthurst, R. A. 1979. The effect of aeration on the growth of *Spartina alterniflora*. *Am. J. Bot.* 66:685–691.

Lissner, J. I., A. Mendelssohn, and C. J. Anastasium. 2003. A method for cultivating plants under controlled redox intensities in hydroponics. *Aquatic Botany.* 76:93–108.

Liu, F., C. Colombo, P. Adams, J. Z. He, and A. Violante. 2002. Trace elements in manganese–iron nodules from a Chinese Alfisol. *Soil Sci. Soc. Am. J.* 66:661–670.

Liu, W. T., T. L. Marsh, H. Cheng, and L. J. Forney. 1997. Characterization of microbial diversity by determining terminal restriction fragment length polymorphisms of genes encoding 16S rRNA. *Appl. Environ. Microbiol.* 63:4516–4522.

Liu, Z. G. 1993. Oxidation–reduction potential and its measurement. In T. R. Yu and G. L. Ji (eds.) *Electrochemical Methods in Soil and Water Research*. Pergamon Press, Tarrytown, NY. pp. 297–313.

Lodge, D. J., W. H. McDowell, and C. P. McSwiney. 1994. The importance of nutrient pulses in tropical forests. *Tree* 9:384–387.

Loeb, L. B. 1934. Kinetic theory of gases. McGraw-Hill, New York.

Lomminicki, A., E. Bandola, and J. Jankowaska. 1968. Modification of the Wiegert-Evans method for estimation of net primary production. *Ecology* 49:147–149.

Lookman, R., D. Freese, R. Merck, K. Vlassak, and W. H. van Riemsdijk. 1995. Long-term kinetics of phosphate release from soil. *Environ. Sci. Technol.* 29:1569–1575.

Loveless, C. M. 1959. A study of the vegetation of the Florida Everglades. *Ecology* 40:1–9.

Lovley, D. R. 1987. Organic matter mineralization with reduction of ferric iron: a review. *Geomicrobiol. J.* 5:375–399.

Lovley, D. R. 1991. Dissimilatory Fe(III) and Mn(IV) reduction. *Microbiol. Rev.* 55:259–287.

Lovley, D. R. 2004. Dissimilatory Fe(III) and Mn(IV) reduction. *Adv. Microbiol. Physiol.* 49:219–286.

Lovley, D. R. 2006. Bug juice: harvesting electricity with microorganisms. *Nat. Rev.* 4:497–508.

Lovley, D. R., J. D. Coates, E. L. Blunt-Harris, E. J. P. Phillips, and J. C. Woodward. 1996. Humic substances as electron acceptors for microbial respiration. *Nature* 382:445–448.

Lovley, D. R., D. E. Holmes, and K. P. Nevin. 2004. Dissimilatory Fe(III) and Mn(IV) reduction. *Adv. Microb. Physiol.* 49:220–286.

Lovley, D. R. and D. J. Lonergan. 1990. Anaerobic oxidation of toluene, phenol and *p*-cresol by the dissimilatory iron reducing organism GS-15. *Appl. Eng. Microbiol.* 56:1858–1864.

Lovley, D. R. and E. J. P. Phillips. 1987. Competitive mechanism for inhibition of sulfate reduction and methane production in the zone of ferric iron reduction in sediments. *Appl. Environ. Microbiol.* 53:2636–2641.

Lovley, D. R., and E. J. P. Phillips. 1987. Rapid Assay for Microbially Reduced Ferric Iron in Aquatic Sediments. *Appl. Environ. Microbiol.* 53(7):1536–1540.

Lovley, D. R., E. J. P. Phillips, and D. J. Lonergan. 1991. Enzymatic versus monenzymatic mechanisms for Fe(III) reduction in aquatic sediments. *Environ. Sci. Technol.* 25:1062–1067.

Lu, Y. H., W. R. Abraham, and R. Conrad. 2007. Spatial variation of active microbiota in the rice rhizosphere revealed by in situ stable isotope probing of phospholipid fatty acids. *Environ. Microbiol.* 9:474–481.

Lu, Y. H., T. Lueders, M. W. Friedrich, and R. Conrad. 2005. Detecting active methanogenic populations on rice roots using stable isotope probing. *Environ. Microbiol.* 7:326–336.

Lundgren, D. G. and W. Dean. 1979. Biogeochemistry of iron. In P. A. Trundinger and D. J. Swaine (eds.) *Biogeochemical Cycling of Mineral-Forming Elements*. Elsevier, Amsterdam. pp. 211–251.

Luther, G. W. III. 1991. Pyrite synthesis via polysulfide compounds. *Geochim. Cosmochim. Acta* 55:2839–2849.

Luther, G. W. III and T. M. Church. 1988. Seasonal cycling of sulfur and iron in pore waters of a Delaware salt marsh. *Mar. Chem.* 23:295–309.

Macur, R. E., C. R. Jackson, L. M. Botero, T. R. McDermott, and W. P. Inskeep. 2004. Bacterial populations associated with the oxidation and reduction of arsenic in unsaturated soil. *Environ. Sci. Technol.* 38:104–111.

Madigan, M. and J. Martinko. 2006. *Brock Biology of Microorganisms*. 11th Edition. Pearson Prentice Hall, Upper Saddle River, NJ.

Madigan, M. T., J. M. Martinko, and J. Parker. 2000. *Brock Biology of Microorganisms*. 9th Edition. Prentice Hall, Upper Saddle River, NJ.

Magid, J., H. Tiessen, and L. M. Condron. 1996. Dynamics of organic phosphorus in soils under natural and agricultural ecosystems. In A. Piccolo (ed.) *Humic Substances in Terrestrial Ecosystems*. Elsevier Science, Oxford. pp. 429–466.

Maie, N., K. J. Parish, A. Watanabe, H. Knicker, R. Benner, T. Abe, K. Kaiser, and R. Jaffe. 2006. Chemical characteristics of dissolved organic nitrogen in an oligotrophic subtropical coastal ecosystem. *Geochim. Cosmochim. Acta* 70:4491–4506.

Malecki, L. M., J. R. White, and K. R. Reddy. 2004. Nitrogen and phosphorus flux rates from sediment in the lower St. Johns River estuary. *J.Environ. Qual.* 33:1545–1555.

Maltby, E. and P. Immirzi. 1993. Carbon dynamics in peatlands and other wetland soils: regional and global perspectives. *Chemosphere* 27:999–1023.

Manceau, A., N. Tamura, R. S. Celestre, A. A. Macdowell, N. Geoffroy, G. Sposito, and H. A. Padmore. 2003. Molecular-scale speciation of Zn and Ni in soil ferromanganese nodules from loess soils of the Mississippi River Basin. *Environ. Sci. Technol.* 37:75–80.

Manning, B. A. and D. L. Suarez. 2000. Modeling arsenic adsorption and heterogeneous oxidation kinetics in soils. *Soil Sci. Am. J.* 64:128–137.

Maquenne, L. 1882. Sur la reduction des nitrates dans la terre ara le. *C. R. Hebd. Seances. Acad. Sci.* 95:690–693.

Marshall, K. C. 1989. Cyanobacterial-heterotrophic bacterial interaction. In Y. Cohen and E. Rosenberg (eds.) *Microbial Mats: Physiological Ecology of Benthic Microbial Communities*. American Society for Microbiology. pp. 239–245.

Marshall, A. G. and S. Guan. 1996. *Rapid Commun. Mass Spectrom.* 10:1819–1823.

Martin, J. B., K. M. Hartl, D. R. Corbett, P. W. Swarzenski, and J. E. Cable. 2003. A multi-level pore-water sampler for permeable sediments. *J. Sedimen. Res.* 73:128–132.

Martin, J. F. and K. R. Reddy. 1997. Interaction and spatial distribution of wetland nitrogen processes. *Ecol. Model.* 105:1–21.

Masscheleyn, P. H., R. D. DeLaune, and W. H. Patrick, Jr. 1990. Transformations of selenium as affected by sediment oxidation–reduction potential and pH. *Environ. Sci. Technol.* 24:91–97.

Masscheleyn, P. H., R. D. DeLaune, and W. H. Patrick, Jr. 1991. Effect of redox potential and pH on arsenic speciation and solubility in a contaminated soil. *Environ. Sci. Technol.* 25(8):1414–1419.

Masscheleyn, P. H., R. D. DeLaune, and W. H. Patrick, Jr. 1991. Biogeochemical behavior of selenium in anoxic soils and sediments: an equilibrium thermodynamics approach. *Environ. Sci. Health* A26(4):555–573.

Masscheleyn, P. H., J. H. Pardue, R. D. DeLaune, and W. H. Patrick, Jr. 1992. Chromium redox chemistry in a lower Mississippi Valley bottomland hardwood wetland. *Environ. Sci. Technol.* 26:1217–1226.

Matheron, G. 1963. Principles in geostatistics. *Econ. Geol.* 58:1246–1266.

Mathews, C. K. and K. E. van Holde. 1990. *Biochemistry*. The Benjamin/Cummings Publishing Company Inc., Redwood, CA. 1097 pp.

Matsumura, F. and H. J. Benezet. 1978. Microbial degradation of insecticides. In I. R. Hill and S. J. L. Wright (eds.) *Pesticide Microbiology*. Academic Press, London. pp. 678–667.

Matthews, E. and I. Fung. 1987. Methane emissions from natural wetlands: global distribution, area and environmental characteristics of sources. *Global Biogeochem. Cycles* 1:61–86.

McBratney, A. B., I. O. A. Odeh, T. F. A. Bishop, M. S. Dunbar, and T. M. Shatar. 2000. An overview of pedometric techniques for use in soil survey. *Geoderma* 97:293–327.

McBride, M. B. 1994. *Environmental Chemistry of Soils*. Oxford University Press, New York, NY. 406 pp.

McCaffrey, R. J. and J. Thomson. 1980. A record of the accumulation of sediment and trace metals in a Connecticut salt marsh. *Adv. Geophys.* 22:165–236.

McCall, P. L. and M. J. S. Tevesz. 1982. The effects of benthos on physical properties of freshwater sediments. In P. L. McCall and M. J. S. Tevesz (eds.) *Animal–Sediment Relations*. Plenum Press, New York. pp. 105–176.

McCarthy, J. F. and B. D. Jimenez. 1985. Interactions between polycyclic aromatic hydrocarbons and dissolved humic material: binding and dissociation. *Environ. Sci. Technol.* 19:1072–1076.

McCormick, P., M. J. Chimney, and D. Swift. 1996. Diel oxygen profiles and aquatic community metabolism in the Florida Everglades, USA. *Arch. Hydrobiol.* 140(1):117–129.

McCormick, P. V. and J. A. Laing. 2003. Effects of increased phosphorus loading on dissolved oxygen in a subtropical wetland, the Florida Everglades. *Wetlands Ecol. Manage.* 11:199–215.

McCormick, P. V., S. Newman, S. Miao, D. E. Gawlik, D. Marley, K. R. Reddy, and T. D. Fontaine. 2002. Effects of anthropogenic phosphorus inputs on the Everglades. In J. W. Porter and K. G. Porter (eds.) *The Everglades, Florida Bay, and Coral Reefs of the Florida Keys: An Ecosystem Sourcebook*. CRC Press, Boca Raton, FL. pp. 83–126.

McCormick, P. V. and M. B. O'Dell. 1996. Quantifying periphyton responses to phosphorus enrichment in the Florida Everglades: a synoptic-experimental approach. *J. North Am. Bentholog. Soc.* 15:450–468.

McCormick, P. V., M. B. O'Dell, R. B. E. Shuford III, J. G. Backus, and W. C. Kennedy. 2001. Periphyton responses to experimental phosphorus enrichment in a subtropical wetland. *Aquat. Bot.* 71:119–139.

McCormick, P. V., R. B. E. Shuford III, J. B. Backus, and W. C. Kennedy. 1998. Spatial and seasonal patterns of periphyton biomass and productivity in the northern Everglades, Florida, USA. *Hydrobiologia* 362:185–208.

McCormick, P. V. and R. J. Stevenson. 1998. Periphyton as a tool for ecological assessment and management in the Florida Everglades. *J. Phycol.* 34:726–733.

McGeehan, S. L. and D. V. Naylor. 1994. Sorption and redox transformation of arsenite and arsenate in two flooded soils. *Soil Sci. Am. J.* 58:337–342.

McIntyre, D. S. 1970. Characterization of soil aeration with a platinum electrode 1. Response in relation to field moisture conditions and electrode diameter. *Aust. J. Soil Res.* 4:95–102.

McIntyre, D. S. 1970. The platinum microelectrode method for soil aeration measurement. *Adv. Agron.* 22:235–285.

McJannet, C. L., P. A. Keddy, and F. R. Pick. 1995. Nitrogen and phosphorus tissue concentrations in 41 wetland plants: a comparison across habitats and functional groups. *Funct. Ecol.* 9:231–238.

McKee, K. L. and I. A. Mendelssohn. 1987. Root metabolism in black mangrove (*Avicennia germinani* L.): response to hypoxia. *Environ. Exp. Bot.* 27:147–150.

McKenzie, R. M. 1989. Manganese oxides and hydroxides. In J. B. Dixon and S. B. Weed (eds.) *Minerals in Soil Environment.* 2nd Edition. SSSA Book Series 1, SSSA, Madison, WI. pp. 439–466.

McLatchey, G. P. and K. R. Reddy. 1998. Regulation of organic matter decomposition and nutrient release in a wetland soil. *J. Environ. Qual.* 27:1268–1274.

McLaughlin, J. R., J. C. Ryden, and J. K. Syers. 1981. Sorption of inorganic phosphate by iron and aluminum containing components. *J. Soil Sci.* 32:365–375.

McLeod, K. W., L. A. Donovan, and N. J. Stumpff. 1987. Responses of woody seedling to elevated floodwater temperatures. In D. D. Hook, McKee, H. K. Smith, J. Gregory, V. G. Burrell, M. R. Devol, R. E. Sojka, S. Gilbert, R. Banks, L. H. Stolzy, C. Brroks, T. D. Matthews, and T. H. Shear (eds.) *The Ecology and Management of Wetlands.* Timber Press, Portland, OR. pp. 441–451.

McMillin, D. J. and J. C. Means. 1996. Spatial and temporal trends of pesticide residues in water and particulates in the Mississippi River plume and the northwestern Gulf of Mexico. *J. Chromatogr. A* 754:169–185.

Meads, R. H. and R. S. Parker. 1985. *Sediments in Rivers of the United States.* National Water Summary, 1984. U.S. Geological Survey Water Supply Paper, 2275.

Meek, B. D. and L. B. Grass. 1975. Redox potential in irrigated desert soils as an indicator of aeration status. *Soil Sci. Soc. Am. Proc.* 39:870–875.

Megonigal, J. P., M. E. Hines, and P. T. Visscher. 2004. Anaerobic metabolism: linkages to trace gases and aerobic processes. In W. H. Schlesinger (ed.) *Biogeochemistry.* Elsevier-Pergamon, Oxford, UK. pp. 317–424.

Megonigal, J. P., W. H. Patrick, Jr., and S. P. Faulker. 1993. Wetland identification in seasonally flooded soils. Soil morphology and redox dynamics. *Soil Sci. Soc. Am. J.* 57:140–149.

Melillo, J. M., J. D. Aber, A. E. Linkins, A. Ricca, B. Fry, and K. J. Nadelhoffer. 1989. Carbon and nitrogen dynamics along the decay continuum—plant litter to soil. *Plant Soil* 115:189–198.

Melillo, J. M., Aber, J. D., Muratore, J. F. 1982. Nitrogen and lignin control of hardwood leaf litter decomposition dynamics. *Ecology* 63:621–626.

Melillo, J. M., C. B. Field, and B. Moldan. 2003. *Interaction of the Major Biogeochemical Cycles.* SCOPE 61. Island Press, Washington, DC. 357 pp.

Mendelssohn, I. A., B. A. Keiss, and J. S. Wakeley. 1995. Factors controlling the formation of oxidized root channels: a review. *Wetlands* 15:37–46.

Mendelssohn, I. A., K. L. McKee, and W. H. Patrick, Jr. 1981. Oxygen deficiency in *Spartina alterniflora* roots: metabolic adaptation to anoxia. *Science* 214:439–441.

Merget, A. 1874. Sur la reproduction artificielle des phenomenes de thermodiffusion gazeuse des feuilles, par les corps poreaux et pulverulents humides. *C. R. Acad. Sci. Paris* 78:884–886.

Meuleman, A. F. M., J. P. Beekman, and J. T. A. Verhoevan. 2002. Nutrient retention and nutrient-use efficiency in *Phragmites australis* stands after wastewater application. *Wetlands* 22:712–721.

Mevi-Schutz, J. and W. Grosse. 1988. A two-way gas transport system in *Nelumbo necifera. Plant Cell Environ.* 11:27–34.

Mevi-Schutz, J. and W. Grosse. 1988. The importance of water vapor for circulating air-flow through *Nelumbo nicifera. J. Exp. Bot.* 39:1231–1236.

Miao, S. L. and W. DeBusk. 1999. Effects of phosphorus enrichment on structure and function of sawgrass and cattail communities in the Everglades. In K. R. Reddy, G. A. O'Connor, and C. L. Schelske (eds.) *Phosphorus Biogeochemistry in Subtropical Ecosystems.* CRC Press, Boca Raton, FL. pp. 275–299.

Miao, S. L., P. V. McCormick, S. Newman, and S. Rajagopalan. 2001. Interactive effects of seed availability, water depth, and phosphorus enrichment on cattail colonization in an Everglades wetland. *Wetland Ecol. Manage.* 9:39–47.

Miao, S. L., S. Newman, and F. H. Sklar. 2000. Effects of habitat nutrients and seed sources on growth and expansion of *Typha domingensis. Aquat. Bot.* 68:297–311.

Miao, S. L. and F. H. Sklar. 1998. Biomass and nutrient allocation of sawgrass and cattail along a nutrient gradient in the Florida Everglades. *Wetland Ecol. Manage.* 5:245–263.

Middelboe, M., N. Kroer, N. O. G. Jorgensen, and D. Pakulski. 1998. Influence of sediment on pelagic carbon and nitrogen turnover in a shallow Danish estuary. *Aquat. Microb. Ecol.* 14:81–90.

Milich, L. 1999. The role of methane in global warming: where might mitigation strategies be focused? *Global Environ. Change* 9:179–201.

Miller, S. L. and L. E. Orgel. 1974. *The Origins of Life on Earth.* Prentice-Hall, Englewood Cliffs, NJ. 228 pp.

Mitsch, W. J., J. W. Day, L. Zhang, and R. R. Lane. 2005. Nitrate-nitrogen retention in wetlands in the Mississippi River Basin. *Ecol. Eng.* 24:267–278.

Mitsch, W. J. and J. G. Gosselink. 2007. *Wetlands.* 4th Edition, Wiley, New York. 920 pp.

Miyamoto, J., K. Kitagawa, and Y. Sato. 1966. Metabolism of organophosphorus insecticides by *Bacillus subtilis* with special emphasis on sumithion. *Jpn. J. Exp. Med.* 36:211–225.

Moeller, R. E., J. M. Burkholder, and R. G. Wetzel. 1988. Significance of sedimentary phosphorus to a submersed freshwater macrophyte (*Najas flexilis*) and its algal epiphytes. *Aquat. Bot.* 32:261–281.

Molden, F., S. Seitzinger, V. T. Eviner, J. N. Galloway, X. Han, M. Keller, P. Nannipieri, W. O. Smith, Jr., and H. Tiessen. 2003. In J. M. Melillo, C. B. Field, and B. Moldan (eds.) *Interactions of the Major Biogeochemical Cycles.* SCOPE 61. Island Press, Washington, DC. pp. 93–116.

Molongoski, J. J. and M. J. Klug. 1976. Characterization of anaerobic heterotrophic bacteria isolated from freshwater lake sediments. *Appl. Environ. Microbiol.* 31:83–90.

Moodie, A. D. and W. J. Ingledew. 1990. Microbial anaerobic respiration. *Adv. Microb. Ecol.* 31:225–269.

Moore, W. S. 1996. Large ground water inputs to coastal waters revealed by Ra-226 enrichments. *Nature* 380(6575):612–614.

Moore, P. A. and K. R. Reddy. 1994. Role of E_h and pH on phosphorus geochemistry in sediments of Lake Okeechobee, Florida. *J. Environ. Qual.* 23:955–964.

Moore, T. R. and Dalva, M. 1993. The influence of temperature and water table position on carbon dioxide and methane emissions from laboratory columns of peatland soils. *J. Soil Sci.* 44:651–664.

Moore, T. R., A. Heyes, and N. T. Roulet. 1993. Methane flux: water table relations in northern peatlands. *Geophys. Res. Lett.* 20:587–590.

Moore, P. A., K. R. Reddy, and M. M. Fisher. 1998. Phosphorus flux between sediment and overlying water in Lake Okeechbee, Florida: spatial and temporal variations. *J. Environ. Qual.* 27:1428–1439.

Moore, W. S. 2003. Sources and fluxes of submarine groundwater discharge delineated by radium isotopes. *Biogeochemistry* 66:75–93.

Moran, M. A., R. Benner, and R. E. Hudson. 1989. Kinetics of microbial degradation of vascular plant material in two wetland ecosystems. *Oecologia* 79:158–167.

Morris, J. T. 2000. Effects of sea-level anomalies on estuarine processes. In J. E. Hobbie (ed.) *Estuarine Science: A Synthetic Approach to Research and Practice.* pp. 107–127.

Mortimer, C. H. 1941. The exchange of dissolved substances between mud and water in lakes. *J. Ecol.* 29:280–329.

Mortimer, C. H. 1942. The exchange of dissolved substances between mud and water in lakes. *J. Ecol.* 30:147–201.

Morton. S. C. and M. Edwards. 2005. Reduced phosphorurs compounds in the environment. *Crit. Rev. Environ. Sci. Technol.* 35(4):333–364.

Mulder, A., A. A. van de Graaf, L. A. Robertson, and J. G. Kuenen. 1995. Anaerobic ammonium oxidation discovered in a denitrifying fluidized bed reactor. *FEMS Microb. Ecol.* 16:177–184.

Mulholland, P. J., G. R. Best, C. C. Coutant, G. M. Hornberger, J. L. Meyer, P. J. Robinson, J. R. Stenberg, R. E. Turner, F. Vera-Herra, and R. G. Wetzel. 1997. Effects of global change on freshwater ecosystems of the Southeastern United States and the Gulf Coast of Mexico. *Hydrologic. Process.* 11:949–970.

Nakatsu, C. H. 2007. Soil microbial community analysis using denaturing gradient gel electrophoresis. *Soil Sci. Soc. Am. J.* 71:562–571.

Nagase, H., Y. Sato, and T. Ishikawa. 1984. Mercury methylation by compounds in humic material. *Sci. Total Environ.* 32:147.

Nealson, H. N. and C. R. Myers. 1992. Microbial reduction of manganese and iron: new approaches to carbon cycling. *Appl. Environ. Microbiol.* 58:439–443.

Nealson, K. H. 1983. The microbial iron cycle. In W. E. Krumbein (ed.) *Microbial geochemistry.* Blackwell Scientific Publications, Oxford, pp. 159–190.

Nealson, H. S. and C. R. Myers. 1992. Microbial reduction of manganese and iron: new approaches to carbon cycling. *Appl. Environ. Microbiol.* 58:439–443.

Nealson, K. H. and R. Rye. 2005. Evolution of metabolism. In W. H. Schlesinger (ed.) *Biogeochemistry.* Elsevier-Pergamon, Oxford, UK. pp. 41–62.

Neely, C. L., M. H. Beare, W. I. Hargrove, and D. C. Coleman. 1991. Relationships between fungal and bacterial substrate-induced respiration, biomass and plant residue decomposition. *Soil Biol. Biochem.* 23(10):947–954.

Nelson, D. W. 1982. Gaseous losses of nitrogen other than through denitrification. In F. J. Stevenson (ed.) *Nitrogen in Agricultural Soils.* American Society of Agronomy, Madison, WI.

Neue, H. 1993. Methane emission from rice fields. *Bioscience* 43:466–474.

Newman, R. H. and K. R. Tate. 1980. Soil phosphorus characterisation by ^{31}P nuclear magnetic resonance. *Commun. Soil Sci. Plant Anal.* 11:835–842.

Newman, S., J. B.Grace, and J. W. Koebel. 1996. Effects of nutrients and hydroperiod on mixtures of *Typha*, *Cladium*, and *Eleocharis*: implications for Everglades restoration. *Ecol. Appl.* 6:774–783.

Newman, S., H. Kumpf, J. A. Laing, and W. C. Kennedy. 2001. Decomposition responses to phosphorus enrichment in an Everglades (USA) slough. *Biogeochemistry* 54:229–250.

Newman, S., P. V. McCormick, and J. G. Backus. 2003. Phosphatase activity as an early warning indicator of wetland eutrophication: problems and prospects. *J. Appl. Phycol.* 15:45–59.

Newman, S., K. R. Reddy, W. F. DeBusk, Y. Wang, and M. M. Fisher. 1997. Spatial distribution of soil nutrients in a northern Everglades marsh: water conservation area 1. *Soil Sci. Soc. Am. J.* 61:1275–1283.

Newman, S., J. Schuette, J. B. Grace, K. Rutchey, T. Fontaine, K. R. Reddy, and M. Pietrucha. 1998. Factors influencing cattail abundance in the northern Everglades. *Aquat. Bot.* 60:265–280.

Noe, G. B., D. L. Childers, A. L. Edwards, E. E. Gaiser, K. Jayachandran, D. Lee, J. Meeder, J. Richards, L. J. Scinto, J. C. Trexler, and R. D. Jones. 2001. Short-term changes in phosphorus storage in an oligotrophic Everglades wetland ecosystem receiving experimental nutrient enrichment. *Biogeochemistry* 59:239–267.

Noe, G. B., D. L. Childers, and R. D. Jones. 2001. Phosphorus biogeochemistry and the impact of phosphorus enrichment: why is the Everglades so unique? *Ecosystems* 4:603–624.

NRC. 2002. National Research Council, National Academy Press.

Nyman, J. A., R. D. DeLaune, and W. H. Patrick, Jr. 1990. Wetland soil formation in the rapidly subsiding Mississippi River deltaic plain: mineral and organic matter relationships. *Estuar. Coastal Shelf Sci.* 31:57–69.

Obernosterer, I., P. Ruardij, and G. J. Herndl. 2001. Spatial and diurnal dynamics of dissolved organic matter (DOM) fluorescence and H_2O_2 and the photochemical oxygen demand of surface water DOM across the subtropical Atlantic Ocean. *Limnol. Oceanogr.* 46(3):632–643.

Obernosterer, I., R. Sempere, and G. J. Herndl. 2001. Ultraviolet radiation induces reversal of the bioavailability of DOM to marine bacterioplankton. *Aquat. Microb. Ecol.* 24(1):61–68.

Odeh, I. O. A., A. B. McBratney, and D. J. Chittleborough. 1995. Further results on prediction of soils properties from terrain attributes: heterotrophic cokriging and regression-kriging. *Geoderma* 67:215–226.

Odum, E. P. 1969. The strategy of ecosystem development. *Science* 164:262–270.

Odum, E. P. 1985. Trends expected in stressed ecosystem. *Bioscience* 35:419–422.

Oenema, O. and R. D. DeLaune. 1988. Accretion rates in salt marshes in the Eastern Scheldt, south-west Netherlands. *Estuar. Coast. Shelf Sci.* 26:379–394.

Ogram, A., S. Bridgham, R. Corstanje, H. Drake, K. Kosel, A. Mills, S. Newman, K. Portier, and R. Wetzel. 2006. Linkages between microbial community composition and biogeochemical processes across scales. *Ecol. Stud.* 190:239–268.

Orem, W. H. and D. P. Krabbenhoft. 2004. Impacts of sulfate contamination on the Florida Everglades ecosystem. USGS Fact Sheet FS109-3.

Oremland, R. S. 1988. Biogeochemistry of methanogenic bacteria. In A. J. Zehnder (ed.) *Biology of Microorganisms.* Wiley, New York. pp. 641–705.

Oremland, R. S. and C. W. Culbertson. 1992. Importance of methane-oxidizing bacteria in the methane budget as revealed by the use of a specific inhibitor. *Nature* 356:421–422.

Oremland, R. S. and J. F. Stolz. 2003. The ecology of arsenic. *Science* 300:939–944.

Orson, R. A. and B. L. Howes. 1992. Salt marsh development studies at Waquoit Bay, Massachusetts: influence of geomorphology on long-term plant community structure. *Estuar. Coast. Shelf Sci.* 35:453–471.

Orson, R. A., R. S. Warren, and W. A. Niering. 1998. Interpreting sea level rise and rates of vertical marsh accretion in a southern New England tidal salt marsh. *Estuar. Coast. Shelf Sci.* 47:419–429.

Osborne, T. 2005. Characterization, mobility, and fate of dissolved organic carbon in a wetland ecosystem. Ph.D. Dissertation. University of Florida, Gainesville, FL.

Osborne, T. Z., P. W. Inglett, and K. R. Reddy. 2007. The use of senescent plant biomass to investigate relationships between potential particulate and dissolved organic matter in a wetland ecosystem. *Aquat. Bot.* 86:53–61.

Osborne, T. Z., K. R. Reddy, S. Newman, and S. Grunwald. 2008. Spatial distribution of soil properties in the Everglades National Park (submitted for publication).

Osenberg, C. W., R. J. Schmitt, S. J. Holbrook, K. E. Anu-saba, and A. R. Flegal. 1994. Detection of environmental impacts: natural variability, effect size, and power analysis. *Ecol. Appl.* 4:16–30.

Ott, R. L. and M. Longnecker. 2001. *An Introduction to Statistical Methods and Data Analysis.* Wadsworth Group, Pacific Grove, CA.

Ottow, J. C. G. 1968. Evaluation of iron reducing bacteria in soil and the physiological mechanism of iron reduction in *Aerobacter aerogenes. Z. Allg. Mikrobiol.* 8:441–443.

Ottow, J. C. G. and A. von Klopotek. 1969. Enzymatic reduction of iron oxide by fungi. *Appl. Environ. Microbiol.* 18:41–43.

Ozesmi, S. L. and M. E. Bauer. 2002. Satellite remote sensing of wetlands. *Wetlands Ecol. Manage.* 10(5):381–402.

Pagenkopf, G. K. 1978. *Introduction to Natural Water Chemistry.* Marcel Dekker, New York. pp. 272.

Pal, D., F. E. Broadbent, and D. S. Mikkelsen. 1975. Influence of temperature on rice straw decomposition in soils. *Soil Sci.* 120:442–449.

Palm, C. A. and Sanchez, P. A. 1991. Nitrogen release from the leaves of some tropical legumes as affected by their lignin and polyphenolic contents. *Soil Biol. Biochem.* 23:83–88.

Paludan, C. and J. T. Morris. 1999. Distribution and speciation of phosphorus along a salinity gradient in intertidal marsh sediments. *Biogeochemistry* 45:197–221.

Palumbo, B., A. Bellanca, R. Neri, and M. J. Roe. 2001. Trace metal partitioning in Fe–Mn nodules from Sicilian soils. *Chem. Geol.* 173:257–269.

Pant, H. K., K. R. Reddy, and F. E. Dierberg. 2002. Bioavailability of phosphorus in the Everglades surface waters. *J. Environ. Qual.* 31:1748–1756.

Pardue, J. H., R. D. DeLaune, and W. H. Patrick, Jr. 1986. Fate of PCB and dioxin in Louisiana aquatic environment Technical Completion Report. Louisiana Water Resource Research Institute. Louisiana State University, Baton Rouge, LA. 90 pp.

Pardue, J. H., R. D. DeLaune, and W. H. Patrick, Jr. 1988. Effect of sediment pH and oxidation–reduction potential on PCB mineralization. *Water Air Soil Pollut.* 37:439–447.

Paterson, M., J. Rudd, and V. Louis. 1998. Increases in total mercury in zooplankton following flooding of a reservoir. *Environ. Sci. Technol.* 32:3868–3874.

Patrick, W. H., Jr. 1977. Oxygen content of soil air by field method. *Soil Sci. Soc. Am. Proc.* 41:651–652.

Patrick, W. H., Jr. 1981. The role of inorganic redox systems in controlling reduction in paddy soils. In Institute of Soil Science, Academia Sinica (eds.) *Proceedings of Symposium on Paddy Soil.* Springer-Verlag, New York.

Patrick, W. H., Jr. 1960. *Trans. 7th Int. Congress Soil Sci.* 2:494–500.

Patrick, W. H., Jr. 1966. Apparatus for controlling oxidation-reduction potential of waterlogged soils. *Nature* 212:1278–1279.

Patrick, W. H., S. Gotoh, and B. G. Williams. 1973. Strengite dissolutions in flooded soils and sediments. *Science* 179:564–565.

Patrick, W. H. and R. A. Khalid. 1974. Phosphorus release and adsorption by soils and sediments: effects of aerobic and anaerobic conditions. *Science* 186:53–57.

Patrick, W. H. and I. C. Mahapatra. 1968. Transformation and availability to rice of nitrogen and phosphorus in waterlogged soils. *Adv. Agron.* 20:323–359.

Patrick, W. H., Jr. and R. D. DeLaune. 1972. Characterization of the oxidized and reduced zones in flooded soil. *Soil Sci. Soc. Am. Proc.* 36:573–576.

Patrick, W. H., Jr. and R. D. DeLaune. 1990. Subsidence, accretion, and sea level rise in south San Francisco Bay marshes. *Limnol. Oceanogr.* 35:1389–1395.

Patrick, W. H., Jr., R. D. DeLaune, and R. Engler. 1973. Soil oxygen content and root development of cotton in Mississippi River alluvial soils. Louisiana State University, Agricultural Experiment Station Bulletin No. 673. pp. 28.

Patrick, W. H., Jr. and R. E. Henderson. 1981. Reduction and reoxidation cycles of manganese and iron in flooded soil and in water solution. *Soil Sci. Soc. Am. J.* 45:855–859.

Patrick, W. H., Jr. and M. Verloo. 1998. Distribution of soluble heavy metals between ionic and complexed forms in a saturated sediment as affected by pH and redox conditions. *Water Sci. Technol.* 37:165–172.

Patrick, W. H., Jr., B. G. Williams, and J. R. Morgan. 1973. A simple system for controlling redox potential and pH in soil suspensions. *Soil Sci. Soc. Am. Proc.* 37:331–332.

Patterson, J. 1999. A Canadian perspective on Wetlands and carbon sequestration. *Natl. Wetlands Newslett.* 21(2):3–4.

Paul, E. A. and F. E. Clark. 1996. Soil Microbiology and Biochemistry. Academic Press. pp. 339.

Payne, W. J. 1973. Reduction of nitrogenous oxides by microorganisms. *Bacteriol. Rev.* 37:409–452.

Payne, G. G., K. C. Weaver, and S. Xue. 2007. Status of phosphorus and nitrogen in the Everglades Protection Area, Chapter 3C, Volume 1. In *South Florida Environmental Report*. South Florida Water Management District, West Palm Beach, FL. http://www.sfwmd.gov/sfer/SFER_2007/index_draft_07.html

Pearl, H. W., J. L. Pinckeny, and T. F. Steppe. 2000. Cyanobacterial-bacterial mat consortia: examining the functional unit of microbial survival and growth in extreme environments. *Environ. Microbiol.* 2:11–26.

Pearsall, W. H. and C. H. Mortimer. 1939. Oxidation–reduction potentials in waterlogged soils, natural water and mud. *J. Ecol.* 27:483–501.

Pedersen, H. D., D. Postma, and R. Jakobsen. 2006. Release of arsenic associated with the reduction and transformation of iron oxides. *Geochim. Cosmochim. Acta* 70:4116–4129.

Pelegri, S. P. and T. H. Blackburn. 1994. Bioturbation effects of the amphipod *Corophium volutator* on microbial nitrogen transformations. *Marine Biol.* 121(2):253–258.

Penland, S. and K. E. Ramsey. 1989. Relative sea-level rise in Louisiana and the Gulf of Mexico: 1908–1988. *J. Coast. Res.* 6:323–342.

Penland, S. and K. Ramsey. 1997. Relative sea-level rise in Louisiana and the Gulf of Mexico: 1908–1988. *J. Coastal Res.* 6:323–342.

Penning, H., P. Claus, P. Casper, and R. Conrad. 2006. Carbon isotope fractionation during acetoclastic methanogenesis by *Methanosaeta concilii* in culture and a lake sediment. *Appl. Environ. Microbiol.* 72:5648–5652.

Pereira, W. E., C. E. Rostad, and T. J. Leiker. 1990. Distribution of agrochemicals in the lower Mississippi River and its tributaries. *Sci. Total Environ.* 97/98:41–53.

Pethick, J. S. 1992. Saltmarsh geomorphology. In J. Allen and K. Pye (eds.) *Saltmarshes: Morphodynamics, Conservation and Engineering Significance*. Cambridge University Press, Cambridge, UK. pp. 41–62.

Pettersson, K. 1986. The fractional composition of phosphorus in lake sediments of different characteristics. In P. G. Sly (ed.) *Sediment and Water Interactions*. Springer-Verlag, Berlin. pp. 149–155.

Pezeshki, S. R. 2001. Wetland plant responses to flooding. *Environ. Exp. Bot.* 46:299–312.

Pezeshki, S. R. and R. D. DeLaune. 1990. Influence of sediment oxidation–reduction potential on root elongation in *Spartina patens*. *Acta Oecol.* 11:377–383.

Pezeshki, S. R., R. D. DeLaune, and S. Z. Pan. 1991. Relationship of soil hydrogen sulfide level to net carbon assimilation of *Panicum hemitomon* and *Spartina patens*. *Vegetation* 95:159–166.

Pezeshki, S. R., R. D. DeLaune, and W. H. Patrick, Jr. 1988. Effect of fluctuating rhizosphere redox potential on carbon assimilation of *Spartina patens*. *Oecologica* 80:132–135.

Pezeshki, S. R., R. D. DeLaune, and W. H. Patrick, Jr. 1989. Assessment of saltwater intrusion impact on gas exchange behavior of Louisiana Gulf Coast plant species. *Wetland Ecol. Manage.* 101:21–300.

Pezeshki, S. R., S. Z. Pan, R. D. DeLaune, and W. H. Patrick, Jr. 1988. Sulfide-induced toxicity: inhibition of carbon assimilation in *Spartina alterniflora*. *Photosynthetica* 2(3):437–442.

Pezeshki, S. R., S. Z. Pan, R. D. DeLaune, and W. H. Patrick, Jr. 1989. Sulfide induced toxicity: inhibition of carbon assimilation in *Spartina alterniflora*. *Photosynthetica* 22(3):437–442.

Pietro, K., R. Bearzotti, M. Chimney, G. Germain, N. Iricanin, and T. Piccone. 2007. STA performance, compliance and optimization. Chapter 5, Volume 1. In South Florida Environmental Report, South Florida Water Management District, West Palm Beach, FL. http://www.sfwmd.gov/sfer/SFER_2007/index_draft_07.html

Pind, A., C. Freeman, and M. A. Lock. 1994. Enzyme degradation of phenolic materials in peatlands—measurement of phenol oxidase activity. *Plant Soil* 159:227–231.

Poiani, K. A., C. W. Johnson, and T. G. F. Kittel, 1995. Sensitivity of prairie wetland to increased temperature and seasonal precipitation changes. *Water Resour. Bull.* 31:283–294.

Ponnamperuma, F. N. 1955. The chemistry of submerged soils in relation to the growth and yield of rice. Ph.D. Dissertation, Cornell University.

Ponnamperuma, F. N. 1972. The chemistry of submerged soils. *Adv. Agron.* 24:29–96.

Ponnamperuma, F. N. 1981. Some aspects of the physical chemistry of paddy soils. In Institute of Soil Science, Academia Sinica (eds.) *Proceedings of Symposium on Paddy Soil*. Springer-Verlag, New York.

Ponnamperuma, F. N. 1984. Effects of flooding on soils. In T. T. Kozlowski (ed.) *Flooding and Plant Growth*. Academic Press, Inc., New York, NY. pp. 9–45.

Post, W. M., W. R. Emanuel, P. J. Zinke, and A. G. Stangenbeger. 1982. Soil carbon pools and world life zones. *Nature* 298:156–159.

Post, W. M., T. H. Peng, W. R. Emanuel, A. W. King, V. H. Dale, and D. L. DeAngelis. 1990. The global climate cycle. *Am. Scientist* 78:310–326.

Postgate, J. 1987. *Fundamentals of Nitrogen Fixation (New Studies in Biology)*. 2nd edition. Edward Arnold, London.

Pourbaix, M. 1966. *Atlas of Electrochemical Equilibria*. (In French). Pergamon Press, New York.

Pourbaix, M. 1966. *Atlas of Electrochemical Equilibria*. Pergamon Press, Oxford. 645 pp.

Prenger, J. P. and K. R. Reddy. 2004. Microbial enzyme activities in a freshwater marsh after cessation on nutrient loading. *Soil Sci. Soc. Am. J.* 68:1796–1804.

Preston, C. M. 1996. Applications of NMR to soil organic matter analysis: history and prospects. *Soil Sci.* 161:144–166.

Psenner, R. and R. Pucsko. 1988. Phosphorus fractionation: advantages and limits of the method for the study of sediment P origins and interactions. *Arch. Hydrobiol. Beih. Ergebn. Limnol.* 30:43–59.

Qualls, R. G. and C. J. Richardson. 1995. Forms of soil phosphorus along a nutrient enrichment gradient in the Florida Everglades. *Soil Sci.* 160:183–198.

Qualls, R. G. and C. J. Richardson. 2000. Phosphorus enrichment affects litter decomposition, immobilization, and soil microbial phosphorus in wetland mesocosms. *Soil Sci. Soc. Am. J.* 64:799–808.

Qualls, R. G. and C. J. Richardson. 2003. Factors controlling concentration, export, and decomposition of dissolved organic nutrients in the Everglades of Florida. *Biogeochemistry* 62:197–229.

Qualls, R. G., C. J. Richardson, and L. J. Sherwood. 2001. Soil reduction-oxidation potential along a nutrient-enrichment gradient in the Everglades. *Wetlands* 24:403–411.

Qui, S. and A. J. McComb. 1994. Effects of oxygen concentration on phosphorus release from reflooded air-dried wetland sediments. *Aust. J. Mar. Freshw. Res.* 45:1319–1328.

Rabenhorst, M. C. 1995. Carbon storage in tidal marsh soils. In R. Lal (ed.) *Soils and Global Change*. Lewis Publishers, Boca Raton, FL.

Raghoebarsing, A. A., A. Pol, K. T. van de Pas-Schoonen1, A. J. P. Smolders, K. F. Ettwig, W. I. C. Rijpstra, S. Schouten, J. S. S. Damste, H. J. M. Op den Camp, M. S. M. Jetten, and M. Strous. 2006. A microbial consortium couples anaerobic methane oxidation to denitrification. *Nature* 440:918–921.

Rainey, F. A., D. P. Kelly, E. Stackebrandt, J. Burghardt, A. Hiraishi, Y. Katayama, and A. P. Wood. 1999. A re-evaluation of the taxonomy of *Paracoccus denitrificans* and a proposal for the creation of *Paracoccus pantotrophus* comb. *Int. J. Syst. Bacteriol.* 49:645–651.

Raskin, I. and H. Kende. 1985. Mechanism of aeration of rice. *Science* 228:327–329.

Ratering, S. and S. Schnell. 2000. Localization of iron-reducing activity in paddy soil by profile studies. *Biogeochemistry* 48:341–365.

Reddy, K. R. and W. H. Patrick, Jr. 1975. Effect of alternate aerobic and anaerobic conditions on redox potential, organic matter decomposition, and nitrogen loss in a flooded soil. *Soil Biol. Biochem.* 7:87–94.

Reddy, K. R. and W. H. Patrick, Jr. 1976. Yield and nitrogen utilization by rice as affected by method and time of application of labeled nitrogen. *Agron. J.* 68:965–969.

Reddy, C. N. and W. H. Patrick, Jr. 1977. Effects of redox potential on the stability of zinc and copper chelates in flooded soils. *Soil Sci. Soc. Am. Proc.* 38:66–71.

Reddy, K. R., P. S. C. Rao, and W. H. Patrick, Jr. 1980. Factors influencing oxygen consumption rates in flooded soils. *Soil Sci. Soc. Am. J.* 44:741–744.

Reddy, K. R., P. S. C. Rao, and R. E. Jessup. 1982. The effect of carbon mineralization and denitrification kinetics in mineral and organic soils. *Soil Sci. Soc. Am. J.* 46:62–68.

Reddy, K. R. and W. H. Patrick. 1984. Nitrogen transformations and loss in flooded soil and sediments. *Crit. Rev. Environ. Control* 13:273–309.

Reddy, K. R. and W. H. Patrick, Jr. 1984. Nitrogen transformations and its loss in flooded soils and sediments. *CRC Crit. Rev. Environ. Contr.* 13(4):273–309.

Reddy, K. R. and W. H. Patrick, Jr. 1986. Fate of fertilizer nitrogen in the rice root zone. *Soil Sci. Soc. Am. J.* 50:649–651.

Reddy, K. R., T. C. Feijtel, and W. H. Patrick, Jr. 1986. Effect of soil redox conditions on microbial oxidation of organic matter. In Y. Chen and Y. Avnimelech (eds.) *The Role of Organic Matter in Modern Agriculture*. Dev. Plant Sci. Martinus Nijhoff, Dordrecht. pp. 117–148.

Reddy, K. R. and K. M. Portier. 1987. Nitrogen utilization by *Typha latifolia* L. as affected by temperature and rate of nitrogen application. *Aquat. Bot.* 27:127–138.

Reddy, K. R. and W. F. DeBusk. 1987. Nutrient storage capabilities of aquatic and wetland plants. In K. R. Reddy and W. H. Smith (eds.) *Aquatic Plants for Water Treatment and Resource Recovery*. Magnolia Publ. Inc., Orlando, FL. pp. 337–357.

Reddy, K. R., W. H. Patrick, Jr., and C. W. Lindau. 1989. Nitrification–denitrification at the plant–root–sediment interface in wetlands. *Limnol. Oceanogr.* 34:1004–1013.

Reddy, K. R., E. M. D'Angelo, and T. A. DeBusk. 1990. Oxygen transport through aquatic macrophytes: the role in wastewater treatment. *J. Environ. Qual.* 19:261–207.

Reddy, K. R. and M. M. Fisher. 1991. Sediment resuspension effects on phosphorus flux across the sediment–water interface: laboratory microcosm studies. In *Lake Okeechobee Phosphorus Dynamics Study: Biogeochemical Processes in the Sediments. Volume V*. Final Report submitted to South Florida Water Management District (Contract No. 531-M88-0445-A4), West Palm Beach, FL.

Reddy, K. R., R. D. DeLaune, W. F. DeBusk, and M. Koch. 1993. Long-term nutrient accumulation rates in the Everglades wetlands. *Soil Sci. Soc. Am. J.* 57:1145–1155.

Reddy, K. R., O. A. Diaz, L. J. Scinto, and M. Agami. 1995. Phosphorus dynamics in selected wetlands and streams of the Lake Okeechobee Basin. *Ecol. Eng.* 5:183–208.

Reddy, K. R. and E. M. D'Angelo. 1996. Biogeochemical indicators to evaluate pollutant removal efficiency in constructed wetlands. *Water Sci. Tech.* 35:1–10.

Reddy, K. R., M. M. Fisher, and D. Ivanoff. 1996. Resuspension and diffusive flux of nitrogen and phosphorus in a hypereutrophic lake. *J. Environ. Qual.* 25:363–371.

Reddy, K. R., G. A. O'Connor, and P. M. Gale. 1998. Phosphorus sorption capacities of wetland soils and stream sediments impacted by dairy effluent. *J. Environ. Qual.* 27:438–447.

Reddy, K. R., Y. Wang, W. F. DeBusk, M. M. Fisher, and S. Newman. 1998. Forms of soil phosphorus in selected hydrologic units of the Florida Everglades. *Soil Sci. Soc. Am. J.* 62:1134–1147.

Reddy, K. R., J. R. White, A. Wright, and T. Chua. 1999. Influence of phosphorus loading on microbial processes in soil and water column of wetlands. Phosphorus in Florida's Ecosystems: Analysis of Current Issues. In K. R. Reddy, G. A. O'Connor, and C. L. Schelske (eds.) *Phosphorus Biogeochemistry in Subtropical Ecosystems: Florida as a Case Example*. CRC Press/Lewis Publishers, Boca Raton, FL. pp. 249–273.

Reddy, K. R., R. H. Kadlec, E. Flaig, and P. M. Gale. 1999. Phosphorus retention in streams and wetlands: a review. *Crit. Rev. Environ. Sci. Technol.* 29(1):83–146.

Reddy, K. R., R. G. Wetzel, and R. Kadlec. 2005. Biogeochemistry of phosphorus in wetlands. In J. T. Sims and A. N. Sharpley (eds.) *Phosphorus: Agriculture and the Environment*. Soil Science Society of America, Madison, WI. pp. 263–316.

Reddy, K. R., R. H. Kadlec, M. J. Chimney, and W. J. Mitsch. 2006. The Everglades nutrient removal project. Special Issue. *Ecol. Eng.* 27:1–379.

Reddy, K. R., T. Z. Osborne, K. S. Inglett, and R. Corstanje, 2006. Influence of water levels on subsidence of organic soils in the Upper St. Johns River Basin. Report submitted to St. John River Water Management District, Palatka, FL.

Redfield, G. and S. Efron. 2007. An integrative perspective on regional water quality and phosphorus. Chapter 1B, Volume 1. In South Florida Environmental Report, South Florida Water Management District, West Palm Beach, FL. http://www.sfwmd.gov/sfer/SFER_2007/index_draft_07.html.

Redfield, G., S. Efron, and K. Burns. 2007. Introduction to the 2007 South Florida Environmental Report, Chapter 1A, Volume 1. In South Florida Environmental Report, South Florida Water Management District, West Palm Beach, FL. http://www.sfwmd.gov/sfer/SFER_2007/index_draft_07.html.

Redman, F. H. and W. H. Patrick, Jr. 1965. *Effect of Submergence on Several Biological and Chemical Properties*. Agricultural Experiment Station, Louisiana State University, Bulletin No. 592, pp. 1–28.

Reed, D. J. 1989. Patterns of sediment deposition in subsiding coastal salt marshes, Terrebonne Bay, Louisiana: the role of winter storms. *Estuaries* 12:222–227.

Reed, D. J. 1989. Patterns of sediment deposition in subsiding coastal salt marshes, Terrebone Bay, Louisiana: the role of winter storms. *Estuaries* 12(4):222–227.

Reed, R. B., Jr. 1988. *National List of Plant Species That Occur in Wetlands: National Summary.* U.S. Fish and Wildlife Service, Washington, DC. *Biol. Rpt.* 88(24):244.

Refsgaard, J. C. and B. Storm. 1995. MIKE SHE. In V. P. Singh (ed.) *Computer Models of Watershed Hydrology.* Water Resources Publishers, Highlands Ranch, CO. pp. 809–846.

Revsbech, N. P. 1989. An oxygen microelectrode with a guard cathode. *Limnol. Oceanogr.* 34:474–476.

Reynolds, C. S. and P. S. Davis. 2001. Sources and bioavailability of phosphorus fractions in freshwaters: a British perspective. *Biol. Rev.* 76:27–64.

Richardson, C. J. 1985. Mechanisms controlling phosphorus retention capacity in freshwater wetlands. *Science* 228:1424–1426.

Richardson, J. L. and M. J. Vepraskas. 2001. *Wetland Soils.* Lewis Publishers, Boca Raton, FL. 417 pp.

Rickard, D. and G. W. Luther. 1997. Kinetics of pyrite formation by the H_2S oxidation on iron (II) monosulfide in aqueous solutions between 25 and 125°C: the mechanism. *Geochim. Cosmochim. Acta* 61:135–147.

Richardson, C. J. and P. Vaithiyanathan. 1995. P sorption characteristics of the Everglades soils along an eutrophication gradient. *Soil Sci. Soc. Am. J.* 59:1782–1788.

Ritchie, J. C. and J. R. McHenry. 1990. Application of radioactive fallout cesium-137 for measuring soil erosion and sediment accumulation rates and patterns: a review. *J. Environ. Qual.* 19:215–233.

Rivero, R. G., S. Grunwald, and G. L. Bruland. 2007. Incorporation of spectral data into multivariate geostatistical models to map soil phosphorus variability in a Florida wetland. *Geoderma* 140: 428–443.

Roberts, H. H., O. K., Huh, S. A. Hsu, and L. J. Rouse, Jr. 1989. Winter storm impacts on the Chenier Plain coast of southwestern Louisiana. *Gulf Coast Assoc. Geol. Soc. Trans.* 39:515–522.

Rochette, E. A., B. C. Bostick, G. Li, and S. Fendorf. 2000. Kinetics of arsenate reduction by dissolved sulfide. *Environ. Sci. Technol.* 34:4714–4720.

Roden, E. E. and J. W. Edmonds. 1997. Phosphate mobilization in iron-rich anaerobic sediments: microbial Fe(III) oxide reduction versus iron-sulfide formation. *Arch. Hydrobiol.* 139:347–378.

Roden, E. E. and R. G. Wetzel. 1996. Organic carbon oxidation and suppression of methane production by microbial Fe(III) reduction in vegetated and unvegetated freshwater wetland sediments. *Limnol. Oceanogr.* 41:1733–1748.

Roden, E. E. and R. G. Wetzel. 2002. Kinetics of microbial Fe(III) oxide reduction in freshwater wetland sediments. *Limnol. Oceanogr.* 47:198–211.

Rodhe, H. 1990. A comparison of the contribution of various gases to the greenhouse effect. *Science* 248:1217–1219.

Rodhe, H. and I. Isaken. 1980. Global distribution of sulfur compounds in the troposphere estimated in a height-latitude transport model. *J. Geophys. Res. Oceans Atmos.* 85:7401–7409.

Roman, C. T., J. A. Peck, J. R. Allen, J. W. King, and P. G. Appleby. 1997. Accretion of a New England (U.S.A.) salt marsh in response to inlet migration, storms, and sea-level rise. *Estuar. Coast. Shelf Sci.* 45:717–727.

Rosenberry, D. O. 2005. Integrating seepage heterogeneity with the use of sanded seepage meter. *Liminol. Oceanogr. Methods* 3:131–138.

Rousseaua, D. P. L., P. A. Vanrolleghemb, and N. De Pauwa. 2004. Model-based design of horizontal subsurface flow constructed treatment wetlands: a review. *Water Res.* 38:1484–1493.

Rowell, D. L. 1981. Oxidation and reduction. In D. J. Greenland and M. H. B. Hayes (eds.) *The Chemistry of Soil Processes.* Wiley, New York. pp. 401–461.

Rutchey, K. and L. Vilchek. 1994. Development of an Everglades vegetation map using a SPOT image and the Global Positioning System. *Photogramm. Eng. Remote Sensing* 60:767–775.

Ruttenberg, K. C. 1992. Development of sequential extraction method for different forms of phosphorus in marine sediments. *Limnol. Oceanogr.* 37:1460–1482.

Ruttenberg, K. C. and M. A. Goni. 1997. Phosphorus distribution, C:N:P ratios, and $^{13}C_{oc}$ in arctic, temperate, and tropical coastal sediments: tools for characterizing bulk sedimentary organic matter. *Mar. Geol.* 139:123–145.

Sah, R. N., D. S. Mikkelsen, and A. A. Hafez. 1989. Phosphorus behavior in flooded-drained soils. I. Effects on phosphorus sorption. *Soil Sci. Soc. Am. J.* 53:1718–1723.

Sah, R. N., D. S. Mikkelsen, and A. A. Hafez. 1989. Phosphorus behavior in flooded-drained soils. II. Iron transformations and phosphorus sorption. *Soil Sci. Soc. Am. J.* 53:1723–1729.

References

Sah, R. N., D. S. Mikkelsen, and A. A. Hafez. 1989. Phosphorus behavior in flooded-drained soils. III. Phosphorus desorption and availability. *Soil Sci. Soc. Am. J.* 53:1729–1732.

Sahrawat, K. L. 2004. Ammonium production in submerged soils and sediments: the role of reducible iron. *Commun. Soil Sci. Plant Anal.* 35:399–411.

Saito, M., H. Wada, and Y. Takai. 1990. Development of microbial community on cellulose buried in waterlogged soil. *Biol. Fert. Soils.* 9:301–305.

Saleque, M. A. and G. J. D. Kirk. 1995. Root-induced solubilization of phosphate in the rhizosphere of lowland rice. *New Physiol.* 129(2):325–336.

Salinger, Y., Y. Geifman, and M. Aronowich. 1993. Orthophosphate and calcium carbonate solubilities in the Upper Jordan watershed basin. *J. Environ. Qual.* 22:672–677.

Sanchez-Carillo, S., M. Alvarez-Cobelas, and D. G. Angeler. 2001. Sedimentation in the semi-arid freshwater wetland Las Tablas de Daimiel (Spain). *Wetlands* 21:112–124.

Sass, R. L., F. M. Fisher, P. A. Harcombe, and F. T. Ferner. 1990. Methane production and emission from a Texas rice field. *Global Biogeochem. Cycles* 4:47–68.

Sasser, C. E. and J. G. Gosselink. 1984. Vegetation and primary production in a floating freshwater marsh in Louisiana. *Aquat. Bot.* 20:245–255.

Savant, N. K. and S. K. DeDatta. 1982. Nitrogen transformations in wetland rice soils. *Adv. Agron.* 35:241–302.

Schabenberger, O. and F. J. Pierce. 2002. *Contemporary Statistical Models for the Plant and Soil Sciences.* CRC Press, New York.

Schipper, L. A. and K. R. Reddy. 1994. In situ determination of plant detritus breakdown in a wetland soil–floodwater profile. *Soil Sci. Soc. Am. J.* 59:565–568.

Schipper, L. A. and K. R. Reddy. 1996. Methane oxidation in the rhizosphere of *Sagittaria lancifolia*. *Soil Sci. Soc. Am. J.* 60:611–616.

Schlesinger, W. H. 1997. *Biogeochemistry: An Analysis of Global Change.* 2nd Edition. Academic Press, San Diego, CA.

Schlesinger, W. H. 2005. *Biogeochemistry.* In H. D. Holland and K. K. Turekian (eds.) *Treatise in Geochemistry.* Vol. 8. Elsevier-Pergamon, Oxford. 702 pp.

Schmidt, E. L. 1982. Nitrification in soil. In F. J. Stevenson (ed.) *Nitrogen in Agricultural Soils. Agronomy* 22:253–288.

Schmidt, K. S. and A. K. Skidmore. 2003. Spectral discrimination of vegetation types in a coastal wetland. *Rem. Sens. Environ.* 5:1–17.

Schmidt, I., R. J. van Spanning, and M. S. Jetten. 2004. Denitrification and ammonia oxidation by *Nitrosomonas europaea* wild-type, and NirK- and NorB-deficient mutants. *Microbiology* 2:495–505.

Scholander, P. F., L. van Dam, and S. I. Scholander. 1955. Gas exchange in the roots of mangroves. *Am. J. Bot.* 42:92–98.

Schroder, P., W. Grosse, and D. Woermann. 1986. Localization of thermoosmotically active partitions in young leaves of *Nuphar lutea*. *J. Exp. Bot.* 37:1450–1461.

Schulten, H. R. and M. Schnitzer. 1998. The chemistry of soil organic nitrogen: a review. *Biol. Fertil. Soils* 26:1–15.

Schutz, H., W. Seiler, and R. Conrad. 1989. Processes involved in formation and emission of methane in rice paddies. *Biogeochemistry* 7:33–53.

Schutz, H., W. Seiler, and R. Conrad. 1990. Influence of soil temperature on methane emission from rice paddy fields. *Biogeochemistry* 11:77–95.

Schwarzenbach, R. P., P. M. Gschwend, and D. M. Imboden. 1993. *Environmental Organic Chemistry.* Wiley, New York. 681 pp.

Schwertmann, U. 1993. Relations between iron oxides, soil color, and soil formation. In J. M. Bigham and E. J. Ciolkosz (eds.) *Soil Color.* Soil Science Society Of America, Madison, WI. pp. 51–70.

Science Coordination Team (SCT). 2003. The role of flow in the Everglades ridge and slough landscape. Online at http://www.sfrestore.org/sct/doc.

Scinto, L. J. 1990. Seasonal variation in soil phosphorus distribution in two wetlands of south Florida, M.S. Thesis, University of Florida.

Scinto, L. J. and K. R. Reddy. 2003. Biotic and abiotic uptake of phosphorus by periphyton in a sub-tropical freshwater wetland. *Aquat. Bot.* 77:202–222.

Sebacher, D. L., R. C. Harris, and K. B. Bartlett. 1985. Methane emissions to the atmosphere through aquatic plants. *J. Environ. Qual.* 14:40–46.

Segel, I. H. 1976. *Biochemical Calculations.* 2nd Edition. Wiley, New York.

Segers, R. 1998. Methane production and methane consumption: a review of processes underlying wetland methane fluxes. *Biogeochemistry* 41:23–51.

Seitzinger, S., J. A. Harrison, J. K. Bohlke, A. F. Bouwman, R. Lowrance, B. Peterson, C. Tobias, and G. Van Drecht. 2006. Denitrification across landscapes and waterscapes: a synthesis. *Ecol. Appl.* 16:2064–2090.

Seitzinger, S., S. Nixon, M. E. W. Pilson, and S. Burke. 1980. Denitrification and N_2O production in near shore marine sediments. *Geochim. Cosmochim. Acta* 44:1853–1860.

Seitzinger, S. P. 1988. Denitrification in freshwater and coastal marine ecosystems: ecological and geochemical significance. *Limnol. Oceanogr.* 33:702–724.

Seitzinger, S. P. and R. W. Sanders. 1997. Contribution of dissolved organic nitrogen from rivers to estuarine eutrophication. *Mar. Ecol. Prog. Ser.* 59:1–12.

Senko, M. W., C. L. Hendrickson, L. Pasa-Tolic, J. A. Marto, F. M. White, S. Guan, and A. G. Marshall. 1996. External accumulation of ions for enhanced electrospray ionization fourier transform ion cyclotron resonance mass spectrometry. *J. Am. Soc. Mass Spectrom.* 8:970–977.

Sethunathan, N. 1973. Degradation of parathion in flooded acid soils. *J. Agric. Food Chem.* 21:602–604.

Seurant, L. and P. G. Strutton (eds.). 2004. *Scaling Methods in Aquatic Ecology—Measurement, Analysis, Simulation.* CRC Press, New York.

Sexstone, A. J., N. P. Revsbech, T. B. P. Arkin, and J. M. Tiedje. 1985. Direct measurement of oxygen profiles and denitrification rates in soil aggregates. *Soil Sci. Soc. Am. J.* 49:645–651.

Shannon, R. D. and J. R. White. 1991. The selectivity of a sequential extraction procedure for the determination of iron oxyhydroxides and iron sulfides in lake sediments. *Biogeochemistry* 14:193–208.

Sharitz, R. R. and S. C. Pennings. 2006. Development of wetland plant communities. In D. P. Batzer and R. R. Sharitz (eds.) *Freshwater and Estuarine Wetlands.* University of California Press, Berkley/Los Angeles, CA. pp. 177–241.

Sharma, K., P. W. Inglett, K. R. Reddy, and A. V. Ogram. 2005. Microscopic examination of photoautotrophic and phosphatase-producing bacteria in phosphorus-limited Everglades periphyton mats. *Limnol. Oceanogr.* 50:2057–2062.

Shaver, G. R. and J. M. Melillo. 1984. Nutrient budgets of marsh plants: efficiency concepts and relation to availability. *Ecology* 65:1491–1510.

Shepherd, K. D. and M. G. Walsh. 2002. Development of reflectance spectral libraries for characterization of soil properties. *Soil Sci. Soc. Am. J.* 66:988–998.

Sillen, L. G. and A. E. Martell. 1964. *Stability Constants of Metal-Ion Complexes.* 2nd Edition. Special Publication No. 17. The Chemical Society, London.

Simon, N. S. 1988. Nitrogen cycling between sediment and the shallow-water column in the transition zone of the Potomac River and estuary. I. Nitrate and ammonium flux. *Estuar. Coast. Shelf Sci.* 26:483–497.

Simon, N. S. 1988. Nitrogen cycling between sediment and the shallow-water column in the transition zone of the Potomac River and estuary. II. The role of wind-driven resuspension and adsorbed ammonium. *Estuar. Coast. Shelf Sci.* 26:483–497.

Sinsabaugh, R. L., R. K. Antibus, and A. E. Linkins. 1991. An enzymatic approach to the analysis of microbial activity during plant litter decomposition Agric. *Ecosyst. Environ.* 34:43–54.

Sinsabaugh, R. L., R. K. Antibus, A. E. Linkins, C. A. McClaugherty, L. Rayburn, D. Repert, and T. Weiland. 1993. Wood decomposition: nitrogen and phosphorus dynamics in relation to extracellular enzyme activity. *Ecology* 74:1586–1593.

Sinsabaugh, R. L., M. M. Carriero, and D. A. Repert. 2002. Allocation of extracellular enzymatic activity in relation to litter composition, N deposition, and mass loss. *Biogeochemistry* 60:1–24.

Sjoblad, R. D. and J. M. Bollag. 1981. Oxidation coupling of aromatic compounds by enzymes from soil microorganisms. In E. A. Paul and J. N. Ladd (eds.) *Soil Biochemistry.* Vol. 5. Marcel Dekker, New York. pp. 113–152.

Sklar, F. H., M. J. Chimney, S. Newman, P. McCormick, D. Gawlik, S. Miao, C. McVoy, W. Said, J. Newman, C. Coronado, G. Cozier, M. Korvela, and K. Rutchey. 2005. The ecological-societal underpinnings of Everglades restoration. *Frontiers Ecol. Environ.* 3(3):161–169.

Smalley, A. E. 1958. *The Role of Two Invertebrate Populations, Littorina Irrorata and Orchelium Fidicenium in the Energy Flow of a Salt Marsh Ecosystem.* Doctoral Thesis. University of Georgia, Athens, Georgia.

Smedley, P. L. and D. G. Kinniburgh. 2002. A review of the source, behavior and distribution of arsenic in natural waters. *Appl. Geochem.* 17:517–568.

Smirnoff, N. and R. M. M. Crawford. 1983. Variation in the structure and response to flooding of root aerenchyma in some wetland plants. *Ann. Bot.* 51(2):237–249.

References

Smith, C. J. and R. D. DeLaune. 1983. Gaseous nitrogen losses from gulf coast marshes. *Northeast Gulf Sci.* 6(1):1–8.

Smith, C. J., R. D. DeLaune, and W. H. Patrick, Jr. 1983. Nitrous oxide emission from Gulf Coast Wetlands. *Geochem. Cosmochim. Acta* 47:1805–1814.

Smith, C. J., R. D. DeLaune, and W. H. Patrick, Jr. 1983. Carbon dioxide emission and carbon accumulation in coastal wetlands. *Est. Coast. Shelf Sci.* 17:21–29.

Smith, E., R. M. Naidu, and A. M. Olston. 1998. Arsenic in the soil environment: a review. *Adv. Agron.* 64:149–195.

Smith, E. P. 2002. BACI design. In A. H. El-Shaarawi and W. W. Piegorsch (eds.) *Encyclopedia of Environmetrics.* Vol. 1. Wiley, Chichester. pp. 141–148.

Smith, S. M., S. Newman, P. B. Garrett, and J. A. Leeds. 2001. Differential effects of surface and peat fire on soil constituents in a degraded wetland of the Northern Florida Everglades. *J. Environ. Qual.* 30:1998–2005.

Soil Survey Staff. 1999. *Soil Taxonomy: A Basic System of Soil Classification for Making and Interpreting Soil Surveys.* 2nd Edition. USDA NRCS, Agriculture Handbook No. 436. U.S. Government Printing Office, Washington, DC.

South Florida Water Management District. 2007. Phosphorus source controls for the basins tributary to the Everglades Protection Area. In South Florida Environmental Report, South Florida Water Management District, West Palm Beach, Fl. http://www.sfwmd.gov/sfer/SFER_2007/index_draft_07.html.

Sowden, F. J., Y. Chen, and M. Schnitzer. 1977. The nitrogen distribution in soils formed under widely differing climatic conditions. *Geochim. Cosmochim. Acta* 41:1524–1526.

Sparks, D. L., A. L. Page, P. A. Helmke, R. H. Loeppert, P. N. Soltanpour, M. A. Tabatabai, C. T. Johnson, and M. E. Sumner. 1996. *Methods of Soil Analysis: Part 3, Chemical Methods.* SSSA and ASA Series No. 5. Soil Science Society of America, Madison, WI. 1358 pp.

Sparling, G. P. 1992. Ratio of microbial biomass carbon to soil organic-carbon as a sensitive indicator of changes in soil organic-matter. *Aust. J. Soil Res.* 30:195–207.

Spratt, H. G. and M. D. Morgan. 1990. Sulfur cycling in a cedar-dominated freshwater wetland. *Limnol. Oceanogr.* 35:1586–1593.

Stal, L. J. 1988. Nitrogen fixation in cyanobacterial mats. In L. Packer and A. N. Glazer (eds.). Methods in *Enzymology Cyanobacteria.* Vol 167. Academic Press, New York. pp. 474–484.

Stal, L. J. 2000. *Cyanobacterial Mats and Stromatolites.* Kluwer Academic Publishers, Dordrecht, The Netherlands.

Steinmann, C. R., E. Veit, S. Grosser, and A. Melzer. 2001. An in-situ mobile interstitial water and sediment sampler (MISS). *Hydrobiologia* 459:173–176.

Sterner, R. W. and J. L. Elser. 2002. *Ecological Stoichiometry.* Princeton University Press, New Jersey. 439 pp.

Steudler, P. A. and B. J. Peterson. 1984. Contribution of gaseous sulfur from salt marshes to the global sulfur cycle. *Nature* 311:455–457.

Stevenson, F. J. 1982. *Humus Chemistry Genesis, Composition, Reactions.* Wiley, New York.

Stevenson, F. J. 1985. *Cycles of Soil: Carbon, Nitrogen, Phosphorus, Sulfur, Micronutrients.* Wiley, New York, NY. 380 pp.

Stevenson, J. C., M. S. Kearney, and E. C. Pendleton. 1985. Sedimentation and erosion in a Chesapeake Bay brackish marsh system. *Mar. Geol.* 67:213–235.

Stevenson, F. J. 1986. *Cycles of Soil: Carbon, Nitrogen, Phosphorus, Sulfur, Micro Nutrients.* Wiley, New York. 380 pp.

Stevenson, F. J. 1994. *Humus Chemistry.* 2nd Edition. Wiley, New York. 496 pp.

Stoddart, D. R., D. J. Reed, and J. R. French. 1989. Understanding salt-marsh accretion, Scolt Head Island, Norfolk, England. *Estuaries* 12:228–236.

Stow, C. A., R. D. DeLaune, and W. H. Patrick, Jr. 1985. Nutrient fluxes in a eutrophic coastal Louisiana freshwater lake. *Environ. Manage.* 9:243.

Straub, K. L., M. Benz, and B. Schink. 2001. Iron metabolism in anoxic environments at near neutral pH. *FEMS Microbiol. Ecol.* 34:181–186.

Straub, K. L., M. Benz, B. Schink, and F. Widdel. 1996. Anaerobic, nitrate-dependent microbial oxidation of ferrous iron. *Appl. Environ. Microbiol.* 62:1458–1460.

Stumm, W. and J. J. Morgan. 1981. *Aquatic Chemistry.* 2nd Edition. Wiley, New York. 780 pp.

Sundareshwar, P. V., J. T. Morris, E. K. Koepfler, and B. Fornwalt. 2003. Phosphorus limitation of coastal ecosystem processes. *Science* 299:563–565.

Suter II, G. W. 1990. Endpoints for regional ecological risk assessment. *Environ. Manage.* 14:19–23.

Sutka, R. L., N. E. Ostrom, P. H. Ostrom, J. A. Breznak, H. Gandhi, A. J. Pitt, and F. Li. 2006. Distinguishing nitrous oxide production from nitrification and denitrification on the basis of isotopomer abundances. *Appl. Environ. Microbiol.* 72:638–644.

Sweerts, J., R. C. Kelly, J. Rudd, R. Hesslein, and T. Cappenberg. 1991. Similarity of whole-sediment diffusion coefficients in freshwater sediments of low and high porosity. *Limnol. Oceanogr.* 36:335–342.

Swift, M. J. 1982. Microbial succession during the decomposition of organic matter. In R. G. Burns and J. H. Slater (eds.) *Experimental Microbial Ecology*. Blackwell Scientific, Oxford. pp. 164–177.

Syers, J. K., R. F. Harris, and D. E. Armstrong. 1973. Phosphate chemistry in lake sediments. *J. Environ. Qual.* 2:1–14.

Takahashi, Y., R. Minamikawa, K. H. Hattori, K. Kurishma, N. Kihou, and K. Yuita. 2004. Arsenic behavior in paddy fields during the cycle of flooded and non-flooded periods. *Environ. Sci. Technol.* 38:1038–1044.

Takai, Y. 1970. The mechanism of methane fermentation in flooded paddy soil. *Soil Sci. Plant Nutr.* 16:238–344.

Thamdrup, B. and T. Dalsgaard. 2002. Production of N_2 through anaerobic ammonium oxidation coupled to nitrate reduction in marine sediments. *Appl. Environ. Microbiol.* 68:1312–1318.

Thaur, R. K., K. Jungerman, and K. Decker. 1977. Energy conservation in chemotrophic anaerobic bacteria. *Bacteriolog. Rev.* 41:100–180.

Thibodeaux, L. G. 1979. *Chemodynamics: Environmental Movement of Chemicals in Air, Water, and Soil.* Wiley, New York.

Thies, J. E. 2007. Soil microbial community analysis using terminal restriction fragment length polymorphisms. *Soil Sci. Soc. Am. J.* 71:579–591.

Thom, R. M. 1992. Accretion rates of low intertidal salt marshes in the Pacific Northwest. *Wetlands* 12:147–156.

Tiedje, J. M. 1988. Ecology of denitrification and dissimilatory nitrate reduction to ammonia. In A. J. B. Zehnder (ed.) *Biology of Anaerobic Microorganisms.* Wiley, New York. pp. 197–244.

Tiedje, J. M. 1994. Denitrifiers. In R. W. Weaver, J. S. Angle, and P. S. Bottomley (eds.) *Methods of Soil Analysis, Part 2: Microbiological and Biochemical Properties.* Soil Science Society of America, Madison, WI. pp. 245–267.

Tilley, D. R. and M. T. Brown. 2006. Dynamic energy accounting for assessing the environmental benefits of subtropical wetland storm water management systems. *Ecol. Model.* 192:327–361.

Torstensson, L., M. Pell, and B. Stenberg. 1998. Need of a strategy for evaluation of arable soil quality. *Ambio* 27:4–8.

Trimmer, M., J. C. Nicholls, and B. Deflandre. 2003. Anaerobic ammonium oxidation measured in sediments along the Thames estuary, United Kingdom. *Appl. Environ. Microbiol.* 69:6446–6454.

Trouwborst, R. E., B. G. Clement, B. M. Tebo, B. T. Glazer, and G. E. Luther. 2006. Soluble Mn(III) suboxic zones, *Science* 1955–1957.

Truesdall, A. 1969. The advantage of using pe rather than Eh in redox equilibrium calculations. *J. Geol. Ed.* 17:17–20.

Turner, B. L., E. Frossard, and D. D. Baldwin. 2004. *Organic Phosphorus in the Environment*. CAB Publishing, Cambridge, MA. 389 pp.

Turner, B. L., N. Mahieu, and L. M. Condron. 2003a. Phosphorus-31 nuclear magnetic resonance spectral assignments of phosphorus compounds in soil NaOH-EDTA extracts. *Soil Sci. Soc. Am. J.* 67:497–510.

Turner, B. L., N. Mahieu, and L. M. Condron. 2003b. The phosphorus composition of temperate pasture soils determined by NaOH-EDTA extraction and solution ^{31}P NMR spectroscopy. *Org. Geochem.* 34:1199–1210.

Turner, B. L. and S. Newman. 2005. Phosphorus cycling in wetland soils: the importance of phosphate diesters. *J. Environ. Qual.* 34:1921–1929.

Turner, B. L., S. Newman, and J. M. Newman. 2006. Organic phosphorus sequestration in sub-tropical treatment wetlands. *Environ. Sci. Technol.* 40:727–733.

Turner, B. L., M. J. Paphazy, P. M. Haygarth, and I. D. McKelvie. 2002. Inositol phosphates in the environment. *Phil. Trans. R. Soc. Long.* 357:449–469.

Turner, R. E., J. J. Baustran, E. M. Swenson, and J. S. Spicer. 2006. Wetland sedimentation from hurricane Katrina and Rita. *Science* 314:449–452.

Tusneem, M. E. and W. H. Patrick, Jr. 1971. Nitrogen transformations in waterlogged soils. Bulletin No. 657. Agricultural Experiment Station, Louisiana State University. pp. 73.

Twinch, A. J. 1987. Phosphate exchange characteristics of wet and dried sediment samples from a hypertrophic reservoir: implications for the measurement of sediment phosphorus status. *Water Res.* 21:1225–1230.

U.S. Environmental Protection Agency (USEPA). 1990. The Quality of Our Nation's Water, EPA 440/4-90-005, USEPA, Washington, DC.

USEPA (U.S. Environmental Protection Agency). 1992. *Framework for Ecological Risk Assessment.* EPA/630/R-92/001. U.S. Environmental Protection Agency, Risk Assessment Forum, Washington, DC.

USEPA (U.S. Environmental Protection Agency). 1993. Methods for determination of inorganic substances in environmental samples. U.S. Government Printing Office, Washington, DC.

USEPA (U.S. Environmental Protection Agency). 1994. National Water Quality Inventory-1992 Report to Congress. U.S. Environmental Protection Agency, 841-R-94-001, Washington, DC.

USEPA (U.S. Environmental Protection Agency). 1996. Drinking water regulations and health advisories. Office of Water, EPA 822-B-96-002, USEPA, Washington, DC.

USEPA (U.S. Environmental Protection Agency). 1997. Pesticides industry sales and usage. 1994 and 1995 Market Estimates. USEPA/Office of Prevention, Pesticides and Toxic Substances 733-R-97-002, Washington, DC.

U.S. EPA. 1993. Methods for the determination of inorganic substances in environmental samples. U.S. Environmental Protection Agency. EPA/600/R-93/100.

USEPA (U.S. Environmental Protection Agency). 2006. *Draft Nutrient Criteria Technical Guidance Manual: Wetlands.* EPA-823-B-05-003. Washington, DC. 179 pp.

Ullirich, S. M., T. W. Tanton, and S. A. Abdrashitova. 2001. Mercury in the aquatic environment. A review of factors affecting methylation. *Crit. Rev. Environ. Sci. Technol.* 31:241–293.

Underwood, A. J. 1994. On beyond BACI: sampling designs that might reliably detect environmental disturbances. *Ecol. Appl.* 4:3–15.

Urban, N. H., S. M. Davis, and N. G. Aumen. 1993. Fluctuations in sawgrass and cattail densities in Everglades water conservation area 2A under varying nutrient, hydrologic and fire regimes. *Aquat. Bot.* 46:203–223.

USDA NRCS. 2006. In G. W. Hurt and L. M. Vasilas (eds.) *Field Indicators of Hydric Soils in the United States, Version 6.0.* USDA NRCS in cooperation with the National Technical Committee for Hydric Soils, Ft. Worth, TX.

USGS. 1994. U.S. Geological Survey Open File Report. 94.376.

Vaithiyanathan, P. and C. J. Richardson. 1997. Nutrient profiles in the everglades: examination along the eutrophication gradient. *Sci. Total Environ.* 205:81–95.

Valiela, I. and J. M. Teal. 1974. Nutrient limitation in salt marsh vegetation. In R. J. Welmore and W. H. Queen (eds.) *Ecology of Halophytes.* Academic Press, New York, pp. 547–563.

Van Bodegom, P. M., J. C. M. Scholter, and A. J. M. Stams. 2004. Direct inhibition of methanogenesis by ferric iron. *FEMS Microbiol. Ecol.* 49:261–268.

van Cleemput, O. and L. Baert. 1984. Nitrite: a key compound in N loss processes under acid conditions? *Plant Soil* 76:233–241.

van Eck, G. T. M. 1982. Forms of phosphorus in particulate matter from the Hollands Diep/Haringvliet, the Netherlands. *Hydrobiologia* 92:665–681.

van Meirvenne, M. and I. van Cleemput. 2006. Pedometrical techniques for soil texture mapping at different scales. In S. Grunwald (ed.) *Environmental Soil-Landscape Modeling: Geographic Information Technologies and Pedometrics.* CRC Press, New York. pp. 323–341.

Van der Zee, C. and W. van Raaphorst. 2004. Manganese oxide reactivity in North Sea sediments. *J. Sea Res.* 52:73–85.

Van Ranst, E. and F. De Coninck. 2002. Evaluation of ferrolysis in soil formation. *Eur. J. Soil Sci.* 53:513–519.

Van Rees, K. C. J., K. R. Reddy, and P. S. C. Rao. 1996. Influence of benthic organisms on solute transport in lake sediments. *Hydrobiologia* 317:31–40.

Van Rees, C. J., E. A. Sudicky, P. S. C. Rao, and K. R. Reddy. 1991. Evaluation of laboratory techniques for measuring diffusion coefficients in sediments. *Environ. Sci. Technol.* 25:1605–1611.

Veldkamp, E., A. M. Weitz, and M. Keller. 2001. Management effects on methane fluxes in humid tropical pasture soils. *Soil Biol. Biochem.* 33:1493–1489.

Vepraskas, M. J. 1992. *Redoximorphic Features for Identifying Aquic Conditions.* Tech. Bull. 301. North Carolina Agricultural Research Service, N.C. State University, Raleigh, NC. 33 pp.

Verhoeven, J. T. A., W. Koerselman, and A. F. M. Meuleman. 1996. Nitrogen or phosphorus limited growth in herbaceous wet vegetation: relations with atmospheric inputs and management regimes. *Trends Ecol. Evol.* 11:494–497.

Villarrubia, C. R. 1998. Ecosystem response to a freshwater diversion: the Caernarvon experience. Abstract. Symposium on recent research in Coastal Louisiana: Natural System Function and Response to Human Influences. February 3–5, Lafayette, Louisiana.

Volk, B. G. 1973. Everglades histosol subsidence 1: CO2 evolution as affected by soil type, temperature, and moisture. *Soil Crop Sci. Soc. Fla. Proc.* 32:132–135.

Wackernagel, H. 2003. *Multivariate Geostatistics*. Springer, Berlin.

Wahlen, M., N. Tanaka, R. Henry, B. Deck, J. Zeglen, J. S. Vogel, J. Southon, A. Shemesh, R. Fairbanks, and W. Broecker. 1989. C-14 in methane sources and in atmospheric methane: the contribution from fossil carbon. *Science* 245:286–290.

Walbridge, M. R., C. J. Richardson, and W. T. Swank. 1991. Vertical distribution of biological and geochemical subcycles in two southern Appalachian forest soils. *Biogeochemistry* 213:61–85.

Walbridge, M. R. and J. P. Struthers. 1993. Phosphorus retention in non-tidal palustrine forested wetlands of the mid-Atlantic region. *Wetlands* 13:84–94.

Wang, T. and J. H. Peverly. 1999. Iron oxidation states on root surfaces of a wetland plant (*Phragmites australis*). *Soil Sci. Soc. Am. J.* 63:247–252.

Wang, W. J., P. M. Chalk, D. Chen, and C. J. Smith. 2001. Nitrogen mineralisation, immobilisation and loss, and their role in determining differences in net nitrogen production during waterlogged and aerobic incubation of soils. *Soil Biol. Biochem.* 33:1305–1315.

Wang, Z. P., R. D. DeLaune, P. H. Masscheleyn, and W. H. Patrick, Jr. 1993. Soil redox and pH effects on methane production in a flooded rice soil. *Soil Sci. Soc. Am. J.* 57:382–385.

Warneck, P. 1988. *Chemistry of The Natural Atmosphere*. San Diego, New York. 499 pp.

Watson, P. G. and T. E. Frickers. 1990. A multilevel, in situ pore-water sampler for use in intertidal sediments and laboratory microcosms. *Limnol. Oceanogr.* 35:1381–1389.

Wassen, M. J., H. O. Venterink, E. D. Lapshina, and F. Tanneberger. 2005. Endangered plants persist under phosphorus limitation. *Nature* 437:547–550.

Weaver, K., G. Payne, and S. Xue. 2007. Status of water quality in the Everglades Protection Area. Chapter 3A, Volume 1. In South Florida Environmental Report, South Florida Water Management District, West Palm Beach, FL. http://www.sfwmd.gov/sfer/SFER_2007/index_draft_07.html.

Weaver, R. W., S. Angle, P. Bottomley, D. Bezdicek, S. Smith, A. Tabatabai, and A. Wollum. 1994. *Methods of Soil Analysis: Part 2-Microbiological and Biochemical Properties*. SSSA Series No. 5. Soil Science Society of America, Madison, WI. 1121 pp.

Weber, K. A., L. A. Achenbach, and J. D. Coates. 2006. Microorganisms pumping iron: anaerobic microbial iron oxidation and reduction. *Nat. Rev.* 4:732–764.

Webster, J. R. and E. F. Benfield. 1986. Vascular plant breakdown in freshwater ecosystems. *Annu. Rev. Ecol. Syst.* 17:567–594.

Webster, R. and M. A. Oliver. 2001. *Geostatistics for Environmental Scientists*. Wiley, New York. 271 pp.

Weider, R. K., J. B. Yavitt, and G. E. Lang. 1990. Methane production and sulfate reduction in two Appalachian peatlands. *Biogeochemistry* 10:81–104.

Weiss, M. S., U. Abele, J. Weckesser, W. Walte, E. Schiltz, and G. E. Schulz. 1991. Molecular architecture and electrostatic properties of a bacterial porin. *Science* 254:1627–1630.

Westermann, P. 1993. Temperature regulation of methanogenesis in wetlands. *Chemosphere* 26:321–328.

Westermann, P. 1993. Wetland and swamp microbiology. In T. E. Ford (ed.) *Aquatic Microbiology*. Blackwell Scientific, Oxford, pp. 215–238.

Wetzel, R. G. 1990. Land-water interfaces: metabolic and limnological regulators. *Verhand. Int. Verein. Limnol.* 24:6–24.

Wetzel, R. G. 1993. Humic compounds from wetlands: complexation, inactivation, and reactivation of surface-bound and extracellular enzymes. *Verh. Int. Verein. Limnol.* 25:122–128.

Wetzel, R. G. 1993. Microcommunities and microgradients: linking nutrient regeneration, microbial mutualism, and high sustained aquatic primary production. *Neth. J. Aquat. Ecol.* 27:3–9.

Wetzel, R. G. 2001. *Limnology: Lake and River Ecosystems*. 3rd Edition. Academic Press, New York. 1006 pp.

Wetzel, R. G. 2002. Dissolved organic carbon: detrital energetics, metabolic regulators, and drivers of ecosystem stability of aquatic ecosystems. In S. Findlay and R. Sinsabaugh (eds.) *Aquatic Ecosystems: Interactivity of Dissolved organic Matter*. Academic Press, San Diego. pp. 455–475.

Wetzel, R. G. 2003. Solar radiation as an ecosystem modulator. In H. Zagarese and W. Helbling (eds.) *Ultraviolet Radiation*. Vol. 7, *Encyclopedia of Solar Radiation*, Oxford University Press, Oxford. pp. 3–18.

Wetzel, R. G. and G. E. Likens. 1990. *Limnological Analysis*. Springer-Verlag, New York. 391 pp.

Wetzel, R. H. 1991. Extracellular enzymatic interactions: storage, redistribution, and interspecific communication. In R. J. Chrost (ed.) *Microbial Enzymes in Aquatic Environments*, Springer-Verlag, New York. pp. 6–28.

Whigham, D. F. 1999. Plant mediated controls on nutrient cycling in temperate fens and bogs. *Ecology* 80:2170–2181.

Whitcomb, J. H., R. D. DeLaune, and W. H. Patrick, Jr. 1989. Chemical oxidation of sulfide to elemental sulfur: its possible role in marsh energy flow. *Marine Chem.* 26:205–214.

White, D. A., T. E. Weiss, J. M. Trapani, and L. B. Thein. 1978. Productivity and decomposition of the dominant salt marsh plants in Louisiana. *Ecology* 59:751–759.

White, D. S. and B. L. Howes. 1994. Long-term ^{15}N-nitrogen retention in the vegetated sediments of a New England salt marsh. *Limnol. Oceanogr.* 39:1878–1892.

White, G. N. and J. B. Dixon. 1996. Iron and manganese distribution in nodules from a young Texas vertisols. *Soil Sci. Soc. Am. J.* 60:1254–1262.

White, J. R., M. A. Belmont, and C. D. Metcalfe. 2006. Pharmaceuticals compounds in wastewater wetland treatment are a potential solution. *Sci. World J.* 6:1731–1736.

White, J. R. and K. R. Reddy. 1999. Influence of nitrate and phosphorus loading on denitrification enzyme activity in Everglades wetland soils. *Soil Sci. Soc. Am. J.* 63:1945–1954.

White, J. R. and K. R. Reddy. 2000. Influence of phosphorus loading on organic nitrogen mineralization of northern everglades soils. *Soil Sci. Soc. Am. J.* 64:1525–1534.

White, J. R. and K. R. Reddy. 2001. Influence of selected inorganic electron acceptors on organic nitrogen mineralization in Everglades soils. *Soil Sci. Soc. Am. J.* 65:941–948.

White, J. R. and K. R. Reddy. 2003. Nitrification and denitrification rates of everglades wetland soil along a phosphorus-impacted gradient. *J. Environ. Qual.* 32:2436–2443.

Whiting, G. J. and J. P. Chanton. 1993. Primary production control of methane emission from wetlands. *Nature* 364:794–795.

Widdel, F. 1988. Microbiology and ecology of sulfate- and sulfur-reducing bacteria. In A. J. B. Zehnder (ed.) *Biology of Anaerobic Microorganisms.* John Wiley & Sons, Inc., New York, NY. pp. 469–585.

Widdel, F., S. Schnell, S. Heising, A. Ehrenreich, B. Assmus, and B. Schink. 1993. Ferrous iron oxidation by anoxygenic phototrophic bacteria. *Nature* 362:834–836.

Wieder, R. K. and G. E. Lang. 1988. Cycling of inorganic and organic sulfur in peat from Big Run Bog, West Virginia. *Biogeochemistry* 5:221–242.

Wiegert, R. G. and F. C. Evans. 1964. Primary production and the disappearance of dead vegetation in an old field in southwestern Michigan. *Ecology* 45:49–63.

Wood, M. E., J. T. Kelley, and D. F. Belknap. 1989. Patterns of sediment accumulation in the tidal marshes of Maine. *Estuaries* 12:237–246.

Worral, F. and T. P. Burt. 2004. Time series analysis of long term river dissolved organic carbon records. *Hydrol. Process.* 18:893–911.

Worral, F., T. P. Burt, and J. Adamson. 2005. Fluxes of dissolved carbon dioxide and inorganic carbon from an upland peat catchment: implications for soil respiration. *Biogeochemistry* 73:515–539.

Worral, F., T. P. Burt, and R. Shedden. 2003. Long term records of riverine carbon flux. *Biogeochemistry* 64:165–178.

Wrage, N., G. L. Velthof, M. L. van Beusichem, and O. Oenema. 2001. Role of nitrifier denitrification in the production of nitrous oxide. *Soil Biol. Biochem.* 33:1723–1732.

Wright, A. 2001. Microbial processes in Everglades wetland soils. Ph.D. Dissertation. University of Florida, Gainesville, FL.

Wright, A. L. and K. R. Reddy. 2001a. Phosphorus loading effects on extracellular enzyme activity in Everglades wetland soil. *Soil Sci. Soc. Am. J.* 65:588–595.

Wright, A. L. and K. R. Reddy. 2001b. Heterotrophic microbial activities in Northern Everglades Wetland. *Soil Sci. Soc. Am. J.* 65:1856–1864.

Wright, A. L. and K. R. Reddy. 2007. Substrate-induced respiration for phosphorus-enriched and oligotrophic peat soils in an Everglades wetland. *Soil Sci. Soc. Am. J.* 71:1579–1583.

Wright, A., K. R. Reddy, and S. Newman, 2008. Biogeochemical indicators of phosphorus loading in select hydrologic units of the Everglades (submitted for publication).

Yu, K. W., R. D. DeLaune, and P. Boecks, 2006a. Direct measurement of denitrification activity in a gulf coast freshwater marsh receiving directed Mississippi River water. *Chemosphere* 65:2449–2455.

Yu, K. W., S. P. Faulkner, and W. H. Patrick. 2006b. Redox potential characterization and soil greenhouse gas concentration across a hydrological gradient in a Gulf coast forest. *Chemosphere* 62(6):905–914.

Yu, K. W., Z. P. Wang, and G. X. Chen. 1997. Nitrous oxide and methane transport through rice plants. *Biol. Fertility Soils* 24:341–343.

Yu, T.-R. 1985. *Physical Chemistry of Paddy Soils*. Springer-Verlag, New York. 217 pp.

Zak, D. R., W. E. Holmes, D. C. White, A. D. Peacock, and D. Tilman. 2003. Plant diversity, soil microbial communities, and ecosystem function: are there any links? *Ecology* 84(8):2042–2050.

Zaman, M., H. J. Di, K. C. Cameron, and C. M. Frampton. 1999. Gross nitrogen mineralization and nitrification rates and their relationships to enzyme activities and the soil microbial biomass in soils treated with dairy shed effluent and ammonium fertilizer at different water potentials. *Biol. Fertil. Soils.* 29:178–186.

Zang, X., D. H. Jasper, J. D. H. van Heemst, K. J. Daria, and P. G. Hatcher. 2000. Encapsulation of protein in humic acid from a Histosol as an explanation for the occurrence of organic nitrogen in soil and sediment. *Org. Geochem.* 31:679–695.

Zehnder, A. J. B. and K. Wuhrman. 1976. Titanium (III) citrate as a non-toxic, oxidation–reduction buffering system for the culture of obligate anaerobes. *Science* 194:1165–1166.

Zehr, J. P. and B. B. Ward. 2002. Nitrogen cycling in the ocean: new perspectives on processes and paradigms. *Appl. Environ. Microbiol.* 68:1015–1024.

Zeikus, J. G. 1981. Lignin metabolism and the carbon cycle. In M. Alexander (ed.) *Advances in Microbial Ecology*, Vol. 5. Plenum Press, New York, pp. 211–243.

Zuberer, D. A. 1999. Recovery and enumeration of viable bacteria. In R. W. Weaver (ed.) *Methods of Soil Analysis Part 2. Microbiological and Biochemical Properties*. SSSA Book Series 5. Soil Science Society of America, Madison, WI. pp. 119–144.

Zwolinski, M. D. 2007. DNA sequencing: strategies for soil microbiology. *Soil Sci. Soc. Am. J.* 71:592–600.

Index

A

A horizon, 19
Abiotic decomposition, 151–153
Abiotic enzymes, 130
Abiotic retention, 343
Absorption, 343
ACC (1-aminocyclopropane-1-carboxylic acid), 226
Acclimation, 513, 514
Accretion balance, 672
Accumulation, 513, 514
Acetaldehyde, 219
Acetate
  oxidation, 145
  utilizers, 148
Acetylene
  reduction, 310
  reduction assay, 313
Achromobacter spp, 292
Acidophilic bacteria, 428
Acids, 8
Acid-volatile sulfur, 693
Activation energy, 18
Activity, 8
Adenosine triphosphate (ATP), 139
Adenylate energy charge ratio, 221
ADH activity, 252
Adsorption, 340, 343
  equilibrium, 344
  isotherm, 344
Advection, 396, 539
Advective flux
  measurement, 540, 541
  processes, 539, 540
  sediment flux, 539
Advective transport, 539
Adventitious roots, 221, 250
AEC ratio, 239
Aerenchyma, 46, 47
  carbon transport, 174, 175
  facilitated oxygen pathway, 239
  lysigenous, 225
  methane emission, 604–606
  morphological adaptations, 221–226, 237, 605
  nitrogen gases, 611
  schizogenous, 225, 226
  sulfur, 473
Aerobic
  catabolism, 138, 139
  decomposition, 265
  denitrification, 301
  metabolic activities, 710
  microbes, 40, 42
  respiration, 17
  soil layer, 201

surface soil layer, 43
zones, 322
Aerobic–anaerobic interface
  exchanges, 45
  global change, 601, 609
  iron and manganese respiration, 406
  oxygen, 200–204
  plant roots, 246
  toxic organic compounds, 509, 512
Air–water exchange, 569, 570
Alachlor, 509
Alcohol dehydrogenase, 220
Alfisols, 22
Alkaline phosphatase, 380, 647
Alkalinity and buffer capacity, 287
Altered roots, 221
Alternaria spp, 415
Alternate electron acceptors, 41
Amino acids, 125, 258
Amino acids in soils, 270
Amino sugars 258, 271
Ammonia
  adsorption–desorption, 280
  equilibrium with ammonium, 285
  fixation, 283
  formation, 286
  volatilization, 261, 263, 284, 286
    physicochemical reaction, 284–286
    regulators, 286–289
Ammonification, 262, 264, 265, 273, 275
Ammonification–immobilization cycle, 264
Ammonium, 257
  concentration, 287, 293
  consumption rates, 318
  equilibrium with ammonia, 285
  exchangeable in soils, 283
  flux 318–320, 322, 545, 546
  in wetland soils, 280
  net release, 275
  nonexchangeable, 284
  oxalate, 353
  oxidation, 707, 292, 294–296
  partition coefficients, 283
  production, 275
  profile in soil pore water, 318
Amorphous iron sulfides, 471
Anabaeana spp, 263
Anabolic nitrate reduction, 296, 297
Anabolism, 16, 136
Anaerobic
  ammonium oxidation, 294–296
  bacterial degradation of organic compounds, 275
  decomposition, 141, 265
  fermentation, 215, 238
  microbes, 40, 42
  respiration, 17

757

Anaerobic (*contd.*)
  root metabolism, 219
  root respiration, 220
  soil layer, 201
  soil volume, 191
  zones, 322
Anaerobiosis, 46
Anammox reactions 294, 320
Andisols 22
Anion exchange, 24, 340
Anions, 24
Anoxic conditions, 220
Anthropogenic nutrients, 581
Antibiotics, 511
Antimony–phospho–molybdate complex, 333
Apatite, 352
Aquic soils, 37, 48
Archaea, 15, 310, 427, 643, 710
Aridisols, 22
Aromatic compound–oxidizing iron reducers, 413
Arrhenius equation, 18
Arsenic, 481
  biotransformation, 486
  chemical oxidation and reduction, 489
  competition with other anions, 488, 489
  coprecipitation with metal oxides and sulfide, 489
  dissolution of primary minerals, 485, 486
  importance of speciation, 487, 488
  kinetics of oxidation–reduction in soils, 487
  oxidation–reduction, 486, 487
  reductive dissolution of metal oxides, 487
  sources, 485
  thermodynamics, 486
Artificial marker horizons, 562, 563
Artificial redox buffer, 247
Arylsulfatases, 456
Ashing, 334
Aspergillis spp, 292
Assimilatory
  iron reduction, 412
  manganese reduction, 412
  nitrate reduction, 303
  sulfate reduction, 453
Atmosphere, 111
Atoms, 7
ATP, 139
ATP sulfurylase, 454
Atrazine, 509, 700
Autochthonous biological nitrogen fixation, 178, 310
Autotrophic methanogens, 149
Autotrophs, 16
Availability of electron acceptors, 164–167
Average lifetime, 18, 19

B

B horizon, 19
BACI design, 585
Bacillus spp, 132, 145, 241, 263, 299, 413, 428, 466
Bacteria, 15, 125, 710
  acidophilic, 428
  ammonia formation and oxidation, 269, 293, 294
  autotrophic methanogens, 149
  autotrophs, 16
  carbon:nitrogen ratio, 265
  chemoautotrophic, 289
  chemoorganotrophs, 16
  decomposition, 129, 130, 132, 135–138, 152, 154
  electrochemical properties, 67
  Everglades, 643, 644, 647, 648, 653, 658, 660, 663
  exchange between soil and floodwater, 548
  fermenting, 142–144
  global change, 604, 605, 608, 611
  gram-negative, 15, 16
  gram-positive, 15, 16
  heterotrophic, 289, 290, 292
  hydrogen-producing acetogenic, 143
  indicators, 575
  inorganic sulfur-oxidizing, 466
  iron and manganese
    ecological significance, 437, 442
    mobile and immobile pools, 426, 427
    oxidation, 427–432
    reduction, 411–414, 416–418, 420
    storage and distribution, 406
  iron reduction with phosphorus, 356
  Louisiana, 675, 682, 700, 701
  metals, 146, 484–486, 491
  methane-oxidizing bacteria, 291, 292
  methanogens, 148–150, 204, 604, 644
  methanotrophs, 291
  methylotrophic bacteria, 291, 292
  neutrophilic bacteria, 428
  nitrate
    flux, 320
    reduction, 296–306
    respiration, 144, 145
  nitrifying chemoautotrophs, 291
  nitrogen-fixing bacteria, 310–313
  nitrogen redox, 262, 263, 268
  organic matter, 119, 125, 175, 178
  organic phosphorus, 333, 379
  oxygen, 200, 201, 203–205
  phosphorus
    mineralization, 387
    mobilization, 389, 390
    release, 382, 386
    uptake and storage, 370–372
  phototrophs, 16, 466
  plants, effects on, 241, 245
  soil aeration, 162
  sulfate reducers, 147, 148, 332
  sulfur, 451, 466
    emission of gases, 470
    mineralization, 455
    oxidation, 466–470
    oxidation–reduction, 451, 453, 454
    reduction, 454, 457–461, 463–465
  syntrophic, 150, 460, 644
  toxic organic compounds, 513, 516, 518, 519, 521, 528, 529
  uses in biogeochemical advances, 703, 707–710
  wetland soils, 38, 40, 41, 52, 53
Bases, 9
Batch incubation experiments, 275

Index

Beggiatoa spp, 310, 454, 466, 467, 473
Benthic
    chambers, 552, 553
    invertebrates, 208, 549, 550
    nitrogen cycle, 418
    unattended generator, 106
Bernoulli's equation, 234
Beryllium-7 dating, 564, 565
β-Glucosidases, 132, 647
β-Xylosidases, 132
Bioaccumulation, 477, 483, 707
Biochemical pathways, 16
Bioconcentration factor, 521
Biodegradation, 513, 514
Biodiffusion, 547, 551
Biogenic trace gases, 599
Biogeochemical cycles, 4, 643
    carbon, 649–653
    enzymes, 647–649
    mechanistic and statistical models, 712–716
    nitrogen, 653–657
    phosphorus, 657–660
    sulfur, 660–663
    tools to study, 710–712
Biogeochemical processes, 705–707
Biogeochemistry, 2, 3, 5, 30, 33, 703–706
    algal and microbial interactions, 708, 709
    diffuse reflectance spectroscopy, 711
    Everglades, 625
    geospatial models, 715, 716
    global change, 601
    iron, 405
    manganese, 415
    mechanistic models, 712, 713
    metals, 496
    microbial communities and diversity, 710, 711
    nuclear magnetic resonance spectroscopy, 711
    phosphorus, 401
    stable isotopes, 711, 712
    stochastic models, 713–715
    sulfur, 447
    vegetation and microbial interactions, 709
BIOLOG plates, 710
Biosphere, 113
Biotic decomposition, 153–157
Bioturbation, 208, 396, 547–550
Bogs, 32
Bomb-produced carbon, 567
Bonds, 7
Boundary layer, 570
Brackish marsh, 696
Browning reactions, 127
Buffering of redox potential, 84, 85

C

C horizon, 20
C:N ratio, 168, 265–267, 278
C:P ratio, 168, 382
Cadmium, 499–501
Caernarvon diversion, 678, 679
Calcite, 391

Calcium phosphate, 394
Calories, 136
Capacity, 100, 101, 343, 351
Carbohydrates, 16
Carbon
    atmospheric, 111–113
    biospheric, 113, 114
    disulfide, 470
    fixation, 111
    hydrospheric, 113
    isotopes, 111
    loss from wetland deterioration, 686, 687
    methane, 111, 112
    monoxide, 111, 112
    net assimilation, 249
    reservoirs, 111
    sequestration, 615, 616
    sinks, 686
Carbon cycle, 649–653, 684–687
Carbon cycle in wetlands
    dissolved organic matter, 117, 118
    gaseous forms, 118, 119
    microbial biomass, 116, 117
    particulate organic matter, 115, 116
    plant biomass, 114, 115
Carbon-14 dating, 567
Carbon dioxide, 111, 112, 704, 706, 707, 709
    carbon cycle, role in, 116, 118, 119
    decomposition, 137–139, 141–143, 145–151
    ecological significance, 173–177
    electrochemical properties, 67, 95, 96, 99,
        100, 105, 106
    Everglades, 659
    exchange processes between soil and floodwater, 545,
        552, 567, 569, 570
    global change, 599–604, 615
    in floodwater, 96
    iron role, 413, 415–417, 426, 428, 431, 435, 437
    Louisiana, 684–686
    manganese role, 413, 415–417, 426, 435, 437
    metals, 493
    nitrogen role, 264, 266, 275, 287, 289, 291–293
    organic matter, 119, 151–154, 162–165, 171–173
    oxygen role, 185–188, 190–192, 201, 207
    phosphorus role, 356, 382, 383, 387, 391, 392
    plant adaptations to anaerobiosis, 219, 225, 228,
        229, 232–234
    soil reduction and plant functions, 242, 247, 249, 250
    solubilization, 233
    substrate utilizers, 148
    sulfur, 457, 459, 466
    toxic organic compounds, 509, 514, 515, 517, 521
    wetland soils, 36, 39, 43
Carbon:nitrogen ratio, 168, 265–267, 278
Carbon:phosphorus ratio, 168, 382
Carbon:sulfur ratio, 456
Carbonyl sulfide, 470
Carex spp
    decomposition, 153–155
    nitrogen, 314
    organic matter, 175
    phosphorus, 373
    soil anaerobiosis, 232–234, 238, 242, 246

Catabolic
  nitrate reduction, 296, 307
  reduction of iron and manganese, 412
  sulfur reduction, 457
Catabolism, 16, 136
Catalase, 205
Cat clays, 97
Cation exchange capacity (CEC), 24, 288, 293
Cations, 24
Cell potential, 72
Cellulase enzyme complex, 132
Cellulose decomposition, 132
Cesium-137 dating, 566, 567
Chemical oxidation of reactants, 206
Chemiosmotic theory, 139
Chemoautotrophic bacteria, 289
Chemodenitrification, 301, 302
Chemolithic oxidation, 466
Chemolithotrophic bacteria, 16, 469
Chemoorganotrophs, 16
Chironomid larvae, 548
Chitin, 258
Chloroform fumigation techniques, 454
Chroma, 20
Chromium, 480, 496–499
Chromophoric dissolved organic matter (CDOM), 152
Cladium spp
  Everglades, 626, 629, 630, 641, 652, 655
  Louisisana, 681
  organic matter, 121, 156, 157
  phosphorus, 374–376
  soil anaerobiosis, 218
Clays
  cat, 97
  fixation by silicate, 340
  silicate, 283
Climate change, 3, 174, 470, 599–601, 616, 709
Clostridium spp, 132, 143, 145, 263, 310–313, 371, 390
Coallocated variables, 591
Coastal marshes
  plant stress factors, 680
  stability, 672, 673
Coastal wetlands, 31, 32, 616–618
Cokriging, 592
Cometabolism, 513, 514
Compaction of sediments, 672
Compliance monitoring, 584
Concentration
  dissolved gas in water, 569
  reductants in root zone, 246
Concretions, 50
Constructed wetlands, 33, 665
Continuous internal cycle, 265
Convective flow, 229, 235
Convenience sampling, 585
Copper, 489–493
Cortical intercellular airspaces, 215
Covalent bonds, 7
Critical oxygen pressure, 227
Current intensity, 196
Cyanobacteria
  algal microbial interactions, 708
  Everglades, 641, 642, 648, 657, 658
  nitrogen, 263, 310–313
  phosphorus, 371, 390
  sulfur, 454, 469
Cysteine, 454
Cytochrome b, 147
Cytoplasmic acidosis, 219

D

Dairy rule, 327
Darcy's law, 540
Dating
  Beryllium-7, 564, 565
  Carbon-14, 567
  Cesium-137, 566, 567
  Lead-210, 565, 566
  sediment, 568
DDT, 532, 699
Decay continuum, 118, 158
Decomposition, 111, 151, 260
  abiotic, 151–153
  aerobic catabolism, 138–141
  aerobic vs. anaerobic catabolism, 150, 151
  anaerobic catabolism, 141, 142, 265
    fermentation, 142–144
    manganese and iron respiration, 145–147
    methanogenesis, 148–150
    nitrate respiration, 144, 145
    sulfate respiration, 147, 148
  biotic, 153–157
  catabolic activity, 136–138
  extracellular enzyme hydrolysis, 130–136
  leaching and fragmentation, 129, 130
  rate, 167
  rate constants, 157
Degraded wetlands, 583
Demethylation, 485
Denaturation, 16
Denitrification, 144, 257, 258, 261, 262, 297, 298
Denitrification enzyme activity, 305, 655
Denitrifying organisms, 298, 528, 529
Depositional fluxes, 556
Depressional wetland, 35
Desorption, 340
Desulfobacter spp, 148, 310, 454, 459,
Desulfobacterium spp, 148, 459
Desulfomonas spp, 147, 459
Desulfonema spp, 148, 459
Desulfotomaculum spp, 147, 241, 454, 459
Desulfovibrio spp
  dinitrogen fixation, 310
  iron respiration, 413
  manganese respiration, 413
  mercury methylation, 485, 701
  sulfate reducers, 147, 241, 451, 454, 458–460
Desulfuration, 453
Detrital
  export, 262
  matter, 274
  organic matter, 121, 261
  plant matter, 122
Diagenesis, 471

Index

Diazotrophs, 310
Diesters, 358
Diffuse reflectance spectroscopy, 711
Diffusion, 46, 396, 605
    coefficients, 192, 544
    due to irrigation, 551
    effective, 551
    gases into soil, 191
    Knudsen, 229, 231
Diffusive flux, 525, 543–547
Dimethyl disulfide, 470
Dimethyl sulfide, 470, 696
Dinitrogen, 257
    fixation, 263
    fixation regulators, 310–313
    gas, 15N-labeled, 313
Disintegration constant, 19
Dispersion, 396
Dissimilatory
    iron reduction, 412, 707
    manganese reduction, 412, 707
    nitrate reductase, 305
    nitrate reduction to ammonia, 144, 263, 302, 303
    sulfate reduction, 453
Dissolution, 341, 351
Dissolved oxygen, 42, 185, 199
Dissolved prokaryotes, 332
Dissolved solutes exchange, 551–555
Distribution of wetland plants, 217
Diversion projects, 676–680
DNA sequencing analysis, 710
DOC, 178
DOM, 178
Drought stress, 249
Dyes, 543

E

$E°$, 14
E horizon, 19
$E°$ vs. $\log K$, 76, 77
Ebullition, 604
Ecophysiology, 217
Ectoenzymes, 130
Effective diffusion, 551
Efficiency factor, 368
Eh, 24
    eh–pH relationship,
        electrochemistry, 69, 81–84, 92, 104, 481, 482
        iron, 407, 408, 425, 428, 429
        manganese, 407, 408, 425, 428, 429
        metals, 486, 490, 484, 495, 497, 500, 503, 504
        oxygen, 187
        sulfur, 451, 452
    gradients, 101
    measurement, 79–81
    scale, 67
Electrical potential, 75
Electrochemical properties, 67
Electrodes
    construction of, 85, 86
    inert, 79
    laboratory, 85
    oxygen microelectrode, 239
    oxygen-permeable membrane–covered, 198
    Platinum (Pt), 86, 195–197
    polarographic membrane–covered, 198
    reference, 86
    saturated calomel reference, 79
    standard hydrogen electrodes (SHE), 70
    standardization of, 86–88
Electron
    acceptors, 14, 137
        alternate, 41
        availability, 164–167
        oxygen as, 42
    affinity, 70
    donors, 14
    pressure, 67, 331, 706
    transport system, 205, 291
Eleocharis spp, 639
EMF, 71
Endergonic, 10
Endocellulases, 132
Energy, 136
Energy source, 310
Enthalpy, 9
Entisols, 22
Entropy, 9
Enzymes, 16–18
    abiotic, 130
    activity, 17, 278
    assays, 135, 136
    biogeochemistry, 43, 703, 705, 706, 710, 711
    carbon
        cycle, 116
        decomposition, 138–140, 144, 145, 147, 149, 150
        environmental significance, 178, 180
        organic matter, 119, 127, 129, 130–136
        regulation of decomposition, 162, 173
    electrochemical properties, 73, 100
    Everglades, 647, 648, 653, 655, 659, 664
    extracellular, 378
    global change, 612, 615
    hydrolytic activity, 705
    indicators, 581, 594
    iron, 406, 415, 416
    latch mechanism, 178
    manganese, 406, 415, 416
    metals, 477, 489
    nitrogen
        fixation, 309–313
        mineralization of, 268, 269, 271, 274, 276, 278
        nitrification of, 290
        redox transformations of, 263
        reduction of nitrate reduction, 296, 298, 302–307
        storage compartments of, 260, 261
    oxygen, 205
    phosphorous, 366, 369, 378–380, 383, 386, 387
    plant adaptations to anaerobiosis, 219, 220
    sulfur, 454–458
    toxic organic compounds, 507, 510, 514–516, 519, 521
Enzyme–substrate complex, 17, 379
Epipedons, 19
Epipelon, 371, 392

Epiphyton, 371
Equilibrium
  ammonium concentration, 281
  constants, 9
  phosphorus concentration, 345, 356
Equivalents of solute, 8
Eriophorum spp, 227
Ethanol, 219, 220
Ethylene-induced cellulase activity, 225
Eukaryotic cells, 15
Eustatic sea-level rise, 672
Eutrophic, 325, 326
Eutrophication, 3
  biogeochemical indicator, 576, 577
  enzymes, 647
  Louisiana, 691
  microbial communities, 643
  minimum monitoring requirements, 590
  modeling, 714
  phosphorus memory, 397
  source of oxygen, 199
  vegetation, 641
  wind mixing, 551
Everglades, 576, 623
  Agricultural Area, 625, 627–629
  biogeochemical characteristics of wetlands, 37, 40, 45, 51
  biological nitrogen fixation, 656, 657
  C-139 basin, 627–629
  carbon role, 117, 121, 157, 161–163, 166–168, 177
  decomposition of organic matter, 649
  electrochemical properties, 90
  Everglades Forever Act, 631
  Everglades National Park, 625, 630
  exchange processes, 538, 561, 572
  floc, 624, 637, 638, 645, 646, 650, 652, 654, 659
  global change, 576, 592, 596
  historical perspective, 626, 627
  Holeyland management area, 630
  hydrologic units, 627–629
  metals, 483
  methane emissions, 651–653
  methylmercury–sulfate link, 663
  microbial biomass, 645–647
  microbial communities, 643–645
  microbial respiration, 649–651
  nitrification–denitrification, 655, 656
  nitrogen roles, 278, 280, 305, 306, 312, 313
  nutrient loads and ecological alternations, 630, 631
  organic nitrogen mineralization, 653–655
  oxygen role, 199, 211
  periphyton, 641–643
  phosphorus, 325
    abiotic processes, 659, 660
    accumulation in wetlands, 328–330
    biotic processes, 658, 659
    exchange, 395, 396
    inorganic, 336, 337
    mineralization, 380, 381, 385, 386
    mobilization, 391–393, 395
    sorption by soils, 346, 348, 353,
    uptake and storage, 370, 374–376
  research, 705, 710–712, 715, 716
  restoration and recovery, 663–666
  Rotenberger management area, 630
  soil nutrient distribution and storage, 633–639
  stormwater treatment areas, 629
  sulfur cycling, 451, 463, 660–663
  surface water quality and loads, 631–633
  vegetation, 639–641
  water conservation areas, 629, 630
  wetlands, 625–630
Exchange
  mud–water interface, 45, 46
  processes in biogeochemical regulation, 570–572
  rates, 46
Exergonic, 10
Exocellulases, 132
Exoenzymes, 130
Exoxylanases, 132
Extracellular enzymes, 130–136, 274

F

Facultative
  anaerobes, 40, 42, 145, 297
  plants, 41, 217
Faraday's constant, 75
Fatty acids, 125
Fens, 33
Fermentation, 17, 141, 150
  anaerobic, 215, 238
  bacteria, 142
  glucose, 143
Fermentative
  iron reducers, 413
  manganese reducers, 413
  step, 416
Ferric oxyhydroxides, 48
Ferric phosphate, 331
Ferrolysis, 441
Ferromanganese nodules, 439, 440
Ferrous iron, 241
Fertilizer, 257, 327
Fibrists, 52
Fick's first law, 191, 525, 543, 551
Field electrodes, 85
Field indicators, 54
Filter pad traps, 562
Fire, 337
First order decay models, 156
Fixation by silicate clays, 340
Flavins, 205
Flavoproteins, 205
Floc, 47, 381, 537, 554, 587, 595, 596, 637, 638
Flooded soils, 200
Flood tolerance, 218
  aerenchyma formation, 223–226
  intercellular oxygen concentration, 226, 227
  metabolic adaptations, 219–221
  morphological/anatomical adaptations, 221–223
  plant species, 217
Floodwater
  depth, 287
  oxygen concentration, 203

Index

Florida Bay, 625
Flow velocity, 539
Flux
    equation, 525
    reduced sulfur gases, 694–696
    solute, 539
Fragmentation, 130
Free energy formation, 10–12
Freshwater
    marsh, 32, 697
    marsh soils, 52
    surface wetlands, 33
    swamps, 32
Freundlich adsorpotion equation, 522
Fuel cells, 106, 707
Fulvic acids, 127
Fungi
    decomposition, 130, 132, 139, 142, 154, 160, 162
    indicators, 575
    iron, 415
    manganese, 415
    metals, 486
    nitrogen, 265, 289, 292, 296, 297
    organic matter, 119
    phosphorus, 366, 370, 379, 382, 386
    sulfur, 453–455
    toxic organic compounds, 513, 516, 521
    wetland soils, 41
Fusarium spp, 415

G

ΔG, 10, 136
$\Delta G°r$, 11, 72
Gallionella spp, 431
Gas chromatography, 198, 369
Gas exchange, 190, 244
Gelisols, 22
Geobacter spp, 413–415
Geostatistics, 592
Gibbs free energy change, 10, 136
Gibbs–Helmholtz equation, 10
Gleyed soils, 48
Global change
    marsh accretion, 619, 620
    wetlands, 601, 602
Global warming, 599, 600
Glucose oxidation
    during nitrate reduction, 144
    during sulfate reduction, 148
Glycolysis, 138, 139
Gradient-based measurements, 552
Granular D-glucose 248
Greenhouse
    effect, 111, 599
    gases, 119, 599
Gregite, 471
Gross mineralization, 264
Groundwater, 34
Groundwater transport of xenobiotics, 528

H

Haber–Bosch process, 257, 310
Hard pan, 53
Heavy metals, transport, 478
Hemists, 52
Henry's law, 188, 569, 570
Heterotrophs, 16
    bacteria, 292
    fungi, 292
    methanogens, 149
    microbial respiration, 206
High-performance liquid chromatography (HPLC), 369
Histosols, 22, 51
Holeyland Wildlife Management Area, 630
Horizons
    A, 19
    Artificial marker, 562, 563
    B, 19
    biogeochemical characteristics, 46, 48, 50–54, 59
    C, 20
    E, 19
    Everglades, 635
    exchange processes, 562, 563
    iron, 441
    Louisiana, 679, 680
    manganese, 441
    metals, 478
    O, 19
    plant adaptations to soil anaerobiosis, 250
    R, 20
    soil, 19, 20, 22, 23
    sulfur, 474
Hue, 20
Humic
    acids, 127, 418, 706
    enzyme complexes, 387
    substances, 127–129
Humidity-induced diffusion, 229, 232
Humidity-induced pressurization, 232, 235
Humification, 179
Humified substances, 179
Humin, 127,129
Humus, 179, 478
Hydric
    soil field indicators, 54–58
        loamy and clayey soils, 60–63
        sandy soils, 58, 59
    soils, 27, 30, 46, 54
    soil types, 46–54
        marsh soils, 52
        organic soils, 51, 52
        paddy soils, 53
        subaqueous soils, 53
        waterlogged mineral soils, 46–50
Hydrodynamics, 34
Hydrogen
    bonds, 7
    oxidizing iron and manganese reducers, 413
    producing acetogenic bacteria, 143
    sulfide, 241, 470
Hydrologic regime, 185
Hydrology, 27, 279

Hydrolysis, 130
Hydroperiod, 34, 162, 185, 320
Hydrophytic vegetation, 27, 46
Hydrosphere, 113
Hydrotrolite, 471
Hydrous oxides, 340
Hydroxyapatite, 335
Hydroxyl radical, 204
Hypereutrophic, 326
Hypoxia, 215, 220
Hysteresis, 347

I

Immobilization, 262, 275
Impact index, 594, 595
Inceptisols, 22
Incubations, overlying water, 552–555
Indicators, 577
    analysis, multivariate method, 591, 592
    assessing nutrient impacts in wetlands, 581
    data analysis, 590–594
    development guidelines, 578, 579
    evaluation, 580
    hydric soil, 38
    impact/recovery indices, 594–597
    levels, 579–581
    multivariate spatial method analysis, 592
    sampling protocol and design, 582
    soil quality, 588, 589
    univariate and multivariate method analysis, 593
    univariate and multivariate spatiotemporal method analysis, 593
    univariate method analysis, 591
    univariate spatial method analysis, 592
    water quality, 588
    wetland ecosystem reference conditions, 581, 582
Industrialization, 599
Inert electrodes, 79
Inland wetlands, 32–34
Inositol phosphates, 366, 381
Intact cores, 554, 555, 637
Integrators of nutrient impacts, 583
Intensity, 99, 100
Intensity factor, 343, 351
Intercellular air spaces, 222
Interspecies hydrogen transfer, 150
Ion activity product, 351
Ion electrodes, 85
Ionic bonds, 7
Iron
    abiotic reduction, 417–421
    amorphous forms, 420, 471
    bacterial reduction, 416
    biotic and abiotic oxidation, 429–432
    biotic reduction, 415–417
    complexation with dissolved organic matter, 425
    coprecipitation of trace elements, 439
    crystalline forms, 420
    eh–pH relationships, 407–410
    electron acceptors, 416
    electron donors, 417
    ferrolysis, 441
    ferromanganese nodules, 439, 440
    ferrous, 241
    forms, 421–427
    immobile pools, 425
    impact of reduction on wetland soil chemistry, 421
    methane emissions, 441–443
    microbial communities and oxidation, 428, 429
    mobile pools, 425
    mobility, 432–435
    nonenzymatic oxidation, 431
    nonenzymatic reduction, 417
    nutrient regeneration/immobilization, 435
    organic matter decomposition and nutrient release, 435–438
    oxidation, 427–432
    oxidation states, 408
    oxides in soil profile, 426
    reductase, 147, 415
    reduction, 411–413
        aromatic compound–oxidizing, 413
        assimilatory, 412
        by microbial communities, 413–415
        phosphorus release or retention, 438
    regulation
        oxidation, 429
        reduction, 425
    respiration, 145–147
    root plaque formation, 246, 440, 441
    storage and distribution, 405–407
    sulfur proteins, 205
Irrigation, 551
Isomerases, 17
Isomorphic substitution, 283
Isotope dilution technique, 275
Isotopes, 18

K

K_{eq}, 9, 74
K_m, 17
Knudsen diffusion, 229, 231
Knudsen regime, 236
Kriging, 592, 593, 715
K_w, 92

L

Labile components, 332
Laboratory electrodes, 85
Lactic acid, 219
Lacunae, 223, 225
Lake Okeechobee, 551
    carbon role, 156
    electrochemical properties, 91
    Everglades, 625–627, 630
    exchange processes, 551, 572
    phosphorus, 327, 336–339
Land Resource Regions, 54
Langmuir adsorptive equation, 523
Leaching, 129, 152

Index

Lead, 501–503
Lead-210 dating, 565, 566
Leaf
  internal pressure, 232
  surface area, 235
  temperature, 235
Legacy phosphorus, 397
Length of root channel aerated, 229
Lenticels, 222, 227
Leptothrix spp, 299, 428, 429, 431
Level I indicators, 579–581
Level II indicators, 579–581
Level III indicators, 579–581
Ligand exchange, 340
Ligases, 17
Light intensity, 235, 312
Lignins, 125, 126
Lignocellulose, 158
Lignocellulose index, 158, 649
Liming soils, 400
Lipids, 16
Lithotrophic oxidation, 206
Litterbag method, 154–156, 275, 384
Litter fall, 121, 260
Louisiana, 558
  carbon role, 121
  coastal marshes, 669
  coastal wetlands
    biogeography and geology, 669, 670
    carbon cycling, 684–687
    carbon losses and wetland deterioration, 686, 687
    carbon sinks, 686
    case studies, 672–680, 697–701
    coastal wetland loss, 670–672
    decomposition of surface peat, 686
    emissions along salinity gradient, 685
    flux of reduced sulfur gases, 694–697
    impacts of flooding and saltwater intrusion on vegetation, 680–683
    mercury, 700, 701
    nitrogen
      budget, 690, 691
      cycle, 687–693
      inputs, 688
      losses, 689, 690
      processing capacity, 691–693
      regeneration and uptake, 688
    primary productivity, 684, 685
    sulfur cycle, 693–697
    toxic organic compounds, 697–700
    vertical accretion comparisons, 673–676
  electrochemical properties, 94
  exchange processes, 558–560
  global change, 602, 607, 612, 616–618, 620
  iron, 407, 410, 423
  manganese, 407, 410, 423
  marsh soils, 694
  metals, 500, 502
  nitrogen role, 273, 274
  oxygen role, 189, 190
  phosphorus role, 395
  plant adaptations to soil anaerobiosis, 242
  salt marsh, 676
  sulfur role, 449–451, 462, 470, 473, 474
  toxic organic compounds, 509, 530–533
Lowland marshes, 52
Low-molecular weight organic matter, 133–136
Low-risk wetlands, 584
Luxury uptake, 327
Lyases, 17
Lysigenous aerenchyma, 225

M

Mackinawite, 471
Macrobenthos communities, 548, 549
Macrophytes, 121, 372, 639
  adaptations pf plants to anaerobiosis, 230, 239, 245, 246
  biogeochemical characteristics of wetlands, 28, 31, 32, 709
  carbon cycle, 115, 116, 118
  decomposition, 129, 136, 141,
  electrochemical properties, 96, 97
  exchange processes, 550
  Everglades
    biogeochemical cycles, 653, 654, 657, 658, 662
    nutrient loads, 631, 635, 639–641, 646
    restoration, 666
  global change, 605, 607, 611
  indicators, 583
  Louisiana, 682, 684, 687, 700
  nitrogen, 260, 287, 313, 317
  organic matter, 121, 156, 157, 178
  oxygen, 185, 200, 206
  phosphorus, 357, 366, 371–377, 383, 394
Major Land Resource Areas, 54
Manganese
  abiotic oxidation, 429–432
  abiotic reduction, 417–421
  amorphous forms, 420
  assimilatory reduction, 412
  bacterial reduction, 416
  biotic oxidation, 429–432
  biotic reduction, 415–417
  complexation with dissolved organic matter, 425
  coprecipitation of trace elements, 439
  crystalline forms, 420
  eh–pH relationships, 407–411
  electron acceptor, 416
  electron donors, 417
  forms, 421–427
  immobile pools, 425
  impact of reduction on wetland soil chemistry, 421
  manganous, 241
  microbial communities and oxidation, 428, 429
  mobile pools, 425
  mobility, 432–435
  nonenzymatic reduction, 417
  nutrient regeneration/immobilization, 435
  organic matter decomposition and nutrient release, 435–438
  oxidation, 427–432
  oxidation states, 408
  oxides in soil profile, 426

Manganese (*contd.*)
    oxyhydroxides, 48
    reductase, 415
    reduction, 411–413
    reduction by microbial communities, 413–415
    regulation of oxidation, 429
    regulation of reduction, 425
    respiration, 145–147
    root plaque formation, 440, 441
    storage and distribution, 405–407
Mangroves, 31, 32
Manning equation, 556
Mantel tests, 714
Marshes
    accretion, 619, 670, 672
    coastal, 672, 673, 680
    salt, 52, 696
    saw grass, 639
    soils, 52
    submergence, 670
    tidal, 31
    upland, 52
Mass flow, 46
Masses, 50
Mass transfer coefficient, 527
Mean residence time, 527, 539
Mechanistic models, 712, 713
Mercury, 700, 701
    cycles, 484
    methyl, 482–485
    organo-mercurial compounds, 482
Mesotrophic, 326
Metabolic phosphates, 366
Metabolic quotient, 649
Metachlor, 509
Metals
    complexes, 477
    factors in availability and transformation, 477–480
    free cations, 477
    soil/sediment redox and pH, 480–482
    toxic, 477
Metaphyton, 371
Methane, 111, 112, 599, 706
    emissions, 441–443
        from soils, 174
        global, 706
        regulators, 607, 608
        wetlands, 604–607
    fluxes, 176
    oxidation, 291, 608, 609
    production in wetlands, 603, 604
    sinks, 608, 609
    wetlands as a source, 602, 603
Methanethiol, 470, 696
Methanobacteriacae, 604
Methanobacterium spp, 148, 604
Methanococcus spp, 148, 310, 604
Methanogenesis, 461, 706
Methanogens, 148, 149, 204, 604, 644
Methanosarcina spp, 148, 310, 604
Methanothrix spp, 148

Methanotrophs, 291
Methionine, 454
Methyl mercaptane, 470
Methyl mercury, 707, 663
Methyl substrate utilizers, 148
Methylotrophic bacteria, 291, 292
Michaelis–Menten equations, 17, 461
Microbes, 40, 42
Microbial
    biomass, 160, 161, 278
    biomass nitrogen, 645
    communities, 160, 161, 310
    fuel cells, 106
    mats, 121
    sulfate reduction, 707
Micronutrients, 168
Microorganism oxygenases, 140
Mineralization, 260, 273, 275
    organic nitrogen, 264
        C:N ratio, 265–267
    rate, 275, 276
Mississippi River
    biogeochemistry, 36
    carbon role, 164, 684–687
    deltaic plain, 669–675
    diversion, 676–680
    electrochemical properties, 94
    exchange processes, 557, 558, 560
    global change, 617–619
    impacts on vegetation, 681, 682
    iron, 407, 422, 425, 440, 442
    manganese, 407, 422, 425, 440, 442
    metals, 493, 494, 500–502
    nitrogen role, 689, 691, 692,
    oxygen role, 189, 210
    sulfur, 694
    toxic organic compounds, 509, 700
Mixing, due to currents and waves, 551
Molality, 7
Molarity, 7
Molecular diffusion, 551
Molecular methods, 710
Molecules, 7
Mole fractions, 8
Moles, 7
Moles of solute, 7
Mollisols, 23
Monitoring, 584, 585, 590
Monoesterases, 379–381
Monoesters, 358
Monomer, 269
Mononucleotides, 258
Mottles, 48
Muck, 52
    Everglades Agricultural Area, 628
    histosols, 22
    indicators of hydric soils, 54, 55
    nitrous oxide emissions, 300
    soils, 55, 56, 58, 60, 62
Mucopolysaccharides, 258
Multisamplers, 552, 553
Multivariate techniques, 714
Munsell color system, 20

Index

N

NAD+/NADH, 139
Negative feedback, 4
Negative potential, 71
Nernst equation, 76, 186, 452
   global change, 613
   iron, 409
   manganese, 411
   oxygen, 166
   redox reactions, 76, 78, 79, 82, 85, 91, 92, 98
   sulfur, 422
Net carbon assimilation, 249
Neutrophilic bacteria, 428
Nickel, 503–505
Nitrate, 258
   flux, 320–322
   reducers, 144
   reduction, 296, 297
      plant roots influence, 304
      rates, 304, 307–309
      regulators, 303–307
   respiration, 145, 299
Nitrate/nitrite reductases, 260, 303
Nitrate/nitrous acid equilibrium, 302
Nitric oxide, 258
Nitric oxide reductase, 305
Nitrification, 258, 262, 282, 289, 293, 322
Nitrification–denitrification, 609, 610
   coupled in wetlands, 320
Nitrifier–denitrification, 298–301
Nitrifying
   chemoautotrophs, 291
   population, 293
Nitrite, 258
Nitrite reductase, 305
Nitrobacter spp, 262, 289, 290, 292, 293, 296
Nitrogen
   ammonia/ammonium, 257
   assimilation
      budget 263, 690, 691
      by vegetation, 314–317
      coefficient, 646
   bioavailability, 260
   concentrations in plant tissue, 315
   cycle, 257, 259, 262, 653–657, 687–693
   dinitrogen, 257
   dissolved organic, 261
   fixation, 178, 257, 258, 309, 310, 313, 314
   forms of, 257, 258, 261
   gaseous end products, 261
   inputs, 258, 688
   isotopes, 257
   losses, 689, 690
   major storage compartments, 258
   microbial biomass, 261
   mineralization, 262, 265
   nitrate, 258
   nitric oxide, 258
   nitrite, 258
   nitrous oxide, 258
   organic, 258, 273, 317, 274–278
   particulate organic, 260
   plant biomass, 260
   potentially mineralizable, 653
   processing by wetlands, 317
   reactions with amino group, 302
   reactions with metal cations, 302
   redox transformations, 262–264
   regeneration and uptake, 688
   substrate-induced mineralization, 653
   utilization efficiency, 314
   wetland reservoirs, 259
Nitrogenase, 310, 312, 313
Nitrogen-fixing bacteria, 310
Nitrogen:nitrous oxide ratio, 321
Nitrogen:phosphorus ratio, 316, 640
Nitroso spp, 289
Nitrosomonas spp, 262, 289–292
Nitrous oxide, 258, 599
   consumption, 614, 615
   emission from wetlands, 611
   evolution, 299, 300
   production in wetlands, 609, 611–613
   reductase, 305–307
   regulators of production and
      emissions, 613, 614
Nodules, 50
Nonexchangeable ammonium, 284
Nonhumic substances
   carbohydrates, 122–125
   lipids, 125
   proteins, 125
Nonliving organic material, 178
Nontidal wetlands, 35
Normality, 8
Nuclear magnetic resonance
      spectroscopy, 369, 711
Nucleic acids, 258, 269, 366
Nucleotides, 16
Nuphar spp, 639
Nutrients, 311
   availability, 167–170
   uptake, 255
   use efficiency, 314
Nymphaea spp, 639

O

O horizon, 19
Obligate anaerobes, 40, 143, 147, 148, 302
Obligate wetland plants, 41, 217
Octahedral layers, 283
Oligotrophic, 325, 326
Ombrotrophic, 52
1-aminocyclopropane-one-carboxylic acid (ACC), 226
Order of reaction, 18
Order of reduction, 43
Organic acids, 241
   oxidizing iron reducers, 413
   oxidizing manganese reducers, 413
Organic compounds
   aromatic ring cleavage, 518
   β-oxidation, 517
   burial, 533

Organic compounds (*contd.*)
  cleavage of ether linkage, 518
  colloidal organic matter, 524
  dealkylation, 517
  decarboxylation, 517, 518
  denitrifying bacteria, 528, 529
  electron acceptors, 528
  epoxidation, 518
  exchange between soil and water column, 525
  heterocyclic ring cleavage, 518
  hydrolysis, 515, 522
  hydroxylation, 516, 517
  oxidation, 516–519
  oxidative coupling, 518
  photolysis, 525, 526
  redox-potential-pH, 521, 522
  reduction, 519, 520
  reductive dehalogenation, 519, 520
  runoff and leaching, 527, 528
  sediment redox–pH conditions, 529–533
  settling and burial of particulate contaminants, 525
  sorption to suspended solids and the substrate bed, 522, 523
  sulfoxidation, 518, 519
  synthesis, 520
  toxic, 507
  volatilization, 526, 527
  wetlands and aquatic ecosystems, 511
Organic mater
  accumulation, 119–121
  accumulation measurements, 120, 121
  burial, 119
  decomposition, 119, 426, 705, 706
  decomposition regulators, 157
    availability of electron acceptors, 164–167
    microbial communities, 160, 161
    nutrient availability, 167–170
    quality and quantity, 158–160
    temperature, 170–173
    soil aeration status, 162, 163
  ecological significance, 173–178
  functions in soils, 178
  humic substances, 122, 127–129
  living, 260
  low-molecular weight, 133–136
  mineralization, 278–280
  nonhumic substances, 122–125
  nonliving, 178, 260
  phenolic substances, 122, 125, 126
  quality, 158–160
  quantity, 158–160
  turnover, 151
    abiotic decomposition, 151–153
    biotic decomposition, 153–157
Organic soils, 51
Organic wastes, 327
Orthophosphate, 330, 332, 333, 358
Orthophosphoric acid, 330
Oryza spp, 243, 249, 251
Overlying water incubations, 552–555
Oxic–anoxic interface, 466
Oxidants, 67

Oxidation
  reduced compounds, 205, 239
  reduction potential, 79
  reduction reactions, 67, 68
  states, 13
Oxidation–reduction reactions, 13–15, 67, 68
Oxidative effect, 205
Oxidative phosphorylation, 17, 139
Oxidized soil–floodwater interface, 43, 44
Oxidized surface soil layer, 43
Oxidizing power of plant roots, 245, 246, 247
Oxidoreductases, 17
Oxisols, 23
Oxygen, 138
  availability, 293
  concentration in solution adjacent to roots, 239
  consumption, 204
    by anaerobic soils, 209
    oxygen as a reactant, 204, 205
    oxygen as an electron acceptor, 205–211
  content and Eh, 186
  content of soils, 191, 303
  convective flow, 232, 233
  depletion, 42
  diffusion, 185, 192
  electron acceptor, 185
  flux (ODR), 196
  gradient, 202
  in restricted zones, 42
  in wetland soil, 36
  leakage from roots, 239
  microelectrodes, 239
  movement in wetland plants, 227–229
  peak production, 200
  profiles, 199, 201, 202
  reactant, 185
  reduction, 204
  reduction in ETS, 140
  release by plants, 237
  release capacity of plant roots, 246
  solubilization, 233
  sources, 199
  status of sediment, 697
  status of soil, 191
  supersaturation, 200
  transport, 46, 228
Oxygen-containing floodwater, 201
Oxygen-free soil zone, 201
Oxygen-permeable membrane-covered electrode, 198

P

Paddy fields, 33, 53
Pancium spp, 639
Paracoccus spp, 299, 301
Particulate materials, 121
Particulate organic matter, 358
Pasteur effect, 221
pe vs. Eh, 77–79
Peak oxygen production, 200
Peat, 22
  accretion, 329

Index

biogeochemical characteristics, 31–33, 42, 47, 51, 52, 55, 58
carbon cycle, 114
decomposition of organic matter, 158–160, 162, 163, 169, 171, 172
decomposition of surface, 686
Everglades, 623, 627, 628, 639, 651, 653, 655, 664
exchange processes, 559
global change, 602, 607, 620
iron, 429, 434
manganese, 429, 434
nitrogen, 260, 265, 266, 273, 300, 313
organic matter, 120–122, 176, 178, 179
oxygen, 197, 199
peatlands 51, 602, 607
phosphorus, 329, 332, 336–338, 347, 354, 357
sulfur, 451, 463, 474
Pedons, 19
Pentachlorophenol (PCP), 698
Percent air-spaces, 189
Perchloric acid digestion, 334
Periphyton, 263, 371, 391–393, 641–643
biogeochemical characteristics, 705, 708, 716
carbon, 129, 136
electrochemical properties, 96
Everglades
biogeochemical cycles, 647, 648, 657, 658, 660, 662
nutrient load, 631, 633, 635, 637, 638, 641–643
restoration, 665, 666
wetland, 623–625
exchange processes, 538, 550
indicators, 577, 580, 581, 583, 584, 588
iron, 438
manganese, 438
nitrogen, 263, 280, 286, 293, 309, 312, 317, 320
oxygen, 199, 200, 206
phosphorus
accumulation, 328
forms, 332, 333
memory, 401
mineralization, 377, 380, 381
mobilization, 391–394
uptake and storage, 371, 372
sulfur, 454, 468
Permafrost, 601, 602
Permanent charge, 341
Peroxidase, 205
Peroxide, 204
Pesticides,
arsenic, 458, 486
Everglades, 631
Louisiana, 697
organic phosphorus, 358
toxic organic compounds, 507, 509, 514, 518–522, 526, 529
water quality indicator, 588
Petroleum hydrocarbons, 509
pH, 12, 13
alkalinity, and carbon dioxide, 293
arsenic, 486–489
biogeochemical characteristics, 39–41, 43, 46, 52
cadmium, 500, 501

carbon, 18, 134, 179–181
chromium, 497–499
copper, 490, 492
effects on electrochemical properties, 67, 69, 76, 79, 81–84, 90, 92, 97, 98
flooded soils, 93–95
floodwater, 96, 97
iron, 405, 407–411, 415, 417, 419, 421, 422
lead, 502, 503
manganese, 405, 407–411, 415, 417, 419, 421, 422
mercury, 482–484
metals, 477, 480–482
nickel, 504, 506
nitrogen, 263, 278, 279, 284, 285–287, 293, 306
oxygen, 186, 187, 193, 195
pH dependent charge on solid phase, 341
phosphorus, 330, 331, 341, 342, 353, 387, 389
plant anaerobiosis, 247
rhizosphere, 247
selenium, 495, 496
soil, 93–96
sulfur, 451–453, 458, 463, 467, 468
zinc, 493, 494
Pharmaceuticals, 511, 512
Phenolic compounds, 125, 134
Phenol oxidase, 133
Phosphatases, 379–381, 385, 706
Phosphate
esters, noncatalyzed hydrolysis of, 377, 378
flux, 546
interactions with periphyton, 391–393
ion exchange in soils, 341
metabolic, 366
reactions in acid soils, 343
reactions in alkaline soils, 343
Phosphine gas, 331, 395
Phosphodiesterases, 381
Phospholipid fatty acid analysis, 710
Phospholipids, 366
Phosphomonoesterases, 387
Phosphorus
abiotic retention, 353
accumulation in wetlands, 328–330
acid hydrolysable, 367
adsorption–desorption, 343–346
anthropogenic loading, 334
assimilation, 375, 646
bioavailable, 336
biotic and abiotic interactions of iron and calcium with, 394
cycle, 332, 657–660
deficient soils, 380
dissolution, 350–353
exchange
between soil and water column, 395–397
processes affecting soil/sediment–water interface, 397
extraction procedure, 369
flux, 325, 397
forms in water column and soil, 330–335
fulvic acid–bound, 367
gaseous loss of, 395
gradients, 631

Phosphorus (contd.)
  humic acid–bound, 367
  immobilization by microorganisms, 382
  inorganic, 333, 335–340
    containing aluminum, 338
    containing calcium, 338
    containing iron, 338
    dissolved, 332, 333, 379
    particulate, 332
  inorganic soil, 334
  inputs, 325
  interactions with iron and
      sulfur, 388–390
  ion speciation, 330
  labile, 367
  leaching, 377, 383
  legacy, 397
  loading, 325, 345
  long term storage in trees, 376
  memory, 328, 397–401
  microorganisms uptake and storage, 370, 371
  mineralization, 382, 384, 385, 386, 659
  mobilization, 388
  modified simple sequential inorganic fractionation
      schemes, 336
  occluded, 336
  organic, 333, 334, 357, 366
    associated with microbial pool, 367
    chemical characterization, 367–370
    complexation, 378
    cycling, 382
    easily mineralizable, 367
    enzymatic hydrolysis, 378–381
    forms, 357–367
    heterotrophic degradation, 383
    highly resistant, 368
    identifying, 369
    mineralization, 376, 383–385
    moderately labile, 368
    moderately resistant, 368
    nonreactive, 368
    particulate, 332
    residual, 367
    slowly mineralizable, 367
    soluble, 332
    stabilization of organic, 378
  oxidation state, 331
  periphyton uptake and storage,
      371, 372
  plant contribution to soluble, 374
  potentially mineralizable, 384, 385, 658
  precipitation, 350–353, 438
  readily available, 338
  redox-sensitive forms, 336
  reductant-soluble, 336
  regulation
    organic mineralization, 387, 388
    retention and release, 353–357
  release by microbial activity, 381
  reservoirs of sediments, 336
  retention, 327
    capacity of soils, 398
    maximum of soils, 354
  slowly available, 340
  soil sequential fractionation, 369
  solubility, 331, 355–357, 386, 387, 400
  solubilization by microorganisms, 382
  sorption
    capacity, 353
    Freundlich equation, 348, 349
    indices, 349, 353
    isotherm equations, 346–350
    Langmuir equation, 349
    linear equation, 347, 348
    quantity (Q)/ intensity (i) relationships, 350
    single point isotherms, 349, 350
    soils, 340–343
  storage, 328, 375, 376
  total
    labile organic, 367
    organic, 334, 358
    particulate, 332, 333
    profiles, 329
  transfer, 327
  uptake, 370–376
Photochemical degradation, 378
Photochemical reactions, 152
Photolysis, 152, 378
Photolytic oxidation, 152
Photomineralization, 378
Photooxidation, 378
Photosynthesis, 708, 709
  adaptations of plants to anaerobiosis, 228, 233,
      249, 252, 255
  biogeochemical characteristics, 41
  carbon role, 111, 114, 118, 135
  electrochemistry, 67, 95
  Everglades, 642, 659
  exchange processes, 554
  iron, 438
  Louisiana, 681, 683, 684
  manganese, 438
  nitrogen role, 263, 284, 286, 287, 293, 310
  oxygen role, 185, 188, 199, 200, 203, 207
  phosphorus role, 353, 356, 392
  sulfur role, 469
Photosynthetically active radiation, 235
Phototrophs, 16, 466
Phragmites spp, 155, 229, 235–238, 242, 317, 372, 374
Phytomass, 113
Phytotoxins, 241
Piezometer, 542
Plants
  density, 288
  flood-tolerant species, 217
  gas exchange by wetland plants, 244
  hydrophytic vegetation, 27, 46
  impacts of flooding on wetland plants, 215
  mechanisms of oxygen movement, 227
    diffusion, 228, 229
    mass flow, 229
  root respiration, 226
  shoot turnover times, 375, 376
  tissue phosphorus concentrations, 372
  nutrient uptake, 252
Platinum electrodes (Pt), 86, 195–197

Index

Playas, 32
Plug flow, 328
Pneumatophores, 31, 221, 222
Poise, 84, 85
Polarographic membrane-covered electrode, 198
Polarography, 239
Polychlorobiphenyl, 698
Polymer, 268
Polymerization of conjugation, 513, 514
Polypedons, 19
Polyphenols, 127
POM, 178
Ponding, 50
Pore diameter, 237
Pore linings, 50
Pore water equilibrators, 552, 553
Porosity, 24, 235, 237, 544
    exchange processes, 543–547
    global change, 608
    nitrogen, 283, 318
    oxygen, 190–192, 201
    plant
        mechanisms of flood tolerance, 222, 223, 227
        oxygen movement, 229, 235, 237
        radial oxygen loss, 239, 250–252, 254, 255
    toxic organic compounds, 527
Positive feedback, 4
Positive potential, 71
Prairie potholes, 32
Precipitation, 34, 340
Primary production, 684, 685
Priming effect, 158
Probability sampling, 585
Prokaryotic cells, 15
Protease, 275, 279
Proteins, 16, 258
Pseudomonas spp, 132, 145, 241, 263, 299, 413, 414, 428, 466
Purposive sampling, 585
Pyrite, 469, 471, 474, 693
Pyruvate decarboxylase, 220

Q

Q_{10}, 18, 170–172
Quinones, 205

R

R horizon, 20
Radial oxygen loss, 237, 239, 250–252, 255
Radioactive
    decay constant, 18
    half-life, 18
    isotopes, 18
Radionuclides, 18
Radium/radon isotopes, 542, 543
Rate of settling particles, 556
Ratio microbial biomass carbon to total organic carbon, 645
Ratio of proton to electron consumption, 94
Reaction quotients, 12
Recently accreted soil, 537

Redfield ratios, 437
Redox
    artificial buffers, 247
    concentrations, 50
    couples in wetlands, 98–101
    depletion, 50
    gradients in soils, 101–105
    gradients in wetlands, 433
    matrix, 50
    potential, 24, 39
        arsenic, 485–489
        buffering, 84, 85
        cadmium, 499–501
        carbon, 142, 165
        chromium, 496, 497, 499
        copper, 489–493
        electrochemical properties, 67–69, 71, 83–85, 88–92, 99–105
        iron, 408–411, 419–423, 429, 431, 432
        lead, 501–503
        manganese, 408–411, 419–423, 429, 431, 432
        measuring
            construction of platinum electrodes, 85, 86
            redox potentials in soils, 88–92
            standardization of electrodes, 86–88
        metals, 481, 482
        mercury, 482–485
        nickel, 503, 504
        nitrogen, 275, 278, 279, 293, 302, 312
        oxygen, 192–195, 206, 208
        pH, 312
        phosphorus
            environmental, 331
            inorganic, 338, 340
            mineralization, 382, 387, 390
            sorption, 350, 353, 355
            uptake and storage, 370, 371
        plant adaptations to anaerobiosis, 217, 239–241, 247, 249–251, 253
        selenium, 494–496
        sulfur, 452, 457, 458
        zinc, 493, 494
    reactions
        balancing, 14, 15
        electrochemical properties, 72, 77, 82, 99, 105
        global change, 601
        metals, 487, 497
        sulfur, 451
        toxic organic compounds, 521, 522
Redoximorphic features, 50, 54, 440
Reduced anaerobic soil layer, 43
Reductants, 67
Reduction, 13
    capacity, 248
    hydroxyl radical, 204
    intensity, 67, 250–252
    kinetics of, 420
    oxygen, 195
    potentials, 14, 69
Reference electrodes, 86
Reference wetlands, 584
Refractory components, 332
Regression kriging, 593

Relative humidity, 235
Remineralization, 261
Respiration, 17, 67, 220
    aerobic, 14, 17, 42
        carbon role, 137, 145, 148, 150, 161, 162, 167
        Everglades, 649
        indicators, 595
        iron and manganese, 430
        plant adaptations to anaerobiosis, 215, 219–221, 225, 238, 245
        sulfur, 464
    anaerobic, 17, 42, 101
        carbon role, 167
        Everglades, 649
        indicators, 595
        plant adaptations to anaerobiosis, 221
        toxic organic compounds, 528
Restricted root elongation, 250
Resuspension, 396
Reverse electron transport, 431
Rheotrophic, 52
Rhizobium spp, 263, 299
Rhizophora spp, 153, 154
Rhizosphere, 46, 239, 245
Rhynchospora spp, 639
16S ribosomal RNA gene sequences, 710
Riparian wetlands, 32
Roots
    altered, 221
    branching, 250
    development, 255
    elongation, 250
    energy status, 221
    exudation, 121
    length of aerated channel, 229
    metabolism, 219, 255
    oxygenation, 46, 238
    plaques, 246, 440, 441
    porosity, 250, 251
    regeneration, 221
    respiration, 220
    turnover times, 376
Rotenberger Wildlife Management Area, 630

S

Sagittaria spp, 173, 238, 245, 246, 694, 697
Salinity/conductivity, 542
Salt marsh, 52, 696
Sampling designs, 584, 585
Saprists, 52
Saturated calomel reference electrodes, 79
Saturation index, 351
Saw grass marshes, 639
Scale, 714
Schizogenous aerenchyma, 225, 226
Sea-level rise, 601
Seasonal response, 373, 374
Second law of thermodynamics, 9
Secondary effects of microbial activity, 513, 514
Sediment
    dating, 568
    transport processes, 556, 557
        measurement of sedimentation or accretion rates, 560–568
        sediment/organic matter accretion in wetlands, 557–560
Sedimentation, 329, 396
Sedimentation–erosion table, 563, 564
Seepage, 396, 541
Seepage meters, 541
Selenium, 481, 494–496
Self-decomposition of nitrous acid, 302
Semivariogram, 592
Sequential extraction procedure, 336
Sequential reduction, 44, 105, 389
Settling particles, rate, 556
Shewanella spp, 413, 414
Shoot turnover times, 375, 376
Siderite, 394
Silicate clays, 283
Sinks, 2
Sloughs, 639
Soil
    aeration status, 162, 163, 192
    air, 24
    anaerobiosis, 252
    bulk density, 23
    capacity of reduction and wetland plant functions, 247
    cation exchange complex, 280
    color, 20
    definitions, 35
    floodwater exchange processes, 537
    gases, 187
        oxygen diffusion rate, 195–198
        redox potential, 192–195
        soil oxygen content, 198, 199
    horizons, 19
    interface with floodwater, 201
    intensity of reduction and wetland plant functions, 247
    microbial biomass, 161
    microorganisms, 357
    orders, 21–23
    organic matter, 173, 179
    oxygen, 192
    oxygen demand (SOD), 153, 211
    particle density, 23
    pH, 24, 279
    phosphorus gradients, 328
    phytotoxic accumulation effects on plant growth, 241
    pore dissolved oxygen, 42
    porosity, 24
    redox potential, 217
    reduction, 248, 249, 250, 252, 253–255
    subaqueous, 53
    subsidence, 51, 52
    taxonomy, 21
    textures, 20, 21
    volume, 36
    water, 24, 221, 288
Solubility product, 351
Solubilization of respiratory carbon dioxide, 229
Solute fluxes, 555
Sorption, 343, 344
Sources, 2